D0786210

Eleventh Edition

Principles of Information Systems

Ralph M. Stair
Professor Emeritus, Florida State University

George W. Reynolds
Instructor, Strayer University

Australia • Brazil • Japan • Korea • Mexico • Singapore • Spain • United Kingdom • United States

COURSE TECHNOLOGY
CENGAGE Learning

Principles of Information Systems, Eleventh Edition
Ralph M. Stair & George W. Reynolds

Editorial Director: Erin Joyner

Editor-in-Chief: Joe Sabatino

Sr. Acquisitions Editor: Charles McCormick, Jr.

Sr. Product Manager: Kate Mason

Development Editor: Lisa Ruffolo, The Software Resource Publications, Inc.

Editorial Assistant: Anne Merrill

Sr. Brand Manager: Robin LeFevre

Market Development Manager: Jon Monahan

Marketing Coordinator: Mike Saver

Media Editor: Chris Valentine

Art and Cover Direction, Production Management, and Composition: PreMediaGlobal

Manufacturing Coordinator: Julio Esperas

Cover Credits:

People Walking: © SVLuma/Shutterstock

Globe and Swish: © iStockphoto/Thinkstock

For product information and technology assistance, contact us at
Cengage Learning Customer & Sales Support, 1-800-354-9706

For permission to use material from this text or product,
submit all requests online at **www.cengage.com/permissions**

Further permissions questions can be emailed to
permissionrequest@cengage.com

Library of Congress Control Number: 2012946958

Student Edition:

ISBN-13: 978-1-133-62966-5

ISBN-10: 1-133-62966-0

Instructor's Edition:

ISBN-13: 978-1-133-95351-7

ISBN-10: 1-133-95351-4

Course Technology
20 Channel Center Street
Boston, MA 02210
USA

Printed in the United States of America
1 2 3 4 5 6 7 16 15 14 13 12

For Lila and Leslie

—RMS

To my grandchildren: Michael, Jacob, Jared, Fievel, Aubrey, Elijah, Abrielle, Sofia, Elliot, and Serena

—GWR

Brief Contents

Contents

PART 3 Business Information Systems 353

8 Electronic and Mobile Commerce 354

Preface

As organizations and entrepreneurs continue to operate in an increasingly competitive and global marketplace, workers in all business areas including accounting, customer service, distribution, finance, human resources, information systems (IS), logistics, marketing, manufacturing, and research and development must be well prepared to make the significant contributions required for success. Regardless of your future role, even if you are an entrepreneur, you need to understand what information systems can and cannot do and be able to use them to help you accomplish your work. You will be expected to discover opportunities to use information systems and to participate in the design of solutions to business problems employing information systems. You will be challenged to identify and evaluate information systems options. To be successful, you must be able to view information systems from the perspective of business and organizational needs. For your solutions to be accepted, you must recognize and address their impact on co-workers, customers, suppliers, and other key business partners. For these reasons, a course in information systems is essential for students in today's high-tech world.

Principles of Information Systems, Eleventh Edition, continues the tradition and approach of the previous editions. Our primary objective is to provide the best information systems text and accompanying materials for the first information systems course required of all business students. We want you to learn to use information systems to ensure your personal success in your current or future job and to improve the success of your organization. Through surveys, questionnaires, focus groups, and feedback that we have received from current and past adopters, as well as others who teach in the field, we have been able to develop the highest-quality set of teaching materials available to help you achieve these goals.

Principles of Information Systems, Eleventh Edition, stands proudly at the beginning of the IS curriculum and remains unchallenged in its position as the only IS principles text offering the basic IS concepts that every business student must learn to be successful. In the past, instructors of the introductory course faced a dilemma. On one hand, experience in business organizations allows students to grasp the complexities underlying important IS concepts. For this reason, many schools delayed presenting these concepts until students completed a large portion of the core business requirements. On the other hand, delaying the presentation of IS concepts until students have matured within the business curriculum often forces the one or two required introductory IS courses to focus only on personal computing software tools and, at best, merely to introduce computer concepts.

This text has been written specifically for the introductory course in the IS curriculum. *Principles of Information Systems, Eleventh Edition*, treats the appropriate computer and IS concepts together with a strong managerial emphasis on meeting business and organizational needs.

APPROACH OF THE TEXT

Principles of Information Systems, Eleventh Edition, offers the traditional coverage of computer concepts, but it places the material within the context of meeting business and organizational needs. Placing information system concepts in this context and taking a general management perspective has always set the text apart from general computer books, thus making it appealing not only to MIS majors but also to students from other fields of study. The text isn't overly technical, but rather deals with the role that information systems play in an organization and the key principles a manager needs to grasp to be successful. These principles of IS are brought together and presented in a way that is both understandable and relevant. In addition, this book offers an overview of the entire IS discipline, while giving students a solid foundation for further study in advanced IS courses such as programming, systems analysis and design, project management, database management, data communications, Web site and systems development, electronic commerce and mobile commerce applications, and decision support. As such, it serves the needs of both general business students and those who will become IS professionals.

The overall vision, framework, and pedagogy that made the previous editions so popular have been retained in the eleventh edition, offering a number of benefits to students. We continue to present IS concepts with a managerial emphasis. While the fundamental vision of this market-leading text remains unchanged, the eleventh edition more clearly highlights established principles and draws out new ones that have emerged as a result of business, organizational, and technological change.

IS Principles First, Where They Belong

Exposing students to fundamental IS principles is an advantage for students who do not later return to the discipline for advanced courses. Since most functional areas in business rely on information systems, an understanding of IS principles helps students in other course work. In addition, introducing students to the principles of information systems helps future business managers and entrepreneurs employ information systems successfully and avoid mishaps that often result in unfortunate consequences. Furthermore, presenting IS concepts at the introductory level creates interest among general business students who may later choose information systems as a field of concentration.

Author Team

Ralph Stair and George Reynolds have teamed up again for the eleventh edition. Together, they have decades of academic and industrial experience. Ralph Stair brings years of writing, teaching, and academic experience to this text. He wrote numerous books and a large number of articles while at Florida State University. George Reynolds brings a wealth of information systems and industrial experience to the project, with more than thirty years of experience working in government, institutional, and commercial IS organizations. He has written numerous college IS texts and has taught the introductory IS course at the University of Cincinnati, the College of Mount St. Joseph, and Strayer University. The Stair and Reynolds team brings a solid conceptual foundation and practical IS experience to students.

GOALS OF THIS TEXT

Because *Principles of Information Systems, Eleventh Edition,* is written for all business majors, we believe it is important not only to present a realistic perspective on IS in business but also to provide students with the skills they can use to be effective business leaders in their organization. To that end, *Principles of Information Systems, Eleventh Edition,* has four main goals:

1. To provide a core of IS principles with which every business student should be familiar
2. To offer a survey of the IS discipline that will enable all business students to understand the relationship of IS courses to their curriculum as a whole
3. To present the changing role of the IS professional
4. To show the value of the discipline as an attractive field of specialization

By achieving these goals, *Principles of Information Systems, Eleventh Edition,* will enable students, regardless of their major, to understand and use fundamental information systems principles so that they can function more efficiently and effectively as workers, managers, decision makers, and organizational leaders.

IS Principles

Principles of Information Systems, Eleventh Edition, although comprehensive, cannot cover every aspect of the rapidly changing IS discipline. The authors, having recognized this, provide students an essential core of guiding IS principles to use as they face the career challenges ahead. Think of principles as basic truths or rules that remain constant regardless of the situation. As such, they provide strong guidance in the face of tough decisions. A set of IS principles is highlighted at the beginning of each chapter. The application of these principles to solve real-world problems is driven home from the opening vignettes to the end-of-chapter material. The ultimate goal of *Principles of Information Systems* is to develop effective, thinking, action-oriented employees by instilling them with principles to help guide their decision making and actions.

Survey of the IS Discipline

This text not only offers the traditional coverage of computer concepts but also provides a broad framework to impart students with a solid grounding in the business uses of technology. In addition to serving general business students, this book offers an overview of the entire IS discipline and solidly prepares future IS professionals for advanced IS courses and their careers in the rapidly changing IS discipline.

Changing Role of the IS Professional

As business and the IS discipline have changed, so too has the role of the IS professional. Once considered a technical specialist, today the IS professional operates as an internal consultant to all functional areas of the organization, being knowledgeable about their needs and competent in bringing the power of information systems to bear throughout the organization. The IS professional views issues through a global perspective that encompasses the entire organization and the broader industry and business environment in which it operates.

The scope of responsibilities of an IS professional today is not confined to just his or her employer but encompasses the entire interconnected network of employees, suppliers, customers, competitors, regulatory agencies, and other entities, no matter where they are located. This broad scope of responsibilities creates a new challenge: how to help an organization survive in a highly interconnected, highly competitive global environment. In accepting that challenge, the IS professional plays a pivotal role in shaping the business itself and ensuring its success. To survive, businesses must now strive for the highest level of customer satisfaction and loyalty through innovative products and services, competitive prices, and ever- improving product and service quality. The IS professional assumes the critical responsibility of determining the organization's approach to both overall cost and quality performance and therefore plays an important role in the ongoing survival of the organization. This new duality in the role of the IS employee—a professional who exercises a specialist's skills with a generalist's perspective—is reflected throughout the book.

IS as a Field for Further Study

Despite the effects of recession and outsourcing, a survey of human resources professionals still puts technology and health care among the top fields of employment opportunity. And business administration and computer science remain among the most sought after majors by employers. Indeed, the long-term job prospects for skilled and business-savvy information systems professionals is optimistic. Employment of such workers is expected to grow faster than the average for all occupations through the year 2020.

A career in IS can be exciting, challenging, and rewarding! It is important to show the value of the discipline as an appealing field of study and that the IS graduate is no longer a technical recluse. Today, perhaps more than ever before, the IS professional must be able to align IS and organizational goals and to ensure that IS investments are justified from a business perspective. The need to draw bright and interested students into the IS discipline is part of our ongoing responsibility. Upon graduation, IS graduates at many schools are among the highest paid of all business graduates. Throughout this text, the many challenges and opportunities available to IS professionals are highlighted and emphasized.

CHANGES IN THE ELEVENTH EDITION

We have implemented a number of exciting changes to the text based on user feedback on how to align the text even more closely with current IS principles and concepts courses. The following list summarizes these changes:

- **All new opening vignettes.** All of the chapter-opening vignettes are new and continue to raise actual issues from foreign-based or multinational companies.
- **All new Information Systems @ Work special interest boxes.** Highlighting current topics and trends in today's headlines, these boxes show how information systems are used in a variety of business career areas.
- **All new Ethical and Societal Issues special interest boxes.** Focusing on ethical issues today's professionals face, these boxes illustrate how information systems professionals confront and react to ethical dilemmas.
- **All new case studies.** Two new end-of-chapter cases for each chapter provide a wealth of practical in-formation for students and instructors. Each case explores a chapter concept or problem that a real-world company or organization has faced. The cases can be assigned as individual homework exercises or serve as a basis for class discussion.
- **Extensive changes and updates in each chapter.** The authors worked hard to provide the most current information available in this latest edition. Over 1,200 new references and all new examples of organizations using information systems are included in this new edition. The full extent of these updates makes it impractical to cover them completely in this forward; however, the following list summarizes these changes:

Chapter 1, An Introduction to Information Systems includes over 100 new or updated references, examples, and material. It serves as an introduction to major topics covered throughout the text. This chapter stresses the importance of having timely information. FedEx, for example, is able to sort more than 3 million packages every day because of timely information on each package. To get timely information, FedEx often scans a package more than 10 times from its origin to its final destination. We also discuss how the Internet has been used to start protests around the world, and how some countries have attempted to censor or control what information is available to their citizens. According to a popular research and consulting company, the total market for Internet-based cloud computing is expected approach $150 billion annually by 2014. A worldwide Hilton Hotel survey indicated that a fast, reliable connection to the Internet was the second most important factor in overall satisfaction, just behind a clean room. We also discuss emerging Internet sites like Groupon. This chapter covers the importance of smartphones and other mobile devices. BMW, the German sports car company, has invested about $100 million in developing mobile applications for its cars and other products. We also introduce augmented reality, a newer form of virtual reality that has the potential to superimpose digital data over real photos or images. The positive impact of information systems is stressed throughout this chapter. DHL, a large shipping company, used information systems to improve its advertising efforts, which helped the company increase its stock value by more than $1 billion during a five year period. Sasol, a South African energy and chemical company, used an information system to streamline its production facilities through better information and control. This helped the company increase its stock value by more than $200 million. The Shinpuku Seika farm used feedback from its fields to help determine when to plant crops and what crops to plant. Not all information systems produce positive results, however. Faulty computer analysis may have caused a multimillion dollar fine for mistakes and possible fraud. In another case, computer criminals stole carbon credits worth about $40 million from a European carbon credit market.

Chapter 2, Information Systems in Organizations includes more than 100 new or updated references, examples, and material on the use of information systems in today's business organizations to deliver the right information to the right person at the right time. This chapter includes a new section on innovation. The founder of Medical Data Services, which provides software for the health care community, says, "Bring innovation in everything you do, not just the big projects. Scan the market externally in your own field, but also in other businesses. Opportunities can come from the strangest places." This chapter also stresses how information systems can be used to increase profits and reduce costs. Procter & Gamble, the large consumer products company, was able to streamline its supply chain by decreasing inventory levels, reducing costs, and making its operations more efficient. Tidewell, a hospice that serves about 8,000 Florida families, acquired software to save money and streamline its operations. According to the chief information officer for FedEx ground, "Over the last five years, we have been on a mission to get faster and faster. We re-engineer and speed up our lanes an average of twice a year, and sometimes more frequently than that." The section on competitive advantage has been fully updated. With today's innovative smartphones and tablet computers, such as Apple's iPad and others, executives are looking for new ways to gain a competitive advantage developing unique and powerful applications for these newer devices. In one case, a Washington library system was able to save $400,000 by using less expensive phone lines for communications and Internet connections. The German-based DHL shipping company analyzed and streamlined its marketing efforts in over 20 countries. The result was an increase in its corporate value of about $1.3 billion over a five year period, representing an ROI of more than 30 percent. Finally, the section on careers has been fully updated. According to the CIO of Sensis, a defense and airline services company, "Being CIO is not just about running the shop anymore, but different ways you can help your business." We discuss that getting certified from a software, database, or network company can result in increased pay of about 7 percent on average. Some Internet sites, like www.freelancer.com, post projects online and offer information and advice for people working on their own. This chapter warns students to be careful of what they post on social media sites, like Facebook. Employers often search the Internet to get information about potential employees before they make hiring decisions.

Chapter 3, Hardware: Input, Processing, Output, and Storage Devices covers the latest in hardware developments from a discussion of the newest devices, such as the Microsoft Surface platform designed to help people learn, collaborate, and make decisions, to the world's fastest supercomputers. In the next few years there will be an increasing amount of data processing, e-mail, Web surfing, and database lookup done on smartphone-like devices. Several examples of smartphones being used to meet business objectives have been added. The chapter expands its coverage of the issue of increased heat with faster CPU speeds and how manufacturers are addressing this problem. The chapter also explains the need for lighter, smaller, and longer life batteries for our mobile computing devices. A current cost per gigabyte comparison is presented for many of the most popular storage devices. Also discussed are new mobile printing solutions that enable print to be sent from a mobile device to a mobile-enabled printer anywhere in the world. The information on various computer types has all been updated. The chapter presents 40 new examples of organizations and individuals using the computer hardware discussed in the chapter.

Chapter 4, Software: Systems and Application Software includes over 100 new and updated references, examples, and material on system and application software. Software is a growing and dynamic industry. In 2011 China's software industry grew almost 30 percent. The material on systems software has been completely updated. OS X Lion, Apple's latest operating system, offers multitouch, full screen applications, mission control features, and other innovations. Today, more than 100 million people are using Google's Android operating system in smartphones and mobile devices. Windows Home Server allows individuals to connect multiple PCs, storage devices, printers, and other devices into a home network. We also discuss Microsoft Windows 8. Red Hat's newest version of Red Hat Enterprise Virtualization (RHEV) software no longer requires Windows Server software to operate. Alibaba Cloud Computing, a part of the Chinese Alibaba Group, has developed an operating system for smartphones and mobile devices. We have also included new information on user interfaces. Today's mobile devices and some PCs, for example, use a touch user interface also called a natural user interface (NUI) or multitouch interface by some. Speech recognition is also available with some operating systems. Sight interfaces use a camera on the computer to determine where a person is looking on the screen and to perform the appropriate command or operation. Some companies are also experimenting with sensors attached to the human brain (brain interfaces) that can detect brain waves and control a computer as a result. Sight and brain interfaces can be very helpful to disabled individuals. We include many new examples of the successful use of application software. One debt collection agency, for example, was able to save more than $250,000 annually by using application software from Latitude to monitor people not paying their bills on time. A Boise, Idaho, architectural firm used ProjectDox software to streamline the paperwork required to get approval and permits for building projects. Software from Amcom allows companies like Eddie Bauer to provide the exact location of someone calling from an Eddie Bauer retail location to emergency 911 call centers. Absolute software, which uses GPS technology, helps people and organizations retrieve stolen computers. The company has recovered almost 10,000 devices worth over $10 million dollars. The city of Winston-Salem,

North Carolina, for example, uses the Microsoft Office 365 suite over the Internet to save money and place software applications on the Internet. The material on programming languages and software issues and trends has also been updated. In one survey, over 70 percent of information systems managers reported that they negotiate for lower licensing prices most of the time or some of the time. The Code for America (CFA) organization used open-source software in Boston and other American cities to help cities and municipalities solve some of their traffic problems, such as locating fire hydrants that could be completely covered with snow in the winter. CFA made its efforts free to other cities and municipalities.

Chapter 5, Database Systems, Data Centers, and Business Intelligence has 100 new or updated references, including many new, exciting examples and quotes from managers and executives that use database systems to their advantage. The chapter has a new section on big data applications. We have also included new information on database administrators and data scientists who can help analyze what is stored in vast corporate databases. Data centers have been stressed to a greater extent, and the term has been added to the title on data modeling and database characteristics to reflect its increased importance. This section has new material and examples. IBM, for example, is helping to construct a huge data center complex in China involving at least seven separate data centers in multiple buildings totaling more than 6 million square feet. Because of lower energy and land costs, rural North Carolina is becoming popular for large data centers. Apple's $1 billion data center, Google's $600 million data center, and Facebook's $450 million data center are located in North Carolina. Data center concerns, including costs and storage capacity, has been stressed. The U.S. Library of Congress, for example, has about 450 billion objects stored on almost 20,000 disk drives attached to about 600 servers, and its storage requirements are growing. According to one study, about one-third to one-half of all data centers will run out of space in the next several years. Data center security and backup are also discussed. After a hurricane disrupted its data center, the Situs Companies decided to back up its data center on the Internet using EVault, a subsidiary of Seagate. Most Japanese data centers survived the devastating earthquake that shook Japan in 2011 by having enough fuel for electric generators and building data centers using advanced construction techniques. This chapter also has a new section on database virtualization, which discusses security concerns in using the database virtualization approach. In one case, a fired employee was able to gain access to a virtual database and delete important applications, e-mail, and other documents, costing the company about $800,000 in losses, according to the FBI. The material on OLAP has been integrated into the section on business intelligence. According to the IDC Digital Universe Study, only about a third of digital information has at least a minimum level of security. We also include new references, examples, and material on new database applications, including relational database management techniques and vendors, popular database management systems, open-source database systems, databases as a service (DaaS), and a variety of database applications for mobile devices, PCs, and tablet computers.

Chapter 6, Telecommunications and Networks has been updated to cover the latest developments in telecommunications and networks. It discusses FiOS, a bundled set of communications services from Verizon. The role of the IEEE and the IEEE 802.1 family of standards has been added. Coverage of 3G and 4G standards has been simplified and clearly explains the role of UMTS, CDMA, TD-SCDMA, LTE, and HPSA+. Material on Wi-Fi and WiMAX technologies has been updated. The rapid growth in wireless data traffic and the significance of this growth are discussed. The coverage of smartphones has been expanded and includes a discussion of their operating systems and available applications. The growing use of Femtocels, small cellular base station devices used to boost cell phone reception, is covered. The chapter also discusses important trends in the location of call centers, with the Philippines providing more call centers than any other country and with many call centers for U.S. firms moving back to the U.S.. Quick response codes, tags that can scanned by smartphone users to view additional information about a product or an exhibit, are covered. Overall, the chapter presents three new tables, one new figure, and 39 new examples of organizations and individuals using the telecommunications technology discussed in the chapter.

Chapter 7: The Internet, Web, Intranets, and Extranets includes over 100 new or updated references, examples, and material on the Internet, Web, intranets, and extranets. Internet sites can have a profound impact on world politics. Some countries try to control the content and services provided by search engines and social networking sites. To make room for more Web addresses, there are efforts to increase the number of available domain names. Today, the .com domain has more than 90 million Web addresses, .net has more than 10 million addresses, and .org has about 9 million Web addresses. The material on cloud computing has been updated. Apple Computer has developed a service called iCloud to allow people to store their music and other documents on its Internet site. With its Office 365 software product, Microsoft is emphasizing cloud computing to a greater extent today. The New York Stock Exchange (NYSE) is starting to offer cloud computing applications that let customers pay for the services and data they use on Euronext, a European market for stocks, bonds, and other investments. The section on Internet and Web applications has been completely updated. Using the Internet, entrepreneurs can start online companies and thrive. Graduates of the University of Pennsylvania's Wharton School, for example, started an Internet prescription eyeglass company. Internet companies like www.frelancer.com and www.livework.com can help entrepreneurs prosper on the Internet. Ratiophram Canada, a pharmaceutical company, used the Internet to help solve a drug distribution problem, where demand for its generic drugs varied considerably. The Internet was used to allow its employees to share information and collaborate about the varying demand. As a result, the percentage of orders filled on time went from below 90 percent to above 95 percent. The popular enterprise resource planning (ERP) company SAP is teaming up with Google to mashup or integrate enterprise data from SAP with Internet geographic data from Google. The result will be creative graphical reports, such as sales by region, loan defaults by neighborhood, and similar reports placed on Google maps. Grady Health Systems upgraded its e-mail service to Microsoft's Exchange Online e-mail system based on cloud computing. The new e-mail system is more stable and less expensive. Pandora, Napster, and Grooveshark are just a few examples of free Internet music sites. Other Internet music sites charge a fee for music. Rhapsody has about 800,000 paid listeners, Slacker Radio has about 300,000 paid listeners, and Spotify has about 1.5 million paid listeners. Internet music has even helped sales of classical music by Mozart, Beethoven, and others. Walmart's acquisition of Vudu has allowed the big discount retailer to successfully get into the Internet movie business. Some TV networks, such as CNN and HLN, are streaming more programming over the Internet. Increasingly, TV networks have iPad and other mobile applications (apps) that stream TV content to tablet computers and other mobile devices. Other TV networks are starting to charge viewers to watch their episodes on the Internet. Some only allow free viewing of an episode a week or longer after the episode first appears. Video games have become a huge industry. Games can generate over $20 billion annually, more than Hollywood movies. Zynga, a fast growing Internet company, sells virtual horses and other virtual items for games, such as Farm Ville. The company, for example, sells a clown pony with colorful clothes for about $5. Zynga has a VIP club for people that spend a lot on virtual items it offers for sale. Some Internet companies also sell food for virtual animals. People can feed and breed virtual animals and sell their offspring. The section on Internet issues has also been updated. Privacy invasion can be a potential problem with Internet and social networking sites. A number of Internet sites, for example, collect personal and financial information about people who visit their sites without their knowledge or consent. Some Internet companies, however, are now starting to allow people to select a "do-not-track" feature that prevents personal and financial information from being gathered and stored. Some people fear that new facial recognition software used by some Internet companies could be an invasion of privacy. Facial recognition software, for example, could be used to identify people in photos on social networking sites and other Internet sites. Some workers have been fired by their employers when they criticized them or their companies using Facebook, Twitter, and other social networking sites. Some fired employees are fighting back by suing their employers. Because of the increased importance of social Web sites and newer Web 2.0 technologies, this material has been moved to the beginning of the section on Web and Internet applications.

Chapter 8, Electronic and Mobile Commerce has been updated to cover the latest developments in e-commerce and m-commerce. A new table providing a forecast of global B2C e-commerce spending is provided. Information on the largest U.S. B2C retailers has been updated. New examples of major data security breaches related to e-commerce are provided. Examples of companies taking action to avoid such security breaches are also provided. Updated forecasts are presented for the volume of mobile commerce. Interesting new examples of companies using e-commerce to reduce costs, speed the flow of goods and services, increase accuracy, and improve customer service are offered. A discussion of both bartering and retargeting becoming e-commerce applications has been added. Information about the best smartphone applications for price comparison is provided. Data on the growth of electronic couponing and new couponing approaches has been added. Overall, the chapter provides two new tables and 43 new examples of organizations and individuals using e-commerce. In addition, the chapter has 66 new references.

Chapter 9, Enterprise Systems has been updated to present the latest developments in enterprise systems. A new table has been added showing the most popular ERP systems for large, medium, and small organizations. The table of the highest rated CRM systems has been updated to include the most recent ratings. The chapter mentions that retailers are using CRM systems to keep aware of what people are saying about their products and services on social networks. Also discussed is the connection between CRM systems and customer loyalty programs. A section has been added to cover Product Lifecycle Management as a type of enterprise system used to manage all the data associated with the product development, engineering design, production, support, and disposal of manufactured products. Two new examples of companies using PLM systems are presented. Overall, the chapter presents three new or updated tables and 44 new examples of organizations using transaction processing systems and enterprise systems to operate their business. In addition, the chapter has 54 new references.

Chapter 10, Information and Decision Support Systems includes more than 100 new or updated references, references, and material. The material on decision making, problem solving, optimization, and other decision making techniques has been updated with new references and examples. Companies such as Chevron are starting to use the latest technology like new tablet computers to help managers make better decisions by allowing them to connect to corporate computer systems and applications. In another example, American Airlines monitored its decision to use probability analysis to reduce inventory levels and shipping costs for airline maintenance equipment and in-flight service items. The value of this inventory can be over $1 billion a year on average. American Airlines used a decision making technique called decision tree analysis that diagrammed major decisions and possible outcomes from these decisions. Shermag, a Canadian furniture manufacturing company, used an optimization program to reduce raw materials costs, including wood, in its manufacturing operations. The optimization program, which used the C++ programming language and CPLEX optimization software, helped the company reduce total costs by more than 20 percent. The material on management information systems and decision support systems, including group support systems, has also been updated. Providence Washington Insurance Company used ReportNet from Cognos, an IBM company, to reduce the number of paper reports they produce and the associated costs. The new reporting system creates an executive dashboard that shows current data, graphs, and tables to help managers make better real-time decisions. Some companies, like Sprint, Levi Strauss, and Mattel, are using college students to help them with their marketing research. The schools are often paid for the work, and their participation can help students get a job after graduation. BMW, the German luxury car maker, performs marketing research using search engines to determine customer preferences and to target ads to people who might want to buy one of its cars. Many small businesses are effectively advertising their products and services using Internet sites like Groupon. Shopkick, Inc., makes smartphone applications that offer discounts to customers for entering a store. According to the company, the smartphone application has drawn about 750,000 customers into stores. Target, Best Buy, and other stores have used this service to their advantage. TurboRouter is a decision support system developed in Norway to reduce shipping costs and cut emissions of merchant ships. Operating a single ship can cost over $10,000 every day. TurboRouter schedules and manages the use of ships to transport oil and other products to locations around the world. Jeppesen, a supplier of charts and navigational products to hundreds of airlines and thousands of pilots, needed a flexible decision support system to monitor and control its operations. With constantly changing navigational charts and documents, Jeppesen found it difficult to ship accurate products in a timely fashion. As a result of its computerized decision support system, Jeppesen was able to reduce its late shipping percentage from 35 percent to almost 0 percent. Yum Brands, owner of Kentucky Fried Chicken (KFC), Taco Bell, and Pizza Hut, uses a group support teleconferencing system by Tanberg to let employees have virtual meetings and make group decisions, reducing travel time and costs. The Tandberg group support system uses high definition videos and group support software to help employees at distant locations collaborate and make group decisions.

Chapter 11, Knowledge Management and Specialized Information Systems has over 100 new or updated references, examples, and material on knowledge management and specialized business information systems, artificial intelligence, expert systems, multimedia, virtual reality, and many other specialized systems. For example, we have included new material and examples on augmented reality. Some luxury car manufacturers, for example, display dashboard information, such as speed and remaining fuel, on windshields. The application is used in some military aircraft and is often called heads-up display. Advent, a San Francisco company that develops investment applications for hedge funds and financial services companies, used a knowledge management system to help its employees locate and use critical information. In one study, workers with more knowledge management experience were able to benefit from knowledge management faster and to a greater extent than workers with less experience. Yum Brands connects its 1.6 million employees around the world to help them share and use knowledge. The vice president of global IT for Yum Brands believes that this type knowledge sharing can help employees "break out of silos and share know-how." This knowledge sharing approach uses an internal social network called iChing, an enterprise search facility developed by Coveco, an online learning system by Saba, and a high-definition videoconferencing system from Tandberg. Sonia Schulenburg, an ex-bodybuilder who also holds a doctorate in artificial intelligence, started a company called Level E Capital that uses artificial intelligence to pick and trade stocks. Her trading system makes as many as 1,000 trades a day, and her company often outperforms popular stock indexes, such as the FTSE 100. Honda Motors has developed a brain computer interface system that allows a person to complete certain operations, like bending a leg, with 90 percent accuracy. The new system uses a special helmet that can measure and transmit brain activity to a computer. The field of robotics has many applications, and research into these unique devices continues. The Robonaut, also called R2, is a human-like robot used on the International Space Station. The Porter Adventist Hospital in Denver, Colorado, uses a $1.2 million Da Vinci Surgical System to perform surgery on prostate cancer patients. The Lantek expert system can be used to cut and fabricate metal into finished products for the automotive, construction, and mining industries. The expert system can help reduce raw material waste and increase profits. The material on multimedia and virtual reality has been updated. Geico uses animation in some of its TV ads. Animation Internet sites like Xtranormal and GoAnimate can help individuals and corporations develop these types of animations. Pixar uses sophisticated animation software to create dazzling 3D movies. The exact process can be seen at Pixar's Web site. Barbara Rothbaum, the director of the Trauma and Recovery Program at Emory University School of Medicine and cofounder of Virtually Better, uses an immersive virtual reality system to help in the treatment of anxiety disorders. Boeing used virtual reality to help design and manufacture airplane parts and new planes, including the 787 Dreamliner. IBM researchers, along with researchers from the Institute of Bioengineering and Nanotechnology based in Singapore, have developed a nanoparticle 50,000 times smaller than the thickness of a human hair. If successful, the nanoparticle could destroy bacteria, which threatens human health and life. Today, more hospitals and healthcare facilities are using the Internet to connect doctors to patients in distant locations. In one case, a physician used Internet video to check on the treatment of a stroke patient located 15 miles away to make sure the drugs being used weren't increasing the chance of bleeding in the brain. After reviewing CT scans and the behavior of the patient, the doctor made specific drug recommendations.

Chapter 12, Systems Development: Investigation and Analysis includes 80 new references, examples, and material on systems development. We discuss that systems development expenditures are expected to soar in the next few years, according to a CIO economic impact survey. IS departments and systems developers will concentrate on creating more mobile applications for their businesses and organizations. Revenues from mobile applications from all sources are expected to be over $15 billion annually according to Gartner, Inc. An individual systems designer, for example, created an application called Word Lens that uses optical character recognition to read text by holding a smartphone or other mobile device up to restaurant menus, books, signs, and other text and taking a picture of the text. In a few seconds, the application can translate the camera image from one language to another, such from German to English. We also stress that companies seek members of the systems development team with training in mobile devices, Internet applications, and social networking. Today, companies are setting up their own internal app stores. We also discuss that mobile devices can pose serious security risks to businesses and nonprofit organizations that allow their workers and managers to use these devices at work. This chapter includes many new examples of the use of systems development. Hallmark, for example, successfully used systems development to create a new Web site to advertise its greeting cards and related products. The new Web site was 300 percent faster than its old one and resulted in increased sales. Federal, state, and local tax breaks have resulted in new systems development efforts. A depreciation tax benefit enacted by the U.S. Congress in 2010, for example, has caused some companies to purchase hardware and related computer equipment in 2011. We include new examples of how specific systems development tools can be used to achieve organizational goals. The CIO of Hewlett-Packard has slashed project completion time to just six months using rapid application development (RAD). The FBI used agile development to save time and reduce the number of people needed for a project called Sentinel. Agile development was able to reduce the number of people involved from about 200 to 50. We introduce a new approach to systems development called Scrum, which stresses agile, incremental development. The use of outsourcing is also stressed. Steel Technologies outsourced many of its computer operations and infrastructure to ERP Suites instead of spending almost $1 million on computers, storage devices, and power supplies. This approach is often called Infrastructure as a Service (IaaS.) Tata, a large outsourcing firm based in India, is now targeting smaller businesses by using cloud computing and a new service is called iON. With cloud computing, smaller businesses can download software that allows Tata to manage their client's programs and processes from a remote location. We have a new section on mobile application development. The sections on systems investigation and analysis have also been updated. A number of hospitals, for example, are scrambling to meet federally mandated systems analysis deadlines. A billion-dollar program called SBInet failed to build a high-tech fence on the border of Arizona and Mexico. As a result, the Department of Homeland Security halted the project. New York awarded a $63 million systems development project to an outside firm. When the project ran into large cost overruns, the city sued the outside firm for $600 million. For one company implementing a new customer relationship management system, the cost of the systems development effort was twice what was projected.

Chapter 13, Systems Development: Design, Implementation, Maintenance, and Review includes more than 80 new or updated references, examples, and material on systems design and implementation. We stress the importance of security in design and implementation. Designing security controls and procedures into the use of smartphones and other mobile devices can be a challenge for many organizations. Employees want to be able to do their work using their smartphones, tablet computers, and other mobile devices at work and while traveling. Corporations want to make sure that the usage of these devices is secure. Stolen laptop computers and other mobile devices have been a major cause of identity theft for individuals. Device theft has also resulted in the loss of corporate secrets and procedures. According to one survey, over 50 percent of respondents indicated that potential security problems with mobile devices have prevented them from using the devices to a greater extent to perform corporate work. To combat these problems, systems developers are installing software to encrypt the data on mobile devices and require IDs and passwords to gain access to them. Companies are also protecting their computers with software and firewalls to block unauthorized people from gaining access to corporate data and programs using corporate-issued mobile devices. The material on disaster planning and recovery has also been updated. The earthquake, tsunami, and resulting problems with several Japans nuclear plants were devastating, but technology helped people stay in touch and deal with the crisis. Keeping its Ginza store open, Apple Computer was able to provide critical emergency communications to employees, customers, and others in the computer store using e-mail, Facebook, Twitter, and other Internet sites. Some people actually used the store as an emergency shelter. Alabama's Troy University upgraded its disaster recovery system to help prevent and recover from a potential disaster, including hurricanes. We also mention that many of today's organizations have compliance departments to make sure the IS department is adhering to its systems controls along with all local, state, and federal laws and regulations. The section on environmental design has been fully updated. Facebook, with the help of its business partners, developed a data center in rural Prineville, Oregon to be more energy efficient by using the latest computer chips and servers that are lighter and easier to maintain. The servers were also less expensive to purchase than the previous ones. Today, solar panels, gas turbines, and fuel cells are making IS departments and data centers more energy efficient. One Japanese firm is designing a new computer chip that could potentially reduce power consumption by 50 percent and almost double battery life for mobile devices, including laptop computers, tablet computers, and smartphones. In the section on systems design, we emphasize the importance of having a good contract. Organizations that use the cloud computing approach need to take special precautions in signing contracts with cloud computing providers, examining how privacy is protected, how the organization can comply with various laws and regulations when using cloud computing, where the cloud computing servers and computers are located in the world, how discovery is handled if there is a lawsuit, and the security of the data stored on cloud computers. The section on systems implementation has been revised with new material and examples. According to a survey of CIOs, not being able to implement new or modified systems was the most important concern for today's IS departments. We stress that implementation is not complete when the software code is finished. Some companies, such as Secure by Design, offer automated software installation and updating services. Using these tools, companies can easily install and update software designed to run on Windows, Linux, and other operating systems. The challenge of older legacy systems is also covered. Like many organizations, Crescent Healthcare, which provides drug treatments for cancer and other life-threatening diseases, has a large investment in older legacy applications. It is a challenge knowing which legacy systems to keep and which ones to replace with newer Internet or cloud applications. British Airways had about 60 legacy systems that were becoming increasingly difficult to update and maintain. The company hopes replacing these older legacy systems will make their applications easier to update and maintain. The material on systems review has also been updated. After reviewing its Virtual Case File System, which some believe was over budget and didn't perform as expected, the FBI initiated a new systems development effort to create Sentinel, hardware and software used to store and analyze important information on its many cases. The events that can trigger systems review can be highly complex or as simple as a broken cable, as was the case when a 75-year old woman broke a fiber-optic cable with her shovel when digging for scrap metal in Armenia. The U.S. State Department reviewed its computer system used to run its annual lottery for visas and discovered an internal programming error. The lottery was supposed to pick 15,000 people for visas at random from the list of 15 million people who applied. As a result, the State Department was forced to fix the programming error and rerun the lottery for visas. The names of people that received a visa under the flawed lottery were put back into the lottery for a second chance.

Chapter 14, The Personal and Social Impact of Computers has been updated to present the most current issues and developments associated with the personal and social impact of computers. The use of nonintegrated information systems that make it difficult to collaborate and share information is discussed as a cause of computer waste and illustrated by Casa Oliveria, a Venezuelan importer. New data from the Government Accounting Office is presented that concludes that 37 of its sample of 810 investments in information systems (amounting to $1.2 billion) were potential duplicates. Procter & Gamble is used as an example of a company whose employees waste valuable information system resources by playing computer games, sending personal e-mail, or browsing the Internet. Many other new examples of computer-related waste are presented. Emerson College, Colorado, Her Majesty's Revenue and Customs organization, and the California Department of Corrections are used to illustrate the importance of careful implementation of new policies and procedures. The role of the Internet Crime Computer Center, an alliance between the White Collar Crime Center and the Federal Bureau of Investigation is discussed. Many new examples of computer crime, fraud, and hacking are covered. The threat of cyberterrorism is discussed and several recent examples of cyberterrorism are provided. The information on Internet gambling and the legality of using and running such Web sites is updated and illustrated with new examples. New examples of using the computer to fight crime are covered. The chapter introduces JusticeXchange, a Web-based data sharing system that provides law enforcement officials with information about current and former offenders held in participating jails. The chapter also discusses new security threats, particularly from employees who use smartphones such as the iPhone and Google's Android on BlackBerry or tablet computers. New data on the volume of software purchased worldwide (about $95 billion) and the volume of software pirated ($59 billion) is presented. New examples of major lawsuits over the alleged illegal downloading of software and movies as well as alleged patent infringement are provided. The Stop Online Piracy Act (SOPA) and Preventing Real Online Threats to Economic Creativity and Theft of Intellectual Property Act (Protect Intellectual Property Act or PIPA) are discussed. Many new examples of computer scams are covered. Separation of duties, a fundamental concept of good internal controls, is introduced and illustrated with an example from the Medicaid Operations Division of the Utah Department of Health. Current highly rated antivirus and Web filtering software is identified. Computer privacy and Internet libel are discussed and several current cases are covered. A model corporate privacy policy from the Better Business Bureau is provided. Information about the potential health effects of heavy computer usage and how to reduce health risks is provided. Overall, the chapter has 48 new examples of organizations dealing with the personal and social impact of computers and 80 new references.

WHAT WE HAVE RETAINED FROM THE TENTH EDITION

The eleventh edition builds on what has worked well in the past; it retains the focus on IS principles and strives to be the most current text on the market.

- **Overall principle.** This book continues to stress a single all-encompassing theme: The right information, if it is delivered to the right person, in the right fashion, and at the right time, can improve and ensure organizational effectiveness and efficiency.
- **Information systems principles.** Information systems principles summarize key concepts that every student should know. These principles are highlighted at the start of each chapter and covered thoroughly in the text.
- **Global perspective.** We stress the global aspects of information systems as a major theme.
- **Learning objectives linked to principles.** Carefully crafted learning objectives are included with every chapter. The learning objectives are linked to the information systems principles and reflect what a student should be able to accomplish after completing a chapter.
- **Opening vignettes emphasize international aspects.** All of the chapter-opening vignettes raise actual issues from foreign-based or multinational companies.
- **Why Learn About features.** Each chapter has a "Why Learn About" section at the beginning of the chapter to pique student interest. The section

sets the stage for students by briefly describing the importance of the chapter's material to the students—whatever their chosen field.

- **Information Systems @ Work special interest boxes.** Each chapter has an entirely new "Information Systems @ Work" box that shows how information systems are used in a variety of business career areas.
- **Ethical and Societal Issues special interest boxes.** Each chapter includes an "Ethical and Societal Issues" box that presents a timely look at the ethical challenges and the societal impact of information systems.
- **Current examples, boxes, cases, and references.** As in each edition, we take great pride in presenting the most recent examples, boxes, cases, and references throughout the text. Some of these were developed at the last possible moment, literally weeks before the book went into publication. Information on new hardware and software, the latest operating systems, mobile commerce, the Internet, electronic commerce, ethical and societal issues, and many other current developments can be found throughout the text. Our adopters have come to expect the best and most recent material. We have done everything we can to meet or exceed these expectations.
- **Summary linked to principles.** Each chapter includes a detailed summary, with each section of the summary tied to an associated information systems principle.
- **Self-assessment tests.** This popular feature helps students review and test their understanding of key chapter concepts.
- **Career exercises.** End-of-chapter career exercises ask students to research how a topic discussed in the chapter relates to a business area of their choice. Students are encouraged to use the Internet, the college library, or interviews to collect information about business careers.
- **End-of-chapter cases.** Two new end-of-chapter cases provide students with an opportunity to apply the principles covered to real-world problems from actual organizations. The cases can be assigned as individual homework exercises or serve as a basis for class discussion.
- **Integrated, comprehensive, Web case.** The Altitude Online case at the end of each chapter provides an integrated and comprehensive case that runs throughout the text. The cases follow the activities of two individuals employed at the fictitious Altitude Online consulting firm as they are challenged to complete various IS-related projects. The cases provide a realistic fictional work environment in which students may imagine themselves in the role of systems analyst. Information systems problems are addressed using the state-of-the-art techniques discussed in the chapters.

STUDENT RESOURCES

PowerPoint Slides

Direct access is offered to the book's PowerPoint presentations that cover the key points from each chapter. These presentations are a useful study tool.

Classic Cases

A frequent request from adopters is that they'd like a broader selection of cases to choose from. To meet this need, a set of over 200 cases from the sixth, seventh, eighth, ninth, and tenth editions of the text are included here. These are the authors' choices of the "best cases" from these editions and span a broad range of companies and industries.

Links to Useful Web Sites

Chapters in *Principles of Information Systems, Eleventh Edition* reference many interesting Web sites. This resource takes you to links you can follow directly to the home pages of those sites so that you can explore them. There are additional links to Web sites that the authors think you would be interested in checking out.

Hands-On Activities

Use these hands-on activities to test your comprehension of IS topics and enhance your skills using Microsoft® Office applications and the Internet. Using these links, you can access three critical-thinking exercises per chapter; each activity asks you to work with an Office tool or do some research on the Internet.

Glossary of Key Terms

The glossary of key terms from the text is available to search.

Online Readings

This feature allows you to access a computer database that contains articles relating to hot topics in information systems.

INSTRUCTOR RESOURCES

The teaching tools that accompany this text offer many options for enhancing a course. And, as always, we are committed to providing one of the best teaching resource packages available in this market.

Instructor's Manual

An all-new *Instructor's Manual* provides valuable chapter overviews; highlights key principles and critical concepts; offers sample syllabi, learning objectives, and discussion topics; and features possible essay topics, further readings and cases, and solutions to all of the end-of-chapter questions and problems, as well as suggestions for conducting the team activities. Additional end-of-chapter questions are also included. As always, we are committed to providing the best teaching resource packages available in this market.

Sample Syllabus

A sample syllabus for both a quarter and semester-length course are provided with sample course outlines to make planning your course that much easier.

Solutions

Solutions to all end-of-chapter material are provided in a separate document for your convenience.

Test Bank and Test Generator

ExamView® is a powerful objective-based test generator that enables instructors to create paper-, LAN- or Web-based tests from test banks designed specifically for their Course Technology text. Instructors can utilize the ultra-efficient QuickTest Wizard to create tests in less than five minutes by taking advantage of Course Technology's question banks or customizing their own exams from scratch. Page references for all questions are provided so you can cross-reference test results with the book.

PowerPoint Presentations

A set of impressive Microsoft PowerPoint slides is available for each chapter. These slides are included to serve as a teaching aid for classroom presentation, to make available to students on the network for chapter review, or to be printed for classroom distribution. Our presentations help students focus on the main topics of each chapter, take better notes, and prepare for examinations. Instructors can also add their own slides for additional topics they introduce to the class.

Figure Files

Figure files allow instructors to create their own presentations using figures taken directly from the text.

ACKNOWLEDGMENTS

A book of this scope and undertaking requires a strong team effort. We would like to thank all of our fellow teammates at Course Technology for their dedication and hard work. We would like to thank Charles McCormick, our Sr. Acquisitions Editor, for his overall leadership and guidance on this effort. Special thanks to Kate Mason, our Product Manager. We would also like to thank Aimee Poirier for filling the Product Manager role during Kate's absence at the start of the project. Our appreciation goes out to all the many people who worked behind the scenes to bring this effort to fruition including Abigail Reip, our photo researcher. We would like to acknowledge and thank Lisa Ruffolo, our development editor, who deserves special recognition for her tireless effort and help in all stages of this project. Thanks also to Arul Joseph Raj and Jennifer Feltri-George, our Content Project Managers, who shepherded the book through the production process.

We are grateful to the salesforce at Cengage Learning whose efforts make this all possible. You helped to get valuable feedback from current and future adopters. As Cengage Learning product users, we know how important you are.

We would especially like to thank Efrem Mallach for his excellent help in writing the vignettes, IS @ Work and Ethical & Societal Issues boxes, and cases for this edition.

Ralph Stair would like to thank the Department of Management and its faculty members in the College of Business Administration at Florida State University for their support and encouragement. He would also like to thank his family, Lila and Leslie, for their support.

George Reynolds would like to thank his wife, Ginnie, for her patience and support in this major project.

To Our Previous Adopters and Potential New Users

We sincerely appreciate our loyal adopters of the previous editions and welcome new users of *Principles of Information Systems, Eleventh Edition.* As in the past, we truly value your needs and feedback. We can only hope the eleventh edition continues to meet your high expectations.

OUR COMMITMENT

We are committed to listening to our adopters and readers and to developing creative solutions to meet their needs. The field of IS continually evolves, and we strongly encourage your participation in helping us provide the freshest, most relevant information possible.

We welcome your input and feedback. If you have any questions or comments regarding *Principles of Information Systems, Eleventh Edition,* please contact us through Course Technology or your local representative.

PART 1 | PART 2 | PART 3 | PART 4 | PART 5

Overview

CHAPTERS

1 An Introduction to Information Systems

Principles	Learning Objectives
• The value of information is directly linked to how it helps decision makers achieve the organization's goals.	• Discuss why it is important to study and understand information systems. • Distinguish data from information and describe the characteristics used to evaluate the quality of data.
• Computers and information systems help make it possible for organizations to improve the way they conduct business.	• Name the components of an information system and describe several system characteristics.
• Knowing the potential impact of information systems and having the ability to put this knowledge to work can result in a successful personal career and in organizations that reach their goals.	• List the components of a computer-based information system. • Identify the basic types of business information systems and discuss who uses them, how they are used, and what kinds of benefits they deliver.
• System users, business managers, and information systems professionals must work together to build a successful information system.	• Identify the major steps of the systems development process and state the goal of each.
• Information systems must be applied thoughtfully and carefully so that society, businesses, and industries can reap their enormous benefits.	• Describe some of the threats that information systems and the Internet can pose to security and privacy. • Discuss the expanding role and benefits of information systems in business and industry.

Information Systems in the Global Economy
CASTILLA-LA MANCHA, SPAIN

Using Cloud Computing to Provide Public Services and Transform Education

The region of Castilla-La Mancha in central Spain covers over 30,000 square miles, or almost 80,000 square kilometers. With a population of over two million people spread over such a large region, Castilla-La Mancha has the lowest population density of any Spanish region. More than half of its 919 communities have fewer than 500 inhabitants. As Pedro-Jesus Rodriguez Gonzalez, head of Information Technology (IT) and Internet for the regional government of Castilla-La Mancha, puts it, "This environment presents one of Spain's most challenging demographics for delivering public services. Although much of the population lives in five major cities, a significant portion of its citizens are widely dispersed."

The government of Castilla-La Mancha has used computers to help it deal with the challenges of delivering public services such as access to social benefits for many years, although limited by the finite resources available to any government agency. Recently, the regional government modernized its technological infrastructure to save money and improve its responsiveness to citizens. It adopted the approach of cloud computing: central applications and data accessed over the Internet, much as people access Web pages. By using a cloud-computing approach, Castilla-La Mancha could centralize its data centers, reducing 18 main sites and 30 smaller facilities to two centers. Direct savings as a result of centralizing the data centers are more than half a million dollars.

To develop its new infrastructure, Castilla-La Mancha chose the Vblock system from the Virtual Computing Environment (VCE) Company, LLC. VCE is a joint venture of networking company Cisco and storage supplier EMC, with additional investment by VMware and Intel. Using the combined strengths of their sponsors, VCE can provide solutions that handle all aspects of creating a cloud platform while avoiding the need for users to deal with multiple suppliers.

The first application Castilla-La Mancha developed to take advantage of the new cloud system was Papás 2.0 (Parents 2.0), a program that enables collaboration among parents, teachers, and pupils and facilitates daily work in twenty-first-century classrooms equipped with information systems. Papás 2.0 was introduced to users in November 2010. "Papás 2.0 Virtual Classroom [gives] teachers and pupils the opportunity to incorporate an online collaborative working environment into the school's daily dynamics," says Tomás Hervás, general secretary of the Council of Education, Science, and Culture. Teachers can follow pupils, set tasks, and send messages to parents; families can access data on their children's performance through an Internet connection. Gonzalez adds, "IT is no longer just a subject, but a main part of the student's daily routine. The students are being educated with the tools they will be using in their future workplace." When fully rolled out, Papás 2.0 will support 345,000 pupils along with their families and teachers.

The new infrastructure also provides considerable cost savings by consolidating what were previously separate systems into a shared data center. This data center will eventually replace about 130 server computers, reducing energy consumption along with space and cooling requirements. Castilla-La Mancha forecasts savings of 20 percent within the first year, with savings continuing to grow as more of the infrastructure is used to replace older, outmoded computers.

Agustina Piedrabuena, Castilla-La Mancha's chief information officer, summarizes: "The project has not only helped us to consolidate and simplify our data center, it also allows us to be completely independent of which department uses the service, where the application is hosted, or what resources it consumes; we are simply automating the provision of applications by means of the cloud."

As you read this chapter, consider the following:

- How does the Castilla-La Mancha information system depend on the various components of any computer-based information system: hardware, software, databases, telecommunications, people, and procedures?
- How do computer-based information systems, such as Papás 2.0 for education, help Castilla-La Mancha deliver services to the people of the region?

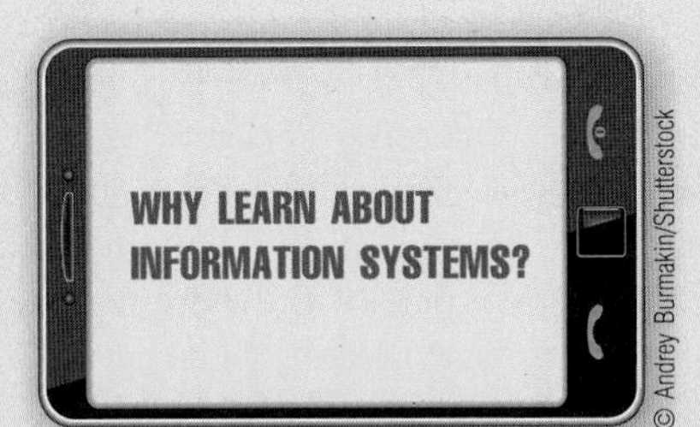

© Andrey Burmakin/Shutterstock

Information systems are used in almost every imaginable profession. Entrepreneurs and small business owners use information systems to reach customers around the world. Sales representatives use information systems to advertise products, communicate with customers, and analyze sales trends. Managers use them to make multimillion-dollar decisions, such as whether to build a manufacturing plant or research a cancer drug. Financial advisors use information systems to advise their clients to help them save for retirement or their children's education. From a small music store to huge multinational companies, businesses of all sizes could not survive without information systems to perform accounting, marketing, management, finance, and similar operations. Regardless of your college major or chosen career, information systems are indispensable tools to help you achieve your career goals. Learning about information systems can help you land your first job, earn promotions, and advance your career.

This chapter presents an overview of information systems, with each section getting full treatment in subsequent chapters. We start by exploring the basics of information systems.

information system (IS): A set of interrelated components that collect, manipulate, store, and disseminate data and information and provide a feedback mechanism to meet an objective.

People and organizations use information every day. The components that are used are often called an information system. An **information system (IS)** is a set of interrelated components that collect, manipulate, store, and disseminate data and information and provide a feedback mechanism to meet an objective. It is the feedback mechanism that helps organizations achieve their goals, such as increasing profits or improving customer service. This book emphasizes the benefits of an information system, including speed, accuracy, increased revenues, and reduced costs. For example, Groupon, an Internet company that offers online coupons for local stores and businesses to customers, uses its information system to generate hundreds of millions of dollars annually.[1]

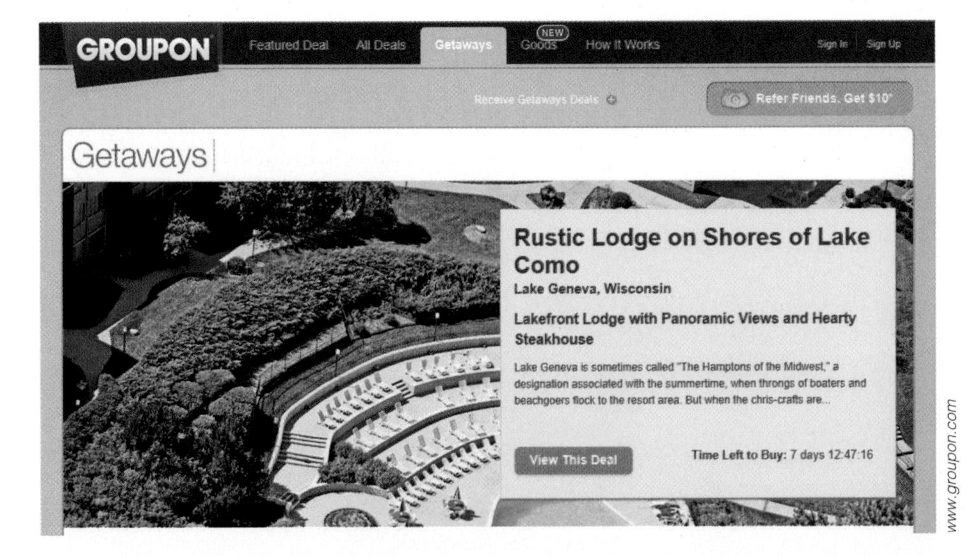

www.groupon.com

Information systems are everywhere. A business offers a discount on a product or service, and an information system sends the offer as a digital coupon to consumers in the area. The system tracks the number of offers accepted. If enough people sign up for the offer, they can use the coupon to apply the discount. Consumers enjoy the savings, while the information system finds enough customers to make the discount worthwhile to the business.

We interact with information systems every day, both personally and professionally. We use automated teller machines at banks, access information over the Internet, select information from kiosks with touch screens, and scan the barcodes on our purchases at self-checkout lanes. Knowing the potential of information systems and putting this knowledge to work can help individuals enjoy a successful career and help organizations reach their goals.

Today, we live in an information economy. Information itself has value, and commerce often involves the exchange of information rather than tangible goods. Systems based on computers are increasingly being used to create, store, and transfer information. Using information systems, investors make multimillion-dollar decisions, financial institutions transfer billions of dollars around the world electronically, and manufacturers order supplies and distribute goods faster than ever before. Computers and information systems will continue to change businesses and the way we live. To prepare for these innovations, you need to be familiar with fundamental information concepts.

INFORMATION CONCEPTS

Information is a central concept of this book. The term is used in the title of the book, in this section, and in almost every chapter. To be an effective manager in any area of business, you need to understand that information is one of an organization's most valuable resources. This term, however, is often confused with *data*.

Data, Information, and Knowledge

data: Raw facts, such as an employee number, total hours worked in a week, inventory part numbers, or sales orders.

information: A collection of facts organized and processed so that they have additional value beyond the value of the individual facts.

Data consists of raw facts, such as an employee number, total hours worked in a week, inventory part numbers, or sales orders. As shown in Table 1.1, several types of data can represent these facts. When facts are arranged in a meaningful manner, they become information. **Information** is a collection of facts organized and processed so that they have additional value beyond the value of the individual facts. For example, sales managers might find that knowing the total monthly sales suits their purpose more (i.e., is more valuable) than knowing the number of sales for each sales representative. Providing information to customers can also help companies increase revenues and profits. Many universities place course information and content on the Internet. Using the Open Course Ware program, the Massachusetts Institute of Technology (MIT) places class notes and contents on the Internet for many of its courses.[2] Some countries, however, have tried to censor or control what information is available to their citizens, especially through the Internet and social media.[3]

Data represents real-world things. Hospitals and health care organizations, for example, maintain patient medical data, which represents actual patients with specific health situations. However, data—raw facts—has little value beyond its existence. Today, hospitals and other health care organizations are investing millions of dollars into developing medical records programs to store and use the vast amount of medical data that is generated each year.

TABLE **1.1** Types of data

Data	Represented by
Alphanumeric data	Numbers, letters, and other characters
Audio data	Sounds, noises, or tones
Image data	Graphic images and pictures
Video data	Moving images or pictures

Medical records systems can be used to generate critical health-related information, which in turn can save money and lives. In addition, integrating information from different sources is an important capability for most travel organizations. According to Kathryn Akerman, CIO of Hurley Travel Experts, "We've had clients who went out, tried to do it themselves and came back to us because they realize that travel booking is complicated and their time is a valuable asset. They're looking to us to put those pieces together for them rather than going to different sites to put together their own itinerary."[4]

Here is another example of the difference between data and information. Consider data as pieces of railroad track in a model railroad kit. Each piece of track has limited inherent value as a single object. However, if you define a relationship among the pieces of the track, they gain value. By arranging the pieces in a certain way, a railroad layout begins to emerge (see Figure 1.1a, top). Data and information work the same way. Rules and relationships can be set up to organize data into useful, valuable information.

The type of information created depends on the relationships defined among existing data. For example, you could rearrange the pieces of track to form different layouts. Adding new or different data means you can redefine relationships and create new information. For instance, adding new pieces to the track can greatly increase the value—in this case, variety and fun—of the final product. You can now create a more elaborate railroad layout (see Figure 1.1b, bottom). Likewise, a sales manager could add specific product data to sales data to create monthly sales information organized by product line. The manager could use this information to determine which product lines are the most popular and profitable.

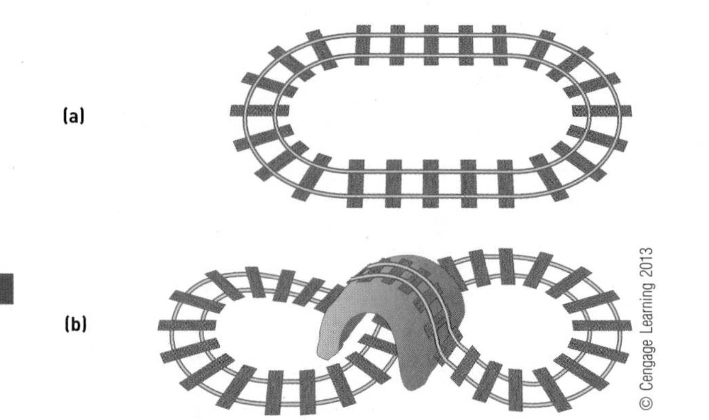

FIGURE 1.1

Defining and organizing relationships among data creates information

process: A set of logically related tasks performed to achieve a defined outcome.

knowledge: The awareness and understanding of a set of information and the ways that information can be made useful to support a specific task or reach a decision.

Turning data into information is a **process**, or a set of logically related tasks performed to achieve a defined outcome. The process of defining relationships among data to create useful information requires knowledge. **Knowledge** is the awareness and understanding of a set of information and the ways that information can be made useful to support a specific task or reach a decision.[5] Having knowledge means understanding relationships in information. Part of the knowledge you need to build a railroad layout, for instance, is the understanding of how much space you have for the layout, how many trains will run on the track, and how fast they will travel. Selecting or rejecting facts according to their relevancy to particular tasks is based on the knowledge used in the process of converting data into information. Therefore, you can also think of information as data made more useful through the application of knowledge. *Knowledge workers (KWs)* are people who create, use, and disseminate knowledge and are usually professionals in science, engineering, business, and other areas.[6] A *knowledge management system (KMS)* is an organized collection of people, procedures, software, databases, and devices used to create, store, and use the organization's knowledge and experience.[7]

In some cases, people organize or process data mentally or manually. In other cases, they use a computer. Where the data comes from or how it

FIGURE 1.2
Process of transforming data into information

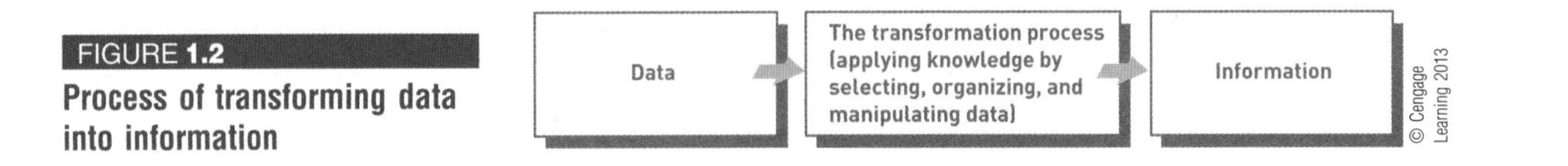

is processed is less important than whether the data is transformed into results that are useful and valuable. This transformation process is shown in Figure 1.2.

The Characteristics of Valuable Information

To be valuable to managers and decision makers, information should have the characteristics described in Table 1.2. These characteristics make the information more valuable to an organization. In contrast, if an organization's information is not accurate or complete, people can make poor decisions, costing thousands or even millions of dollars. If an inaccurate forecast of future demand indicates that sales will be very high when the opposite is true, an organization can invest millions of dollars in a new plant that is not needed. Furthermore, if information is not relevant, not delivered to decision makers in a timely fashion, or too complex to understand, it can be of little value to the organization.

TABLE **1.2** Characteristics of valuable information

Characteristics	Definitions
Accessible	Information should be easily accessible by authorized users so they can obtain it in the right format and at the right time to meet their needs.
Accurate	Accurate information is error free. In some cases, inaccurate information is generated because inaccurate data is fed into the transformation process. This is commonly called garbage in, garbage out (GIGO).
Complete	Complete information contains all the important facts. For example, an investment report that does not include all important costs is not complete.
Economical	Information should also be relatively economical to produce. Decision makers must always balance the value of information with the cost of producing it.
Flexible	Flexible information can be used for a variety of purposes. For example, information on how much inventory is on hand for a particular part can be used by a sales representative in closing a sale, by a production manager to determine whether more inventory is needed, and by a financial executive to determine the total value the company has invested in inventory.
Relevant	Relevant information is important to the decision maker. Information showing that lumber prices might drop might not be relevant to a computer chip manufacturer.
Reliable	Reliable information can be trusted by users. In many cases, the reliability of the information depends on the reliability of the data-collection method. In other instances, reliability depends on the source of the information. A rumor from an unknown source that oil prices might go up might not be reliable.
Secure	Information should be secure from access by unauthorized users.
Simple	Information should be simple, not complex. Sophisticated and detailed information might not be needed. In fact, too much information can cause information overload, whereby a decision maker has too much information and is unable to determine what is really important.
Timely	Timely information is delivered when it is needed. Knowing last week's weather conditions will not help when trying to decide what coat to wear today.
Verifiable	Information should be verifiable. This means that you can check it to make sure it is correct, perhaps by checking many sources for the same information.

Depending on the type of data you need, some of these attributes are more important than others. For example, having timely information is a key for many organizations. FedEx, for example, can sort more than 3 million packages every day because of timely information on each package.[8] To get timely information, FedEx often scans a package more than 10 times from its origin to its final destination. Verifiability and completeness are critical for data used in accounting to manage company assets such as cash, inventory, and equipment.

The Value of Information

The value of information is directly linked to how it helps decision makers achieve their organization's goals. Valuable information can help people in their organizations perform tasks more efficiently and effectively. Consider a market forecast that predicts a high demand for a new product. If you use this information to develop the new product and your company makes an additional profit of $10,000, the value of this information to the company is $10,000 minus the cost of the information. Valuable information can also help managers decide whether to invest in additional information systems and technology. A new computerized ordering system might cost $30,000 but generate an additional $50,000 in sales. The *value added* by the new system is the additional revenue from the increased sales of $20,000. Most corporations have cost reduction as a primary goal. Using information systems, some manufacturing companies have slashed inventory costs by millions of dollars. The value of information can also be measured by what people or organizations are willing to pay for it. One company, for example, offered a $3-million prize for the individual or group that could most accurately predict when patients go to hospitals for medical procedures and care.[9]

SYSTEM CONCEPTS

system: A set of elements or components that interact to accomplish goals.

Like information, another central concept of this book is that of a system. A **system** is a set of elements or components that interact to accomplish goals. Systems have inputs, processing mechanisms, outputs, and feedback (see Figure 1.3). For example, consider an automatic car wash. Tangible *inputs* for the process are a dirty car, water, and various cleaning ingredients. Time, energy, skill, and knowledge also serve as inputs to the system because they are needed to operate it. Skill is the ability to successfully operate the liquid sprayer, foaming brush, and air dryer devices. Knowledge is used to define the steps in the car wash operation and the order in which the steps are executed.

FIGURE **1.3**

Components of a system

A system's four components consist of input, processing, output, and feedback.

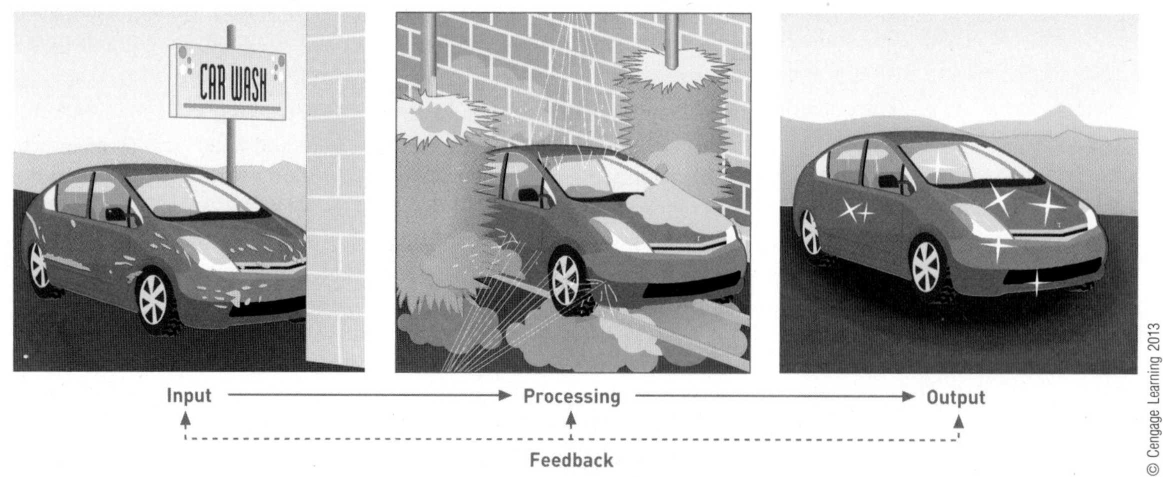

The *processing mechanisms* consist of first selecting which cleaning option you want (wash only, wash with wax, wash with wax and hand dry, etc.) and communicating that to the operator of the car wash. A *feedback mechanism* is your assessment of how clean the car is. Liquid sprayers shoot clear water, liquid soap, or car wax depending on where your car is in the process and which options you selected. The *output* is a clean car. As in all systems, independent elements or components (the liquid sprayer, foaming brush, and air dryer) interact to create a clean car.

System Performance and Standards

efficiency: A measure of what is produced divided by what is consumed.

System performance can be measured in various ways. **Efficiency** is a measure of what is produced divided by what is consumed. It can range from 0 to 100 percent.[10] For example, the efficiency of a motor is the energy produced (in terms of work done) divided by the energy consumed (in terms of electricity or fuel). Some motors have an efficiency of 50 percent or less because of the energy lost to friction and heat generation.

Efficiency is a relative term used to compare systems. For example, a hybrid gasoline engine for an automobile or truck can be more efficient than a traditional gasoline engine because, for the equivalent amount of fuel consumed, the hybrid engine travels more miles and gets better gas mileage. Many organizations can reduce their energy usage by investing in more energy-efficient computer systems.

effectiveness: A measure of the extent to which a system achieves its goals; it can be computed by dividing the goals actually achieved by the total of the stated goals.

Effectiveness is a measure of the extent to which a system achieves its goals. It can be computed by dividing the goals actually achieved by the total of the stated goals. For example, a company might want to achieve a net profit of $100 million for the year using a new information system. Actual profits, however, might only be $85 million for the year. In this case, the effectiveness is 85 percent (85/100 = 85%). Of course, companies measure effectiveness using different measures.

system performance standard: A specific objective of the system.

Evaluating system performance also calls for using performance standards. A **system performance standard** is a specific objective of the system. For example, a system performance standard for a marketing campaign might be to have each sales representative sell $100,000 of a certain type of product each year (see Figure 1.4a). A system performance standard for a manufacturing process might be to provide no more than 1 percent defective parts (see Figure 1.4b). After standards are established, system performance is measured and compared with the standard. Variances from the standard are determinants of system performance.

WHAT IS AN INFORMATION SYSTEM?

As mentioned previously, an information system (IS) is a set of interrelated elements or components that collect (input), manipulate (process), store, and disseminate (output) data and information and that provide a corrective reaction (feedback mechanism) to meet an objective (see Figure 1.5). The feedback mechanism is the component that helps organizations achieve their goals, such as increasing profits or improving customer service.

Input, Processing, Output, Feedback

Input

input: The activity of gathering and capturing raw data.

In information systems, **input** is the activity of gathering and capturing raw data. In producing paychecks, for example, the number of hours every employee works must be collected before paychecks can be calculated or printed. In a university grading system, instructors must submit student grades before a summary of grades can be compiled and sent to students.

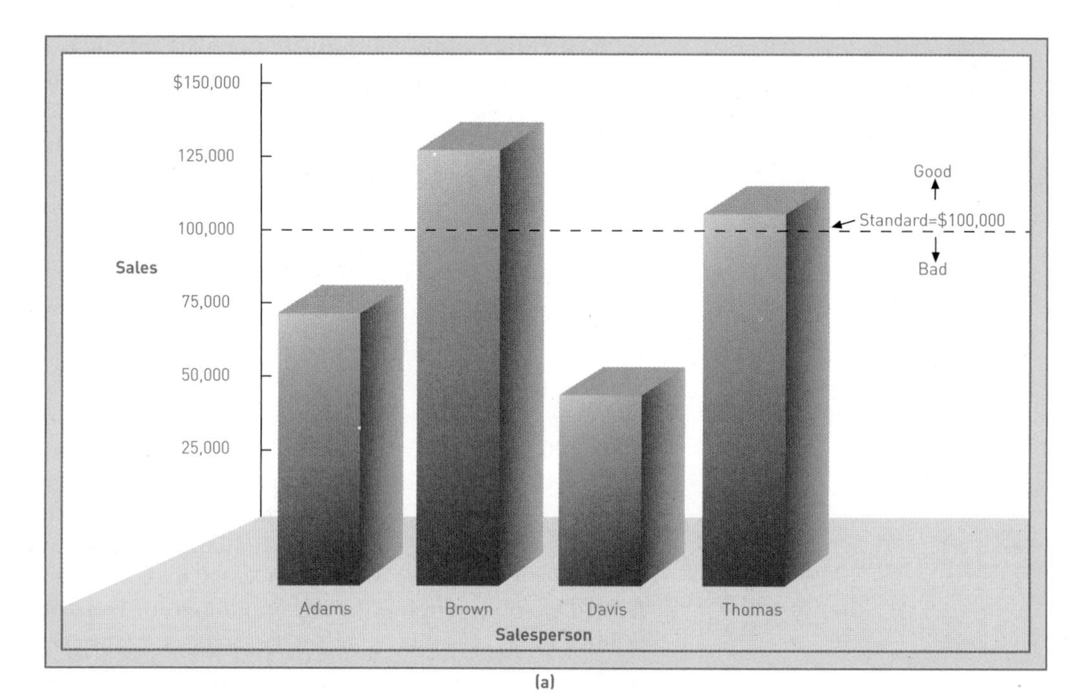

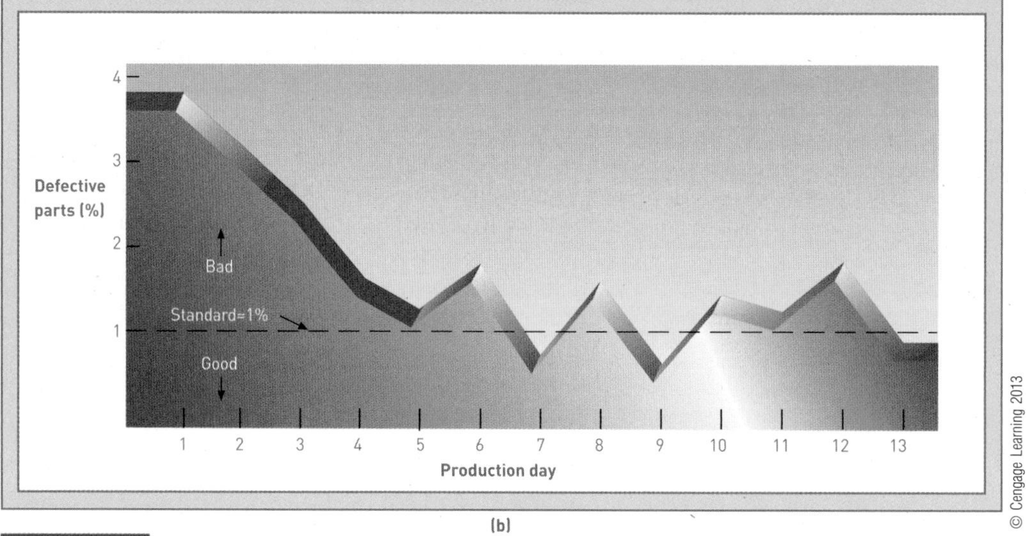

FIGURE 1.4
System performance standards

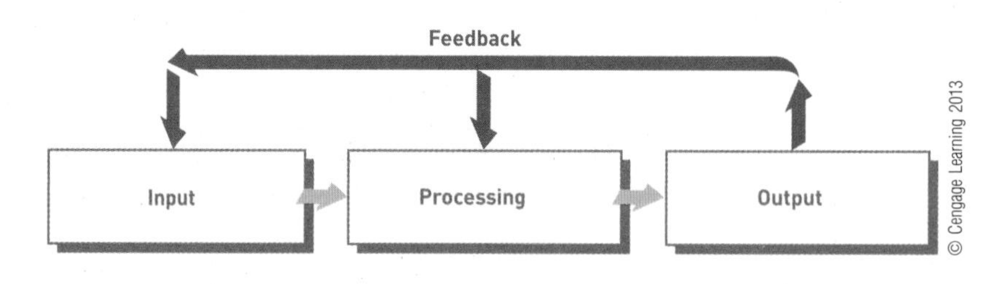

FIGURE 1.5
Components of an information system
Feedback is critical to the successful operation of a system.

Processing

processing: Converting or transforming data into useful outputs.

In information systems, **processing** means converting or transforming data into useful outputs. Processing can involve making calculations, comparing data and taking alternative actions, and storing data for future use. Processing data into useful information is critical in business settings.

Processing can be done manually or with computer assistance. In a payroll application, the number of hours each employee worked must be converted into net, or take-home, pay. Other inputs often include employee ID number and department. The processing can first involve multiplying the number of hours worked by the employee's hourly pay rate to get gross pay. If weekly hours worked exceed 40, overtime pay might also be included. Then deductions—for example, federal and state taxes or contributions to insurance or savings plans—are subtracted from gross pay to get net pay.

After these calculations and comparisons are performed, the results are typically stored. *Storage* involves keeping data and information available for future use, including output, discussed next.

Output

output: Production of useful information, usually in the form of documents and reports.

In information systems, **output** involves producing useful information, usually in the form of documents and reports. Outputs can include paychecks for employees, reports for managers, and information supplied to stockholders, banks, government agencies, and other groups. In some cases, output from one system can become input for another. For example, output from a system that processes sales orders can be used as input to a customer billing system. Not all output, however, is accurate or useful. Faulty mathematical models used by one investment firm resulted in a multimillion dollar fine for mistakes and possible fraud.[11]

Feedback

feedback: Information from the system that is used to make changes to input or processing activities.

In information systems, **feedback** is information from the system that is used to make changes to input or processing activities.[12] For example, errors or problems might make it necessary to correct input data or change a process. Consider a payroll example. Perhaps the number of hours an employee worked was entered as 400 instead of 40. Fortunately, most information systems check to make sure that data falls within certain ranges. For number of hours worked, the range might be from 0 to 100 because it is unlikely that an employee would work more than 100 hours in a week. The information system would determine that 400 hours is out of range and provide feedback. The feedback is used to check and correct the input on the number of hours worked to 40. If undetected, this error would result in a very high net pay!

The Shinpuku Seika farm used feedback from its fields to help determine when to plant crops and what crops to plant.[13] The farm placed sensors in its fields to measure soil temperatures and moisture to determine the best time to plant and the best crop to plant to maximize profits. This feedback loop can be more profitable than using past experience and hunches to determine what and when to plant. Royal Caribbean Cruises used smartphones and other mobile devices to get feedback on where people and items are located onboard. If necessary, corrective action can be taken to make sure everything is in its proper location, ensuring the safety and comfort of the passengers.[14] The system can even be used to locate children wearing special bracelets. According to one cruise customer, "My daughter wore one onboard last year. From a couple of decks away, we could tell where she was standing in the arcade."

forecasting: Predicting future events to avoid problems.

In addition to feedback, a computer system can predict future events to avoid problems. This concept, often called **forecasting**, can be used to estimate future sales and order more inventory before a shortage occurs. Forecasting is also used to predict the strength and landfall sites of hurricanes, future stock market values, and the winner of a political election.

© Caro/Alamy

Forecasting systems can help economists predict the strengths and weaknesses in the global economy.

Computer-Based Information Systems

computer-based information system (CBIS): A single set of hardware, software, databases, telecommunications, people, and procedures that are configured to collect, manipulate, store, and process data into information.

As discussed earlier, an information system can be manual or computerized. A **computer-based information system (CBIS)** is a single set of hardware, software, databases, telecommunications, people, and procedures that are configured to collect, manipulate, store, and process data into information. Increasingly, companies are incorporating computer-based information systems into their products and services. The chief information officer for Volkswagen Group of America, for example, calls future Volkswagens "rolling computers."[15] The computers integrated into Volkswagen vehicles will be able to determine if something is wrong with the car, recommend the needed repair work, check on the available parts to make the repair, and schedule a service appointment with the local VW dealer. Lloyd's Insurance in London used a CBIS to reduce paper transactions and convert to an electronic insurance system. The CBIS allows Lloyd's to insure people and property more efficiently and effectively. Lloyd's often insures the unusual, including actress Betty Grable's legs, Rolling Stone Keith Richard's hands, and a possible appearance of the Lock Ness Monster (Nessie) in Scotland, which would result in a large payment for the person first seeing the monster. In addition to supporting the organization, computer-based information systems often become part of the product or service. A worldwide Hilton Hotel survey indicated that a fast, reliable connection to the Internet was the second most important factor in overall customer satisfaction, just behind a clean room.[16] Today, many excellent computer-based information systems follow stock indexes and markets and suggest when large blocks of stocks should be purchased or sold (called "program trading") to take advantage of market discrepancies.[17]

technology infrastructure: All the hardware, software, databases, telecommunications, people, and procedures that are configured to collect, manipulate, store, and process data into information.

The components of a CBIS are illustrated in Figure 1.6. "Information technology (IT)" refers to hardware, software, databases, and telecommunications. A business's **technology infrastructure** includes all the hardware, software, databases, telecommunications, people, and procedures that are configured to collect, manipulate, store, and process data into information. The technology infrastructure is a set of shared IS resources that form the foundation of each computer-based information system.

Hardware

hardware: Computer equipment used to perform input, processing, storage, and output activities.

Hardware consists of computer equipment used to perform input, processing, storage, and output activities. Input devices include keyboards, mice and other

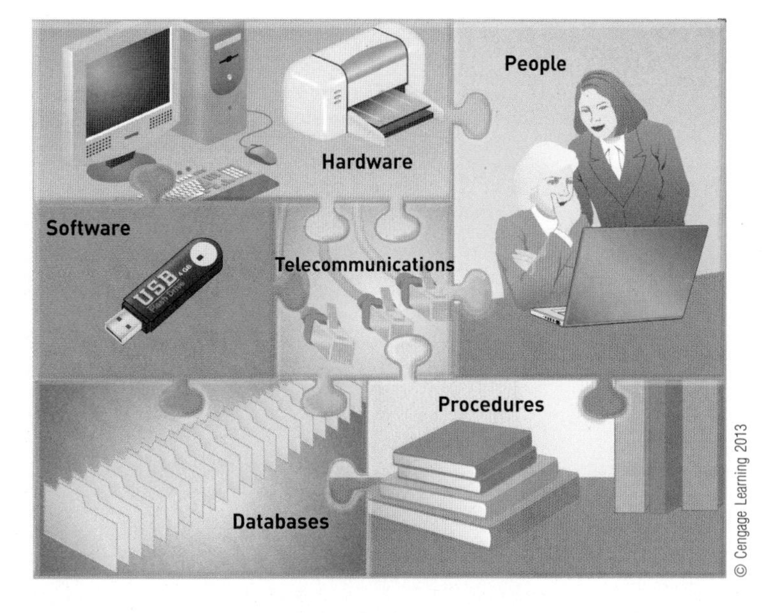

FIGURE 1.6
Components of a computer-based information system

pointing devices, automatic scanning devices, and equipment that can read magnetic ink characters. Processing devices include computer chips that contain the central processing unit and main memory. Advances in chip design allow faster speeds, less power consumption, and larger storage capacity. A university professor, for example, has designed a chip that takes shortcuts and is not completely accurate. However, this "slightly inaccurate" chip could be up to 100,000 times faster than the typical chip.[18] SanDisk and other companies make small, portable chips that are used to conveniently store programs, data files, and more.[19] The publisher of this book, for example, has used this type of chip storage device to send promotional material for this book to professors and instructors.

Processor speed is also important. Today's more advanced processor chips have the same power as 1990s-era supercomputers that occupied a room measuring 10 feet by 40 feet. Today, a large IBM computer used by U.S. Livermore National Laboratories to analyze nuclear explosions is one of the fastest computers in the world at up to 300 teraflops—300 trillion operations per second.[20] The super-fast computer, called Blue Gene, costs about $40 million.[21] It received the *National Medal of Technology and Innovation* award from President Obama. Small, inexpensive computers and handheld devices are also becoming popular. In addition, the iPhone by Apple Computer can perform many functions that can be done on a desktop or laptop computer.[22] The One Laptop Per Child computer costs under $200.[23] The Classmate PC by Intel will cost about $300 and include some educational software. Both computers are intended for regions of the world that can't afford traditional personal computers.

The many types of output devices include printers and computer screens. Some touch-sensitive computer screens, for example, can be used to execute functions or complete programs, such as connecting to the Internet or running a new computer game or word-processing program. Many special-purpose hardware devices have also been developed. Computerized event data recorders (EDRs) are now being placed into vehicles. Like an airplane's black box, EDRs record vehicle speed, possible engine problems, driver performance, and more.

The use of tablet computers, including Apple's iPad and Motorola's Xoom, has soared for individuals and corporations.[24] These devices can use tens of thousands of applications specifically designed for tablet computers.[25] Many magazine and book publishers, for instance, have developed applications that allow their publications to be downloaded and read on these new tablet devices. Chevron and many other companies are investing in new tablet computers, such as Apple's iPad, to connect managers and corporate workers to corporate computer systems.[26] Today, many companies are letting their

Hardware consists of computer equipment used to perform input, processing, and output activities. The trend in the computer industry is to produce smaller, faster, and more mobile hardware, such as tablet computers.

employees use personal hardware devices, such as smartphones and other mobile devices, to work from home or while traveling.[27] According to one study, about 90 percent of businesses will allow and support corporate applications on personal devices by 2014.[28] These personal hardware devices are convenient for employees, but there can be security threats when the organization doesn't completely control the devices that access corporate data and programs.

Software

software: The computer programs that govern the operation of the computer.

Software consists of the computer programs that govern the operation of the computer. These programs allow a computer to process payroll, send bills to customers, and provide managers with information to increase profits, reduce costs, and provide better customer service. Fab Lab software, for example, controls tools such as cutters, milling machines, and other devices.[29] One Fab Lab system, which costs about $20,000, has been used to make radio frequency tags to track animals in Norway, engine parts to allow tractors to run on processed castor beans in India, and many other fabrication applications. Salesforce sells software to help companies manage their sales force and help improve customer satisfaction.[30]

The two types of software are system software, such as Microsoft Windows, which controls basic computer operations including start-up and printing, and applications software, such as Microsoft Office, which allows you to accomplish specific tasks including word processing or tabulating numbers.[31] Software is needed for computers of all sizes, from small handheld computers to large supercomputers. The Android operating system by Google and Microsoft's Phone, for example, are operating system for cell phones and small portable devices. Although most software can be installed from CDs, many of today's software packages can be downloaded through the Internet.

Sophisticated application software, such as Adobe Creative Suite, can be used to design, develop, print, and place professional-quality advertising, brochures, posters, prints, and videos on the Internet.[32] Nvidia's GeForce 3D is software that can display three-dimensional images on a computer screen when using special glasses.[33]

Windows is systems software that controls basic computer operations including start-up and printing.

Databases

database: An organized collection of facts and information, typically consisting of two or more related data files.

A **database** is an organized collection of facts and information, typically consisting of two or more related data files. An organization's database can contain facts and information on customers, employees, inventory, competitors' sales, online purchases, and much more.

Most managers and executives consider a database to be one of the most valuable parts of a computer-based information system.[34] People can analyze databases, such as Medicare's claims database, to find potential fraud and abuse.[35] Data can be stored in large data centers, within computers of all sizes, on the Internet, and in smartphones and small computing devices.[36] The huge increase in database storage requirements, however, often leads to more storage devices, more space to house the additional storage devices, and additional electricity to operate them. Most organizations using database systems have seen storage requirements increase more than 10 percent every year. An important issue for any organization is how to keep a vast database secure and safe from the prying eyes of outside individuals and groups.

Telecommunications, Networks, and the Internet

telecommunications: The electronic transmission of signals for communications that enables organizations to carry out their processes and tasks through effective computer networks.

Telecommunications is the electronic transmission of signals for communications, which enables organizations to carry out their processes and tasks through effective computer networks. Telecommunications can take place through wired, wireless, and satellite transmissions. The Associated Press was one of the first users of telecommunications in the 1920s, sending news over 103,000 miles of wire in the United States and over almost 10,000 miles of cable across the ocean. Recently, a public library in the state of Washington used a telecommunications system to connect branch libraries using cables and existing phone lines.[37] The library was also able to use the telecommunications system to make phone calls. Today, telecommunications is used by people and organizations of all sizes around the world. With telecommunications, people can work at home or while traveling. This approach to work, often called "telecommuting," allows someone living in England to send work to the United States, China, or any location with telecommunications capabilities.

Businesses use cloud computing and other types of Web sites to provide better customer service and reduce costs.

network: Computers and equipment that are connected in a building, around the country, or around the world to enable electronic communications.

Internet: The world's largest computer network, consisting of thousands of interconnected networks, all freely exchanging information.

Networks connect computers and equipment in a building, around the country, or around the world to enable electronic communication. Wireless transmission allows aircraft drones, such as Boeing's Scan Eagle, to fly using a remote control system that can monitor commercial buildings or enemy positions.[38] The drones are smaller and less-expensive versions of the Predator and Global Hawk drones that the U.S. military has used in the Afghanistan and Iraq conflicts.

The **Internet** is the world's largest computer network, consisting of thousands of interconnected networks, all freely exchanging information. Research firms, colleges, universities, high schools, hospitals, and media companies are just a few examples of organizations using the Internet.[39] Increasingly, businesses and people are using the Internet to run and deliver important applications, such as those accessing vast databases, performing sophisticated business analyses, and getting a variety of reports.[40] This concept, called *cloud computing*, allows people to get the information they need from the Internet (the cloud) instead of from desktop or corporate computers. According to a popular research and consulting company, the total market for cloud computing is expected to approach $150 billion annually by 2014.[41] Reliability and security, however, remain top concerns for using the cloud-computing approach.[42]

People use the Internet to research information, buy and sell products and services, make travel arrangements, conduct banking, download music and videos, read books, and listen to radio programs, among other activities.[43] Amazon's Web Services (AWS) allows businesses of all sizes to use Amazon's computer power and only pay for the services and computer resources used.[44] According to an Amazon vice president, "This completely levels the playing field." Bank of America allows people to check their bank balances and pay their bills on the Internet using Apple's iPhone and other handheld devices.[45] Internet sites such as Facebook (*www.facebook.com*) have become popular places to connect with friends and colleagues. People can also send short messages up to 140 characters using Twitter (*www.twitter.com*) over the Internet.[46] In early 2011, social networking sites such as Facebook and Twitter were used to help organize massive protests in Egypt, Libya, and other places.[47] Some countries try to block Internet traffic on these sites to stop or retard the protesters from organizing.[48]

This increased use of the Internet is not without its risks. Some people fear that this increased usage can lead to problems, including criminals hacking into the Internet and gaining access to sensitive personal information.[49]

The World Wide Web (WWW), or the Web, is a network of links on the Internet to documents containing text, graphics, video, and sound. Information about the documents and access to them are controlled and provided by tens of thousands of special computers called Web servers. The Web is one of

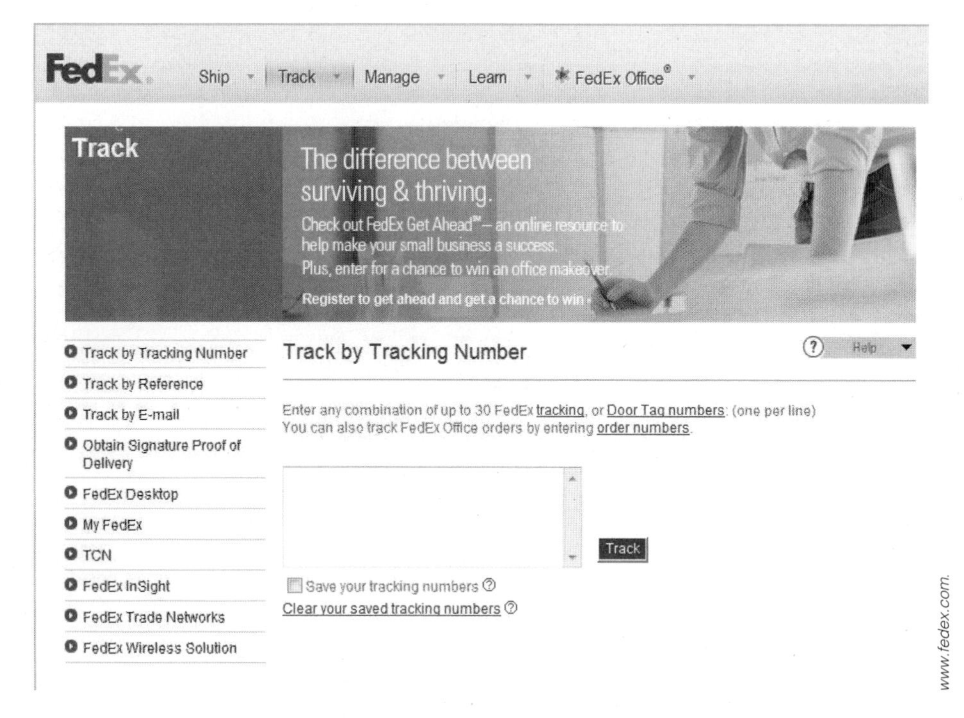

www.fedex.com.

When you log on to the FedEx site (*www.fedex.com*) to check the status of a package, you are using an extranet.

many services available over the Internet and provides access to millions of documents. New Internet technologies and increased Internet communications and collaboration are collectively called Web 2.0.

intranet: An internal network based on Web technologies that allows people within an organization to exchange information and work on projects.

extranet: A network based on Web technologies that allows selected outsiders, such as business partners and customers, to access authorized resources of a company's intranet.

The technology used to create the Internet is also being applied within companies and organizations to create **intranets**, which allow people in an organization to exchange information and work on projects. ING DIRECT Canada, for example, uses its intranet to get ideas from its employees.[50] Companies often use intranets to connect their employees around the globe. An **extranet** is a network based on Web technologies that allows selected outsiders, such as business partners and customers, to access authorized resources of a company's intranet. Many people use extranets every day without realizing it—to track shipped goods, order products from their suppliers, or access customer assistance from other companies. Penske Truck Leasing, for instance, uses an extranet for Penske leasing companies and its customers.[51] The extranet site allows customers to schedule maintenance, find Penske fuel stops, receive emergency roadside assistance, participate in driver training programs, and more. Likewise, if you log on to the FedEx site to check the status of a package, you are using an extranet.[52]

People

People are the most important element in most computer-based information systems. They make the difference between success and failure for most organizations. Information systems personnel include all the people who manage, run, program, and maintain the system, including the chief information officer (CIO), who manages the IS department. According to Dennis Strong, senior vice president and chief information officer of McCoy's Building Supply company, "Know your team and those you interact with on a personal level. Be able to share their celebrations and support them during their life struggles."[53]

Other people are users who work with information systems to get results. Users include financial executives, marketing representatives, manufacturing operators, and many others. Certain computer users are also IS personnel.

Procedures

procedures: The strategies, policies, methods, and rules for using a CBIS.

Procedures include the strategies, policies, methods, and rules for using the CBIS, including the operation, maintenance, and security of the computer. For example, some procedures describe when each program should be run. Others describe who can access facts in the database or what to do if a

The chief information officer (CIO) manages the Information Systems department, which includes all the people who manage, run, program, and maintain a computer-based information system.

disaster, such as a fire, earthquake, or hurricane, renders the CBIS unusable. Good procedures can help companies take advantage of new opportunities and avoid potential disasters. Poorly developed and inadequately implemented procedures, however, can cause people to waste their time on useless rules or result in inadequate responses to disasters.

Now that we have looked at computer-based information systems in general, we will briefly examine the most common types used in business today. These IS types are covered in more detail in Part 3.

BUSINESS INFORMATION SYSTEMS

The most common types of information systems used in business organizations are those designed for electronic and mobile commerce, transaction processing, management information, and decision support.[54] In addition, some organizations employ special-purpose systems, such as virtual reality, that not every organization uses. Although these systems are discussed in separate sections in this chapter and explained in more detail later, they are often integrated in one product and delivered by the same software package (see Figure 1.7). For example, some business information systems process

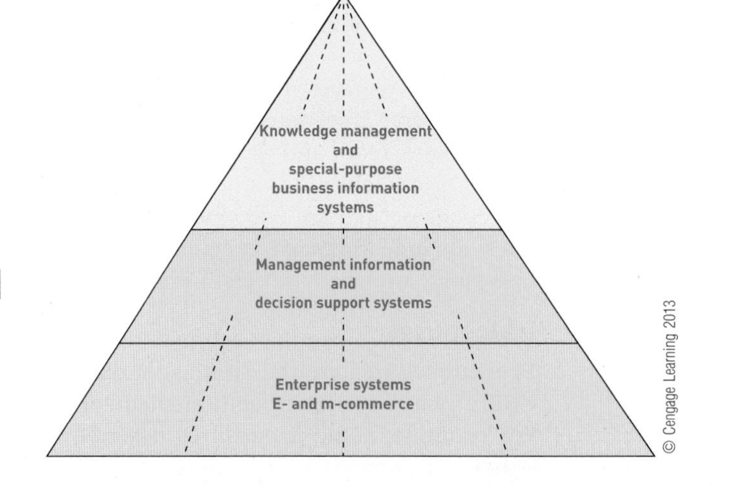

FIGURE 1.7

Business information systems

Business information systems are often integrated in one product and can be delivered by the same software package.

INFORMATION SYSTEMS @ WORK

CIBS: Maintaining Quality While Growing

You probably don't think of cleaning washrooms and high-tech software at the same time. That means you don't usually think of CIBS.

CIBS, a division of CI (originally Clean Interiors) Business Services, provides a complete range of washroom and pest control services. Founded more than 20 years ago in the United Kingdom, CIBS is now an award-winning cleaning and hygiene services provider.

In its early years, managers at CIBS scheduled services using a combination of spreadsheets, paper files, and small business software. As the company grew, however, the managers reevaluated this method. General Manager Julia Kulinski explains that "because data was scattered across spreadsheets and paper files, it was difficult for us to get an integrated view of our customers, which we needed to service them properly. For example, when customers called in regarding errors or other service-related issues, service representatives couldn't find the information needed to resolve the issues on the first call."

Other problems affected revenue and expenses. Because invoicing was a manual process, the staff prepared and sent invoices only once a month, and invoices based on paper records often had errors. Management wasn't aware of cost overruns until it was too late to correct them. Further, it wasn't practical to motivate employees by moving to performance-based pay methods, which was a business objective. Finally and most importantly, CIBS couldn't grow.

CIBS evaluated its options and selected an integrated ERP (enterprise resource planning) system from the German firm SAP. This chapter defines an ERP system as "a set of integrated programs that manages the vital business operations for an entire multisite, global organization." Although most ERP users are large organizations, smaller companies are also taking advantage of ERP systems. With 200 employees, CIBS is a medium-sized company. It too wanted the benefits that an ERP system offers, including a single shared database to store its information and coordinate its operations. As Kulinski puts it, they wouldn't be able to grow otherwise: "For the business to scale, we needed a centralized database, automated processes, and real-time reporting." Small companies need these as much as large ones do.

In fact, ERP use is growing more quickly among small companies than large ones. According to Albert Pang in the *Apps Run the World* blog, total annual revenue of all ERP vendors is growing at 5.6 percent per year for customers with 100 or fewer employees, dropping steadily to an annual growth rate of 2.4 percent among companies with 5,000 or more employees. The reason is that early ERP systems required large computers, which only the largest organizations could afford. Knowing this, ERP vendors offered complex programs suited to such users.

Today, computers are far less expensive, and much ERP software is tailored to the needs of smaller organizations. However, many people still think that ERP is only for large companies. It often takes time for the conventional wisdom to catch up with reality, but CIBS didn't fall into this trap.

Kulinski maintains that the move to ERP software solved business problems at CIBS and helped the company grow. She evaluates the financial benefits this way: "I look at the running costs of SAP software—which for us is the equivalent of one full-time employee—and then at the value the software delivers to the business. There's no way one person could deliver this much value to the business."

In the final analysis, delivering value to the business is what information systems are about.

Discussion Questions

1. This case is about an SAP customer and is based in part on SAP materials. However, other software firms also offer ERP software. The two largest vendors in the small company segment are Oracle and Microsoft. Compare their ERP offerings.
2. Before it moved to SAP, CIBS kept and processed its data with a collection of simpler, and in some cases manual, methods. Did these methods constitute an information system? Why or why not?

Critical Thinking Questions

1. List the five problems that CIBS had with its earlier spreadsheet- and paper-based system. Rank them by their importance to CIBS. Justify your rankings. If you disagree with Kulinski's top ranking of limited growth potential, explain why.
2. Does your university use an ERP system or any comparable integrated software with a comprehensive shared database? Try to find out how it chose that system, or if it doesn't have one, how

it might choose one in the future. (This exploration can be done as a class project to avoid asking a few people the same questions over and over again.)

SOURCES: CIBS Web site, *www.ci-bs.co.uk*, and subsidiaries' Web sites, *www.cibshygiene.com* and *www.cibsfacilities.com*, accessed January 17, 2012; SAP, "CIBS: Enabling Growth and Exceptional Service Quality with SAP Software," *http://download.sap.com/uk/download.epd?context=A8700D6A2BB022BCF7C1BA6D4FFF4D1837ACD9F06DE9B34FF8945FE34BEA4AE61EA46CBB6C5EA8DAE21568331B293774E37074837BC3CE5D*, March 2011, downloaded January 16, 2012; Pang, A., "Infor's Daring Move to Buy Lawson, Shake Up ERP MidMarket," *Apps Run the World* blog, *www.appsruntheworld.com/blogs/?p=370*, March 13, 2011, accessed January 17, 2012.

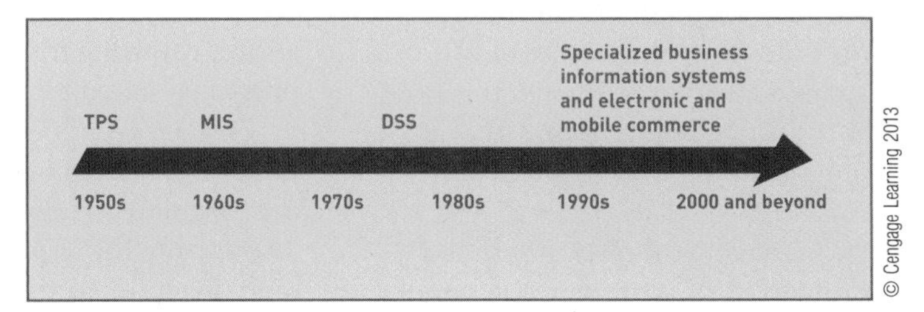

FIGURE **1.8**
Development of important business information systems

transactions, deliver information, and support decisions. Figure 1.8 shows a simple overview of the development of important business information systems discussed in this section. In addition to owning a complete business information system including hardware, software, databases, telecommunications, and Internet capabilities, companies can rent business information systems from others.[55] As mentioned before, Amazon Web Services (AWS) allows people and companies to pay for the business information systems they use. According to the vice president of AWS, "Instead of spending millions on a data center and servers, you can simply pay as you go."

Electronic and Mobile Commerce

electronic commerce (e-commerce): Any business transaction executed electronically between companies (business-to-business), companies and consumers (business-to-consumer), consumers and other consumers (consumer-to-consumer), business and the public sector, and consumers and the public sector.

mobile commerce (m-commerce): The use of mobile, wireless devices to place orders and conduct business.

Electronic commerce (e-commerce) involves any business transaction executed electronically between companies (business-to-business, or B2B), companies and consumers (business-to-consumer, or B2C), consumers and other consumers (consumer-to-consumer, or C2C), business and the public sector, and consumers and the public sector. E-commerce offers opportunities for businesses of all sizes to market and sell at a low cost worldwide, allowing them to enter the global market.

Mobile commerce (m-commerce) is the use of mobile, wireless devices to place orders and conduct business. M-commerce relies on wireless communications that managers and corporations use to place orders and conduct business with handheld computers, portable phones, laptop computers connected to a network, and other mobile devices. BMW, the German sports car company, has invested about $100 million in developing mobile applications for its cars and other products.[56] Today, mobile commerce has exploded in popularity with advances in smartphones, including Apple's iPhone. Customers are using their cell phones to purchase concert tickets from companies such as Ticketmaster Entertainment (*www.ticketmaster.com*) and Tickets (*www.tickets.com*).[57] Phones and other mobile devices can be used to pay for products and services.[58] Google, a popular Internet company, is looking into incorporating a payment feature into its software. As another example, the Square card reader can be attached to some smartphones and computer tablets to collect payments from people and organizations.[59] This small device allows entrepreneurs and small businesses to accept credit card charges at any time and at any place.

With mobile commerce (m-commerce), people can use cell phones to pay for goods and services anywhere, anytime.

FIGURE **1.9**

E-commerce greatly simplifies purchasing

E-commerce offers many advantages for streamlining work activities. Figure 1.9 provides a brief example of how e-commerce can simplify the process of purchasing new office furniture from an office supply company. In the manual system, a corporate office worker must get approval for a purchase that exceeds a certain amount. That request goes to the purchasing department, which generates a formal purchase order to procure the goods from the approved vendor. Business-to-business e-commerce automates the entire process. Employees go directly to the supplier's Web site, find the item in a catalog, and order what they need at a price set by their company. If management approval is required, the manager is notified automatically. As the use of e-commerce systems grows, companies are phasing out their traditional systems. The resulting growth of e-commerce is creating many new business opportunities.

electronic business (e-business): Using information systems and the Internet to perform all business-related tasks and functions.

In addition to e-commerce, business information systems use telecommunications and the Internet to perform many related tasks. Electronic procurement (e-procurement), for example, involves using information systems and the Internet to acquire parts and supplies. **Electronic business (e-business)** goes beyond e-commerce and e-procurement by using information systems and the Internet to perform all business-related tasks and functions, such as accounting, finance, marketing, manufacturing, and human resource activities. E-business also includes working with customers, suppliers, strategic partners, and stakeholders. Compared to traditional business strategy, e-business strategy is flexible and adaptable (see Figure 1.10).

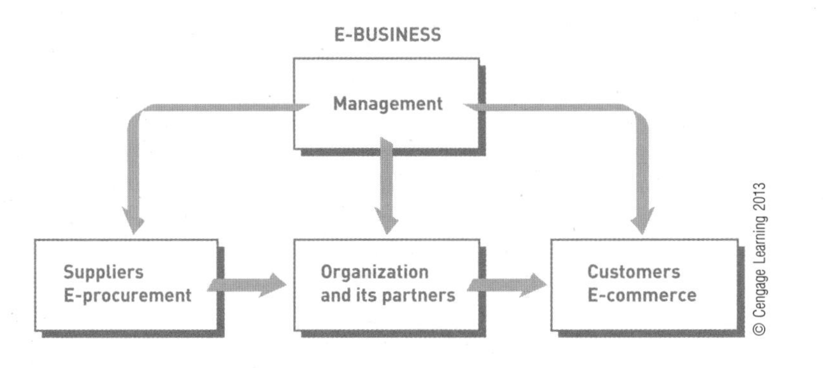

FIGURE **1.10**
Electronic business
E-business goes beyond e-commerce to include using information systems and the Internet to perform all business-related tasks and functions, such as accounting, finance, marketing, manufacturing, and human resources activities.

Enterprise Systems: Transaction Processing Systems and Enterprise Resource Planning

Enterprise systems that process daily transactions have evolved over the years and offer important solutions for businesses of all sizes. Bar codes, often used by enterprise systems in grocery stores and retail outlets, speed checkout and provide a wealth of information for business executives. Some bar codes have been converted into art work with flowers and artistic drawings placed on top, to the side, or below the bar codes to make them more interesting and attractive to customers.[60] In 2011 eBay, the large Internet auction site, acquired GSI to help it develop an enterprise system to compete with other Internet retail operations such as Amazon.[61] The enterprise system will help eBay receive and process online orders. The acquisition will also help eBay process online transactions from a customer's smartphones and other mobile devices. Traditional transaction processing systems are still being used today, but increasingly, companies are turning to enterprise resource planning systems.

Transaction Processing Systems

transaction: Any business-related exchange such as payments to employees, sales to customers, and payments to suppliers.

transaction processing system (TPS): An organized collection of people, procedures, software, databases, and devices used to perform and record business transactions.

Since the 1950s, computers have been used to perform common business applications. Many of these early systems were designed to reduce costs by automating routine, labor-intensive business transactions. A **transaction** is any business-related exchange such as payments to employees, sales to customers, or payments to suppliers. Processing business transactions was the first computer application developed for most organizations. A **transaction processing system (TPS)** is an organized collection of people, procedures, software, databases, and devices used to perform and record business transactions. If you understand a transaction processing system, you understand basic business operations and functions.

One of the first business systems to be computerized was the payroll system (see Figure 1.11). The primary inputs for a payroll TPS are the number of employee hours worked during the week and the pay rate. The primary output consists of paychecks. Early payroll systems produced employee paychecks and related reports required by state and federal agencies, such as the Internal Revenue Service. Other routine applications include sales ordering, customer billing and customer relationship management, and inventory control.

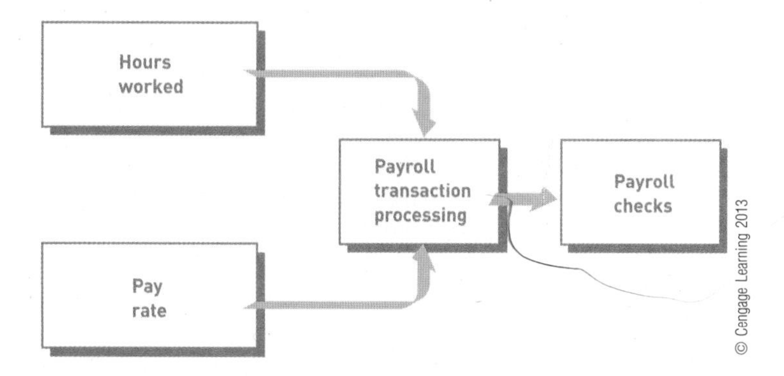

FIGURE **1.11**

Payroll transaction processing system

In a payroll TPS, the inputs (numbers of employee hours worked and pay rates) go through a transformation process to produce outputs (paychecks).

Enterprise systems help organizations perform and integrate important tasks, such as paying employees and suppliers, controlling inventory, sending invoices, and ordering supplies. In the past, companies accomplished these tasks using traditional transaction processing systems. Today, more companies use enterprise resource planning systems for these tasks.

Enterprise Resource Planning

enterprise resource planning (ERP) system: A set of integrated programs that manages the vital business operations for an entire multisite, global organization.

An **enterprise resource planning (ERP) system** is a set of integrated programs that manages the vital business operations for an entire multisite, global organization. An ERP system can replace many applications with one unified set of programs, making the system easier to use and more effective. Today, using ERP systems and getting timely reports from them can be done using cell phones and mobile devices.

Although the scope of an ERP system might vary from company to company, most ERP systems provide integrated software to support manufacturing and finance. Many ERP systems also have a purchasing subsystem that orders the needed items. In addition to these core business processes, some ERP systems can support functions such as customer service, human resources, sales, and distribution. The primary benefits of implementing an ERP system include easing adoption of improved work processes and increasing access to timely data for decision making.

Information and Decision Support Systems

The benefits provided by an effective TPS or ERP, including reduced processing costs and reductions in needed personnel, are substantial and justify their associated costs in computing equipment, computer programs, and specialized personnel and supplies. Companies soon realize that they can use the data stored in these systems to help managers make better decisions, whether in human resource management, marketing, or administration. Satisfying the needs of managers and decision makers continues to be a major factor in developing information systems.

www.sap.com

SAP AG, a German software company, is one of the leading suppliers of ERP software. The company employs more than 50,000 people in more than 120 countries.

Management Information Systems

management information system (MIS): An organized collection of people, procedures, software, databases, and devices that provides routine information to managers and decision makers.

A **management information system (MIS)** is an organized collection of people, procedures, software, databases, and devices that provides routine information to managers and decision makers. An MIS focuses on operational efficiency.[62] Manufacturing, marketing, production, finance, and other functional areas are supported by MISs and are linked through a common database.[63] MISs typically provide standard reports generated with data and information from the TPS or ERP (see Figure 1.12). Dell Computer, for example, used manufacturing MIS software to develop a variety of reports on its manufacturing processes and costs.[64] Dell was able to double its product variety, while saving about $1 million annually in manufacturing costs as a result. As another example, some marketing research companies specialize in finding the best approach for investing in Internet advertising.[65]

MISs were first developed in the 1960s and typically use information systems to produce managerial reports. In many cases, these early reports were produced periodically—daily, weekly, monthly, or yearly. Because of their value to managers, MISs have proliferated throughout the management ranks.

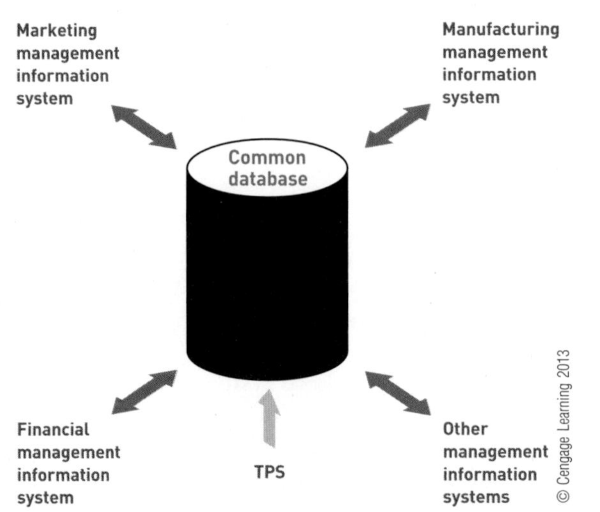

FIGURE **1.12**

Management information system

Functional management information systems draw data from the organization's transaction processing system.

Decision Support Systems

decision support system (DSS): An organized collection of people, procedures, software, databases, and devices used to support problem-specific decision making.

By the 1980s, dramatic improvements in technology resulted in information systems that were less expensive but more powerful than earlier systems.[66] People quickly recognized that computer systems could support additional decision-making activities. A **decision support system (DSS)** is an organized collection of people, procedures, software, databases, and devices that support problem-specific decision making. The focus of a DSS is on making effective decisions. Whereas an MIS helps an organization "do things right," a DSS helps a manager "do the right thing."

A DSS goes beyond a traditional MIS by providing immediate assistance in solving problems. Many of these problems are unique and complex, and key information is often difficult to obtain. For instance, an auto manufacturer might try to determine the best location to build a new manufacturing facility. Some researchers are using DSSs to analyze Medicare's claims databases to search for potential fraud and abuse.[67] The DSS analysis can determine when doctors perform the same or similar procedures on the same patient, revealing potential Medicare abuses. Traditional MISs are seldom used to solve these types of problems; a DSS can help by suggesting alternatives and assisting in final decision making. A DSS recognizes that different managerial styles and decision types require different systems. For example, two production managers in the same position trying to solve the same problem might require different information and support. The overall emphasis is to support, rather than replace, managerial decision making.

A DSS can include a collection of models used to support a decision maker or user (model base), a collection of facts and information to assist in decision making (database), and systems and procedures (user interface or dialogue manager) that help decision makers and other users interact with the DSS (see Figure 1.13). Software called the database management system (DBMS) is often used to manage the database, and software called the model management system (MMS) is used to manage the model base. Not all DSSs have all of these components.

Endeca provides Discovery for Design, decision support software that helps businesspeople assess risk and analyze performance. The data shown here is for electronic component development.

Courtesy of Endeca

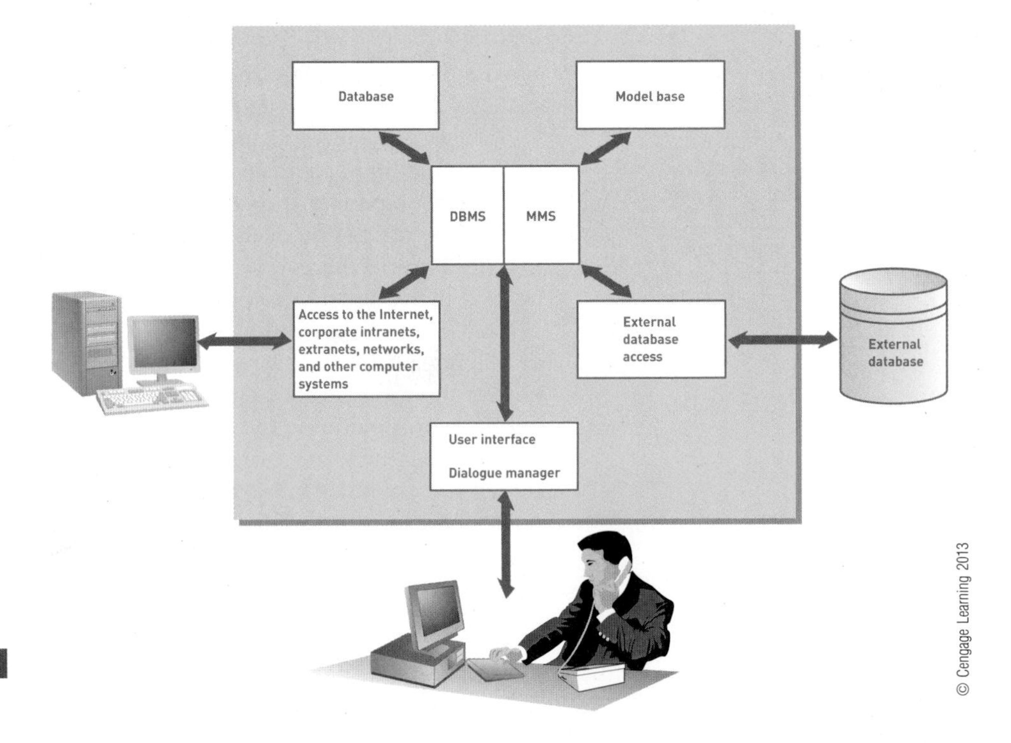

FIGURE 1.13
Essential DSS elements

In addition to DSSs for managers, other systems use the same approach to support groups and executives. A group support system includes the DSS elements just described as well as software, called groupware, to help groups make effective decisions. An executive support system, also called an executive information system, helps top-level managers, including a firm's president, vice presidents, and members of the board of directors, make better decisions. Healthland and Performance Management Institute, a health care company, has developed an executive information system to help small community and rural hospital executives make better decisions about delivering quality health care to patients and increasing the efficient delivery of health care services for hospitals.[68] As an incentive for such efforts, the American Recovery and Reinvestment Act provided funds for qualifying health care companies that invest in better information and decision support systems. An executive support system can assist with strategic planning, top-level organizing and staffing, strategic control, and crisis management.

Specialized Business Information Systems: Knowledge Management, Artificial Intelligence, Expert Systems, and Virtual Reality

In addition to ERPs, MISs, and DSSs, organizations often rely on specialized systems. Many use knowledge management systems (KMSs), an organized collection of people, procedures, software, databases, and devices to create, store, share, and use the organization's knowledge and experience.[69] A shipping company, for example, can use a KMS to streamline its transportation and logistics business.

artificial intelligence (AI): A field in which the computer system takes on the characteristics of human intelligence.

In addition to knowledge management, companies use other types of specialized systems. Some are based on the notion of **artificial intelligence (AI)** in which the computer system takes on the characteristics of human intelligence. Artificial intelligence allows computers to beat human champions in games, helps doctors make medical diagnoses, and allows cars to be driven hundreds of miles without a human behind the wheel.[70] The field of artificial intelligence includes several subfields (see Figure 1.14). Some people predict that in the future we will have nanobots, small molecular-sized robots,

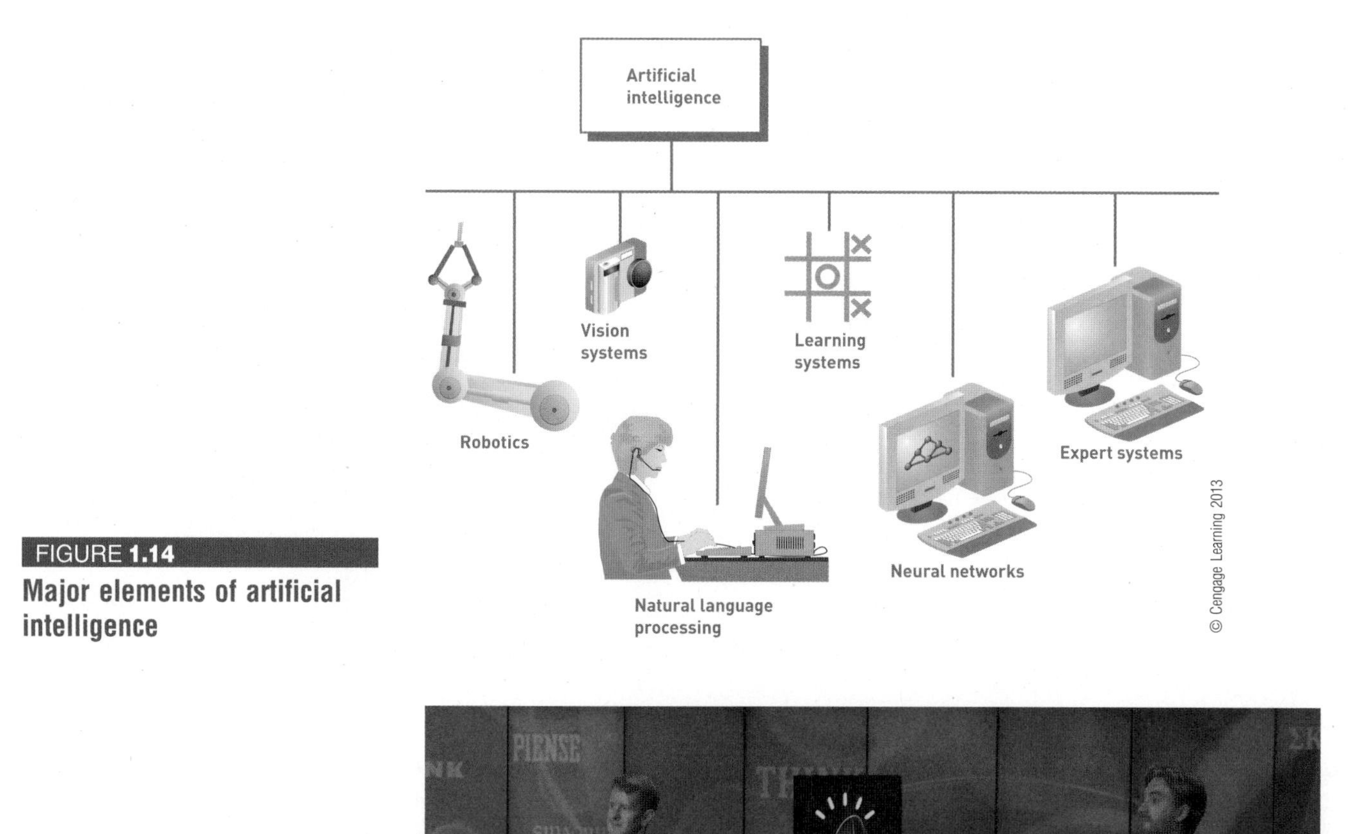

FIGURE 1.14
Major elements of artificial intelligence

IBM's supercomputer Watson used artificial intelligence to compete with and eventually win against Jeopardy champions.

Courtesy of IBM Corporation

traveling throughout our bodies and in our bloodstream, monitoring our health. Other nanobots will be embedded in products and services.

Artificial Intelligence

Robotics is an area of artificial intelligence in which machines take over complex, dangerous, routine, or boring tasks, such as welding car frames or assembling computer systems and components.[71] Honda has spent millions of dollars on advanced robotics that allows a person to give orders to a computer using only his or her thoughts. The new system uses a special helmet that can measure and transmit brain activity to a computer.[72] Vision systems allow robots and other devices to "see," store, and process visual images.[73] Natural language processing involves computers understanding and acting on verbal or written commands in English, Spanish, or other human languages.[74] Learning systems allow computers to learn from past mistakes or experiences, such as playing games or making business decisions. Neural networks is a branch of artificial intelligence that allows computers to recognize and act on patterns or trends. Some successful stock, options, and futures traders use neural networks to spot trends and improve the profitability of their investments.

Industrial robots perform complex, dangerous, routine, or boring tasks, such as milling machine tools or assembling computer systems and components.

Expert Systems

expert system: A system that gives a computer the ability to make suggestions and function like an expert in a particular field.

Expert systems give the computer the ability to make suggestions and function like an expert in a particular field, helping enhance the performance of the novice user.[75] The unique value of expert systems is that they allow organizations to capture and use the wisdom of experts and specialists. Therefore, years of experience and specific skills are not completely lost when a human expert dies, retires, or leaves for another job. Imprint Business Systems, for example, has an expert system that helps printing and packaging companies manage their businesses.[76] The collection of data, rules, procedures, and relationships that must be followed to achieve value or the proper outcome is contained in the expert system's **knowledge base**.

knowledge base: The collection of data, rules, procedures, and relationships that must be followed to achieve value or the proper outcome.

Virtual Reality and Multimedia

Virtual reality and multimedia are specialized systems that are valuable for many businesses and nonprofit organizations. Many imitate or act like real environments. These unique systems are discussed in this section.

virtual reality: The simulation of a real or imagined environment that can be experienced visually in three dimensions.

Virtual reality is the simulation of a real or imagined environment that can be experienced visually in three dimensions.[77] Originally, virtual reality referred to immersive virtual reality, which means the user becomes fully immersed in an artificial, computer-generated 3D world. The virtual world is presented in full scale and relates properly to the human size. Virtual reality can also refer to applications that are not fully immersive, such as mouse-controlled navigation through a 3D environment on a graphics monitor, stereo viewing from the monitor via stereo glasses, stereo projection systems, and others. Boeing, for example, used virtual reality and computer simulation to help design and build its Dreamliner 787.[78] The company used 3D models from Dassault Systems to design and manufacture the new aircraft. *Augmented reality,* a newer form of virtual reality, has the potential to superimpose digital data over real photos or images.[79]

A variety of input devices, such as head-mounted displays (see Figure 1.15), data gloves, joysticks, and handheld wands, allow the user to navigate through a virtual environment and to interact with virtual objects. Directional sound, tactile and force feedback devices, voice recognition, and other technologies enrich the immersive experience. Because several people can share and interact in the same environment, virtual reality can be a powerful medium for communication, entertainment, and learning.

Courtesy of Mechdyne Corporation

The Cave Automatic Virtual Environment (CAVE) is a virtual reality room that allows scientists to interact with 3D models of aerospace systems.

FIGURE **1.15**

Head-mounted display

The head-mounted display (HMD) was the first device to provide the wearer with an immersive experience. A typical HMD houses two miniature display screens and an optical system that channels the images from the screens to the eyes, thereby presenting a stereo view of a virtual world. A motion tracker continuously measures the position and orientation of the user's head and allows the image-generating computer to adjust the scene representation to the current view. As a result, the viewer can look around and walk through the surrounding virtual environment.

Courtesy of 5DT, Inc.

Multimedia is a natural extension of virtual reality. It can include photos and images, the manipulation of sound, and special 3D effects. Once used primarily in movies, 3D technology can be used by companies to design products, such as motorcycles, jet engines, bridges, and more. Autodesk, for instance, makes exciting 3D software that companies can use to design large skyscrapers and other buildings. The software can also be used by Hollywood animators to develop action and animated movies.

SYSTEMS DEVELOPMENT

systems development: The activity of creating or modifying information systems.

Systems development is the activity of creating or modifying information systems. Systems development projects can range from small to very large and are conducted in fields as diverse as stock analysis and video game development. Systems development can be initiated because companies want to take advantage of federal, state, and local tax breaks and related legislation.[80] Other companies are initiating systems development to take advantage of new technologies, such as smartphones and tablet computers, by developing unique and powerful applications for these newer devices.[81] The hope is to reduce costs and achieve a competitive advantage. Budgets for systems development projects are expected to increase in the next few years, according to a CIO economic impact survey.[82] It is expected that IS departments and systems developers will concentrate on creating more mobile applications for their businesses and organizations.

People inside a company can develop systems, or companies can use outsourcing, hiring an outside company to perform some or all of a systems development project. Outsourcing allows a company to focus on what it does best and delegate other functions to companies with expertise in systems development.

One strategy for improving the results of a systems development project is to divide it into several steps, each with a well-defined goal and set of tasks to accomplish (see Figure 1.16). These steps are summarized next.

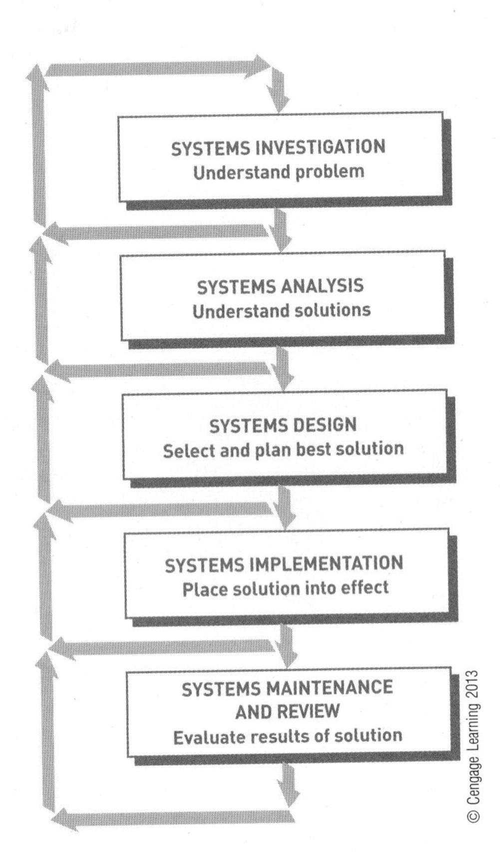

FIGURE 1.16
Overview of systems development

Systems Investigation and Analysis

The first two steps of systems development are systems investigation and analysis. The goal of the systems investigation is to gain a clear understanding of the problem to be solved or the opportunity to be addressed. After an organization understands the problem, the next question is, "Is the problem worth solving?" Given that organizations have limited resources—people and money—this question deserves careful consideration. If the decision is to continue with the solution, the next step, systems analysis, defines the problems and opportunities of the existing system. During systems investigation and analysis, as well as design maintenance and review, discussed next, the project must have the complete support of top-level managers and must focus on developing systems that achieve business goals.

ETHICAL & SOCIETAL ISSUES

How Much Privacy for Your Data?

In 1928, United States Supreme Court Justice Louis Brandeis wrote that "the right to be left alone [is] the most comprehensive of rights." When management consultant Tom Peters framed the privacy policy of his Web site, that concept was uppermost in his mind. Peters's entire policy is 10 lines long, the following being four of those lines:

"What we do with what you tell us

Your information will be used only by tompeterscompany! and its affiliates.

WE WILL NOT RENT OR SELL THIS INFORMATION TO ANY OTHER PARTY.

Period."

Now consider Amazon.com's privacy policy, which is four screens long in small type. Why the difference? Both companies thought about what should be in their privacy notices, but they reached dramatically different conclusions.

The difference is not because Amazon is a large company and Tom Peters is a small one. Some Web sites for large companies have brief policies, while some small Web sites have long ones. For example, the privacy policy on the Web site of fifteen-year-old blogger Harrison Li is about as long as Amazon's.

Part of the reason that the privacy policies for Amazon and Tom Peters are different is the nature of their businesses. Amazon needs credit card numbers, shipping addresses, and other personal information that could be used to harm someone. Peters does not sell products to consumers, so he doesn't need to collect similar information.

However, business objectives don't explain all the differences. Amazon tracks visit patterns via *cookies*, small data files stored by a visitor's Web browser. With cookies, a Web site can customize the content it displays based on a visitor's actions during earlier visits. Peters could also track visitor information with cookies but chooses not to use them. Amazon uses personal information to send offers on behalf of other businesses. Peters could also do so, and might earn some money that way, but doesn't.

Both policies make sense for the two firms. Both are as complete as necessary. Because both sites provide clear and easy access to the privacy policies, they are using them ethically. Why, then, are the policies so different? When a company defines its Web privacy policy, it must consider several factors:

1. *What legal obligations apply?* Legal obligations trump all other considerations.
2. *Are there other business considerations?* In the United States, Web sites are not required to have a privacy policy. Indeed, having one creates a risk of not complying with it. However, including a privacy policy creates some benefits. For example, a firm must have a privacy policy that meets specific requirements if it wants to advertise with Google AdSense, which allows Web site publishers to earn revenue from ads displayed on other sites. Companies with online privacy policies can also take advantage of the "Safe Harbor" agreement with the European Union (EU) that lets U.S. firms share data with EU computers without conforming to all EU privacy regulations.
3. *What does the market call for?* If customers expect a privacy policy, not having one that meets their expectations can hurt business.
4. *What can the firm realistically promise?* The ideal course might be for a company to promise to completely protect its customers' data.

However, it may not be practical to protect data against all forms of attack.

5. *What is a firm willing to commit to?* A privacy policy is, after all, a commitment. A firm should not make promises it doesn't intend to keep.
6. *What are a firm's moral obligations?* A privacy policy is a statement of its sponsor's moral standards. Most companies want public statements to reflect high ethical principles.

The answers to these questions led Tom Peters, Harrison Li, and Amazon in different directions. They will lead other companies in still other directions. As long as a company's management asks these questions and addresses them according to its own situation, the privacy policy will suit the company and its business objectives.

Discussion Questions

1. Visit three Web sites that sell the same type of product. Compare their privacy policies point by point. Did you find a site that you would not do business with now that you have read its privacy policy?
2. Consider your school's Web site privacy policy. How do the six factors for privacy policies apply to the site? As an alternative, discuss how the six factors help explain the differences between Tom Peters's and Amazon's privacy policies.

Critical Thinking Questions

1. Use your browser to visit *adsense.google.com*. Follow the "Program Policies" link at the bottom of the page, and then click "learn more" under "Google advertising cookies." Follow the first two links in the "Google advertising cookies" section. Do you think those policies do a good job of balancing consumers' privacy interests with Google's business objectives? Why or why not?
2. Write the privacy policy for a student musical organization with about 50 members. This Web site stores members' contact information and information about their musical abilities and interests. Who might want to use that information outside of the organization? Should the organization enable them to use the information?

SOURCES: Peters, T., "Tom Peters' Privacy Policy," *www.tompeters.com/privacy_policy.php*, accessed January 16, 2012; Amazon, "Privacy Notice," *www.amazon.com/gp/help/customer/display.html/ref=footer_privacy?nodeId=468496*, accessed January 16, 2012; Li, H., "Privacy Policy," *Blog Lectures, www.bloglectures.com/privacy-policy*, April 17, 2011, accessed January 16, 2012.

Systems Design, Implementation, and Maintenance and Review

Systems design determines how the new system should be developed to meet the business needs defined during systems analysis. Firefox, for example, used systems design to include a "do-not-track" feature into its Internet browser.[83] Some systems development projects find western states in the United States attractive because cooler temperatures can mean low cooling costs.[84] Implementing systems development projects in these states can save money and help protect the environment because less fuel is burned to generate electricity and cool buildings. Systems implementation involves creating or acquiring the various system components (such as hardware, software, and databases) defined in the design step, assembling them, and putting the new system into operation. For many organizations, this process includes purchasing software, hardware, databases, and other IS components. The purpose of systems maintenance and review is to check and modify the system so that it continues to meet changing business needs. Companies often hire outside companies to do their design, implementation, maintenance, and review functions.

INFORMATION SYSTEMS IN SOCIETY, BUSINESS, AND INDUSTRY

Information systems have been developed to meet the needs of all types of organizations and people. The speed and widespread use of information systems, however, opens users to a variety of threats from unethical people. Computer criminals and terrorists, for example, have used the Internet to steal millions of dollars or promote terrorism and violence. Some studies report that most of corporate security attacks come from people inside the company. Computer-related attacks can come from individuals, groups, companies, and even countries.

Security, Privacy, and Ethical Issues in Information Systems and the Internet

Although information systems can provide enormous benefits, they do have drawbacks. In one case, computer criminals stole carbon credits worth about $40 million from a European carbon credit market.[85] According to the president of the International Emissions Trading Association, "If you make sure that it's transferred to an account that you own and you sell it very quickly, then you've essentially got something for nothing, sold it for a lot, and you get out of town with all the dollars in your bag." According to one study, critical infrastructure for a large developed nation in the G20 will be attacked and damaged by 2015.[86]

Computer-related mistakes and waste are also a concern. In Japan, a financial services firm had trading losses of $335 million due to a typing mistake in entering a trade. Unwanted e-mail, called spam, can also be a huge

Computer criminals stole carbon credits worth about $40 million from a European carbon credit market.

waste of people's time. Many individuals and organizations are trying to find better ways to block spam. A technical problem at BlackBerry, a popular smartphone company, resulted in its customers not being able to access their e-mail or the Internet for days and consequently the company's stock value to plummet.[87]

Ethical issues concern what is generally considered right or wrong. Some IS professionals believe that computers may create new opportunities for unethical behavior. For example, unethical investors have placed false rumors or incorrect information about a company on the Internet and tried to influence its stock price to make money.

Individual privacy is also an important social issue.[88] People can inadvertently disclose personal information while using the Internet, and once private information or photos have been placed on the Internet, it can be very difficult to remove them. A number of Internet sites, for instance, collect personal and financial information about site visitors without their knowledge or consent. Some Internet companies, such as Firefox mentioned previously, are now starting to allow people to select a "do-not-track" feature that prevents personal and financial information from being gathered and stored while visiting their Internet sites.[89] While social networks provide convenient connections among friends and family, they can cause problems at the workplace.[90] Some workers have been fired when they criticized their employers or their companies using Facebook, Twitter, and other social networking sites. Some fired employees are fighting back by suing their employers.

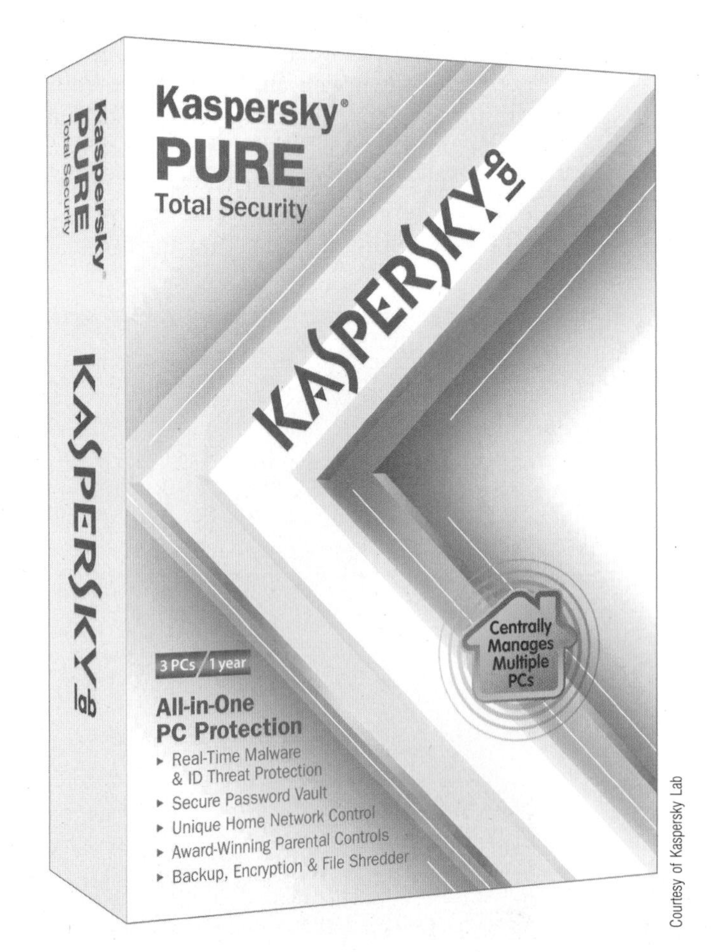

Courtesy of Kaspersky Lab

Kaspersky is a popular virus-protection program.

You can install firewalls (software and hardware that protect a computer system or network from outside attacks) to avoid viruses and prevent unauthorized people from gaining access to your computer system. You can also use identification numbers and passwords. In response to possible abuses, a number of laws have been passed to protect people from invasion of their privacy, including The Privacy Act, enacted in the 1970s.

Use of information systems also raises work concerns, including job loss through increased efficiency and potential health problems from making repetitive motions. Ergonomics, the study of designing and positioning workplace equipment, can help you avoid health-related problems of using computer systems.

Computer and Information Systems Literacy

Whatever your college major or career path, understanding computers and information systems will help you cope, adapt, and prosper in this challenging environment. Some colleges are requiring a certain level of computer and information systems literacy before students are admitted or accepted into the college.

Knowledge of information systems will help you make a significant contribution on the job. It will also help you advance in your chosen career or field. Managers are expected to identify opportunities to implement information systems to improve their business. They are also expected to lead IS projects in their areas of expertise. To meet these personal and organizational goals, you must acquire both computer literacy and information systems literacy. **Computer literacy** is the knowledge of computer systems and equipment and the ways they function. It includes the knowledge of equipment and devices (hardware), programs and instructions (software), databases, and telecommunications.

computer literacy: Knowledge of computer systems and equipment and the ways they function; it includes the knowledge of equipment and devices (hardware), programs and instructions (software), databases, and telecommunications.

information systems literacy: Knowledge of how data and information are used by individuals, groups, and organizations.

Information systems literacy goes beyond knowing the fundamentals of computer systems and equipment. **Information systems literacy** is the knowledge of how data and information are used by individuals, groups, and organizations. It includes knowledge of computer technology and the broader range of information systems. Most important, however, it encompasses *how* and *why* this technology is applied in business. Knowing about various types of hardware and software is an example of computer literacy. Knowing how

People connect with friends and other contacts using a social networking site, such as Facebook (*www.facebook.com*) or Twitter (*www.twitter.com*).

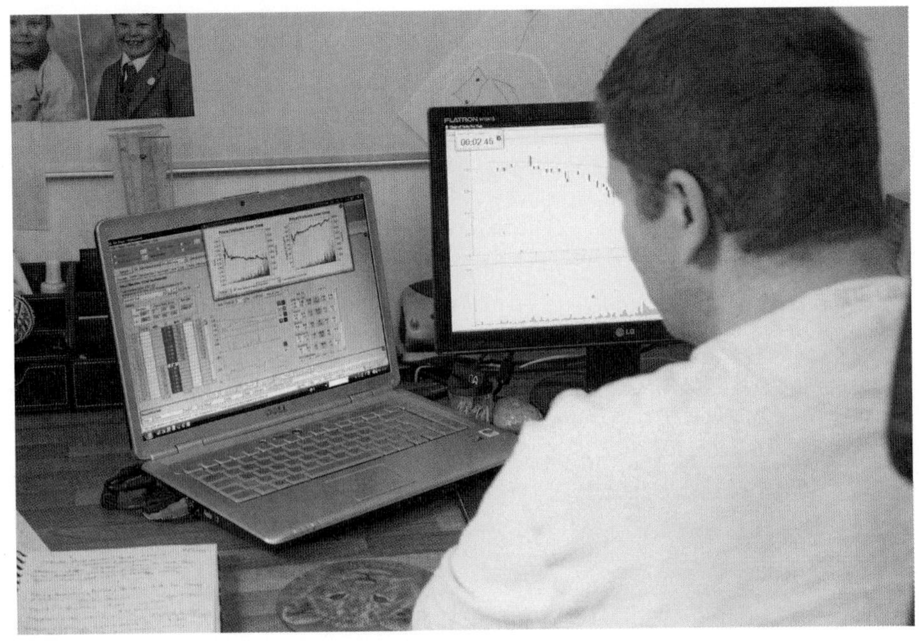

© David J. Green–lifestyle themes/Alamy

Stock traders use information systems to track stock performance.

to use hardware and software to increase profits, cut costs, improve productivity, and increase customer satisfaction is an example of information systems literacy. Information systems literacy often leads to more career opportunities and higher paying jobs.[91]

Information Systems in the Functional Areas of Business

Information systems are used in all functional areas and operating divisions of business. In finance and accounting, information systems forecast revenues and business activity, determine the best sources and uses of funds, manage cash and other financial resources, analyze investments, and perform audits to make sure that the organization is financially sound and that all financial reports and documents are accurate. Sales and marketing professionals use information systems develop new goods and services (product analysis), select the best location for production and distribution facilities (place or site analysis), determine the best advertising and sales approaches (promotion analysis), and set product prices to get the highest total revenues (price analysis). DHL, a large shipping company, for example, used information systems to improve its advertising efforts, which helped the company increase its stock value by more than $1 billion during a five-year period.[92] In manufacturing, information systems process customer orders, develop production schedules, control inventory levels, and monitor product quality. Procter & Gamble, for instance, was able to use information systems to control manufacturing processes and help reduce inventory levels and costs.[93] In addition, information systems help to design products (computer-assisted design, or CAD), manufacture items (computer-assisted manufacturing, or CAM), and integrate machines or pieces of equipment (computer-integrated manufacturing, or CIM). Human resource management uses information systems to screen applicants, administer performance tests to employees, monitor employee productivity, and more. Legal information systems analyze product liability and warranties and help to develop important legal documents and reports.

Information Systems in Industry

In addition to being used in every department in a company, information systems are used in almost every industry or field in business. The airline industry develops Internet auction sites to offer discount fares and increase revenue. Many managers in the automotive industry, for example, are increasingly using smartphones and mobile devices to do their work.[94] Sasol, a South African energy and chemical company, used an information system to streamline its production facilities through better information and control. This streamlining helped the company increase its stock value by more than $200 million.[95] Even art collectors are starting to buy hard-to-find works of art using online auctions instead of using traditional methods.[96] A Shang Dynasty vessel, for example, was purchased online for more than $3 million. Other artwork has been purchased online for thousands or hundreds of thousands of dollars. Investment firms use information systems to analyze stocks, bonds, options, the futures market, and other financial instruments, and provide improved services to their customers. Banks use information systems to help make sound loans and good investments as well as to provide online check payment for account holders. The transportation industry uses information systems to schedule trucks and trains to deliver goods and services at the lowest cost. Publishing companies use information systems to analyze markets and to develop and publish newspapers, magazines, and books. Health care organizations use information systems to diagnose illnesses, plan medical treatment, track patient records, and bill patients. Health maintenance organizations (HMOs) use Web technology to access patients' insurance eligibility and other information stored in data bases to cut patient costs. Retail companies are using the Web to take orders and provide customer service support. Retail companies also use information systems to help market products and services, manage inventory levels, control the supply chain, and forecast demand. Power management and utility companies use information systems to monitor and control power generation and usage. Professional services firms have information systems to improve the speed and quality of the services they provide to customers. Management consulting firms use intranets and extranets to offer information on products, services, skill levels, and past engagements to their consultants. These industries are discussed in more detail as we continue through the book.

GLOBAL CHALLENGES IN INFORMATION SYSTEMS

Having a global presence is a key to financial success for many organizations.[97] In his book *The World Is Flat*, Thomas Friedman describes three eras of globalization.[98] (See Table 1.3.) According to Friedman, we have progressed from the globalization of countries (Globalization 1.0) to the globalization of multinational corporations (Globalization 2.0) and individuals (Globalization 3.0). Today, people in remote areas can use the Internet to compete with and contribute to other people, the largest corporations, and entire countries. These workers are empowered by high-speed Internet access, making the world flatter. In the Globalization 3.0 era, designing a new airplane or computer can be separated into smaller subtasks and then completed by a

TABLE 1.3 Eras of globalization

Era	Dates	Characterized by
Globalization 1.0	Late 1400–1880	Countries with the power to explore and influence the world
Globalization 2.0	1800–2000	Multinational corporations that have plants, warehouses, and offices around the world
Globalization 3.0	2000–today	Individuals from around the world who can compete and influence other people, corporations, and countries by using the Internet and powerful technology tools

person or small group that can do the best job. These workers can be located in India, China, Russia, Europe, and other areas of the world. The subtasks can then be combined or reassembled into the complete design. This approach can be used to prepare tax returns, diagnose a patient's medical condition, fix a broken computer, and many other tasks.

Global markets have become more important today. Western Union, for instance, has recognized the importance of globalization in generating creative and innovative ideas for future products and services.[99] According to the chief information officer for the company, "We are undergoing a radical shift as more of our innovation is coming from remote points. There is a leapfrog effect taking place where developing countries are leveraging how cheap technology has become." People and companies can get products and services from around the world, instead of around the corner or across town. Fundamental ways of doing business can be very different for different parts of the world.[100] According to the chief information officer for Coca-Cola's Bottling Investments Group, "Building a sales force solution in Germany is very different than in China. In Germany, bottlers drive the truck right up to the retail outlet, but in China they work through distributors. You need to deliver a technology solution that provides value to each country's unique business needs now, but you must also build toward a common footprint." Global opportunities, however, introduce numerous obstacles and issues, including challenges involving culture, language, and many others:

- **Cultural challenges.** Countries and regional areas have their own cultures and customs that can significantly affect individuals and organizations involved in global trade.
- **Language challenges.** Language differences can make it difficult to translate exact meanings from one language to another.
- **Time and distance challenges.** Time and distance issues can be difficult to overcome for individuals and organizations involved with global trade in remote locations. Large time differences make it difficult to talk to people on the other side of the world. With long distance, it can take days to get a product or part from one location to another.
- **Infrastructure challenges.** High-quality electricity and water might not be available in certain parts of the world. Telephone services, Internet connections, and skilled employees might be expensive or not readily available.
- **Currency challenges.** The value of various currencies can vary significantly over time, making international trade more difficult and complex.
- **Product and service challenges.** Traditional products that are physical or tangible, such as an automobile or a bicycle, can be difficult to deliver to the global market. However, electronic products (e-products) and electronic services (e-services) can be delivered to customers electronically over the phone, through networks, through the Internet, or via other electronic means. Software, music, books, manuals, and advice can all be delivered globally and over the Internet.
- **Technology transfer issues.** Most governments don't allow certain military-related equipment and systems to be sold to some countries. Even so, some believe that foreign companies are stealing intellectual property, trade secrets, and copyrighted materials and counterfeiting products and services.
- **State, regional, and national laws.** Each state, region, and country has a set of laws that must be obeyed by citizens and organizations operating in the country. These laws can deal with a variety of issues, including trade secrets, patents, copyrights, protection of personal or financial data, privacy, and much more. Laws restricting how data enters or exits a country are often called transborder data-flow laws. Keeping track of these laws

and incorporating them into the procedures and computer systems of multinational and transnational organizations can be very difficult and time consuming, requiring expert legal advice.

- **Trade agreements.** Countries often enter into trade agreements with each other. The North American Free Trade Agreement (NAFTA) and the Central American Free Trade Agreement (CAFTA) are examples. The European Union (EU) is another example of a group of countries with an international trade agreement.[101] The EU is a collection of mostly European countries that have joined together for peace and prosperity. Additional trade agreements include the Australia-United States Free Trade Agreement (AUSFTA), signed into law in 2005, and the Korean-United States Free Trade Agreement (KORUS-FTA), signed into law in 2007. Free trade agreements have been established between Bolivia and Mexico, Canada and Costa Rica, Canada and Israel, Chile and Korea, Mexico and Japan, the United States and Jordan, and many others.[102]

SUMMARY

Principle:

The value of information is directly linked to how it helps decision makers achieve the organization's goals.

Information systems are used in almost every imaginable career area. Regardless of your college major or chosen career, you will find that information systems are indispensable tools to help you achieve your career goals. Learning about information systems can help you get your first job, earn promotions, and advance your career.

Data consists of raw facts; information is data transformed into a meaningful form. The process of defining relationships among data requires knowledge. Knowledge is an awareness and understanding of a set of information and the way that information can support a specific task. To be valuable, information must have several characteristics: It should be accurate, complete, economical to produce, flexible, reliable, relevant, simple to understand, timely, verifiable, accessible, and secure. The value of information is directly linked to how it helps people achieve their organizations' goals.

Principle:

Computers and information systems help make it possible for organizations to improve the way they conduct business.

A system is a set of elements that interact to accomplish a goal or set of objectives. The components of a system include inputs, processing mechanisms, and outputs. A system uses feedback to monitor and control its operation to make sure that it continues to meet its goals and objectives.

System performance is measured by its efficiency and effectiveness. Efficiency is a measure of what is produced divided by what is consumed; effectiveness measures the extent to which a system achieves its goals. A systems performance standard is a specific objective.

Principle:

Knowing the potential impact of information systems and having the ability to put this knowledge to work can result in a successful personal career and in organizations that reach their goals.

Information systems are sets of interrelated elements that collect (input), manipulate and store (process), and disseminate (output) data and information. Input is the activity of capturing and gathering new data, processing involves converting or transforming data into useful outputs, and output involves producing useful information. Feedback is the output that is used to make adjustments or changes to input or processing activities.

The components of a computer-based information system (CBIS) include hardware, software, databases, telecommunications and the Internet, people, and procedures. The types of CBISs that organizations use can be classified into four basic groups: (1) e-commerce and m-commerce, (2) TPS and ERP systems, (3) MIS and DSS, and (4) knowledge management and specialized business information systems. The key to understanding these types of systems begins with learning their fundamentals.

E-commerce involves any business transaction executed electronically between parties such as companies (business-to-business), companies and consumers (business-to-consumer), business and the public sector, and consumers and the public sector. The major volume of e-commerce and its fastest-growing segment is business-to-business transactions that make purchasing easier for big corporations. E-commerce also offers opportunities for small businesses to market and sell at a low cost worldwide, thus allowing them to enter the global market right from start-up. M-commerce involves anytime, anywhere computing that relies on wireless networks and systems.

The most fundamental system is the transaction processing system (TPS). A transaction is any business-related exchange. The TPS handles the large volume of business transactions that occur daily within an organization. An enterprise resource planning (ERP) system is a set of integrated programs that can manage the vital business operations for an entire multisite, global organization. A management information system (MIS) uses the information from a TPS to generate information useful for management decision making.

A decision support system (DSS) is an organized collection of people, procedures, databases, and devices that help make problem-specific decisions. A DSS differs from an MIS in the support given to users, the emphasis on decisions, the development and approach, and the system components, speed, and output.

Specialized business information systems include knowledge management, artificial intelligence, expert systems, multimedia, and virtual reality systems. Knowledge management systems are organized collections of people, procedures, software, databases, and devices used to create, store, share, and use the organization's knowledge and experience. Artificial intelligence (AI) includes a wide range of systems in which the computer takes on the characteristics of human intelligence. Robotics is an area of artificial intelligence in which machines perform complex, dangerous, routine, or boring tasks, such as welding car frames or assembling computer systems and components. Vision systems allow robots and other devices to have "sight" and to store and process visual images. Natural language processing involves computers interpreting and acting on verbal or written commands in English, Spanish, or other human languages. Learning systems let computers "learn" from past mistakes or experiences, such as playing games or making business decisions, while neural networks is a branch of artificial intelligence that allows computers to recognize and act on patterns or trends. An expert system (ES) is designed to act as an expert consultant to a user who is seeking advice about a specific situation. Originally, the term "virtual reality" referred to immersive virtual reality in which the user becomes fully immersed in an artificial, computer-generated 3D world. Virtual reality can also refer to applications that are not fully immersive, such as mouse-controlled navigation through a 3D environment on a graphics monitor, stereo viewing from the monitor via stereo glasses, and stereo projection systems. Augmented reality, a newer form of virtual reality, has the potential to superimpose digital data over real photos or images. Multimedia is a natural extension of virtual reality. It can include photos and images, the manipulation of sound, and special 3D effects.

Principle:

System users, business managers, and information systems professionals must work together to build a successful information system.

Systems development involves creating or modifying existing business systems. The major steps of this process and their goals include systems investigation (gain a clear understanding of what the problem is), systems analysis (define what the system must do to solve the problem), systems design (determine exactly how the system will work to meet the business needs), systems implementation (create or acquire the various system components defined in the design step), and systems maintenance and review (maintain and then modify the system so that it continues to meet changing business needs).

Principle:

Information systems must be applied thoughtfully and carefully so that society, business, and industry around the globe can reap their enormous benefits.

Information systems play a fundamental and ever-expanding role in society, business, and industry. But their use can also raise serious security, privacy, and ethical issues. Effective information systems can have a major impact on corporate strategy and organizational success. Businesses around the globe are enjoying better safety and service, greater efficiency and effectiveness, reduced expenses, and improved decision making and control because of information systems. Individuals who can help their businesses realize these benefits will be in demand well into the future.

Computer and information systems literacy are prerequisites for numerous job opportunities, and not only in the IS field. Computer literacy is knowledge of computer systems, software, and equipment; information systems literacy is knowledge of how data and information are used by individuals, groups, and organizations. Today, information systems are used in all the functional areas of business, including accounting, finance, sales, marketing, manufacturing, human resource management, and legal. Information systems are also used in every industry, such as airlines, investment firms, banks, transportation companies, publishing companies, health care, retail, power management, professional services, and more.

Changes in society as a result of increased international trade and cultural exchange, often called globalization, has always had a significant impact on organizations and their information systems. In his book *The World Is Flat,* Thomas Friedman describes three eras of globalization, spanning the globalization of countries to the globalization of multinational corporations and individuals. Today, people in remote areas can use the Internet to compete with and contribute to other people, the largest corporations, and entire countries. People and companies can get products and services from around the world, instead of around the corner or across town. These opportunities, however, introduce numerous obstacles and issues, including challenges involving culture and language.

CHAPTER 1: SELF-ASSESSMENT TEST

The value of information is directly linked to how it helps decision makers achieve the organization's goals.

1. A(n) ________________ is a set of interrelated components that collect, manipulate, and disseminate data and information and provide a feedback mechanism to meet an objective.
2. What consists of raw facts, such as an employee number?
   a. bytes
   b. data
   c. information
   d. knowledge

3. Knowledge is the awareness and understanding of a set of information and the ways that information can be made useful to support a specific task or reach a decision. True or False?

Computers and information systems help make it possible for organizations to improve the way they conduct business.

4. A(n) ________________ is a set of elements or components that interact to accomplish a goal.
5. A measure of what is produced divided by what is consumed is known as ________________.
   a. efficiency
   b. effectiveness
   c. performance
   d. productivity
6. A specific objective of a system is called a performance standard. True or False?

Knowing the potential impact of information systems and having the ability to put this knowledge to work can result in a successful personal career and in organizations that reach their goals.

7. A(n) ________________ consists of hardware, software, databases, telecommunications, people, and procedures.
8. Computer programs that govern the operation of a computer system are called ________________.
   a. feedback
   b. feedforward
   c. software
   d. transaction processing systems
9. Augmented reality is a newer form of virtual reality and has the potential to superimpose digital data over real photos or images. True or False?
10. What is an organized collection of people, procedures, software, databases, and devices used to create, store, share, and use the organization's experience and knowledge?
    a. TPS (transaction processing system)
    b. MIS (management information system)
    c. DSS (decision support system)
    d. KM (knowledge management)
11. ________________ is a set of integrated programs that manage vital business operations.

System users, business managers, and information systems professionals must work together to build a successful information system.

12. What defines the problems and opportunities of an existing system?
    a. systems analysis
    b. systems review
    c. systems development
    d. systems design

Information systems must be applied thoughtfully and carefully so that society, business, and industry around the globe can reap their enormous benefits.

13. ________________ literacy is knowledge of how data and information are used by individuals, groups, and organizations.

CHAPTER 1: SELF-ASSESSMENT TEST ANSWERS

1. information system
2. b
3. True
4. system
5. a
6. True
7. computer-based information system (CBIS)
8. c
9. True
10. d
11. Enterprise resource planning (ERP)
12. a
13. Information systems

REVIEW QUESTIONS

1. What is an information system? What are some of the ways information systems are changing our lives?
2. How is data different from information? Information from knowledge?
3. Describe the various types of data.
4. What is the difference between efficiency and effectiveness?
5. What are the components of any information system?

6. What is feedback? What are possible consequences of inadequate feedback?
7. How is system performance measured?
8. What is the difference between data, information, and knowledge? Give examples of each.
9. What is a computer-based information system? What are its components?
10. Describe the characteristics of a decision support system.
11. What is the difference between an intranet and an extranet?
12. What is m-commerce? Describe how it can be used.
13. What are the most common types of computer-based information systems used in business organizations today? Give an example of each.
14. What is the difference between virtual reality and augmented reality?
15. What are computer literacy and information systems literacy? Why are they important?
16. What are some of the benefits organizations seek to achieve through using information systems?
17. Identify the steps in the systems development process and state the goal of each.

DISCUSSION QUESTIONS

1. Why is the study of information systems important to you? What do you hope to learn from this course to make it worthwhile?
2. List the ways an information system can be used in a career area of interest to you.
3. What is the value of software? Give several examples of software you use at school or home.
4. Why is a database an important part of a computer-based information system?
5. What is the difference between e-commerce and m-commerce?
6. What is the difference between DSS and knowledge management?
7. Describe how tablet computers can be used in a DSS.
8. Describe the "ideal" automated class registration system for a college or university. Compare this "ideal" system with what is available at your college or university.
9. What computer application needs the most improvement at your college or university? Describe how systems development could be used to develop it.
10. Discuss how information systems are linked to the business objectives of an organization.
11. For an industry of your choice, describe how a CBIS could be used to reduce costs or increase profits.

PROBLEM-SOLVING EXERCISES

1. Prepare a data disk and a backup disk (or USB flash drives) for the problem-solving exercises and other computer-based assignments you will complete in this class. Create one folder for each chapter in the textbook (you should have 14 folders). As you work through the problem-solving exercises and complete other work using the computer, save your assignments for each chapter in the appropriate folder. On the label of each disk or USB flash drive, be sure to include your name, course, and section. On one disk write "Working Copy"; on the other write "Backup."
2. Search through several business magazines (*Bloomberg Businessweek*, *Computerworld*, *PC Week*, etc.) or use an Internet search engine to find recent articles that describe potential social or ethical issues related to the use of an information system. Use word-processing software to write a one-page report summarizing what you discovered.
3. Create a table that lists 10 or more possible career areas, annual salaries, and brief job descriptions, and rate how much you would like the career area on a scale from 1 (don't like) to 10 (like the most). Print the results. Sort the table according to annual salaries from high to low and print the resulting table. Sort the table from the most liked to least liked and print the results.
4. Use a graphics program to create a diagram showing a billing transaction processing system, similar to the payroll transaction processing

figure shown in this chapter. Your diagram should show how a company collects sales data and sends out bills to customers. Share your findings with the class.

TEAM ACTIVITIES

1. Before you can do a team activity, you need a team! As a class member, you might create your own team, or your instructor might assign members to groups. After your group has been formed, meet and introduce yourselves to each other. Find out the first name, hometown, major, e-mail address, and phone number of each member. Find out one interesting fact about each member of your team as well. Brainstorm a name for your team. Put the information on each team member into a database and print enough copies for each team member and your instructor.
2. With the other members of your group, use word-processing software to write a summary of the members of your team, the courses each team member has taken, and the expected graduation date of each team member. Send the report to your instructor via e-mail.

WEB EXERCISES

1. Throughout this book, you will see how the Internet provides a vast amount of information to individuals and organizations. We will stress the World Wide Web, or simply the Web, which is an important part of the Internet. Most large universities and organizations have an address on the Internet, called a Web site or home page. The address of the Web site for this publisher is *www.cengage.com*. You can gain access to the Internet through a browser, such as Microsoft Internet Explorer or Safari. Using an Internet browser, go to the Web site for this publisher. What did you find? Try to obtain information on this book. You might be asked to develop a report or send an e-mail message to your instructor about what you found.
2. Go to an Internet search engine, such as *www.google.com* or *www.bing.com*, and search for information about a specialized business information system, such as robotics, an expert system, or multimedia. Write a brief report that summarizes what you found.
3. Using the Internet, search for information on the use of knowledge management in a company or organization that interests you. Write a brief report that summarizes what you found.

CAREER EXERCISES

1. In the Career Exercises found at the end of every chapter, you will explore how material in the chapter can help you excel in your college major or chosen career. Write a brief report on the career that appeals to you the most. Do the same for two other careers that interest you.
2. Research two or three possible careers that interest you. Describe the job opportunities, job duties, and the possible starting salaries for each career area in a report.

CASE STUDIES

Case One

Business before Technology at Campbell's Soup

Soup. Canned soup. It's been around for ages. What could be said about it that hasn't been said? What could be done to improve soup that hasn't already been done—probably when your grandparents were in school?

Soup companies still find room for improvement. Your grandparents may not have cared about using less salt, but you do. Other opportunities for improvement, however, may not involve the product itself but the business processes used to produce and distribute it. Information systems can do a great deal to improve business processes. The challenge a company faces today is to find those opportunities for improvement. These opportunities are not in the same places that senior managers looked for them a decade or two ago.

Joseph Spagnoletti, senior vice president and CIO of Campbell's Soup, is the person who must figure out how to improve business processes at Campbell's. His challenge is to find opportunities for improvement and take advantage of them. In a recent interview, he expressed his approach to improving business processes this way: "You focus on the enabling strategies at a company level, and those are … the primary driver. Second, you look at value. And there's economic, noneconomic, and strategic values, and again you're making trade-offs between them. So when you're trying to decide, you ask how it fits with our [company's] strategies and, second, how [it creates value] economically and foundationally."

In other words, Spagnoletti puts business ahead of technology and uses technology to support Campbell's business. When asked what technologies are the most important for giving companies a competitive edge, Spagnoletti said, "The big shifts are mobility and having the ability to be out in front of the consumers with information.… It's about shoppers' behaviors and patterns, local demographics and data.… How do you bring your brand to the world, and how do you represent your company in mobile and social media?" As business managers now recognize, addressing these behaviors has more impact on a company's bottom line than using spreadsheets and word processors or processing payroll and accounts receivable.

To gain a competitive edge as Spagnoletti suggests, a company can't focus on technology. As he put it at the Enterprise CIO Forum in March 2011, "It's less about what device and what specific applications. It's all about the information and the delivery mechanism."

However, Spagnoletti says it would also be a mistake to let business completely drive technology. They're interdependent. Earlier in that interview, he said, "I don't think you can separate the two [business and technology]. You have to think about them as being intertwined or interwoven. It's all about making information more visible to suppliers or consumers." Spagnoletti's focus is on information and getting it to those who can use it.

What does Spagnoletti's boss think of this focus? Campbell CEO-elect Denise Morrison says, "I'm a firm believer in cross-functional teamwork, and IT needs to be at that table." Clearly, Campbell's sees information systems as vital to its future.

Discussion Questions

1. Spagnoletti received his undergraduate degree in computer science and spent his entire career in information systems before being appointed Campbell's CIO in August 2008. Given his present responsibilities, do you think that career path is still appropriate for someone who wants to be a CIO today?
2. In two different interviews, Spagnoletti discussed (a) making information more visible to consumers and (b) using social media such as Facebook and Twitter. How do you think a company can use social media to make information visible to consumers?

Critical Thinking Questions

1. Campbell's is a large company, with 2011 revenues of over $7.5 billion. How can Spagnoletti's focus on delivering information to consumers and suppliers work for smaller organizations? Consider three types of organizations: a college with about 2,000 students, the police department of a city with a population of about 250,000, and a family-owned chain of five car dealerships in the same region.
2. Spagnoletti's focus on information suggests that you can use any device you want as long as it can handle and deliver the necessary information. Still, many companies settle on one kind of computer, one operating system, and one application for a specific purpose. Do you think he means that this standardization is no longer necessary, or should companies continue to standardize for other reasons?

SOURCES: Campbell's Web site, "Executive Team: Joseph C. Spagnoletti," 2011, *www.campbellsoupcompany.com/bio_spagnoletti.asp*, accessed November 30, 2011; EnterpriseCIOForum, "Business Transformation through IT at the Campbell Soup Co., Part 1," (video), June 14, 2011, *www .youtube.com/watch?v= hxpuJ7QRz4g*, accessed November 30, 2011; EnterpriseCIOForum, "Joseph Spagnoletti on Business Transformation" (video), March 3, 2011, *www.youtube.com/watch?v=AVGqlt8pKbs*, accessed November 30, 2011; Pratt, M., "The Grill: Joseph Spagnoletti," *Computerworld*, November 21, 2011, *www.computerworld.com/s/article/ 359405/Joe_Spagnoletti*, accessed November 30, 2011.

Case Two

Shoes Plus Information: A Winning Formula

Skechers USA, Inc., a $2-billion-a-year company, describes itself as "an award-winning global leader in the lifestyle footwear industry, [and] designs, develops and markets lifestyle footwear that appeals to men, women, and children of all ages.… With more than 3,000 styles, Skechers meets the needs of male and female consumers across every age and demographic."

Any shoe company could say something similar. What separates one from another? Increasingly, it isn't the shoes. It's the *information*.

Information systems are woven into every part of Skechers's business. Its recent investment in Oracle applications, including cloud computing (introduced in the "Telecommunications, Networks, and the Internet" section of this chapter) demonstrates the company's commitment to information systems. Mark Bravo, Skechers senior vice president of finance, says, "As we manage growth, we are establishing a business structure that lowers costs and creates more value and flexibility across the business. The … cloud services help us to lighten our IT overhead and enable us to respond more quickly to market opportunities." Therefore, it was natural that Skechers would turn to information systems to help with customer retention.

In a fast-moving consumer product category like shoes, using information to understand, attract, and retain customers is even more important than having the latest technology. Many companies use loyalty programs to help retain customers. A pizza shop might give its customers a card that is punched every time they buy a pizza. When the card has 10 punches, the customer can order a free medium pizza with two toppings. Loyalty programs reduce the chances of a regular customer switching suppliers even if another shop sells pizza for less during a promotion or offers a different advantage.

After Skechers decided to offer a loyalty program, their challenge was this: How to design the program for greatest sales impact? The company had to balance ease of earning rewards, the value of the rewards, and other factors so they gave away as little as possible while retaining as many loyal customers as possible. In the pizza shop, a free pizza after buying five might cost too much revenue; a free pizza after twenty might put the rewards too far out in the future to be attractive. Ten is a good middle ground.

The loyalty program that Skechers designed, planned jointly by their marketing and information systems departments, is called Skechers Elite. Members earn free merchandise ($10 credit for every $150 spent), get free shipping, and enjoy special promotions. In addition, Gold members (who spend at least $750 on Skechers shoes in a calendar year) and Platinum members ($1,000) get higher merchandise credits, sneak peeks at future products, and earn other higher benefits.

Skechers couldn't operate Skechers Elite without information systems. The system that supports this loyalty program records information about members, their purchases, and the rewards they're entitled to, so members can track their participation online. In addition, the system provides Skechers's management with information about the purchase patterns of regular customers, such as shoe designs that appeal to them. The system also lets Skechers send targeted promotional materials to its best customers.

Does this use of information technology pay off? According to analyst Peter Chu, it does. He found on November 2, 2011, that Skechers (SKX) stock performance outpaced that of the other shoe manufacturers he tracked. He considers that performance "a bullish sign of underlying fundamental and technical strength."

Discussion Questions

1. Which information systems applications described in the case are unique to Skechers and would not benefit other shoe manufacturers? Which aspects of their loyalty program could other firms duplicate and quickly benefit from? Which would take competitors longer to use or offer?
2. What kind of information does the Skechers Elite program use? Aside from its direct benefit in increasing customer loyalty, what other benefits might the program have? How could Skechers use the information in its planning and sales activities?

Critical Thinking Questions

1. In the five years from 2005 to 2010, Skechers approximately doubled its revenue: from about $1 billion to about $2 billion. This is a high growth rate. How do you think this growth affected their spending on information technology?
2. In its 2011 annual report, Skechers warns investors that "Many of our competitors are larger, have been in existence for a longer period of time, have achieved greater recognition for their brand names, have captured greater market share and/or have substantially greater financial, distribution, marketing, and other resources than we do." Part of this wording is required by financial regulations, but it is nonetheless true. How can intelligent use of information systems help Skechers overcome the difficulties described in its annual report?

SOURCES: Chu, Peter, "Skechers U.S.A. Has the Best Relative Performance in the Footwear Industry," Financial News Network Online, November 2, 2011, www.fnno.com/story/fast-lane/331-skechers-usa-has-best-relative-performance-footwear-industry-skx-crox-shoo-icon-nke-fast-lane, accessed November 6, 2011; Oracle, "Skechers Leverages Oracle Applications, Business Intelligence, and On Demand Offerings to Drive Long-Term Growth," News Release, June 6, 2011, http://emeapressoffice.oracle.com/Press-Releases/Skechers-Leverages-Oracle-Applications-Business-Intelligence-and-On-Demand-Offerings-to-Drive-Long-Term-Growth-1e2d.aspx, accessed November 6, 2011; Skechers corporate Web sites, *www.skechers.com* and *www.skx.com*, accessed November 6, 2011; 2011 annual report downloaded from *http://skx.com/investor.jsp?p=2*.

Questions for Web Case

See the Web site for this book to read about the Altitude Online case for this chapter. The following questions cover this Web case.

Altitude Online: Outgrowing Systems

Discussion Questions

1. Why do you think it's a problem for Altitude Online to use different information systems in its branch locations?
2. What information do you think Jon should collect from the branch offices to plan the new centralized information system?

Critical Thinking Questions

1. With Jon's education and experience, he could design and implement a new information system for Altitude Online himself. What would be the benefits and drawbacks of doing the job himself compared to contracting with an information systems contractor?
2. While Jon is visiting the branch offices, how might he prepare them for the inevitable upheaval caused by the upcoming overhaul to the information system?

NOTES

Sources for the opening vignette: Government site of Castilla-La Mancha, *www.jccm.es* (in Spanish), accessed October 29, 2011; Government of Castilla-La Mancha, "Case Study: Castilla-La Mancha Government," submitted as *Computerworld* case study, *https://www.eiseverywhere.com/file_uploads/eedce33bed14338d4f98066e49364b82_Castila_la_Mancha_Government_-_Vblock_Castilla_la_Mancha.pdf*, accessed October 27, 2011; VCE, "Regional Government Creates New Collaborative Cloud Model," *www.vce.com/pdf/solutions/vce-case-study-castilla-la-mancha.pdf*, accessed October 27, 2011; VCE, the Virtual Computing Environment Company Web site, *http://vce.com*, accessed October 29, 2011; VCE, Castilla-La Mancha project video (partly in English, partly in Spanish), *www.vce.com/media/videos/vce-customer-castilla-la-mancha.htm*, accessed October 27, 2011.

1. Hickins, M., "Groupon Revenue Hit $760 Million, CEO Memo Shows," *The Wall Street Journal*, February 26, 2001, p. B3.
2. MIT Open Course Ware home page, *http://ocw.mit.edu/OcwWeb/web/home/home/index.htm*, accessed June 27, 2011.
3. Chao, Loretta and Solomon, Jay, "U.S. Steps Up Web Freedom Efforts," *The Wall Street Journal*, February 17, 2011, p. A15.
4. Pratt, Mary, "The Grill," *Computerworld*, January 24, 2011, p. 12.
5. Ravishankar, M., et al, "Examining the Strategic Alignment and Implementation Success of a KMS," *Information Systems Research*, March 2011, p. 39.
6. Betts, Mitch, "IT Leader Builds a Know-how Network," *Computerworld*, April 18, 2011, p. 4.
7. Ko, D. and Dennis, A., "Profiting from Knowledge Management: The Impact of Time and Experience," *Information Systems Research*, March 2011, p. 134.
8. King, Julia, "Extreme Automation," *Computerworld*, June 6, 2011, p. 16.
9. DeVries, J., "May the Best Algorithm Win ..." *The Wall Street Journal*, May 16, 2011, p. B4.
10. Schultz, B., "Florida Hospice Saves with SaaS-based CRM," *Network World*, June 6, 2011, p. 24.
11. Eaglesham, J. and Strasburg, J., "Big Fine over Bug in Quant Program," *The Wall Street Journal*, February 4, 2011, p. C1.
12. Sanna, N., "Getting Predictive About IT," *CIO*, February 1, 2011, p. 39.
13. Wakabayashi, D., "Japanese Farms Look to the Cloud," *The Wall Street Journal*, January 18, 2011, p. B5.
14. Staff, "Cruise Control," *CIO*, April 1, 2011, p. 22.
15. Thibodeau, Patrick, "VW Reorganizes IT to Help Develop 'Rolling Computers,'" *Computerworld*, June 6, 2011, p. 6.
16. Murphy, Chris, "Create," *Information Week*, March 14, 2011, p. 23.
17. Munoz, D., et al, "INDEVAL Develops a New Operating and Settlement System Using Operations Research," *Interfaces*, January-February 2011, p. 8.
18. Bennett, D., "Innovator," *Bloomberg Businessweek*, January 31, 2011, p. 40.
19. SanDisk Web site, *www.sandisk.com*, accessed June 28, 2011.
20. Lawrence Livermore National Laboratory home page, *www.llnl.gov*, accessed June 28, 2011.
21. IBM Web site, *www.ibm.com/systems/deepcomputing/bluegene*, accessed June 28, 2011.
22. Apple Web site, *www.apple.com*, accessed June 28, 2011.
23. One Laptop per Child Web site, *http://one.laptop.org*, accessed June 28, 2011.
24. Staff, "A Quick Tablet Tumble," *Network World*, March 7, 2011, p. 14.
25. Peters, Jeremy, "A Technological Divide," *The Tampa Tribune*, January 24, 2011, p. 6.
26. Hamblen, Matt, "Chevron, TD Bank Hope to Tap Tablets' Potential," *Computerworld*, February 21, 2011, p. 6.
27. Cheng, Roger, "So You Want to Use your iPhone for Work?" *The Wall Street Journal*, April 25, 2011, p. R1.
28. Gartner Webinar, *www.gartner.com/it/content/1462300/1462334/december_15_top_predictions_for_2011_dplummer.pdf*, accessed June 7, 2011.
29. Fab Lab, *http://fab.cba.mit.edu*, accessed June 28, 2011.
30. Salesforce Web site, *www.salesforce.com*, accessed June 28, 2011.
31. Microsoft Web site, *www.microsoft.com*, accessed July 24, 2011.
32. "Adobe Creative Suite 5.5," *www.adobe.com/products/creativesuite*, accessed June 28, 2011.
33. Nvidia Web site, *www.nvidia.com/object/3d-vision-main.html*, accessed June 28, 2011.
34. Howson, C., "Predictions of What's in Store for the BI Market," *Information Week*, January 31, 2011, p. 22.
35. Carreyrou, J. and McGinty, T., "Medicare Records Reveal Troubling Trail of Surgeries," *The Wall Street Journal*, March 29, 2011, p. A1.
36. Kane, Y. and Smith, E., "Apple Readies iCloud Service," *The Wall Street Journal*, June 1, 2011, p. B1.
37. Greene, Tim, "Library System Shushes MPLS for Cheaper DSL," *Network World*, January 24, 2011, p. 14.
38. Boeing Web site, *www.boeing.com/defense-space/military/scaneagle/index.html*, accessed June 28, 2011.
39. Schechner, S. and Vascellaro, J., "Hulu Reworks Its Script," *The Wall Street Journal*, January 27, 2011, p. A1.
40. Sheridan, B., "The Apps Class of 2010," *Bloomberg Businessweek*, January 3, 2011, p. 80.
41. Vance, Ashlee, "The Power of Cloud Computing," *Bloomberg Businessweek*, March 3, 2011.
42. Niccolai, J. and Thibodeau, P., "Top Tech Vendors Renew Cloud Push," *Computerworld*, April 18, 2011, p. 6.
43. Sheridan, B., "The Apps Class of 2010," *Bloomberg Businessweek*, January 3, 2011, p. 80.

44. Vance, Ashlee, "The Power of Cloud Computing," *Bloomberg Businessweek*, March 3, 2011.
45. Bank of America Web site, *www.bankofamerica.com*, accessed June 28, 2011.
46. Twitter Web site, *www.twitter.com*, accessed June 28, 2011.
47. Crovitz, L. Gordon, "Tunisia and Cyber Utopia," *The Wall Street Journal*, January 24, 2001, p. A15.
48. Page, Jeremy, "China Co-Opts Social Media to Head Off Unrest," *The Wall Street Journal*, February 22, 2011, p. A8.
49. Angwin, Julia, "Web Tool on Firefox to Deter Tracking," *The Wall Street Journal*, January 24, 2011, p. B1.
50. ING Direct Web site, *www.ingdirect.ca/en*, accessed June 28, 2011.
51. Penske Web site, *www.MyFleetAtPenske.com*, accessed July 24, 2011.
52. FedEx Web site, *www.fedex.com*, accessed July 24, 2011.
53. Strong, Dennis, "CIO Profiles," *Information Week*, March 14, 2011, p. 12.
54. Pratt, Mary, "Turning Business Objectives into IT Opportunities," *Computerworld*, February 21, 2011, p. 56.
55. Gohring, Nancy, "Seeding the Cloud," *CIO*, April 1, 2011, p. 27.
56. Murphy, Chris, "Create," *InformationWeek*, March 14, 2011, p. 23.
57. Ticketmaster Web site, *www.ticketmaster.com*, accessed June 28, 2011.
58. Efrati, A. and Sidel, R., "Google Sets Role in Mobile Payment," *The Wall Street Journal*, March 28, 2011, p. B1.
59. Square Web site, *https://squareup.com*, accessed March 22, 2011.
60. Nassauer, Sarah, "Art in Aisle 5," *The Wall Street Journal*, June 22, 2011, p. D1.
61. Morrison, S. and Fowler, G., "eBay Pushes into Amazon Turf," *The Wall Street Journal*, March 29, 2011, p. B1.
62. Munoz, D., et al, "INDEVAL Develops a New Operating and Settlement System Using Operations Research," *Interfaces*, January-February 2011, p. 8.
63. Tam, Pui-Wing, et al, "A Venture-Capital Newbie Shakes up Silicon Valley," *The Wall Street Journal*, May 10, 2011, p. A1.
64. Dell Web site, *www.dell.com*, accessed June 28, 2011.
65. Steel, E. and Fowler, S., "Facebook Gets New Friends," *The Wall Street Journal*, April 4, 2011, p. B4.
66. Wakabayashi, D., "Japanese Farms Look to the Cloud," *The Wall Street Journal*, January 18, 2011, p. B5.
67. Carreyrou, J. and McGinty, T., "Medicare Records Reveal Troubling Trail of Surgeries," *The Wall Street Journal*, March 29, 2011, p. A1.
68. Healthland Web site, *www.healthland.com*, accessed June 28, 2011.
69. Chen, A., et al, "Knowledge Life Cycle, Knowledge Inventory, and Knowledge Acquisition," *Decision Sciences*, February 2010, p. 21.
70. Kurzweil, R., "When Computers Beat Humans on Jeopardy," *The Wall Street Journal*, February 17, 2011, p. A19.
71. Staff, "Space Station Assistant Unpacked," *The Tampa Tribune*, March 17, 2011, p. 2.
72. Honda Web site, *www.honda.com*, accessed June 28, 2011.
73. Staff, "SA Photonics Develops an Advanced Digital Night Vision System," *Business Wire*, April 19, 2001.
74. Staff, "AlchemyAPI Announces Major Updates to Natural Language Processing Service," *PR Newswire*, May 23, 2011.
75. EZ-Xpert Expert System Web site, *www.ez-xpert.com*, accessed June 6, 2011.
76. Imprint Web site, *www.imprint-mis.co.uk*, accessed June 16, 2011.
77. Gamerman, E., "Animation Nation," *The Wall Street Journal*, February 11, 2011, p. D1.
78. Boeing: Commercial Airplanes—747 home page, *www.boeing.com/commercial/787family*, accessed June 28, 2011.
79. Boehret, K., "Why Smartphones Can See More Than We Can," *The Wall Street Journal*, May 4, 2011, p. D3.
80. Thibodeau, P., "Tax Law May Accelerate IT Purchases," *Computerworld*, February 7, 2011, p. 4.
81. Evans, Bob, "Why CIOs Must Have a Tablet Strategy," *Information Week*, March 14, 2001, p. 10.
82. Brousell, Lauren, "IT Budgets Bulge in 2011," *CIO*, February 1, 2011, p. 8.
83. Angwin, Julia, "Web Tool on Firefox to Deter Tracking," *The Wall Street Journal*, January 24, 2011, p. B1.
84. Simon, S., "Wyoming Plays It Cool," *The Wall Street Journal*, March 8, 2011, p. A3.
85. Gjelten, Tom, "Cyberthieves Target European Carbon Credit Market," *National Public Radio, www.npr.gov*, accessed January 22, 2011.
86. Gartner Webinar, *www.gartner.com/it/content/1462300/1462334/december_15_top_predictions_for_2011_dplummer.pdf*, accessed June 7, 2011.
87. Connors, Will, et al, "For BlackBerry Maker, Crisis Mounts," *The Wall Street Journal*, October 13, 2011, p. A1.
88. Angwin, J. and Steel, E., "Web's Hot New Commodity: Privacy," *The Wall Street Journal*, February 28, 2011, p. A1.
89. Miller, John, "Yahoo Cookie Plan in Place," *The Wall Street Journal*, March 19, 2011, p. B3.
90. Borzo, Jeanette, "Employers Tread a Minefield," *The Wall Street Journal*, January 21, 2011, p. B6.
91. King, Julia, "Premier 100 IT Leaders of 2011," *Computerworld*, February 21, 2011, p. 16.
92. Fischer, M., et al, "Managing Global Brand Investments at DHL," *Interfaces*, January-February 2011, p. 35.
93. Farasyn, I., et al, "Inventory Optimization at Procter & Gamble," *Interfaces*, January-February 2011, p. 66.
94. Hamblen, Matt, "Chevron, TD Bank Hope to Tap Tablets' Potential," *Computerworld*, February 21, 2011, p. 6.
95. Meer, M., et al, "Innovative Decision Support in a Petrochemical Production Environment," *Interfaces*, January–February 2011, p. 79.
96. Gamerman, E. and Crow, K., "Clicking on a Masterpiece," *The Wall Street Journal*, January 14, 2011, p. C1.

97. Evans, Bob, "The Top 10 CIO Priorities and Issues for 2011," *InformationWeek*, January 31, 2011, p. 12.
98. Friedman, Thomas, *The World Is Flat*, New York: Farrar, Straus and Giroux, 2005, p. 488.
99. Heller, Martha, "Global Strategy, Local Tactics," *CIO*, April 1, 2011, p. 40.
100. Ibid.
101. European Commission Web site, *http://ec.europa.eu/trade/index_en.htm*, accessed June 28, 2011.
102. Foreign Trade Information Web site, *www.sice.oas.org/agreements_e.asp*, accessed June 28, 2011.

2 Information Systems in Organizations

Principles	Learning Objectives
• The use of information systems to add value to the organization is strongly influenced by organizational structure, culture, and change.	• Identify the value-added processes in the supply chain and describe the role of information systems within them. • Provide a clear definition of the terms "organizational structure," "culture," and "change" and discuss how they affect the type of information systems that the organization implements.
• Because information systems are so important, businesses need to be sure that improvements or completely new systems help lower costs, increase profits, improve service, or achieve a competitive advantage.	• Define the term "competitive advantage" and identify the factors that lead firms to seek competitive advantage. • Discuss strategic planning for competitive advantage. • Describe how the performance of an information system can be measured.
• IS personnel is a key to unlocking the potential of any new or modified system.	• Define the types of roles, functions, and careers available in the field of information systems.

Information Systems in the Global Economy
TESCO, UNITED KINGDOM

Connecting with Customers through Information

Tesco has come a long way since it began as a market stall selling surplus groceries in London's East End in 1919. It is now the U.K.'s largest food seller, though the company has expanded into general merchandise as well. It operates in 14 countries across Europe, Asia, and North America, has over 5,000 stores (about half outside the U.K.), and has annual revenues of £67.6 billion for fiscal 2011, equivalent to US $107 billion at February 2012 exchange rates.

Despite its history of nearly a century, Tesco is up to date with today's information systems. One way it uses these systems is to understand its customers better. As former CEO Sir Terry Leahy put it in April 2011, "The hardest thing to know is where you stand relative to your customers, your suppliers and your competitors. Collecting, analyzing, and acting on the insights revealed by customer behavior, at the [cash register] and online, allowed Tesco to find the truth." He added, "Customers [are] the best guide. They have no axe to grind. You have to follow the customers."

To track and analyze customer information, Tesco invested in a data warehousing system from Teradata along with reporting software from Business Objects. A data warehouse is a large collection of historical data to use for analysis and decision making. At Tesco, "large" is no exaggeration: its data warehouse contains over 100 TB (terabytes) of data. By comparison, a high-end personal computer in 2011 might have a total storage of 1 TB.

Connecting with customers, though, isn't a one-way process of collecting data about them. Connecting also means reaching out to customers, allowing them to interact in new ways. Tesco is doing that, too. Using augmented reality technology from Kishino AR, Tesco lets customers see products online almost as if they were physically in a store. (You can see this in action in the Kishino AR video listed under Sources.) Tesco will also put computers in eight of its U.K. stores that allow customers to check out more products than a store can stock and look at heavy, bulky items from all angles. In Korea, Tesco has opened a complete virtual store: customers can view over 500 items, scan their bar codes using a special smartphone app, and order products. The products can be delivered later that same day if they order by 1 pm.

Recognizing that many of the customers it wants to connect with are members of social networking sites, Tesco has also developed a Facebook application in which Clubcard holders (or most of its regular customers, 16 million in the U.K. alone) can vote on products they want added to its Big Price Drop promotion. Richard Brasher, CEO of Tesco U.K., explains, "We are committed to doing all we can to help our customers, and our new Facebook application will enable them to tell us directly where they most value reduced prices." Aside from the benefits of lower prices, voting on which prices should be lowered gives customers a feeling of being connected with the store and participating in decisions.

Tesco's applications require modern information systems. More importantly, however, they require the ability to see the value of information and conceive of innovative ways to use it. In this chapter, you'll see how that can happen.

As you read this chapter, consider the following:

- How does Tesco's Teradata database add value to the organization?
- How do Tesco's use of augmented reality and its Facebook application give Tesco a potential competitive advantage?

WHY LEARN ABOUT INFORMATION SYSTEMS IN ORGANIZATIONS?

© Andrey Burmakin/Shutterstock

Organizations of all types use information systems to cut costs and increase profits. After graduating, a management major might be hired by a shipping company to help design a computerized system to improve employee productivity. A marketing major might work for a national retailer using a network to analyze customer needs in different areas of the country. An accounting major might work for an accounting or consulting firm using an information system to audit other companies' financial records. A real estate major might use the Internet and work in a loose organizational structure with clients, builders, and a legal team located around the world. A biochemist might conduct research for a drug company and use a computer to evaluate the potential of a new cancer treatment. An entrepreneur might use information systems to advertise and sell products and bill customers.

Although your career might be different from those of your classmates, you will almost certainly work with computers and information systems to help you and your company or organization become more efficient, effective, productive, and competitive in its industry. In this chapter, you will see how information systems can help organizations produce higher-quality products and increase their return on investment. We begin by investigating organizations and information systems.

Information systems have changed the way organizations work in recent years. While information systems were once used primarily to automate manual processes, they have transformed the nature of work and the shape of organizations themselves. In this chapter and throughout the book, you will explore the benefits and drawbacks of information systems in today's organizations around the globe.

ORGANIZATIONS AND INFORMATION SYSTEMS

organization: A formal collection of people and other resources established to accomplish a set of goals.

An **organization** is a formal collection of people and other resources established to accomplish a set of goals. It constantly uses money, people, materials, machines and other equipment, data, information, and decisions. As shown in Figure 2.1, resources such as materials, people, and money serve as inputs to the organizational system from the environment; they go through a transformation mechanism; and then outputs are produced to the environment. The outputs from the transformation mechanism are usually goods or services, which are of higher relative value than the inputs alone. Through adding value or worth, organizations attempt to increase performance and achieve their goals.

How does the organizational system increase the value of resources? In the transformation mechanism, subsystems contain processes that help turn inputs into goods or services of increased value. These processes increase the relative worth of the combined inputs on their way to becoming final outputs. Let's reconsider the simple car wash example from Chapter 1 (see Figure 1.3). The first process is washing the car. The output of this system—a clean but wet car—is worth more than the mere collection of ingredients (soap and water), as evidenced by the popularity of automatic car washes. Consumers are willing to pay for the skill, knowledge, time, and energy required to wash their cars. The second process is drying—transforming the wet car into a dry one with no water spotting. Again, consumers are willing to pay for the additional skill, knowledge, time, and energy required to accomplish this transformation.

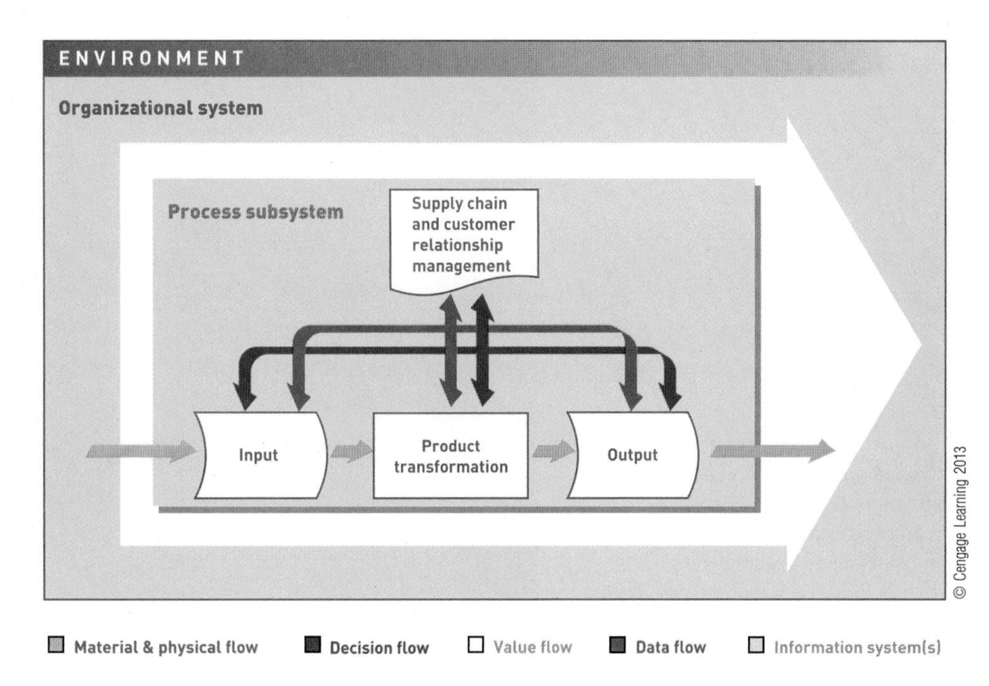

FIGURE 2.1
General model of an organization
Information systems support and work within all parts of an organizational process. Although not shown in this simple model, input to the process subsystem can come from internal and external sources. Just prior to entering the subsystem, data is external. After entering the subsystem, it becomes internal. Likewise, goods and services can be output to either internal or external systems.

Providing value to a stakeholder—customer, supplier, manager, shareholder, or employee—is the primary goal of any organization. The value chain, first described by Michael Porter in a 1985 *Harvard Business Review* article titled "How Information Gives You Competitive Advantage," reveals how organizations can add value to their products and services. The **value chain** is a series (chain) of activities that includes inbound logistics, warehouse and storage, production and manufacturing, finished product storage, outbound logistics, marketing and sales, and customer service. See Figure 2.2. You investigate each activity in the chain to determine how to increase the value perceived by a customer. The value chain is just as important to companies that don't manufacture products, such as tax preparers, retail stores, legal firms, and other service providers. By adding a significant amount of value to their products and services, companies ensure success.

value chain: A series (chain) of activities that includes inbound logistics, warehouse and storage, production and manufacturing, finished product storage, outbound logistics, marketing and sales, and customer service.

Managing the supply chain and customer relationships are two key elements of managing the value chain. *Supply chain management (SCM)* helps determine what supplies are required for the value chain, what quantities are

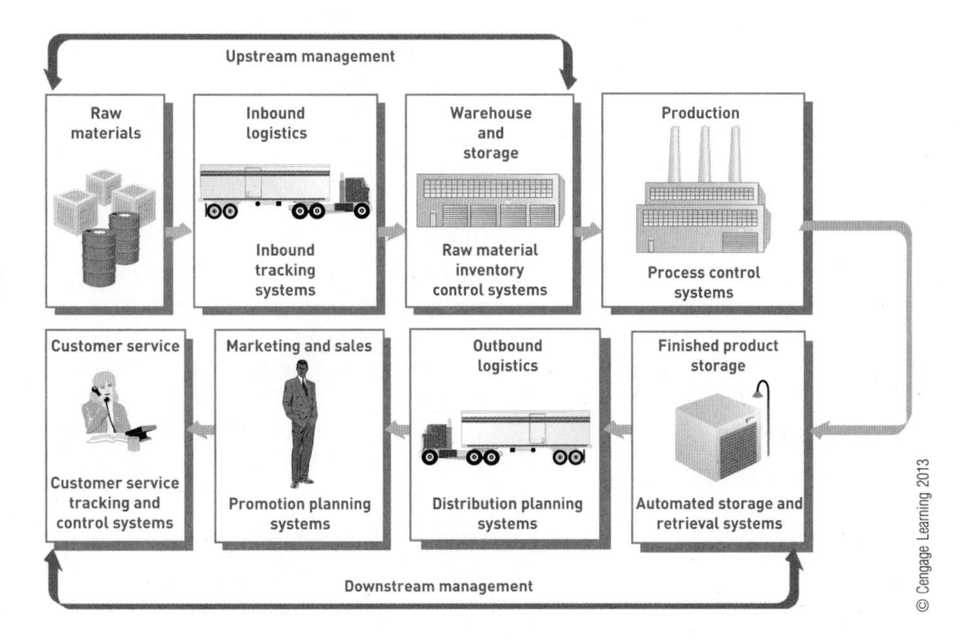

FIGURE 2.2
Value chain of a manufacturing company
Managing raw materials, inbound logistics, and warehouse and storage facilities is called *upstream management*. Managing finished product storage, outbound logistics, marketing and sales, and customer service is called *downstream management*.

Combining a value chain with just-in-time (JIT) inventory means companies can deliver materials or parts when they are needed. General Mills uses JIT to help reduce inventory costs and enhance customer satisfaction.

needed to meet customer demand, how the supplies should be processed (manufactured) into finished goods and services, and how the shipment of supplies and products to customers should be scheduled, monitored, and controlled.[1] Waste Management, for example, works with customers to improve its supply chain.[2] According to the CEO of Waste Management, "The best thing we can do for our customers is to give them full visibility into the data along the supply chain." Suppliers can also help a business improve its operations.[3] Suppliers want the businesses they supply to succeed because the practice also means their success. Procter & Gamble, the large consumer products company, was able to streamline its supply chain by reducing inventory levels, reducing costs, and making its operations more efficient.[4] The company hopes to use inventory control methods on more than 50 percent of its supply chain.

Ford Motor Company's use of information systems is an integral part of its operation. The company gives suppliers access to its inventory system so that the suppliers can monitor the database and automatically send another shipment when stocks are low, such as for engine parts, eliminating the need for purchase orders. This procedure speeds delivery and assembly time and lowers Ford's inventory-carrying costs.

Delivering products and services to customers is the end of most organization's supply chain. Many companies are increasing their use of free shipping to customers in hopes of increasing sales and profits. Amazon, the popular online shopping company, is experimenting with Amazon Tote, a Web site that offers fast, free delivery of groceries and other products in the Seattle area.[5] Customers can also get actively involved in having products and services delivered to them. Hard Rock's Las Vegas hotels and casinos, for instance, have a smartphone application that customers can download and use to browse menus and order food and drinks.[6] The application uses GPS technology to locate customers at a specific restaurant or in their rooms. Increasingly, small and medium-sized businesses are hiring outside companies, often called fulfillment companies, to store and deliver their products.[7] Fulfillment companies include Amazon, Shipwire, Webgistics, and others. Companies also involve their customers to improve supply chain operations.

Traditionally, companies determined the best supply chain to deliver products and services to customers. With the Internet, customers today are starting to determine the best delivery of products and services to them.[8] They research the speed, cost, and convenience of the delivery system that best meets their needs. According to an executive of an Internet company, "People now control distribution on the Web."

As seen in Japan in 2011, a natural disaster can disrupt or destroy a company's supply chain.[9] The tsunami and resulting nuclear power plant problems in Japan have not only disrupted many Japanese manufacturing companies but also manufacturing companies around the world that use raw materials and supplies from Japan.

Customer relationship management (CRM) programs by SAP and others help companies of all sizes manage all aspects of customer encounters, including marketing and advertising, sales, customer service after the sale, and programs to retain loyal customers (see Figure 2.3). To be most beneficial, CRM programs must be tailored for each company or organization. Tidewell, a hospice that serves about 8,000 Florida families, acquired CRM software from Salesforce.com to save money and streamline its operations.[10] According to the chief information officer (CIO) of Tidewell, "While I can't talk in hard, fast dollars, I can tell you that the cost to run Salesforce for a year, with about

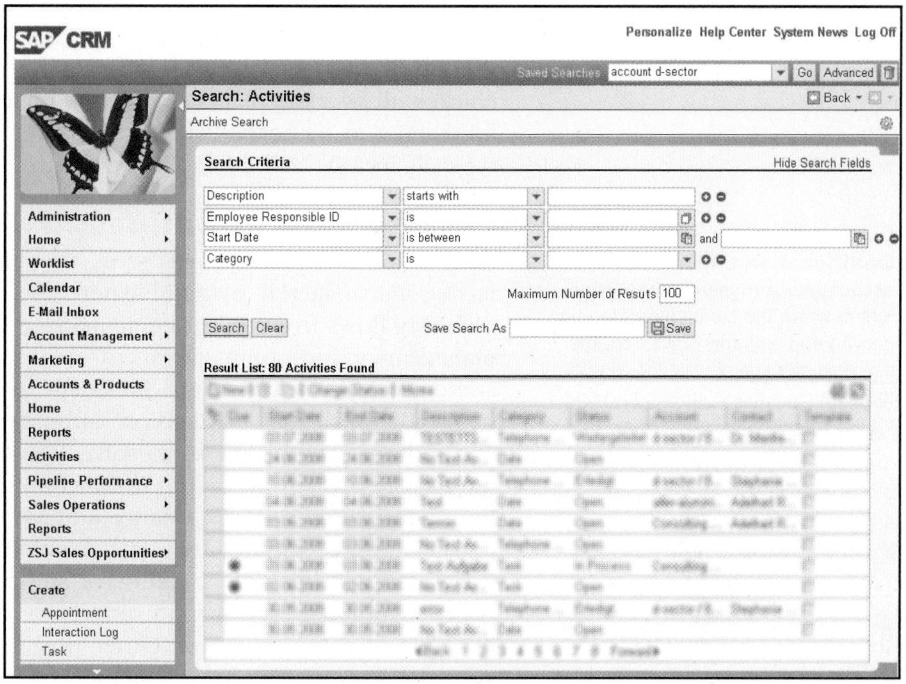

FIGURE **2.3**

SAP CRM

Companies in more than 25 industries use SAP CRM to reduce cost and increase decision-making ability in all aspects of their customer relationship management.

www.sap.com

70 or so users today, is probably a little over a third the cost of some of the typical hardware/software solutions we had quotes on." Duke Energy, an energy holding company, uses Convergys (*www.convergys.com*) to provide CRM software that is specifically configured to help the energy company manage its customer's use of energy grids and energy services.[11] Oracle, Salesforce, SAP, and other companies develop and sell CRM software.[12] Microsoft's Dynamics CRM application is available in over 40 languages.[13] Pricing on popular CRM applications can range from about $25 per user per month to over $100 per user per month. CRM software can also be purchased as a service and delivered over the Internet instead of being installed on corporate computers.

What role does an information system play in these processes? A traditional view of information systems holds that organizations use them to control and monitor processes and ensure effectiveness and efficiency. In this view, the information system is external to the process and serves to monitor or control it.

A more contemporary view, however, holds that information systems are often so intimately involved that they are *part of* the process itself. From this perspective, the information system plays an integral role in the process, whether providing input, aiding product transformation, or producing output. Consider a phone directory business that creates phone books for international corporations. A corporate customer requests a phone directory listing all steel suppliers in Western Europe. Using its information system, the directory business can sort files to find the suppliers' names and phone numbers and organize them into an alphabetical list. The information system itself is an integral part of this process. It does not just monitor the process externally but works as part of the process to transform raw data into a product. In this example, the information system turns input (names and phone numbers) into a salable output (a phone directory). The same system might also provide the input (data files) and output (printed pages for the directory).

Organizational Structures

organizational structure: Organizational subunits and the way they relate to the overall organization.

Organizational structure refers to organizational subunits and the way they relate to the overall organization. An organization's structure depends on its goals and its approach to management. Organization structure can affect how a company views and uses information systems. According to the vice president of information technology at Yum Brands, owner of Pizza Hut, KFC, Taco Bell, and other restaurant chains, "We're making pizza, tacos, and chicken all over the planet. We need to have a structure to allow our collective know-how to be seamlessly shared."[14] The types of organizational structures typically include traditional, project, team, and virtual.

Traditional Organizational Structure

traditional organizational structure: An organizational structure in which the hierarchy of decision making and authority flows from the strategic management at the top down to operational management and nonmanagement employees.

A **traditional organizational structure**, also called a "hierarchical structure," is like a managerial pyramid where the hierarchy of decision making and authority flows from the strategic management at the top down to operational management and nonmanagement employees. Compared to lower levels, the strategic level, including the president of the company and vice presidents, has a higher degree of decision authority, more impact on corporate goals, and more unique problems to solve (see Figure 2.4). The major departments are usually divided according to function and can include marketing, production, information systems, finance and accounting, research and development, and so on (see Figure 2.5). The positions or departments that are directly associated with making, packing, or shipping goods are called line positions. A production supervisor who reports to a vice president of production is an example of a line position. Other positions might not be directly involved with the formal chain of command but instead assist a department or area.

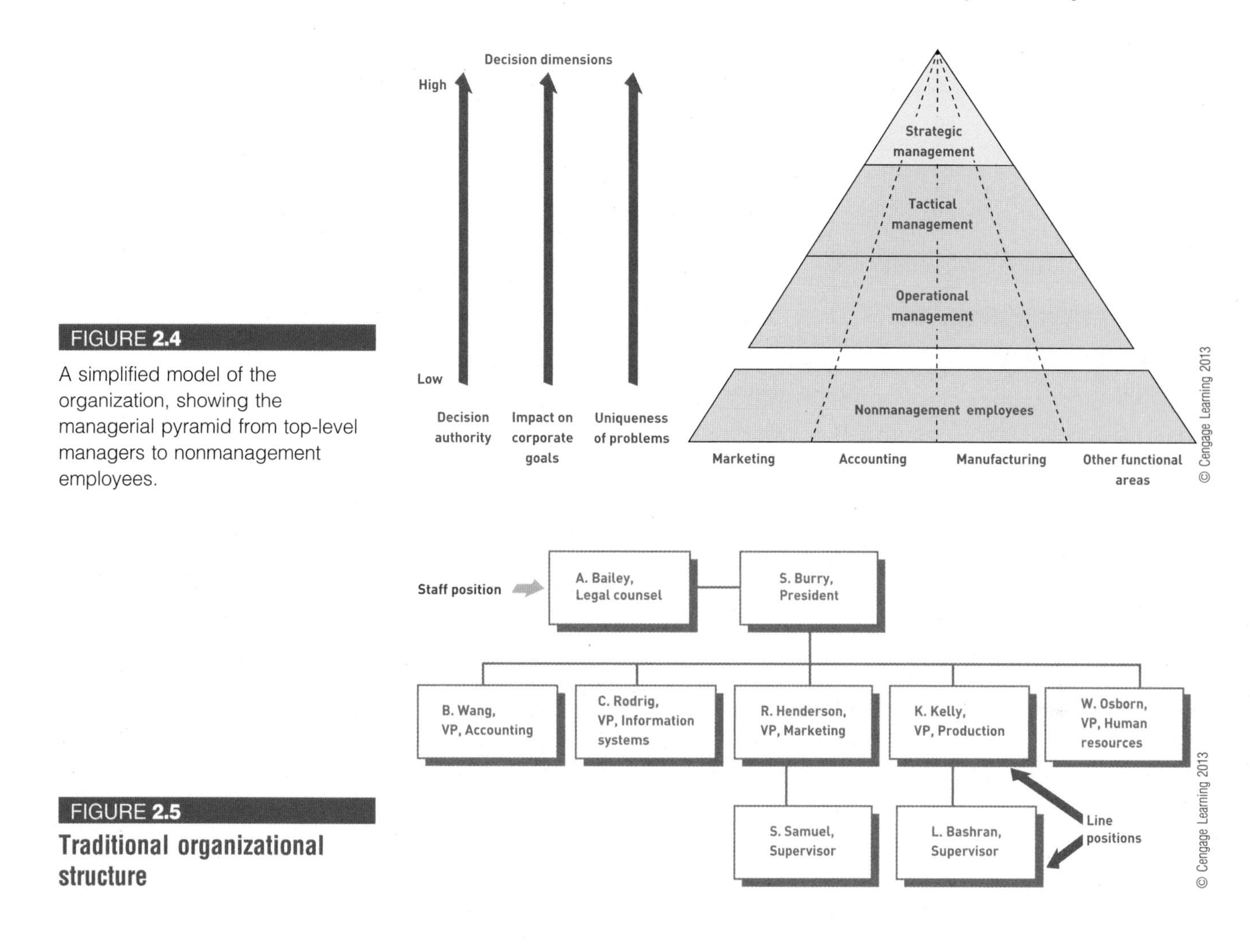

FIGURE 2.4

A simplified model of the organization, showing the managerial pyramid from top-level managers to nonmanagement employees.

FIGURE 2.5

Traditional organizational structure

These are called staff positions, such as a legal counsel reporting to the president.

Today, the trend is to reduce the number of management levels, or layers, in the traditional organizational structure. This type of structure, often called a **flat organizational structure**, empowers employees at lower levels to make decisions and solve problems without needing permission from midlevel managers. **Empowerment** gives employees and their managers more responsibility and authority to make decisions, take action, and have more control over their jobs. For example, an empowered sales clerk could respond to certain customer requests or problems without needing permission from a supervisor.

flat organizational structure: An organizational structure with a reduced number of management layers.

empowerment: Giving employees and their managers more responsibility and authority to make decisions, take action, and have more control over their jobs.

Project and Team Organizational Structures

A **project organizational structure** is centered on major products or services. For example, in a manufacturing firm that produces baby food and other baby products, each line is produced by a separate unit. Traditional functions such as marketing, finance, and production are positioned within these major units (see Figure 2.6). Many project teams are temporary: When the project is complete, the members go on to new teams formed for another project.

project organizational structure: A structure centered on major products or services.

The **team organizational structure** is centered on work teams or groups. In some cases, these teams are small; in others, they are very large. Typically, each team has a leader who reports to an upper-level manager, and depending on its tasks, the team can be temporary or permanent. Many organizations use the team organizational structure. According to Dennis Strong, senior vice president and CIO of McCoy's building supply company, "Know your team

team organizational structure: A structure centered on work teams or groups.

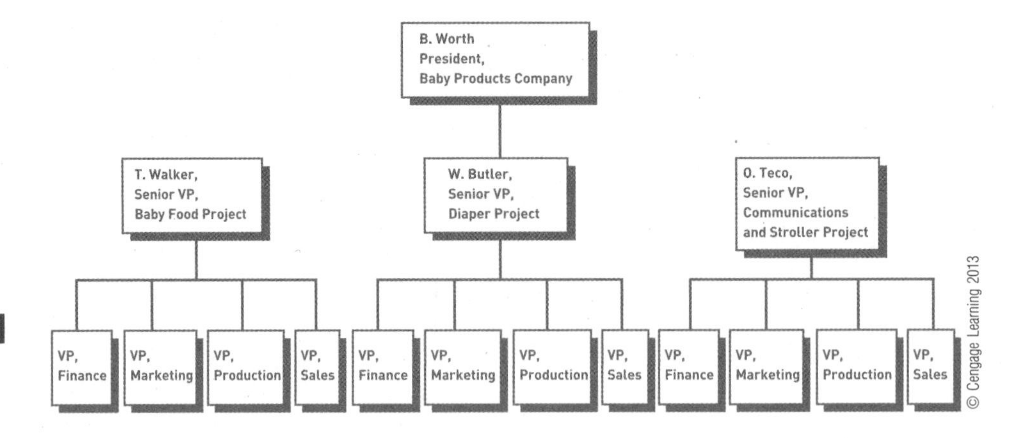

FIGURE 2.6
Project organizational structure

and those you interact with on a personal level. Be able to share their celebrations and support them during their life struggles."[15]

Virtual Organizational Structure and Collaborative Work

virtual organizational structure: A structure that uses individuals, groups, or complete business units in geographically dispersed areas; these groups can last for a few weeks or years, often requiring telecommunications and the Internet.

A **virtual organizational structure** uses individuals, groups, or complete business units in geographically dispersed areas. These groups can last for a few weeks or years, often requiring telecommunications and the Internet. Work can be done anywhere, anytime. Virtual teams help ensure the participation of the best available people to solve important organizational problems. Virtual teams are also used in education, where teams of students are increasingly connecting online. Even middle schools and high schools are starting to use online education and virtual teams to a greater extent.[16] Florida, for example, has the *Florida Virtual School*, which allows students from kindergarten through eighth grade to take online courses.[17] According to the staff director for a large Florida county, "It's a really good idea for a teacher to have some experience in virtual education. The future of education is going to push a lot more in that direction." The authors and publishing team for this book used a virtual team structure, consisting of over a dozen people from across the United States who worked over a year to complete this textbook project. They used the Internet to send chapter files and related documents to each other in developing the textbook and related materials you are using.

Successful virtual organizational structures share key characteristics. One strategy is to have in-house employees concentrate on the firm's core businesses and use virtual employees, groups, or businesses to do other business tasks. Even with sophisticated IS tools, teams still benefit from face-to-face meetings, especially at the beginning of new projects.

To stay connected, some virtual teams use smartphones and mobile devices that connect to the Internet. Social networking sites such as Facebook can be used to help virtual teams work together.[18]

Innovation

Organizations are continuously trying to improve their operations by looking for fresh, innovative ideas.[19] Innovative, cutting-edge products and services can create new revenue streams.[20] In some cases, innovation can help companies explore new markets and business approaches. The founder of Medical Data Services, which provides software for the health care community, says, "Bring innovation in everything you do, not just the big projects. Scan the market externally in your own field, but also in other businesses. Opportunities can come from the strangest places."[21] The London Business School houses the Institute of Innovation and Entrepreneurship. According to the dean of the school, "Innovation as far as we're concerned, is one of the key drivers of everything that goes on in business."[22] America Online (AOL) leased a 200,000 square foot office building in Palo Alto, California, to help it

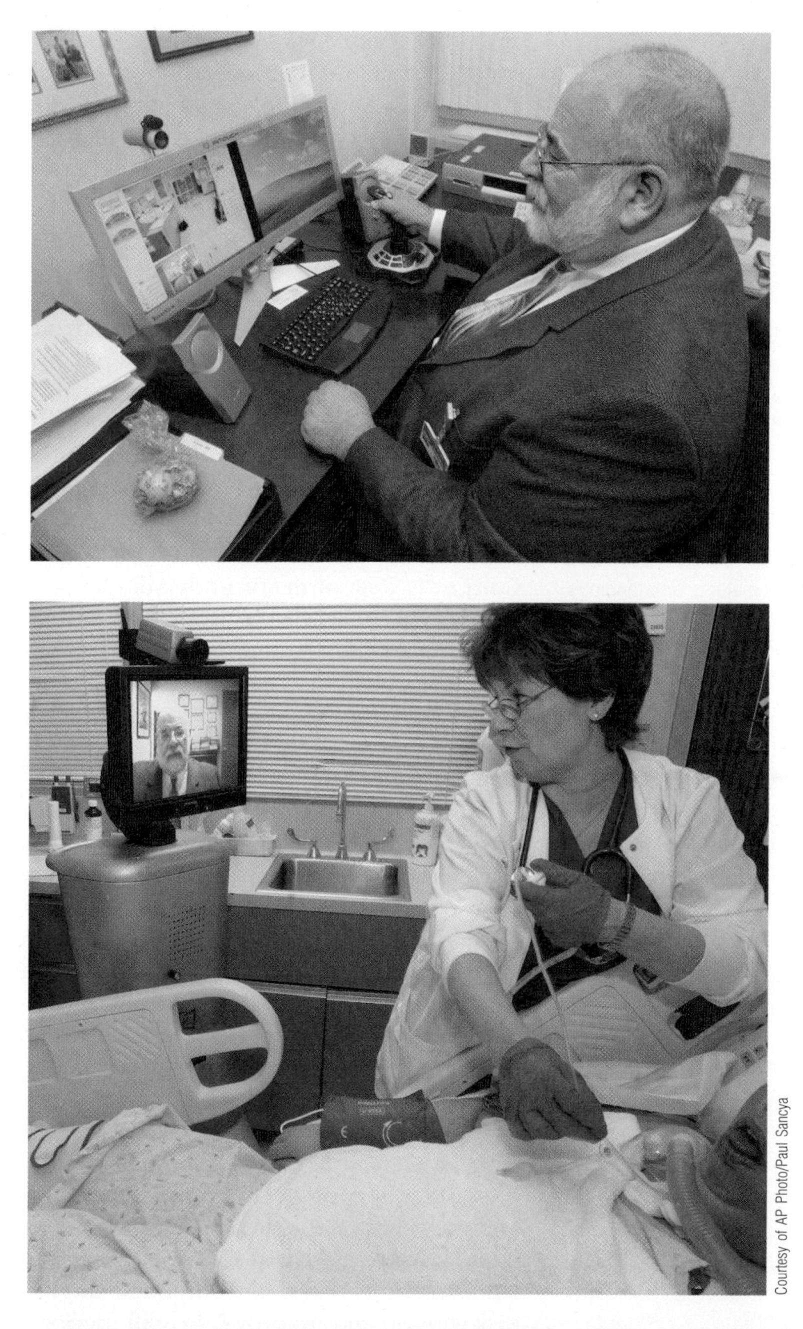

Virtual teams let people consult with experts no matter their physical location; they are especially useful in the health care industry.

A virtual organizational structure allows collaborative work in which managers and employees can effectively work in groups, even those composed of members from around the world.

and other companies become more innovative.[23] AOL is hoping its new office complex will be an incubator for new ideas and business ventures. Many believe that Steve Jobs, who died in 2011, was a real innovator, creating Pixar, the iPod, the iPhone, the iPad, and many other products.[24] According to Bill Gates, founder of Microsoft, "Steve Jobs' ability to focus on a few things that count ... is an amazing thing."[25]

Using IS personnel to increase innovation can lead to increased revenue streams and profitability in the future. Some IS departments are creating separate groups that explore new, innovative ideas. According to the vice president of Diversey, a provider of hygiene products, "We've created an IT innovation team of four people, separate from the operations group, whose sole mission is innovation."[26] Increasingly, IS departments are motivating their employees to be more innovative.[27] According to the CIO of Jet Blue, "I want my IT staff to understand innovation as making something better without waiting for me or another business leader to ask them to."

Organizational Culture and Change

culture: A set of major understandings and assumptions shared by a group, such as within an ethnic group or a country.

organizational culture: The major understandings and assumptions for a business, corporation, or other organization.

organizational change: How for-profit and nonprofit organizations plan for, implement, and handle change.

Culture is a set of major understandings and assumptions shared by a group, such as within an ethnic group or a country. **Organizational culture** consists of the major understandings and assumptions for a business, corporation, or other organization. The understandings, which can include common beliefs, values, and approaches to decision making, are often not stated or documented as goals or formal policies. For example, employees might be expected to be clean-cut, wear conservative outfits, and be courteous in dealing with all customers. Sometimes organizational culture is formed over years.

Organizational change deals with how for-profit and nonprofit organizations plan for, implement, and handle change. Change can be caused by internal factors, such as those initiated by employees at all levels, or by external factors, such as those wrought by competitors, stockholders, federal and state laws, community regulations, natural occurrences (such as hurricanes), and general economic conditions. Organizational change also occurs when two or more organizations merge. When organizations merge, integrating their information systems can be critical to future success. Unfortunately, many organizations consider the integration of their various information systems too late in the merger process.

Change can be sustaining or disruptive.[28] *Sustaining change* can help an organization improve the supply of raw materials, the production process, and the products and services it offers. Developing new manufacturing equipment to make disk drives is an example of a sustaining change for a computer manufacturer. The new equipment might reduce the costs of producing the disk drives and improve overall performance. *Disruptive change*, on the other hand, can completely transform an industry or create new ones, which can harm an organization's performance or even put it out of business. In general, disruptive technologies might not originally have good performance, low cost, or even strong demand. Over time, however, they often replace existing technologies. Cloud computing will likely cause disruptive change for many businesses and industries in the future. According to a researcher at the Deloitte Center for the Edge, "Cloud computing has the potential to generate a series of disruptions that will ripple out from the tech industry and ultimately transform many industries around the world."[29] On a positive note, disruptive change often results in new, successful companies and offers consumers the potential of new products and services at reduced costs and superior performance. An institute called Singularity University, located at the NASA Ames Research Center, Moffett Field, California, offers workshops on how to deal with disruptive change.[30] The purpose of the institute is to prepare managers and executives for the fast, ever-changing nature of information systems.

Singularity University offers workshops on how organizations can deal with disruptive change.

change model: A representation of change theories that identifies the phases of change and the best way to implement them.

The dynamics of change can be viewed in terms of a change model. A **change model** represents change theories by identifying the phases of change and the best way to implement them. Kurt Lewin and Edgar Schein propose a three-stage approach for change (see Figure 2.7). The first stage, unfreezing, is ceasing old habits and creating a climate that is receptive to change. Moving, the second stage, is learning new work methods, behaviors, and systems. The final stage, refreezing, involves reinforcing changes to make the new process second nature, accepted, and part of the job.[31] In addition to the Lewin-Schein change model discussed above, other change models, publications, and courses can help organizations manage change, including *Managing at the Speed of Change* and *Project Change Management* by Conner Partners (*www.connerpartners.com*), *Leading Change, The Heart Of Change* by John Kotter (*www.theheartofchange.com*), and many others.

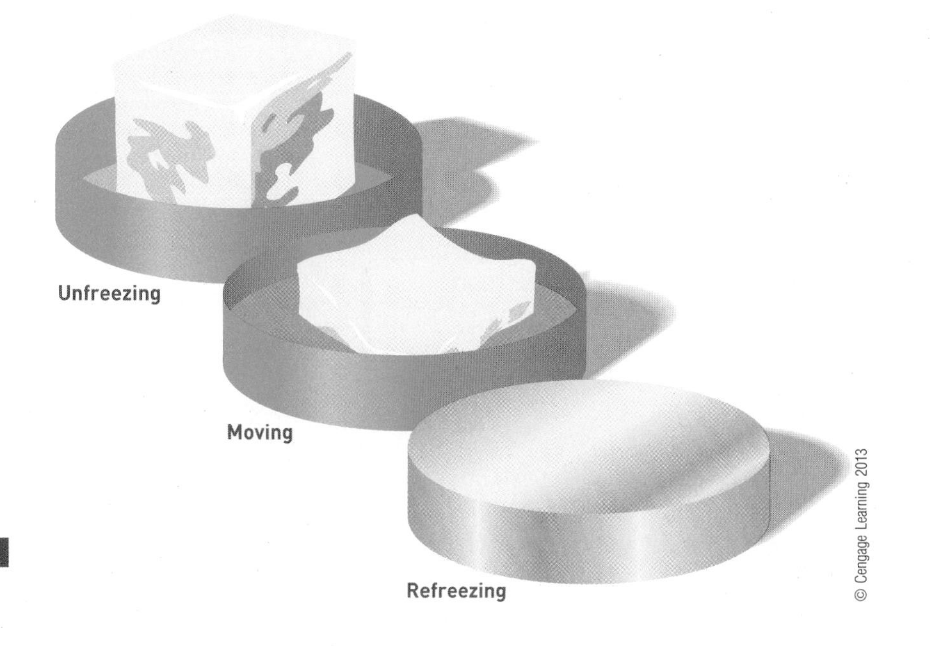

FIGURE **2.7**
Change model

organizational learning: The adaptations and adjustments based on experience and ideas over time.

Organizational learning is closely related to organizational change. All organizations adapt to new conditions or alter their practices over time—some better than others. Collectively, these adaptations and adjustments based on experience and ideas are called **organizational learning**. Assembly-line workers, secretaries, clerks, managers, and executives learn better ways of doing business and then incorporate them into their day-to-day activities. In some cases, the adjustments can be a radical redesign of business processes, often called reengineering. In other cases, adjustments can be more incremental, a concept called continuous improvement. Both adjustments reflect an organization's strategy, the long-term plan of action for achieving its goals.

Reengineering and Continuous Improvement

reengineering (process redesign): The radical redesign of business processes, organizational structures, information systems, and values of the organization to achieve a breakthrough in business results.

To stay competitive, organizations must occasionally make fundamental changes in the way they do business. In other words, they must change the activities, tasks, or processes they use to achieve their goals. **Reengineering**, also called **process redesign** and business process reengineering (BPR), involves the radical redesign of business processes, organizational structures, information systems, and values of the organization to achieve a breakthrough in business results. BBVA, a large financial services firm with offices around the world, used reengineering to streamline its operations and save about $2 million in its Madrid and New York offices.[32] See Figure 2.8. Reengineering can reduce delivery time, increase product and service quality, enhance customer satisfaction, and increase revenues and profitability. According to the CIO for FedEx ground, "Over the last five years, we have been on a mission to get faster and faster. We reengineer and speed up our lanes an average of twice a year, and sometimes more frequently than that."[33]

continuous improvement: Constantly seeking ways to improve business processes and add value to products and services.

In contrast to reengineering, the idea of **continuous improvement** is to constantly seek ways to improve business processes and add value to products and services. This continual change will increase customer satisfaction and loyalty and ensure long-term profitability. Manufacturing companies make continual product changes and improvements. Service organizations regularly find ways to provide faster and more effective assistance to customers. By doing so, these companies increase customer loyalty, minimize the chance of customer dissatisfaction, and diminish the opportunity for competitive inroads. Table 2.1 compares the two strategies of business process reengineering and continuous improvement.

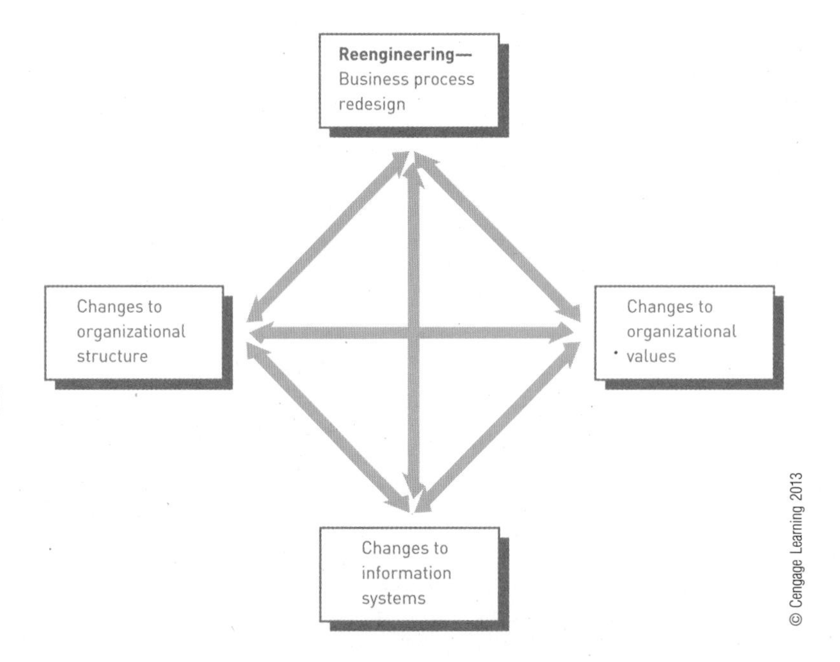

FIGURE 2.8

Reengineering

Reengineering involves the radical redesign of business processes, organizational structure, information systems, and values of the organization to achieve a breakthrough in business results.

FRED PROUSER/Reuters/Landov

FedEx used reengineering to streamline its distribution centers around the world.

TABLE 2.1 Comparing business process reengineering to continuous improvement

Business Process Reengineering	Continuous Improvement
Strong action taken to solve serious problem	Routine action taken to make minor improvements
Top-down change driven by senior executives	Bottom-up change driven by workers
Broad in scope; cuts across departments	Narrow in scope; focuses on tasks in a given area
Goal is to achieve a major breakthrough	Goal is continuous, gradual improvements
Often led by outsiders	Usually led by workers close to the business
Information system integral to the solution	Information systems provide data to guide the improvement team

User Satisfaction and Technology Acceptance

To be effective, reengineering and continuous improvement efforts must result in satisfied users and be accepted and used throughout the organization. Over the years, IS researchers have studied user satisfaction and technology acceptance as they relate to IS attitudes and usage.[34] Although user satisfaction and technology acceptance started as two separate theories, some believe that they are related concepts.[35]

User satisfaction with a computer system and the information it generates often depend on the quality of the system and the value of the information it delivers to users.[36] A quality information system is usually flexible, efficient, accessible, and timely. Recall from Chapter 1 that quality information is accurate, reliable, current, complete, and delivered in the proper format.[37]

Low user satisfaction often means that fewer people use the computer-based information system. This low satisfaction is true for employees and customers. Boring and uninteresting Internet sites, for example, attract fewer visitors, which usually translates into lower sales and profits.[38] As a result, consultants are often hired to make Internet sites more interesting and fun to visit to increase user satisfaction and enjoyment. In addition to traditional content, the Internet site makeovers can include exciting video games and other forms of entertainment.

technology acceptance model (TAM): A model that specifies the factors that can lead to better attitudes about an information system, along with higher acceptance and usage of it.

The **technology acceptance model (TAM)** specifies the factors that can lead to better attitudes about the information system, along with higher acceptance and usage of it.[39] These factors include the perceived usefulness of the technology, the ease of its use, its quality, and the degree to which the organization supports its use.[40] Studies have shown that user satisfaction and

ETHICAL & SOCIETAL ISSUES

Electronic Medical Records

Doctors have been recording information about their patients since the days of ancient Egypt. Written records served well when medicine was less specialized than it is today, people were less mobile, and patients were (for better or worse) more tolerant of physician errors.

Today, the world is moving to electronic medical records (EMRs) to consolidate medical information about a patient in a central place. According to Trisha Torrey, a patient advocate, an EMR is "a digital record kept by your doctor's office, your insurance company or a facility where you are a patient." She goes on to say that "EMR systems are intended to keep track of a patient's entire health and medical history in a computerized, electronic format. By keeping these potentially vast records in this manner, they are more easily retrievable, and can make a patient's navigation through the health care system much safer and more efficient."

In the United States, the federal government has thrown its weight behind EMRs. It will pay physicians up to $18,000 each for their use. To earn this incentive, physicians must reach certain stages of "meaningful use" at milestones from 2011 through 2014. Similar incentives, with different definitions of meaningful use, apply to hospitals and other health care organizations.

Brighton Hospital is at the forefront of EMR adoption. Located in Brighton, Michigan, it is the second oldest alcohol and substance abuse treatment center in the United States. Its goal for EMRs was "to increase [the hospital's] efficiency and patient safety without dramatically changing its workflow. Its staff also wanted to be able to more easily collect data and process a multitude of reports from that data." Since adopting EMRs in 2010, Brighton Hospital has achieved several specific benefits. One was a reduction of 2.5 (full-time equivalent) nurses, releasing them to areas that had no funds to hire new employees. The following are other benefits the hospital realized:

- 80 percent reduction in medication errors
- Increased patient safety and compliance
- Around the clock access to patient records
- Enhanced decision making using EMR data

Yet EMRs are not without concerns. Some are technical: will hardware and software be sufficiently reliable so that EMRs remain accessible and will electronic records be safe from intrusion? Other concerns, however, relate to the human side of health care. Dr. Danielle Ofri writes, "In the old days, when a patient arrived in my office ... I looked directly at the patient. As we spoke, I would briefly drop my eyes to jot a note on the page, and then look right up to continue our conversation. My gaze and my body language remained oriented toward the patient.... In the current computerized medical world, this is impossible. I have to be turned toward the computer screen." She summarizes: "The computer has much to offer, but I mourn the loss of intimacy that it has engendered."

As with so many other advances, innovations often involve trade-offs as an organization gains one benefit while losing another. Information system professionals can help to optimize the benefits while minimizing the losses.

Discussion Questions

1. Consider Dr. Ofri's comments about the trade-offs of using EMRs in the context of your most recent visit to a health care professional. Did he or she use a computer? If so, did you feel that it interfered with your discussion? Compare answers with your classmates.

2. Suppose your family lives in Vermont, you attend school in Texas, and you break your leg while skiing in Colorado. Describe how EMRs could help in that situation.

Critical Thinking Questions

1a. (For U.S. students) The United States has been criticized for having excellent health care but no health care system. Do you feel that criticism is justified? What, if anything, can EMRs do to address this concern?

1b. (For students outside the United States) Compare what your country is doing with EMRs with the situation in any other country at approximately the same economic level. Is your country ahead of the other in its adoption or behind? Why is this so? In your opinion, is that a problem?

2. Find the definition of "meaningful use" of EMRs on the Web. Do you feel that the meaningful use standards for physicians force them to move too fast? Let them move too slowly?

SOURCES: Dollinger, A., "Ancient Egyptian Medicine," *www.reshafim.org.il/ad/egypt/timelines/topics/medicine.htm*, updated November 2010, accessed December 19, 2011; Torrey, T., "What is an EMR (Electronic Medical Record) or EHR (Electronic Health Record)?" *About.com, http://patients.about.com/od/electronicpatientrecords/a/emr.htm*, April 11, 2011, accessed December 19, 2011; Fiegl, C., "Early EMR adopters get a break; tougher criteria delayed to 2014," *American Medical News, www.ama-assn.org/amednews/2011/12/12/gvl11212.htm*, December 12, 2011, accessed December 19, 2011; iPatientCare, "iPatientCare Helps Brighton Hospital Fulfill Its Passion for Paperless," *www.ipatientcare.com/KnowledgeCenter.aspx*, July 2011, downloaded December 19, 2011; Ofri, D., "When Computers Come between Doctors and Patients," *The New York Times, http://well.blogs.nytimes.com/2011/09/08/when-computers-come-between-doctors-and-patients*, September 8, 2011, accessed December 19, 2011.

technology acceptance are critical in health care.[41] Doctors and other health care professionals need training and time to accept and use new technology and databases before it can help them reduce medical errors and save lives. To increase user satisfaction and technology acceptance, more companies are allowing their employees to use smartphones and other mobile devices to get their work done and get access to corporate databases.[42]

technology diffusion: A measure of how widely technology is spread throughout the organization.

technology infusion: The extent to which technology permeates an area or department.

You can determine the actual usage of an information system by the amount of technology diffusion and infusion.[43] **Technology diffusion** is a measure of how widely technology is spread throughout an organization. An organization in which computers and information systems are located in most departments and areas has a high level of technology diffusion.[44] Some online merchants such as Amazon.com have a high level of diffusion and use computer systems to perform most of their business functions, including marketing, purchasing, and billing. **Technology infusion**, on the other hand, is the extent to which technology permeates an area or department. In other words, it is a measure of how deeply embedded technology is in an area of the organization. Some architectural firms, for example, use computers in all aspects of designing a building, from drafting to final blueprints (see Figure 2.9). The design area, thus, has a high level of infusion. Of course, a firm can have a high level of infusion in one part of its operations and a low level of diffusion overall. The architectural firm might use computers in all aspects of design (high infusion in the design area) but not to perform other business functions, including billing, purchasing, and marketing (low diffusion overall). Diffusion and infusion often depend on the technology available now and in the future, the size and type of the organization, and the environmental factors that include the competition, government regulations, suppliers, and so on. This is often called the technology, organization, and environment (TOE) framework.[45]

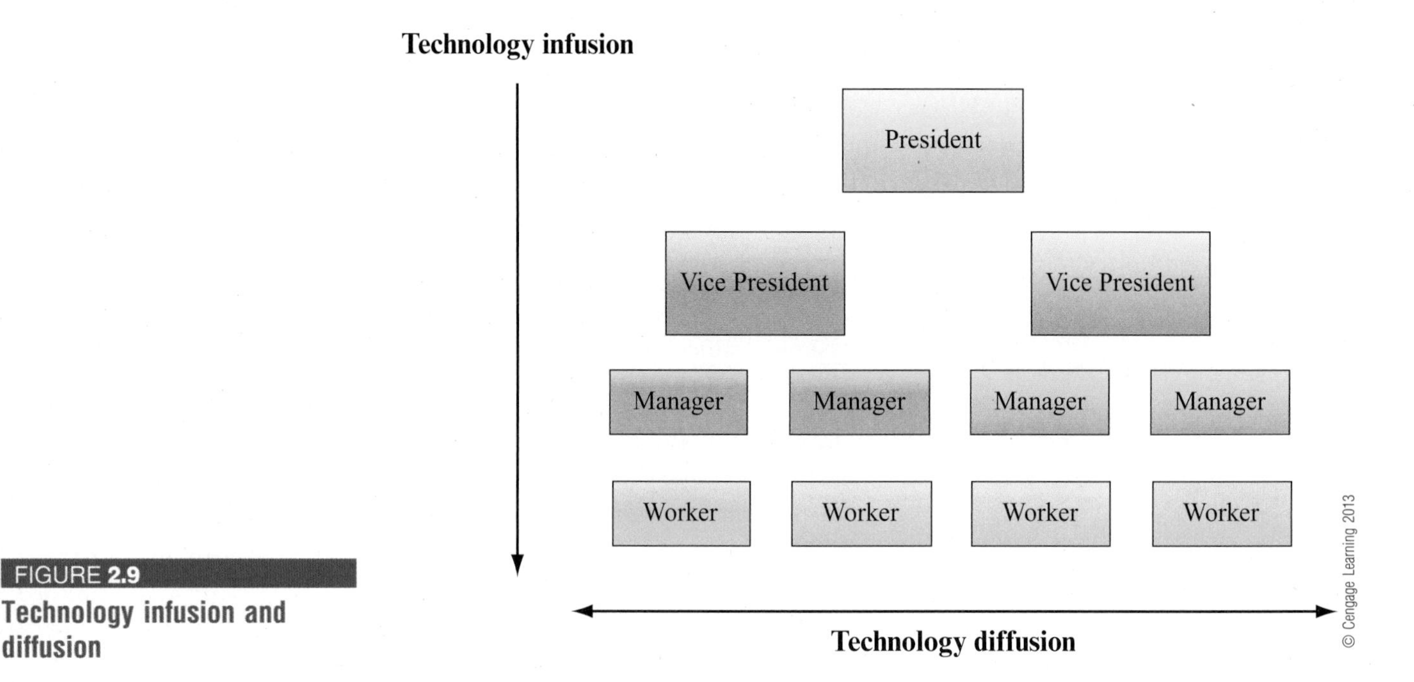

FIGURE 2.9
Technology infusion and diffusion

Although an organization might have a high level of diffusion and infusion with computers throughout the organization, this does not necessarily mean that information systems are being used to their full potential. In fact, the assimilation and use of expensive computer technology throughout organizations varies greatly.[46] Providing training and support to employees usually increases the use of a new information system.[47] Companies also hope that a high level of diffusion, infusion, satisfaction, and acceptance will lead to greater performance and profitability.[48] How appropriate and useful the information system is to the tasks or activities being performed, often called task-technology fit (TTF), can also lead to greater performance and profitability.[49]

Quality

quality: The ability of a product or service to meet or exceed customer expectations.

The definition of the term "quality" has evolved over the years. In the early years of quality control, firms were concerned with meeting design specifications, that is, conforming to standards. If a product performed as designed, it was considered a high-quality product. A product can perform its intended function, however, and still not satisfy customer needs. Today, **quality** means the ability of a product or service to meet or exceed customer expectations. For example, a computer that users say performs well and is easy to maintain and repair would be considered a high-quality product. This view of quality is completely customer oriented. Laps Care, an information system that helps assign medical personnel to home-health care patients in Sweden, has been able to improve the quality of medical care delivered to the elderly.[50] The system has also improved the efficiency of health care for the elderly by 10 to 15 percent and lowered costs by more than 20 million Euros. Organizations now use techniques to ensure quality, including total quality management and Six Sigma (see Table 2.2).

Outsourcing, On-Demand Computing, and Downsizing

A significant portion of an organization's expenses are used to hire, train, and compensate employees. Naturally, organizations try to control costs by determining the number of employees they need to maintain high-quality goods and services. Strategies to contain these personnel costs include outsourcing, on-demand computing, and downsizing.

TABLE **2.2** Total Quality Management and Six Sigma

Technique	Description	Examples
Total Quality Management (TQM)	Involves developing a keen awareness of customer needs, adopting a strategic vision for quality, empowering employees, and rewarding employees and managers for producing high-quality products.	The Ghana Investment Promotion Center, an organization that promotes investment and businesses in Ghana, South Africa, won a world quality award based on TQM.
Six Sigma	A statistical term that means products and services will meet quality standards 99.9997% of the time. In a normal distribution curve used in statistics, six standard deviations (six sigma) is 99.9997% of the area under the curve. Six Sigma was developed at Motorola, Inc., in the mid-1980s.[51]	Transplace, a $57 million trucking and logistics company, uses Six Sigma to improve quality by eliminating waste and unneeded steps. There are a number of training and certification programs for Six Sigma.[52]

outsourcing: Contracting with outside professional services to meet specific business needs.

Outsourcing involves contracting with outside professional services to meet specific business needs. Organizations often outsource a process to focus more closely on their core business—and target limited resources to meet strategic goals. Australian airline Virgin Australia, as an example, outsourced many of its routine business applications to Verizon, a business communications company, to free up its IS staff for important strategic projects.[53]

Companies that are considering outsourcing to cut the cost of their IS operations need to make this decision carefully. A growing number of organizations are finding that outsourcing does not necessarily lead to reduced costs. One of the primary reasons for cost increases is poorly written contracts that allow vendors to tack on unexpected charges. Other potential drawbacks of outsourcing include loss of control and flexibility, overlooked opportunities to strengthen core competency, and low employee morale.

on-demand computing: Contracting for computer resources to rapidly respond to an organization's flow of work as the need for computer resources arises. Also called on-demand business and utility computing.

On-demand computing is an extension of the outsourcing approach, and many companies offer it to business clients and customers. **On-demand computing**, also called on-demand business and utility computing, involves rapidly responding to the organization's flow of work as the need for computer resources varies. It is often called utility computing because the organization pays for computing resources from a computer or consulting company, just as it pays for electricity from a utility company.

This approach treats the information system—including hardware, software, databases, telecommunications, personnel, and other components—more as a service than as separate products. This approach can save money because the

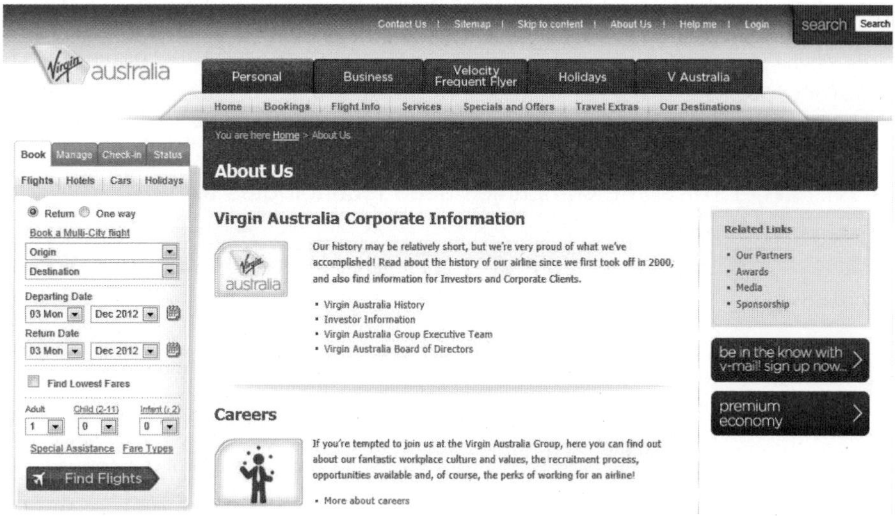

Virgin Australia outsourced many of its routine business applications to free up its IS staff for important strategic IS projects.

www.virginaustralia.com

organization does not pay for systems that it doesn't routinely need. It also allows the organization's IS staff to concentrate on more strategic issues.

downsizing: Reducing the number of employees to cut costs.

Downsizing involves reducing the number of employees to cut costs. The term "rightsizing" is also used. Rather than pick a specific business process to downsize, companies usually look to downsize across the entire company. Downsizing clearly reduces total payroll costs, though the quality of products and services and employee morale can suffer.

COMPETITIVE ADVANTAGE

competitive advantage: A significant and ideally long-term benefit to a company over its competition.

A **competitive advantage** is a significant and ideally long-term benefit to a company over its competition and can result in higher-quality products, better customer service, and lower costs.[54] An organization often uses its information system to help gain a competitive advantage.[55] According to the director of technology for the city of Berkeley, California, "We worked for so many years to get a seat at the table, and when we did we focused on selling the need for IT. Now, we're past that, and we've become strategic partners, figuring out how to get things done better." In addition, many companies consider their IS staff a key competitive weapon against other companies in the marketplace, especially if they have employees with training in the development and use of mobile devices, Internet applications, social networks, and collaborative tools.[56] Firms that gain a competitive advantage often emphasize the alignment of organizational goals and IS goals.[57] In other words, these organizations make sure that their IS departments are totally supportive of the broader goals and strategies of the organization.

To help achieve a competitive advantage, Apple, Inc., requires that companies selling music, books, and other content on Apple's devices, such as iPhones and iPads, give Apple customers the best deals and prices offered.[58] In other words, these companies cannot give customers using other devices better deals and prices than Apple customers can get. Some people, however, believe this policy might violate U.S. antitrust regulations.

In his book *Good to Great,* Jim Collins outlines how technology can be used to accelerate companies to greatness.[59] Table 2.3 shows how a few companies accomplished this. Ultimately, it is not how much a company spends on information systems but how it makes and manages investments in technology. Companies can spend less and get more value.

Factors That Lead Firms to Seek Competitive Advantage

five-forces model: A widely accepted model that identifies five key factors that can lead to attainment of competitive advantage, including (1) the rivalry among existing competitors, (2) the threat of new entrants, (3) the threat of substitute products and services, (4) the bargaining power of buyers, and (5) the bargaining power of suppliers.

A number of factors can lead to attaining a competitive advantage.[60] Michael Porter, a prominent management theorist, proposed a now widely accepted competitive forces model, also called the **five-forces model**. The five forces include (1) the rivalry among existing competitors, (2) the threat of new entrants, (3) the threat of substitute products and services, (4) the bargaining power of buyers, and (5) the bargaining power of suppliers. The more these forces combine in any instance, the more likely firms will seek competitive advantage and the more dramatic the results of such an advantage will be.

TABLE 2.3 How some companies used technologies to move from good to great

Company	Business	Competitive Use of Information Systems
Gillette	Shaving products	Developed advanced computerized manufacturing systems to produce high-quality products at low cost
Walgreens	Drug and convenience stores	Developed satellite communication systems to link local stores to centralized computer systems
Wells Fargo	Financial services	Developed 24-hour banking, ATMs, investments, and increased customer service using information systems

(Source: Data from Jim Collins, *Good to Great*, New York: Harper Collins Books, 2001, p. 300.)

Rivalry among Existing Competitors

Typically, highly competitive industries are characterized by high fixed costs of entering or leaving the industry, low degrees of product differentiation, and many competitors. To gain an advantage over competitors, companies constantly analyze how they use their resources and assets. This resource-based view is an approach to acquiring and controlling assets or resources that can help the company achieve a competitive advantage. For example, a transportation company might decide to invest in radio-frequency technology to tag and trace products as they move from one location to another.

Threat of New Entrants

A threat appears when entry and exit costs to an industry are low and the technology needed to start and maintain a business is commonly available. For example, a small restaurant is threatened by new competitors. Owners of small restaurants do not require millions of dollars to start the business, food costs do not decline substantially for large volumes, and food processing and preparation equipment is easily available. When the threat of new market entrants is high, the desire to seek and maintain competitive advantage to dissuade new entrants is also usually high.

Threat of Substitute Products and Services

Companies that offer one type of goods or services are threatened by other companies that offer similar goods or services. The more consumers can obtain similar products and services that satisfy their needs, the more likely firms are to try to establish competitive advantage. For example, consider the photographic industry. When digital cameras became popular, traditional film companies had to respond to try to stay competitive and profitable.

Bargaining Power of Customers and Suppliers

Large customers tend to influence a firm, and this influence can increase significantly if the customers threaten to switch to rival companies. When customers have a lot of bargaining power, companies increase their competitive advantage to retain their customers. Similarly, when the bargaining power of suppliers is strong, companies need to improve their competitive advantage to maintain their bargaining position. Suppliers can also help an organization gain a competitive advantage. Some suppliers enter into strategic alliances with firms and eventually act as a part of the company.

In the restaurant industry, competition is fierce because entry costs are low. Therefore, a small restaurant that enters the market can be a threat to existing restaurants.

Strategic Planning for Competitive Advantage

To be competitive, a company must be fast, nimble, flexible, innovative, productive, economical, and customer oriented. It must also align its IS strategy with general business strategies and objectives.[61] Given the five market forces previously mentioned, Porter and others have proposed a number of strategies to attain competitive advantage, including cost leadership, differentiation, niche strategy, altering the industry structure, creating new products and services, and improving existing product lines and services[62]:

- **Cost leadership.** Deliver the lowest possible cost for products and services. Walmart, Costco, and other discount retailers have used this strategy for years. Cost leadership is often achieved by reducing the costs of raw materials through aggressive negotiations with suppliers, becoming more efficient with production and manufacturing processes, and reducing warehousing and shipping costs. Some companies use outsourcing to cut costs when making products or completing services.
- **Differentiation.** Deliver different products and services. This strategy can involve producing a variety of products, giving customers more choices, or delivering higher-quality products and services. Many car companies make different models that use the same basic parts and components, giving customers more options. Other car companies attempt to increase perceived quality and safety to differentiate their products and appeal to consumers who are willing to pay higher prices for these features. Companies that try to differentiate their products often strive to uncover and eliminate counterfeit products produced and delivered by others.
- **Niche strategy.** Deliver to only a small, niche market. Porsche, for example, doesn't produce inexpensive economy cars, but rather, it makes high-performance sports cars and SUVs. Rolex only makes high-quality, expensive watches; it doesn't make inexpensive, plastic watches.
- **Altering the industry structure.** Change the industry to become more favorable to the company or organization. The introduction of low-fare airline carriers, such as Southwest Airlines, has forever changed the airline industry, making it difficult for traditional airlines to make high profit margins. Creating strategic alliances can also alter the industry structure. A **strategic alliance**, also called a **strategic partnership**, is an agreement between two or more companies that involves the joint production and distribution of goods and services.

strategic alliance (or strategic partnership): An agreement between two or more companies that involves the joint production and distribution of goods and services.

Costco and other discount retailers have used a cost leadership strategy to deliver the lowest possible price for products and services.

© iStockphoto/slobo

Porsche is an example of a company with a niche strategy, producing only high-performance sports cars and SUVs.

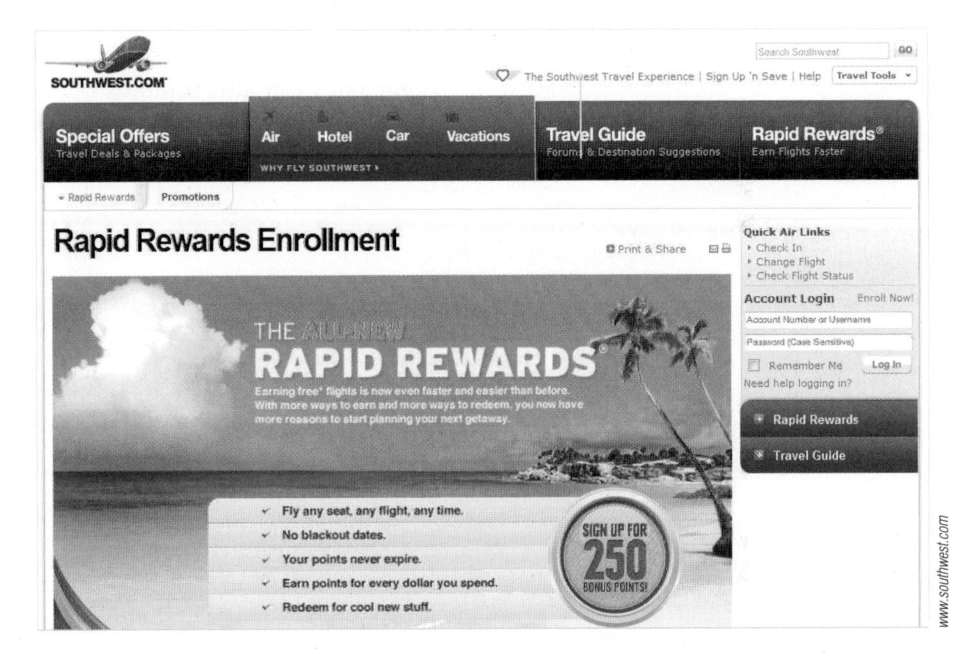

Low-fare airline carriers such as Southwest Airlines altered the structure of the airline industry.

- **Creating new products and services.** Introduce new products and services periodically or frequently. This strategy always helps a firm gain a competitive advantage, especially in the computer industry and for other high-tech businesses. If an organization does not introduce new products and services every few months, the company can quickly stagnate, lose market share, and decline. Companies that stay on top are constantly developing new products and services. Apple Computer, for example, introduced the iPod, iPhone, and iPad as new products.
- **Improving existing product lines and services.** Make real or perceived improvements to existing product lines and services. Manufacturers of household products are always advertising new and improved products. In some cases, the improvements are more perceived than actual refinements; usually, only minor changes are made to the existing product, such as reducing the amount of sugar in a breakfast cereal.
- **Other strategies.** Some companies seek strong *growth in sales*, hoping that it can increase profits in the long run due to increased sales. Being the *first to market* is another competitive strategy. Apple Computer, for instance, was one of the first companies to offer complete and ready-to-use personal computers. Some companies offer *customized* products and services to achieve a competitive advantage. Dell, for example, builds

custom PCs for consumers. *Hire the best people* is another example of a competitive strategy. The assumption is that the best people will determine the best products and services to deliver to the market and the best approach to deliver these products and services. Having *agile* information systems that can rapidly change with changing conditions and environments can be a key to information systems success and a competitive advantage.

Innovation is another competitive strategy. With today's innovative smartphones and tablet computers, such as Apple's iPad and others, executives are looking for new ways to gain a competitive advantage developing unique and powerful applications for these newer devices.[63] Innovation led Natural Selection, a San Diego company, to develop a computer program that attempted to analyze past inventions and suggest future ones.[64] Although the original program was not an immediate success, the approach has been used by General Electric, the U.S. Air Force, and others to cut costs and streamline delivery routes of products. According to one expert, "Successful innovations are often built on the back of failed ones."

Companies can also combine one or more of these strategies. In addition to customization, Dell attempts to offer low-cost computers (cost leadership) and top-notch service (differentiation).

PERFORMANCE-BASED INFORMATION SYSTEMS

Businesses have passed through at least three major stages in their use of information systems. In the first stage, organizations focused on using information systems to reduce costs and improve productivity. In one case, a Washington library system was able to save $400,000 by using less expensive phone lines for communications and Internet connections.[65] According to the cofounder of Silver Lake, a private equity and investment firm, "In the technology business, you're trying to drive adoption by making your product cheaper."[66] Companies can also use software tools, such as Apptio's IT Cost Optimization Solutions, to cut the costs of computer upgrades, reduce the number of computers, and help determine what to charge business units for providing computer services and equipment.[67]

The second stage was defined by Porter and others. It was oriented toward gaining a competitive advantage. In many cases, companies spent large amounts on information systems and downplayed the costs.

Today, companies are shifting from strategic management to performance-based management of their information systems. In this third stage, companies carefully consider both strategic advantage and costs. They use productivity, return on investment (ROI), net present value, and other measures of performance to evaluate the contributions their information systems make to their businesses. Figure 2.10 illustrates these three stages. This balanced approach attempts to reduce costs and increase revenues.

Productivity

productivity: A measure of the output achieved divided by the input required.

Developing information systems that measure and control productivity is a key element for most organizations. **Productivity** is a measure of the output achieved divided by the input required. In one survey, almost 70 percent of responding IS executives said that improving workforce productivity is a top priority.[68] A higher level of output for a given level of input means greater productivity; a lower level of output for a given level of input means lower productivity. The numbers assigned to productivity levels are not always based on labor hours: Productivity can be based on factors such as the amount of raw materials used, resulting quality, or time to

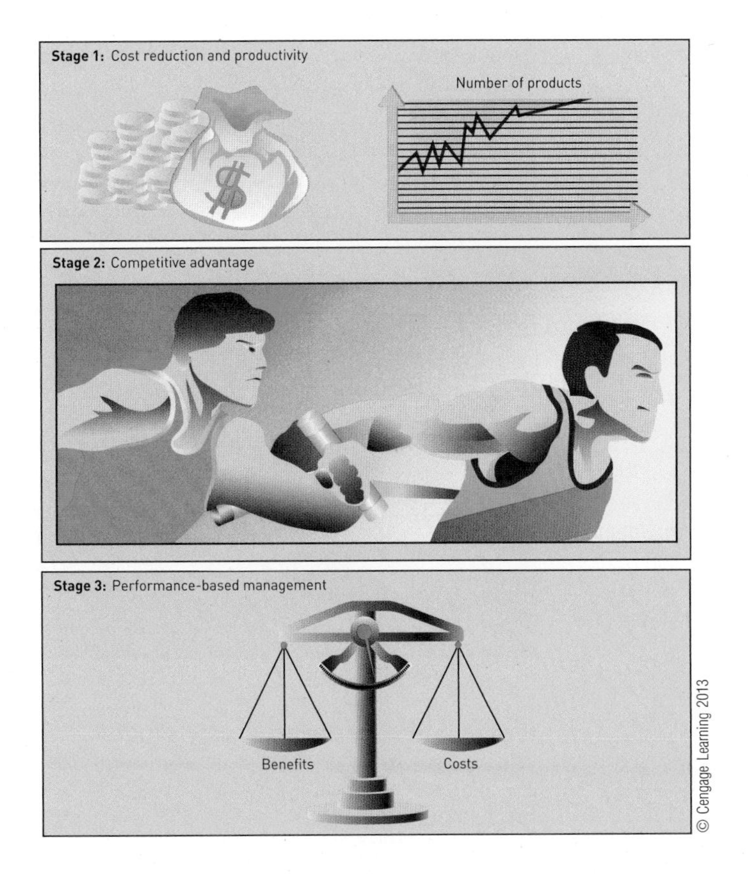

FIGURE 2.10
Three stages in the business use of information systems

produce the goods or service. The value of the productivity number is not as significant as how it compares with other time periods, settings, and organizations.

$$\text{Productivity} = (\text{Output}/\text{Input}) \times 100\%$$

Return on Investment and the Value of Information Systems

return on investment (ROI): One measure of IS value that investigates the additional profits or benefits that are generated as a percentage of the investment in IS technology.

One measure of IS value is **return on investment (ROI)**. This measure investigates the additional profits or benefits that are generated as a percentage of the investment in IS technology on an after-tax basis.[69] A small business that generates an additional profit of $20,000 for the year as a result of an investment of $100,000 for additional computer equipment and software would have a return on investment of 20 percent ($20,000/$100,000). ROI calculations can be complex, including investment returns over multiple years and the impact of the time value of money. The German-based DHL shipping company, for example, analyzed and streamlined its marketing efforts in more than 20 countries. The result was an increase in its corporate value of about $1.3 billion over a five-year period, representing an ROI of more than 30 percent.[70] Today, companies are demanding a higher return from their investments in information systems.[71] The chief financial officer of Tibco Software in Palo Alto, California asks, "The recession has focused us more on the fact that we've made investments. Are we really getting all we can from them?" Many companies rely on ROI calculations to make decisions about which IS projects to fund.[72] According to the chief information officer of a health care company, "We move quickly if we think there's a strong, quick ROI. If it's not obvious, we're probably not going to do it." ROI calculators are typically provided on a vendor's Web site and can be used to estimate returns.[73] Of course, computing ROI can be difficult. It can depend on the assumptions that are made for cost and revenue projections. According to the chief financial officer of Tatum, LLC, an executive consulting company, "I see ROIs all

Shipping company DHL analyzed and streamlined its marketing efforts, increasing its corporate value by about $1.3 billion over a five-year period, representing an ROI of more than 30 percent.

© iStockphoto/Wittelsbach bernd

the time that can have a wide range of values depending on how you work your assumptions."[74]

Earnings Growth

Another measure of IS value is the increase in profit, or earnings growth, the system brings. For instance, a mail-order company might install an order-processing system that generates a 7 percent earnings growth compared with the previous year.

Market Share and Speed to Market

Market share is the percentage of sales that a product or service has in relation to the total market. If installing a new online catalog increases sales, it might help a company boost its market share by 20 percent. Information systems can also help organizations bring new products and services to customers in less time. This is often called speed to market. Speed can also be a critical performance objective for many organizations.

Customer Awareness and Satisfaction

Although customer satisfaction can be difficult to quantify, about half of today's best global companies measure the performance of their information systems based on feedback from internal and external users. Some companies and nonprofit organizations use surveys and questionnaires to determine whether the IS investment has increased customer awareness and satisfaction.

Total Cost of Ownership

total cost of ownership (TCO): The sum of all costs over the life of an information system, including the costs to acquire components such as the technology, technical support, administrative costs, and end-user operations.

Another way to measure the value of information systems was developed by the Gartner Group and is called the **total cost of ownership (TCO)**. TCO is the sum of all costs over the life of the information system, including the costs to acquire components such as the technology, technical support, administrative costs, and end-user operations. Hitachi uses TCO to promote its projectors to businesses and consumers.[75] TCO is also used by many other companies to rate and select hardware, software, databases, and other computer-related components.

ROI, earnings growth, market share, customer satisfaction, and TCO are only a few measures that companies use to plan for and maximize the value

Hitachi manufactures audiovisual equipment and computer storage products and uses TCO to promote its projectors to businesses and consumers.

of their IS investments. Regardless of the difficulties, organizations must attempt to evaluate the contributions that information systems make to assess organizational progress and plan for the future. Information systems and personnel are too important to leave to chance.

Risk

In addition to the ROI measures of a new or modified system discussed in Chapter 1 and this chapter, managers must also consider the risks of designing, developing, and implementing these systems. Information systems can sometimes be costly failures. There is always a risk that any IS project will not be successful.[76] According to one corporate executive, "If there's a $500,000 expenditure, you have to consider the magnitude of success, the probability of success, and the risk if you don't succeed."

CAREERS IN INFORMATION SYSTEMS

As the economy starts to rebound, career opportunities in information systems are expected to increase.[77] While some CIOs cut IS budgets during an economic downturn, others see it as an opportunity to invest for the future.[78] According to the vice president of information systems and CIO of Verizon Wireless, "If you wait until the economy picks up steam to prepare for future growth, you'll find yourself behind the curve and constantly working to catch up. Now is the time to optimize operations, eliminate the clutter, and get better focused for the road ahead." The budget for the typical IS department is expected to surge in the next few years, according to a CIO economic impact survey.[79] Mobile applications will be a key factor for this economic surge.[80] As a result, some companies are concerned that their top IS talent will be recruited by other IS companies. According to the managing director of Dice.com, a job-posting company, "You've got a lot of people who aren't interested in staying where they are and employers who are saying it's time to get aggressive and poach talent from bigger players." IS jobs posted at Dice.com increased by about 30 percent from 2010 to 2011.[81] Some states, such as California, are seeing growth in jobs in general and IS jobs in particular.[82] California added more jobs in the first few months of 2011 than it added in 2010. In addition, some traditional "rust belt" cities in the United States, such as Detroit, Cincinnati, and Cleveland, have high IS job growth rates, exceeding 50 percent in some cases.[83] Even with the economic downturn in the last few years, IS employees have seen modest gains in salary and benefits, although some IS employees feel under more pressure to get more done in less time.[84] In one survey, salary was the most important job factor, followed by job and company stability, benefits, and flexible work schedules.[85]

Numerous schools have degree programs with such titles as information systems, computer information systems, and management information systems. These programs are typically offered by information schools, by business schools, and within computer science departments. Information systems skills can also help people start their own companies. When David Ulevitch and Daniel Kaminsky originally got together, it was to discuss Internet vulnerabilities at a computer conference.[86] Later, they started an Internet business, along with some investors, called OpenDNS that helps people filter Internet content and make surfing the Web safer. OpenDNS can also help organizations protect their Internet sites from attacks. Today, OpenDNS has more than 10 million users.

Skills that some experts believe are important for IS workers to have include the following, all of which are discussed in the chapters throughout this book:

- Mobile applications for smartphones, tablet computers, and other mobile devices
- Program and application development
- Help desk and technical support
- Project management
- Networking
- Business intelligence
- Security
- Web 2.0
- Data center
- Telecommunications

Nontechnical skills are also important for IS personnel, including communication skills, a detailed knowledge of the organization, and how information systems can help the organization achieve its goals.

The Internet giant Google, as an example, is looking for IS workers with experience in developing mobile applications for smartphones and other mobile devices.[87] Companies such as Google and Apple have hundreds of thousands of applications available that can be used on their systems, and they are looking for more systems developers to create even more mobile applications. Companies often rely on in-house IS personnel for developing their mobile applications.[88] Companies can also get mobile applications from application vendors, mobile device makers, and third-party developers. Demand for careers in developing applications for mobile devices and the Internet will likely explode in the future as more people purchase and use these devices.[89]

The U.S. Department of Labor's Bureau of Labor Statistics (*www.bls.gov*) publishes the fastest-growing occupations and predicts that many technology jobs will increase through 2012 or beyond. Table 2.4 summarizes some of the best places to work as an IS professional. Career development opportunities,

TABLE 2.4 Best places to work as an IS professional

Company	Additional Benefits
Booz Allen Hamilton	Good training and benefits
Chesapeake Energy Corporation	Retention of workers
General Mills	Good training programs
Genetech, Inc.	A mission to improve the lives of patients
Quicken Loans, Inc.	Good training
Salesforce.com, Inc.	Good training and retention.
Securian Financial Group	Pay, perks, and promotions
University of Pennsylvania	Employee benefits and diversity
USAA	Innovation and Service
Verizon Wireless	Good training programs

(Source: "100 Best Places to Work in IT," *Computerworld*, June 20, 2011, p. 15).

training, benefits, retention, diversity, and the nature of the work itself are just a few of the qualities these top employers offer. SAS, a popular software company, for example, offers its employees a good salary, generous benefits, a health center, a hotel, and other perks.[90] The company provides excellent job security and has very low employee turnover.

Opportunities in information systems are also available to people from foreign countries. The U.S. H-1B and L-1 visa programs seek to allow skilled employees from foreign lands into the United States. These programs, however, are limited and are usually in high demand. In 2011 U.S. companies made substantially fewer applications for H1-B visas than they did in 2009 or 2010.[91] The L-1 visa program is often used for intracompany transfers for multinational companies. The H-1B program can be used for new employees. The number of H-1B visas offered annually can be political and controversial, with some fearing that the program is being abused to replace high-paid U.S. workers with less expensive foreign workers. Indeed, some believe that companies pretend to seek U.S. workers while actually seeking less expensive foreign workers. In fact, in 2011 the United States accused a large foreign consulting company with improperly using the U.S. visa program to place its workers in U.S. companies.[92] Some foreign consulting companies get more than half of their revenues from U.S. clients.[93] Others, however, believe the H-1B program and similar programs are invaluable to the U.S. economy and its competitiveness.

Roles, Functions, and Careers in IS

IS offers many exciting and rewarding careers. Professionals with careers in information systems can work in an IS department or outside a traditional IS department as Web developers, computer programmers, systems analysts, computer operators, and many other positions. There are also opportunities for IS professionals in the public sector. In addition to technical skills, IS professionals need skills in written and verbal communication, an understanding of organizations and the way they operate, and the ability to work with people and in groups. Today, many good schools in information, business, and computer science require these skills of their graduates. At the end of every chapter, you will find career exercises that will help you explore careers in IS and career areas that interest you.

Most medium to large organizations manage information resources through an IS department. In smaller businesses, one or more people might manage information resources, with support from outsourced services. (Recall that outsourcing is also popular with some organizations.) As shown in Figure 2.11, the IS organization has three primary responsibilities: operations, systems development, and support.

Operations

System operators primarily run and maintain IS equipment and are typically trained at technical schools or through on-the-job experience. They are responsible for efficiently starting, stopping, and correctly operating mainframe systems, networks, tape drives, disk devices, printers, and so on. Other operations include scheduling, hardware maintenance, and preparing input and output. Data-entry operators convert data into a form the computer system can use, using terminals or other devices to enter business transactions, such as sales orders and payroll data. In addition, companies might have local area network and Web operators who run the local network and any Web sites the company has.

Systems Development

The systems development component of a typical IS department focuses on specific development projects and ongoing maintenance and review. Systems analysts and programmers, for example, address these concerns to achieve and maintain IS effectiveness. The role of a systems analyst is multifaceted. *Systems*

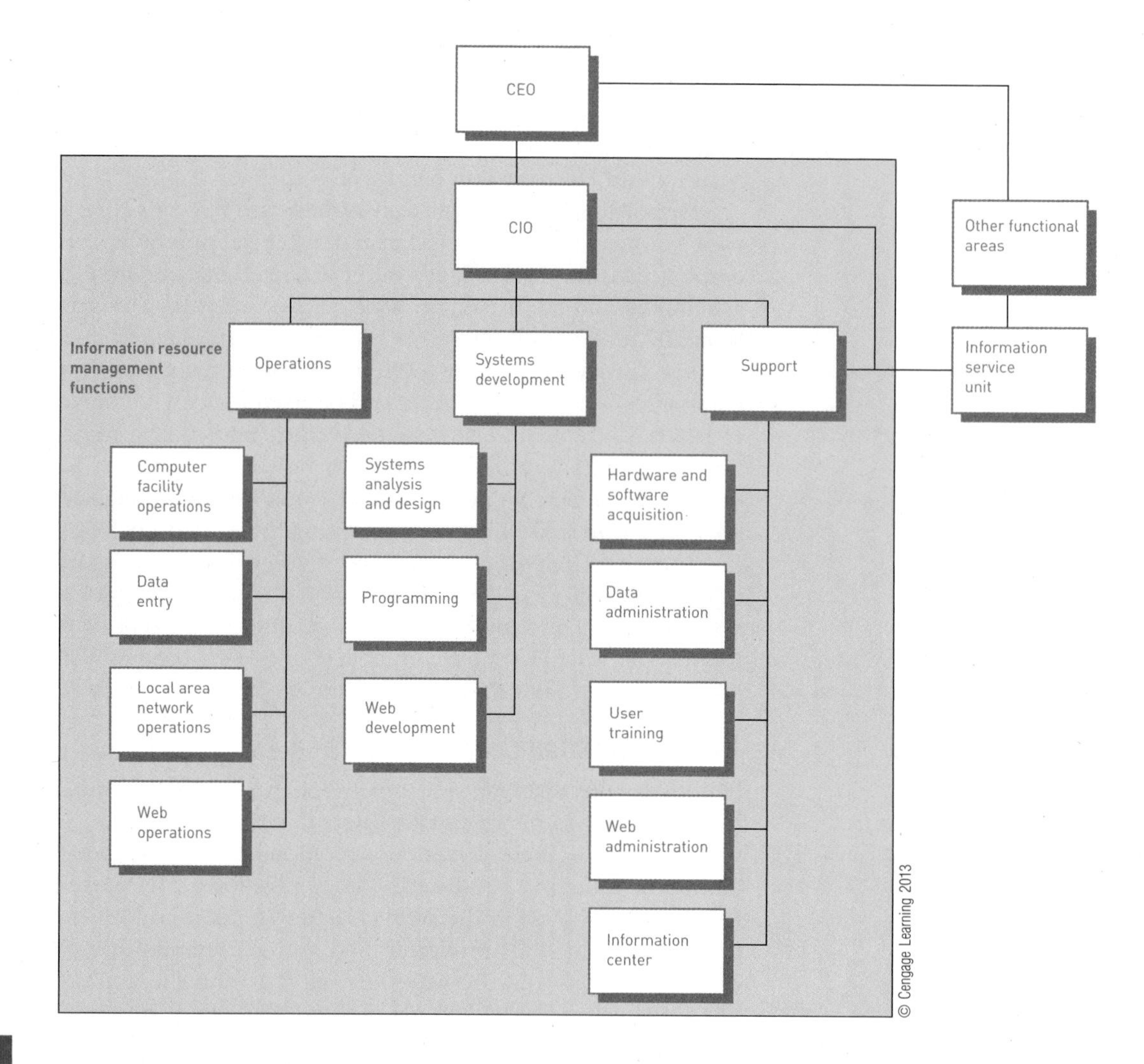

FIGURE 2.11

Three primary responsibilities of information systems

Each of these elements—operations, systems development, and support—contains subelements that are critical to the efficient and effective performance of the organization.

Web developers create and maintain company Web sites.

analysts help users determine what outputs they need from the system and help construct plans for developing the necessary programs that produce these outputs. Systems analysts then work with one or more programmers to make sure that the appropriate programs are purchased, modified from existing programs, or developed. A *computer programmer* uses the plans created by the systems analyst to develop or adapt one or more computer programs that produce the desired outputs. To help businesses select the best analysts and programmers, companies such as TopCoder offer tests to evaluate the proficiency and competence of current IS employees or job candidates. TopCoder Collegiate Challenge allows programming students to compete with other programmers around the world.[94] Some companies, however, are skeptical of the usefulness of these types of tests.[95] In addition, with the dramatic increase in the use of the Internet, intranets, and extranets, many companies have Web or Internet developers who create effective and attractive Web sites for customers, internal personnel, suppliers, stockholders, and others who have a business relationship with the company.

Support

The support component of a typical IS department provides user assistance in hardware and software acquisition and use, data administration, user training and assistance, and Web administration. Often training is done using the Internet.

Because IS hardware and software are costly, a specialized support group often manages computer hardware and software acquisitions. This group sets guidelines and standards for the rest of the organization to follow in making purchases. A database administrator focuses on planning, policies, and procedures regarding the use of corporate data and information. Web administration is another key area for support staff. With the increased use of the Internet and corporate Web sites, Web administrators are sometimes asked to regulate and monitor Internet use by employees and managers to make sure that it is authorized and appropriate. User training is a key to get the most from any information system, and the support area ensures that appropriate training is available. Training can be provided by internal staff or from external sources.

information center: A support function that provides users with assistance, training, application development, documentation, equipment selection and setup, standards, technical assistance, and troubleshooting.

The support component typically operates the information center. An **information center** provides users with assistance, training, application development, documentation, equipment selection and setup, standards, technical assistance, and troubleshooting. Although many firms have attempted to phase out information centers, others have changed the focus of these centers

IS personnel provide assistance in hardware and software acquisition, data administration, user training and assistance, and Web administration.

from technical training to helping users find ways to maximize the benefits of the information resource.

Information Service Units

information service unit: A miniature IS department attached and directly reporting to a functional area in a large organization.

An **information service unit** is basically a miniature IS department attached and directly reporting to a functional area in a large organization. Notice the information service unit shown in Figure 2.11. Even though this unit is usually staffed by IS professionals, the project assignments and the resources necessary to accomplish these projects are provided by the functional area to which it reports. Depending on the policies of the organization, the salaries of IS professionals staffing the information service unit might be budgeted to either the IS department or the functional area.

Typical IS Titles and Functions

The organizational chart shown in Figure 2.11 is a simplified model of an IS department in a typical medium-sized or large organization. Many organizations have even larger departments, with specialized positions such as librarian or quality assurance manager. Smaller firms often combine the roles shown in Figure 2.11 into fewer formal positions.

Chief Information Officer

The role of the CIO is to employ an IS department's equipment and personnel to help the organization attain its goals. The CIO is usually a vice president concerned with the overall needs of the organization and sets corporate-wide policies and plans, manages, and acquires information systems.[96] According to one survey, almost 80 percent of CIOs are actively involved in or consulted on most major decisions. Good CIOs know the top-level executives who can make the difference in the performance of the organization.[97] They also understand the importance of finance, accounting, and return on investment. CIOs can also help companies avoid damaging ethical challenges by monitoring how their firms are complying with a large number of laws and regulations. According to the CIO of Sensis, a defense and airline services company, "Being CIO is not just about running the shop anymore, but different ways you can help your business."[98] The CIO of Case Western Reserve University observes, "Tactical people are providing a tool to answer a question. But broader-thinking CIOs want to know what variety of questions employees have and the many ways to answer them.[99] *Computerworld's* top CIOs include Jan Marshall of Southwest Airlines, Craig Young of Verizon, Joseph AbiDaoud of HudBay Minerals, Tom Amburgey of the City of Wellington, Florida, James Attardi of Medidata Solutions, and Kate Bass of Valspar.[100]

Courtesy of Southwest Airlines

Jan Marshall of Southwest Airlines is one of *Computerworld*'s top CIOs.

The high level of the CIO position reflects the fact that information is one of an organization's most important resources. A good CIO is typically a visionary who provides leadership and direction to the IS department to help an organization achieve its goals. CIOs need technical, business, and personal skills. For federal agencies, the Clinger-Cohen Act of 1996 requires that a CIO coordinate the purchase and management of information systems.[101] The U.S. federal government has also instituted a CIO position to manage federal IS projects, including budgets and deadlines. Vivek Kundra was the first person appointed to this new position—CIO of the United States. Some colleges and universities provide educational programs for CIOs. CIO University, for example, consists of a group of several universities that offer specific training for future and existing CIOs in a variety of technical and organizational topics.[102] Students attending one of the participating universities receive a CIO University Certificate in areas such as Master of Science in Information Systems Technology and Master of Science in Information Management.

Senior IS Managers

Depending on the size of the IS department, several people might work in senior IS managerial levels.[103] Some job titles associated with IS management are the vice president of information systems, manager of information systems, and chief technology officer (CTO). A central role of all these people is to communicate with other areas of the organization to determine changing needs. Often these employees are part of an advisory or steering committee that helps the CIO and other IS managers make decisions about the use of information systems. Together, they can best decide what information systems will support corporate goals. The CTO, for example, typically works under a CIO and specializes in networks and related equipment and technology.

LAN Administrators

Local area network (LAN) administrators set up and manage the network hardware, software, and security processes. They manage the addition of new users, software, and devices to the network. They also isolate and fix operations problems. Solving both technical and nontechnical problems, LAN administrators are in high demand.

Internet Careers

The use of the Internet to conduct business continues to grow and has stimulated a steady need for skilled personnel to develop and coordinate Internet usage. As shown in Figure 2.11, these careers are in the areas of Web operations, Web development, and Web administration. As with other areas in IS, many top-level administrative jobs are related to the Internet. These career opportunities are found in both traditional companies and those that specialize in the Internet, such as Google, Amazon.com, Yahoo!, eBay, and many others.

Certification

certification: A process for testing skills and knowledge, which results in a statement by the certifying authority that confirms an individual is capable of performing particular tasks.

Often, the people filling IS roles have completed some form of certification. **Certification** is a process for testing skills and knowledge resulting in an endorsement by the certifying authority that an individual is capable of performing particular tasks or jobs. Certification frequently involves specific, vendor-provided or vendor-endorsed coursework. Popular certification programs include Microsoft Certified Systems Engineer, Certified Information Systems Security Professional (CISSP), Oracle Certified Professional, Cisco Certified Security Professional (CCSP), and many others. Getting certified from a software, database, or network company can result in increased pay of about 7 percent on average.[104] Some certifications result in even bigger pay increases. Not all certifications, however, provide this financial incentive. Still,

INFORMATION SYSTEMS @ WORK

Profile of a CIO: Bringing Technology to Health Care

A CIO, or chief information officer, is the top manager responsible for how an organization uses information systems to advance the purpose of the organization. Aside from supervising programmers and network technicians, what do these people really think about? What's on the mind of a CIO?

Dr. John Halamka is CIO of Beth Israel Deaconess Medical Center (BIDMC) and an emergency room physician. He summarizes his job as follows in a post on his blog of November 1, 2011: "The modern CIO is no longer a technologist or evangelist for innovation. The modern CIO is a customer relationship manager, a strategic communicator, and a project manager, delicately balancing project portfolios, available resources, and governance."

Dr. Alan Shark is the author of *CIO Leadership for Cities and Counties, Emerging Trends and Practices* (Washington, DC: Public Technology Institute, 2009). As he puts it, "The new CIO has to be a leader, not a dictator; a technologist, not a technician; a business person, not an accountant; and finally, a diplomat, not a politician." A CIO must lead with "vision, knowledge, and team-building." In that context, BIDMC has been an early adopter of "electronic health records, patient portals, and clinical decision support tools. [It] began offering subsidized, hosted EHRs to its 300 affiliated doctors more than a year before the American Recovery and Reinvestment Act's HITECH provision provided financial incentives for hospitals to roll out Web-based EHRs to their affiliated physicians in the effort to get more of these doctors wired."

To lead with "vision, knowledge, and team-building," on what should CIOs focus? Gary Beach, publisher emeritus of *CIO* magazine, surveyed CIOs to learn how they spend their time now and how they want to spend time in the future. He distills their answers as follows:

> In the "how they currently spend their time" category, the top three vote-getters are: 1) aligning IT and business goals, 2) implementing new architectures, and 3) managing cost control. For "where they want to spend more time in the future," the list looks like this: 1) developing new go-to-market strategies and technologies, 2) studying market trends for commercial opportunities, and 3) identifying opportunities for competitive differentiation.

Only two of these six items have anything to do with technology. Four, and part of a fifth, are about business. According to the CIOs surveyed, CIOs are primarily managers.

As managers, CIOs are responsible for developing their staff. BIDMC follows this model. Halamka says, "The only way I am able to succeed is by hiring people smarter than me." He goes on to explain that it's the CIO's job to turn those people into a *team*.

Halamka also plays an important role outside of the medical center, a role he takes seriously. He speaks at conferences, writes a blog, and gives interviews. He is a forthright advocate for using technology to improve health care and spends about one day a week in Washington, D.C., advising legislators on how to accomplish that goal. A recent *InformationWeek* article described him as "the hardest-working man in health IT." Not all CIOs are as visible as Halamka, but many are. Because CIOs focus on "aligning IT and business goals," as Gary Beach found, part of the job often involves gathering and communicating business information.

Developing a diverse set of skills that allows a CIO to manage people effectively and keep up with innovations in technology can be challenging. In a recent blog post, Halamka quotes Meg Aranow, the CIO of Boston Medical Center: "The content of our jobs is great, the context is really challenging." Halamka calls that "a profound observation." He goes on to list other challenges health care CIOs reported in 2011, including:

- You don't receive credit for everything that works. Instead, you are held accountable for the .01% that doesn't work.
- Demand always exceeds supply. Success is finishing half the projects you are asked to complete.
- The pace of change in consumer information systems creates expectations that far exceed the abilities of a thinly staffed IS organization.
- Regulatory burdens will increase exponentially. Compliance is a must-do, though your "customers" do not want their projects or services postponed while you work on compliance.

Why does Halamka keep this job when he could return to the practice of emergency room medicine or move into a different management position? In his blog post of February 24, 2011, he wrote, "The organizations in which I work will last for generations. Their reputations transcend anything I will ever do personally. My role is to champion, support, and publicize a few key innovations every year that will keep the organizations highly visible. That visibility will attract smart people and retain the best employees who want to work for a place on a rising trajectory." These reasons make the work worthwhile.

Discussion Questions

1. Do you think you would enjoy the job of a CIO? Do you think you would be successful as a CIO? Explain why you feel this way.
2. What should someone study in school if his or her eventual career objective is a CIO's job? Compare your answers with those of your classmates. Discuss any differences.

Critical Thinking Questions

1. An organization's CFO (chief financial officer) does not manage an organization's assets. If a company owns a truck, the CFO doesn't decide where it should go, what it should carry, or when to replace its tires. A CFO must, however, budget for its cost, pay taxes on it, choose a depreciation method, and handle capital gains or losses when it is disposed of. How do a CIO's responsibilities in the information area parallel these? How are they conceptually different?
2. Contact your school's CIO or that of a nearby organization. Find out how the CIO typically allocates time to the following major categories of activity:
   - Meetings with top management and people in other departments
   - Meetings with others in the IS department
   - Meetings with people outside the organization
   - Keeping current: reading, Web research, seminars
   - Individual work: budgeting, writing memos/ reports, planning, HR work
   - Other technical tasks (ask for examples)
   - Other nontechnical tasks (ask for examples)

   Also find out the total hours the CIO works in a week. Combine your results with those of your classmates to assemble a composite picture of the CIO's job.

SOURCES: Beach, Gary, "Time Is Money: What CIOs Should Know About How They Spend Their Time," *CIO, www.cio.com/article/693037/Time_Is_Money_What_CIOs_Should_Know_about_How_they_Spend_Their_Time*, November 2, 2011, accessed November 4, 2011; Halamka, John D., "Life as a Healthcare CIO," blog, *http://geekdoctor.blogspot.com*, accessed November 1, 2011; McGee, M.K., Mitchell, R.N., and Versel, N., "Healthcare CIO 25: The Leaders behind the Healthcare IT Revolution," *InformationWeek, http://reports.informationweek.com/abstract/105/5954/Healthcare/research-healthcare-cio-25-the-leaders-behind-the-healthcare-it-revolution.html* (free registration required), March 18, 2011, downloaded December 19, 2011; Motorola, "The Evolving Role of the CIO" (interview with Dr. Alan Shark), Enterprise, *http://ezine.motorola.com/enterprise?a=443*, October 2009, accessed November 5, 2011; Opensource.com (under screen name "opensourceway"), "Dr. John Halamka on Openness and Privacy in Medicine" (video), *www.youtube.com/watch?v=4zn_9eiLfcA*, July 6, 2010, accessed December 15, 2011; Versel, Neil, "Halamka to Leave Harvard Med School CIO Post," *InformationWeek, www.informationweek.com/news/healthcare/leadership/231002441*, July 22, 2011, accessed December 18, 2011.

as certification becomes more important in finding a good IS job, it is becoming increasingly tempting for some unemployed individuals to attempt to cheat on certification exams.[105] Some try to pay others to take certification exams and others illegally purchase stolen tests from the Internet, called braindump materials.

Other IS Careers

To respond to the increase in attacks on computers, new and exciting careers have developed in security and fraud detection and prevention. Today, many organizations have IS security positions, such as a chief information security officer or a chief privacy officer. Some universities offer degree programs in security or privacy.

In addition to working for an IS department in an organization, IS personnel can work for large consulting firms, such as Accenture, IBM, Hewlett-Packard, and others.[106] Some consulting jobs can entail frequent travel because consultants are assigned to work on various projects wherever the client is. Such jobs require excellent project management and people skills in addition to IS technical skills. Related career opportunities include computer training, computer and computer-equipment sales, computer repair and maintenance, and many others.

Other IS career opportunities include being employed by technology companies, such as Microsoft (*www.microsoft.com*), Google (*www.google.com*), and Dell (*www.dell.com*), among others. Such a career enables an individual to work on the cutting edge of technology, which can be extremely challenging and exciting.

As some computer companies cut their services to customers, new companies are being formed to fill the need. With names such as Speak with a Geek and Geek Squad, these companies are helping people and organizations with their computer-related problems that computer vendors are no longer solving.

Some people start their own IS businesses from scratch, such as Craig Newmark, founder of craigslist.[107] In the mid-1990s, Newmark was working for a large financial services firm and wanted to give something back to society by developing an e-mail list for arts and technology events in the San Francisco area. This early e-mail list turned into craigslist. According to Newmark, to run a successful business, you should "Treat people like you want to be treated, including providing good customer service. Listening skills and effective communication are essential." Other people are becoming IS entrepreneurs or freelancers, working from home writing programs, working on IS projects with larger businesses, or developing new applications for the iPhone or similar devices.[108] Some Internet sites, such as *www.freelancer.com*, post projects online and offer information and advice for people working on their own. Many freelancers work for small to medium-sized enterprises in the U.S. market. If you are thinking of freelance or consulting work, be creative and protect yourself. Aggressively market your talents and make sure you are paid by having some or all of your fees put into an escrow account.

Working in Teams

Most IS careers involve working in project teams that can consist of many of the positions and roles discussed above. Thus, it is always good for IS professionals to have good communications skills and the ability to work with other people. Many colleges and universities have courses in information systems and related areas that require students to work in project teams. At the end of every chapter in this book, we have "team activities" that require teamwork to complete a project. You may be required to complete one or more of these team-oriented assignments.

Finding a Job in IS

Traditional approaches to finding a job in the information systems area include attending on-campus visits from recruiters and referrals from professors, friends, and family members. Many colleges and universities have excellent programs to help students develop résumés and conduct job interviews. Developing an online résumé can be critical to finding a good job. Many companies accept résumés only online and use software to search for key words and skills used to screen job candidates. Consequently, having the right key words and skills can mean the difference between getting or not getting a job interview. Some corporate recruiters, however, are starting to actively search for employees rather than sifting through thousands of online resumes or posting jobs on their Web sites.[109] Instead, these corporate recruiters do their own Internet searches and check with professional job sites such as *www.linkedin.com*, *www.branchout.com*, and others.[110] Other companies hire college students to help them market products and services to students.[111] In addition to being paid, students can get invaluable career experience. In some cases, it can help them get jobs after graduation. Increasingly, CIOs are becoming actively involved in hiring employees for their IS departments.[112] In the past, many CIOs relied on the company's human resources (HR) department to fill key IS jobs.

Students who use the Internet and other nontraditional sources to find IS jobs have more opportunities to land a job. Many Web sites, such as Dice.com, CareerBuilder.com, TheLadders.com, LinkedIn.com, Computerjobs.com, and Monster.com, post job opportunities for Internet careers and more traditional careers. Most large companies list job opportunities on their Web sites. These sites allow prospective job hunters to browse job opportunities, locations, salaries, benefits, and other factors. In addition, some sites allow job hunters to post their résumés. Many people use social networking sites such as Facebook to help get job leads. Corporate recruiters also use the Internet or Web logs (blogs) to gather information on existing job candidates or to locate new job

© iStockphoto/kristian sekulic

As with other areas in IS, many top-level administrative jobs, such as Internet systems developers and Internet programmers, are related to the Internet.

candidates. Students are often warned to be careful of what they post on social media sites, including Facebook. Employers often search the Internet to get information about potential employees before they make hiring decisions. (Even law firms are using the Internet and social media sites to try to determine how a potential juror might vote during jury selection for important legal cases.[113]) In addition, many professional organizations and user groups can be helpful in finding a job, staying current once employed, and seeking new career opportunities. These groups include the Association for Computer Machinery (ACM: *www.acm.org*), the Association of Information Technology Professionals (AITP: *www.aitp.org*), Apple User Groups (*www.apple.com/usergroups*), and Linux users groups located around the world. Many companies, including Microsoft and Viacom, among others, use Twitter, an Internet site that allows short messages of 140 characters or less, to advertise job openings. People who have quit jobs or been laid off often use informal networks of colleagues or business acquaintances from their previous jobs to help find new jobs.

SUMMARY

Principle:

The use of information systems to add value to the organization is strongly influenced by organizational structure, culture, and change.

Organizations use information systems to support their goals. Because information systems typically are designed to improve productivity, organizations should devise methods for measuring a system's impact on productivity.

An organization is a formal collection of people and other resources established to accomplish a set of goals. The primary goal of a for-profit organization is to maximize shareholder value. Nonprofit organizations include social groups, religious groups, universities, and other organizations that do not have profit as the primary goal.

Organizations are systems with inputs, transformation mechanisms, and outputs. Value-added processes increase the relative worth of the combined inputs on their way to becoming final outputs of the organization. The value chain is a series (chain) of activities that includes (1) inbound logistics, (2) warehouse and storage, (3) production, (4) finished product storage, (5) outbound logistics, (6) marketing and sales, and (7) customer service.

Organizational structure refers to how organizational subunits relate to the overall organization. Several basic organizational structures include traditional, project, team, and virtual. A virtual organizational structure employs individuals, groups, or complete business units in geographically dispersed areas. These can involve people in various countries operating in different time zones and different cultures from each other.

Organizational culture consists of the major understandings and assumptions for a business, corporation, or organization. Organizational change deals with how profit and nonprofit organizations plan for, implement, and handle change. Change can be caused by internal or external factors. The stages of the change model are unfreezing, moving, and refreezing. According to the concept of organizational learning, organizations adapt to new conditions or alter practices over time.

Principle:

Because information systems are so important, businesses need to be sure that improvements or completely new systems help lower costs, increase profits, improve service, or achieve a competitive advantage.

Business process reengineering involves the radical redesign of business processes, organizational structures, information systems, and values of the organization to achieve a breakthrough in results. Continuous improvement to business processes can add value to products and services.

The extent to which technology is used throughout an organization can be a function of technology diffusion, infusion, and acceptance. Technology diffusion is a measure of how widely technology is in place throughout an organization. Technology infusion is the extent to which technology permeates an area or department. User satisfaction with a computer system and the information it generates depends on the quality of the system and the resulting information. The technology acceptance model (TAM) investigates factors, such as the perceived usefulness of the technology, the ease of use of the technology, the quality of the information system, and the degree to which the organization supports the use of the information system, to predict IS usage and performance.

Total quality management (TQM) consists of a collection of approaches, tools, and techniques that fosters a commitment to quality throughout the organization. Six Sigma is often used in quality control and is based on a statistical term that means 99.9997 percent of the time, products and services will meet quality standards.

Outsourcing involves contracting with outside professional services to meet specific business needs. This approach allows the company to focus more closely on its core business and to target its limited resources to meet strategic goals. Downsizing involves reducing the number of employees to reduce payroll costs; however, it can lead to unwanted side effects.

Competitive advantage is usually embodied in either a product or service that has the most added value to consumers and that is unavailable from the competition or in an internal system that delivers benefits to a firm not enjoyed by its competition. A five-forces model covers factors that lead firms to seek competitive advantage: the rivalry among existing competitors, the threat of new market entrants, the threat of substitute products and services, the bargaining power of buyers, and the bargaining power of suppliers. Strategies to address these factors and to attain competitive advantage include cost leadership, differentiation, niche strategy, altering the industry structure, creating new products and services, and improving existing product lines and services, as well as other strategies.

Developing information systems that measure and control productivity is a key element for most organizations. A useful measure of the value of an IS project is return on investment (ROI). This measure investigates the additional profits or benefits that are generated as a percentage of the investment in IS technology. Total cost of ownership (TCO) can also be a useful measure.

Principle:

IS personnel is a key to unlocking the potential of any new or modified system.

Information systems personnel typically work in an IS department that employs a chief information officer, chief technology officer, systems analysts, computer programmers, computer operators, and other personnel. The chief information officer (CIO) employs an IS department's equipment and personnel to help the organization attain its goals. The chief technology officer (CTO) typically works under a CIO and specializes in hardware and related equipment and technology. Systems analysts help users determine what outputs they need from the system, and they help construct the plans needed to develop the necessary programs that produce these outputs. Systems analysts then work with one or more programmers to make sure that the appropriate programs are purchased, modified from existing programs, or developed. The major responsibility of a computer programmer is to use the plans developed by the systems analyst to build or adapt one or more computer programs that produce the desired outputs.

Computer operators are responsible for starting, stopping, and correctly operating mainframe systems, networks, tape drives, disk devices, printers, and so on. LAN administrators set up and manage the network hardware, software, and security processes. Trained personnel are also needed to set up and manage a company's Internet site, including Internet strategists, Internet systems developers, Internet programmers, and Web site operators. Information systems personnel can also support other functional departments or areas.

In addition to technical skills, IS personnel need skills in written and verbal communication, an understanding of organizations and the way they operate, and the ability to work with people. In general, IS personnel are charged with maintaining the broadest enterprise-wide perspective.

Besides working for an IS department in an organization, IS personnel can work for a large consulting firm, such as Accenture, IBM, Hewlett-Packard, and others. Developing or selling products for a hardware or software vendor is another IS career opportunity.

CHAPTER 2: SELF-ASSESSMENT TEST

The use of information systems to add value to the organization is strongly influenced by organizational structure, culture, and change.

1. A value chain is a series of activities that includes inbound logistics, warehouse and storage, production and manufacturing, finished product storage, outbound logistics, marketing and sales, and customer service. True or False?
2. A(n) ________________ is a formal collection of people and other resources established to accomplish a set of goals.
3. User satisfaction with a computer system and the information it generates often depends on the quality of the system and the resulting information. True or False?
4. A ________________ employs individuals, groups, or complete business units in geographically dispersed areas that can last for a few weeks or years, often requiring telecommunications or the Internet.
   a. learning structure
   b. virtual structure
   c. continuous improvement plan
   d. reengineering project

Because information systems are so important, businesses need to be sure that improvements or completely new systems help lower costs, increase profits, improve service, or achieve a competitive advantage.

5. ________________ involves contracting with outside professional services to meet specific business needs.
6. Today, quality means ________________.
   a. achieving production standards
   b. meeting or exceeding customer expectations
   c. maximizing total profits
   d. meeting or achieving design specifications
7. TCO is the sum of all costs over the life of the information system, including the costs to acquire components such as the technology, technical support, administrative costs, and end-user operations. True or False?
8. Reengineering is also called ________________.
9. What is a measure of the output achieved divided by the input required?
   a. efficiency
   b. effectiveness
   c. productivity
   d. return on investment
10. ________________ is a measure of the additional profits or benefits generated as a percentage of the investment in IS technology.

IS personnel is a key to unlocking the potential of any new or modified system.

11. Who is involved in helping users determine what outputs they need and constructing the plans required to produce these outputs?
    a. CIO
    b. applications programmer
    c. systems programmer
    d. systems analyst
12. An information center provides users with assistance, training, and application development. True or False?
13. The ________________ is typically in charge of the IS department or area in a company.

CHAPTER 2: SELF-ASSESSMENT TEST ANSWERS

1. True
2. organization
3. True
4. b
5. Outsourcing
6. b
7. True
8. process redesign
9. c
10. Return on investment
11. d
12. True
13. chief information officer (CIO)

REVIEW QUESTIONS

1. What is the difference between a value chain and a supply chain?
2. What is supply chain management?
3. What role does an information system play in today's organizations?
4. What is reengineering? What are the potential benefits of performing a process redesign?
5. What is technology diffusion?
6. What is the difference between reengineering and continuous improvement?
7. What is the difference between technology infusion and technology diffusion?
8. What is quality? What is total quality management (TQM)? What is Six Sigma?
9. What are organizational culture and change?
10. List and define the basic organizational structures.
11. Sketch and briefly describe the three-stage organizational change model.
12. What is downsizing? How is it different from outsourcing?
13. What are some general strategies employed by organizations to achieve competitive advantage?
14. List and describe popular job finding Internet sites.
15. Define the term "productivity." How can a company best use productivity measurements?
16. What is on-demand computing? What two advantages does it offer to a company?
17. What is the total cost of ownership?
18. Describe the role of the CIO.

DISCUSSION QUESTIONS

1. You have been hired to work in the IS area of a manufacturing company that is starting to use the Internet to order parts from its suppliers and to offer sales and support to its customers. What types of Internet positions would you expect to see at the company?
2. You have decided to open an Internet site to buy and sell used music CDs. Describe your approach to customer relationship management for your new business.
3. What sort of IS career would be most appealing to you: working as a member of an IS organization, consulting, or working for an IT hardware or software vendor? Why?
4. How can a company encourage innovation? Give several examples of companies that have been innovative.
5. How would you measure technology diffusion and infusion?
6. You have been asked to participate in preparing your company's strategic plan. Specifically, your task is to analyze the competitive marketplace using Porter's five-forces model. Prepare your analysis, using your knowledge of a business you have worked for or have an interest in working for.
7. Based on the analysis you performed in Discussion Question 6, what possible strategies could your organization adopt to address these challenges? What role could information systems play in these strategies? Use Porter's strategies as a guide.
8. Describe the advantages and disadvantages of using the Internet to search for a job.
9. Assume you are the manager of a retail store and need to hire a CIO to run your new computer system. What characteristics would you want in a new CIO?

PROBLEM-SOLVING EXERCISES

1. Identify three companies that make the highest-quality products or services for an industry of your choice. Find the number of employees, total sales, total profits, and earnings growth rate for these three firms. Using a database program, enter this information for the last year. Use the database to generate a report of the three companies with the highest earnings growth rate. Use your word processor to create a document that describes these firms and why you believe they make the highest-quality products and services. What other measures would you use to determine which is the best company in terms of future profit potential? Does high quality always mean high profits?
2. A new IS project has been proposed that is expected to produce not only cost savings but also an increase in revenue. The initial costs to establish the system are estimated to be $500,000. The remaining cash flow data is presented in the following table.

	Year 1	Year 2	Year 3	Year 4	Year 5
Increased revenue	$0	$100	$150	$200	$250
Cost savings	$0	$50	$50	$50	$50
Depreciation	$0	$75	$75	$75	$75
Initial expense	$500				

Note: All amounts are in hundreds of thousands.

   a. Using a spreadsheet program, calculate the return on investment (ROI) for this project. Assume that the cost of capital is 7 percent.
   b. How would the rate of return change if the project delivered $50,000 in additional revenue and generated cost savings of $25,000 in the first year?

3. Using a database program, develop a table listing five popular Internet sites for a job search. The table should include columns on any costs of using the site, any requirements such as salary and job type, important features, advantages, and disadvantages.

TEAM ACTIVITIES

1. With your team, interview one or more instructors or professors about the organizational culture at your college or university.
2. With your team, research a firm that has achieved a competitive advantage. Write a brief report that describes how the company was able to achieve its competitive advantage.

WEB EXERCISES

1. This book emphasizes the importance of information. You can get information from the Internet by going to a specific address, such as *www.ibm.com*, *www.whitehouse.gov*, or *www.fsu.edu*. Going to these sites will give you access to the home page of the IBM corporation, the White House, or Florida State University, respectively. Note that "com" is used for businesses or commercial operations, "gov" for governmental offices, and "edu" for educational institutions. Another approach is to use a search engine, which is a Web site that allows you to enter key words or phrases to find information. Yahoo!, developed by two Tulane University students, was one of the first search engines on the Internet. You can also locate information through lists or menus. The search engine will return other Web sites (hits) that correspond to a search request. Using Yahoo! at *www.yahoo.com* or Google at *www.google.com*, search for information about a company discussed in Chapter 1 or 2. You might be asked to develop a report or send an e-mail message to your instructor about the company and its products and services.
2. Use the Internet to search for information about the total cost of ownership (TCO). What companies use this approach? Describe any Web sites you found that measure TCO.

CAREER EXERCISES

1. Assume that you have decided to become an entrepreneur in a technology-related business. Describe the business, including its products and services. How would you organize your business? Write a brief business plan on how you would make your business a success.
2. Analyze several Internet sites that can be used to find a job. Write a report describing the features of each. Which job-related Internet site would you recommend to a friend as being the best?

CASE STUDIES

Case One

Gaining the Edge

To succeed, a business needs an edge over its competitors: a *competitive advantage*. A big part of creating a competitive advantage is using information systems effectively, meaning a business can't simply buy computers and expect good results. As Oscar Berg puts it in his blog *The Content Economy*, "What [creates] competitive advantage is how we use technologies, how we let them affect our practices and behaviors.... If technologies are carefully selected and applied, they can help to create competitive advantage."

This chapter discusses the five forces that define any competitive situation: rivalry among existing firms in an industry, the threats of new competitors and of substitute products/services, and a firm's relationships with suppliers and customers. Firms use these forces to achieve a *sustainable* competitive advantage, which is one that others cannot copy immediately to eliminate the edge an innovator can have.

TUI Deutschland is Germany's leading tour operator. Targeted pricing is vital in its market, with the travel company that sets prices to accommodate customers' preferences and habits gaining a competitive advantage. For a large tour operator like TUI, setting optimal prices is not easy. Each season, the employee responsible for a particular tour must set around 100,000 prices for each destination region. The factors that affect the final price of hotel rooms, for example, include facilities, types of rooms, arrival dates, and expected demand.

"In the past, decision-making processes were not clear," explains Matthias Wunderlich, head of Business Intelligence at TUI Deutschland GmbH. "There were too many gaps in the system, since the information needed to make pricing decisions was hidden in different places. The result was a pricing process that was complex, laborious, time-consuming, and occasionally inconsistent."

Wunderlich's team developed a new information system to make this process more effective. Used for the first time for the destination of Tenerife in 2010, it organizes historical booking data, making relevant information available to pricing specialists. They define the desired margin for a destination and specify parameters for results. The system calculates combinations and dependencies until the optimum result is achieved. It forecasts which group of customers will drive demand for particular accommodations at each point of the season, from coastal hotels for families during the school holidays to luxury hotels with first-class amenities for premium customers during the low season.

"We have to ensure that a four-star hotel, for example, is always cheaper on a given date than a five-star hotel in the same customer segment," explains Wunderlich. "With the new solution, this is guaranteed. There is no need for a time-consuming manual procedure to ensure it is done correctly."

Because the new pricing process is based on customer data, it reflects the needs and habits of customers. A pricing specialist can set prices that are attractive to customers while still achieving desired margins.

"Traditional pricing methods are no longer appropriate for today's travel and tourism market," says Wunderlich. "In the past it was practically impossible to set prices in a way that was flexible and customer-focused. This has all changed. The pricing specialist in effect becomes an expert in a particular customer group and knows exactly what a certain customer is prepared to pay for a certain travel service. This increases profits, but not at the expense of our customers."

Discussion Questions

1. Are there any losers with the TUI system described in this case? If so, who are they? Is this system justified despite creating losers?
2. Of the five competitive forces discussed in this case, which do you think TUI's system affects?

Critical Thinking Questions

1. Visit the Web page at *www-01.ibm.com/software/success/cssdb.nsf/CS/JHUN-8N748A* to read about a new information system for MediaMath. According to this case, the project can be justified on purely financial grounds. Although cost savings could justify this project, do you think they are the main reason for pursuing it? Should MediaMath go ahead with the system even if it does not promise direct savings? Why or why not?
2. Consider a book store that gives customers a card to be punched for each book they buy. With ten punches, they get a free paperback of their choice. This low-tech system leverages the force of customer power: By promising customers future benefits, it reduces their motivation to switch suppliers even if another store sells books for less. How could a book store use technology to make this loyalty program more effective in retaining customers?

SOURCES: Berg, Oscar, "Creating Competitive Advantage with Social Software," *The Content Economy* (blog), *www.thecontenteconomy.com/2011/06/creating-competitive-advantage-with.html*, June 9, 2011, accessed November 6, 2011; IBM: "Getting the Price Right," IBM Success Stories, *www-01.ibm.com/software/success/cssdb.nsf/CS/STRD-8MQLX4*, October 31, 2011, accessed November 6, 2011; IBM: "Netezza Mediamath—A Nucleus ROI Case Study," IBM Success Stories, *www-01.ibm.com/software/success/cssdb.nsf/CS/JHUN-8N748A*, October 31, 2011, accessed November 6, 2011; Porter, Michael E., "How Competitive Forces Shape Strategy," *Harvard Business Review*, *http://hbr.org/1979/03/how-competitive-forces-shape-strategy/ar/1* (free registration required to read beyond the first page), March/April 1979; accessed November 6, 2011.

Case Two

Listen to Customers with Voice of the Customer Programs

Recall that quality is "the ability of a product or service to meet or exceed customer expectations." In other words, it depends on what customers want or expect. But what *do* customers want or expect? Many companies are launching voice of the customer (VOC) programs to find out.

Consider women's clothing retailer Charming Shoppes, which does business as Lane Bryant, Fashion Bug, and Catherine's Plus Sizes. According to a recent article in *Computerworld*, in the past "various departments and brand groups received input from customer e-mails and online product reviews, and store personnel received verbal comments from shoppers." Anything deemed relevant was "passed up the command chain" to top executives via email distribution lists," says Jeffrey H. Liss, senior VP of corporate strategy. As a result, "We had a lot of anecdotal information floating around," and executives had no way to distinguish important data from rumor.

Charming Shoppes hopes to change this with VOC. In December 2010, Charming Shoppes signed up for VOC services from Clarabridge, Inc. The firm now provides its data to Clarabridge's VOC service, which Clarabridge keeps secure and separate from their other clients' data. In July 2011, Liss reported that implementation is ongoing: "It takes time to learn how to harness the power of this tool." However, Charming Shoppes is one of a growing number of firms that find VOC services worthwhile.

Hotel chain Gaylord Entertainment is further along with its VOC program. The company uses the program to analyze customer comments on social media sites such as Yelp and TripAdvisor as well as from its own surveys. Gaylord Entertainment found two related bits of information:

- Guests who have a good check-in experience are less likely to complain
- Satisfaction with check-in drops off dramatically if it takes over five minutes

As a result, Gaylord created incentives for staff to speed up check-in activities. Shawn Madden, executive director of operations analysis, reports that in the 15 months from late 2009 to early 2011, check-in times under five minutes rose from 39 percent to 49 percent of the total, while the average customer rating for check-in ease rose from 2.4 to 4 on a 1 to 5 scale.

Are VOC programs perfect? Andrew McInnes, a Forrester Research analyst who follows VOC, finds that many are flawed. These programs may not provide a systematic way to take action based on customer insights or to make those insights easy to access. These shortcomings mean potential participants within companies may not receive the information they need in order to take action, or they may not be able to take action based on the information they receive.

These problems can be overcome, and so success stories are not hard to come by. A Forrester survey of 118 customer experience professionals found that 52 percent had a VOC program in place and 29 percent were actively considering one. "Big companies have finally embraced the link between customer experience, loyalty, and long-term financial success," says McInnes. "Investing in voice-of-the-customer programs is the next logical step."

Discussion Questions

1. Charming Shoppes's Jeffrey Liss gives the following example of VOC input: "Suppose a customer says 'I really love going to Fashion Bug, but I don't like sorting through all of the jeans to find the ones that fit me well.'" Is this customer's experience good, bad, or some of each? What were the words in the Liss quotation that led you to this conclusion? If Fashion Bug hears the same or similar reaction from many other customers, what should it do with this information?
2. A major theme of this chapter is the use of information to obtain a competitive advantage. Do VOC programs help to provide a competitive advantage? Why or why not?

Critical Thinking Questions

1. Suppose your school wanted to launch a VOC program for its food service operations. Suggest types of information they could gather from students, other types of information they could get from their own records, and how they could combine these two types of information to get better insight into changes they should make. Be as specific as possible.
2. Do a Web search for "voice of the customer." (Use quotation marks to search for the exact phrase.) Identify three vendors who provide a VOC program. Suppose you had to choose among them for a chain of electronics stores. Which program would you choose? Why?

SOURCES: Horwitt, Elisabeth, "BI: 'Voice of the Customer' Programs Combine Feedback in One Place," *Computerworld*, *www.computerworld.com/s/article/9218483/BI_Voice_of_the_customer_programs_combine_feedback_in_one_place*, July 20, 2011, accessed December 11, 2011; Charming Shoppes Web site, *www.charmingshoppes.com*, accessed December 11, 2011; Clarabridge Web site, *www.clarabridge.com*, accessed December 11, 2011; McInnes, Andrew, "Taking VoC Programs to the Next Level," Forrester Research, *http://blogs.forrester.com/andrew_mcinnes/11-05-16-taking_voc_programs_to_the_next_level*, May 16, 2011, accessed December 11, 2011.

Questions for Web Case

See the Web site for this book to read about the Altitude Online case for this chapter. The following questions cover this Web case.

Altitude Online: Addressing the Needs of the Organization

Discussion Questions

1. What are the advantages of Altitude Online adopting a new ERP system compared to simply connecting existing corporate systems?
2. Why isn't an out-of-the-box ERP system enough for Altitude Online? What additional needs does the company have? Is this the case for businesses in other industries as well?

Critical Thinking Questions

1. Why do you think Jon is taking weeks to directly communicate with stakeholders about the new system?
2. Why do you think Jon and the system administrators decided to outsource the software for this system to an ERP company rather than developing it from scratch themselves?

NOTES

Sources for the opening vignette: Grant, I., "Tesco Uses Customer Data to Stride ahead of Competition," *ComputerWeekly*, *www.computerweekly.com/news/1280095684/Tesco-uses-customer-data-to-stride-ahead-of-competition*, April 12, 2011, accessed December 2, 2011; Kishino AR, "Kishino Augmented Reality for Tesco" (video), *www.youtube.com/watch?v=S5QDRoxuHtk*, November 15, 2011, accessed December 2, 2011; Sillitoe, B., "Tesco Trials Virtual Store in South Korea," *Retail Gazette*, *www.retailgazette.co.uk/articles/43224-tesco-trials-virtual-store-in-south-korea*, August 26, 2011, accessed December 2, 2011; Taylor, G., "Tesco Launches Big Price Drop Facebook App," *Retail Gazette*, *www.retailgazette.co.uk/articles/41023-tesco-launches-big-price-drop-facebook-app*, October 19, 2011, accessed December 2, 2011; Tesco PLC, "Interim Results 2011/12," October 5, 2011, downloaded from *www.tescoplc.com/investors/results-and-events*; Whiteaker, J., "Tesco to Trial Augmented Reality In-store," *Retail Gazette*, *www.retailgazette.co.uk/articles/44432-tesco-to-trial-augmented-reality-instore*, November 17, 2011, accessed December 2, 2011.

1. Kim, D. and Lee, R., "Systems Collaboration and Strategic Collaboration: Their Impacts on Supply Chain Responsiveness and Market Performance," *Decision Sciences*, November 2010, p. 955.
2. Bhasin, P., "Keep on Tracking," *CIO*, February 1, 2011, p. 43.
3. Perkins, Bart, "What Suppliers Can Tell You about Your Own Business," *Computerworld*, May 9, 2011, p. 48.
4. Farasyn, I., et al, "Inventory Optimization at Procter & Gamble," *Interfaces*, January-February 2011, p. 66.
5. Woo, Stu, "Amazon May Expand Free Home Delivery," *The Wall Street Journal*, January 25, 2011, p. B5.

6. Villano, Matt, "Liquor Is Quicker," *CIO*, May 1, 2011, p. 15.
7. Woo, Stu, "Fulfilling Work," *The Wall Street Journal*, June 23, 2011, p. B6.
8. Vascellaro, J., "Building Loyalty on the Web," *The Wall Street Journal*, March 28, 2011, p. B6.
9. Einhorn, B., et al, "Now, a Weak Link in the Global Supply Chain," *Bloomberg Businessweek*, March 27, 2011, p. 18.
10. Schultz, B., "Florida Hospice Saves with SaaS-based CRM," *Network World*, June 6, 2011, p. 24.
11. Duke Energy Web site, *www.duke-energy.com*, accessed June 28, 2011.
12. Oracle Web site, *www.oracle.com*, accessed June 28, 2011.
13. Henschen, D., "Microsoft Goes After Salesforce, Oracle," *Information Week*, January 31, 2011, p. 30.
14. Nash, K., "Recipe for Collaboration," *CIO*, June 1, 2011, p. 17.
15. Strong, Dennis, "CIO Profiles," *Information Week*, March 14, 2011, p. 12.
16. Silvestrini, E., "More Online Education is a Virtual Certainty," *The Tampa Tribune*, March 7, 2011, p. 1.
17. Florida Virtual Web site, *http://melearning.flvs.net*, accessed June 28, 2011.
18. Facebook Web site, *www.facebook.com*, accessed June 28, 2011.
19. Swanborg, Rick, "How to Find Fresh Ideas," *CIO*, February 1, 2011, p. 16.
20. Silverman, R., "Allowing Innovation to Bubble Up," *The Wall Street Journal*, August 29, 2011, p. B4.
21. Elfering, Ingo, "The Grill," *Computerworld*, April 18, 2011, p. 10.
22. Korn, Melissa, "Dean in London Champions Innovation," *The Wall Street Journal*, May 5, 2011, p. B7.
23. Aley, Jim, "AOL Tries for Something New," *Bloomberg Businessweek*, March 28, 2011, p. 43.
24. Keizer, G. and Shah, A., "Steve Jobs Remembered," *Computerworld*, October 10, 2011, p. 6.
25. Isaacson, W., "Book Excerpt: Steve Jobs The Biography," *Fortune*, November 7, 2011, p. 97.
26. Hoag, Brent, "Innovation Rejuvenation," *CIO*, January 1, 2001, p. 50.
27. Eng, Joe, "How to Inspire Everyone," *CIO*, April 1, 2011, p. 18.
28. Christensen, Clayton, *The Innovator's Dilemma*, Boston: Harvard Business School Press, 1997, p. 225 and *The Inventor's Solution*, Boston: Harvard Business School Press, 2003.
29. Mullich, Joe, "16 Ways Cloud Computing Will Change Our Lives," *The Wall Street Journal*, January 7, 2011, p. B4.
30. Singularity University Web site, *http://singularityu.org*, accessed June 28, 2011.
31. Schein, E. H., *Process Consultation: Its Role in Organizational Development*, Reading, MA: Addison-Wesley, 1969. *See also* Keen, Peter G. W., "Information Systems and Organizational Change," *Communications of the ACM*, vol. 24, no. 1, January 1981, pp. 24–33.
32. BBVA Web site, *www.bbva.com*, accessed June 28, 2011.
33. King, Julia, "Extreme Automation," *Computerworld*, June 6, 2011, p. 16.
34. Conry-Murray, Andrew, "A Measure of Satisfaction," *InformationWeek*, January 26, 2009, p. 19.
35. Wixom, Barbara and Todd, Peter, "A Theoretical Integration of User Satisfaction and Technology Acceptance," *Information Systems Research*, March 2005, p. 85.
36. Bailey, J. and Pearson, W., "Development of a Tool for Measuring and Analyzing Computer User Satisfaction," *Management Science*, vol. 29, no. 5, 1983, p. 530.
37. Chaparro, Barbara, et al, "Using the End-User Computing Satisfaction Instrument to Measure Satisfaction with a Web Site," *Decision Sciences*, May 2005, p. 341.
38. MacMillian, D., "Creating Web Addicts for $10,000 a Month," *Bloomberg Businessweek*, January 24, 2011, p. 35.
39. Schwarz, A. and Chin, W., "Toward an Understanding of the Nature and Definition of IT Acceptance," *Journal of the Association for Information Systems*, April 2007, p. 230.
40. Davis, F., "Perceived Usefulness, Perceived Ease of Use, and User Acceptance of Information Technology," *MIS Quarterly*, vol. 13, no. 3, 1989, p. 319. Kwon, et al, "A Test of the Technology Acceptance Model," *Proceedings of the Hawaii International Conference on System Sciences*, January 4–7, 2000.
41. Ilie, V., et al, "Paper versus Electronic Medical Records," *Decision Sciences*, May 2009, p. 213.
42. Hartung, Adam, "Tools They Want to Use," *CIO*, May 1, 2001, p. 18.
43. Barki, H., et al, "Information System Use-Related Activity," *Information Systems Research*, June 2007, p. 173.
44. Loch, Christoph and Huberman, Bernardo, "A Punctuated-Equilibrium Model of Technology Diffusion," *Management Science*, February 1999, p. 160.
45. Tornatzky, L. and Fleischer, M., "The Process of Technological Innovation," *Lexington Books*, Lexington, MA, 1990; Zhu, K. and Kraemer, K., "Post-Adoption Variations in Usage and Value of E-Business by Organizations," *Information Systems Research*, March 2005, p. 61.
46. Armstrong, Curtis and Sambamurthy, V., "Information Technology Assimilation in Firms," *Information Systems Research*, April 1999, p. 304.
47. Sykes, T. and Venkatesh, V., "Model of Acceptance with Peer Support," *MIS Quarterly*, June 2009, p. 371.
48. Agarwal, Ritu and Prasad, Jayesh, "Are Individual Differences Germane to the Acceptance of New Information Technology?" *Decision Sciences*, Spring 1999, p. 361.
49. Fuller, R. and Denis, A., "Does Fit Matter?" *Information Systems Research*, March 2009, p. 2.
50. Tieto Web site, *www.tieto.com/healthcare*, accessed December 5, 2012.
51. Richardson, Karen, "The Six Sigma Factor for Home Depot," *The Wall Street Journal*, January 4, 2007, p. C3.
52. Six Sigma Web sites, *www.sixsigma.com* and *www.6sigma.us*, accessed November 15, 2009.
53. Virgin Australia Web site, *www.virginaustralia.com*, accessed December 5, 2012.

54. Ferrier, W., et al, "Digital Systems and Competition," *Information Systems Research*, September 2010, p. 413.
55. Lasala, Donna, "Outlook 2011," *Network World*, January 10, 2011, p. 8.
56. Nash, Kim, "The Talent Advantage," *CIO*, March 1, 2011, p. 32.
57. Tanriverdi, H., et al, "Reframing the Dominant Quests of Information Systems Strategy Research for Complex Adaptive Business Systems," *Information Systems Research*, December 2010, p. 822.
58. Catan, Thomas and Koppel, Nathan, "Regulators Eye Apple Anew," *The Wall Street Journal*, February 18, 2011, p. B1.
59. Collins, Jim, *Good to Great*, New York: Harper Collins, 2001, p. 300.
60. Pavlou, P., and Sawy, O., "The Third Hand: IT-Enabled Competitive Advantage in Turbulence through Improvisational Capabilities," *Information Systems Research*, September 2010, p. 443.
61. Porter, M. E., *Competitive Advantage: Creating and Sustaining Superior Performance*, New York: Free Press, 1985; *Competitive Strategy: Techniques for Analyzing Industries and Competitors*, New York: Free Press, 1980; and *Competitive Advantage of Nations*, New York: Free Press, 1990.
62. Porter, M. E. and Millar, V., "How Information Systems Give You Competitive Advantage," *Journal of Business Strategy*, Winter 1985. *See also* Porter, M. E., *Competitive Advantage*, New York: Free Press, 1985.
63. Evans, Bob, "Why CIOs Must Have a Tablet Strategy," *Information Week*, March 14, 2011, p. 10.
64. Natural Selection Web site, *www.natural-selection.com*, accessed June 29, 2011.
65. Greene, Tim, "Library System Shushes MPLS for Cheaper DSL," *Network World*, January 24, 2011, p. 14.
66. Staff, "The Innovator's Solution," *The Wall Street Journal*, June 27, 2011, p. C9.
67. Apptio Web site, *www.apptio.com*, accessed June 29, 2011.
68. Nash, Kim, "IT Leaders Fueling Growth with Strategic Investments," *CIO*, January 1, 2011, p. 34.
69. Thibodeau, P., "Tax Law May Accelerate IT Purchases," *Computerworld*, February 7, 2011, p. 4.
70. Fischer, M., et al, "Managing Global Brand Investments at DHL," *Interfaces*, January-February 2011, p. 35.
71. Pratt, Mary, "What CFOs Want from IT," *Computerworld*, January 24, 2011, p. 19.
72. Ibid.
73. Kodak Web site, *www.kodak.com*, accessed June 29, 2011.
74. Pratt, Mary, "What CFOs Want from IT," *Computerworld*, January 24, 2011, p. 19.
75. Hitachi Data Center Web site, *www.hds.com/assets/pdf/itcentrix_report_may_2006.pdf*, accessed June 29, 2011.
76. Pratt, Mary, "What CFOs Want from IT," *Computerworld*, January 24, 2011, p. 19.
77. Heylar, J. and MacMillan, D., "In Tech, Poaching Is the Sincerest Form of Flattery," *Bloomberg Businessweek*, March 7, 2011, p. 17.
78. Waghray, A., "CIO Profiles," *Information Week*, February 14, 2011, p. 10.
79. Brousell, Lauren, "IT Budgets Bulge in 2011," *CIO*, February 1, 2011, p. 8.
80. Evans, Bob, "The Top 10 CIO Priorities and Issues for 2011," *Information Week*, January 31, 2011, p. 12.
81. Bednaza, Ann, "Prepare for Talent Wars and IT Poaching," *Network World*, March 7, 2011, p. 19.
82. Carlton, Jim, "California Economy Gets a Jolt from Tech Hiring," *The Wall Street Journal*, May 10, 2011, p. A3.
83. Staff, "Career Watch, *Computerworld*, May 9, 2011, p. 50.
84. Staff, "2011 Salary Survey," *Computerworld*, April 4, 2011, p. 13.
85. Murphy, Chris, "Long, Hard Road," *Information Week*, April 25, 2011, p. 21.
86. OpenDNS Web site, *www.opendns.com*, accessed June 29, 2011.
87. Nash, Kim, "The Talent Advantage," *CIO*, March 1, 2011, p. 32.
88. Henschen, D., "They're Ready," *Information Week*, February 14, 2001, p. 21.
89. Light, J., "How's Your HTML5? App Skills in Demand," *The Wall Street Journal*, January 31, 2001, p. B7.
90. Flint, J., "Analyze This," *Bloomberg Business*, February 21, 2011, p. 82.
91. Jordan, Miriam, "Long-Prized Tech Visas Lose Cachet," *The Wall Street Journal*, May 7, 2011, p. A3.
92. Jordan, Miriam, "U.S. Probes Infosys Over Visas," *The Wall Street Journal*, May 25, 2011, p. B1.
93. Bahree, M. and Sharma, A., "Tighter U.S. Visa Scrutiny Alters Landscape for Indian Tech Firms," *The Wall Street Journal*, May 26, 2011, p. B3.
94. Top Coder Collegiate Challenge Web site, *www.topcoder.com*, accessed June 29, 2011.
95. Ibid.
96. Zetlin, Minda, "Surviving CIO Regime Change," *Computerworld*, March 7, 2011, p. 12.
97. McGreavy, John, "Meet the Manager Who Will Replace Me," *Information Week*, February 14, 2011, p. 12.
98. Nash, Kim, "Market Makers," *CIO*, June 1, 2011, p. 37.
99. Nash, Kim, "The Thought That Counts," *CIO*, April 1, 2011, p. 28.
100. King, Julia, "Premier 100 IT Leaders of 2011," *Computerworld*, February 21, 2011, p. 16.
101. Department of Defense Web site, *http://cio-nii.defense.gov/docs/ciodesrefvolone.pdf*, accessed June 29, 2011.
102. CIO University Web site, *www.cio.gov/admin-pages.cfm/page/cio-university*, accessed April 2, 2011.
103. Perkins, Bart, "Disappearing CIOs," *Computerworld*, January 10, 2011, p. 18.
104. Brodkin, Jon, "Microsoft Certifications Won't Boost Your Pay," *Network World*, February 21, 2011, p. 16.
105. Duffy-Marsan, Carolyn, "Survey Shows Increase in Cert Cheating," *Network World*, February 21, 2011, p. 22.
106. IBM Web site, *www.ibm.com/services*; Hewlett-Packard Web site, *http://www8.hp.com/us/en/services/it-services.html*; and Accenture Web site, *www.accenture.com*, all accessed June 29, 2011.

107. Craigslist Web site, *www.craigslist.org*, accessed June 29, 2011.
108. Elance Web site, *www.elance.com*, accessed December 5, 2011.
109. Light, Joe, "Recruiters Rethink Online Playbook," *The Wall Street Journal*, January 18, 2011, p. B7.
110. Berfield, Susan, "Dueling Your Facebook Friends for a New Job," *Bloomberg Businessweek*, March 7, 2011, p. 35.
111. Rosman, Katherine, "Here, Tweeting Is a Class Requirement," *The Wall Street Journal*, March 9, 2011, p. D1.
112. Lamoreaux, Kristen, "Rethinking the Talent Search," *CIO*, May 1, 2011, p. 30.
113. Campoy, A. and Jones, A., "Searching for Details Online," *The Wall Street Journal*, February 22, 2011, p. A2.

PART 1 PART 2 PART 3 PART 4 PART 5

Information Technology Concepts

CHAPTERS

3 Hardware: Input, Processing, Output, and Storage Devices

Principles	Learning Objectives
• Computer hardware must be carefully selected to meet the evolving needs of the organization and of its supporting information systems.	• Describe the role of the central processing unit and main memory. • State the advantages of multiprocessing and parallel computing systems and provide examples of the types of problems they address. • Describe the access methods, capacity, and portability of various secondary storage devices. • Identify and discuss the speed, functionality, and importance of various input and output devices. • Identify the characteristics of and discuss the usage of various classes of single-user and multiuser computer systems.
• The computer hardware industry is rapidly changing and highly competitive, creating an environment ripe for technological breakthroughs.	• Describe Moore's Law and discuss its implications for future computer hardware developments. • Give an example of recent innovations in computer CPU chips, memory devices, and input/output devices.
• The computer hardware industry and users are implementing green computing designs and products.	• Define the term green computing and identify the primary goals of this program. • Identify several benefits of green computing initiatives that have been broadly adopted.

Information Systems in the Global Economy

DEUTSCHE BANK, GERMANY

Sending Computers into the Cloud

Since the modern electronic computer was invented in the 1940s, the trend has been toward reducing the size of the computer while increasing its capability. The logical end of this trend is to remove the physical computer altogether. While that isn't likely to happen in business, companies have found ways to make their central computers disappear.

Central computers still exist, of course, but if you look around business offices, follow the cables from a desktop or wireless router through walls and down halls, you may not find a central computer. What you'll find instead in more and more organizations are signals going "into the cloud." That saying refers to *cloud computing*, which provides computing services and database access over the Internet that are accessible from anywhere in the world rather than from a specific computer in a specific location.

Deutsche Bank (DB), the German financial services firm, made a decision to send its computers into the cloud. As Alistair McLaurin of its Global Technology Engineering group put it, the bank "wanted to create something radically different," to "challenge assumptions around what centrally provided IT services could be and how much they must cost." DB created a system in which computing is done by *virtual machines (VMs)*: software-managed "slices" of real computers that behave in every respect like a full computer but that share the hardware of one real computer with many other VMs. A virtual machine is an extension of the familiar concept of running more than one program at a time. In a VM, you run more than one operating system at a time, with each completely isolated from the others. The result is substantial savings in hardware cost and everything that goes with it, such as space and electricity. By putting the computers that host their virtual machines in the cloud, DB freed themselves from the constraints of being at a particular physical location. DB can thus optimize the use of these virtual computers across the entire company.

Another advantage of the virtual approach is that someone who needs a new computer doesn't have to purchase one. Instead, they can use a virtual computer inside a real computer that the company already has; such a VM is easier to set up than a new system. In fact, "a user who is a permanent employee, who wants a new Virtual Machine for their own use only, can do it by visiting one Web site, selecting an operating system [Windows, Solaris, or Linux] and clicking three buttons. The new VM will be ready and available for them within an hour."

The Open Data Center Alliance recently chose DB as the grand prize winner of its Conquering the Cloud Challenge. The specific basis for the award was the way DB's cloud-based system manages user identities. When a user requests a virtual machine, the system already knows who has to approve the request (if anyone), where its cost should be billed, and who should be allowed to administer the machine. The cloud-based system means users don't have to worry about how virtual machines are created, making it more practical to use them. Because a virtual machine is less expensive than a new desktop computer, DB management wanted to encourage employees to use the virtual machines. Removing barriers to their adoption was important, which is why they designed the cloud-based system to manage user identities.

Currently, programmers and other system developers use DB's cloud system for application development and testing. If a developer is working with a computer that runs Solaris and wants to test an application under Windows 7 or Windows Vista, he or she can do so using a virtual machine quickly and efficiently. The cloud system will be used next for DB production applications, except for those that need 100 percent uptime (such as the one that operates a network of ATMs). After that? Who knows?

As you read this chapter, consider the following:

- How does the *location independence* of cloud computing help Deutsche Bank or any other organization?
- Would cloud computing be useful to your school? To a specific small business you can think of?
- How does being aware of technology trends help a business become and stay successful?

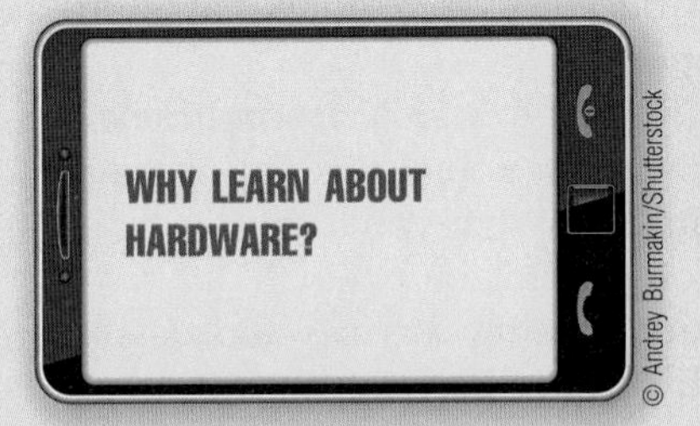
WHY LEARN ABOUT
HARDWARE?
© Andrey Burmakin/Shutterstock

Organizations invest in computer hardware to improve worker productivity, increase revenue, reduce costs, provide better customer service, speed up time-to-market, and enable collaboration among employees. Organizations that don't make wise hardware investments are often stuck with outdated equipment that is unreliable and that cannot take advantage of the latest software advances. Such obsolete hardware can place an organization at a competitive disadvantage. Managers, no matter what their career field and educational background, are expected to help define the business needs that the hardware must support. In addition, managers must be able to ask good questions and evaluate options when considering hardware investments for their areas of the business. This need is especially true in small organizations, which might not have information system specialists. Managers in marketing, sales, and human resources often help IS specialists assess opportunities to apply computer hardware and evaluate the options and features specified for the hardware. Managers in finance and accounting especially must keep an eye on the bottom line, guarding against overspending, yet be willing to invest in computer hardware when and where business conditions warrant it.

Today's use of technology is practical—it's intended to yield real business benefits, as demonstrated by Deutsche Bank. Using the latest information technology and providing additional processing capabilities can increase employee productivity, expand business opportunities, and allow for more flexibility. This chapter concentrates on the hardware component of a computer-based information system (CBIS). Recall that hardware refers to the physical components of a computer that perform the input, processing, output, and storage activities of the computer. When making hardware decisions, the overriding consideration of a business should be how hardware can support the objectives of the information system and the goals of the organization.

COMPUTER SYSTEMS: INTEGRATING THE POWER OF TECHNOLOGY

People involved in selecting their organization's computer hardware must clearly understand current and future business requirements so that they can make informed acquisition decisions. Consider the following examples of applying business knowledge to reach sound decisions on acquiring hardware:

- When Facebook needed to add two new data centers, it elected to build its own custom computers rather than buy off-the-shelf computers from traditional manufacturers such as Dell or HP. The computers installed had only the minimum components needed to perform their specific task and did not include expensive manufacturer upgrade and backup services. As a result, the cost of building and outfitting the data center was reduced by 24 percent.[1]
- The Air Force Information Technology Commodity Council, composed of top USAF officials, selects the computer hardware vendors that provide wares to the service agency. Recently this group chose a vendor to provide new workstations and desktop personal computers based on a number of

criteria but primarily on their performance in environmental extremes of heat, humidity, cold, dryness, and air pollutants including dust and sand. The equipment has to perform reliably in the extremes in which the USAF must carry out its missions.[2]

- Russell's Convenience Stores operates 24 convenience stores across the U.S. western states and Hawaii. Russell's converted to cloud computing technology to improve how its employees collaborate with one another and with their licensees, vendors, and other business partners. It also believes that cloud computing will provide more flexibility and reduce computing costs over a five-year hardware planning horizon. According to Raymond Huff, president of HJB Convenience Corporation/Russell's Convenience, cloud computing "is helping our licensees to operate as one business—one that is connected, informed, and cohesive."[3]

As these examples demonstrate, choosing the right computer hardware requires understanding its relationship to the information systems and the needs of an organization.

central processing unit (CPU): The part of the computer that consists of three associated elements: the arithmetic/logic unit, the control unit, and the register areas.

arithmetic/logic unit (ALU): The part of the CPU that performs mathematical calculations and makes logical comparisons.

control unit: The part of the CPU that sequentially accesses program instructions, decodes them, and coordinates the flow of data in and out of the ALU, the registers, the primary storage, and even secondary storage and various output devices.

register: A high-speed storage area in the CPU used to temporarily hold small units of program instructions and data immediately before, during, and after execution by the CPU.

primary storage (main memory; memory): The part of the computer that holds program instructions and data.

Hardware Components

Computer system hardware components include devices that perform input, processing, data storage, and output, as shown in Figure 3.1.

Recall that any system must be able to process (organize and manipulate) data, and a computer system does so through an interplay between one or more central processing units and primary storage. Each **central processing unit (CPU)** consists of three associated elements: the arithmetic/logic unit, the control unit, and the register areas. The **arithmetic/logic unit (ALU)** performs mathematical calculations and makes logical comparisons. The **control unit** sequentially accesses program instructions, decodes them, and coordinates the flow of data in and out of the ALU, the registers, the primary storage, and even secondary storage and various output devices. **Registers** are high-speed storage areas used to temporarily hold small units of program instructions and data immediately before, during, and after execution by the CPU.

Primary storage, also called **main memory** or **memory**, is closely associated with the CPU. Memory holds program instructions and data immediately before or after the registers. To understand the function of processing and the interplay between the CPU and memory, let's examine the way a typical computer executes a program instruction.

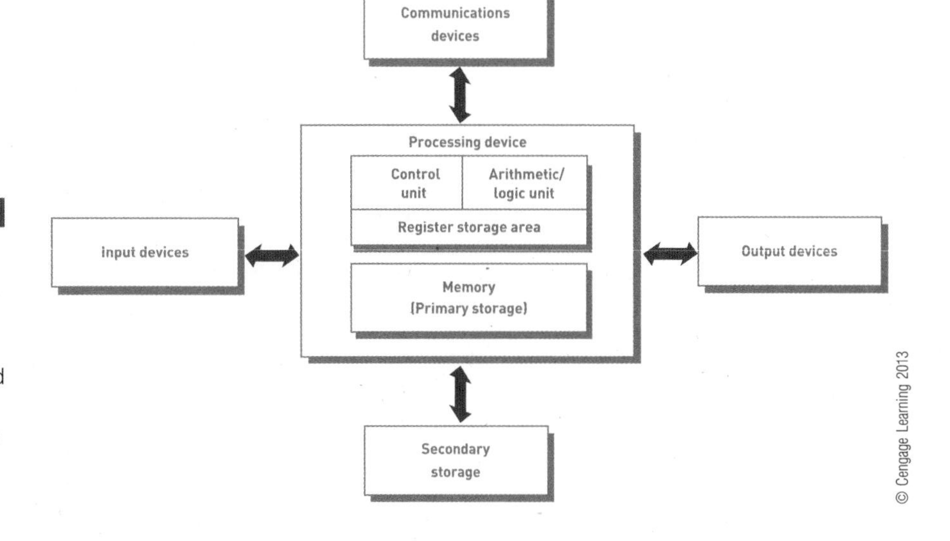

Hardware components

These components include the input devices, output devices, communications devices, primary and secondary storage devices, and the central processing unit (CPU). The control unit, the arithmetic/logic unit (ALU), and the register storage areas constitute the CPU.

Hardware Components in Action

Executing any machine-level instruction involves two phases: instruction and execution. During the instruction phase, a computer performs the following steps:

- Step 1: Fetch instruction. The computer reads the next program instruction to be executed and any necessary data into the processor.
- Step 2: Decode instruction. The instruction is decoded and passed to the appropriate processor execution unit. Each execution unit plays a different role. The arithmetic/logic unit performs all arithmetic operations; the floating-point unit deals with noninteger operations; the load/store unit manages the instructions that read or write to memory; the branch processing unit predicts the outcome of a branch instruction in an attempt to reduce disruptions in the flow of instructions and data into the processor; the memory-management unit translates an application's addresses into physical memory addresses; and the vector-processing unit handles vector-based instructions that accelerate graphics operations.

The time it takes to perform the instruction phase (Steps 1 and 2) is called the **instruction time (I-time)**.

instruction time (I-time): The time it takes to perform the fetch instruction and decode instruction steps of the instruction phase.

The second phase is execution. During the execution phase, a computer performs the following steps:

- Step 3: Execute instruction. The hardware element, now freshly fed with an instruction and data, carries out the instruction. This process could involve making an arithmetic computation, logical comparison, bit shift, or vector operation.
- Step 4: Store results. The results are stored in registers or memory.

The time it takes to complete the execution phase (Steps 3 and 4) is called the **execution time (E-time)**.

execution time (E-time): The time it takes to execute an instruction and store the results.

After both phases have been completed for one instruction, they are performed again for the second instruction, and so on. Completing the instruction phase followed by the execution phase is called a **machine cycle**, as shown in Figure 3.2. Some processing units can speed processing by using **pipelining**, whereby the processing unit gets one instruction, decodes another, and executes a third at the same time. The Pentium 4 processor, for example, uses two execution unit pipelines. This feature means the processing unit can execute two instructions in a single machine cycle.

machine cycle: The instruction phase followed by the execution phase.

pipelining: A form of CPU operation in which multiple execution phases are performed in a single machine cycle.

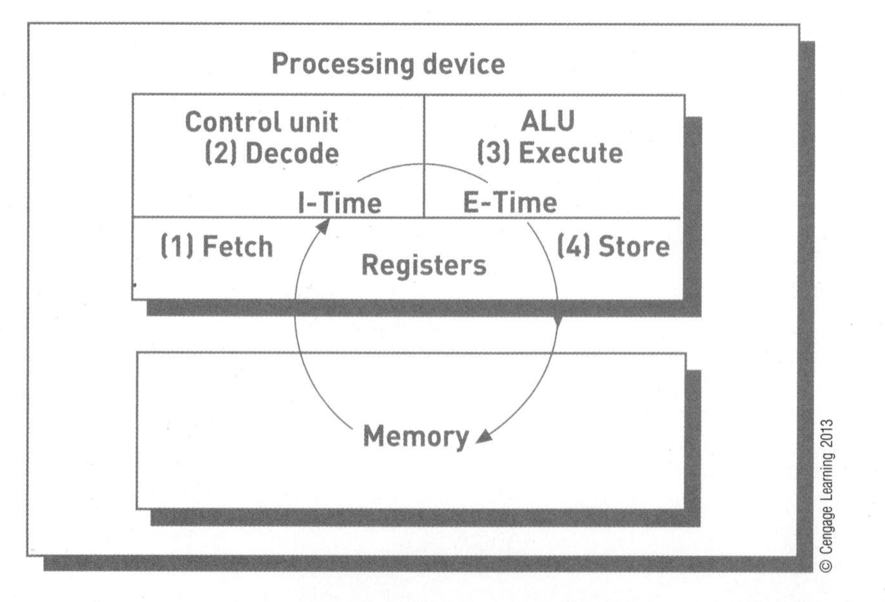

FIGURE 3.2

Execution of an instruction

In the instruction phase, a program's instructions and any necessary data are read into the processor (1). Then the instruction is decoded so that the central processor can understand what to do (2). In the execution phase, the ALU does what it is instructed to do, making either an arithmetic computation or a logical comparison (3). Then the results are stored in the registers or in memory (4). The instruction and execution phases together make up one machine cycle.

PROCESSING AND MEMORY DEVICES: POWER, SPEED, AND CAPACITY

The components responsible for processing—the CPU and memory—are housed together in the same box or cabinet, called the *system unit*. All other computer system devices, such as the monitor, secondary storage, and keyboard, are linked directly or indirectly into the system unit housing. In this section, we investigate the characteristics of these important devices.

Processing Characteristics and Functions

Because organizations want efficient processing and timely output, they use a variety of measures to gauge processing speed. These measures include the time it takes to complete a machine cycle and clock speed.

Machine Cycle Time

As you've seen, a computer executes an instruction during a machine cycle. The time in which a machine cycle occurs is measured in *nanoseconds* (one-billionth of one second) and *picoseconds* (one-trillionth of one second). Machine cycle time also can be measured by how many instructions are executed in one second. This measure, called **MIPS**, stands for millions of instructions per second. MIPS is another measure of speed for computer systems of all sizes.

MIPS: Millions of instructions per second, a measure of machine cycle time.

Clock Speed

Each CPU produces a series of electronic pulses at a predetermined rate, called the **clock speed**, which affects machine cycle time. The control unit executes instructions in accordance with the electronic cycle, or pulses of the CPU "clock." Each instruction takes at least the same amount of time as the interval between pulses. The shorter the interval between pulses, the faster each instruction can be executed.

clock speed: A series of electronic pulses produced at a predetermined rate that affects machine cycle time.

Clock speed is often measured in **megahertz** (MHz, millions of cycles per second) or **gigahertz** (GHz, billions of cycles per second). Unfortunately, the faster the clock speed of the CPU, the more heat the processor generates. This heat must be dissipated to avoid corrupting the data and instructions the computer is trying to process. Also, chips that run at higher temperatures need bigger heat sinks, fans, and other components to eliminate the excess heat. This increases the size of the computing device whether it is a desktop computer, tablet computer, or smartphone, which increases the cost of materials and makes the device heavier—counter to what manufacturers and customers desire.

megahertz (MHz): Millions of cycles per second, a measure of clock speed.

gigahertz (GHz): Billions of cycles per second, a measure of clock speed.

Chip designers and manufacturers are exploring various means to avoid heat problems in their new designs. ARM is a computer chip design company whose energy-efficient chip architecture is broadly used in smartphones and tablet computers. Its Cortex-A7 chip design is expected to lead to much less expensive smartphones with a battery life five times longer than in current devices. Its more powerful Cortex-A15 processor can be used for processing-intensive tasks such as navigation or video playback.[4] Intel expects to begin producing computer processor chips based on a new 3D technology that it has been developing for over a decade. Traditionally, transistors, the basic elements of computer chips, are produced in flat two-dimensional structures. The new 3D transistors will cut chip power consumption in half, making the chips ideal for use in the rapidly growing smartphone and tablet computer market.[5]

Manufacturers are also seeking more effective sources of energy as portable devices grow increasingly power hungry. A number of companies are exploring the substitution of fuel cells for lithium ion batteries to provide additional, longer-lasting power. Fuel cells generate electricity by consuming fuel (often methanol), while traditional batteries store electricity and release it through a chemical reaction. A spent fuel cell is replenished in moments by simply refilling its reservoir or by replacing the spent fuel cartridge with a fresh one.

Physical Characteristics of the CPU

Most CPUs are collections of digital circuits imprinted on silicon wafers, or chips, each no bigger than the tip of a pencil eraser. To turn a digital circuit on or off within the CPU, electrical current must flow through a medium (usually silicon) from point A to point B. The speed the current travels between points can be increased by either reducing the distance between the points or reducing the resistance of the medium to the electrical current.

Reducing the distance between points has resulted in ever smaller chips, with the circuits packed closer together. Gordon Moore, who would cofound Intel (the largest maker of microprocessor chips) and become its chairman of the board, hypothesized that progress in chip manufacturing ought to make it possible to double the number of transistors (the microscopic on/off switches) on a single chip every two years. The hypothesis became known as **Moore's Law**, and this "rule of thumb" has become a goal that chip manufacturers have met more or less for more than four decades.

Moore's Law: A hypothesis stating that transistor densities on a single chip will double every two years.

Chip makers have been able to improve productivity and performance by putting more transistors on the same size chip while reducing the amount of power required to perform tasks. Furthermore, because the chips are smaller, chip manufacturers can cut more chips from a single silicon wafer and thus reduce the cost per chip. As silicon-based components and computers perform better, they become cheaper to produce and therefore more plentiful, more powerful, and more a part of our everyday lives. This process makes computing devices affordable for an increasing number of people around the world and makes it practical to pack tremendous computing power into the tiniest of devices.

Memory Characteristics and Functions

Main memory is located physically close to the CPU, although not on the CPU chip itself. It provides the CPU with a working storage area for program instructions and data. The chief feature of memory is that it rapidly provides the data and instructions to the CPU.

Storage Capacity

Like the CPU, memory devices contain thousands of circuits imprinted on a silicon chip. Each circuit is either conducting electrical current (on) or not conducting current (off). Data is stored in memory as a combination of on or off circuit states. Usually, 8 bits are used to represent a character, such as the letter *A*. Eight bits together form a **byte (B)**. In most cases, storage capacity is measured in bytes, with 1 byte equivalent to one character of data. The contents of the Library of Congress, with over 126 million items and 530 miles of bookshelves, would require about 20 petabytes of digital storage. It is estimated that all the words ever spoken represented in text form would equal about 5 exabytes of information.[6] Table 3.1 lists units for measuring computer storage.

byte (B): Eight bits that together represent a single character of data.

Types of Memory

Computer memory can take several forms. Instructions or data can be temporarily stored in and read from **random access memory (RAM)**. As currently designed, RAM chips are volatile storage devices, meaning they lose their contents if the current is turned off or disrupted (as happens in a power surge, brownout, or electrical noise generated by lightning or nearby machines). RAM chips are mounted directly on the computer's main circuit board or in other chips mounted on peripheral cards that plug into the main circuit

random access memory (RAM): A form of memory in which instructions or data can be temporarily stored.

TABLE 3.1 Computer storage units

Name	Abbreviation	Number of Bytes
Byte	B	1
Kilobyte	KB	2^{10} or approximately 1,024 bytes
Megabyte	MB	2^{20} or 1,024 kilobytes (about 1 million)
Gigabyte	GB	2^{30} or 1,024 megabytes (about 1 billion)
Terabyte	TB	2^{40} or 1,024 gigabytes (about 1 trillion)
Petabyte	PB	2^{50} or 1,024 terabytes (about 1 quadrillion)
Exabyte	EB	2^{60} or 1,024 petabytes (about 1 quintillion)

board. These RAM chips consist of millions of switches that are sensitive to changes in electric current.

RAM comes in many varieties: static random access memory (SRAM) is byte-addressable storage used for high-speed registers and caches; dynamic random access memory (DRAM) is byte-addressable storage used for the main memory in a computer; and double data rate synchronous dynamic random access memory (DDR SDRAM) is an improved form of DRAM that effectively doubles the rate at which data can be moved in and out of main memory. Other forms of RAM memory include DDR2 SDRAM and DDR3 SDRAM.

read-only memory (ROM): A nonvolatile form of memory.

Read-only memory (ROM), another type of memory, is nonvolatile, meaning that its contents are not lost if the power is turned off or interrupted. ROM provides permanent storage for data and instructions that do not change, such as programs and data from the computer manufacturer, including the instructions that tell the computer how to start up when power is turned on. ROM memory also comes in many varieties: programmable read-only memory (PROM), which is used to hold data and instructions that can never be changed; erasable programmable read-only memory (EPROM), which is programmable ROM that can be erased and reused; and electrically erasable programmable read-only memory (EEPROM), which is user-modifiable read-only memory that can be erased and reprogrammed repeatedly through the application of higher than normal electrical voltage.

Chip manufacturers are competing to develop a nonvolatile memory chip that requires minimal power, offers extremely fast write speed, and can store data accurately even after a large number of write-erase cycles. Such a chip could eliminate the need for RAM and simplify and speed up memory processing. Phase change memory (PCM) is one potential approach to provide such a memory device. PCM employs a specialized glass-like material that can change its physical state, shifting between a low-resistance crystalline state to a high-resistance gaseous state by applying voltage to rearrange the atoms of the material. This technology is expected to perform 100 times faster than flash memory and may be used by server computers by 2016.[7]

cache memory: A type of high-speed memory that a processor can access more rapidly than main memory.

Although microprocessor speed has roughly doubled every 24 months over the past decades, memory performance has not kept pace. In effect, memory has become the principal bottleneck to system performance. **Cache memory** is a type of high-speed memory that a processor can access more rapidly than main memory to help to ease this bottleneck (see Figure 3.3). Frequently used data is stored in easily accessible cache memory instead of slower memory such as RAM. Because cache memory holds less data, the CPU can access the desired data and instructions more quickly than when selecting from the larger set in main memory. Thus, the CPU can execute instructions faster, improving the overall performance of the computer system. Cache memory is available in three forms. The level 1 (L1) cache is on the CPU chip. The level 2 (L2) cache memory can be accessed by the CPU over a high-speed

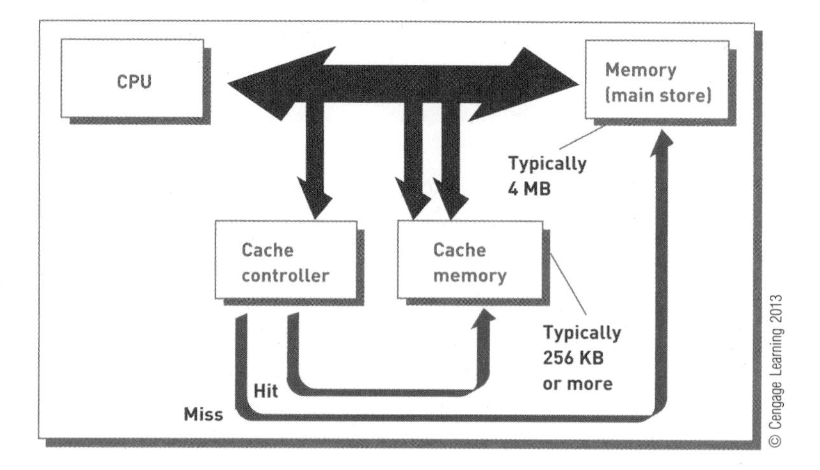

FIGURE 3.3

Cache memory

Processors can access this type of high-speed memory faster than main memory. Located on or near the CPU chip, cache memory works with main memory. A cache controller determines how often the data is used, transfers frequently used data to cache memory, and then deletes the data when it goes out of use.

dedicated interface. The latest processors go a step further and place the L2 cache directly on the CPU chip itself and provide high-speed support for a tertiary level 3 (L3) external cache.

Memory capacity contributes to the effectiveness of a computer. For example, complex processing problems, such as computer-assisted product design, require more memory than simpler tasks such as word processing. Also, because computer systems have different types of memory, they might need other programs to control how memory is accessed and used. In other cases, the computer system can be configured to maximize memory usage. Before purchasing additional memory, an organization should address all these considerations.

Multiprocessing

multiprocessing: The simultaneous execution of two or more instructions at the same time.

coprocessor: The part of the computer that speeds processing by executing specific types of instructions while the CPU works on another processing activity.

multicore microprocessor: A microprocessor that combines two or more independent processors into a single computer so that they share the workload and improve processing capacity.

Generally, **multiprocessing** involves the simultaneous execution of two or more instructions at the same time. One form of multiprocessing uses coprocessors. A **coprocessor** speeds processing by executing specific types of instructions while the CPU works on another processing activity. Coprocessors can be internal or external to the CPU and can have different clock speeds than the CPU. Each type of coprocessor performs a specific function. For example, a math coprocessor chip speeds mathematical calculations, while a graphics coprocessor chip decreases the time it takes to manipulate graphics.

A **multicore microprocessor** combines two or more independent processors into a single computer so that they share the workload and boost processing capacity. In addition, a dual-core processor enables people to perform multiple tasks simultaneously, such as playing a game and burning a CD.

AMD and Intel are battling for leadership in the multicore processor marketplace with both companies offering quad-core, six-core, and eight-core CPU chips that can be used to build powerful desktop computers. AMD announced that its eight-core desktop processors based on its so-called Bulldozer architecture broke the world's record for clock speed by running at 8.5 GHz.[8]

You need to use multithreaded software designed to take full advantage of the computing power of multicore processors. Scientists at the Jet Propulsion Laboratory are planning to take advantage of the full capabilities of multicore processors to perform image analysis and other tasks on future space missions.[9]

When selecting a CPU, organizations must balance the benefits of processing speed with energy requirements and cost. CPUs with faster clock speeds and shorter machine cycle times require more energy to dissipate the heat generated by the CPU and are bulkier and more expensive than slower ones.

Parallel Computing

parallel computing: The simultaneous execution of the same task on multiple processors to obtain results faster.

massively parallel processing systems: A form of multiprocessing that speeds processing by linking hundreds or thousands of processors to operate at the same time, or in parallel, with each processor having its own bus, memory, disks, copy of the operating system, and applications.

Parallel computing is the simultaneous execution of the same task on multiple processors to obtain results faster. Systems with thousands of such processors are known as **massively parallel processing systems**, a form of multiprocessing that speeds processing by linking hundreds or thousands of processors to operate at the same time, or in parallel, with each processor having its own bus, memory, disks, copy of the operating system, and applications. The processors might communicate with one another to coordinate when executing a computer program, or they might run independently of one another but under the direction of another processor that distributes the work to the other processors and collects their results. The dual-core processors mentioned earlier are a simple form of parallel computing.

The most frequent uses for parallel computing include modeling, simulation, and analyzing large amounts of data. Parallel computing is used in medicine to develop new imaging systems to complete ultrasound scans in less time with greater accuracy, enabling doctors to provide better diagnosis to patients, for example. Instead of building physical models of new products, engineers can create a virtual model of them and use parallel computing to test how the products work and then change design elements and materials as needed. Chevron uses seismic imaging to bounce sound waves off underground rock formations to detect potential oil- and gas-bearing formations. The resulting echoes create a vast amount of data that powerful parallel processing computers process to boost the company's exploration discovery rate to 57 percent.[10]

grid computing: The use of a collection of computers, often owned by multiple individuals or organizations, to work in a coordinated manner to solve a common problem.

Grid computing is the use of a collection of computers, often owned by multiple individuals or organizations, to work in a coordinated manner to solve a common problem. Grid computing is a low-cost approach to parallel computing. The grid can include dozens, hundreds, or even thousands of computers that run collectively to solve extremely large processing problems. Key to the success of grid computing is a central server that acts as the grid leader and traffic monitor. This controlling server divides the computing task into subtasks and assigns the work to computers on the grid that have (at least temporarily) surplus processing power. The central server also monitors the processing, and if a member of the grid fails to complete a subtask, it restarts or reassigns the task. When all the subtasks are completed, the controlling server combines the results and advances to the next task until the whole job is completed.

CERN is the European Organization for Nuclear Research and its main area of research is the study of the fundamental constituents of matter and the forces acting between them.[11] CERN uses grid computing with the processing power of over 300,000 high-end personal computers. This computing power is needed to process some 25 petabytes of data generated each year by the Large Hadron Collider (LHC) particle accelerator looking for evidence of new particles that can provide clues to the origins of our universe.[12]

SECONDARY STORAGE

Storing data safely and effectively is critical to an organization's success. Driven by many factors—such as needing to retain more data longer to meet government regulatory concerns, storing new forms of digital data such as audio and video, and keeping systems running under the onslaught of increasing volumes of e-mail—the estimated amount of data that companies store digitally is growing so rapidly that by 2020 storage requirements will be 44 times what they were in 2009 (an average compounded growth rate of 42 percent per year).[13] International Data Corporation (IDC) estimates that more than 1.8 zettabytes (10^{21} bytes) of information was created and stored in 2011 alone.[14]

The Indian government is undertaking a massive effort (estimated cost is more than $4 billion) to register its 1.2 billion residents in a universal citizen ID system. When complete, the result will be the world's largest database of biometric data including retina scans, fingerprints, and multiple facial images of each individual. The database will have many applications including use at India's borders to recognize travelers and to identify people who should not be in controlled areas such as the hangar area of an airport. The system can also be used in crowd control to recognize the gender and age of a crowd of people and identify where security personnel might be most needed.[15] Advanced data storage technologies will be required to store the large quantity of data and enable users to gain quick access to the information.

secondary storage: Devices that store large amounts of data, instructions, and information more permanently than allowed with main memory.

For most organizations, the best overall data storage solution is likely a combination of different **secondary storage** options that can store large amounts of data, instructions, and information more permanently than allowed with main memory. Compared with memory, secondary storage offers the advantages of nonvolatility, greater capacity, and greater economy. On a cost-per-megabyte basis, secondary storage is considerably less expensive than primary memory (see Table 3.2). The selection of secondary storage media and devices requires understanding their primary characteristics: access method, capacity, and portability.

As with other computer system components, the access methods, storage capacities, and portability required of secondary storage media are determined by the information system's objectives. An objective of a credit card company's information system might be to rapidly retrieve stored customer data to approve customer purchases. In this case, a fast access method is critical. In other cases, such as equipping the Coca-Cola field sales force with pocket-sized personal computers, portability and storage capacity might be major considerations in selecting and using secondary storage media and devices.

In addition to cost, capacity, and portability, organizations must address security issues to allow only authorized people to access sensitive data and critical programs. Because the data and programs kept on secondary storage devices are so critical to most organizations, all of these issues merit careful consideration.

sequential access: A retrieval method in which data must be accessed in the order in which it is stored.

Access Methods

Data and information access can be either sequential or direct. **Sequential access** means that data must be accessed in the order in which it is stored.

TABLE 3.2 Cost comparison for various forms of storage

All forms of secondary storage cost considerably less per gigabyte of capacity than SDRAM (synchronous dynamic random access memory), although they have slower access times. A data cartridge costs about $0.02 per gigabyte, while SDRAM can cost around $16 per gigabyte—800 times more expensive.

Description	Cost	Storage Capacity (GB)	Cost Per GB
1.6 TB 4 mm backup data tape cartridge	$39.95	1,600	$0.02
1 TB desktop external hard drive	$90.99	1000	$0.09
25 GB rewritable Blu-ray disk	$2.88	25	$0.11
500 GB portable hard drive	$77.99	500	$0.15
72 GB DAT 72 data cartridge	$16.95	72	$0.24
50 4.7 GB DVD+R disks	$74.95	235	$0.31
4 GB flash drive	$9.95	4	$2.48
9.1 GB write-once, read-many optical disk	$73.95	9.1	$8.12
1 GB DDR2 SDRAM computer memory upgrade	$15.95	1	$15.95

(Source: Office Depot Web site, *www.officedepot.com*, October 2011.)

For example, inventory data might be stored sequentially by part number, such as 100, 101, 102, and so on. If you want to retrieve information on part number 125, you must read and discard all the data relating to parts 001 through 124.

direct access: A retrieval method in which data can be retrieved without the need to read and discard other data.

sequential access storage device (SASD): A device used to sequentially access secondary storage data.

direct access storage device (DASD): A device used for direct access of secondary storage data.

Direct access means that data can be retrieved directly without the need to pass by other data in sequence. With direct access, it is possible to go directly to and access the needed data—for example, part number 125—without having to read through parts 001 through 124. For this reason, direct access is usually faster than sequential access. The devices used only to access secondary storage data sequentially are called **sequential access storage devices (SASDs)**; those used for direct access are called **direct access storage devices (DASDs)**.

Secondary Storage Devices

Secondary data storage is not directly accessible by the CPU. Instead, computers usually use input/output channels to access secondary storage and transfer the desired data using intermediate areas in primary storage. The most common forms of secondary storage devices are magnetic, optical, and solid state.

Magnetic Secondary Storage Devices

magnetic tape: A type of sequential secondary storage medium, now used primarily for storing backups of critical organizational data in the event of a disaster.

Magnetic storage uses tape or disk devices covered with a thin magnetic coating that enables data to be stored as magnetic particles. **Magnetic tape** is a type of sequential secondary storage medium, which is now used primarily for storing backups of critical organizational data in the event of a disaster. Examples of tape storage devices include cassettes and cartridges measuring a few millimeters in diameter, requiring very little storage space. Magnetic tape has been used as storage media since the time of the earliest computers, such as the 1951 Univac computer.[16]

A robotic tape backup system is used at the National Center for Atmospheric Research where several supercomputers are used to solve the world's most computationally intensive climate modeling problems.[17]

magnetic disk: A direct access storage device with bits represented by magnetized areas.

A **magnetic disk** is a direct access storage device that represents bits using small magnetized areas and uses a read/write head to go directly to the desired piece of data. Because direct access allows fast data retrieval, this type of storage is ideal for companies that need to respond quickly to customer requests, such as airlines and credit card firms. For example, if a manager

The National Center for Atmospheric Research uses a robotic tape backup system to back up the world's most computationally intensive climate modeling problems.

Courtesy of Deutsches Klimarechenzentrum GmbH

FIGURE 3.4

Hard disk

Hard disks provide direct access to stored data. The read/write head can move directly to the location of a desired piece of data, dramatically reducing access times compared to magnetic tape.

needs information on the credit history of a customer or the seat availability on a particular flight, the information can be obtained in seconds if the data is stored on a direct access storage device. Magnetic disk storage varies widely in capacity and portability. Hard disks, though more costly and less portable, are more popular because of their greater storage capacity and quicker access time (see Figure 3.4).

IBM is building a data repository nearly 10 times larger than anything in existence. This huge storehouse consists of 200,000 conventional hard disks working together to provide a storage capacity of 120 petabytes—large enough to hold 60 copies of the 150 billion pages needed to back up the Web. An unnamed client will use the storage device with a supercomputer to perform detailed simulations of real-world events such as weather forecasts, seismic processing for the petroleum industry, or molecular studies of genomes or proteins.[18]

Putting an organization's data online involves a serious business risk—the loss of critical data can put a corporation out of business. The concern is that the most critical mechanical components inside a magnetic disk storage device—the disk drives, the fans, and other input/output devices—can fail. Thus organizations now require that their data storage devices be fault tolerant, that is, they can continue with little or no loss of performance if one or more key components fail.

redundant array of independent/inexpensive disks (RAID): A method of storing data that generates extra bits of data from existing data, allowing the system to create a "reconstruction map" so that if a hard drive fails, the system can rebuild lost data.

A **redundant array of independent/inexpensive disks (RAID)** is a method of storing data that generates extra bits of data from existing data, allowing the system to create a "reconstruction map" so that if a hard drive fails, it can rebuild lost data. With this approach, data is split and stored on different physical disk drives using a technique called *striping* to evenly distribute the data. RAID technology has been applied to storage systems to improve system performance and reliability.

disk mirroring: A process of storing data that provides an exact copy that protects users fully in the event of data loss.

RAID can be implemented in several ways. In the simplest form, RAID subsystems duplicate data on drives. This process, called **disk mirroring**, provides an exact copy that protects users fully in the event of data loss. However, to keep complete duplicates of current backups, organizations need to double the amount of their storage capacity. Other RAID methods are less expensive because they only duplicate part of the data, allowing storage managers to minimize the amount of extra disk space they must purchase to protect data. Optional second drives for personal computer users who need to mirror critical data are available for less than $100.

Advanced Audio Rentals produced the soundtrack for the movie *Avatar*. They used 3 GB and 6 GB RAID storage devices to store the film's soundtracks and to ensure smooth, efficient data transfers to guarantee high, consistent performance.[19]

virtual tape: A storage device for less frequently needed data so that it appears to be stored entirely on tape cartridges, although some parts of it might actually be located on faster hard disks.

Virtual tape is a storage technology for less frequently needed data so that it appears to be stored entirely on tape cartridges, although some parts might actually be located on faster hard disks. The software associated with a virtual tape system is sometimes called a *virtual tape server*. Virtual tape can be used with a sophisticated storage-management system that moves data to slower but less costly forms of storage media as people use the data less often. Virtual tape technology can decrease data access time, lower the total cost of ownership, and reduce the amount of floor space consumed by tape operations.

Baldor Electric Company designs, manufactures, and markets industrial electric motors, transmission products, drives, and generators. The firm implemented a virtual tape system to replace its tape-based storage system consisting of thousands of magnetic tapes. Baldor uses the new virtual tape system to back up its five production databases twice a day and stores the data for 14 days. The time to create backups has been cut by 40 percent, and the new system takes up about 100 square feet less of data center floor space.[20]

Optical Secondary Storage Devices

optical storage device: A form of data storage that uses lasers to read and write data.

An **optical storage device** uses special lasers to read and write data. The lasers record data by physically burning pits in the disk. Data is directly accessed from the disc by an optical disc device, which operates much like a compact disc player. This optical disc device uses a low-power laser that measures the difference in reflected light caused by a pit (or lack thereof) on the disc.

compact disc read-only memory (CD-ROM): A common form of optical disc on which data cannot be modified once it has been recorded.

A common optical storage device is the **compact disc read-only memory (CD-ROM)** with a storage capacity of 740 MB of data. After data is recorded on a CD-ROM, it cannot be modified—the disc is "read-only." A CD burner, the informal name for a CD recorder, is a device that can record data to a compact disc. *CD-recordable* (*CD-R*) and *CD-rewritable* (*CD-RW*) are the two most common types of drives that can write CDs, either once (in the case of CD-R) or repeatedly (in the case of CD-RW). CD-rewritable (CD-RW) technology allows PC users to back up data on CDs.

digital video disc (DVD): A storage medium used to store software, video games, and movies.

A **digital video disc (DVD)** looks like a CD but can store about 135 minutes of digital video or several gigabytes of data (see Figure 3.5). Software, video games, and movies are often stored and distributed on DVDs. At a data transfer rate of 1.352 MB per second, the access speed of a DVD drive is faster than that of the typical CD-ROM drive.

DVDs have replaced recordable and rewritable CD discs (CD-R and CD-RW) as the preferred format for sharing movies and photos. Whereas a CD can hold about 740 MB of data, a single-sided DVD can hold 4.7 GB, with double-sided DVDs having a capacity of 9.4 GB. Several types of recorders and discs are currently in use. Recordings can be made on record-once discs (DVD-R and DVD+R) or on rewritable discs (DVD-RW, DVD+RW, and DVD-RAM). Not all types of rewritable DVDs are compatible with other types.

The U.S. Naval Air Warfare Center Weapons Division at China Lake, California, is testing various brands of archival-quality DVDs for longevity and reliability. They have a need to store large volumes of data for generations.[21]

The Blu-ray high-definition video disc format based on blue laser technology stores at least three times as much data as a DVD. The primary use for this new format is in home entertainment equipment to store high-definition video, though this format can also store computer data. A dual-layer Blu-ray disk can store 50 GB of data.

The Holographic Versatile Disc (HVD) is an advanced optical disc technology still in the development stage that will store more data than even the Blu-ray optical disc system. HVD is the same size and shape as a regular DVD

Courtesy of LaCie USA

FIGURE 3.5

Digital video disc and player

DVDs look like CDs but have a greater storage capacity and can transfer data at a faster rate.

but can hold 1 terabyte (or more) of information. One HVD approach records data through the depth of the storage media in three dimensions by splitting a laser beam in two—the signal beam carries the data, and the reference beam positions where the data is written and read. HVD will make it possible to view 3D visuals on home TVs.[22]

Solid State Secondary Storage Devices

Solid state storage devices (SSDs) store data in memory chips rather than magnetic or optical media. These memory chips require less power and provide faster data access than magnetic data storage devices. While hard drives can provide 250 to 350 IOPS (input/output operations per second or read/write operations per second), advanced SSDs can perform at the rate of one-half million IOPS.[23]

In addition, SSDs have few moving parts, so they are less fragile than hard disk drives. All these factors make the SSD a preferred choice for portable computers. Two current disadvantages of SSD are their high cost per GB of data storage (roughly a 5:1 disadvantage compared to hard disks) and lower capacity compared to current hard drives. However, SSD is a rapidly developing technology, and future improvements will lower its cost and increase its capacity.

A Universal Serial Bus (USB) flash drive is one example of a commonly used SSD (see Figure 3.6). USB flash drives are external to the computer and are removable and rewritable. Most weigh less than an ounce and can provide storage of 1 to 64 GB. For example, SanDisk manufactures its Ultra Backup USB Flash Drive with a storage capacity of 64 GB for around $200, which includes password protection and hardware encryption.[24]

Qualcomm is a U.S. global communications company that designs, manufactures, and markets wireless communications products and services. Its information systems organization used SSD to reduce the time required to boot up its employees' notebook computers, speed up overall system performance, and reduce downtime caused by hard disk failures.[25]

The Vaillant Group and its over 12,000 employees provide heating, ventilation, and air conditioning (HVAC) products and systems worldwide. Vaillant converted from the use of traditional hard drive storage systems to SSD to improve the performance of its key business systems. Critical business interactive transactions as well as batch processing jobs were dramatically improved, taking one-tenth or less of the time required using the old technology.[26]

Enterprise Storage Options

Businesses need to store large amounts of data created throughout an organization. Such large secondary storage is called *enterprise storage* and comes in three forms: attached storage, network-attached storage (NAS), and storage area networks (SANs).

Attached Storage

Attached storage methods include the tape, hard disks, and optical devices discussed previously, which are connected directly to a single computer.

FIGURE **3.6**

Flash drive

Flash drives are solid state storage devices.

Attached storage methods, though simple and cost effective for single users and small groups, do not allow systems to share storage, and they make it difficult to back up data.

Because of the limitations of attached storage, firms are turning to network-attached storage (NAS) and storage area networks (SANs). These alternatives enable an organization to share data storage resources among a much larger number of computers and users, resulting in improved storage efficiency and greater cost effectiveness. In addition, they simplify data backup and reduce the risk of downtime. Nearly one-third of system downtime is a direct result of data storage failures, so eliminating storage problems as a cause of downtime is a major advantage.

Network-Attached Storage

network-attached storage (NAS): Hard disk storage that is set up with its own network address rather than being attached to a computer.

Network-attached storage (NAS) is hard disk storage that is set up with its own network address rather than being attached to a computer. Figure 3.7 shows a NAS storage device. NAS includes software to manage storage access and file management, relieving the users' computers of those tasks. The result is that both application software and files can be served faster because they are not competing for the same processor resources. Computer users can share and access the same information, even if they are using different types of computers. Common applications for NAS include consolidated storage, Internet and e-commerce applications, and digital media.

One of the most popular Swiss skiing destinations is the Davos Klosters resort with more than 300 kilometers of ski slopes, five mountain railways, and 22 hotels with 1,700 beds. Resort guests expect hassle-free hotel check-ins, an always available online ticket shop, reliable information display boards, and efficient and on-time mountain railways. It takes powerful information systems to meet these expectations. The resort decided to implement NAS storage devices to make sure its information systems are reliable and provide fast access to data, dependable backups of operational data, and easy expansion of storage capacity.[27]

Storage Area Network

storage area network (SAN): A special-purpose, high-speed network that provides high-speed connections among data storage devices and computers over a network.

A **storage area network (SAN)** is a special-purpose, high-speed network that provides direct connections among data storage devices and computers across

Courtesy of Seagate Technology LLC.

FIGURE **3.7**

NAS storage device

The Seagate BlackArmor NAS 440 has a capacity of 4 to 12 Terabytes at a cost of less than $.27 per GB.

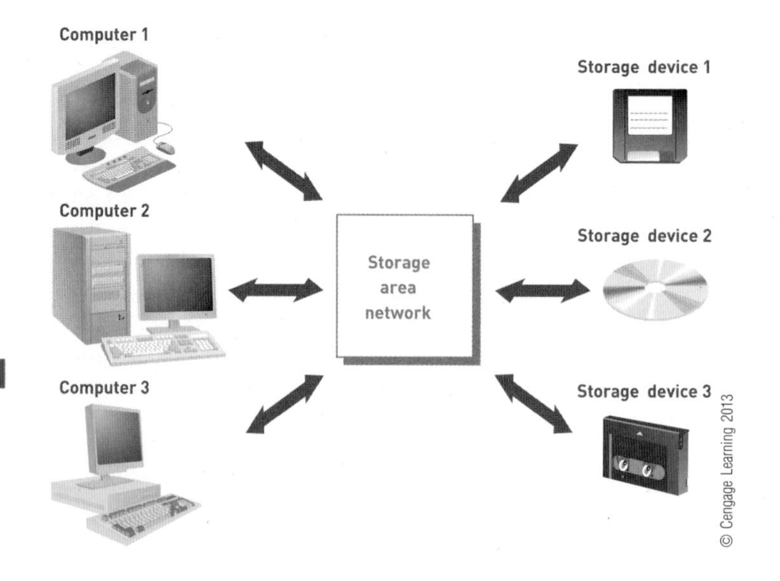

FIGURE 3.8
Storage area network
A SAN provides high-speed connections among data-storage devices and computers over a network.

the enterprise (see Figure 3.8). A SAN also integrates different types of storage subsystems, such as multiple RAID storage devices and magnetic tape backup systems, into a single storage system. Use of a SAN offloads the network traffic associated with storage onto a separate network. The data can then be copied to a remote location, making it easier for companies to create backups and implement disaster recovery policies.

Using a SAN, an organization can centralize the people, policies, procedures, and practices for managing storage, and a data storage manager can apply the data consistently across an enterprise. This centralization eliminates inconsistent treatment of data by different system administrators and users, providing efficient and cost-effective data storage practices.

NorthgateArinso is a global human resources services provider that equips its clients with HR solutions using advanced technology, outsourcing, and consulting. The firm's systems support multicountry payroll, training, recruiting, and talent management.[28] NorthgateArinso implemented two integrated data centers, one in London and one in Brussels, with an information systems architecture based on standard servers from a single supplier and data storage provided by SAN hardware and software. The SAN makes the total data stored available to all users. The company's prior collection of separate servers, applications, and databases is now integrated into an infrastructure that is easier to manage and can more flexibly meet the challenges of a highly fluctuating workload.[29]

A fundamental difference between NAS and SAN is that NAS uses file input/output, which defines data as complete containers of information, while SAN deals with block input/output, which is based on subsets of data smaller than a file. SAN manufacturers include EMC, Hitachi Data Systems Corporation, Xiotech, and IBM.

As organizations set up large-scale SANs, they use more computers and network connections than in a NAS environment, and consequently, the network becomes difficult to manage. In response, software tools designed to automate storage using previously defined policies are finding a place in the enterprise. Known as **policy-based storage management**, the software products from industry leaders such as Veritas Software Corporation, Legato Systems, EMC, and IBM automatically allocate storage space to users, balance the loads on servers and disks, and reroute network traffic when systems go down—all based on policies set up by system administrators.

policy-based storage management: Automation of storage using previously defined policies.

The trend in secondary storage is toward higher capacity, increased portability, and automated storage management. Organizations should select a type of storage based on their needs and resources. In general, storing large amounts of data and information and providing users with quick access makes an organization more efficient.

Storage as a Service

storage as a service: Storage as a service is a data storage model where a data storage service provider rents space to individuals and organizations.

Storage as a service is a data storage model in which a data storage service provider rents space to people and organizations. Users access their rented data storage via the Internet. Such a service enables the users to store and back up their data without requiring a major investment to create and maintain their own data storage infrastructure. Businesses can also choose pay-per-use services, where they rent space on massive storage devices housed either at a service provider (such as Hewlett-Packard or IBM) or on the customer's premises, paying only for the amount of storage they use. This approach is sensible for organizations with wildly fluctuating storage needs, such as those involved in the testing of new drugs or in developing software.

AT&T, Aviva, Amazon.com, EMC, Google, Microsoft, and ParaScale are a few of the storage-as-a-service providers used by organizations. Amazon.com's Simple Storage Service (S3) provides storage as a service with a monthly cost of roughly $0.15 per GB stored and $0.10 per GB of data transferred into the Amazon.com storage.

Box.net, Carbonite, SugarSynch, Symantec, and Mozy are a few of the storage-as-a-service providers used by individuals. This set of providers all charge less than $8.00 per month for up to 5 GB of storage.

A Mozy customer who had his laptop stolen was able to provide police with photos of the thief because Mozy continued to back up data after the laptop was stolen, including the thief's photos and documents. The customer accessed the photos from his online storage site, and police captured the thief and returned the laptop.[30]

INPUT AND OUTPUT DEVICES: THE GATEWAY TO COMPUTER SYSTEMS

Your first experience with computers is usually through input and output devices. These devices are the gateways to the computer system—you use them to provide data and instructions to the computer and receive results from it. Input and output devices are part of a computer's user interface, which includes other hardware devices and software that allow you to interact with a computer system.

As with other computer system components, an organization should keep its business goals in mind when selecting input and output devices. For example, many restaurant chains use handheld input devices or computerized terminals that let food servers enter orders efficiently and accurately. These systems have also cut costs by helping to track inventory and market to customers.

Characteristics and Functionality

In general, businesses want input devices that let them rapidly enter data into a computer system, and they want output devices that let them produce timely results. When selecting input and output devices, businesses also need to consider the form of the output they want, the nature of the data required to generate this output, and the speed and accuracy they need for both. Some organizations have very specific needs for output and input, requiring devices that perform specific functions. The more specialized the application, the more specialized the associated system input and output devices.

The speed and functions of input and output devices should be balanced with their cost, control, and complexity. More specialized devices might make it easier to enter data or output information, but they are generally more costly, less flexible, and more susceptible to malfunction.

The Nature of Data

Getting data into the computer—input—often requires transferring human-readable data, such as a sales order, into the computer system. "Human-readable" means data that people can read and understand. A sheet of paper containing inventory adjustments is an example of human-readable data. In contrast, machine-readable data can be read by computer devices (such as the universal barcode on many grocery and retail items) and is typically stored as bits or bytes. Inventory changes stored on a disk is an example of machine-readable data.

Some data can be read by people and machines, such as magnetic ink on bank checks. Usually, people begin the input process by organizing human-readable data and transforming it into machine-readable data. Every keystroke on a keyboard, for example, turns a letter symbol of a human language into a digital code that the machine can manipulate.

Data Entry and Input

Getting data into the computer system is a two-stage process. First, the human-readable data is converted into a machine-readable form through **data entry**. The second stage involves transferring the machine-readable data into the system. This is **data input**.

data entry: Converting human-readable data into a machine-readable form.

data input: Transferring machine-readable data into the system.

Today, many companies are using online data entry and input: They communicate and transfer data to computer devices directly connected to the computer system. Online data entry and input place data into the computer system in a matter of seconds. Organizations in many industries require the instantaneous updating offered by this approach. For example, when ticket agents enter a request for concert tickets, they can use online data entry and input to record the request as soon as it is made. Ticket agents at other terminals can then access this data to make a seating check before they process another request.

Source Data Automation

Regardless of how data gets into the computer, it should be captured and edited at its source. **Source data automation** involves capturing and editing data where it is originally created and in a form that can be directly entered into a computer, thus ensuring accuracy and timeliness. For example, using source data automation, salespeople enter sales orders into the computer at the time and place they take the orders. Any errors can be detected and corrected immediately. If an item is temporarily out of stock, the salesperson can discuss options with the customer. Prior to source data automation, orders were written on paper and entered into the computer later (usually by a clerk, not by the person who took the order). Often the handwritten information wasn't legible or, worse yet, got lost. If problems occurred during data entry, the clerk had to contact the salesperson or the customer to "recapture" the data needed for order entry, leading to further delays and customer dissatisfaction.

source data automation: Capturing and editing data where it is initially created and in a form that can be directly entered into a computer, thus ensuring accuracy and timeliness.

Input Devices

Data entry and input devices come in many forms. They range from special-purpose devices that capture specific types of data to more general-purpose input devices. Some of the special-purpose data entry and input devices are discussed later in this chapter. First, we focus on devices used to enter and input general types of data, including text, audio, images, and video for personal computers.

Personal Computer Input Devices

A keyboard and a computer mouse are the most common devices used for entry and input of data such as characters, text, and basic commands. Some

© iStockphoto/300dpi

A keyboard and mouse are two of the most common devices for computer input. Wireless mice and keyboards are now readily available.

companies are developing keyboards that are more comfortable, more easily adjusted, and faster to use than standard keyboards. These ergonomic keyboards, such as the split keyboard offered by Microsoft and others, are designed to avoid wrist and hand injuries caused by hours of typing. Other keyboards include touchpads that let you enter sketches on the touchpad while still using keys to enter text. Other innovations are wireless mice and keyboards, which keep a physical desktop free from clutter.

You use a computer mouse to point to and click symbols, icons, menus, and commands on the screen. The computer takes a number of actions in response, such as placing data into the computer system.

Speech-Recognition Technology

speech-recognition technology: Input devices that recognize human speech.

Using **speech-recognition technology**, a computer equipped with a source of speech input, such as a microphone, can interpret human speech as an alternative means of providing data or instructions to the computer. The most basic systems require you to train the system to recognize your speech patterns or are limited to a small vocabulary of words. More advanced systems can recognize continuous speech without requiring you to break your speech into discrete words. Interactive voice response (IVR) systems allow a computer to recognize both voice and keypad inputs.

Companies that must constantly interact with customers are eager to reduce their customer support costs while improving the quality of their service. Time Warner Cable implemented a speech-recognition application as part of its customer call center. Subscribers who call customer service can speak commands to begin simple processes such as "pay my bill" or "add ShowTime." The voice recognition system saves time and money even though most people would prefer to speak to a live person. "We have roughly 13 million customers, and a few seconds or minutes here or there for each customer can really add up to longer hold times and higher staffing costs—which makes cable rates climb," says Time Warner spokesman Matthew Tremblay.[31]

Digital Cameras

digital camera: An input device used with a PC to record and store images and video in digital form.

Digital cameras record and store images or video in digital form, so when you take pictures, the images are electronically stored in the camera. You can download the images to a computer either directly or by using a flash memory card. After you store the images on the computer's hard disk, you can then edit and print them, send them to another location, or paste them into another application. This digital format saves time and money by eliminating the need to process film in order to share photos. For example, you can download a photo of your project team captured by a digital camera and then post it on a Web site or paste it into a project status report. Digital cameras have eclipsed film cameras used by professional photographers for photo quality and features such as zoom, flash, exposure controls, special effects, and even video-capture capabilities. With the right software, you can add sound and handwriting to the photo. Many computers and smartphones come equipped with a digital camera to enable their users to place video calls and take pictures and videos.

Canon, Casio, Nikon, Olympus, Panasonic, Pentax, Sony, and other camera manufacturers offer full-featured, high-resolution digital camera models at prices ranging from $250 to $3500. Some manufacturers offer pocket-sized camcorders for less than $150.

The police department in Wallis, Mississippi, consists of only five officers but is one of the first departments in the United States to use tiny digital cameras that clip onto the front pocket of the officers' uniforms. The cameras are the size of a pack of gum and come with a memory card capable of holding hours of evidence. The cameras record each police stop in its entirety and provide evidence that supports prosecution of suspects.[32]

Scanning Devices

Scanning devices capture image and character data. A page scanner is like a copy machine. You either insert a page into the scanner or place it face down on the glass plate of the scanner and then scan it. With a handheld scanner, you manually move or roll the scanning device over the image you want to scan. Both page and handheld scanners can convert monochrome or color pictures, forms, text, and other images into machine-readable digits. Considering that U.S. enterprises generate an estimated 1 billion pieces of paper daily, many companies are looking to scanning devices to help them manage their documents and reduce the high cost of using and processing paper.

Silicon Valley Bank (SVB) Financial Group is headquartered in Santa Clara, California, and is surrounded by hundreds of high-tech companies and start-up ventures in the life science, clean technology, venture capital, private equity, and premium wine markets.[33] SVB used to store loan and deposit documents for some 4,000 clients in paper files at its headquarters. The firm received more than 75 requests per day from branches for copies of documents, with each request taking about 15 minutes to process. SVB implemented document-scanning hardware and software that can create a digital, online copy of all documents. Now users can immediately access the documents online, leading to improvements in customer service and reductions in administrative costs.[34]

Optical Data Readers

You can also use a special scanning device called an *optical data reader* to scan documents. The two categories of optical data readers are for optical mark recognition (OMR) and optical character recognition (OCR). You use OMR readers for grading tests and other purposes such as forms. With this technology, you use pencils to fill in bubbles or check boxes on OMR paper, which is also called a "mark sense form." OMR systems are used in standardized tests, including the SAT and GMAT tests, and to record votes in elections. In contrast, most OCR readers use reflected light to recognize and scan various machine-generated characters. With special software, OCR readers can also convert handwritten or typed documents into digital data. After being entered, this data can be shared, modified, and distributed over computer networks to hundreds or thousands of people.

Magnetic Ink Character Recognition (MICR) Devices

In the 1950s, the banking industry became swamped with paper checks, loan applications, bank statements, and so on. The result was the development of magnetic ink character recognition (MICR), a system for reading banking data

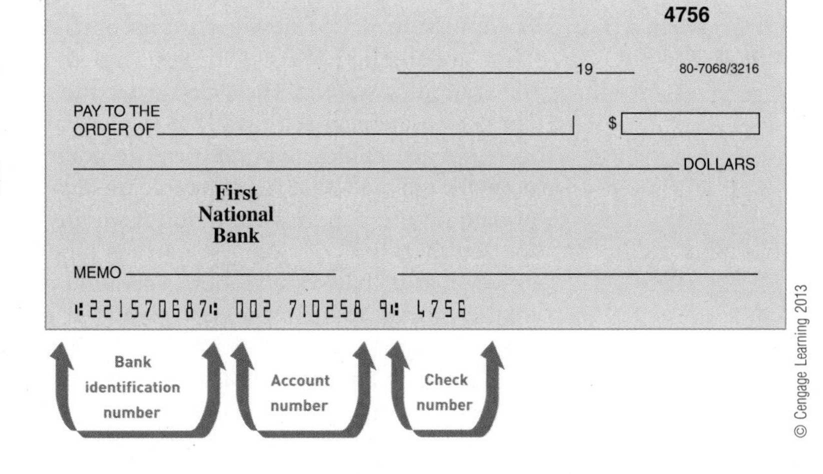

FIGURE 3.9
MICR device
Magnetic ink character recognition technology codes data on the bottom of a check or other form using special magnetic ink, which is readable by people and computers. For an example, look at the bottom of a bank check.

quickly. With MICR, data is placed on the bottom of a check or other form using a special magnetic ink. Using a special character set, data printed with this ink is readable by people and computers (see Figure 3.9).

Magnetic Stripe Cards

magnetic stripe card: A type of card that stores a limited amount of data by modifying the magnetism of tiny iron-based particles contained in a band on the card.

A **magnetic stripe card** stores a limited amount of data by modifying the magnetism of tiny iron-based particles contained in a band on the card. The magnetic stripe is read by physically swiping the card at a terminal. For this reason, such cards are called a contact card. Magnetic stripe cards are commonly used in credit cards, transportation tickets, and driver's licenses.

Chip-and-PIN Cards

Most European countries as well as many countries in Asia and South America are converting to the chip-and-PIN card, which uses "smart card" technology.[35] These cards are also known as EMV-enabled cards because they are named for their developers Europay, MasterCard, and Visa.[36] This technology employs a computer chip that communicates with a card reader using radio frequencies, which means the cards do not need to be swiped at a terminal. **Chip-and-PIN cards** require different terminals from those used for magnetic stripe cards. For security, the card holder is also required to enter a PIN at the point of sale, making such cards more effective at preventing fraud. Although credit card fraud is a problem in the United States, credit card issuers cannot force merchants to invest in the new terminals required for chip-and-PIN cards. As a result, deployment of this technology is lagging in the United States.

chip-and-PIN card: A type of card that employs a computer chip that communicates with a card reader using radio frequencies; it does not need to be swiped at a terminal.

Cardholders of Chase's premium Palladium card and a test market sample of some 15,000 Wells Fargo customers who are frequent overseas travelers will be issued cards with both the magnetic stripe and chip-and-PIN technology.[37]

Contactless Cards

contactless card: A card with an embedded chip that only needs to be held close to a terminal to transfer its data; no PIN number needs to be entered.

A **contactless card** has an embedded chip that only needs to be held close to a terminal to transfer its data; no PIN number needs to be entered. MasterCard is testing the use of contactless-based payments aboard flights of WestJet, a Canadian airline. One problem with the current method of payment using regular credit cards is that these transactions are conducted in an offline mode with no credit check authorization possible. With contactless cards, it will eventually be possible to support online credit card authorization, thus reducing the potential of fraud. This fraud protection will open the door to expanding in-flight sales beyond drinks and refreshments to the sale of wine and apparel that is delivered to the passengers' homes when they land.[38]

Point-of-Sale Devices

point-of-sale (POS) device: A terminal used to enter data into the computer system.

Point-of-sale (POS) devices are terminals used to capture data for data entry. They are frequently used in retail operations to enter sales information into the computer system. The POS device then computes the total charges, including tax. In medical settings, POS devices are often used for remote monitoring in hospitals, clinics, laboratories, doctor's offices, and patients' homes. With network-enabled POS equipment, medical professionals can instantly get an update on the patient's condition from anywhere at any time via a network or the Internet. POS devices use various types of input and output devices, such as keyboards, bar-code readers, scanning devices, printers, and screens. Much of the money that businesses spend on computer technology involves POS devices. Figure 3.10 shows a handheld POS terminal device.

La Paz is an upscale Mexican restaurant in the Vinings Jubilee section of Atlanta, Georgia. However, its POS was outdated, and workers needed hours each night to close out and balance the books. The restaurant replaced its POS with a new system that provides useful payroll and management reporting features such as tracking tips and sales. Workers also like the feature that lets them close books rapidly and accurately.[39]

Special input devices can be attached to smartphones and computers to accept payments from credit and debit cards for goods and services. Intuit Go Payment and Square can provide a small credit card scanner that plugs into your smartphone.[40]

Automated Teller Machine (ATM) Devices

Another type of special-purpose input/output device, the automated teller machine (ATM) is a terminal that bank customers use to perform transactions with their bank accounts. Companies use various ATM devices, sometimes called *kiosks*, to support their business processes. Some can dispense tickets, such as for airlines, concerts, and soccer games. Some colleges use them to produce transcripts.

Pen Input Devices

By touching the screen with a pen input device, you can activate a command or cause the computer to perform a task, enter handwritten notes, and draw objects and figures (see Figure 3.11). Pen input requires special software and hardware. Handwriting recognition software, for example, converts handwriting on the screen into text. The Tablet PC from Microsoft and its hardware partners can transform handwriting into typed text and store the "digital ink" just the way a person writes it. People can use a pen to write and send e-mail, add comments to Word documents, mark up PowerPoint presentations,

FIGURE 3.10

Handheld POS terminal device

Using a wireless, handheld POS device, restaurant staff can take orders and payments on the floor.

FIGURE **3.11**

Using a pen input device

Using a pen input device directly on a 3D display, physicians can precisely trace organs at a very detailed level to prepare for a complex operation.

and even hand draw charts in a document. The data can then be moved, highlighted, searched, and converted into text. If perfected, this interface is likely to become widely used. Pen input is especially attractive if you are uncomfortable using a keyboard. The success of pen input depends on how accurately and at what cost handwriting can be read and translated into digital form.

Audi AG installed interactive white boards at its five technical training locations in Germany. PowerPoint presentations can be projected on the screen, and trainers can draw over the image and highlight features by circling or underlining. Consequently, static presentations have become more interactive.[41]

Touch-Sensitive Screens

Advances in screen technology allow display screens to function as input as well as output devices. By touching certain parts of a touch-sensitive screen, you can start a program or trigger other types of action. Touch-sensitive screens can remove the need for a keyboard, which conserves space and increases portability. Touchscreens are frequently used at gas stations to allow customers to select grades of gas and request a receipt; on photocopy machines for selecting options; at fast-food restaurants for entering customer choices; at information centers for finding facts about local eating and drinking establishments; and at amusement parks to provide directions to patrons. They also are used in kiosks at airports and department stores. Touch-sensitive screens are also being considered for gathering votes in elections.

As touchscreens get smaller, the user's fingers begin to block the information on the display. Nanotouch technology is being explored as a means of overcoming this problem. Using this technology, users control the touchscreen from its backside so that fingers do not block the display. As the user's finger moves on the back of the display, a tiny graphical finger is projected onto the touchscreen. Such displays are useful for mobile audio players the size of a coin.

Bar-Code Scanners

A bar-code scanner employs a laser scanner to read a bar-coded label and pass the data to a computer. The bar code reader may be stationary or handheld to support a wide variety of uses. This form of input is used widely in store checkouts and warehouse inventory control. Bar codes are also used in hospitals, where a nurse scans a patient's wristband and then a bar code on the medication about to be administered to prevent medication errors.

Mobio, an international mobile payments and marketing company, has developed an iPhone application that it is testing with fans of the Jacksonville Jaguars football team. The fans use the application to scan a barcode at their seat and a menu of food and beverage items is displayed on their phone. Fans then select and pay for the items they want using the iPhone application, and the order is then delivered to their seat.[42]

Courtesy of Intermec Technologies

FIGURE 3.12
RFID tag
An RFID tag is small compared to current bar-code labels used to identify items.

Radio Frequency Identification

Radio Frequency Identification (RFID): A technology that employs a microchip with an antenna to broadcast its unique identifier and location to receivers.

Radio Frequency Identification (RFID) is a technology that employs a microchip with an antenna to broadcast its unique identifier and location to receivers. The purpose of an RFID system is to transmit data by a mobile device, called a tag (see Figure 3.12), which is read by an RFID reader and processed according to the needs of a computer program. One popular application of RFID is to place microchips on retail items and install in-store readers that track the inventory on the shelves to determine when shelves should be restocked. The RFID tag chip includes a special form of EPROM memory that holds data about the item to which the tag is attached. A radio frequency signal can update this memory as the status of the item changes. The data transmitted by the tag might provide identification, location information, or details about the product tagged, such as date manufactured, retail price, color, or date of purchase.

The Canadian government is supporting a move to require the sheep industry to use RFID chips to enable a "farm to fork" traceability system.[43]

Output Devices

Computer systems provide output to decision makers at all levels of an organization so they can solve a business problem or capitalize on a competitive opportunity. In addition, output from one computer system can provide input into another computer system. The desired form of this output might be visual, audio, or even digital. Whatever the output's content or form, output devices are designed to provide the right information to the right person in the right format at the right time.

Display Monitors

The display monitor is a device used to display the output from the computer. Because early monitors used a cathode-ray tube to display images, they were sometimes called *CRTs*. The cathode-ray tubes generate one or more electron beams. As the beams strike a phosphorescent compound (phosphor) coated on the inside of the screen, a dot on the screen called a pixel lights up. A **pixel** is a dot of color on a photo image or a point of light on a display screen. It appears in one of two modes: on or off. The electron beam sweeps across the screen so that as the phosphor starts to fade, it is struck and lights up again.

pixel: A dot of color on a photo image or a point of light on a display screen.

plasma display: A type of display using thousands of smart cells (pixels) consisting of electrodes and neon and xenon gases that are electrically turned into plasma (electrically charged atoms and negatively charged particles) to emit light.

A **plasma display** uses thousands of smart cells (pixels) consisting of electrodes and neon and xenon gases that are electrically turned into plasma (electrically charged atoms and negatively charged particles) to emit light. The plasma display lights up the pixels to form an image based on the information in the video signal. Each pixel is made up of three types of light—red,

© Justin Pumfrey/Getty Images

FIGURE 3.13
Four-screenwide display

green, and blue—with the plasma display varying the intensities of the lights to produce a full range of colors. Plasma displays can produce high resolution and accurate representation of colors to create a high-quality image.

LCD display: Flat display that uses liquid crystals—organic, oil-like material placed between two polarizers—to form characters and graphic images on a backlit screen.

LCD displays are flat displays that use liquid crystals—organic, oil-like material placed between two polarizers—to form characters and graphic images on a backlit screen. These displays are easier on your eyes than CRTs because they are flicker-free, are brighter, and do not emit the type of radiation that concerns some CRT users. In addition, LCD monitors take up less space and use less than half of the electricity required to operate a comparably sized CRT monitor. *Thin-film transistor (TFT) LCDs* are a type of liquid crystal display that assigns a transistor to control each pixel, resulting in higher resolution and quicker response to changes on the screen. TFT LCD monitors have displaced the older CRT technology and are available in sizes from 12 to 30 inches. Many companies now provide multimonitor solutions that enable users to see a wealth of related information at a single glance, as shown in Figure 3.13.

organic light-emitting diode (OLED) display: Flat display that uses a layer of organic material sandwiched between two conductors which in turn are sandwiched between a glass top plate and a glass bottom plate so that when electric current is applied to the two conductors, a bright, electroluminescent light is produced directly from the organic material.

Organic light-emitting diode (OLED) uses a layer of organic material sandwiched between two conductors, which in turn are sandwiched between a glass top plate and a glass bottom plate. When electric current is applied to the two conductors, a bright, electroluminescent light is produced directly from the organic material. OLEDs can provide sharper and brighter colors than LCDs and CRTs, and because they do not require a backlight, the displays can be half as thick as LCDs, and they are flexible. Another big advantage is that OLEDs do not break when dropped. OLED technology can also create 3D video displays by taking a traditional LCD monitor and then adding layers of transparent OLED films to create the perception of depth without the need for 3D glasses or laser optics. The iZ3D monitor is capable of displaying in both 2D and 3D mode. The manufacturer offered a 22-inch version of the monitor at a price of $300 to coincide with the debut of *Avatar*, a film directed by James Cameron.

Because most users leave their computers on for hours at a time, power usage is an important factor when deciding which type monitor to purchase. Although the power usage varies from model to model, LCD monitors generally consume between 35 and 50 percent less power than plasma screens.[44] OLED monitors use even less power than LCD monitors.

Aspect ratio and screen size describe the size of the display screen. Aspect ratio is the ratio of the width of the display to its height. An aspect ratio of 4 to 3 is common. For widescreen LCD monitors used for viewing DVD movies in widescreen format, playing games, or displaying multiple screens side-by-side, an aspect ratio of 16 to 10 or 15 to 9 is preferred. The screen

TABLE 3.3 Common display monitor standards and associated resolutions

Standard	Resolution (number of horizontal pixels × vertical pixels)
WSXGA (Wide SGXA plus)	1680 × 1050
UXGA (Ultra XGA)	1600 × 1200
WUXGA (Wide Ultra XGA)	1920 × 1200
QXGA (Quad XGA)	2048 × 1536

size is measured diagonally from the outside of the screen casing for CRT monitors and from the inside of the screen casing for LCD displays.

With today's wide selection of monitors, price and overall quality can vary tremendously. The quality of a screen image is measured by the number of horizontal and vertical pixels used to create it. Resolution is the total number of pixels contained in the display; the more pixels, the clearer and sharper the image. The size of the display monitor also affects the quality of the viewing. The same pixel resolution on a small screen is sharper than on a larger screen, where the same number of pixels is spread out over a larger area. Over the years, display monitor sizes have increased and display standards and resolutions have changed, as shown in Table 3.3.

Another way to measure image quality is the distance between one pixel on the screen and the next nearest pixel, which is known as *dot pitch*. The common range of dot pitch is from 0.25 mm to 0.31 mm. The smaller the dot pitch, the better the picture. A dot pitch of 0.28 mm or smaller is considered good. Greater pixel densities and smaller dot pitches yield sharper images of higher resolution.

The characteristics of screen color depend on the quality of the monitor, the amount of RAM in the computer system, and the monitor's graphics adapter card. Digital Video Interface (DVI) is a video interface standard designed to maximize the visual quality of digital display devices such as flat-panel LCD computer displays.

Companies are competing on the innovation frontier to create thinner display devices for computers, cell phones, and other mobile devices. In its effort to gain an edge, LG Phillips has developed an extremely thin display that is only 0.15 mm thick, or roughly as thick as a human hair. The display is also so flexible that it can be bent or rolled without damage. Nokia has demonstrated a flexible portable computer that you can actually twist and bend to change a music track or adjust the volume.[45] Such screens open possibilities for manufacturers to make cell phones and laptops with significantly larger displays but without increasing the size of the device itself, as the screen could be rolled up or folded and tucked away into a pocket.

The Microsoft Surface platform is designed to help people learn, collaborate, and make decisions. The Surface can be used as a table, mounted on the wall, or embedded in furniture or fixtures. Its large 40-inch screen is an effective way to share photos, maps, modeling, and simulations. The Surface allows a single user or multiple users to manipulate digital content by motion of their hands.

The Bank of Canada is Canada's largest bank and uses the Surface as a component of its Discovery Zone, a unique and digitally interactive approach to engage customers to learn more about the bank, its services, and its employees. Arbie, an animated character, guides people through various applications with the results displayed on the Surface screen. These applications help people learn which bank services can help them meet their financial needs, demonstrates the value of a Tax Free Savings Account compared

The Microsoft Surface 2.0 is a 4-inch thick computer on legs that has an embedded AMD Athlon II X2 2.9GHz processor paired with an AMD Radeon HD 6700M Series GPU. The 40-inch touchscreen offers a 1920 x 1080 (1080p) resolution. The table costs around $7,600.

with a standard savings account, and provides photos and brief profiles of the local branch staff. The software includes an Instant Win application and puzzles for children to work while their parents do their banking.[46]

Printers and Plotters

One of the most useful and common forms of output is called *hard copy*, which is simply paper output from a printer. The two main types of printers are laser printers and inkjet printers, and they are available with different speeds, features, and capabilities. Some can be set up to accommodate paper forms, such as blank check forms and invoice forms. Newer printers allow businesses to create customized printed output using full color for each customer from standard paper and data input. Ticket-receipt printers such as those used in restaurants, ATMs, and point-of-sale systems are in wide-scale use.

The speed of the printer is typically measured by the number of pages printed per minute (ppm). Like a display screen, the quality, or resolution, of a printer's output depends on the number of dots printed per inch (dpi). A 600-dpi printer prints more clearly than a 300-dpi printer. A recurring cost of using a printer is the inkjet or laser cartridge that must be replaced periodically—every few thousand pages for laser printers and every 500 to 900 pages for inkjet printers. Figure 3.14 shows an inkjet printer.

Costing less than $550, laser printers are generally faster than inkjet printers and can handle more volume: 25 to 60 ppm for black and white and 6 to 25 ppm for color. Inkjet printers that can print 12 to 40 ppm for black and white and 5 to 20 ppm for color are available for less than $200. For color printing, inkjet printers print vivid hues with an initial cost much less than color laser printers and can produce high-quality banners, graphics, greeting cards, letters, text, and photo prints.

A number of manufacturers offer multiple-function printers that can copy, print (in color or black and white), fax, and scan. Such multifunctional devices are often used when people need to do a relatively low volume of copying, printing, faxing, and scanning. The typical price of multifunction printers ranges from $100 to $500, depending on features and capabilities. Because these devices take the place of more than one piece of equipment, they are less expensive to acquire and maintain than a standalone fax plus a standalone printer, copier, and so on. Also, eliminating equipment that was once located on a countertop or desktop clears a workspace for other work-related activities. As a result, such devices are popular in homes and small office settings.

FIGURE 3.14
Hewlett-Packard all-in-one 8500 inkjet printer

Courtesy of Z Corporation/3D Systems

FIGURE 3.15
ZPrinter 650 3D printer

3D printers can be used to turn three-dimensional computer models into three-dimensional objects (see Figure 3.15). One form of 3D printer uses an inkjet printing system to print an adhesive in the shape of a cross-section of the model. Next, a fine powder is sprayed onto the adhesive to form one layer of the object. This process is repeated thousands of times until the object is completed. 3D printing is commonly used by aerospace firms, auto manufacturers, and other design-intensive companies. It is especially valuable during the conceptual stage of engineering design when the exact dimensions and material strength of the prototype is not critical.

Some new printers from Hewlett-Packard and others allow printing without requiring the printer to be connected to a computer. A USB storage device such as a thumb drive can be inserted into a slot in the printer, and the user can specify what is to be printed without entering commands to the computer.[47]

Other mobile print solutions enable users to wirelessly send documents, e-mail messages and attachments, presentations, and even boarding passes from any smartphone, tablet computer, or laptop to any mobile printer in the world. For example, mobile users who use the PrinterOn service only need to access a directory of PrinterOn printers and locations and then send to the e-mail address of the printer an e-mail with attachment to be printed. American Airlines Admiral Club, Delta Sky Club, Embassy Suites, and DoubleTree by Hilton have installed PrinterOn printers at many of their locations.[48]

Plotters are a type of hard-copy output device used for general design work. Businesses typically use plotters to generate paper or acetate blueprints, schematics, and drawings of buildings or new products. Standard plot widths are 24 inches and 36 inches, and the length can be whatever meets the need—from a few inches to many feet.

Digital Audio Players

digital audio player: A device that can store, organize, and play digital music files.

MP3: A standard format for compressing a sound sequence into a small file.

A **digital audio player** is a device that can store, organize, and play digital music files. **MP3** (MPEG-1 Audio Layer-3) is a popular format for compressing a sound sequence into a very small file while preserving the original level of sound quality when it is played. By compressing the sound file, it requires less time to download the file and less storage space on a hard drive.

You can use many different music devices smaller than a deck of cards to download music from the Internet and other sources. These devices have no moving parts and can store hours of music. Apple expanded into the digital music market with an MP3 player (the iPod) and the iTunes Music Store, which allows you to find music online, preview it, and download it in a way that is safe, legal, and affordable. Other MP3 manufacturers include Dell, Sony, Samsung, Iomega, Creative, and Motorola, whose Rokr product is the first iTunes-compatible phone.

The Apple iPod Touch, with a 3.5-inch widescreen, is a music player that also plays movies and TV shows, displays photos, and connects to the Internet. You can, therefore, use it to view YouTube videos, buy music online, check e-mail, and more. The display automatically adjusts the view when it is rotated from portrait to landscape. An ambient light sensor adjusts brightness to match the current lighting conditions.

E-book Readers

The digital media equivalent of a conventional printed book is called an e-book (short for electronic book). The Project Gutenberg Online Book Catalog lists over 36,000 free e-books and a total of over 100,000 e-books available. E-books can be downloaded from Project Gutenberg (*www.gutenberg.org*) or many other sites onto personal computers or dedicated hardware devices known as e-book readers. The devices themselves cost from around $200 to $350, and downloads of the bestselling books and new releases cost less than $10.00. The e-book reader has the capacity to store thousands of books. The most current Amazon.com Kindle, Sony PRS, Kobo e-reader, Barnes & Noble's Nook, and iRiver's Story are popular e-readers that use e-paper displays that either look like printed pages or LCD screens that are bright and shiny but can be difficult to read in bright sunlight.[49] E-books weigh less than three-quarters of a pound, are around one-half inch thick, and come with a display screen ranging from 5 to 8 inches. Thus, these readers are more compact than most paperbacks and can be easily held in one hand. Recent e-book readers display content in 16 million colors and high resolution (see Figure 3.16). On many e-readers, the size of the text can be magnified for readers with poor vision.

Apple's iPod Touch

FIGURE **3.16**
Kindle Fire e-book reader

COMPUTER SYSTEM TYPES

In general, computers can be classified as either special purpose or general purpose. *Special-purpose computers* are used for limited applications, for example, by military, government, and scientific research groups such as the CIA and NASA. Other applications include specialized processors found in appliances, cars, and other products. For example, automobile repair shops connect special-purpose computers to your car's engine to identify specific performance problems. As another example, IBM is developing a new generation of computer chips to develop so-called cognitive computers that are designed to mimic the way the human brain works. Rather than being programmed as today's computers are, cognitive computers will be able to learn through experiences and outcomes and mimic human learning patterns.[50]

General-purpose computers are used for a variety of applications and to execute the business applications discussed in this text. General-purpose computer systems can be divided into two major groups: systems used by one user at a time and systems used by multiple concurrent users. Table 3.4 shows the general ranges of capabilities for various types of computer systems. General-purpose computer systems can range from small handheld computers to massive supercomputers that fill an entire room. We will first cover single-user computer systems.

Portable Computers

portable computer: A computer small enough to carry easily.

Many computer manufacturers offer a variety of **portable computers**, those that are small enough to carry easily. Portable computers include handheld computers, laptop computers, notebook computers, netbook computers, and tablet computers.

TABLE 3.4 Types of computer systems

Single-user computer systems can be divided into two groups: portable computers and nonportable computers.

Factor	Single-User Computers				
	Portable Computers				
	Handheld	Laptop	Notebook	Netbook	Tablet
Cost	$150–$400	$500–$2,200	$650–$2,300	$200–$800	$250–$550
Weight (pounds)	<.30	4.0–7.0	<4	<2.5	0.75–2.0
Screen size (inches)	2.4–3.6	11.5–15.5	11.6–14.0	7.0–11.0	5.0–14.0
Typical use	Organize personal data	Improve worker productivity	Improve productivity of mobile worker	Access the Internet and e-mail	Capture data via pen input, improve worker productivity

	Nonportable Computers			
	Thin Client	Desktop	Nettop	Workstation
Cost	$200–$500	$500–$2,500	$150–$350	$750–$5,000
Weight (pounds)	1–3	<30	<5	<35
Screen size (inches)	10.0–15.0	13.0–27.0	Comes with or without attached screen	13.0–27.0
Typical use	Enter data and access applications via the Internet	Improve worker productivity	Replace desktop with small, low-cost, low-energy computer	Perform engineering, CAD, and software development

TABLE 3.4 (*Continued*)

Multiple-user computer systems include servers, mainframes, and supercomputers.

Factor	Multiple-User Computers		
	Server	Mainframe	Supercomputer
Cost	$500–$50,000	>$100,000	>$250,000
Weight (pounds)	>25	>100	>100
Screen size (inches)	n/a	n/a	n/a
Typical use	Perform network and Internet applications	Perform computing tasks for large organizations and provide massive data storage	Run scientific applications; perform intensive number crunching

handheld computer: A single-user computer that provides ease of portability because of its small size.

smartphone: A handheld computer that combines the functionality of a mobile phone, camera, Web browser, e-mail tool, MP3 player, and other devices into a single device.

Handheld computers are single-user computers that provide ease of portability because of their small size—some are as small as a credit card. These systems often include a variety of software and communications capabilities. Most can communicate with desktop computers over wireless networks. Some even add a built-in GPS receiver with software that can integrate location data into the application. For example, if you click an entry in an electronic address book, the device displays a map and directions from your current location. Such a computer can also be mounted in your car and serve as a navigation system. One of the shortcomings of handheld computers is that they require a lot of power relative to their size.

A **smartphone** is a handheld computer that combines the functionality of a mobile phone, camera, Web browser, e-mail tool, MP3 player, and other devices into a single device (see Figure 3.17). BlackBerry was one of the earliest smartphone devices developed by the Canadian company Research in Motion in 1999.

In early 2011, both LG and Motorola announced new smartphones based on dual-core 1 GHz processors that can run in parallel to deliver faster performance, provided the software has been designed to take advantage of the dual-core processing capability. Such chips can enable 3D enabled handsets that would not require users to wear special glasses to view the content.[51]

While Apple with its iPhone dominated the smartphone market for several years, it is now facing stiff competition from Amazon, Samsung, HTC, Motorola, Nokia, Samsung, and others. Apple has sued many of its competitors for allegedly violating its patents and trademarks used in mobile devices. In return, many of these competitors have countersued Apple, with Samsung and Apple alone being engaged in 30 legal battles in nine countries.[52] Industry

FIGURE 3.17
BlackBerry Torch 9810 smartphone
A smartphone combines the features of a mobile phone, camera, Web browser, e-mail tool, MP3 player, and other devices.

observers point out that such patent wars stifle innovation and competition, but unfortunately, they have become a common tactic in dealing with the competition.[53]

In early 2011, Deloitte Canada predicted that worldwide sales of smartphones and tablet computers would exceed sales of computers for the first time during the year. The combined sales of smartphones and tablet computers are expected to be 425 million units compared to 400 million units for traditional computers.[54]

Increasingly consumers and workers alike will perform data processing, e-mail, Web surfing, and database lookup tasks on smartphone-like devices. The number of business applications for smartphones is increasing rapidly to meet this need, especially in the medical field where it is estimated that about 30 percent of doctors have an iPad and more than 80 percent carry a smartphone.[55] ePocrates is a provider of online and mobile information to health care providers. Physicians writing prescriptions use their smartphones to access ePocrates on a daily basis to confirm proper drug dosages and look for potential adverse reactions with other drugs their patients are taking. The firm also offers mobile access to patients' electronic medical records.[56]

AccessReflex is software that enables users to view Microsoft Access databases from a smartphone.[57] EBSCOhost Mobile offers mobile access to a broad range of full text and bibliographic databases for research.[58] Mobile police officers can use their smartphones to connect to national crime databases, and emergency first responders can use theirs to connect to hazardous materials databases to find advice of how to deal with dangerous spills or fires involving such materials.

Laptop Computers

laptop computer: A personal computer designed for use by mobile users, being small and light enough to sit comfortably on a user's lap.

A **laptop computer** is a personal computer designed for use by mobile users, being small and light enough to sit comfortably on a user's lap. Laptop computers use a variety of flat panel technologies to produce a lightweight and thin display screen with good resolution. In terms of computing power, laptop computers can match most desktop computers as they come with powerful CPUs as well as large-capacity primary memory and disk storage. This type of computer is highly popular among students and mobile workers who carry their laptops on trips and to meetings and classes. Many personal computer users now prefer a laptop computer over a desktop because of its portability, lower energy usage, and smaller space requirements.

Notebook Computers

notebook computer: Smaller than a laptop computer, an extremely lightweight computer that weighs less than 4 pounds and can easily fit in a briefcase.

Many highly mobile users prefer notebook computers that weigh less than 4 pounds compared to larger laptops that weigh up to around 7 pounds. However, there are limitations to these small, lighter laptops. They typically come without optical drives for burning DVDs or playing Blu-ray movies. Because they are thinner, they have less room for larger, longer-life batteries. Their thin profile also does not allow for heat sinks and fans to dissipate the heat generated by fast processors, so they typically have less processing power. Finally, few come with a high-power graphics card so these machines are less popular with gamers.

Netbook Computers

netbook computer: A small, light, inexpensive member of the laptop computer family.

Netbook computers are small, light, and inexpensive members of the laptop computer family that are great for tasks that do not require a lot of computing power, such as sending and receiving e-mail, viewing DVDs, playing games, or accessing the Internet. However, netbook computers are not good for users who want to run demanding applications, have many applications open at one time, or need lots of data storage capacity.

Many netbooks use the Intel Atom CPU (the N450), which is specially designed to run on minimal power so that the computer can use small, lightweight batteries and avoid potential overheating problems without the need for fans and large heat sinks. Battery life is a key distinguishing feature when comparing various netbooks with expected operating time varying from 4 hours to nearly 12 hours depending on the manufacturer and model.

All 320 high school students in Bloomingdale, Michigan, were provided their own netbook computers and free wireless Internet access. The goal of this program is to provide students with additional learning resources and improve the level of instruction. This is important for a community where nearly half of the households are without Internet access. Teachers update the system each day to notify students and parents about missing assignments; they can also attach a worksheet if applicable. Students appreciate the ability to track their homework and to watch online videos covering many topics and providing step-by-step directions on how to complete homework assignments.[59]

Tablet Computers

tablet computer: A portable, lightweight computer with no keyboard that allows you to roam the office, home, or factory floor carrying the device like a clipboard.

Tablet computers are portable, lightweight computers with no keyboard that allow you to roam the office, home, or factory floor carrying the device like a clipboard. You can enter text with a writing stylus directly on the screen thanks to built-in handwriting recognition software. Other input methods include an optional keyboard or speech recognition. Tablet PCs that support input only via a writing stylus are called *slate computers*. The *convertible tablet PC* comes with a swivel screen and can be used as a traditional notebook or as a pen-based tablet PC. Most new tablets come with a front-facing camera for videoconferencing and a second camera for snapshot photos and video.[60]

Tablets do not yet have the processing power of desktop computers. They also are limited in displaying some videos because the Flash software does not run at all on an iPad nor does it run reliably on Android tablets.[61] Further, tablet screens of most manufacturers need better antiglare protection before they can be used outside in full sunlight.

Tablet computers are especially popular with students and gamers. They are also frequently used in the health care, retail, insurance, and manufacturing industries because of their versatility. M&D Oral Care and Maxillofacial Surgery is an oral surgery practice in Connecticut that installed iPads and a Motorola Xoom tablet at five patient chairs in its office so that patients can view their CT scans and X-rays as well as educational videos.[62]

De Santos is a high-end Italian-American restaurant in New York's West Village. Its waiters use tablet computers to take orders and swipe credit cards. In addition to displaying the full menu, the tablet displays the restaurant's table and seating chart. Accordingly, the tablets make the whole process of seating customers, taking orders, sending orders to the kitchen, and paying the bill simpler and more time efficient. By using tablets, waiters can serve more tables and provide improved customer service. "Nowadays in New York City, the menus don't list the entire specifications of each dish," says Sebastian Gonella, one of the owners and cofounders of the restaurant. "With this software, you can show them exactly the dish itself and all the specifications for each dish, so people are really buying what they're seeing and there's no more confusion. It's pretty important."[63]

The Apple iPad is a tablet computer capable of running the same software that runs on the older Apple iPhone and iPod Touch devices, giving it a library of over 300,000 applications.[64] It also runs software developed specifically for it. The device has a 9.7-inch screen and an on-screen keypad, weighs 1.5 pounds, and supports Internet access over wireless networks.

A number of computer companies are offering tablet computers to compete with Apple's iPad and iPad II, including the Playbook from BlackBerry,

© iStockphoto/mozcann

Samsung Galaxy Tab 10.1 Android tablet

the TouchPad from Hewlett Packard, the Kindle Fire from Amazon, the Streak by Dell, the Tablet S and Tablet P from Sony, the Thrive by Toshiba, the Galaxy Tab from Samsung, the Xoom from Motorola, and the low-cost (less than $75) Aakash and Ubislate from the India-based company Quad.

Nonportable Single-User Computers

Nonportable single-user computers include thin client computers, desktop computers, nettop computers, and workstations.

Thin Clients

thin client: A low-cost, centrally managed computer with essential but limited capabilities and no extra drives (such as CD or DVD drives) or expansion slots.

A **thin client** is a low-cost, centrally managed computer with no extra drives (such as CD or DVD drives) or expansion slots. These computers have limited capabilities and perform only essential applications, so they remain "thin" in terms of the client applications they include. As stripped-down computers, they do not have the storage capacity or computing power of typical desktop computers, nor do they need it for the role they play. With no hard disk, they never pick up viruses or suffer a hard disk crash. Unlike personal computers, thin clients download data and software from a network when needed, making support, distribution, and updating of software applications much easier and less expensive.[65] Thin clients work well in a cloud-computing environment to enable users to access the computing and data resources available within the cloud.[66]

Jewelry Television is viewed by over 65 million people in the United States and millions more visit its Web site (*www.jtv.com*). As a result of this exposure, the company has become one of the world's largest jewelry retailers. Its 300 call center representatives handle more than 6 million calls per year using thin client computers to access information to respond to customer queries and to place orders.[67]

Desktop Computers

desktop computer: A nonportable computer that fits on a desktop and that provides sufficient computing power, memory, and storage for most business computing tasks.

Desktop computers are single-user computer systems that are highly versatile. Named for their size, desktop computers can provide sufficient computing power, memory, and storage for most business computing tasks.

The Apple iMac is a family of Macintosh desktop computers first introduced in 1998 in which all the components (including the CPU, the disk drives, and so on) fit behind the display screen. The Intel iMac is available with Intel's new core i5 or i7 processors making such machines the first quad-core iMacs.

Nettop Computers

nettop computer: An inexpensive desktop computer designed to be smaller, lighter, and consume much less power than a traditional desktop computer.

A **nettop computer** is an inexpensive (less than $350) desktop computer designed to be smaller and lighter and to consume one-tenth the power of a

Courtesy of ASUS Computers International

FIGURE 3.18

ASUS Eee Box nettop

The ASUS Eee Box looks like an oversized external hard drive, but includes integrated graphics, wireless networking capabilities, solid state media connections, and an HDMI port so it can play HD video.

traditional desktop computer.[68] A nettop is designed to perform basic processing tasks such as exchanging e-mail, Internet surfing, and accessing Web-based applications. This computer can also be used for home theatre activities such as watching video, viewing pictures, listening to music, and playing games. Unlike netbook computers, nettop computers are not designed to be portable, and they come with or without an attached screen. (Nettops with attached screens are called all-in-ones.) A nettop without an attached screen can be connected to an existing monitor or even a TV screen. They also may include an optical drive (CD/DVD). The CPU is typically an Intel Atom or AMD Geode, with a single-core or dual-core processor. Choosing a single-core processor CPU reduces the cost and power consumption but limits the processing power of the computer. A dual-core processor nettop has sufficient processing power to enable you to watch video and do limited processing tasks. Businesses are considering using nettops because they are inexpensive to buy and run, and therefore, these computers can improve an organization's profitability. Figure 3.18 shows the ASUS Eee Box nettop computer.

Workstations

workstation: A more powerful personal computer used for mathematical computing, computer-assisted design, and other high-end processing but still small enough to fit on a desktop.

Workstations are more powerful than personal computers but still small enough to fit on a desktop. They are used to support engineering and technical users who perform heavy mathematical computing, computer-assisted design (CAD), video editing, and other applications requiring a high-end processor. Such users need very powerful CPUs, large amounts of main memory, and extremely high-resolution graphic displays. Workstations are typically more expensive than the average desktop computer. Some computer manufacturers are now providing laptop versions of their powerful desktop workstations.

Sekotec Security and Communication is a small company that provides video surveillance systems to supermarkets, gas stations, retail stores, and hotels throughout Germany. Their customer solutions typically involve clusters of 4, 8, 16, or 32 video cameras linked to a powerful workstation that controls the operation of the cameras and can store multiple terabytes of video data for rapid viewing.[69]

Multiple-User Computer Systems

Multiple-user computers are designed to support workgroups from a small department of two or three workers to large organizations with tens of thousands of employees and millions of customers. Multiple-user systems include servers, mainframe computers, and supercomputers.

Servers

server: A computer employed by many users to perform a specific task, such as running network or Internet applications.

A **server** is a computer employed by many users to perform a specific task, such as running network or Internet applications. Servers typically have large memory and storage capacities, along with fast and efficient communications abilities. A Web server handles Internet traffic and communications. An enterprise server stores and provides access to programs that meet the needs of an entire organization. A file server stores and coordinates program and data files. Server systems consist of multiuser computers, including supercomputers, mainframes, and other servers. Often an organization will house a large number of servers in the same room where access to the machines can be controlled and authorized support personnel can more easily manage and maintain the servers from this single location. Such a facility is called a *server farm.*

A new 360,000-square-foot server farm in Santa Clara, California, built by DuPont Fabros Technology, holds tens of thousands of servers. The facility is powered by 32 diesel generators fed by two 50,000-gallon fuel tanks. In the event of a power outage, 32 rotary power systems are on standby. The cooling system consists of 16 chillers and two 500,000-gallon tanks filled with chilled water.[70]

scalability: The ability to increase the processing capability of a computer system so that it can handle more users, more data, or more transactions in a given period.

Servers offer great **scalability**, the ability to increase the processing capability of a computer system so that it can handle more users, more data, or more transactions in a given period. Scalability is increased by adding more, or more powerful, processors. *Scaling up* adds more powerful processors, and *scaling out* adds many more equal (or even less powerful) processors to increase the total data-processing capacity.

The Intel 10-core Westmere-EX processor is targeted at high-end servers such as those frequently used in data centers to support large databases and other processing intense applications. The processor supports a technology called hyperthreading, which enables each core to conduct two sets of instructions, called threads. Hyperthreading gives each processor the ability to run up to 20 threads simultaneously.[71] Advanced Micro Devices (AMD) has announced the release of a 16-core server chip code-named Interlagos Opteron 6200.[72]

Less powerful servers are often used on a smaller scale to support the needs of many users. For example, the Gashora Girls Academy located in the republic of Rwanda deployed server computers to ensure that its 270 students and 12 teachers could have access to the latest technology.[73] Each server computer supports multiple students, each working independently using their own basic workstation that allows them to run word processing and spreadsheet software, use computer science applications, stream videos, and listen to audio programs. This solution minimized the school's investment in hardware and reduced ongoing power and maintenance costs, making it feasible for students to access computing technology to gain the work skills necessary to be successful in the twenty-first century.[74]

Server manufacturers are also competing heavily to reduce the power required to operate their servers and making "performance per watt" a key part of their product differentiation strategy. Low power usage is a critical factor for organizations that run servers farms of hundreds or even thousands of servers. Typical servers draw up to 220 watts, while new servers based on

INFORMATION SYSTEMS @ WORK

Build Your Own

You've probably heard of people building their own computers, such as hobbyists or a small business that assembles computers for other local small businesses. It can be a fun hobby or a reasonable way to earn a living. A large international company, of course, would never do such a thing.

Unless you are talking about Facebook.

For most of its history, Facebook bought the same servers everyone else bought. The company probably received big discounts from suppliers because it uses a lot of servers, but the computer designs were the same that anyone else could buy. The servers got the job done, but Facebook thought the job could be done better. So, the company managers did what anyone else could do: They hired hardware designers and told them, "Come up with something better."

This does not mean Facebook management thought that engineers at Dell, HP, IBM, and all the other server suppliers are incompetent. Far from it. However, the company recognized that those engineers design for a broad market and must satisfy a wide range of customer needs. The customer can specify some server components, such as the number and size of its drives, but the basic design of the server is fixed. Facebook managers realized that if engineers only need to satisfy the needs of a single company, they can meet those needs more closely. The new hardware designers at Facebook started the Open Compute Project based on the model of open source software projects. Their goal was to create energy-efficient, low-cost servers.

As hardware engineering manager Amir Michael puts it, "Trying to optimize costs, we took out a lot of components you find in a standard server. That made it easier to service. It made the thermals more efficient because you had less obstructions blocking cool air. And it made it ten pounds lighter. That's ten pounds less material you're buying, ten pounds less you have to lift every time you put it into the rack or pull it out, and ten pounds less you have to recycle when you're done with it."

The Open Compute Project built a "vanity-free server"—one free of extras that Facebook wouldn't need. "We didn't pay attention to how the servers looked," Michael says. "There's no paint. There's no buttons on the front. There's no fancy logos or emblems." But this no-frills approach was also part of the effort to significantly reduce the cost of cooling the machine. With computers using an estimated 2 percent of the world's power, with much of that going to cooling, reducing cooling costs is significant. (See the Ethical and Societal Issues box in this chapter, on Green IT, for more on that topic.)

Because of its size, Facebook could do more than just redesign its servers. It also redesigned the building that houses the servers, developing a new data center from scratch in Prineville, Oregon. The data center is close to sources of hydroelectric power, benefits from cool winds that can reduce the need for air conditioning, and is designed for minimum power consumption and environmental impact.

Unlike most other high-tech organizations, which keep their development activities quiet, Facebook doesn't mind if you know about its hardware designs. The firm realizes that it's not its hardware that makes it successful; it's what Facebook does with that hardware. You can download Facebook's design documentation at no charge from the Open Compute Project Web site. If you agree to the terms of its open license, you can even join the community and contribute your own ideas.

Discussion Questions

1. Facebook uses more servers than all but a handful of organizations. Most use far fewer. How can Facebook's experience help smaller organizations? (Use your school as an example if you like.)
2. Visit the Open Compute Project Web site (*opencompute.org*). Browse the pages in the Open Updates section, such as "The OCP Wants You" (November 16, 2011) and "Making It Real: Next Steps for the Open Compute Project" (October 27, 2011), including the comments on these updates. If you were a computer hardware designer, would you want to participate in this project? If your employer didn't want you contributing to the project during working hours, would you contribute to it (with your employer's permission) in your spare time?

Critical Thinking Questions

1. Does Facebook's designing its own computers give it a competitive advantage as described in Chapter 2? If so, via which of the five competitive forces does it gain an advantage and how? If not, why would Facebook design its own servers?
2. Suppose you were about to start a social network site to compete with Facebook. Can you find any

information from the Open Compute Project Web site (*opencompute.org*) to make you more successful than you would be otherwise? Why or why not?

SOURCES: Metz, C., "Facebook Hacks Shipping Dock into World-Class Server Lab," *Wired*, *www.wired.com/wiredenterprise/2012/01/facebook-server-lab*, January 9, 2012; Michael, Amir, "Inside the Open Compute Project Server," Facebook Engineering Notes, *www.facebook.com/notes/facebook-engineering/inside-the-open-compute-project-server/10150144796738920*, April 8, 2011; Chang, E., "Facebook Shares Technology to Build Data Centers," Bloomberg TV, *www.bloomberg.com/video/73907872*, August 12, 2011; Open Compute Project Web site, *http://opencompute.org*, accessed January 9, 2012.

Intel's Atom microprocessor draw 8 or fewer watts. The annual power savings from such low-energy usage servers can amount to tens of thousands of dollars for operators of a large server farm.

A virtual server is a method of logically dividing the resources of a single physical server to create multiple logical servers, each acting as its own dedicated machine. The server administrator uses software to divide one physical server into multiple isolated virtual environments. For example, a single physical Web server might be divided into two virtual private servers. One of the virtual servers hosts the organization's live Web site, while the other hosts a copy of the Web site. The second private virtual server is used to test and verify updates to software before changes are made to the live Web site. The use of virtual servers is growing rapidly. In a typical data center deployment of several hundred servers, companies using virtualization can build 12 virtual machines for every actual server with a resulting savings in capital and operating expenses (including energy costs) of millions of dollars per year.

EZZI.net is a Web hosting service provider for many companies including some of the largest *Fortune* 500 companies. It has data centers located in New York City and Los Angeles that employ virtual servers because they are easy to use, can be supported around the clock, and operate with a 99.7 percent uptime to meet the needs of its many customers.[75]

blade server: A server that houses many individual computer motherboards that include one or more processors, computer memory, computer storage, and computer network connections.

A **blade server** houses many computer motherboards that include one or more processors, computer memory, computer storage, and computer network connections. These all share a common power supply and air-cooling source within a single chassis. By placing many blades into a single chassis,

Rack-mounted system with blade servers.

and then mounting multiple chassis in a single rack, the blade server is more powerful but less expensive than traditional systems based on mainframes or server farms of individual computers. In addition, the blade server approach requires much less physical space than traditional server farms.

Norddeutsche Landesbank (NORD/LB) is a major financial institution with headquarters in Hanover, Germany. The bank was suffering from slow response time for its key systems while trying to meet new business needs. New blade server computers were installed to improve system response time by 40 percent and provide the bank's departments with the data they need on a timely basis so they could operate efficiently.[76]

Mainframe Computers

mainframe computer: A large, powerful computer often shared by hundreds of concurrent users connected to the machine over a network.

A **mainframe computer** is a large, powerful computer shared by dozens or even hundreds of concurrent users connected to the machine over a network. The mainframe computer must reside in a data center with special HVAC equipment to control temperature, humidity, and dust levels. In addition, most mainframes are kept in a secure data center with limited access. The construction and maintenance of a controlled-access room with HVAC can add hundreds of thousands of dollars to the cost of owning and operating a mainframe computer.

The role of the mainframe is undergoing some remarkable changes as lower-cost, single-user computers become increasingly powerful. Many computer jobs that used to run on mainframe computers have migrated onto these smaller, less expensive computers. This information-processing migration is called *computer downsizing*.

The new role of the mainframe is as a large information-processing and data storage utility for a corporation—running jobs too large for other computers, storing files and databases too large to be stored elsewhere, and storing backups of files and databases created elsewhere. For example, the mainframe can handle the millions of daily transactions associated with airline,

Courtesy of IBM Corporation

Mainframe computers have been the workhorses of corporate computing for more than 50 years. They can support hundreds of users simultaneously and can handle all of the core functions of a corporation.

automobile, and hotel/motel reservation systems. It can process the tens of thousands of daily queries necessary to provide data to decision support systems. Its massive storage and input/output capabilities enable it to play the role of a video computer, providing full-motion video to multiple, concurrent users.

Payment Solution Providers (PSP) is a Canadian corporation specializing in e-payment networks and the integration of financial transaction processing systems. PSP recently selected an IBM system z mainframe computer on which to run its credit card processing business. Other alternatives examined lacked the security PSP requires and would make it difficult to meet the banking industry's PCI compliance standards for increasing controls around cardholder data to reduce credit card fraud. In addition, consolidation of operations onto a single mainframe provides a compact, efficient infrastructure that minimizes space requirements and reduces costs for IT management, power and cooling, and software licenses by 35 percent.[77]

Supercomputers

supercomputers: The most powerful computer systems with the fastest processing speeds.

Supercomputers are the most powerful computers with the fastest processing speed and highest performance. They are special-purpose machines designed for applications that require extensive and rapid computational capabilities. Originally, supercomputers were used primarily by government agencies to perform the high-speed number crunching needed in weather forecasting, earthquake simulations, climate modeling, nuclear research, study of the origin of matter and the universe, and weapons development and testing. They are now used more broadly for commercial purposes in the life sciences and the manufacture of drugs and new materials. For example, Procter & Gamble uses supercomputers in the research and development of many of its leading commercial brands such as Tide and Pampers to help develop detergent with more soap suds and improve the quality of its diapers.[78]

A Japanese supercomputer built by Fujitsu named simply "K" and capable of making 8.2 quadrillion calculations per second (8.2 petaflops, where flop is floating point operation per second) was identified as the world's fastest computer in June 2011. This machine has the computing power of one million laptop computers working in tandem. Although considered energy efficient,

AFP PHOTO/HO/RIKEN

Japan's K supercomputer, the world's fastest computer.

the computer requires enough electricity to operate 10,000 homes at a cost of $10 million per year. This computer is three times faster than the previous speed champion, the Tianhe-1A supercomputer at the National Supercomputing Center in Tianjin, China.[79]

graphics processing unit (GPU): A specialized circuit that is very efficient at manipulating computer graphics and is much faster than the typical CPU chip at performing floating point operations and executing algorithms for which processing of large blocks of data is done in parallel.

The K and the Tianhe-1A supercomputers are based on a new architecture that employs both **graphics processing unit (GPU)** chips to perform high-speed processing. The GPU chip is a specialized circuit that is very efficient at manipulating computer graphics and is much faster than the typical CPU chip at performing floating point operations and executing algorithms for which processing of large blocks of data is done in parallel. This math is precisely the type performed by supercomputers.[80]

The fastest operational U.S. supercomputer as of this writing is the XT-5 Jaguar built by Cray and residing at the Department of Energy's Oak Ridge Laboratory in Tennessee. This computer was recently used to conduct simulations of the airflow around 18-wheeler tractor trailers to prove that outfitting these trailers with a set of integrated aerodynamic truck parts would substantially reduce gas consumption and the production of carbon dioxide. Use of the XT-5 shortened the computing time required from days to a few hours and avoided the need to build time-consuming and expensive physical prototypes. All told, running the simulations on the XT-5 cut the time required to go from concept to design of the necessary truck parts from 3.5 years to just 1.5 years.[81] Table 3.5 lists the three most powerful supercomputers in use as of June 2011.

Watson, an IBM supercomputer, is best known for defeating former *Jeopardy!* quiz champions. The contest was a means to demonstrate the various systems, data management, and analytics technology that can be applied in business and across different industries.[82] Watson is being "trained" by the insurance company WellPoint to diagnose treatment options for patients.[83]

There are plans to develop a new supercomputer (called Titan) with a computing power of 20 petaflops at the Oak Ridge Laboratory. This machine will employ the new architecture using a combination of GPU chips from Nvidia and CPU chips from Intel and Advanced Micro Devices.[84] Table 3.6 lists the processing speeds of supercomputers.

TABLE 3.5 The three most powerful operational supercomputers (June 2011)

Rank	System	Manufacturer	Research Center	Location	Speed (petaflops)
1	K	Fujitsu	Riken Advanced Institute for Computational Science	Kobe, Japan	8.2
2	Tianhe-1A	National University of Defense Technology	National Supercomputing Center	Tianjin, China	2.57
3	XT-5 Jaguar	Cray	Department of Energy's Oak Ridge Laboratory	United States	1.76

TABLE 3.6 Supercomputer processing speeds

Speed	Meaning
GigaFLOPS	1×10^{9} FLOPS
TeraFLOPS	1×10^{12} FLOPS
PetaFLOPS	1×10^{15} FLOPS
ExaFLOPS	1×10^{18} FLOPS

GREEN COMPUTING

green computing: A program concerned with the efficient and environmentally responsible design, manufacture, operation, and disposal of IS-related products.

Green computing is concerned with the efficient and environmentally responsible design, manufacture, operation, and disposal of IS-related products, including all types of computers, printers, and printer materials such as cartridges and toner. Business organizations recognize that going green is in their best interests in terms of public relations, safety of employees, and the community at large. They also recognize that green computing presents an opportunity to substantially reduce total costs over the life cycle of their IS equipment. Green computing has three goals: reduce the use of hazardous material, allow companies to lower their power-related costs (including potential cap and trade fees), and enable the safe disposal or recycling of computers and computer-related equipment. According to Greenpeace, 50 million tons of computers, monitors, laptops, printers, disk drives, cell phones, DVDs, and CDs are discarded worldwide each year.[85]

Computers contain many toxic substances including beryllium, brominated flame retardants, cadmium, lead, mercury, polyvinyl chloride, and selenium. As a result, electronic manufacturing employees and suppliers at all steps along the supply chain and in the manufacturing process are at risk of unhealthy exposure. Computer users can also be exposed to these substances when using poorly designed or damaged devices.

Because it is impossible to ensure safe recycling or disposal, the best practice is to eliminate the use of toxic substances, particularly since recycling of used computers, monitors, and printers has raised concerns about toxicity and carcinogenicity of some of the substances. Safe disposal and reclamation operations must be extremely careful to avoid exposure in recycling operations and leaching of materials such as heavy metals from landfills and incinerator ashes. In many cases, recycling companies export large quantities of used electronics to companies in undeveloped countries. Unfortunately, many of these countries do not have strong environmental laws, and they sometimes fail to recognize the potential dangers of dealing with hazardous materials. In their defense, these countries point out that the United States and other first-world countries were allowed to develop robust economies and rise up out of poverty without the restrictions of strict environmental policies.

One of the earliest initiatives toward green computing in the United States was the voluntary labeling program known as Energy Star. It was conceived of by the Environmental Protection Agency in 1992 to promote energy efficiency in hardware of all kinds. This program resulted in the widespread adoption of sleep mode for electronic products. For example, Energy Star monitors have the capability to power down into two levels of "sleep." In the first level, the monitor energy consumption is less than or equal to 15 Watts, and in the second, power consumption reduces to 8 Watts, which is less than 10 percent of its operating power consumption.[86]

The European Union Directive 2002/95/EC required that as of July 2006, new electrical and electronic equipment cannot contain any of six banned substances in quantities exceeding certain maximum concentration values. The six banned substances are lead, mercury, cadmium, hexavalent chromium, polybrominated biphenyls, and polybrominated diphenylethers. The directive was modified in September 2011 to exclude lead and cadmium because of the impracticality of finding suitable substitutes for these materials.[87] This directive applies to U.S. organizations selling equipment to members of the European Union and has encouraged U.S. manufacturers to meet the standards as well.

The Green Electronics Council manages the Electronic Product Environment Assessment Tool (EPEAT) to assist in the evaluation and purchase of green computing systems. The EPEAT assesses products against 51 lifecycle environmental criteria developed by representatives of the environmental community, manufacturers, private and public purchasers, resellers, recyclers,

So Many Servers: Reducing Energy Consumption in Information Systems

According to a recent article in *CNN Money,* information technology accounts for over 2 percent of the world's energy use. Google alone uses as much electricity as all the homes in San Diego, which has a population of over 1.3 million people.

Energy consumption is one of the secret downsides of the computer revolution. The computing devices we've come to depend on require energy resources to build, need electricity to operate, and ultimately create waste, which requires still more resources to handle.

Organizations are taking steps to deal with this problem. Some of the most promising solutions come from organizations that use many computers internally to carry out their business. The newspaper *Computerworld* selected the top 12 "Green IT" organizations, including Allstate Insurance, Northrop Grumman, Prudential Insurance, and Kaiser Permanente, from a field of 70 applicants. All twelve are striving to find new ways to use less energy for information technology.

Allstate Insurance eliminated over 3,000 computers and reduced its energy usage by about 40 percent over 18 months. The company followed a variety of strategies to accomplish that energy reduction, including upgrading to more efficient types of hardware for processing and storage. Anthony Abbattista, senior vice president of technology solutions for Allstate, says "We are a collection of socially conscious engineers and business people who have passion for doing things green." When deciding which hardware to purchase for its data centers, Allstate considers energy efficiency along with price and compatibility. The company also monitors its energy consumption and plans to continue replacing older, inefficient hardware with less energy-intensive alternatives.

Other companies use similar approaches. Aerospace and defense company Northrop Grumman is using virtualization, as discussed in this chapter, to eliminate 3,000 of its servers—about 80 percent of the total. Eliminating those servers is a major step toward its corporate goal of reducing greenhouse gas emissions by 25 percent from 2010 to 2015. "IT is a large consumer of power, equipment, and consumables, and as such, IT will continue to be a focus in our company's sustainability [efforts]," says vice president and chief technology officer Brad Furukawa. Virtualization will let Northrop Grumman consolidate 100 data centers and server rooms in three data centers, ultimately eliminating over 13,000 tons of CO_2 emissions every year.

Prudential Insurance is also looking to virtualization to deal with the problem of energy consumption. The company shifted 1,000 servers to a virtual environment in 2010. It's installing solar panels to power its data center in Roseland, New Jersey, and pushing hard to minimize the number of pages that employees print. "You have to base decisions on economics and functionality to be green," says Michael Mandelbaum, a vice president in Prudential's IT group. "That's how we approach everything, which means we can continue to do better because we can point to a track record of serving the shareholders while serving the broader community through sustainability."

Health care company Kaiser Permanente earned the top spot on *Computerworld's* list. By focusing on the areas with the biggest payoff, such as a Unix system that consumed up to 100 kilowatts per cabinet, Kaiser Permanente reduced its power consumption by 7.2 million kilowatt-hours in 2011. That is equivalent to the power consumption of 800 family homes, and it cut the company's power budget by over $770,000. Major savings came from balancing cooling in its data centers, so the firm doesn't have to freeze some areas in a room to control the temperature in the warmest parts. Like Northrop Grumman and Prudential, Kaiser Permanente also used virtualization: About 65 percent of its servers that run the Unix operating system and 20 percent of those that run Windows were virtual in late 2011. Even so, vice president of data center services Laz Garcia says "There's still a lot of gains to be had!"

Discussion Questions

1. Large computer installations use power for two purposes: to operate equipment and to cool it, essentially counteracting each other. They are using approaches such as solar power, mentioned in this sidebar, to generate this power. Can they also reduce the need for power by reducing the need for air conditioning? How?
2. How does virtualization allow companies to reduce their energy consumption?

Critical Thinking Questions

1. The four companies profiled here and the other winners in *Computerworld's* study are large organizations. Should small businesses care about the energy consumption of their computers? Medium-sized ones? Why or why not? If you were a manager in a small or medium business, what factors would you consider in deciding if it's important to spend time to reduce energy consumption?
2. Consider your home, apartment or dorm room. Do you leave printers switched on when you're not printing? Digital photo frames when you're asleep? TV in "standby," with cable box on rather than off even when you'll be away all day or for several days? (Don't just use these examples; look around and see what you have.) See how much power each consumes when not in use, assume 10 watts for those you don't have data for and compute the total. Multiply by 0.024 for kilowatt-hours (kWh) per day. Figure the annual cost, using the cost of electricity per kWh in your area if you know it, $0.25/kWh if you don't. Is the convenience of having these devices on every instant, compared to switching them on when you need them, worth the cost?

SOURCES: Fanning, Ellen, "The Top Green-IT organizations: Hard-wired to Be Green," *Computerworld, www.computerworld.com/s/article/359173/The_top_Green_IT_organizations_Hard_wired_to_be_green*, October 24, 2011; Hargreaves, Steve, "The Internet: One Big Power Suck," *CNN Money, http://money.cnn.com/2011/05/03/technology/internet_electricity/index.htm*, May 9, 2011; Mitchell, Robert L., "Kaiser Permanente: Slashes Data Center Power by 7.2 Million Kilowatt-Hours," *Computerworld, www.computerworld.com/s/article/358844/Kaiser_Permanente_Slashes_data_center_power*, October 24, 2011; Pratt, Mary K., "Allstate: Reduces Nearly 3,000 Servers or Devices in 18 Months," *Computerworld, www.computerworld.com/s/article/358932/Allstate_Reduces_nearly_3_000_servers_devices*, October 24, 2011; Pratt, Mary K., "Northrop Grumman: Virtualizing or Retiring Approximately 3,000 Servers," *Computerworld, www.computerworld.com/s/article/359031/Northrop_Grumman_Virtualizing_3_000_servers*, October 24, 20111; Pratt, Mary K., "Prudential Financial: 1,000 Servers Moved into a Virtual Environment," *Computerworld, www.computerworld.com/s/article/359041/Prudential_Financial_Moving_to_a_virtual_environment*, October 24, 2011.

and other interested parties. These criteria are documented in IEEE Standard 1680 and have to do with the reduction of hazardous materials, the use of recycled materials, the design for recovery through recycling systems, product longevity, energy conservation, end of life management, the manufacturer's corporate environmental policy, and packaging.[88] The products evaluated

TABLE 3.7 EPEAT product tiers

Tier	Number of Required Criteria That Must Be Met	Number of Optional Criteria That Must Be Met
Bronze	All 23	None
Silver	All 23	At least 50%
Gold	All 23	At least 75%

against the EPEAT criteria are placed into one of three tiers based on their rating, as shown in Table 3.7.

Computer manufacturers such as Apple, Dell, and Hewlett-Packard have long competed on the basis of price and performance. As the difference among the manufacturers in these two arenas narrows, support for green computing is emerging as a new business strategy for these companies to distinguish themselves from the competition. Apple claims to have the "greenest lineup of notebooks" and is making progress at removing toxic chemicals. Dell's new mantra is to become "the greenest technology company on Earth." Hewlett-Packard highlights its long tradition of environmentalism and is improving its packaging to reduce the use of materials. It is also urging computer users around the world to shut down their computers at the end of the day to save energy and reduce carbon emissions.

SUMMARY

Principle:

Computer hardware must be carefully selected to meet the evolving needs of the organization and its supporting information systems.

Computer hardware should be selected to meet specific user and business requirements. These requirements can evolve and change over time.

The central processing unit (CPU) and memory cooperate to execute data processing. The CPU has three main components: the arithmetic/logic unit (ALU), the control unit, and the register areas. Instructions are executed in a two-phase process called a machine cycle, which includes the instruction phase and the execution phase.

Computer system processing speed is affected by clock speed, which is measured in gigahertz (GHz). As the clock speed of the CPU increases, heat is generated that can corrupt the data and instructions the computer is trying to process. Bigger heat sinks, fans, and other components are required to eliminate the excess heat. This excess heat can also raise safety issues. Chip designers and manufacturers are exploring various means to avoid heat problems in their new designs.

Primary storage, or memory, provides working storage for program instructions and data to be processed and provides them to the CPU. Storage capacity is measured in bytes.

A common form of memory is random access memory (RAM). RAM is volatile; loss of power to the computer erases its contents. RAM comes in many different varieties including dynamic RAM (DRAM), synchronous DRAM (SDRAM), Double Data Rate SDRAM, and DDR2 SDRAM.

Read-only memory (ROM) is nonvolatile and contains permanent program instructions for execution by the CPU. Other nonvolatile memory types include programmable read-only memory (PROM), erasable programmable read-only memory (EPROM), electrically erasable PROM (EEPROM), and flash memory.

Cache memory is a type of high-speed memory that CPUs can access more rapidly than RAM.

A multicore microprocessor is one that combines two or more independent processors into a single computer so that they can share the workload. Intel and AMD have introduced eight-core processors that are effective in working on problems involving large databases and multimedia.

Parallel computing is the simultaneous execution of the same task on multiple processors to obtain results faster. Massively parallel processing involves linking many processors to work together to solve complex problems.

Grid computing is the use of a collection of computers, often owned by multiple individuals or organizations, to work in a coordinated manner to solve a common problem.

Computer systems can store larger amounts of data and instructions in secondary storage, which is less volatile and has greater capacity than memory. The primary characteristics of secondary storage media and devices include access method, capacity, portability, and cost. Storage media can implement either sequential access or direct access. Common forms of secondary storage include magnetic storage devices such as tape, magnetic disk, virtual tape; optical storage devices such as optical disc, digital video disc (DVD), and holographic versatile disc (HVD); and solid state storage devices such as flash drives.

Redundant array of independent/inexpensive disks (RAID) is a method of storing data that generates extra bits of data from existing data, allowing the system to more easily recover data in the event of a hardware failure.

Network-attached storage (NAS) and storage area networks (SAN) are alternative forms of data storage that enable an organization to share data resources among a much larger number of computers and users for improved storage efficiency and greater cost effectiveness.

Storage as a service is a data storage model in which a data storage service provider rents space to people and organizations.

Input and output devices allow users to provide data and instructions to the computer for processing and allow subsequent storage and output. These devices are part of a user interface through which human beings interact with computer systems.

Data is placed in a computer system in a two-stage process: Data entry converts human-readable data into machine-readable form; data input then transfers it to the computer. Common input devices include a keyboard, a mouse, speech recognition, digital cameras, terminals, scanning devices, optical data readers, magnetic ink character recognition devices, magnetic stripe cards, chip-and-PIN cards, contactless cards, point-of-sale devices, automated teller machines, pen input devices, touch-sensitive screens, bar-code scanners, and Radio Frequency Identification tags.

Display monitor quality is determined by aspect ratio, size, color, and resolution. Liquid crystal display and organic light-emitting diode technology is enabling improvements in the resolution and size of computer monitors. Other output devices include printers, plotters, Surface touch tables, digital audio players, and e-book readers.

Computer systems are generally divided into two categories: single user and multiple users. Single-user systems include portable computers such as handheld, laptop, notebook, netbook, and tablet computers. Nonportable single-user systems include thin client, desktop, nettop, and workstation computers.

Multiuser systems include servers, blade servers, mainframes, and supercomputers.

Principle:

The computer hardware industry is rapidly changing and is highly competitive, creating an environment ripe for technological breakthroughs.

CPU processing speed is limited by physical constraints such as the distance between circuitry points and circuitry materials. Moore's Law is a hypothesis stating that the number of transistors on a single chip doubles every two years. This hypothesis has been accurate since it was introduced in 1970.

Manufacturers are competing to develop a nonvolatile memory chip that requires minimal power, offers extremely fast write speed, and can store data accurately even after it has been stored and written over many times. Such a chip could eliminate the need for RAM forms of memory.

Principle:

The computer hardware industry and users are implementing green computing designs and products.

Green computing is concerned with the efficient and environmentally responsible design, manufacture, operation, and disposal of IT related products.

Business organizations recognize that going green can reduce costs and is in their best interests in terms of public relations, safety of employees, and the community at large.

Three specific goals of green computing are reduce the use of hazardous material, lower power related costs, and enable the safe disposal and/or recycling of IT products.

Three key green computing initiatives are the Energy Star program to promote energy efficiency, the European Union Directive 2002/95/EC to reduce the use of hazardous materials, and the use of the EPEAT tool to evaluate and purchase green computing systems.

CHAPTER 3: SELF-ASSESSMENT TEST

Computer hardware must be carefully selected to meet the evolving needs of the organization and its supporting information systems.

1. All organizations require the most powerful and most current software to remain competitive. True or False?
2. The faster the clock speed of the CPU, the more heat the processor generates. True or False?
3. The overriding consideration for a business in making hardware decisions should be how the hardware supports the ________________ of the information system and goals of the organization.
4. Which represents the largest amount of data—an exabyte, terabyte, or gigabyte?
5. Which of the following CPU components provides a high-speed storage area to temporarily hold small units of program instruction and data immediately before, during, and after execution by the CPU?
   a. control unit
   b. register
   c. ALU
   d. main memory
6. Executing an instruction by the CPU involves two phases: the ________________ phase and the execution phase.
7. ________________ involves capturing and editing data when it is originally created and in a form that can be directly entered into a computer, thus ensuring accuracy and timeliness.

The computer hardware industry is rapidly changing and highly competitive, creating an environment ripe for technological breakthroughs.

8. Many computer jobs that used to run on mainframe computers have migrated onto smaller, less expensive computers. This information-processing migration is called ________________.
9. The transistor densities on a single chip double every two years. True or False?

The computer hardware industry and users are implementing green computing designs and products.

10. Green computing is about saving the environment; there are no real business benefits associated with this program. True or False?
11. The disposal and reclamation operations for IT equipment must be careful to avoid unsafe exposure to ________________.

CHAPTER 3: SELF-ASSESSMENT TEST ANSWERS

1. False
2. True
3. objectives
4. exabyte
5. b
6. instruction
7. Source data automation
8. computer downsizing
9. True
10. False
11. hazardous materials

REVIEW QUESTIONS

1. What is a virtual tape system and what is it used for?
2. How does the role of primary storage differ from secondary storage?
3. Identify and briefly discuss the fundamental characteristic that distinguishes RAM from ROM memory.
4. What is RFID technology? Identify three practical uses for this technology.
5. What is a fuel cell? What advantages do fuel cells offer over batteries for use in portable electronic devices? Do they have any disadvantages?
6. What is the difference between a CPU chip and a GPU chip?
7. What is RAID storage technology?
8. Outline and briefly explain the two-phase process for executing machine-level instructions.
9. What is a massively parallel processing computer system? How is grid computing different from such a system? How is it similar?
10. Identify the three components of the CPU and explain the role of each.
11. Distinguish between a netbook computer and a laptop computer. Distinguish between a nettop and desktop computer.
12. Identify and briefly describe the various classes of single-user, portable computers.
13. What is a solid state storage device?
14. Define the term green computing and state the primary goals of this program.
15. What is the EPEAT tool? How is it used?

DISCUSSION QUESTIONS

1. Discuss the role of the business manager in helping to determine the computer hardware to be used by the organization.
2. Explain why the clock speed is not directly related to the true processing speed of a computer.
3. Briefly describe the concept of multiprocessing. How does parallel processing differ from multiprocessing?
4. What issues can arise when the CPU runs at a very fast clock speed? What measures are manufacturers taking to deal with this problem?
5. Briefly discuss the advantages and disadvantages of installing thin clients for use in a university computer lab versus desktop computers.
6. What is an eight-core processor? What advantages does it offer users over a single-core processor? Are there any potential disadvantages?
7. Outline the Electronic Product Environment Assessment for rating computer products.
8. Briefly describe Moore's Law. What are the implications of this law? Are there any practical limitations to Moore's Law?
9. Identify and briefly discuss the advantages and disadvantages of solid state secondary storage devices compared to magnetic secondary storage devices.
10. Briefly discuss the advantages and disadvantages of attached storage, network-attached storage, and storage area networks in meeting enterprise data storage challenges.
11. If cost were not an issue, describe the characteristics of your ideal computer. What would you use it for? Would you choose a handheld, portable, desktop, or workstation computer? Why?
12. Briefly explain the differences between the magnetic stripe card and the chip-and-PIN card.
13. Discuss potential issues that can arise if an organization is not careful in selecting a reputable service organization to recycle or dispose of its IS equipment.

PROBLEM-SOLVING EXERCISES

1. Use word-processing software to document what your needs are as a computer user and your justification for selecting either a desktop or portable computer. Find a Web site that allows you to order and customize a computer and select those options that best meet your needs in a cost-effective manner. Assume that you have a budget of $650. Enter the computer specifications and associated costs from the Web site into an Excel spreadsheet that you cut and paste into the document defining your needs. E-mail the document to your instructor.
2. Develop a spreadsheet that compares the features, initial purchase price, and a two-year estimate of operating costs (paper, cartridges, and toner) for three color laser printers. Assume that you will print 100 color pages and 200 black and white pages each month. Now do the same for three inkjet printers. Write a brief memo on which printer you would choose and why. Cut and paste the spreadsheet into a document.
3. Use a database program to document the features and capabilities of three different e-book readers.

TEAM ACTIVITIES

1. With one or two of your classmates, visit a retail store that employs Radio Frequency Identification chips to track inventory. Interview an employee involved in inventory control and document the advantages and disadvantages they see in this technology. Go online and identify what concerns have been expressed about the use of RFID by privacy advocates. Write a summary of your findings.
2. With two or three of your classmates, visit three different retail stores in search of your ideal e-book reader. Document the costs, features, advantages, and disadvantages of three different readers using a spreadsheet program. Be sure to consider the cost and ease with which e-books can be purchased for each reader. Analyze your data and make a recommendation on which one you would buy.

WEB EXERCISES

1. There is great competition among countries and manufacturers to develop the fastest supercomputer. Do research on the Web to identify the current three fastest supercomputers and how they are being used. Write a brief report summarizing your findings.
2. Do research on the Web to learn more about the Electronic Product Environment Assessment Tool and the criteria it uses to evaluate products. Use the EPEAT tool to find out the rating for your current computer. Which company seems to have the greenest notebook computers? Write a brief report summarizing your findings.

CAREER EXERCISES

1. Imagine that you are going to buy a smartphone to improve your communication capabilities and organizational abilities. What tasks do you need it to perform? What features would you look for in this device? Visit a phone store or a consumer electronics store and identify the specific device and manufacturer that comes closest to meeting your needs at a cost under $225.
2. Your $50 million in annual sales organization of 500 employees plans to acquire 100 new portable computers. The chief financial officer has asked you to lead a project team assigned to define users' computer hardware needs and recommend the most cost-effective solution for meeting those needs. Who else (role, department) and how many people would you select to be a member of the team? How would you go about defining users' needs? Do you think that only one kind of portable computer will meet everyone's needs? Should you define multiple portable computers based on the needs of various classes of end user? What business justification can you define to justify this expenditure of roughly $50,000?

CASE STUDIES

Case One

Extreme Storage

The amount of data stored in the world's computers is growing faster than drive capacities are growing to hold it.

The United States Library of Congress processes about 40 TB of data each week. That's about 20 times the capacity of the largest commonly available disk drives. The library uses nearly 20,000 disk drives to store over 3 petabytes (PB) of data in total. Thomas Youkel, group chief of enterprise systems engineering at the library, estimates that the library's data load will quadruple in the next few years. The library realized that it had to manage this data intelligently to cope with its size. One method the library used was to separate the actual content, which is usually needed only at the end of a search, from the information about the content (*metadata*) used in searching. Only the metadata is kept in high-speed storage, with the content kept offline or on low-speed, less expensive storage media.

On a smaller (but still large) scale, Mazda Motor Corporation has "only" about 90 TB of data. To trade off access time and cost, the firm divides its data into four levels. The most frequently needed data is kept in solid state storage. The next level of data is stored on high-performance 15,000-RPM drives; next, on less expensive 7,200-RPM drives; and finally, on magnetic tape. The top and bottom tiers each store about 20 percent of the total, with the remaining 60 percent on the middle magnetic-disk tiers.

Whether an organization has extraordinary storage requirements like the Library of Congress or more manageable requirements like Mazda, the amount of data organizations need to store is exploding for several reasons. One is the accumulation of years of data, which companies are reluctant to discard since they may be of value in the future. Another is the proliferation of storage-intensive data types such as photos and video, which occupy more space than letters and numbers. Advanced data analysis methods, which help extract value from this data, are another business reason to collect data. As a report from the Data Warehousing Institute puts it, "The fastest growing use case for big data analytics is advanced data visualization. A growing number of companies are running sophisticated analytics tools on big data sets in order to build highly complex visual representations of their data." This representation helps managers make better decisions but uses a lot of storage space.

Storage vendors compete to meet the demand for greater storage capacity by using the enterprise storage system types discussed earlier in this chapter. A recent article by Desmond Fuller in *Computerworld* discusses how businesses use NAS to handle data storage needs such as those of the Library of Congress and Mazda:

> A larger business might use a low-cost NAS box to offload stagnant, rarely used data from more expensive, high-performance storage. Or it might place one alongside a virtual server farm to store virtual machine images or ship one to a satellite office to serve as low-cost file storage. For a small to medium-size business, one of these NAS boxes would serve the needs for daily file storage, with the bonus that nontechnical staff could set it up and start using it without professional IT help.

Fuller goes on to review five NAS offerings, noting, "I found the richest sets of business features—straightforward setup, easy remote access, plentiful backup options—at the higher end of the scale." It's important for a business to analyze its needs carefully before looking at what's available. Some features will only come at a price, and the business shouldn't pay that price if it doesn't need the features.

The range of storage area network (SAN) offerings is equally wide. Blackpool and The Fylde College, north of Liverpool in England, chose a NetApp SAN to support its faculty, staff, and 20,000+ students. Network manager Simon Bailey says, "Previously, we had over 100 servers with attached storage to manage. These were constantly running out of disk space and we were forever trying to cram in extra disk capacity. To be honest, the situation had got past manageable." Today, after expanding its initial SAN, the college has centralized 99 percent of its storage. "The centralized NetApp storage system has allowed us to implement a scalable data management strategy in line with organizational growth," says Bailey. "It's flexible, scalable, and affordable—we wouldn't be without it."

Discussion Questions

1. Suppose you use about 500 GB (0.5 TB) of storage in your personal system, filling about two-thirds of its 750 GB disk. Most of that storage is used by your video library. Photos, music, and software (including the operating system) use about 20 GB. Traditional data files, such as word processing files, spreadsheets, and e-mail, require less than 1 GB. Which of the technologies and strategies discussed in this case might be of practical value to you? Why?
2. Both SAN and NAS systems can use RAID internally (see discussion earlier in this chapter). Why might this be a good idea?

Critical Thinking Questions

1. Consider the Library of Congress's 15,000 to 18,000 disk drives. The mean time between failures (MTBF) of a typical disk drive has been estimated at about 50 years. If that figure is correct, the library will have 300 to 360 drive failures every year, an average of about one a day. What does that mean for managing its hardware? How is that different from the way a small business with five or six disk drives should manage its hardware?
2. Both Mazda and the Library of Congress divide their data into multiple tiers to manage it effectively and cost effectively. SAN systems can support this strategy by incorporating several types of storage into one SAN.

How could your college or university take advantage of this concept? Outline three tiers of data it could have, with two specific examples of data that would be in each tier. (For example, records of students who graduated or left at least ten years ago could be in the lowest tier.)

SOURCES: Brandon, J., "Storage Tips from Heavy-Duty Users," *Computerworld, www.computerworld.com/s/article/358624/Extreme_Storage*, October 10, 2011; Fuller, D., "NAS Shoot-out: 5 Storage Servers Battle for Business," *Computerworld, www.computerworld.com/s/article/9220996/NAS_shoot_out_5_storage_servers_battle_for_business*, October 19, 2011; Vijayan, J., "New Tools Driving Big Data Analytics, Survey Finds," *Computerworld, www.computerworld.com/s/article/9219487/New_tools_driving_big_data_analytics_survey_finds*, August 25, 2011; Williams, N., "Blackpool and The Fylde College Achieves Business Continuity and Saves Cost with NetApp SAN Storage," *www.computerweekly.com/Articles/2011/01/24/245031/Case-study-NetApp-storage-solution-gains-top-results-for-Blackpool-and-the-Fylde-College.htm* (free registration required to download full paper), January 24, 2011.

Case Two

Serving Up Football

When you visit the Web site of the Dallas Cowboys professional football team, you have a good chance of seeing a banner ad for the team's stadium with the headline "It's a data center with 80,000 seats." Calling a stadium a data center may be an exaggeration, but today's professional sports arenas aren't just seats around a grass field. For at least a few hours a week, they are in effect good-sized cities. It would be impossible to manage such stadiums without technology.

Professional sports teams are also businesses. Making a profit is close to the top of their agendas. Costs must be controlled, expenditures minimized. These needs apply to computer hardware as much as it does anywhere else in the Cowboys organization.

Therefore, when the team designed its new stadium for the 2009–10 NFL season, technology was part of the picture. As well as handling stadium operations, the customer experience (such as ticketing, running the huge screens along the sidelines and in the end zone, and operating 2,900 other screens in the stadium) and other businesses of Cowboys owner Jerry Jones in over 90 locations, the infrastructure had to handle growth for 15 to 20 years. "At our previous home, 30-year old Texas Stadium, all you really did was turn on the lights and bring out the football," says Bill Haggard, director of enterprise infrastructure for the Cowboys. "This is different."

"If we had gone all physical with the servers, we would have in the neighborhood of 500 physical servers," Haggard observes. Aside from the cost of the servers and the space they'd occupy, they would have cost about $2.2 million per year just in power and cooling. That expense was too high for the organization. By using two technologies, blade servers and virtualization, the Cowboys reduced these numbers: from 500 to 130 servers and from $2.2 million to $365,000 for power and cooling.

First, the team used virtualization to reduce computing costs. Virtualization let the Cowboys operate one server as if it were several individual ones. For example, the team's concession stand application required each stand to have its own server. Using virtualization, the Cowboys could create the 212 virtual servers the team needed by using only 16 physical ones. Applied across the board, that approach reduced the number of required servers from 500 to 130.

However, a 74 percent reduction in server count doesn't explain an 83 percent drop in power and cooling requirements. Using blade servers accounts for the rest of the energy saving. As you read, blade technology lets many servers share common components—power supplies, cooling fans, network connections, and more. Using blade technology reduced the power consumption of each server by about a third and cut the cooling requirements to remove that power by a similar amount. Putting it all together, the team estimates savings at over $8.2 million in total cost of ownership (TCO) over five years.

Discussion Questions

1. Consider Jerry Jones's other 30+ businesses in other locations. Chances are they do most of their computing during working hours, while football games are usually scheduled for evenings and weekends. How does this schedule make virtualization more desirable than it might be otherwise?
2. One clear benefit of the combination of virtualization and blade servers to the Dallas Cowboys is the financial savings. What other benefits did they obtain by using these technologies?

Critical Thinking Questions

1. The Cowboys estimate that the team will save a total of $8.2 million over five years by using the approaches outlined in this case rather than by using separate servers. What do you think will happen at the end of this period?
2. Can you think of any reasons why a company would *not* use blade servers? Virtualization?

SOURCES: Dallas Cowboys Web site, *www.dallascowboys.com*, accessed January 2, 2012; Dallas Cowboys stadium fact sheet (no date), downloaded January 2, 2012, from *http://stadium.dallascowboys.com/assets/pdf/mediaArchitectureFactSheet.pdf*; Hewlett-Packard, "Dallas Cowboys Generate 30% More Revenue with New Stadium and HP Solutions," *www.convergedinfrastructure.com/document/1319477924_337* (free registration required), October 27, 2011.

Questions for Web Case

See the Web site for this book to read about the Altitude Online case for this chapter. Following are questions concerning this Web case.

Altitude Online: Choosing Hardware

Discussion Questions

1. How might Altitude Online determine what new hardware devices it requires to support the service that its employees use?
2. How will Altitude Online determine the computing power and storage requirements of the new system?

Critical Thinking Questions

1. What should Altitude Online do with its old computer hardware as it is replaced with new hardware?
2. Why do you think Altitude Online decided to phase in new desktop computers, but replace mobile devices all at once?

NOTES

Sources for the opening vignette: King, L., "Deutsche Bank Completes Cloud Computing Overhaul," *Computerworld UK*, *www.itworld.com/it-managementstrategy/229793/deutsche-bank-completes-cloud-computing-overhaul*, December 2, 2011; McLaurin, A., "Identity Management in the New Hybrid Cloud World," Deutsche Bank's entry in the Open Data Center Alliance 2011 Conquering the Cloud Challenge, downloaded December 19, 2011, from *www.opendatacenteralliance.org/contest;* Morgan, G., "Deutsche Bank Lifts the Hood on Cloud Transition," *Computing*, *www.computing.co.uk/ctg/news/2128892/deutsche-bank-lifts-hood-cloud-transition*, November 30, 2011.

1. King, Ian and Bass, Dina, "Dell Loses Orders as Facebook Do-It-Yourself Servers Gain: Tech," *Bloomberg News*, *www.bloomberg.com/news/2011-09-12/dell-loses-orders-as-facebook-do-it-yourself-servers-gain-tech.html*, September 12, 2011.
2. Staff, "U.S. Air Force Chooses HP as Key Technology Provider," *Market Watch*, August 29, 2011.
3. Staff, "Convenience Store Chain Chooses IBM to Collaborate in the Cloud," IBM Press Release, August 16, 2011.
4. Arthur, Charles, "ARM Chip Offers Cheaper Smartphones with Longer Battery Life by 2013," *The Guardian*, October 20, 2011.
5. D'Altorio, Tony, "Intel's 3D Chip Challenge," *www.investmentu.com/2011/May/intel-chip-challenges-arm.html*, May 12, 2011.
6. Seubert, Curtis, "How Many Bytes Is an Exabyte?" *www.ehow.com/about_6370860_many-bytes-exabyte_.html*, accessed October 21, 2011.
7. Lee, Kevin, "IBM's Next-Gen Memory Is 100 Times Faster Than Flash," *PC World*, June 30, 2011.
8. Shah, Agam, "AMD's First Eight-Core Desktop Processors Detailed," *PC World*, October 6, 2011.
9. Bronstein, Benjamin, et al, "Using a Multicore Processor for Rover Autonomous Science," IEEEAC paper, no. 1104, version 1, *http://ase.jpl.nasa.gov/public/papers/bornstein_ieeeaero2011_using.pdf*, January 10, 2011.
10. Chevron Web site, "Seismic Imaging," *www.chevron.com/deliveringenergy/oil/seismicimaging*, June 2011.
11. European Organization for Nuclear Research Web site, "The Name CERN," *http://public.web.cern.ch/public/en/About/Name-en.html*, accessed October 21, 2011.
12. The Best Physics Videos Web site, "What's New @ CERN?" *http://bestphysicsvideos.blogspot.com/2011/12/whats-new-cern-n3-grid-computing.html*, December 5, 2011.
13. Cariaga, Vance, "Teradata Helps Corporations Save Vast Amounts of Data," *Investor's Business Daily*, October 21, 2011.
14. Kerschberg, Ben, "Names to Know in Big Data: Hitachi Data Systems," *Forbes*, October 25, 2011.
15. Jellinek, Dan, "India Builds World's Largest Biometric ID Database," *www.ukauthority.com/Headlines/tabid/36/NewsArticle/tabid/64/Default.aspx?id=3371*, accessed October 17, 2011.
16. Mims, Christopher, "And the Longest Running Digital Storage Media is ...," *www.technologyreview.com/blog/mimssbits/26990*, July 13, 2011.
17. Ibid.
18. Simonite, Tom, "IBM Builds Biggest Data Drive Ever," *Technology Review*, August 25, 2011.
19. ATTO Technology Web site, "Avatar 'Sounds Off' Using ATTO's Technology," *www.attotech.com/pdfs/Advanced-Audio Rentals.pdf*, accessed October 27, 2011.
20. IBM Success Stories, "Baldor Electric Opts for IBM ProtecTIER and IBM XIV," *www-01.ibm.com/software/success/cssdb.nsf/CS/DLAS-8JYRFM?OpenDocument&Site=corp&cty=en_us*, accessed July 26, 2011.
21. Millenniata Web site, "Millenniata Partners with Hitachi-LG Data Storages," *http://millenniata.com/2011/08/24/millenniata-partners-with-hitachi-lg-data-storage*, August 15, 2011.
22. Holographics Projectors Web site, Holographic Video Disk page, *www.holographicprojectors.com.au/hvd*, October 27, 2011.
23. Staff, "Vaillant Accelerates SAP Environment through the Use of SSD Technology," *www.ramsan.com/resources/successStories/91*, October 10, 2011.
24. SanDisk Web site, "SanDisk Ultra Backup USB Flash Drive," *www.sandisk.com/products/usb-flash-drives/sandisk-ultra-backup-usb-flash-drive*, accessed October 28, 2011.
25. Kingston Technology Web site, "SSD Performance and Productivity Revitalize Notebook Assets," *http://media.kingston.com/images/branded/MKF_357_Qualcomm_Case_study.pdf*, accessed December 19, 2011.
26. Staff, "Vaillant Accelerates SAP Environment through the Use of SSD Technology," *www.ramsan.com/resources/successStories/91*, October 10, 2011.
27. HP Web site, "High-availability IT for Swiss Holiday Destination, Davos Klosters, HP Customer Case Study," h20195.*www2.hp.com/v2/GetPDF.aspx/4AA3-4360EEW.pdf*, accessed October 29, 2011.
28. NorthgateArinso Web site, "Delivering HR Excellence," *http://ngahr.com*, accessed October 30, 2011.
29. IBM Success Stories, "NorthgateArinso Builds Compact, Energy-efficient and Scalable Solution,"

www-01.ibm.com/software/success/cssdb.nsf/CS/STRD-8GXE3S?OpenDocument&Site=corp&cty=en_us, May 18, 2011.

30. Mozy Web site, "I Found My Stolen Laptop," *http://mozy.com/home/reviews*, accessed October 30, 2011.
31. Spangler, Todd, "Time Warner Cable Adds Speech Recognition to Customer Service," *www.multichannel.com/article/464064-Time_Warner_Cable_Adds_Speech_Recognition_To_Customer_Service_Line.php*, February 16, 2011.
32. Prann, Elizabeth, "Police Officers Find Tiny Pocket Cams Are Silent Partners," *www.foxnews.com/scitech/2011/07/04/police-officers-find-tiny-pocket-cams-are-silent-partners*, July 4, 2011.
33. Silicon Valley Bank Web site, "About SVB Financial Group," *www.svb.com/Company/About-SVB-Financial-Group*, accessed October 29, 2011.
34. PSIGEN Software Web site, "Silicon Valley Bank," *www.psigen.com/industry_solutions/banking_finanical_services_scanning_capture_imaging_software_industry_solution.aspx*, accessed October 29, 2011.
35. Perkins, Ed, "Travel-Friendly Chip-and-Pin Credit Cards Coming to U.S., *USA Today*, April 4, 2011.
36. Herron, Janna, "U.S. Credit Cards Add Chip and PIN Security," *www.foxbusiness.com/personal-finance/2011/10/20/us-credit-cards-add-chip-and-pin-security*, October 21, 2011.
37. Perkins, Ed, "Travel-Friendly Chip-and-Pin Credit Cards Coming to U.S.," *USA Today*, April 21, 2011.
38. Fitzgerald, Kate, "MasterCard Testing Contactless Payments for In-Flight Purchases," *www.paymentssource.com/news/mastercard-testing-contactless-payments-in-flight-guestlogix-3008292-1.html?zkPrintable=1&nopagination=1*, October 25, 2011.
39. First Data Web site, "A First Data Customer Success Story: First Data's POS Restaurant Solution Helps La Paz Restaurant Improve Operations and Offer Gift Cards," *www.firstdata.com/en_us/insights/lapaz-case-study.html?cat=Success+Stories&tag=POS+Payments+%26+Customer+Contact*, accessed December 20, 2011.
40. Mastin, Michelle, "Square vs. Intuit Go Payment: Mobile Credit Card Systems Compared," *PC World*, *www.pcworld.com/businesscenter/article/239250/square_vs_intuit_gopayment_mobile_credit_card_systems_compared.html#tk.mod_rel*, September 6, 2011.
41. SMART Customer Stories, "Audi AG Advancement through Technology at Audi AG," *www.smarttech.com/us/Customer%20Stories/Browse%20Stories?q=1&business=1&allstories=1*, accessed December 27, 2011.
42. Barcode.com Web site, "Mobio Uses Barcode to Enhance Game Day," *http://barcode.com/mobio-uses-barcodes-to-enhance-game-day.html*, accessed October 31, 2011.
43. Canadian Sheep Federation, "Sheep Industry Continues toward Mandatory RFID Tags," *www.seregonmap.com/repository/wool/scm/uploads/CSF-Mandatory-RFID-Tags-English.pdf*, June 20, 2011.
44. Kondolojy, Amanda, "LCD vs Plasma Monitors," *www.ehow.com/about_4778386_lcd-vs-plasma-monitors.html*, accessed November 1, 2011.
45. Grubb, Ben, "Nokia Demos Flexible Mobile and Smudge-Free Screen Technology," *www.theage.com.au/digital-life/mobiles/nokia-demos-flexible-mobile-and-smudgefree-screen-technology-20111028-1mmhe.html*, October 28, 2011.
46. Microsoft Case Studies, "Royal Bank of Canada Delights Customers with Innovative Microsoft Surface Experience," *www.microsoft.com/casestudies/Microsoft-Surface/Royal-Bank-of-Canada/Royal-Bank-of-Canada-delights-customers-with-innovative-Microsoft-Surface-experience/4000011029*, August 25, 2011.
47. HP Web site, "HP ePrint Enterprise Mobile Printing," *http://h71028.www7.hp.com/enterprise/us/en/ipg/HPe-print-solution.html*, accessed November 6, 2011.
48. PrinterOn Web site, "Mobile Printing Solutions," *www.printeron.com*, accessed November 6, 2011.
49. Kozlowski, Michael, "Inspiring Technologies in e-Readers during 2011," *http://goodereader.com/blog/electronic-readers/inspiring-technologies-in-e-readers-during-2011*, November 7, 2011.
50. IBM Web site, "IBM Unveils Cognitive Computing Chips," *www-03.ibm.com/press/us/en/pressrelease/35251.wss*, August 18, 2011.
51. Lomas, Natasha, "Dual Core Smartphones: The Next Mobile Arms Race," *www.silicon.com/technology/mobile/2011/01/12/dual-core-smartphones-the-next-mobile-arms-race*, January 11, 2011.
52. Ji-hyun, Cho, "Samsung, Google Gain in Apple Patent Fight," *Asia News Network*, *www.asianewsnet.net/home/news.php?id=23574*, July 11, 2011.
53. Chen, Brian X., "Samsung Wins a Round in Patent Fight with Apple," *New York Times*, December 9, 2011.
54. Hogg, Chris, "Study: Smartphones and Tablets to Outsell Computers in 2011," *www.futureofmediaevents.com/2011/01/19/study-smartphones-and-tablets-to-out-sell-computers-in-2011*, January 18, 2011.
55. Fuquay, Jim, "Doctors Using Smartphones, Tablets to Access Medical Data," *Star-Telegram*, July 6, 2011.
56. MedHealthWorld Web site, "ePocrates Survey Confirms Expanded Role for Smartphones and Electronic Records," *www.medhealthworld.com/?p=814*, February 6, 2011.
57. AccessReflex Web site, "Stay Connected with Your MS Access," *http://access.mobilereflex.com/features.html*, accessed November 7, 2011.
58. EBSCO Publishing Web site, *www.ebscohost.com/schools/mobile-access*, accessed November 8, 2011.
59. Mack, Julie, "Bloomingdale High School Students Each Given Netbook Computer, Internet Access," *Kalamazoo Gazette*, October 12, 2011.
60. Strohmeyer, Robert and Perenson, Melissa, J., "Why Your Next PC Will Be A Tablet," *PC World*, January 2011.
61. Wenzel, Elsa, "Slates Enable a Surgery Practice to Improve Patient Care," *PC World*, October 1, 2011.
62. Wenzel, Elsa, "Tablets Help a Business Stand Out, Improve Client Care," *PC World*, July 24, 2011.
63. Smith, Dave, "The Birth of the iRestaurant," *Inc.*, August 16, 2011.
64. Strohmeyer, Robert and Perenson, Melissa, J., "Why Your Next PC Will Be A Tablet," *PC World*, January 2011.
65. Wyse Web site, "Wyse Thin Clients," *http://wyse.com/products/hardware/thinclients/index.asp*, October 19, 2011.
66. Deboosere, Lien, et al, "Cloud-based Desktop Services for Thin Clients," *http://doi.ieeecomputersociety.org/10.1109/MIC.2011.139*, October 3, 2011.

67. Wyse Web site, "Wyse TV Jewelry Case Study," *www.wyse.com/resources/casestudies/CS-JTV-register.asp*, October 19, 2011.
68. Wiesen, G., "What Is a Nettop," *www.wisegeek.com/what-is-a-nettop.htm*, accessed October 19, 2011.
69. Dell Web site, "Zooming in on Time-Savings," *http://i.dell.com/sites/content/corporate/case-studies/en/Documents/2010-sekotec-10008226.pdf*, accessed April 1, 2010.
70. Carey, Peter, "Server Farms Sprouting in Silicon Valley," *San Jose Mercury News*, October 6, 2011.
71. Parrish, Kevin, "Intel to Ship 10-Core CPUs in the First Half of 2011," *IDG New Service*, February 11, 2011.
72. Morgan, Timothy Prickett, "AMD Shoots Lower with Opteron 3000 Chips," *The Register*, *www.theregister.co.uk/2011/11/14/amd_opteron_3000_server_chip*, November 14, 2011.
73. Microsoft Case Studies, "Gashora Girls Academy," *www.microsoft.com/casestudies/Case_Study_Detail.aspx?CaseStudyID=4000010078*, May 26, 2011.
74. North, Jeffrey, "The Total Economic Impact of Microsoft Windows MultiPoint Server 2011," Forrester Consulting Project, April 2011.
75. PRWeb, "Companies on Tight Budgets Thrive with Low Prices on EZZI.net Dedicated Servers, Dedicated Cloud Servers, and Virtual Private Servers," *http://news.yahoo.com/companies-tight-budgets-thrive-low-prices-ezzi-net-110250858.html*, October 20, 2011, and EZZI.net Web site, "Virtual Private Servers," *www.ezzi.net/vps.php*, accessed October 21, 2011.
76. IBM Success Stories, "Top German Bank NORD/LB Improves Business and Technical Efficiencies Using IBM and SAP," *www-01.ibm.com/software/success/cssdb.nsf/CS/STRD-8M4GN6?OpenDocument&Site=default&cty=en_us*, September 27, 2011.
77. Taft, Darryl K., "IBM Mainframe Replaces HP, Oracle Systems for Payment Solutions," *eWeek*, April 19, 2011.
78. ChemInfo Web site, "Supercomputer Center Partners with P&G on Simulation," *www.chem.info/News/2011/02/Software-Supercomputer-Center-Partners-with-PG-on-Simulation*, February 28, 2011.
79. Staff, "Japanese Supercomputer Is the Most Powerful," *Thaindian News*, June 20, 2011.
80. Hesseldahl, Arik, "Nvidia Chips to Power World's Most Powerful Computer," *http://allthingsd.com/?p=130810&ak_action=printable*, October 11, 2011.
81. Staff, "Big Rigs Go Aerodynamic, Thanks to Supercomputers," *TechNewsDaily*, February 9, 2011.
82. IBM Web site, "IBM's Watson, What is Watson," *www-03.ibm.com/innovation/us/watson/what-is-watson/index.html*, accessed December 23, 2011.
83. Technology staff, "IBM's Watson Supercomputer to Give Instant Medical Diagnosis," *Los Angeles Times*, September 12, 2011.
84. Hesseldahl, Arik, "Nvidia Chips to Power World's Most Powerful Computer," *http://allthingsd.com/?p=130810&ak_action=printable*, October 11, 2011.
85. Parsons, June and Oja, Dan, *New Perspectives on Computer Concepts 2011*, Boston: Cengage Learning, 2011, p. 106.
86. U.S. Department of Energy Web site, "Energy Savers–When to Turn off Your Computer," *www.energysavers.gov/your_home/appliances/index.cfm/mytopic=10070*, accessed November 8, 2011.
87. Staff, "Commission Decision of September 8, 2011," *Official Journal of the European Union*, September 10, 2011.
88. EPEAT Web site, "Welcome to EPEAT," *www.epeat.net*, accessed November 8, 2011.

4 Software: Systems and Application Software

Principles	Learning Objectives
• Systems and application software are critical in helping individuals and organizations achieve their goals.	• Identify and briefly describe the functions of the two basic kinds of software. • Outline the role of the operating system and identify the features of several popular operating systems.
• Organizations use off-the-shelf application software for common business needs and proprietary application software to meet unique business needs and provide a competitive advantage.	• Discuss how application software can support personal, workgroup, and enterprise business objectives. • Identify three basic approaches to developing application software and discuss the pros and cons of each.
• Organizations should choose programming languages with functional characteristics that are appropriate for the task at hand and well suited to the skills and experience of the programming staff.	• Outline the overall evolution and importance of programming languages and clearly differentiate among the generations of programming languages.
• The software industry continues to undergo constant change; users need to be aware of recent trends and issues to be effective in their business and personal life.	• Identify several key software issues and trends that have an impact on organizations and individuals.

Information Systems in the Global Economy

EQUITAS MICRO FINANCE, INDIA

Microfinance Needs Software

India may be the most entrepreneurial country in the world. However, since most of its new enterprises are small, business owners often don't qualify for conventional banking services. The *microfinance* system has grown up to meet their needs. According to Consultative Group to Assist the Poor (CGAP), which provides microfinance information and services to "governments, financial service providers, donors and investors," microfinance "offers poor people access to basic financial services such as loans, savings, money transfer services and insurance." Unfortunately, as CGAP points out, "the administrative cost of making tiny loans is much higher in percentage terms than the c-ost of making a large loan." Besides, "inefficient operations can make them higher than necessary."

Efficiency is a concern at Equitas Micro Finance India, Pvt. Ltd., perhaps the fastest-growing start-up microfinance institution in the world. "We've been growing very quickly now for the last two years," says chief technology officer Hariharan Mahalingam. "This pace of growth poses challenges for the IT department, ... with new branches opening, as many as nearly 30 per month." To keep pace with the growth, Equitas purchased new hardware, but perhaps importantly, the company also purchased new application software to run on that hardware. After considering its options, Equitas selected the T24 for Microfinance and Community Banking (T24 MCB) application package from Swiss software firm Temenos. T24 MCB is a subset of Temenos's full T24 application, preconfigured for small retail financial institutions such as microfinance firms. Equitas runs T24 MCB on IBM pSeries servers under the UNIX-based AIX operating system.

T24 MCB automates most processes so that Equitas can now scale up without extra manpower. For example, when a loan is taken out, the loan form is completed at the branch, and the forms are delivered to Chennai for processing via automatic scanning and optical-mark reading. Manual data entry is required for only about 20 percent of the form. Without T24 MCB, form processing would require more people. Equitas has not needed to increase its back office staff, despite the customer base growing from 500,000 in July 2009 to 1.3 million at the end of November 2011—passing one million in May 2010, two years and five months after the company's founding.

Equitas branches cover a large area. When branch managers and collection officers record payments toward outstanding loans and other information, they send the data from their mobile devices to headquarters in Chennai via SMS. In the other direction, Chennai sends reports to the branch managers, such as information about payments due. Currently, Equitas sends details of 132,000 customers every day.

To sum up, T24 MCB enables Equitas Micro Finance to support rapid expansion, to improve its operational efficiency, and to reduce transaction processing costs. These benefits are necessities in its world, although not only there as these are necessities for just about any business. It is *application software* that makes these benefits possible.

As you read this chapter, consider the following:

- Why is selecting the right software more important to Equitas than choosing its hardware?
- If packaged applications such as T24 MCB did not exist, would it have been practical for Equitas to automate its processes? Why or why not?

WHY LEARN ABOUT SYSTEMS AND APPLICATION SOFTWARE?

© Andrey Burmakin/Shutterstock

Software is indispensable for any computer system and the people using it. In this chapter, you will learn about systems and application software. Without systems software, computers would not be able to accept data input from a keyboard, process data, or display results. Application software is one of the keys to helping you achieve your career goals. Sales representatives use software on their smart phones and tablet computers to enter sales orders and help their customers get what they want. Stock and bond traders use software to make split-second decisions involving millions of dollars. Scientists use software to analyze the threat of climate change. Regardless of your job, you most likely will use software to help you advance in your career and earn higher wages. You can also use software to help you prepare your personal income taxes, keep a budget, and keep in contact with friends and family online. Software can truly advance your career and enrich your life. We begin with an overview of software.

Software has a profound impact on individuals and organizations. It can make the difference between profits and losses and between financial health and bankruptcy. As Figure 4.1 shows, companies recognize this impact, spending more on software than on computer hardware.

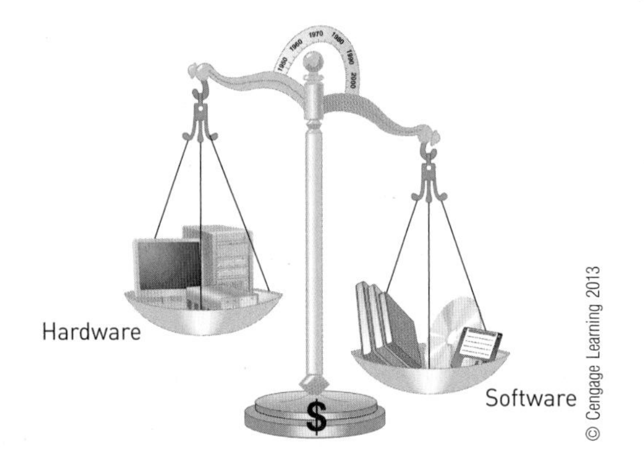

FIGURE 4.1
Importance of software in business
Since the 1950s, businesses have greatly increased their expenditures on software compared with hardware.

© Cengage Learning 2013

AN OVERVIEW OF SOFTWARE

computer programs: Sequences of instructions for the computer.

documentation: Text that describes a program's functions to help the user operate the computer system.

As you learned in Chapter 1, software consists of computer programs that control the workings of computer hardware. **Computer programs** are sequences of instructions for the computer. **Documentation** describes the program functions to help the user operate the computer system. Some documentation is given on-screen or online, while other forms appear in external resources, such as printed manuals. Software is a growing and dynamic industry. In 2011, for example, China's software industry grew almost 30 percent.[1] Some believe that software development and sales have more growth potential than hardware.[2] According to an early Internet pioneer and board of directors member of Hewlett-Packard, "This week, Hewlett-Packard (where I am on the board) announced that it is exploring jettisoning its struggling PC business in favor of investing more heavily in software, where it sees better potential for growth."

Systems Software

Systems software is the set of programs that coordinates the activities and functions of the hardware and other programs throughout the computer system. Each type of systems software is designed for a specific CPU and class of hardware. The combination of a hardware configuration and systems software is known as a computer system platform.

© iStockphoto/Avatar_023

Application software has the greatest potential to affect processes that add value to a business because it is designed for specific organizational activities and functions.

Application Software

Application software consists of programs that help users solve particular computing problems.[3] An architectural firm in Boise, Idaho, for example, used ProjectDox software to streamline the paperwork required to get approval and permits for building projects.[4] According to one architect from the firm, "The nice thing is that most files, whether it be a PDF or Word document, can be dropped into different folders online and sent. It's not a big deal like it was before." Software from Amcom allows companies such as Eddie Bauer to provide the exact location of someone who calls from an Eddie Bauer retail location to emergency 911 call centers.[5] According to a company technical analyst, "We take communications and security very seriously. The Amcom system is a perfect communications safety net in case someone dials 911 and can't explain where they are."

In most cases, application software resides on the computer's hard disk before it is brought into the computer's memory and run. Application software can also be stored on CDs, DVDs, and even USB flash drives. An increasing amount of application software is available on the Web. Sometimes referred to as a *rich Internet application* (*RIA*), a Web-delivered application combines hardware resources of the Web server and the PC to deliver valuable software services through a Web browser interface. Before a person, group, or enterprise decides on the best approach for acquiring application software, they should analyze their goals and needs carefully.

Supporting Individual, Group, and Organizational Goals

Every organization relies on the contributions of people, groups, and the entire enterprise to achieve its business objectives. One useful way of classifying the many potential uses of information systems is to identify the scope of the problems and opportunities that an organization addresses. This scope is called the *sphere of influence*. For most companies, the spheres of influence are personal, workgroup, and enterprise. Table 4.1 shows how various kinds of software support these three spheres.

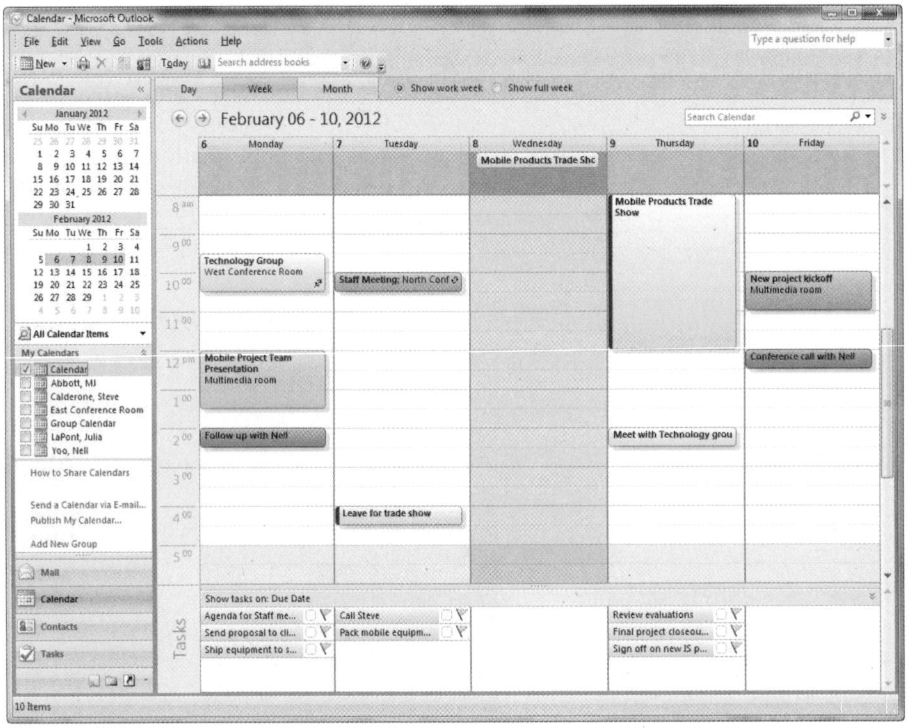

Microsoft Outlook is an application that workgroups can use to schedule meetings and coordinate activities.

TABLE 4.1 Software supporting individuals, workgroups, and enterprises

Software	Personal	Workgroup	Enterprise
Systems software	Smartphone, tablet computer, personal computer, and workstation operating systems	Network operating systems	Server and mainframe operating systems
Application software	Word processing, spreadsheet, database, and graphics	Electronic mail, group scheduling, shared work, and collaboration	General ledger, order entry, payroll, and human resources

personal sphere of influence: The sphere of influence that serves the needs of an individual user.

personal productivity software: The software that enables users to improve their personal effectiveness, increasing the amount of work and quality of work they can do.

workgroup: Two or more people who work together to achieve a common goal.

workgroup sphere of influence: The sphere of influence that helps workgroup members attain their common goals.

enterprise sphere of influence: The sphere of influence that serves the needs of the firm in its interaction with its environment.

Information systems that operate within the **personal sphere of influence** serve the needs of individual users. These information systems help users improve their personal effectiveness, increasing the amount and quality of work they can do. Such software is often called **personal productivity software**. For example, MindManager software from Mindjet provides tools to help people diagram complex ideas and projects using an intuitive graphic interface.[6]

When two or more people work together to achieve a common goal, they form a **workgroup**. A workgroup might be a large formal, permanent organizational entity, such as a section or department, or a temporary group formed to complete a specific project. An information system in the **workgroup sphere of influence** helps workgroup members attain their common goals. Often, software designed for the personal sphere of influence can extend into the workgroup sphere. For example, people can use online calendar software such as Google Calendar to store personal appointments but also to schedule meetings with others.

Information systems that operate within the **enterprise sphere of influence** support the firm in its interaction with its environment, which includes customers, suppliers, shareholders, competitors, special-interest groups, the financial community, and government agencies. This means the enterprise sphere of influence includes business partners, such as suppliers that provide raw materials; retail companies that store and sell a company's products; and shipping companies that transport raw materials to the plant and finished goods to retail outlets. For example, many enterprises use IBM Cognos software as a centralized Web-based system where employees, partners, and stakeholders can report and analyze corporate financial data.[7]

SYSTEMS SOFTWARE

Controlling the operations of computer hardware is one of the most critical functions of systems software. Systems software also supports the application programs' problem-solving capabilities. Types of systems software include operating systems, utility programs, and middleware.

Operating Systems

operating system (OS): A set of computer programs that controls the computer hardware and acts as an interface with applications.

An operating system (OS) is a set of programs that controls the computer hardware and acts as an interface with applications (see Figure 4.2). Operating systems can control one or more computers, or they can allow multiple users to interact with one computer. The various combinations of OSs, computers, and users include the following:

- **Single computer with a single user.** This system is commonly used in a personal computer, tablet computer, or a smart phone that supports one user at a time. Examples of OSs for this setup include Microsoft Windows, Mac OS X, and Google Android.
- **Single computer with multiple simultaneous users.** This system is typical of larger server or mainframe computers that can support hundreds or thousands of people, all using the computer at the same time. Examples of OSs that support this kind of system include UNIX, z/OS, and HP UX.
- **Multiple computers with multiple users.** This type of system is typical of a network of computers, such as a home network with several computers attached or a large computer network with hundreds of computers attached supporting many users, sometimes located around the world. Most PC operating systems double as network operating systems. Network server OSs include Red Hat Linux, Windows Server, and Mac OS X Server.
- **Special-purpose computers.** This type of system is typical of a number of computers with specialized functions, such as those that control sophisticated military aircraft, space shuttles, digital cameras, or home appliances. Examples of OSs for these purposes include Windows Embedded, Symbian, and some distributions of Linux.

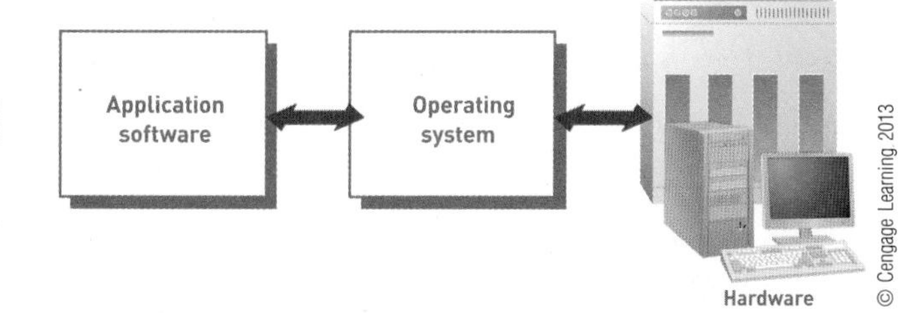

FIGURE 4.2
Role of operating systems
The role of the operating system is to act as an interface between application software and hardware.

The OS, which plays a central role in the functioning of the complete computer system, is usually stored on disk on general-purpose computers and in solid state memory on special-purpose computers such as cell phones and smartphones. After you start, or "boot up," a computer system, portions of the OS are transferred to memory as the system needs them. This process can take anywhere from a split second on a smartphone, to a few minutes on a desktop PC, to hours on a large mainframe or distributed computer systems. OS developers are continuously working to shorten the time required to boot devices from being shut down and wake devices from sleep mode.

You can also boot a computer from a CD, DVD, or even a USB flash drive. A storage device that contains some or all of the OS is often called a *rescue disk* because you can use it to start the computer if you have problems with the primary hard disk.

The set of programs that make up the OS performs a variety of activities, including the following:

- Performing common computer hardware functions
- Providing a user interface and input/output management
- Providing a degree of hardware independence
- Managing system memory
- Managing processing tasks
- Sometimes providing networking capability
- Controlling access to system resources
- Managing files

kernel: The heart of the operating system and controls its most critical processes.

The **kernel**, as its name suggests, is the heart of the OS and controls its most critical processes. The kernel ties all of the OS components together and regulates other programs.

Common Hardware Functions

All applications must perform certain hardware-related tasks, such as the following:

- Get input from the keyboard or another input device
- Retrieve data from disks
- Store data on disks
- Display information on a monitor or printer

Each of these tasks requires a detailed set of instructions. The OS converts a basic request into the instructions that the hardware requires. In effect, the OS acts as an intermediary between the application and the hardware. The OS uses special software provided by device manufacturers, called device drivers, to communicate with and control a device. Device drivers are installed when a device is initially connected to the computer system.

User Interface and Input/Output Management

user interface: The element of the operating system that allows people to access and interact with the computer system.

command-based user interface: A user interface that requires you to give text commands to the computer to perform basic activities.

One of the most important functions of any OS is providing a **user interface**, which allows people to access and interact with the computer system. The first user interfaces for mainframe and personal computer systems were command based. A **command-based user interface** requires you to give text commands to the computer to perform basic activities. For example, the command ERASE 00TAXRTN would cause the computer to erase a file named 00TAXRTN. RENAME and COPY are other examples of commands used to rename files and copy files from one location to another. Today's systems engineers and administrators often use a command-based user interface to control the low-level functioning of computer systems. Most modern OSs (including popular graphical user interfaces such as Windows) provide a way to interact with the system through a command line (see Figure 4.3).

graphical user interface (GUI): An interface that displays pictures (icons) and menus that people use to send commands to the computer system.

A **graphical user interface (GUI)** displays pictures (called *icons*) and menus that people use to send commands to the computer system. GUIs are more intuitive to use because they anticipate the user's needs and provide easy to recognize options. Microsoft Windows is a popular GUI. As the name

FIGURE **4.3**

Command-based and graphical user interfaces

A Windows file system viewed with a GUI (a) and from the command prompt (b).

suggests, Windows is based on the use of a window, or a portion of the display screen dedicated to a specific application. The screen can display several windows at once.

While GUIs have traditionally been accessed using a keyboard and mouse, more recent technologies allow people to use touch screens and spoken commands. Today's mobile devices and some PCs, for example, use a touch user interface also called a *natural user interface (NUI)* or multi-touch interface by some. Apple's Mountain Lion operating system, for example, uses a touch user interface to allow people to control the personal computer by touching the screen.[8] Speech recognition is also available with some operating systems.[9] By speaking into a microphone, the operating system commands and controls the computer system. Sight interfaces use a camera on the computer to determine where a person is looking on the screen and performs the appropriate command or operation. Some companies are also experimenting with sensors attached to the human brain (brain interfaces) that can detect brain waves and control a computer as a result. Sight and brain interfaces can be very helpful to disabled individuals.

Hardware Independence

application program interface (API): Tools software developers use to build application software without needing to understand the inner workings of the OS and hardware.

Software applications are designed to run on a particular operating system by using the operating system's **application program interface (API)**, which provides software developers with tools they use to build application software without needing to understand the inner workings of the OS and hardware (see Figure 4.4). Being able to develop software without concern for the specific underlying hardware is referred to as *hardware independence*. When new hardware technologies are introduced, the operating system is required to adjust to address those changes, not the application software that runs on the operating system.

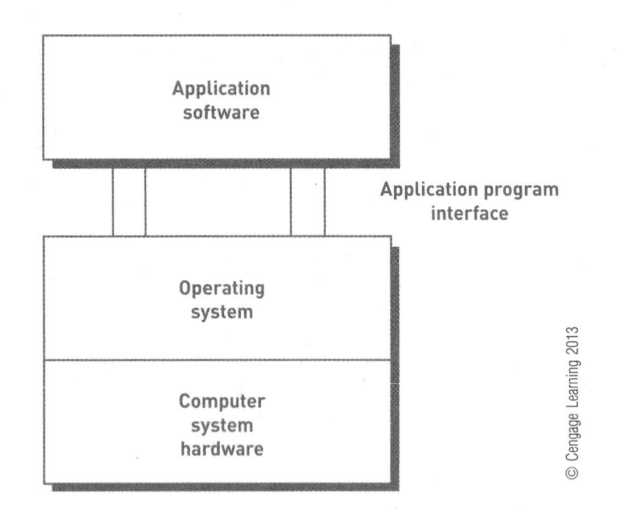

FIGURE 4.4
Application program interface (API)
The API links application software to the operating system, providing hardware independence for software developers.

Memory Management

The OS also controls how memory is accessed, maximizing the use of available memory and storage to provide optimum efficiency. The memory-management feature of many OSs allows the computer to execute program instructions effectively and to speed processing. One way to increase the performance of an old computer is to upgrade to a newer OS and increase the amount of memory.

Most OSs support *virtual memory*, which allocates space on the hard disk to supplement the immediate, functional memory capacity of RAM. Virtual memory works by swapping programs or parts of programs between memory and one or more disk devices—a concept called paging. This procedure reduces CPU idle time and increases the number of jobs that can run in a given time span.

Processing Tasks

The task-management features of today's OSs manage all processing activities. Task management allocates computer resources to make the best use of each system's assets. Task-management software lets one user run several programs or tasks at the same time (multitasking) and allows several users to use the same computer at the same time (time sharing).

An OS with multitasking capabilities allows a user to run more than one application at the same time. While you're working in the *foreground* in one program, one or more other applications can be churning away, unseen, in the *background*. Background activities include tasks such as sorting a database, printing a document, or performing other lengthy operations that otherwise would monopolize your computer and leave you staring at the screen unable to perform other work. Multitasking can save users a considerable amount of time and effort. *Time sharing* allows more than one person to use a computer system at the same time. For example, 15 customer service representatives might enter sales data into a computer system for a mail-order company at the same time. The ability of the computer to handle an increasing number of concurrent users smoothly is called *scalability*. This feature is critical for systems expected to handle a large and possibly fluctuating number of users, such as a mainframe computer or a Web server.

Networking Capability

Most operating systems include networking capabilities so that computers can join together in a network to send and receive data and share computing resources. Operating systems for larger server computers are designed specifically for computer networking environments.

Access to System Resources and Security

Because computers often handle sensitive data that can be accessed over networks, the OS needs to provide a high level of security against unauthorized access to the users' data and programs. Typically, the OS establishes a logon procedure that requires users to enter an identification code, such as a user name, and a matching password. Operating systems may also control what system resources a user may access. When a user successfully logs on to the system, the OS restricts access to only portions of the system for which the user has been cleared. The OS records who is using the system and for how long and reports any attempted breaches of security.

File Management

The OS manages files to ensure that files in secondary storage are available when needed and that they are protected from access by unauthorized users. Many computers support multiple users who store files on centrally located disks or tape drives. The OS keeps track of where each file is stored and who can access them.

Current Operating Systems

Today's operating systems incorporate sophisticated features and impressive graphic effects. Table 4.2 classifies a few current OSs by sphere of influence.

TABLE **4.2** Operating systems serving three spheres of influence

Personal	Workgroup	Enterprise
Microsoft Windows	Microsoft Windows Server	Microsoft Windows Server
Mac OS X, Mac OS X iPhone	Mac OS X Server	
Linux	Linux	Linux
Google Android, Chrome OS		
HP webOS		
	UNIX	UNIX
	IBM i and z/OS	IBM i and z/OS
	HP-UX	HP-UX

Microsoft PC Operating Systems

Since a once small company called Microsoft developed PC-DOS and MS-DOS to support the IBM personal computer introduced in the 1980s, personal computer OSs have steadily evolved. PC-DOS and MS-DOS had command-driven interfaces that were difficult to learn and use. MS-DOS gave way to Windows, which opened the PC market to everyday users. Windows evolved through several versions, including Windows 1.01, 2.03, 3.0, and 3.1, Windows 95, 98, and Me, Windows NT, Windows 2000, Windows XP, Windows Vista, Windows 7, and Windows 8.

Windows XP (XP reportedly stands for the positive e*xp*erience that you will have with your personal computer) was released in fall of 2001. In 2007, Microsoft released *Windows Vista* to the public, introducing it as the most secure version of Windows ever. The next version, *Windows 7*, was released in 2009 with improvements and new features. Many analysts classified Windows 7 as "Vista done right." Windows 7 has strong support for touch displays and netbooks, ushering in a new era of mobile computing devices. Windows 7 is available in configurations designed for 32-bit or 64-bit processors. Users running newer computers are advised to install the 64-bit version, if their computers can support it, to experience faster processor performance.[10]

Microsoft Windows 8, available in 2012, offers a number of enhancements, including features for tablet computers.[11] Windows 8 includes a touch interface and many new features for the consumer market.[12] The home screen displays colorful application "tiles" instead of icons.[13] Windows 8 is available for a number of platforms, including smartphones, tablet computers, PCs, and servers.[14] According to one industry analyst, "They're betting the farm on this one. Microsoft's problem is how do they keep the existing customer base with Windows while addressing touch?" Many smartphone and mobile device makers plan to use Microsoft's Windows operating system in their devices.[15] See Figure 4.5.

FIGURE 4.5
Microsoft Windows 8

Apple Computer Operating Systems

In July 2001, Mac OS X was released as an entirely new OS for the Mac based on the UNIX operating system. It included a new user interface, which provided a new visual appearance for users—including luminous and semitransparent elements, such as buttons, scroll bars, windows, and fluid animation to enhance the user's experience.

Since its first release, Apple has upgraded OS X several times. OS X Mountain Lion is Apple's latest operating system.[16] See Figure 4.6. It offers multitouch, full-screen applications, mission control features, and other innovations. It also incorporates many features of Apple's mobile devices into Apple's desktop and laptop computers.[17] Mountain Lion can also automatically save a document every time a change is made to an application, such as a word-processing or spreadsheet application. In one survey, ease of use, the number of available applications, and overall user appeal for Apple's mobile operating system (iOS) received high ratings.[18]

Because Mac OS X runs on Intel processors, Mac users can set up their computer to run both Windows and Mac OS X and select which platform they want to work with when they boot their computer. Such an arrangement is called *dual booting*. While Macs can dual boot into Windows, the opposite is not true. Apple does not allow OS X to be run on any machine other than an Apple. However, Windows PCs can dual boot with Linux and other OSs.

FIGURE **4.6**
Mac OS X Mountain Lion

Linux

Linux is an OS developed by Linus Torvalds in 1991 as a student in Finland. The OS is distributed under the *GNU General Public License*, and its source code is freely available to everyone. It is, therefore, called an *open-source* operating system. This designation doesn't mean, however, that Linux and its assorted distributions are necessarily free—companies and developers can charge money for a distribution as long as the source code remains available. Linux is actually only the kernel of an OS, the part that controls hardware, manages files, separates processes, and so forth.

Several combinations of Linux are available, with various sets of capabilities and applications to form a complete OS. Each of these combinations is called a *distribution* of Linux. Many distributions are available as free downloads.

Linux is available on the Internet and from other sources. Popular versions include Red Hat Linux, OpenSUSE (see Figure 4.7), and Caldera OpenLinux. Several large computer vendors, including IBM, Hewlett-Packard, and Intel, support the Linux operating system. Although Linux is free software, Red Hat had revenues of about $1 billion in 2011 distributing and servicing the software.[19]

FIGURE **4.7**
OpenSUSE operating system

INFORMATION SYSTEMS @ WORK

Linux in Business

If you use a computer as a university business student, odds are that it runs either Microsoft Windows or Apple Mac OS. Few students outside computer science, and even fewer business school computer labs, use any other platform.

That's not the case in business. In May 2012, about 65 percent of 662 million Web servers surveyed use the Linux operating system, described in this chapter, to run a Web server application called Apache. Apache's market share fluctuates but has been above 40 percent since 1997. Linux is popular among businesses in other application areas as well.

Why do businesses use Linux? Their reasons for choosing it vary. For PrintedArt, an online shop that sells limited editions of fine art photography, the reasons involved the availability of open-source applications developed for Linux. President and CEO of PrintedArt, Klaus Sonnenleiter, explains the company's choice of the open-source package Drupal for managing Web content. "Before settling on Drupal, we went through a major evaluation shoot-out between the different CMS [Content Management System] options. After looking at a fairly large number of options, Joomla, Drupal, Alfresco, and Typo3 became the finalists. Drupal came out on top because of its layered API [Application Program Interface] that lets PrintedArt … create their own integrations and modules."

Ubercart, the free open-source e-commerce shopping cart module, is also a core part of the PrintedArt system. "In addition, we use Capsule running as a Google App as our CRM," Sonnenleiter adds. "We also use MailChimp, and we are evaluating Producteev as our project and to do-list manager."

Gompute of Göteborg, Sweden, is larger than the six-person PrintedArt. Gompute operates a cluster of 336 IBM servers to provide high-performance computing on demand for technical and scientific users. Those people use Gompute's computers in fields such as fluid dynamics, stress analysis, and computational chemistry. Linux gives them the ability to run the variety of applications that the company's customers require. These include proprietary applications such as ANSYS for computer-aided engineering and PERMAS for structural analysis for which users must purchase a license and open-source programs such as OpenFOAM for advanced computation, which are free for anyone to use. If none of these applications meet a user's needs, users can write their own programs and then run them on Gompute's advanced hardware.

The Linux groundswell in business is so strong that not even Microsoft is immune. In June 2012, Microsoft announced that its Azure cloud-computing service will let customers run Linux as well as Windows. Linux provides important features in this environment, such as the ability to retain data even after a virtual machine is rebooted. By offering Linux, Microsoft can pursue customers that need data retention and other Linux capabilities.

Discussion Questions

1. If you use Windows or Mac OS, did you consider Linux as a possible operating system when you bought your present computer? If not, why not? If you did consider Linux, why did you reject it? If you use Linux, why did you choose Linux over Windows and Mac OS? What advantages and disadvantages have you found since you made that choice? Would you make the same choice again?
2. As the case mentions, about two-thirds of all Web servers run Linux but only about one percent of personal computers do. What factors do you think account for this difference?

Critical Thinking Questions

1. Basic applications are available for all operating systems. Beyond those applications, some users depend on packages while others tend to write their own applications. How does application availability affect their choice of an operating system?
2. Microsoft support for Linux in its Azure cloud-computing service could increase Azure revenue but could also decrease Windows revenue. Discuss the pros and cons of offering Azure support from a business standpoint.

SOURCES: Endsley, R., "How Small Business PrintedArt Uses Linux and Open Source," *www.linux.com/learn/tutorials/539523-case-study-how-small-business-printedart-uses-linux-and-open-source*, January 25, 2012; Staff, "Gompute Harnesses Sophisticated IBM High Performance Computing," IBM, *www-01.ibm.com/software/success/cssdb.nsf/CS/STRD-8SYJ2K*, April 3, 2012; Metz, C., "Microsoft Preps for Public Embrace of Linux," *Wired*, *www.wired.com/wiredenterprise/2012/05/microsoft-linux*, May 30, 2012; Meyer, D., "Microsoft Azure Starts Embracing Linux and Python," ZDNet UK, *www.zdnet.co.uk/news/cloud/2012/06/07/microsoft-azure-starts-embracing-linux-and-python-40155346*, June 7, 2012; Staff, May 2012 Web Server Survey, Netcraft, *news.netcraft.com/archives/2012/05/02/may-2012-web-server-survey.html*, May 2, 2012; PrintedArt Web site, *www.printedart.com*, accessed May 31, 2012.

Google

Over the years, Google has extended its reach from providing the most popular search engine to application software (Google Docs), mobile operating system (Android), Web browser (Chrome), and more recently, PC operating system—*Chrome OS*.[20] Today, over 100 million people are using Google's Android operating system in smartphones and mobile devices.[21] This number is up from less than 10 million users in 2009. Some believe that the number of Android users could explode to more than 200 million in a few years or less.

Google's Gingerbread operating system was designed for smartphones and other mobile devices, such as Samsung's Galaxy Note.[22] Chrome OS is a Linux-based operating system for netbooks and nettops, which are notebooks and desktop PCs primarily used to access Web-based information and services such as e-mail, Web browsing, social networks, and Google online applications. The OS is designed to run on inexpensive low-power computers. Chrome OS for personal computers doesn't need application software.[23] All applications can be accessed through the Internet. An open-source version of Chrome OS, named Chromium OS, was made available at the end of 2009. Because it is open-source software, developers can customize the source code to run on different platforms, incorporating unique features.

Workgroup Operating Systems

To keep pace with user demands, the technology of the future must support a world in which network usage, data-storage requirements, and data-processing speeds increase at a dramatic rate. Powerful and sophisticated OSs are needed to run the servers that meet these business needs for workgroups.

Windows Server

Microsoft designed *Windows Server* to perform a host of tasks that are vital for Web sites and corporate Web applications. For example, Microsoft Windows Server can be used to coordinate large data centers. It delivers benefits such as a powerful Web server management system, virtualization tools that allow various operating systems to run on a single server, advanced security features, and robust administrative support. Windows Home Server allows individuals to connect multiple PCs, storage devices, printers, and other devices into a home network.[24] It provides a convenient way to store and manage photos, video, music, and other digital content. It also provides backup and data recovery functions.

UNIX

UNIX is a powerful OS originally developed by AT&T for minicomputers—the predecessors of servers that are larger than PCs and smaller than mainframes. Ken Thompson, one of the creators of the UNIX operating system, was awarded the Japan Prize, a prize for outstanding contribution to science and technology.[25] UNIX can be used on many computer system types and platforms including workstations, servers, and mainframe computers. UNIX also makes it much easier to move programs and data among computers or to connect mainframes and workstations to share resources. There are many variants of UNIX, including HP/UX from Hewlett-Packard, AIX from IBM, and Solaris from Oracle. Oracle's Solaris operating system manages eBay's systems, including database servers, Web servers, tape libraries, and identity management systems. The online auction company found that when it switched to Solaris, system performance increased.[26]

Red Hat Linux

Red Hat Software offers a Linux network OS that taps into the talents of tens of thousands of volunteer programmers who generate a steady stream of improvements for the Linux OS. The *Red Hat Linux* network OS is very efficient at serving Web pages and can manage a cluster of up to eight servers. Distributions such as SuSE and Red Hat have proven Linux to be a very stable and efficient OS. Red Hat's newest version of Red Hat Enterprise Virtualization (RHEV) software no longer requires Windows Server software to operate.[27] According to director of Red Hat's virtualization business, "We're in a really good position to capitalize on the growing demand for alternatives to VMware." RHEV provides virtualization capabilities for servers and desktop computers.[28] Other vendors are also investigating virtualization for open-source software such as Linux.[29]

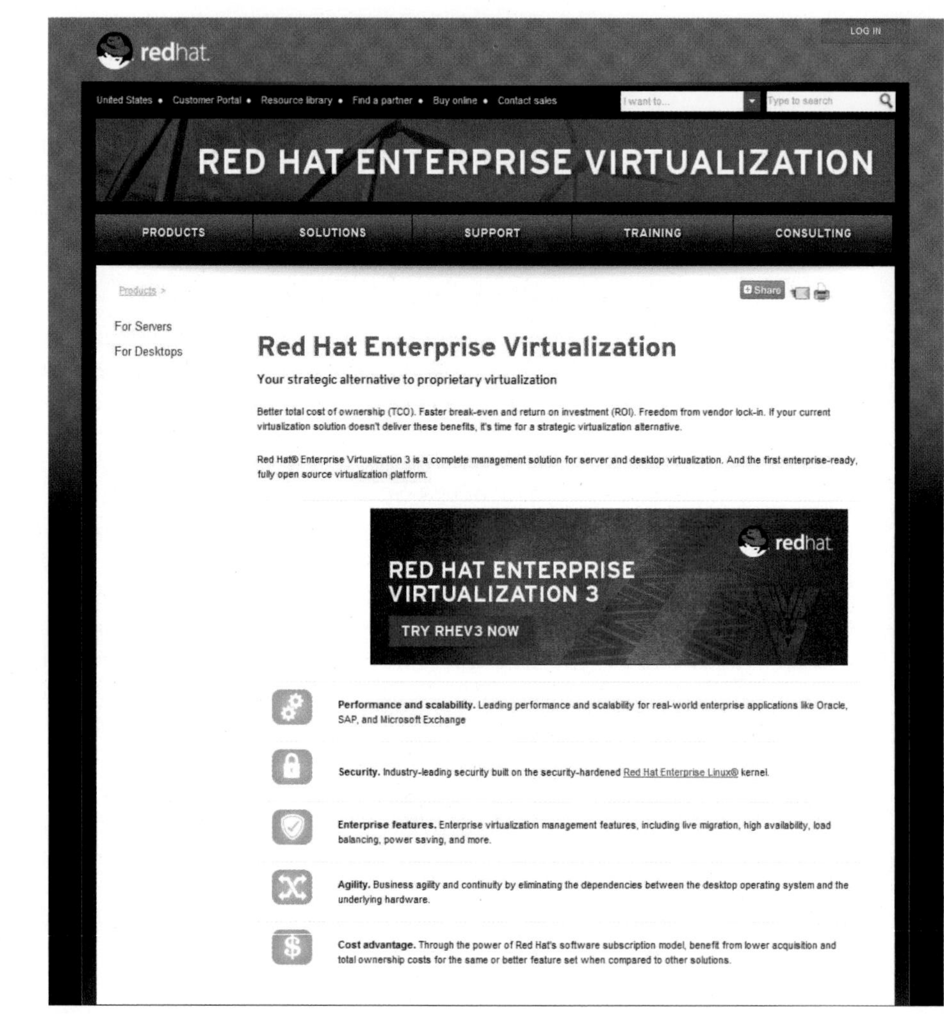

www.redhat.com

Red Hat Enterprise Virtualization (RHEV) software provides virtualization capabilities for servers and desktop computers.

Mac OS X Server

The *Mac OS X Server* is the first modern server OS from Apple Computer and is based on the UNIX OS. The most recent version is OS X Mountain Lion Server. It includes support for 64-bit processing, along with several server functions and features that allow the easy management of network and Internet services such as e-mail, Web site hosting, calendar management and sharing, wikis, and podcasting.

Enterprise Operating Systems

Mainframe computers, often referred to as "Big Iron," provide the computing and storage capacity to meet massive data-processing requirements and offer many users high performance and excellent system availability, strong security, and scalability. In addition, a wide range of application software has been developed to run in the mainframe environment, making it possible to purchase software to address almost any business problem. Examples of mainframe OSs include z/OS from IBM, HP-UX from Hewlett-Packard, and Linux. The *z/OS* is IBM's first 64-bit enterprise OS. It supports IBM's mainframes that can come with up to sixteen 64-bit processors.[30] (The z stands for zero downtime.) The *HP-UX* is a robust UNIX-based OS from Hewlett-Packard designed to handle a variety of business tasks, including online transaction processing and Web applications. HP-UX supports Hewlett-Packard's computers and those designed to run Intel's Itanium processors.

Operating Systems for Small Computers, Embedded Computers, and Special-Purpose Devices

New OSs are changing the way we interact with smartphones, cell phones, digital cameras, TVs, and other digital electronics devices. Companies around the world are developing operating systems for these devices. Alibaba Cloud Computing, a part of the Chinese Alibaba Group, has developed an operating system for smartphones and mobile devices.[31] This operating system will compete with operating systems from Google, Apple, and Microsoft in China.[32] Hewlett-Packard is hoping that car and appliance makers will increasingly use its webOS operating system.[33] The webOS uses a touch interface and allows people to connect to the Internet.[34] According to a company spokesperson, "We're looking at expanding the base and bringing to the webOS an ecosystem that inspires developers."

These OSs are also called *embedded operating systems* or just *embedded systems* because they are typically embedded within a device. Embedded systems are typically designed to perform specialized tasks. For example, an automotive embedded system might be responsible for controlling fuel injection. A digital camera's embedded system supports taking and viewing photos and may include a limited set of editing tools. A GPS device uses an embedded system to help people find their way around town or more remote areas (see Figure 4.8). Some of the more popular OSs for devices are described in the following section.

FIGURE **4.8**

GPS devices use embedded operating systems

A GPS device uses an embedded system to acquire information from satellites, display your current location on a map, and direct you to your destination.

Cell Phone Embedded Systems and Operating Systems

Cell phones have traditionally used embedded systems to provide communication and limited personal information management services to users. *Symbian*, a popular cell phone embedded OS, has traditionally provided voice and text communication, an address book, and a few other basic applications. Nokia has introduced three new cell phones using its updated Symbian operating system.[35] According to the head of sales for Nokia, "We will use Symbian to introduce competitive products that offer more choice at affordable prices to people all over this world." When RIM introduced the BlackBerry smartphone in 2002, the mobile phone's capabilities were vastly expanded.[36] Since then, cell phone embedded systems have transformed into full-fledged personal computer OSs such as the iPhone OS, Google Android, and Microsoft Windows Mobile. Even traditional embedded systems such as Palm OS (now webOS) and Symbian have evolved into PC operating systems with APIs and software development kits that allow developers to design hundreds of applications providing a myriad of mobile services.

Windows Embedded

Windows Embedded is a family of Microsoft OSs included with or embedded into small computer devices.[37] Windows Embedded includes several versions that provide computing power for TV set top boxes, automated industrial machines, media players, medical devices, digital cameras, PDAs, GPS receivers, ATMs, gaming devices, and business devices such as cash registers. Windows Embedded Automotive provides a computing platform for automotive software such as Ford Sync. The Ford Sync system uses an in-dashboard display and wireless networking technologies to link automotive systems with cell phones and portable media players.[38] See Figure 4.9.

Daniel Acker/Bloomberg via Getty Images

FIGURE **4.9**

Microsoft Auto and Ford Sync

The Ford Sync system, developed on the Microsoft Auto operating system, allows drivers to wirelessly connect cell phones and media devices to automotive systems.

Proprietary Linux-Based Systems

Because embedded systems are usually designed for a specific purpose in a specific device, they are usually proprietary or custom-created and owned by the manufacturer. Sony's Wii, for example, uses a custom-designed OS based on the Linux kernel. Linux is a popular choice for embedded systems because it is free and highly configurable. It has been used in many embedded systems, including e-book readers, ATM machines, cell phones, networking devices, and media players.

Utility Programs

utility program: Program that helps to perform maintenance or correct problems with a computer system.

Utility programs help to perform a variety of tasks. For example, some utility programs merge and sort sets of data, keep track of computer jobs being run, compress files of data before they are stored or transmitted over a network (thus saving space and time), and perform other important tasks. Parallels

Desktop is a popular utility that allows Apple Mac computers to run Windows programs.[39] The utility, which costs under $100, creates a virtual Windows machine inside a Mac computer (see Figure 4.10).

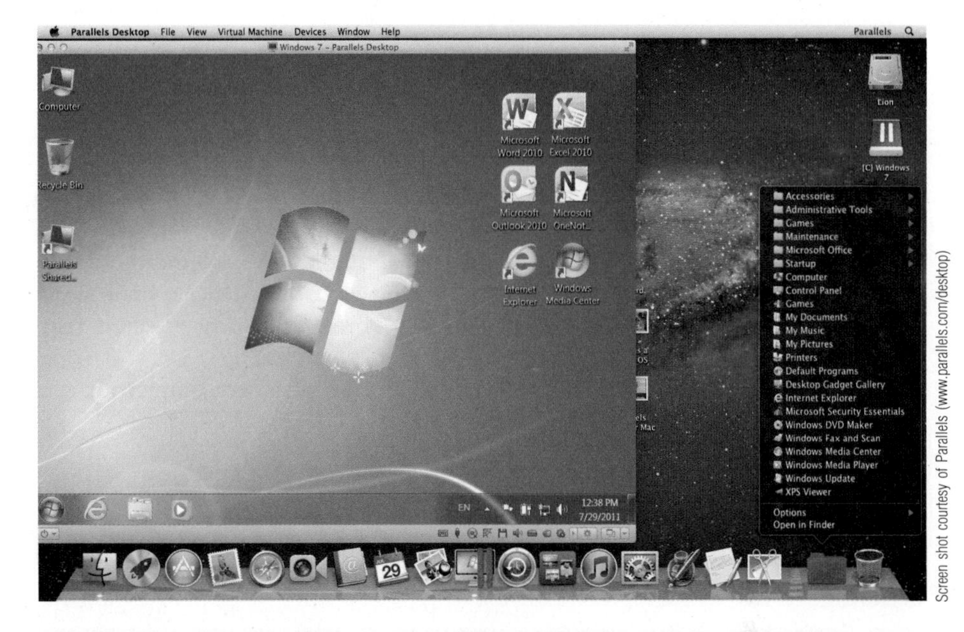

Screen shot courtesy of Parallels (www.parallels.com/desktop)

FIGURE **4.10**

Parallels Desktop

Parallels Desktop for Mac is shown running Windows on a Mac with Windows applications available on the OS X Lion dock.

Another type of utility program allows people and organizations to take advantage of unused computer power over a network. Often called *grid computing*, this approach can be very efficient and less expensive than purchasing additional hardware or computer equipment. CERN, home of the Large Hadron Collider (LHC), the world's largest scientific instrument, is also home to one of the world's largest scientific grid computing and storage systems. The LHC Computing Grid (LCG) project provides scientists around the world with access to shared computer power and storage systems over the Internet.[40] In 2012, this project helped identify a particle that may be the Higgs boson, also called the "God Particle" by some people.

Although many PC utility programs come installed on computers, you can also purchase utility programs separately. The following sections examine some common types of utilities.

Hardware Utilities

Some hardware utilities are available from companies such as Symantec, which produces Norton Utilities. Hardware utilities can check the status of all parts of the PC, including hard disks, memory, modems, speakers, and printers. Disk utilities check the hard disk's boot sector, file allocation tables, and directories and analyze them to ensure that the hard disk is not damaged. Disk utilities can also optimize the placement of files on a crowded disk.

Security Utilities

Computer viruses and spyware from the Internet and other sources can be a nuisance—and sometimes can completely disable a computer. Antivirus and antispyware software can be installed to constantly monitor and protect the computer. If a virus or spyware is found, often times it can be removed. It is also a good idea to protect computer systems with firewall software. Firewall software filters incoming and outgoing packets, making sure that neither hackers nor their tools are attacking the system. Symantec, McAfee, and Microsoft are the most popular providers of security software.

File-Compression Utilities

File-compression programs can reduce the amount of disk space required to store a file or reduce the time it takes to transfer a file over the Internet. Both

Windows and Mac operating systems let you compress or decompress files and folders. A zip file has a .zip extension, and its contents can be easily unzipped to the original size. *MP3 (Motion Pictures Experts Group-Layer 3)* is a popular file-compression format used to store, transfer, and play music and audio files, such as podcasts—audio programs that can be downloaded from the Internet.

Spam-Filtering Utilities

Receiving unwanted e-mail (spam) can be a frustrating waste of time. E-mail software and services include spam-filtering utilities to assist users with these annoyances. E-mail filters identify spam by learning what the user considers spam and routing it to a junk mail folder. However, this method is insufficient for protecting enterprise-level e-mail systems where spam containing viruses is a serious threat. Businesses often use additional spam-filtering software from companies including Cisco, Barracuda Networks, and Google at the enterprise level to intercept dangerous spam as it enters the corporate e-mail system.

Network and Internet Utilities

A broad range of network- and systems-management utility software is available to monitor hardware and network performance and trigger an alert when a server is crashing or a network problem occurs.[41] IBM's Tivoli Netcool and Hewlett-Packard's Automated Network Management Suite can be used to solve computer-network problems and help save money.[42] In one survey, about 60 percent of responding organizations used monitoring software to determine if their Internet sites and Internet applications were running as expected.

Server and Mainframe Utilities

Some utilities enhance the performance of servers and mainframe computers. James River Insurance uses a utility program from Confio to help it monitor the performance of its computer systems and databases.[43] According to a manager for James River, "We take a proactive approach to database management to ensure we maintain high availability and performance in our virtual and physical environments." IBM and other companies have created systems-management software that allows a support person to monitor the growing number of desktop computers attached to a server or mainframe computer. Similar to the virtual machine software discussed earlier, *server virtualization software* allows a server to run more than one operating system at the same time. For example, you could run four different virtual servers simultaneously on one physical server.

Other Utilities

Utility programs are available for almost every conceivable task or function. Managing the vast array of operating systems for smartphones and mobile devices, for example, has been difficult for many companies. In one survey, two-thirds of responding organizations allowed managers and workers to connect to corporate databases using smartphones and mobile devices with very little or no guidance or supervision.[44] Utility programs can help. Research in Motion (RIM) has developed a utility program that helps companies manage cell phones and mobile devices from its company and others.[45] Often called *mobile device management (MDM)*, this type of software should help companies as smartphones and other mobile devices become more popular for managers and workers in a business setting. MDM software helps a company manage security, enforce corporate strategies, and control downloads and content streaming from corporate databases into smartphones and mobile devices. In addition, a number of companies, such as CNET, offer utilities that can be downloaded for most popular operating systems.[46]

Middleware

middleware: Software that allows various systems to communicate and exchange data.

Middleware is software that allows various systems to communicate and exchange data. It is often developed to address situations where a company acquires different types of information systems through mergers, acquisitions, or expansion and wants the systems to share data and interact. Middleware can also serve as an interface between the Internet and private corporate systems. For example, it can be used to transfer a request for information from a corporate customer on the corporate Web site to a traditional database on a mainframe computer and return the results to the customer on the Internet.

service-oriented architecture (SOA): A modular method of developing software and systems that allows users to interact with systems and systems to interact with each other.

The use of middleware to connect disparate systems has evolved into an approach for developing software and systems called SOA. A **service-oriented architecture (SOA)** uses modular application services to allow users to interact with systems and systems to interact with each other. Systems developed with SOA are flexible and ideal for businesses that need a system to expand and evolve over time. SOA modules can be reused for a variety of purposes, thus reducing development time. Because SOA modules are designed using programming standards so that they can interact with other modules, rigid custom-designed middleware software is not needed to connect systems.

APPLICATION SOFTWARE

As discussed earlier in this chapter, the primary function of application software is to apply the power of the computer to give people, workgroups, and the entire enterprise the ability to solve problems and perform specific tasks. One debt collection agency, for example, was able to save more than $250,000 annually by using application software from Latitude to monitor people not paying their bills on time.[47] Applications help you perform common tasks, such as create and format documents, perform calculations, or manage information. Some applications are more specialized. Accenture, for example, offers application software specifically for the property and causality insurance industry.[48] Land O'Lakes, a large food and agricultural cooperative, used application software to help synchronize its supply chain by shipping perishable products such as milk and cheese to customers in a timely fashion.[49] Application software is used throughout the medical profession to save and prolong lives. For example, Swedish Medical Center in Seattle, Washington, uses content management software from Oracle to access patient records when and where they are needed.[50] New passenger-screening software at the Tulsa International Airport has streamlined the check-in process and reduced privacy concerns.[51] The software, called automated target recognition, uses a new full-body scanning technology. The U.S. Army is testing new application software on smart phones and tablet computers in combat zones.[52] The military software will help commanders and combat troops analyze surveillance video and data from battle fields to help them locate and eliminate enemy troops, giving new meaning to the term "killer app."

Overview of Application Software

proprietary software: One-of-a-kind software designed for a specific application and owned by the company, organization, or person that uses it.

off-the-shelf software: Software mass-produced by software vendors to address needs that are common across businesses, organizations, or individuals.

Proprietary software and off-the-shelf software are important types of application software. **Proprietary software** is one-of-a-kind software designed for a specific application and owned by the company, organization, or person that uses it. Proprietary software can give a company a competitive advantage by providing services or solving problems in a unique manner, better than methods used by a competitor. **Off-the-shelf software** is mass-produced by software vendors to address needs that are common across businesses, organizations, or individuals. For example, Amazon.com uses the same off-the-shelf payroll software as many businesses, but the company uses custom-designed proprietary software on its Web site that allows visitors to more easily find items to purchase. The relative advantages and disadvantages of proprietary software and off-the-shelf software are summarized in Table 4.3.

TABLE 4.3 Comparison of proprietary and off-the-shelf software

Proprietary Software		Off-the-Shelf Software	
Advantages	**Disadvantages**	**Advantages**	**Disadvantages**
You can get exactly what you need in terms of features, reports, and so on.	It can take a long time and significant resources to develop required features.	The initial cost is lower because the software firm can spread the development costs over many customers.	An organization might have to pay for features that are not required and never used.
Being involved in the development offers control over the results.	In-house system development staff may be hard pressed to provide the required level of ongoing support and maintenance because of pressure to move on to other new projects.	The software is likely to meet the basic business needs—you can analyze existing features and the performance of the package before purchasing.	The software might lack important features, thus requiring future modification or customization. This lack can be very expensive because users must adopt future releases of the software as well.
You can modify features that you might need to counteract an initiative by competitors or to meet new supplier or customer demands.	The features and performance of software that has yet to be developed presents more potential risk.	The package is likely to be of high quality because many customer firms have tested the software and helped identify its bugs.	The software might not match current work processes and data standards.

Many companies use off-the-shelf software to support business processes. Key questions for selecting off-the-shelf software include the following. First, will the software run on the OS and hardware you have selected? Second, does the software meet the essential business requirements that have been defined? Third, is the software manufacturer financially solvent and reliable? Finally, does the total cost of purchasing, installing, and maintaining the software compare favorably to the expected business benefits?

Some off-the-shelf programs can be modified, in effect blending the off-the-shelf and customized approaches. For example, El Camino Hospital in Mountain View, California, customized Microsoft's e-health management system, Amalga, to track patients with the H1N1 flu and those that may have been exposed to it.[53]

application service provider (ASP): A company that provides the software, support, and computer hardware on which to run the software from the user's facilities over a network.

Software as a Service (SaaS): A service that allows businesses to subscribe to Web-delivered application software.

Another approach to obtaining a customized software package is to use an application service provider. An **application service provider (ASP)** is a company that can provide the software, support, and computer hardware on which to run the software from the user's facilities over a network. Some vendors refer to the service as *on-demand software*.

Today, many companies are running software on the Web. This approach is called **Software as a Service (SaaS)**, which allows businesses to subscribe to Web-delivered application software. In most cases, the company pays a monthly service charge or a per-use fee.[54] Guardian Life Insurance, for example, implemented an actuarial application by using Amazon's Ec2 SaaS approach.[55] According to the CIO of the company, "We don't do anything because it's Cloud. But if the financials look right, if the risk profile looks right, if the richness and robustness look right, we go with that solution." Like ASP, SaaS providers maintain software on their own servers and provide access to it over the Internet. SaaS usually uses a Web browser-based user interface. Many business activities are supported by SaaS. Vendors include Oracle, SAP, Net Suite, Salesforce, and Google. Tidewell, a hospice that serves about 8,000 Florida families, acquired software from Salesforce.com to save money and streamline its operations.[56] SaaS can reduce expenses by sharing its running applications among many businesses. Some people, however, are concerned about the security of data and programs on the Internet using the SaaS approach.[57]

SaaS and new Web development technologies have led to a new paradigm in computing called cloud computing.[58] *Cloud computing* refers to the use of computing resources, including software and data storage, on the Internet (the cloud) rather than on local computers. Google, for example, is launching new personal computers built by Samsung and Acer called Chromebooks that include only an Internet browser. All of the software applications are accessed through an Internet connection.[59] Businesses can get a Chromebook and Chrome OS for under $30 per user.[60] In addition, Google's e-mail and productivity suite can be purchased for about $50 per month per individual. Rather than installing, storing, and running software on your own computer, with cloud computing, you use the Web browser to access software stored and delivered from a Web server. Typically the data generated by the software is also stored on the Web server. For example, Tableau software allows users to import databases or spreadsheet data to create powerful visualizations that provide useful information.[61] Cloud computing also provides the benefit of being able to easily collaborate with others by sharing documents on the Internet.

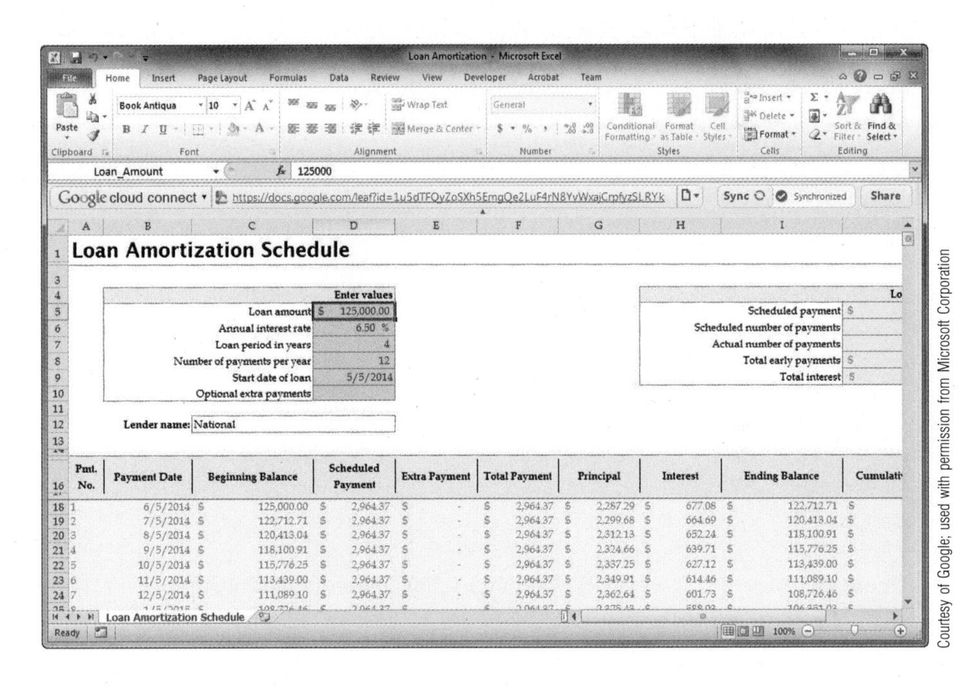

Google has a cloud application called Google Cloud Connect, which lets you share and edit Microsoft Office documents simultaneously with other people in your organization.

ASP, SaaS, and cloud computing, however, involve some risks. For example, sensitive information could be compromised in a number of ways, including unauthorized access by employees or computer hackers; the host might not be able to keep its computers and network up and running as consistently as necessary; or a disaster could disable the host's data center, temporarily putting an organization out of business. In addition, these approaches are not accepted and used by everyone.[62] According to one survey, about 15 percent of enterprises are either using the SaaS approach or plan to use the approach in the next year. It can also be difficult to integrate the SaaS approach with existing software. According to the CIO of Hostess Brands, "Figuring out integration requirements and how providers handle those and getting everything in sync have been among our tougher challenges."

Personal Application Software

Hundreds of computer applications can help people at school, home, and work. New computer software under development and existing GPS technology, for example, will allow people to see 3D views of where they are, along with directions and 3D maps to where they would like to go. Absolute software, which uses GPS technology, helps people and organizations retrieve stolen computers. The company has recovered almost 10,000 devices worth over $10 million dollars.[63] According to a special investigator for the Detroit Public Schools (DPS), "At DPS, we've already seen the effect of these recoveries. We would have never recovered any of the 300 plus laptops stolen from our district without the aid of Absolute Software."

The features of some popular types of personal application software are summarized in Table 4.4. In addition to these general-purpose programs, thousands of other personal computer applications perform specialized tasks that help you do your taxes, get in shape, lose weight, get medical advice, write wills and other legal documents, repair your computer, fix your car, write music, and edit your pictures and videos. This type of software, often called *user software* or *personal productivity software*, includes the general-purpose tools and programs that support individual needs.

TABLE **4.4** Examples of personal application software

Type of Software	Explanation	Example
Word processing	Create, edit, and print text documents	Microsoft Word Google Docs Apple Pages Open Office Writer
Spreadsheet	Provide a wide range of built-in functions for statistical, financial, logical, database, graphics, and date and time calculations	Microsoft Excel IBM Lotus 1-2-3 Google Spreadsheet Apple Numbers Open Office Calc
Database	Store, manipulate, and retrieve data	Microsoft Access IBM Lotus Approach Borland dBASE Google Base Open Office Base
Graphics	Develop graphs, illustrations, and drawings	Adobe Illustrator Adobe FreeHand Microsoft PowerPoint Open Office Impress
Project management	Plan, schedule, allocate, and control people and resources (money, time, and technology) needed to complete a project according to schedule	Microsoft Project Symantec On Target Scitor Project Scheduler Symantec Time Line
Financial management	Provide income and expense tracking and reporting to monitor and plan budgets (some programs have investment portfolio management features)	Intuit Quicken
Desktop publishing (DTP)	Use with personal computers and high-resolution printers to create high-quality printed output, including text and graphics; various styles of pages can be laid out; art and text files from other programs can also be integrated into published pages	Quark Xpress Microsoft Publisher Adobe PageMaker Corel Ventura Publisher Apple Pages

Word Processing

Word-processing applications are installed on most PCs today. These applications come with a vast array of features, including those for checking spelling, creating tables, inserting formulas, creating graphics, and much more (see Figure 4.11). Much of the work required to create this book used the popular word-processing software, Microsoft Word.

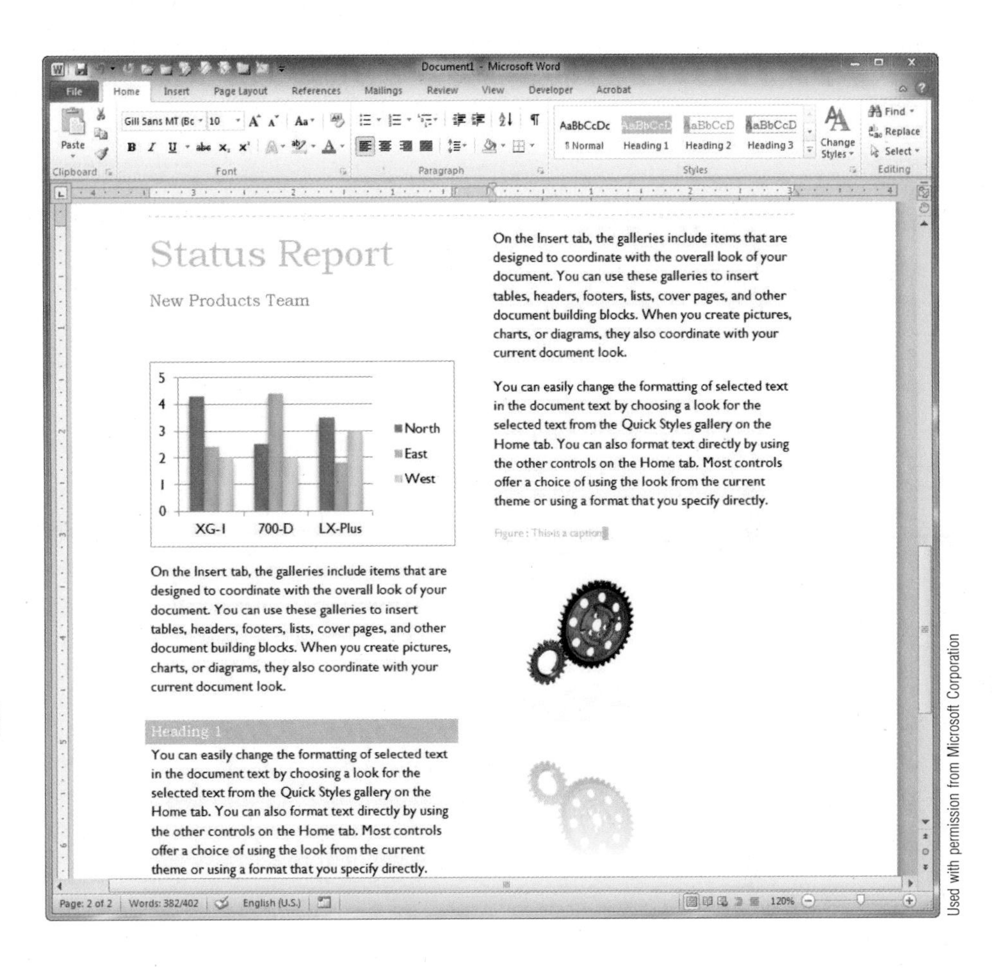

FIGURE 4.11

Word-processing program

Word-processing applications, such as Microsoft Word, can be used to write letters, professional documents, work reports, and term papers.

A team of people can use a word-processing program to collaborate on a project. The authors and editors who developed this book, for example, used the Track Changes and Reviewing features of Microsoft Word to track and make changes to chapter files. With these features, you can add comments or make revisions to a document that a coworker can review and either accept or reject.

Spreadsheet Analysis

Spreadsheets are powerful tools for manipulating and analyzing numbers and alphanumeric data. Individuals and organizations use spreadsheets. Features of spreadsheets include formulas, statistical analysis, built-in business functions, graphics, limited database capabilities, and much more (see Figure 4.12). The business functions include calculation of depreciation, present value, internal rate of return, and the monthly payment on a loan, to name a few. Optimization is another powerful feature of many spreadsheet programs. *Optimization* allows the spreadsheet to maximize or minimize a quantity subject to certain constraints. For example, a small furniture manufacturer that produces chairs and tables might want to maximize its profits. The constraints could be a limited supply of lumber, a limited number of workers who can assemble the chairs and tables, or a limited amount of various hardware fasteners that might

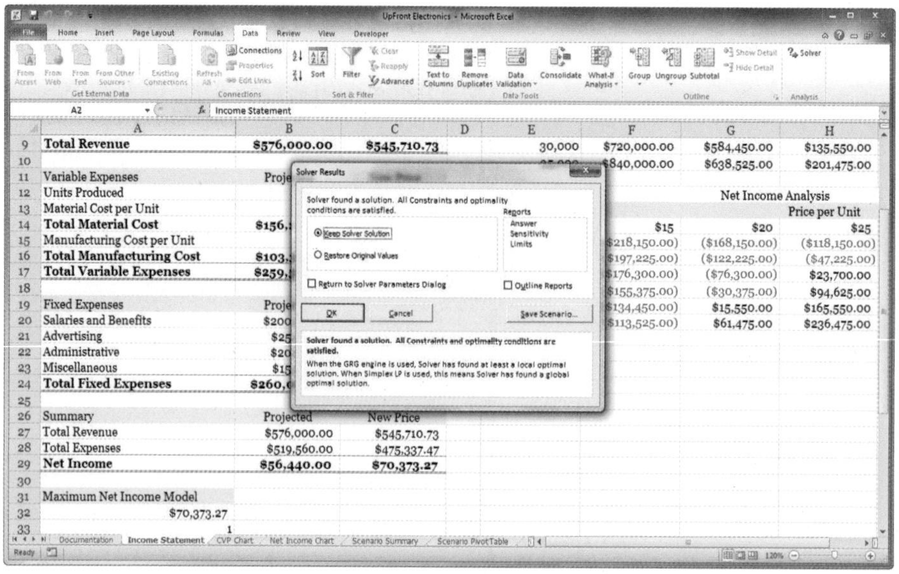

FIGURE 4.12

Spreadsheet program

Consider spreadsheet programs, such as Microsoft Excel, when calculations are required.

be required. Using an optimization feature, such as Solver in Microsoft Excel, the spreadsheet can determine what number of chairs and tables to produce with labor and material constraints to maximize profits.

Database Applications

Database applications are ideal for storing, organizing, and retrieving data. These applications are particularly useful when you need to manipulate a large amount of data and produce reports and documents. Database manipulations include merging, editing, and sorting data. The uses of a database application are varied. You can keep track of a CD collection, the items in your apartment, tax records, and expenses. A student club can use a database to store names, addresses, phone numbers, and dues paid. In business, a database application can help process sales orders, control inventory, order new supplies, send letters to customers, and pay employees. Database management systems can be used to track orders, products, and customers; analyze weather data to make forecasts for the next several days; and summarize medical research results. A database can also be a front end to another application. For example, you can use a database application to enter and store income tax information and then export the stored results to other applications, such as a spreadsheet or tax-preparation application.

Presentation Graphics Program

It is often said that a picture is worth a thousand words. With today's graphics programs, it is easy to develop attractive graphs, illustrations, and drawings that assist in communicating important information (see Figure 4.13). Presentation graphics programs can be used to develop advertising brochures, announcements, and full-color presentations and to organize and edit photographic images. If you need to make a presentation at school or work, you can use a special type of graphics program called a presentation application to develop slides and then display them while you are speaking. Because of their popularity, many colleges and departments require students to become proficient at using presentation graphics programs.

Many graphics programs, including Microsoft PowerPoint, consist of a series of slides. Each slide can be displayed on a computer screen, printed as a handout, or (more commonly) projected onto a large viewing screen for audiences. Powerful built-in features allow you to develop attractive slides and complete presentations. You can select a template for a type of presentation,

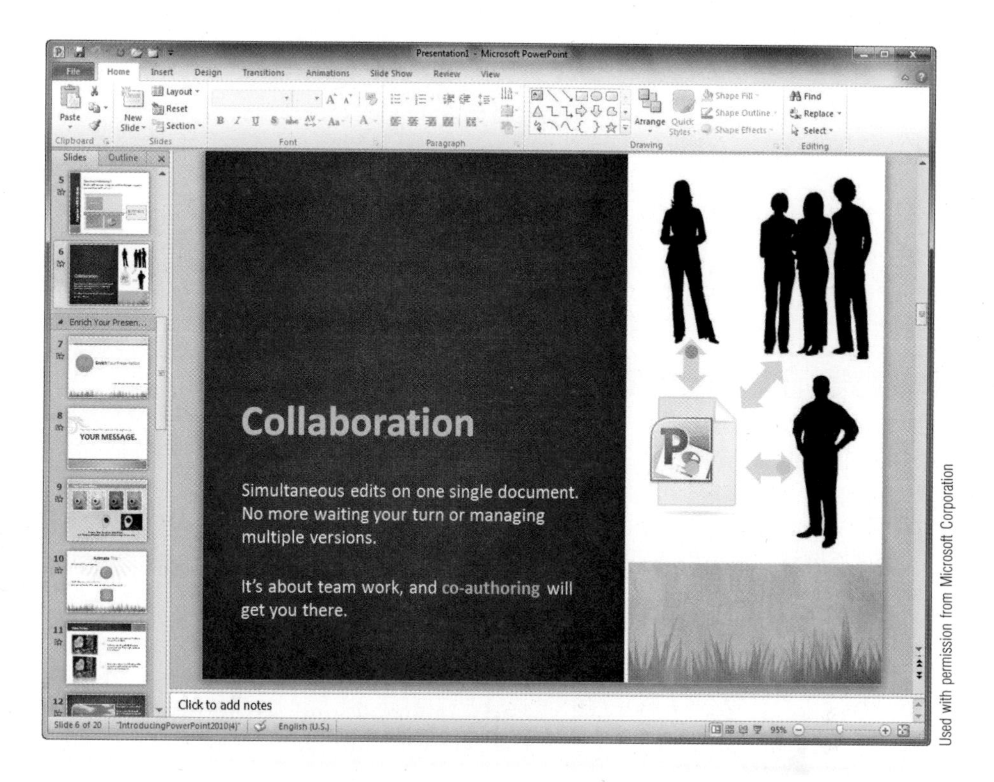

FIGURE 4.13

Presentation graphics program

Presentation graphics programs, such as Microsoft PowerPoint, can help you make a presentation at school or work.

such as recommending a strategy for managers, communicating news to a sales force, giving a training presentation, or facilitating a brainstorming session. The presentation graphics program lets you create a presentation step-by-step, including applying color and attractive formatting. You can also design a custom presentation using the many types of charts, drawings, and formatting available. Most presentation graphics programs come with many pieces of *clip art*, such as drawings and photos of people meeting, medical equipment, telecommunications equipment, entertainment, and much more.

Personal Information Managers

Personal information management (PIM) software helps people, groups, and organizations store useful information, such as a list of tasks to complete or a set of names and addresses. PIM software usually provides an appointment calendar, an address book or contacts list, and a place to take notes. In addition, information in a PIM can be linked. For example, you can link an appointment with a sales manager in the calendar to information on the sales manager in the address book. When you click the appointment in the calendar, a window opens displaying information on the sales manager from the address book. Microsoft Outlook is an example of very popular PIM software. Increasingly, PIM software is moving online where it can be accessed from any Internet-connected device (see Figure 4.14).

Some PIMs allow you to schedule and coordinate group meetings. If a computer or handheld device is connected to a network, you can upload the PIM data and coordinate it with the calendar and schedule of others using the same PIM software on the network. You can also use some PIMs to coordinate e-mails inviting others to meetings. As users receive their invitations, they click a link or button to be automatically added to the guest list.

Software Suites and Integrated Software Packages

software suite: A collection of single programs packaged together in a bundle.

A **software suite** is a collection of single programs packaged together in a bundle. Software suites can include a word processor, spreadsheet program, database management system, graphics program, communications tools, organizers, and more. Some suites support the development of Web pages, note

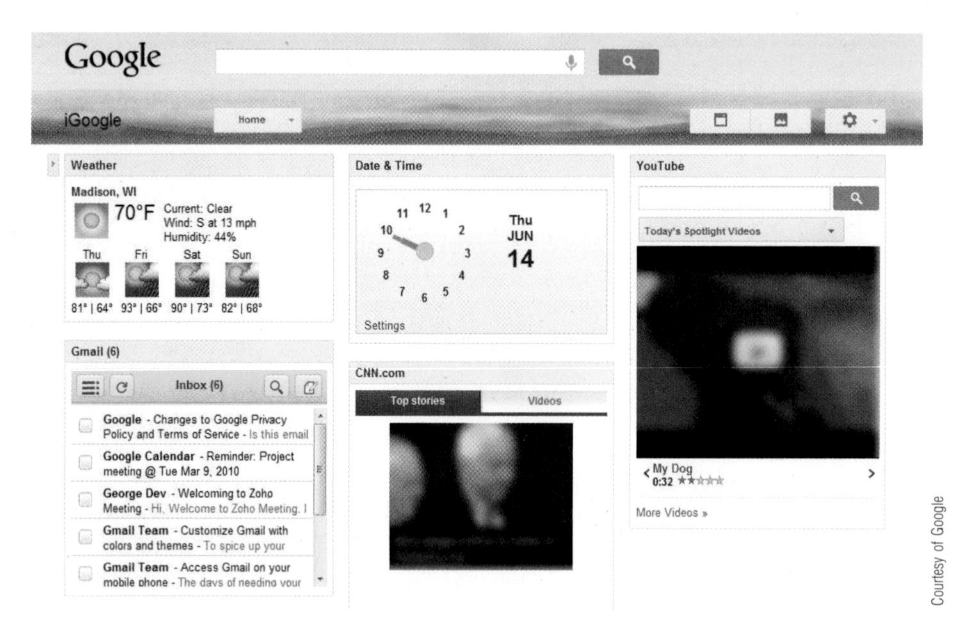

Courtesy of Google

FIGURE **4.14**

Personal information management software

iGoogle and other Web portals support PIM by allowing users to access calendars, to-do lists, e-mail, social networks, contacts, and other information all from one page.

taking, and speech recognition so that applications in the suite can accept voice commands and record dictation. Software suites offer many advantages. The software programs have been designed to work similarly so that, after you learn the basics for one application, the other applications are easy to learn and use. Buying software in a bundled suite is cost effective; the programs usually sell for a fraction of what they would cost individually.

Microsoft Office, Corel WordPerfect Office, Lotus SmartSuite, and Sun Microsystems OpenOffice are examples of popular general-purpose software suites for personal computer users. Microsoft Office has the largest market share. Most of these software suites include a spreadsheet program, word processor, database program, and graphics presentation software. All can exchange documents, data, and diagrams (see Table 4.5). In other words, you can create a spreadsheet and then cut and paste that spreadsheet into a document created using the word-processing application.

TABLE **4.5** Major components of leading software suites

Personal Productivity Function	Microsoft Office	Lotus Symphony	Corel WordPerfect Office	Open Office	AppleiWork	Google
Word Processing	Word	Documents	WordPerfect	Writer	Pages	Docs
Spreadsheet	Excel	Spreadsheets	Quattro Pro	Calc	Numbers	Spreadsheet
Presentation Graphics	PowerPoint	Presentations	Presentations	Impress and Draw	Keynote	Presentation
Database	Access			Base		

In addition to suites, some companies produce *integrated application packages* that contain several programs. For example, Microsoft Works is one program that contains a basic word processor, spreadsheet, database, address book, calendar, and other applications. Although not as powerful as standalone software included in software suites, integrated software packages offer a range of capabilities for less money. QuickOffice can be used on tablet computers and smartphones to read and edit Microsoft Office documents.[64] Onlive can also be used to open and edit Microsoft Office documents on an Apple iPad.[65] Some integrated packages cost about $100.

Some companies offer Web-based productivity software suites that require no installation—only a Web browser. Zoho, Google, and Thinkfree offer free online word processor, spreadsheet, presentation, and other software that require no installation on the PC. Adobe has developed Acrobat.com, a suite of programs that can be used to create and combine Adobe PDF (Portable Document Format) files, convert PDF files to Microsoft Word or Excel files, create Web forms, and more.[66] After observing this trend, Microsoft responded with an online version of some of its popular Office applications. Office 365 offers basic software suite features over the Internet using cloud computing.[67] See Figure 4.15. Microsoft Word, Outlook, Excel, Exchange for messaging, SharePoint for collaboration, and Lync for conferencing can be accessed.[68] The cloud-based applications can cost $10 per user per month depending on the features used.[69] Microsoft offers plans for professionals and small businesses, enterprises, and education. Some believe that Office 365 has advantages over many other online suites.[70] According to the director of online services at Microsoft, "With Office 365, businesses of all sizes can get the same robust capabilities that have given larger businesses an edge for years."[71] The city of Winston-Salem, North Carolina, for example, used Office 365 to save money and place software applications on the Internet. According to the CIO of the city, "I have to improve technology with a constrained budget. Because we were able to package Microsoft cloud and local products in one enterprise agreement, we ended up with more bang for no additional cost." The online versions of Word, Excel, PowerPoint, and OneNote are tightly integrated with Microsoft's desktop Office suite for easy sharing of documents among computers and collaborators.

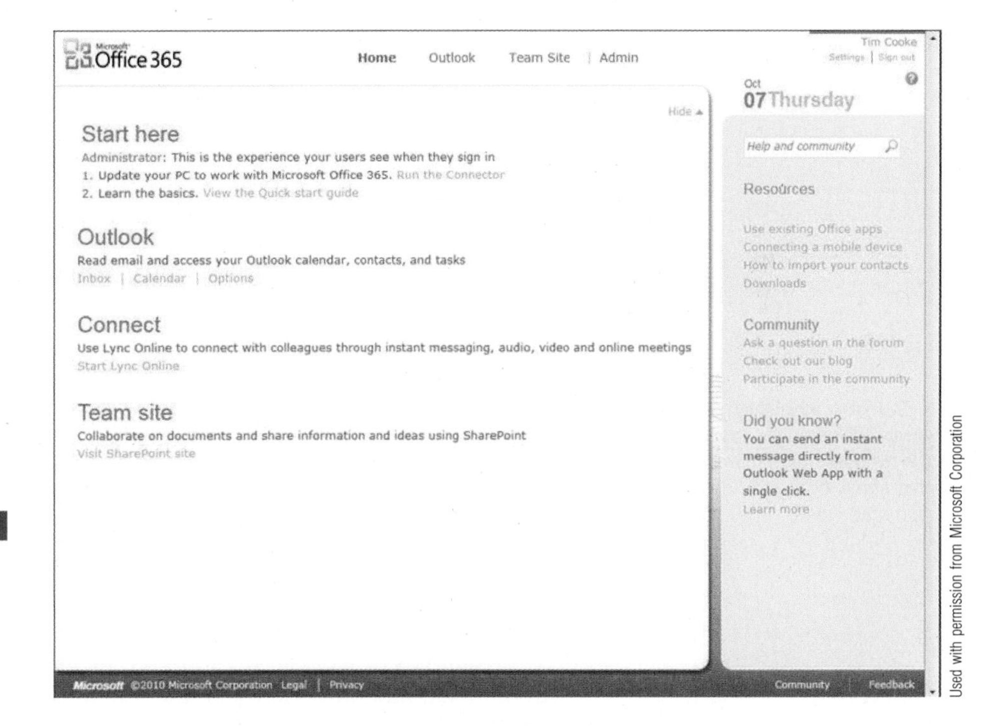

FIGURE **4.15**

Web suites

Microsoft Office 365 is a Web suite that offers basic software suite features over the Internet using cloud computing.

Other Personal Application Software

In addition to the software already discussed, people can use many other interesting and powerful application software tools. In some cases, the features and capabilities of these applications can more than justify the cost of an entire computer system. TurboTax, for example, is a popular tax-preparation program. You can find software for creating Web pages and sites, composing music, and editing photos and videos. Many people use educational and reference software and entertainment, games, and leisure software. Game-playing software is popular and can be very profitable for companies that develop games and various game accessories, including virtual avatars such as colorful animals, fish, and people.[72] Game-playing software has even been used as therapy for young children and adults recovering from cancer and other disease.[73] According to a hospital executive, "It's a very motivating tool for the patients. It's visual, the feedback is instant, and it's fun." Some believe that online game players may have solved an important AIDS research question.[74] Engineers, architects, and designers often use computer-aided design (CAD) software to design and develop buildings, electrical systems, plumbing systems, and more. Autosketch, CorelCAD, and AutoCad are examples of CAD software. Other programs perform a wide array of statistical tests. Colleges and universities often have a number of courses in statistics that use this type of application software. Two popular applications in the social sciences are SPSS and SAS.

Mobile Application Software

The number of applications (apps) for smartphones and other mobile devices has exploded in recent years. Besides the valuable mobile applications that come with these devices, tens of thousands of applications have been developed by third parties. For example, iPhone users can download and install thousands of applications using Apple's App Store.[75] Many iPhone apps are free, while others range in price from 99 cents to hundreds of dollars. Thousands of mobile apps are available in the Android Market for users of Android handsets. Microsoft and other software companies are also investing in mobile applications for devices that run on its software.[76] SceneTap, an application for iPhones and Android devices, can determine the number of people at participating bars, pubs, or similar establishments and the ratio of males to females.[77] This approach uses video cameras and facial recognition software to identify males and females. SocialCamera, an application for Android phones, allows people to take a picture of someone and then search their Facebook friends for a match.[78] New facial-recognition software developed at Carnegie Mellon University was able to correctly identify about one-third of the people tested from a simple photograph from a cell phone or camera.[79] Facial-recognition software, however, could be a potential invasion to privacy.[80] The market for mobile application software for smartphones and mobile devices could reach $80 billion by 2017.[81] Table 4.6 lists a few mobile application categories.

Workgroup Application Software

workgroup application software: Software that supports teamwork, whether team members are in the same location or dispersed around the world.

Workgroup application software is designed to support teamwork, whether team members are in the same location or dispersed around the world. This support can be accomplished with software known as *groupware* that helps groups of people work together effectively. Microsoft Exchange Server, for example, has groupware and e-mail features.[82] Also called *collaborative software*, this approach allows a team of managers to work on the same production problem, letting them share their ideas and work via connected computer systems.

Examples of workgroup software include group-scheduling software, electronic mail, and other software that enables people to share ideas. Lotus Notes and Domino are examples of workgroup software from IBM (see Figure 4.16). Web-based software is ideal for group use. Because documents are stored on an Internet server, anyone with an Internet connection can access them easily. Google provides options in its online applications that allow users to share documents, spreadsheets, presentations, calendars, and notes with other specified users or everyone on the Web. This sharing makes it convenient for several people to contribute to a document without concern for software compatibility or storage. Google also provides a tool for creating Web-based forms and surveys. When invited parties fill out the form, the data is stored in a Google spreadsheet.

TABLE 4.6 Categories of mobile applications for smartphones

Category	Description
Books and reference	Access e-books, subscribe to journals, or look up information in Webster's or Wikipedia
Business and finance	Track expenses, trade stocks, and access corporate information systems
Entertainment	Access all forms of entertainment, including movies, television programs, music videos, and local night life
Games	Play a variety of games, from 2D games such as Pacman and Tetris to 3D games such as Need for Speed, Rock Band, and The Sims
Health and fitness	Track workout and fitness progress, calculate calories, and even monitor your speed and progress from your wirelessly connected Nike shoes
Lifestyle	Find good restaurants, select wine for a meal, and more
Music	Find, listen to, and create music
News and weather	Access major news and weather providers including Reuters, AP, the *New York Times*, and the Weather Channel
Photography	Organize, edit, view, and share photos taken on your camera phone
Productivity and utilities	Create grocery lists, practice PowerPoint presentations, work on spreadsheets, synchronize with PC files, and more
Social networking	Connect with others via major social networks including Facebook, Twitter, and MySpace
Sports	Keep up with your favorite team or track your own golf scores
Travel and navigation	Use the GPS in your smartphone to get turn-by-turn directions, find interesting places to visit, access travel itineraries, and more

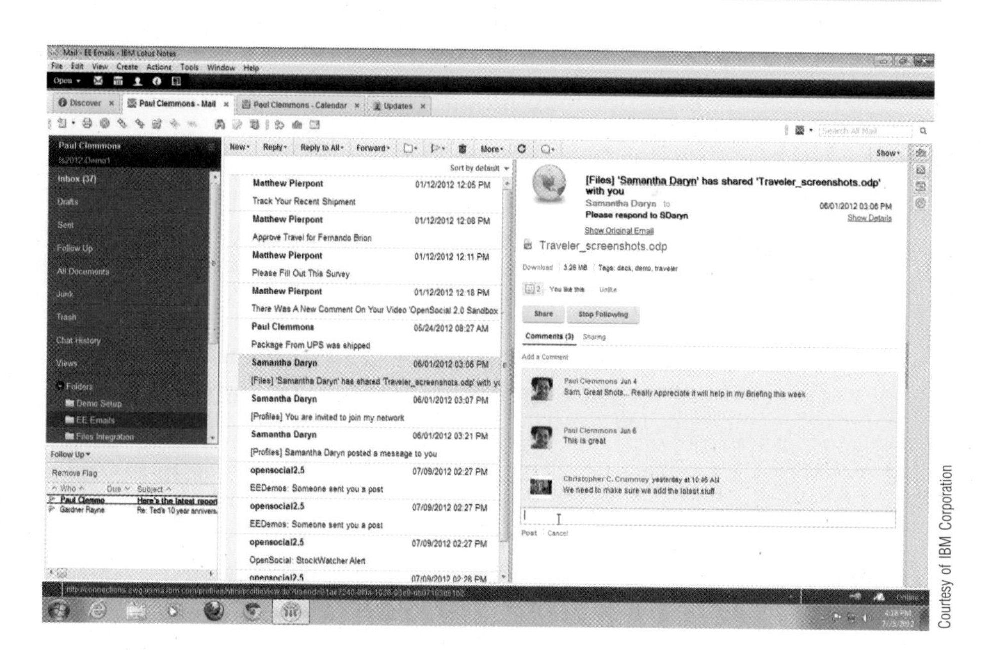

Courtesy of IBM Corporation

FIGURE **4.16**

IBM Lotus Notes Social Edition

IBM Lotus Notes Social Edition is workgroup software.

Enterprise Application Software

Software that benefits an entire organization—enterprise application software—can also be developed specifically for the business or purchased off the shelf. The Copper Mountain Ski Resort used Visual One software from Agilysys to manage condominiums and other real estate holdings.[83] According the information technology director, "We need a dynamic software system that allows us to manage our rather complex condominium lodging model." More CPA firms are acquiring sophisticated tax software for their corporate clients.[84] According to a manager at the Citrin Cooper CPA firm, "The number of available software programs has expanded in recent years. At the same time, more specialized software is available that focuses on individual industries." Verafin has developed specialized software that helps banks find people and organizations that attempt to launder money.[85] The software works by looking for suspicious transactions or patterns in large databases of financial transactions.[86]

Enterprise software also helps managers and workers stay connected. Traditional e-mail might not be the best approach.[87] According to the vice president of a large publishing company, "If you have a really important message you need to get to people; email is where it goes to die. People need a sense of ambient awareness." This type of awareness can come from enterprise software and group support systems, first introduced in Chapter 1. The following are some applications that can be addressed with enterprise software:

Accounts payable	Invoicing
Accounts receivable	Manufacturing control
Airline industry operations	Order entry
Automatic teller systems	Payroll
Cash-flow analysis	Receiving
Check processing	Restaurant management
Credit and charge card administration	Retail operations
Distribution control	Sales ordering
Fixed asset accounting	Savings and time deposits
General ledger	Shipping
Human resource management	Stock and bond management
Inventory control	Tax planning and preparation

According to a survey, cost is the greatest concern for selecting enterprise software.[88] Other factors include the difficulty to install and manage enterprise software and the ability to integrate enterprise software with other software applications. Increasingly, enterprise application software is being found on smartphones and mobile devices. In one survey, over 80 percent of respondents believe that having enterprise application software that can be used on smartphones and mobile devices was an important factor in selecting enterprise software.[89]

Application Software for Information, Decision Support, and Competitive Advantage

Specialized application software for information, decision support, and competitive advantage is available in every industry. For example, many schools and colleges use Blackboard or other learning management software to organize class materials and grades. Genetic researchers, as another example, are using software to visualize and analyze the human genome. Music executives use decision support software to help pick the next hit song. Companies seeking a competitive advantage, first discussed in Chapter 2, are increasingly building or developing their own enterprise software.[90] According to the CIO of the New York Stock Exchange Euronext, "Building is not easy. If it were, everyone would do it and we'd get no edge." But how are all these systems actually developed and built? The answer is through the use of programming languages, discussed next.

PROGRAMMING LANGUAGES

Both system and application software are written in coding schemes called *programming languages*. The primary function of a programming language is to provide instructions to the computer system so that it can perform a

ETHICAL & SOCIETAL ISSUES

Software Controls Nuclear Power Plants

The safety of nuclear power plants has always been an important consideration in their design. In the wake of the Fukushima plant failure after a record tsunami in March 2011, safety has an even higher priority. Using software to control power plants offers the potential of increased safety compared to earlier methods.

Duke Energy's Oconee nuclear power plant on the eastern shore of Lake Keowee near Seneca, South Carolina, was commissioned in 1973. As it entered the twenty-first century, its older analog control systems were showing their age. The plant suffered minor control failures during the 1990s, though no people were injured and no radiation leaked out as a result. Digital controls were added to some parts of the system in the late 1990s and early 2000s to deal with the most acute problems, but it was clear that Oconee's entire control structure needed to be replaced.

The purpose of a reactor protection system (RPS) is to protect the integrity of the plant's nuclear fuel by monitoring inputs from the reactor core. To accomplish this monitoring, application software must check sensors located throughout the reactor. If any safe operating values are exceeded, the software takes action, such as injecting cooling water or shutting the reactor down by inserting control rods.

After reviewing RPS applications, Duke Energy chose the Teleperm XS (TXS) system from Areva of France because TXS is designed to modernize existing analog instrumentation and control systems, and because its design includes features to ensure reliability. TXS is licensed in 11 countries and was already in use in other nuclear reactors outside the United States, thus assuring Duke that Oconee would not be a test site. TXS encompasses three functional systems:

- **Protection**: Monitoring safety parameters, enabling automatic protection and safeguard actions when an initiating event occurs
- **Surveillance**: Monitoring the core, rod control, and reactor coolant system and performing actions to protect reactor thresholds from being breached
- **Priority and actuator control system**: Managing the control and monitoring of operational and safety system actuators

Reactor Unit 1 of the Oconee facility became the first U.S. nuclear power plant to convert to fully digital control in May 2011. Unit 3 was converted in May 2012, with Unit 2 scheduled for May 2013. The first two conversions took place, and the third will take place, during the respective reactors' scheduled refueling shutdowns.

The nuclear power industry has recognized the importance of this instrumentation and control system upgrade. In May 2012, the Nuclear Energy Institute awarded Duke Energy its "Best of the Best" Top Industry Practice award. Speaking at the award ceremony, Preston Gillespie, vice president at the Oconee site, said, "When I look back over the decision of leaders that I worked for ten years ago, who had the vision of what it would take to install a safety-related digital system, I stand very much in respect of what those leaders did. They knew it would be hard; they knew the cost would be great; they knew they had to find the right partner; they knew they had to get it through the licensing process. All of this, they knew, would result in reliable and safe operation of the plant. Because of that vision, the trail is now blazed for the rest of the industry to take advantage of the fruits of their labor."

If the conversion goes well, other nuclear power plants will likely follow Oconee's lead as soon as they can afford it, said David Lochbaum, director of the Nuclear Safety Project for the Union of Concerned Scientists: "There are a lot of eyes on that. If it goes well, you'll probably see many people in the queue making it happen. If it doesn't go well, they are going to wait for Duke Energy to iron out the kinks."

Discussion Questions

1. Does a computer controlling a nuclear power plant need an operating system? Justify your answer in terms of what an operating system does and whether these functions are necessary in an RPS application.
2. Duke Energy selected off-the-shelf software for Oconee rather than writing custom software (or having a software development firm write it for them). Discuss the pros and cons of these two approaches in this situation. Do you think Duke Energy made the correct choice? Why or why not?

Critical Thinking Questions

1. At first glance, you might think a system that uses computers and software to control a nuclear plant has more ways to fail than a system that doesn't use them and is, therefore, at greater risk of failure. Why is a computer-controlled nuclear power system not at a greater risk of failure?
2. Computers are increasingly used to control systems that affect human lives. Besides nuclear power plants, examples include passenger aircraft, elevators, and medical equipment. Should the programmers who write software for those systems be licensed, be certified, or be required to pass standardized official examinations?

SOURCES: Areva Web site, *www.areva.com*, accessed May 31, 2012; Collins, J., "S.C. Nuke Plant First in U.S. to Go Digital," *Herald-Sun* (Durham, N.C.), *www.heraldsun.com/view/full_story/13488870/article-S-C–nuke-plant-first-in-U-S–to-go-digital*, May 29, 2011; Staff, "Oconee Nuclear Station Projects Honored with Three Awards by the Nuclear Energy Institute," Duke Energy, *www.duke-energy.com/news/releases/2012052301.asp*, May 23, 2012; Hashemian, H., "USA's First Fully Digital Station," *Nuclear Engineering International*, *www.neimagazine.com/story.asp?storyCode=2058654*, January 21, 2011; Staff, "Duke Energy Employees Win Top Nuclear Industry Award for Improving Safety With Digital Milestone," Nuclear Energy Institute, *www.nei.org/newsandevents/newsreleases/duke-energy-employees-win-top-nuclear-industry-award-for-improving-safety-with-digital-milestone*, May 23, 2012.

programming languages: Sets of keywords, commands, symbols, and rules for constructing statements by which humans can communicate instructions to a computer.

syntax: A set of rules associated with a programming language.

processing activity. Information systems professionals work with **programming languages**, which are sets of keywords, symbols, and rules for constructing statements that people can use to communicate instructions to a computer. Programming involves translating what a user wants to accomplish into a code that the computer can understand and execute. *Program code* is the set of instructions that signal the CPU to perform circuit-switching operations. In the simplest coding schemes, a line of code typically contains a single instruction such as, "Retrieve the data in memory address X." As discussed in Chapter 3, the instruction is then decoded during the instruction phase of the machine cycle. Like writing a report or a paper in English, writing a computer program in a programming language requires the programmer to follow a set of rules. Each programming language uses symbols, keywords, and commands that have special meanings and usage. Each language also has its own set of rules, called the **syntax** of the language. The language syntax dictates how the symbols, keywords, and commands should be combined into statements capable of conveying meaningful instructions to the CPU. Rules such

as "statements must terminate with a semicolon," and "variable names must begin with a letter," are examples of a language's syntax. A variable is a quantity that can take on different values. Program variable names such as SALES, PAYRATE, and TOTAL follow the syntax because they start with a letter, whereas variables such as %INTEREST, $TOTAL, and #POUNDS do not.

The Evolution of Programming Languages

The desire for faster, more efficient, more powerful information processing has pushed the development of new programming languages. The evolution of programming languages is typically discussed in terms of generations of languages (see Table 4.7).

TABLE 4.7 Evolution of programming languages

Generation	Language	Approximate Development Date	Sample Statement or Action
First	Machine language	1940s	00010101
Second	Assembly language	1950s	MVC
Third	High-level language	1960s	READ SALES
Fourth	Query and database languages	1970s	PRINT EMPLOYEE NUMBER IF GROSS PAY > 1000
Beyond Fourth	Natural and intelligent languages	1980s	IF gross pay is greater than 40, THEN pay the employee overtime pay

Visual, Object-Oriented, and Artificial Intelligence Languages

Today, programmers often use visual and object-oriented languages. In the future, they may be using artificial intelligence languages to a greater extent. In general, these languages are easier for nonprogrammers to use, compared with older generation languages.

Visual programming uses a graphical or "visual" interface combined with text-based commands. Prior to visual programming, programmers were required to describe the windows, buttons, text boxes, and menus that they were creating for an application by using only text-based programming language commands. With visual programming, the software engineer drags and drops graphical objects such as buttons and menus onto the application form. Then, using a programming language, the programmer defines the capabilities of those objects in a separate code window. Visual Basic was one of the first visual programming interfaces. Today, software engineers use Visual Basic .NET, Visual C++, Visual C# (# is pronounced "sharp" as in music), and other visual programming tools.

Many people refer to visual programming interfaces such as Visual C# as "visual programming languages." This custom is fine for casual references, but a lesser-known category of programming language is more truly visual. With a true visual programming language, programmers create software by manipulating programming elements only graphically, without the use of any text-based programming language commands. Examples include Alice, Mindscript, and Microsoft Visual Programming Language (VPL). Visual programming languages are ideal for teaching novices the basics about programming without requiring them to memorize programming language syntax.

Some programming languages separate data elements from the procedures or actions that will be performed on them, but another type of programming language ties them together into units called *objects*. An object consists of data and the actions that can be performed on the data. For example, an object could be data about an employee and all the operations (such as payroll calculations) that might be performed on the data. Programming languages that are based on objects are called *object-oriented programming languages*. C++ and Java are popular general-purpose object-oriented programming languages.[91] Languages used for Web development, such as Javascript and PHP, are also object oriented. In fact, most popular languages in use today take the object-oriented approach—and for good reason.

Using object-oriented programming languages is like constructing a building using prefabricated modules or parts. The object containing the data, instructions, and procedures is a programming building block. The same objects (modules or parts) can be used repeatedly. One of the primary advantages of an object is that it contains reusable code. In other words, the instruction code within that object can be reused in different programs for a variety of applications, just as the same basic prefabricated door can be used in two different houses. An object can relate to data on a product, an input routine, or an order-processing routine. An object can even direct a computer to execute other programs or to retrieve and manipulate data. So, a sorting routine developed for a payroll application could be used in both a billing program and an inventory control program. By reusing program code, programmers can write programs for specific application problems more quickly (see Figure 4.17). By combining existing program objects with new ones, programmers can easily and efficiently develop new object-oriented programs to accomplish organizational goals.

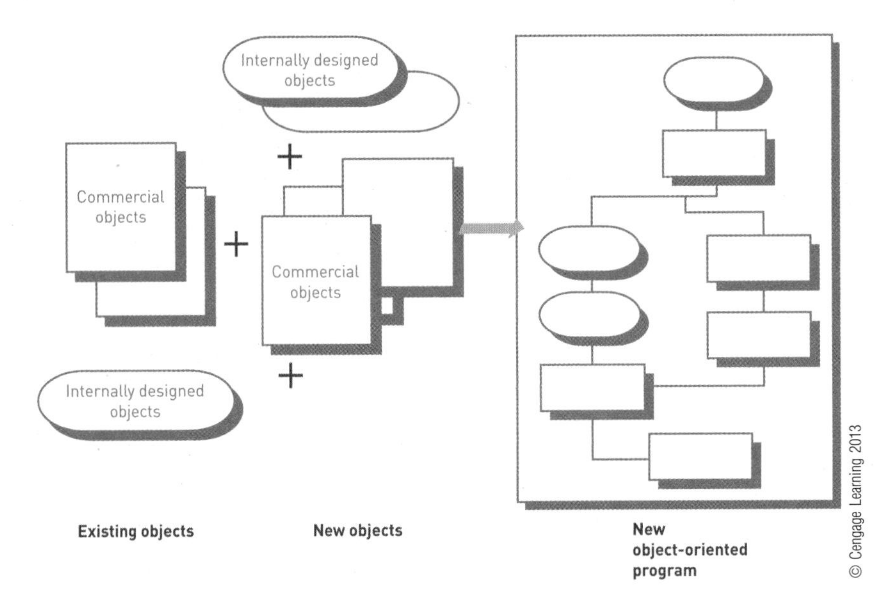

FIGURE 4.17

Reusable code in object-oriented programming

By combining existing program objects with new ones, programmers can easily and efficiently develop new object-oriented programs to accomplish organizational goals. Note that these objects can be either commercially available or designed internally.

Programming languages used to create artificial intelligence or expert systems applications are often called *fifth-generation languages (5GLs)*. Fifth-generation languages are sometimes called *natural languages* because they use even more English-like syntax than 4GLs. They allow programmers to communicate with the computer by using normal sentences. For example, computers programmed in fifth-generation languages can understand queries such as "How many athletic shoes did our company sell last month?"

With third-generation and higher-level programming languages, each statement in the language translates into several instructions in machine language. A special software program called a **compiler** converts the programmer's source code into the machine-language instructions, which consist of binary digits, as shown in Figure 4.18. A compiler creates a two-stage process for program execution. First, the compiler translates the program into a

compiler: A special software program that converts the programmer's source code into the machine-language instructions, which consist of binary digits.

machine language; second, the CPU executes that program. Another approach is to use an *interpreter*, which is a language translator that carries out the operations called for by the source code. An interpreter does not produce a complete machine-language program. After the statement executes, the machine-language statement is discarded, the process continues for the next statement, and so on.

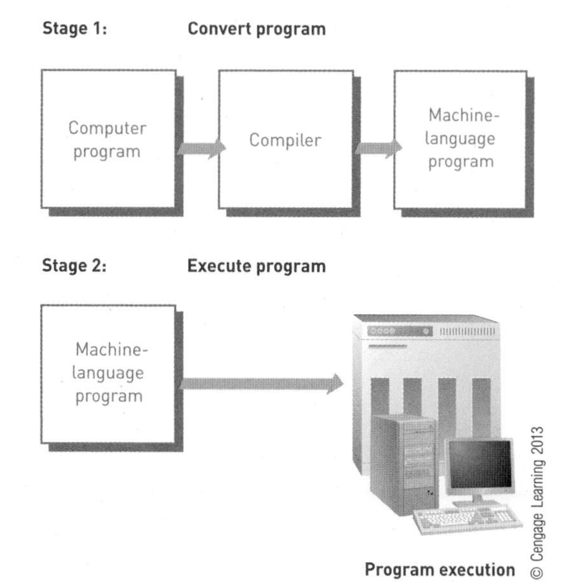

FIGURE 4.18

How a compiler works

A compiler translates a complete program into a complete set of binary instructions (Stage 1). After this is done, the CPU can execute the converted program in its entirety (Stage 2).

The majority of software used today is created using an integrated development environment. An *integrated development environment*, or *IDE*, combines all the tools required for software engineering into one package. For example, the popular IDE Microsoft Visual Studio includes an editor that supports several visual programming interfaces and languages, a compiler and interpreter, programming automation tools, a debugger (a tool for finding errors in the code), and other tools that provide convenience to the developer.[92]

Software Development Kits (SDKs) often serve the purpose of an IDE for a particular platform. For example, software developers for Google's Android smartphone platform use Java (an object-oriented programming language) along with the Eclipse SDK. They use special code libraries provided by Google for Android functionality, and they test out their applications in an Android Emulator.[93] See Figure 4.19.

IDEs and SDKs have made software development easier than ever. Many novice coders and some who might have never considered developing software are publishing applications for popular platforms such as Facebook and the iPhone.

SOFTWARE ISSUES AND TRENDS

Because software is such an important part of today's computer systems, issues such as software bugs, licensing, upgrades, global software support, and taxation have received increased attention. The governor of Colorado and the Colorado General Assembly, for example, repealed a tax on certain types of software.[94] The tax repeal should help Colorado's software industry. Software can also be harmful to companies and countries. For example, a foreign TV program might have shown software to its viewers designed to attack Web sites in the United States.[95]

Courtesy of Google

FIGURE **4.19**

Emulator for Android smartphones

To develop for the Android, you use an SDK with a mobile device emulator so you can prototype, develop, and test Android applications without transferring them to a physical device.

Software Bugs

A software bug is a defect in a computer program that keeps it from performing as it is designed to perform.[96] Some software bugs are obvious and cause the program to terminate unexpectedly. Other bugs are subtler and allow errors to creep into your work. Computer and software vendors say that as long as people design and program hardware and software, bugs are inevitable.[97] The following list summarizes tips for reducing the impact of software bugs:

- Register all software so that you receive bug alerts, fixes, and patches.
- Check the manual or read-me files for solutions to known problems.
- Access the support area of the manufacturer's Web site for patches.
- Install the latest software updates.
- Before reporting a bug, make sure that you can recreate the circumstances under which it occurs.
- After you can recreate the bug, call the manufacturer's tech support line.
- Consider waiting before buying the latest release of software to give the vendor a chance to discover and remove bugs. Many schools and businesses don't purchase software until the first major revision with patches is released.

Copyrights and Licenses

Most companies aggressively guard and protect the source code of their software from competitors, lawsuits, and other predators.[98] As a result, most software products are protected by law using copyright or licensing provisions. Those provisions can vary, however. In some cases, you are given unlimited use of software on one or two computers. This stipulation is typical with many applications developed for personal computers. In other cases, you pay for your usage: If you use the software more, you pay more. This approach is becoming popular with software placed on networks or larger computers. Most of these protections prevent you from copying software and giving it to others. Some software now requires that you *register* or *activate* it before it can be fully used. This requirement is another way software companies prevent illegal distribution of their products.

When people purchase software, they don't actually own the software, but rather they are licensed to use the software on a computer. This is called a single-user license. A **single-user license** permits you to install the software on one or more computers, used by one person. A single-user license does not allow you to copy and share the software with others. Table 4.8 describes different types of software licenses. Licenses that accommodate multiple users are usually provided at a discounted price.

single-user license: A software license that permits you to install the software on one or more computers, used by one person.

TABLE **4.8** Software licenses

License	Description
Single-user license	Permits you to install the software on one computer, or sometimes two computers, used by one person.
Multiuser license	Specifies the number of users allowed to use the software and can be installed on each user's computer. For example, a 20-user license can be installed on 20 computers for 20 users.
Concurrent-user license	Designed for network-distributed software, this license allows any number of users to use the software but only a specific number of users to use it at the same time.
Site license	Permits the software to be used anywhere on a particular site, such as a college campus, by everyone on the site.

Freeware and Open-Source Software

Some software developers are not as interested in profiting from their intellectual property as others and have developed alternative copyrights and licensing agreements. *Freeware* is software that is made available to the public for free.[99] Software developers might give away their product for several reasons. Some want to build customer interest and name recognition. Others simply don't need the money and want to make a valuable donation to society. Still others, such as those associated with the Free Software Foundation (*www.fsf.org*), believe that all software should be free. Some freeware is placed in the public domain where anyone can use the software free of charge. (Creative works that reach the end of their term of copyright revert to the public domain.) Table 4.9 shows some examples of freeware.

TABLE **4.9** Examples of freeware

Software	Description
Thunderbird	E-mail and newsgroup software
Pidgin	Instant messaging software
Adobe Reader	Software for viewing Adobe PDF documents
AVG Anti-Virus	Antivirus security software
WinPatrol	Antispyware software
IrfanView	Photo-editing software

Freeware differs slightly from free software. Freeware simply implies that the software is distributed for free. The term free software was coined by Richard Stallman and the Free Software Foundation and implies that the software is not only freeware, but it is also open source. **Open-source software** is

open-source software: Software that is distributed, typically for free, with the source code also available so that it can be studied, changed, and improved by its users.

distributed, typically for free, with the source code also available so that it can be studied, changed, and improved by its users.[100] Open-source software evolves from the combined contribution of its users. The Code For America (CFA) organization, for example, used open-source software in Boston and other American cities to help cities and municipalities solve some of their traffic problems, such as locating fire hydrants that might be completely covered with snow in the winter.[101] CFA made its efforts free to other cities and municipalities. Table 4.10 provides examples of popular open-source software applications.

TABLE **4.10** Examples of open-source software

Software	Category
Linux	Operating system
Open Office	Application software
MySQL	Database software
Mozilla Firefox	Internet browser
Gimp	Photo editing
OpenProj	Project management
Grisbi	Personal accounting

Open-source software is not completely devoid of restrictions. Much of the popular free software in use today is protected by the GNU General Public License (GPL). The GPL grants you the right to do the following:

- Run the program for any purpose
- Study how the program works and adapt it to your needs
- Redistribute copies so you can help others
- Improve the program and release improvements to the public

Software under the GPL is typically protected by a "copyleft" (a play on the word copyright), which requires that any copies of the work retain the same license. A copyleft work cannot be owned by any one person, and no one is allowed to profit from its distribution. The Free Software Directory (*http://directory.fsf.org*) lists over 5,000 software titles in 22 categories licensed under the GPL.

Why would an organization run its business using software that's free? Can something that's given away over the Internet be stable, reliable, or sufficiently supported to place at the core of a company's day-to-day operations? The answer is surprising—many believe that open-source software is often *more* reliable and secure than commercial software. How can this be? First, because a program's source code is readily available, users can fix any problems they discover. A fix is often available within hours of the problem's discovery. Second, because the source code for a program is accessible to thousands of people, the chances of a bug being discovered and fixed before it does any damage are much greater than with traditional software packages.

However, using open-source software does have some disadvantages. Although open-source systems can be obtained for next to nothing, the upfront costs are only a small piece of the total cost of ownership that accrues over the years that the system is in place. Some claim that open-source systems contain many hidden costs, particularly for user support or solving problems with the software. Licensed software comes with guarantees and support services while open-source software does not. Still, many businesses appreciate the additional freedom that open-source software provides. The question of software support is the biggest stumbling block to the acceptance of open-source software at the corporate level. Getting support for traditional software packages is easy—you call a company's toll-free support number or access its Web site. But how do you get help if an open-source package doesn't work as expected? Because the open-source community lives on the Internet, you look there for help. Through use of Internet discussion areas, you can communicate with others who use the same software, and you might even reach someone who helped develop it. Users of popular open-source packages can get correct answers to their technical questions within a few hours of asking for help on the appropriate Internet forum. Another approach is to contact one of the many companies emerging to support and service such software—for example, Red Hat for Linux and Sendmail, Inc., for Sendmail. These companies offer high-quality, for-pay technical assistance.

Software Upgrades

Software companies revise their programs periodically. Software upgrades vary widely in the benefits that they provide, and what some people call a benefit others might call a drawback. Deciding whether to upgrade to a new version of software can be a challenge for corporations and people with a large investment in software. Should the newest version be purchased when it is released? Some users do not always get the most current software upgrades or versions unless it includes significant improvements or capabilities. Developing an upgrading strategy is important for many businesses. American Express, for example, has standardized its software upgrade process around the world to make installing updated software faster and more efficient.[102] The standardized process also helps the company make sure that updated software is more stable with fewer errors and problems.

Global Software Support

Large global companies have little trouble persuading vendors to sell them software licenses for even the most far-flung outposts of their company. But can those same vendors provide adequate support for their software customers in all locations? Supporting local operations is one of the biggest challenges IS teams face when putting together standardized companywide systems. Slower technology growth markets, such as Eastern Europe and Latin America, might not have any official vendor presence. Instead, large vendors such as Sybase, IBM, and Hewlett-Packard typically contract with local providers to support their software.

One approach that has been gaining acceptance in North America is to outsource global support to one or more third-party distributors. The user company can still negotiate its license with the software vendor directly, but it then hands the global support contract to a third-party supplier. The supplier acts as a middleman between software vendor and user, often providing distribution, support, and invoicing.

In today's computer systems, software is an increasingly critical component. Whatever approach people and organizations take to acquire software, everyone must be aware of the current trends in the industry. Informed users are wise consumers.

SUMMARY

Principle:

Systems and application software are critical in helping individuals and organizations achieve their goals.

Software consists of programs that control the workings of the computer hardware. The two main categories of software are systems software and application software. Systems software is a collection of programs that interacts between hardware and application software and includes operating systems, utility programs, and middleware. Application software can be proprietary or off the shelf and enables people to solve problems and perform specific tasks.

An operating system (OS) is a set of computer programs that controls the computer hardware to support users' computing needs. An OS converts an instruction from an application into a set of instructions needed by the hardware. This intermediary role allows hardware independence. An OS also manages memory, which involves controlling storage access and use by converting logical requests into physical locations and by placing data in the best storage space, including virtual memory.

An OS manages tasks to allocate computer resources through multitasking and time sharing. With multitasking, users can run more than one application at a time. Time sharing allows more than one person to use a computer system at the same time.

The ability of a computer to handle an increasing number of concurrent users smoothly is called *scalability*, a feature critical for systems expected to handle a large number of users.

An OS also provides a user interface, which allows users to access and command the computer. A command-based user interface requires text commands to send instructions. A graphical user interface (GUI), such as Windows, uses icons and menus. Other user interfaces include touch and speech.

Software applications use the OS by requesting services through a defined application program interface (API). Programmers can use APIs to create application software without having to understand the inner workings of the OS. APIs also provide a degree of hardware independence so that the underlying hardware can change without necessarily requiring a rewrite of the software applications.

Over the years, many popular OSs have been developed, including Microsoft Windows, the Mac OS X, and Linux. There are several options for OSs in the enterprise as well, depending on the type server. UNIX is a powerful OS that can be used on many computer system types and platforms, from workstations to mainframe systems. Linux is the kernel of an OS whose source code is freely available to everyone. Some OSs, such as Mac OS X iPhone, Windows Embedded, Symbian, Android, webOS, and variations of Linux, have been developed to support mobile communications and consumer appliances. When an OS is stored in solid state memory, embedded in a device, it is referred to as an embedded operating system or an embedded system for short.

Utility programs can perform many useful tasks and often come installed on computers along with the OS. This software is used to merge and sort sets of data, keep track of computer jobs being run, compress files of data, protect against harmful computer viruses, monitor hardware and network performance, and perform dozens of other important tasks. Virtualization software simulates a computer's hardware architecture in software so that computer systems can run operating systems and software designed for other architectures or run several operating systems simultaneously on one system. Middleware is software that allows different systems to communicate and transfer data back and forth.

Principle:

Organizations use off-the-shelf application software for common business needs and proprietary application software to meet unique business needs and provide a competitive advantage.

Application software applies the power of the computer to solve problems and perform specific tasks. One useful way of classifying the many potential uses of information systems is to identify the scope of problems and opportunities addressed by a particular organization or its sphere of influence. For most companies, the spheres of influence are personal, workgroup, and enterprise.

User software, or personal productivity software, includes general-purpose programs that enable users to improve their personal effectiveness, increasing the quality and amount of work that can be done. Software that helps groups work together is often called workgroup application software. It includes group scheduling software, electronic mail, and other software that enables people to share ideas. Enterprise software that benefits the entire organization, called enterprise resource planning software, is a set of integrated programs that help manage a company's vital business operations for an entire multisite, global organization.

Three approaches to acquiring application software are to build proprietary application software, buy existing programs off the shelf, or use a combination of customized and off-the-shelf application software. Building proprietary software (in-house or on contract) has the following advantages. The organization gets software that more closely matches its needs. Further, by being involved with the development, the organization has further control over the results. Finally, the organization has more flexibility in making changes. The disadvantages include the following. It is likely to take longer and cost more to develop. Additionally, the in-house staff will be hard pressed to provide ongoing support and maintenance. Lastly, there is a greater risk that the software features will not work as expected or that other performance problems will occur.

Some organizations have taken a third approach—customizing software packages. This approach usually involves a mixture of the preceding advantages and disadvantages and must be carefully managed.

An application service provider (ASP) is a company that provides the software, support, and computer hardware on which to run the software from the user's facilities over a network. ASPs customize off-the-shelf software on contract and speed deployment of new applications while helping IS managers avoid implementation headaches. ASPs reduce the need for many skilled IS staff members and also lower a project's start-up expenses. Software as a Service (SaaS) allows businesses to subscribe to Web-delivered business application software by paying a monthly service charge or a per-use fee.

SaaS and recent Web development technologies have led to a new paradigm in computing called cloud computing. Cloud computing refers to the use of computing resources, including software and data storage, on the Internet (the cloud), not on local computers. Rather than installing, storing, and running software on your own computer, with cloud computing, you access software stored on and delivered from a Web server.

Although hundreds of computer applications can help people at school, home, and work, the most popular applications are word processing, spreadsheet analysis, database, graphics, and personal information management. A software suite, such as SmartSuite, WordPerfect, StarOffice, or Microsoft Office, offers a collection of these powerful programs sold as a bundle.

Many thousands of applications are designed for businesses and workgroups. Business software generally falls under the heading of information systems that support common business activities, such as accounts receivable, accounts payable, inventory control, and other management activities.

Principle:

Organizations should choose programming languages with functional characteristics that are appropriate for the task at hand and well suited to the skills and experience of the programming staff.

All software programs are written in coding schemes called *programming languages*, which provide instructions to a computer to perform some processing activity. The several classes of programming languages include machine, assembly, high-level, query and database, object-oriented, and visual programming.

Programming languages have changed since their initial development in the early 1950s. In the first generation, computers were programmed in machine language, and in the second, assembly languages were used. The third generation consists of many high-level programming languages that use Englishlike statements and commands. They must be converted to machine language by special software called a compiler and include BASIC, COBOL, FORTRAN, and others. Fourth-generation languages include database and query languages such as SQL.

Fifth-generation programming languages combine rules-based code generation, component management, visual programming techniques, reuse management, and other advances. Object-oriented programming languages use groups of related data, instructions, and procedures called *objects*, which serve as reusable modules in various programs. These languages can reduce program development and testing time. Java can be used to develop applications on the Internet. Visual programming environments, integrated development environments (IDEs), and software development kits (SDKs) have simplified and streamlined the coding process and have made it easier for more people to develop software.

Principle:

The software industry continues to undergo constant change; users need to be aware of recent trends and issues to be effective in their business and personal life.

Software bugs, software licensing and copyrighting, open-source software, shareware and freeware, multiorganizational software development, software upgrades, and global software support are all important software issues and trends.

A software bug is a defect in a computer program that keeps it from performing in the manner intended. Software bugs are common, even in key pieces of business software.

Freeware is software that is made available to the public for free. Open-source software is freeware that also has its source code available so that others may modify it. Open-source software development and maintenance is a collaborative process, with developers around the world using the Internet to download the software, communicate about it, and submit new versions of it.

Software upgrades are an important source of increased revenue for software manufacturers and can provide useful new functionality and improved quality for software users.

Global software support is an important consideration for large global companies putting together standardized companywide systems. A common solution is outsourcing global support to one or more third-party software distributors.

CHAPTER 4: SELF-ASSESSMENT TEST

Systems and application software are critical in helping individuals and organizations achieve their goals.

1. Which of the following is an example of a command-driven operating system?
   a. XP
   b. Snow Leopard
   c. MS DOS
   d. Windows 7
2. Many of today's mobile devices and some PCs use a touch user interface, also called a *natural user interface (NUI)* or multitouch interface by some. True or False?
3. ______________ is an open-source OS that is used in all computer platforms: PC, server, embedded, smartphones, and other.
4. Spam filtering is a function of the operating system. True or False?
5. Some companies use ______________ to run multiple operating systems on a single computer.
   a. multitasking
   b. middleware

c. service-oriented architecture
d. virtualization

Organizations use off-the-shelf application software for common business needs and proprietary application software to meet unique business needs and provide a competitive advantage.

6. Application software that determines the best shipping option for products from a manufacturing company to a consumer is software for the personal sphere of influence. True or False?
7. Software that enables users to improve their personal effectiveness, increasing the amount of work they can do and its quality, is called ______________.
   a. personal productivity software
   b. operating system software
   c. utility software
   d. graphics software
8. Optimization can be found in which type of application software?
   a. spreadsheets
   b. word-processing programs
   c. personal information management programs
   d. presentation graphics programs
9. ______________ software is one-of-a-kind software designed for a specific application and owned by the company, organization, or person that uses it.
10. ______________ allows businesses to subscribe to Web-delivered business application software by paying a monthly service charge or a per-use fee.
    a. Software as a Service (SaaS)
    b. An application service provider (ASP)
    c. Proprietary software
    d. Off-the-shelf software

Organizations should choose programming languages with functional characteristics that are appropriate for the task at hand and well suited to the skills and experience of the programming staff.

11. Most software purchased to run on a personal computer uses a ______________ license.
    a. site
    b. concurrent-user
    c. multiuser
    d. single-user
12. One of the primary advantages of ______________ programming is the use of reusable code modules that save developers from having to start coding from scratch.
13. Each programming language has its own set of rules, called the ______________ of the language.
14. An object-oriented language converts a programmer's source code into the machine-language instructions consisting of binary digits. True or False?

The software industry continues to undergo constant change; users need to be aware of recent trends and issues to be effective in their business and personal life.

15. ______________ allows users to tweak the software to their own needs.
    a. Freeware
    b. Off-the-shelf software
    c. Open-source software
    d. Software in the public domain
16. What type of license is an enterprise likely to purchase for software that it intends all of its employees to use while on site?

CHAPTER 4: SELF-ASSESSMENT TEST ANSWERS

1. c
2. True
3. Linux
4. False
5. d
6. False
7. a
8. a
9. Proprietary
10. a
11. d
12. object-oriented
13. syntax
14. False
15. c
16. site license

REVIEW QUESTIONS

1. What is the difference between systems software and application software? Give four examples of personal productivity software.
2. What steps can a user take to correct software bugs?
3. Identify and briefly discuss two types of user interfaces provided by an operating system. What are the benefits and drawbacks of each?
4. What is a software suite? Give several examples.

5. Name four operating systems that support the personal sphere of influence.
6. What is Software as a Service (SaaS)?
7. What is multitasking?
8. Define the term *utility software* and give two examples.
9. Identify the two primary sources for acquiring application software.
10. What is cloud computing? What are the pros and cons of cloud computing?
11. What is open-source software? What are the benefits and drawbacks for a business that uses open-source software?
12. What does the acronym API stand for? What is the role of an API?
13. Briefly discuss the advantages and disadvantages of frequent software upgrades.
14. List four application software packages that would be useful for an enterprise.
15. What is the difference between freeware and open-source software?

DISCUSSION QUESTIONS

1. Assume that you must take a computer-programming course next semester. What language do you think would be best for you to study? Why? Do you think that a professional programmer needs to know more than one programming language? Why or why not?
2. You are going to buy a personal computer. What operating system features are important to you? What operating system would you select and why?
3. You have been asked to develop a user interface for someone with limited sight—someone without the ability to recognize shapes on a computer screen. Describe the user interface you would recommend.
4. You are using a new release of an application software package. You think that you have discovered a bug. Outline the approach that you would take to confirm that it is indeed a bug. What actions would you take if it truly were a bug?
5. For a company of your choice, describe the three most important application software packages you would recommend for the company's profitability and success.
6. Define the term *Software as a Service (SaaS)*. What are some of the advantages and disadvantages of employing a SaaS? What precautions might you take to minimize the risk of using one?
7. Describe three personal productivity software packages you are likely to use the most. What personal productivity software packages would you select for your use?
8. Describe the most important features of an operating system for a smartphone.
9. If you were the IS manager for a large manufacturing company, what issues might you have with the use of open-source software? What advantages might there be for use of such software?
10. Identify four types of software licenses frequently used. Which approach does the best job of ensuring a steady, predictable stream of revenue from customers? Which approach is most fair for the small company that makes infrequent use of the software?
11. How have software development kits (SDKs) influenced software development?
12. How can virtualization save a company a lot of money?

PROBLEM-SOLVING EXERCISES

1. Develop a ten-slide presentation that compares the advantages and disadvantages of proprietary software versus off-the shelf software.
2. Use a spreadsheet package to prepare a simple monthly budget and forecast your cash flow—both income and expenses—for the next six months (make up numbers rather than using actual ones). Now use a graph to plot the total monthly income and monthly expenses for six months. Cut and paste both the spreadsheet and the graph into a word-processing document that summarizes your (fictitious) financial situation.
3. Use a database program to enter five software products you are likely to use at work. List the name, vendor or manufacturer, cost, and features in the columns of a database table. Use a word processor to write a report on the software. Copy the database table into the word-processing program.

TEAM ACTIVITIES

1. Form a group of three or four classmates. Find articles from business periodicals, search the Internet, or interview people on the topic of cloud computing. Make sure to analyze the advantages and disadvantages of the cloud-computing approach. Compile your results for an in-class presentation or a written report.
2. Form a group of three or four classmates. Identify and contact an employee of any local business or organization. Interview the individual and describe the application software the company uses and discuss the importance of the software to the organization. Write a brief report summarizing your findings.
3. Team members should learn how to use a PC operating system with which they are unfamiliar. Explore how to launch applications, minimize and maximize windows, close applications, view files on the system, and change system settings such as the wallpaper. Team members should collaborate on a report, using the track changes features of Word or collaborative features of Google Docs to summarize findings and opinions on at least three PC OSs.

WEB EXERCISES

1. Use the Web to research four productivity software suites from various vendors (see *http://en.wikipedia.org/wiki/Office_Suite*). Create a table in a word-processing document to show what applications are provided by the competing suites. Write a few paragraphs on which suite you think best matches your needs and why.
2. Use the Internet to search for three popular freeware utilities that you would find useful. Write a report that describes the features of these three utility programs.
3. Use the Internet to search for information on embedded operating systems. Describe how embedded operating systems can be used in vehicles, home appliances, TVs, and other devices.
4. Do research on the Web about application software that is used in an industry and is of interest to you. Write a brief report describing how the application software can be used to increase profits or reduce costs.

CAREER EXERCISES

1. What applications for a smartphone or other mobile device would help you the most in your next professional job? Why? What features are the most important to you?
2. Think of your ideal job. Describe five application software packages that could help you advance in your career. If the software package doesn't exist, describe the kinds of software packages that could help you in your career.

CASE STUDIES

Case One

Tendring District Council: Open for Business Online

Tendring District Council, with a population of about 150,000, is located in the county of Essex in southeast England. Every year the council receives thousands of applications for building permits and other items that could be affected by regulations or could affect other people.

Applications to Tendring District Council span a wide range of requests. Applications received during the week ending May 25, 2012, for example, ranged from a request by Mr. A. Maloney of Frinton & Walton to prune a cherry tree in his front garden to a request by Mr. T. Munson of Wix to install two wind turbines, 50-feet tall from ground to hub, with 18-foot blades. Installing two wind turbines needs more consideration than pruning a cherry tree, but Tendring District Council must process both requests and render decisions according to the established rules.

When making decisions, the council solicits the opinions of neighbors, neighborhood organizations, the Essex County highway department, National Heritage (for buildings or locations of historic importance), and gas, electricity, and water companies. In a typical year, Tendring issues about 9,000 requests for comments—*consultations*, in the official terminology—and receives about 8,000 replies.

Traditionally, the council sent all consultations by mail, which posed three problems:

1. Significant costs are associated with producing multiple copies of the documents and for postage.
2. Sending and receiving paper files delayed planning activities.
3. The consultation process generated a great deal of paper, all of which had to be stored (requiring space) or discarded (having an environmental impact, even with recycling).

To reduce or eliminate those problems, Tendring decided to invest in an electronic document management system (EDMS) using software from Idox. This system generates electronic consultations where the consultee—that is, one with whom Tendring District Council consults—has an e-mail address. The consultation e-mail contains all relevant details of the planning proposal plus a hyperlink the consultee can use to view and comment on the proposal on Tendring's Web site.

Tendring recognized the importance of adapting the system to its users, not forcing users to adapt to the system. The council knew that instead of using the EDMS, some consultees might prefer to reply by e-mail, while others would prefer to submit hard copy documents. Tendring, therefore, left open the traditional e-mail and hard copy routes for responding but found that few consultees used them. Two reasons might be that the EDMS also allows comments to be public or private and that it maintains appropriate security to ensure the privacy of comments that have been submitted as private.

Today, 99 percent of planning consultations are handled electronically. (This figure includes both Web site and e-mail replies.) The district council's finance department calculated total savings at £150,000 (about U.S. $230,000) per year, including £8,000 (about U.S. $12,500) on postage alone. Consultees like the new system, too. Vicky Presland, district manager of the Essex County East Area Highways Office, says "Consultee Access has saved considerable time in producing our responses to the local Planning Authority, and has allowed us to reduce our own filing systems due to the easy access to ours and other consultees' responses." This system truly has no losers.

Discussion Questions

1. Tendring District Council selected an off-the-shelf package for its document management system rather than developing custom software or having custom software developed for them. Do you agree with this decision? Justify your answer.
2. What is the sphere of influence of the Tendring District Council EDMS and online consultee response application: personal, workgroup, or enterprise? Indicate whether the system has aspects of more than one sphere of influence, being as specific as possible.

Critical Thinking Questions

1. Tendring District Council is responsible for a relatively small area. Larger regional and municipal government agencies have used electronic document systems for much longer than Tendring has. Why do you think other agencies use electronic document systems? Consider both business and technical factors.
2. Suppose Tendring evaluated EDMS systems and found that the best one for its needs was not compatible with its existing operating system. What would you suggest Tendring do? Justify your answer.

SOURCES: Staff, "Tendring Saves £150,000 per Annum with Idox e-Planning," *www.idoxgroup.com/downloads/news/Idox_case_study_Tendring_e-Planning.pdf*, May 17, 2011; Idox Group Web site, *www.idoxgroup.com*, accessed May 31, 2012; Tendring District Council Web site, *www.tendringdc.gov.uk*, accessed May 31, 2012.

Case Two

Your Next Car on a Tablet

Like most companies today, automotive information publisher Edmunds collects and analyzes statistical data about visitors to its Web site. When chief operating officer Seth Berkowitz saw a spike in the percentage of mobile page views that came from iPads in the summer of 2010, he knew something was up.

In April 2011, Edmunds launched an iPhone app (application program) for its Inside Line car fan site, followed by an iPad version in June. An Android version (for the other major smartphone and tablet OS) is "in the long-range plan."

However, for a business, the decision is never as simple as "Let's develop an iPhone app." It's not enough to know that visitors to its Web site use certain mobile devices. A business must also know *why*. To develop an app without that knowledge is to risk developing the wrong app. That would be worse than not having one at all. However, that information, the *why*, cannot be obtained from statistical analysis of Web-visit data. Coming up with an app that would earn a five-star rating on the iTunes store requires more than programming skills. It calls for a deep understanding of why mobile users access a site and how their needs differ from those of people who use other devices. For example, knowing that users of smartphones have small screens but may still need to see details, Edmunds included high-quality car photos that users can zoom to enlarge.

To Edmunds users, the Inside Line iPhone app can be the most convenient way to access the Edmunds site. Not all car buyers prefer or even use tablets (any more than people in any other group), but some do. In a competitive market, being accessible to all potential customers, not just some, can help companies gain an advantage over competitors. As one blogger put it, "I'll be in the market for another car soon. Used, of course. Which is where the value of the Edmunds site really comes into play, since owners are free to review these cars and post them on the site."

But to Edmunds, tablets such as the iPad offer more than a way to connect with its customers. They also support valuable business analysis tools. Using an app that Edmunds.com developed with tools from software company MicroStrategy, Edmunds management can sort and analyze huge volumes of data about auto sales. President Avi Steinlauf refers to this app when he says that, "as a result of having more data at his fingertips [a manager] can ask more specific questions of his staff and make quicker decisions."

The analytics software indicates such tendencies as the propensity of a consumer to consider other models when researching a particular vehicle. Automakers can review and interact with the data on their iPads to help them with marketing and advertising decisions, MicroStrategy said. Solomon Kang, director of client analytic services at Edmunds.com, agrees: "Our new iPad app is particularly critical for executives who are always on the go and need to be able to react quickly."

Discussion Questions

1. Why did Edmunds develop the Inside Line iPhone app? It's nice to provide Inside Line as a free news service and discussion forum for car fans, but companies that provide free services for no reason don't stay in business. How does Edmunds expect to earn enough money with this app to justify its development and support costs?
2. Companies must plan their software carefully before developing it. Think of a situation where a mobile app for accessing information about cars would be more useful than a site that required a full-size computer. What features would such an app need that a fixed-location Web site would not need?

Critical Thinking Questions

1. Gulliver International sells about $2 billion worth of used cars per year in Japan. Compare its iPad app (*www.apple.com/ipad/business/profiles/gulliver*) with that of Edmunds. Start by considering the two firms' objectives with their software packages: how are their objectives the same and how are they different?
2. This case study says that Edmunds's Inside Line app includes full-resolution photos that users can zoom. The other possible decision would have been to include low-resolution photos that load faster and use fewer megabytes of a smartphone's data cap but do not offer the zoom option. For what market or application might that software design decision have been the better choice? What differences between the two make you think so?

SOURCES: Campbell, J., "Edmunds Has Their Own App (Finally!)," Apple Thoughts, *forums.thoughtsmedia.com/f387/edmunds-has-their-own-app-finally-124564.html*, December 5, 2011; Taylor, P., "iPad Case Study: Edmunds.com," *Financial Times*, *www.ft.com/intl/cms/s/0/d8e5eda6-613c-11e0-ab25-00144feab49a.html#axzz1b6VLxELL* (free registration required), April 8, 2011; Edmunds, "*Inside Line* App Comes to iPad," Inside Line, *www.insideline.com/car-news/inside-line-app-comes-to-ipad.html*, June 3, 2011; Staff, "*Inside Line* Launches iPhone App," Edmunds, *Inside Line*, *www.insideline.com/car-news/inside-line-launches-iphone-app.html*, April 14, 2011; Staff, "Edmunds' InsideLine iPhone and iPad Apps Earn Five-Star Ratings in iTunes Store," Edmunds, *Business Wire*, *www.businesswire.com/news/home/20110826005691/en/Edmunds'-InsideLine-iPhone-iPad-Apps-Earn-Five-Star*, August 26, 2011; Konrad, A., "Tablets Storm the Corner Office," *Fortune*, *tech.fortune.cnn.com/2011/10/13/ipad-executives-managing*, October 17, 2011; Moore, C. W., "Edmunds InsideLine iPhone and iPad Automotive Enthusiast App," PowerBook Central, *www.pbcentral.com/blog/2011/08/29/edmunds-insideline-iphone-and-ipad-automotive-enthusiast-app*, August 29, 2011.

Questions for Web Case

See the Web site for this book to read about the Altitude Online case for this chapter. Following are questions concerning this Web case.

Altitude Online: Systems and Application Software

Discussion Questions

1. Why do you think Altitude Online uses two PC platforms—Windows and Mac—rather than standardizing on one? What are the benefits and drawbacks of this decision?
2. Why do you think a business is required to keep copies of all of its software licenses?

Critical Thinking Questions

1. How much freedom should a company like Altitude Online allow for its employees to choose their own personal application software? Why might a company prefer to standardize around specific software packages?
2. What benefits might be provided to an advertising media company like Altitude Online by upgrading to the latest media development and production software? How might upgrading provide the company with a competitive advantage?

NOTES

Sources for the opening vignette: Staff, "What Is Microfinance?," CGAP, *www.cgap.org/p/site/c/template.rc/1.26.1302*, accessed June 14, 2012; Staff, "Microfinance Gateway," CGAP, *www.microfinancegateway.org/p/site/m/template.rc/1.26.12263*, 2012; Equitas Web site, *www.equitas.in*, accessed January 3, 2012; Staff, "Equitas Micro—Case Study," Temenos, *www.temenos.com/Equitas-Micro—Case-Study* (free registration required), 2011.

1. "Chinese Software Revenues in H1 2011 Surged 29.3%," *Asia Pulse*, July 29, 2011.
2. Andreessen, Marc, "Why Software Is Eating the World," *The Wall Street Journal*, August 20, 2011, p. C2.
3. Jones, S., "Microsoft Challenges Itself in the Clouds," *The Wall Street Journal*, June 27, 2011, p. B2.
4. Gonzalez, J., "Boise Planners Adopt New Software to Speed Up Permitting Process," *The Idaho Business Review*," August 3, 2011.

5. "Eddie Bauer Protects Staff with Amcom Software," *Business Wire*, August 4, 2011.
6. MindJet Web site, *info.mindjet.com*, accessed September 12, 2011.
7. IBM Web site, *www-01.ibm.com/software/data/cognos*, accessed September 19, 2011.
8. Mossberg, W., "Apple's Lion Brings PCs into the Tablet Era," *The Wall Street Journal*, July 21, 2011, p. D1.
9. Microsoft Web site, *windows.microsoft.com/en-US/windows7/What-can-I-do-with-Speech-Recognition*, accessed September 18, 2011.
10. Microsoft Web site, *technet.microsoft.com/en-us/windows/dd320286*, accessed September 12, 2011.
11. Wingfield, N. and Tibken, S., "Microsoft to Limit Tablets," *The Wall Street Journal*, June 2, 2011, p. B4.
12. Keizer, G., "Microsoft Gambles with Windows 8," *Computerworld*, June 20, 2011, p. 6.
13. Grundberg, S. and Ovide, S., "A Test Ride for Windows 8," *The Wall Street Journal*, March 1, 2012, p. B4.
14. Henderson, Tom, "Windows 8 Breaks New Ground," *Network World*, January 23, 2012, p. 28.
15. Ramstad, E., "Samsung Plans to Expand Tablet Line to Use Windows," *The Wall Street Journal*, September 9, 2011, p. B4.
16. Apple Web site, *www.apple.com/macosx*, accessed September 18, 2011.
17. Vascellaro, Jessica, "Apple's Mac Makeover," *The Wall Street Journal*, February 17, 2012, B1.
18. Nelson, F., "IT Pro Ranking," *InformationWeek*, September 5, 2011, p. 16.
19. Vance, Ashlee, "Red Hat Sees Lots of Green," *Bloomberg Businessweek*, April 2, 2012, p. 41.
20. "And Now, Google's Other Operating System," *Bloomberg Businessweek*, June 13, 2011, p. 42.
21. Kowitt, B., "One Hundred Million Android Fans Can't Be Wrong," *Fortune*, July 4, 2011, p. 93.
22. Mossberg, Walter, "Mobile Device That's Better for Jotter than a Talker," *The Wall Street Journal*, February 16, 2012, p. D1.
23. Clayburn, T., "Google Gambles on Chromebooks," *InformationWeek*, May 30, 2011, p. 18.
24. Microsoft Web site, *www.microsoft.com/windows/products/winfamily/windowshomeserver/default.mspx*, accessed September 20, 2011.
25. Binstock, Andrew, "Q&A: Ken Thompson, Creator of Unix," *InformationWeek*, June 27, 2011, p. 45.
26. Sun Web site, *www.sun.com/customers/index.xml?c=ebay.xml&submit=Find*, accessed September 12, 2011.
27. "Red Hat RHEV Freed from Windows Fetters," *Network World*, August 22, 2011, p. 8.
28. Red Hat Web site, *www.redhat.com/virtualization/rhev*, accessed September 12, 2011.
29. Dornan, A., "Linux Virtualization Finds Some Rich Uncles," *InformationWeek*, June 13, 2011, p. 21.
30. IBM Web site, *www-03.ibm.com/systems/z/os/zos*, accessed September 20, 2011.
31. Fletcher, O., "Alibaba Develops Cloud Mobile Operating System," *The Wall Street Journal*, July 5, 2011, p. B5.
32. Apple Web site, *www.apple.com/ios/ios5*, accessed September 17, 2011.
33. Sherr, I., "H-P Looks to Kitchens, Cars," *The Wall Street Journal*, August 16, 2011, p. B5.
34. HP Web site, *www.hpwebos.com/us/products/software/webos2*, accessed September 15, 2011.
35. Lawton, C. and Kim, Y., "Nokia Updates Smart Phones," *The Wall Street Journal*, August 25, 2011, p. B5.
36. King, C., "RIM, Dolby Settle Dispute," *The Wall Street Journal*, September 13, 2011, p. B9.
37. Microsoft Web site, *www.microsoft.com/windowsembedded/en-us/windows-embedded.aspx*, accessed September 20, 2011.
38. Ford Motors Web site, *www.ford.com/technology/sync*, accessed September 20, 2011.
39. Mossberg, W., "A Parallels World Where Windows Zips on Macs," *The Wall Street Journal*, September 1, 2011, p. D1.
40. CERN Web site, *http://wlcg.web.cern.ch*, accessed September 12, 2011.
41. Babcock, C., "What You Can't See," *InformationWeek*, September 5, 2011, p. 18.
42. Nance, Barry, "HP, IBM, CA Deliver Powerful Toolkits," *Network World*, March 12, 2012, p. 26.
43. "James River Insurance Selects Confio Software," *Business Wire*, August 2, 2011.
44. Healy, M., "The OS Mess," *InformationWeek*, July 11, 2011, p. 21.
45. Murphy, C., "Is Management Software RIM's Secret Weapon?" *InformationWeek*, September 5, 2011, p. 6.
46. CNET Web site, *www.cnet.com*, accessed September 20, 2011.
47. "First Financial Asset Management Deploys Debt Collection Solution from Latitude Software," *Business Wire*, June 30, 2011.
48. "Accenture to Expand Property and Casualty Insurance Software," *Business Wire*, July 2011.
49. Williams, J., "Advanced Analytics at Land O'Lakes," *OR/MS Today*, August 2011, p. 18.
50. Swedish Medical Center Web site, *www.swedish.org*, accessed September 12, 2011.
51. McClatchy, S., "New Software Installed to Help Speed Airport Screening," *Tribune Business News*, August 11, 2011.
52. Hodge, N., "Killer App," *The Wall Street Journal*, June 3, 2011, p. A2.
53. El Camino Hospital Web site, *www.elcaminohospital.org/Locations/El_Camino_Hospital_Mountain_View*, accessed September 12, 2011.
54. "Globus Online to Provide Software-as-a-Service for NSF," *PR Newswire*, September 2, 2011.
55. "Big SaaS Done Right," *Computerworld*, February 13, 2012, p. 13.
56. Schultz, B., "Florida Hospice Saves with SaaS," *Network World*, June 6, 2011, p. 24.
57. Thurman, Mathias, "Plugging a SaaS Access Hole," *Computerworld*, March 12, 2012, p. 33.
58. Mossberg, W., "Google Unveils a Laptop with Its Brain in the Cloud," *The Wall Street Journal*, June 23, 2011, p. D1.
59. "And Now, Google's Other Operating System," *Bloomberg Businessweek*, June 13, 2011, p. 42.

60. Clayburn, T., "Google Gambles on Chromebooks," *InformationWeek*, May 30, 2011, p. 18.
61. Tableau Software Web site, *www.tableausoftware.com*, accessed September 20, 2011.
62. Biddick, M., "IT Management Goes SaaS," *Information-Week*, September 5, 2011, p. 33.
63. "Absolute Software Helps Recover 20,000th Stolen Computer," *PR Wire*, June 2011.
64. Burrows, Peter, "It Looks Like You're Trying to Use Word on an iPad," *Bloomberg Businessweek*, January 23, 2012, p. 35.
65. Mossberg, W., "Working in Word, Excel, PowerPoint On an iPad," *The Wall Street Journal*, January 12, 2012, p. D1.
66. Adobe Acrobat Web site, *https://www.acrobat.com/welcome/en/home.html*, accessed March 7, 2012.
67. Henschen, Doug, "Microsoft Places Bigger Bet on Cloud Apps," *InformationWeek*, July 11, 2011, p. 10.
68. Wingfield, N., "Microsoft Sets Rival to Google Apps," *The Wall Street Journal*, June 28, 2011, p. B6.
69. Microsoft Web site, *www.microsoft.com/en-us/office365*, accessed September 15, 2011.
70. "Office 365 vs. Google: Advantage Microsoft," *InformationWeek*, July 11, 2011, p. 10.
71. Rizzo, Tom, "Office 365: Best of Both Worlds," *Network World*, August 22, 2011, p. 20.
72. Wingfield, N., "Virtual Products, Real Profits," *The Wall Street Journal*, September 9, 2011, p. A1.
73. Ramachandran, S., "Playing on a Tablet at Therapy," *The Wall Street Journal*, July 26, 2011, p. D1.
74. Horn, Leslie, "Gamers Unlock Protein Mystery That Baffled AIDS Researchers for Years" *www.pcmag.com/article2/0,2817,2393200,00.asp*, accessed September 20, 2011.
75. Satariano, A and MacMillan, D., "Anarchy in the App Store," *Bloomberg Businessweek*, March 19, 2012, p. 47,
76. Ovide, Shira and Sherr, Ian, "Microsoft Banks on Mobile Apps," *The Wall Street Journal*, April 6, 2012, p. B1.
77. Steel, E., "A Face Launches 1,000 Apps," *The Wall Street Journal*, August 5, 2011, p. B5.
78. Android Web site, *https://market.android.com/details?id=com.viewdle.socialcamera&hl=en*, accessed September 15, 2011.
79. Angwin, J., "Face-ID Tools Pose New Risk," *The Wall Street Journal*, August 1, 2011, p. B1.
80. Fowler, G. and Lawton, C., "Facebook Again in Spotlight on Privacy," *The Wall Street Journal*, June 9, 2011, p. B1.
81. "Mobile Software Market to Reach $80 Billion by 2017," *Business Wire*, June 16, 2011.
82. Microsoft Exchange Server Web site, *microsoft.com/exchange*, accessed September 20, 2011.
83. "Copper Mountain Resort Selects Agilysys Visual One," *PR Newswire*, August 2011.
84. Martin, D., "Software Creates Less-Taxing Environment," *NJ Biz*, August 1, 2011, p. 18.
85. Tozzi, J., "Bank Data Miner," *Bloomberg Businessweek*, July 3, 2011, p. 41.
86. Verafin Web site, *verafin.com*, accessed September 15, 2011.
87. Burnham, K., "Spreading the Word," *CIO*, September 1, 2011, p. 11.
88. Biddick, M., "IT Management Goes SaaS," *Information-Week*, September 5, 2011, p. 33.
89. "81 Percent Find Mobile ERP Software Interface Important," *Business Wire*, July 12, 2011.
90. Nash, K., "Do It Yourself," *CIO*, September 1, 2011, p. 28.
91. C++ Web site, *www.cplusplus.com*, accessed September 25, 2011.
92. Microsoft Web site, *www.microsoft.com/visualstudio/en-us*, accessed September 25, 2011.
93. Android Web site, *developer.android.com/guide/developing/tools/emulator.html*, accessed September 25, 2011.
94. "Hickenlooper Merits Praise for Repealing Software Tax," *Boulder County Business Report*, June 24, p. 30.
95. Page, Jeremy, "Chinese State TV Alludes to U.S. Website Attacks," *The Wall Street Journal*, August 25, 2011, p. A8.
96. "Malware in Android Apps Rises," *The Tampa Tribune*, March 26, 2012, p. 3.
97. Babcock, Charles, "Leap Day Bug Caused Azure Outage," *InformationWeek*, March 26, 2012, p. 14.
98. Searcey, D., "Toyota Maneuvers to Protect Crown Jewels," *The Wall Street Journal*, March 22, 2011, p. B1.
99. Freeware Web site, *freewarehome.com*, accessed September 25, 2011.
100. Binstock, A., ".NET Alternative in Transition," *InformationWeek*, June 13, 2011, p. 42.
101. Matlin, C., "Innovator," *Bloomberg Businessweek*, April 11, 2011, p. 34.
102. Nash, K., "Discipline for Unruly Updates," *CIO*, July 1, 2011, p. 14.

5 Database Systems, Data Centers, and Business Intelligence

Principles	Learning Objectives
• Data management and modeling are key aspects of organizing data and information.	• Define general data management concepts and terms, highlighting the advantages of the database approach to data management. • Describe logical and physical database design considerations, the function of data centers, and the relational database model.
• A well-designed and well-managed database is an extremely valuable tool in supporting decision making.	• Identify the common functions performed by all database management systems, and identify popular database management systems.
• The number and types of database applications will continue to evolve and yield real business benefits.	• Identify and briefly discuss business intelligence, data mining, and other database applications.

Information Systems in the Global Economy

MEDIHELP, SOUTH AFRICA

From a Traditional Database to Business Intelligence

Medihelp is South Africa's third largest health insurance company. It covers about 350,000 people with plans ranging from R744 to R4,278 (about U.S. $95 to $545) per month per person. Medihelp needed a better way to access and analyze data on customers, claims, and third-party providers in order to monitor the effectiveness of its insurance products and to fine-tune and create new products as needed.

Medihelp's problem was that its data was stored in a traditional database. As discussed in this chapter, traditional databases are not designed to support decision making. They're not efficient at the types of information retrieval that decision making uses. With Medihelp's existing database, reports took unacceptably long times to run. Reports based on the content of the full database couldn't be run at all. This inefficiency detracted from Medihelp's ability to make informed business decisions.

For example, Medihelp's claim file had about 55 million rows. Each row contained about 35 data values describing medical conditions, treatments, and payments. Each row was associated with one of 15 million rows of historical member data, which held another 15 or so data values describing the member and his or her coverage. For claim processing, accessing the data for a single claim was fast and efficient. Combining data from thousands of rows, as decision support calls for, was not.

"Logging into our [traditional] database ... was not providing us the information we needed to make the best business decisions," explains Jan Steyl, senior manager of business intelligence at Medihelp. "We needed a dedicated, high-performance data warehouse."

After looking at several options, Medihelp made a preliminary selection of Sybase IQ as the basis for its data warehouse. Working with B. I. Practice, a Sybase subsidiary in South Africa, Medihelp carried out a proof of concept to confirm that Sybase IQ could deliver the needed performance at an acceptable cost. This procedure involved loading a subset of the tables from the operational database into Sybase IQ, executing queries that used only that subset of the data, and evaluating the results.

Because these queries now used a database designed for queries, performance improved dramatically. Response time was reduced by an average of 71.5 percent. Response time for ad hoc queries, those which were not programmed into the database system ahead of time, was reduced by an average of 74.1 percent. One query's response time dropped by 92.8 percent.

Theo Els, Medihelp's senior manager of client relations, likes the new system. "Health insurers supply data to employer groups. These demographic and claims profiles are essential for employer groups seeking to understand their employees' health risks ... and the consequent impact that risk can have on business productivity. Brokers and healthcare consultants also use this information in their annual client reviews to ensure that employees receive the most suitable coverage. It is imperative that the ... data warehouse provide accurate information in a format that is easily understood."

The biggest beneficiary of Medihelp's data warehouse is its product development team. The team uses data from the data warehouse to understand trends in claims by benefit code, condition, area, age group, and other factors. Medihelp also uses its data warehouse to determine what financial effects changes to a benefit in a specific product will have. This provides the sales force with the right offering at the right price for specific target markets in South Africa.

As you read this chapter, consider the following:

- How does the structure of a database affect what an organization can do with the information in it?
- How can businesses use the information in their databases to be more effective?

WHY LEARN ABOUT DATABASE SYSTEMS, DATA CENTERS, AND BUSINESS INTELLIGENCE?

© Andrey Burmakin/Shutterstock

A huge amount of data is entered into computer systems every day. Where does all this data go, and how is it used? How can it help you on the job? In this chapter, you will learn about database systems and business intelligence tools that can help you make the most effective use of information. If you become a marketing manager, you can access a vast store of data on existing and potential customers from surveys, their Web habits, and their past purchases. This information can help you sell products and services. If you become a corporate lawyer, you will have access to past cases and legal opinions from sophisticated legal databases. This information can help you win cases and protect your organization legally. If you become a human resource (HR) manager, you will be able to use databases and business intelligence tools to analyze the impact of raises, employee insurance benefits, and retirement contributions on long-term costs to your company. Regardless of your field of study in school, using database systems and business intelligence tools will likely be a critical part of your job. In this chapter, you will see how you can use data mining to extract valuable information to help you succeed. This chapter starts by introducing basic concepts of database management systems.

A database is an organized collection of data. Like other components of an information system, a database should help an organization achieve its goals. A database can contribute to organizational success by providing managers and decision makers with timely, accurate, and relevant information built on data. Databases also help companies generate information to reduce costs, increase profits, track past business activities, and open new market opportunities. According to an executive of SAP, "We need to get data into the hands of users in departments like sales ... and give them tools so they could analyze data themselves."[1]

database management system (DBMS): A group of programs that manipulate the database and provide an interface between the database and the user of the database and other application programs.

A **database management system (DBMS)** consists of a group of programs that manipulate the database and provide an interface between the database and its users and other application programs. Usually purchased from a database company, a DBMS provides a single point of management and control over data resources, which can be critical to maintaining the integrity and security of the data. Oracle's DBMS, for example, has released a database firewall to help secure the databases of its customers.[2] A database, a DBMS, and the application programs that use the data make up a database environment. A **database administrator (DBA)** is a skilled and trained IS professional who directs all activities related to an organization's database, including providing security from intruders. Some companies also hire a *data scientist* who can help analyze what is stored in vast corporate databases. According to a vice president at EMC, "A data scientist doesn't only look at one data set and then stop digging. They need to find nuggets of truth in data and then explain it to the business leaders."[3]

database administrator (DBA): A skilled IS professional who directs all activities related to an organization's database.

Databases and database management systems are becoming even more important to businesses as they deal with increasing amounts of digital information. A report from IDC called "The Digital Universe Decade," estimates the size of the digital universe to be 1.8 zettabytes, or 1.8 trillion gigabytes.[4] If a tennis ball were one byte of information, a zettabyte-sized ball would be about the size of a million earths.

DATA MANAGEMENT

Without data and the ability to process it, an organization could not successfully complete most business activities. It could not pay employees, send out bills, order new inventory, or produce information to assist managers in decision making. As you recall, data consists of raw facts, such as employee numbers and sales figures. For data to be transformed into useful information, it must first be organized in a meaningful way.

The Hierarchy of Data

Data is generally organized in a hierarchy that begins with the smallest piece of data used by computers (a bit) and progresses through the hierarchy to a database. A bit (a binary digit) represents a circuit that is either on or off. Bits can be organized into units called *bytes*. A byte is typically eight bits. Each byte represents a **character**, which is the basic building block of most information. A character can be an uppercase letter (A, B, C ... Z), lowercase letter (a, b, c ... z), numeric digit (0, 1, 2 ... 9), or special symbol (., !, +, -, /, ...).

character: A basic building block of most information, consisting of uppercase letters, lowercase letters, numeric digits, or special symbols.

Characters are put together to form a field. A **field** is typically a name, number, or combination of characters that describes an aspect of a business object (such as an employee, a location, or a truck) or activity (such as a sale). In addition to being entered into a database, fields can be computed from other fields. *Computed fields* include the total, average, maximum, and minimum value. A collection of data fields all related to one object, activity, or individual is called a **record**. By combining descriptions of the characteristics of an object, activity, or individual, a record can provide a complete description of it. For instance, an employee record is a collection of fields about one employee. One field includes the employee's name, another field contains the address, and still others the phone number, pay rate, earnings made to date, and so forth. A collection of related records is a **file**—for example, an employee file is a collection of all company employee records. Likewise, an inventory file is a collection of all inventory records for a particular company or organization. Some database software refers to files as tables.

field: Typically a name, number, or combination of characters that describes an aspect of a business object or activity.

record: A collection of data fields all related to one object, activity, or individual.

file: A collection of related records.

At the highest level of this hierarchy is a *database*, a collection of integrated and related files. Together, bits, characters, fields, records, files, and databases form the **hierarchy of data** (see Figure 5.1). Characters are combined to make a field, fields are combined to make a record, records are combined to make a file, and files are combined to make a database. A database houses not only all these levels of data but also the relationships among them.

hierarchy of data: Bits, characters, fields, records, files, and databases.

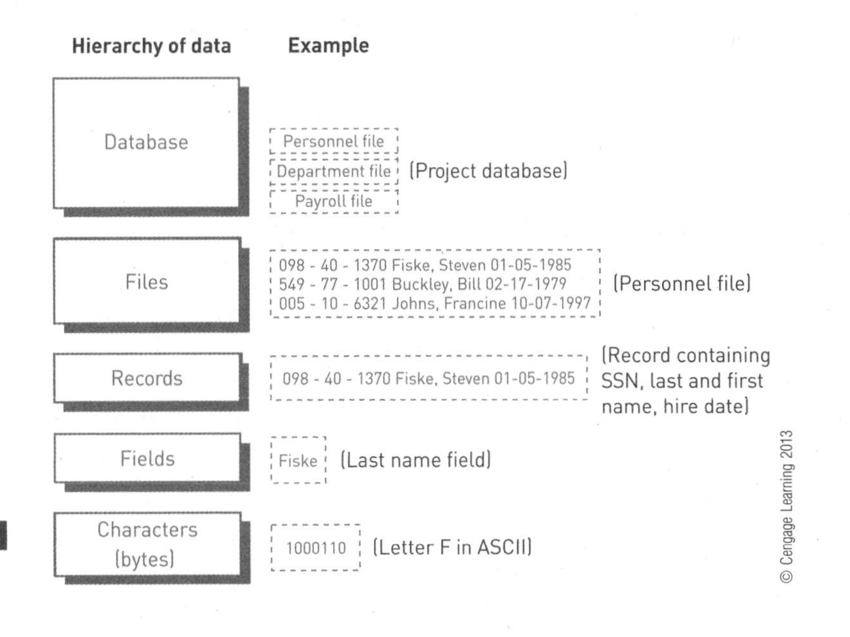

FIGURE 5.1
Hierarchy of data

Data Entities, Attributes, and Keys

entity: A general class of people, places, or things for which data is collected, stored, and maintained.

Entities, attributes, and keys are important database concepts. An **entity** is a general class of people, places, or things (objects) for which data is collected, stored, and maintained. Examples of entities include employees, inventory, and customers. Most organizations organize and store data as entities.

attribute: A characteristic of an entity.

An **attribute** is a characteristic of an entity. For example, employee number, last name, first name, hire date, and department number are attributes for an employee (see Figure 5.2). The inventory number, description, number of units on hand, and location of the inventory item in the warehouse are attributes for items in inventory. Customer number, name, address, phone number, credit rating, and contact person are attributes for customers. Attributes are usually selected to reflect the relevant characteristics of entities such as employees or customers. The specific value of an attribute, called a **data item**, can be found in the fields of the record describing an entity.

data item: The specific value of an attribute.

Employee #	Last name	First name	Hire date	Dept. number
005-10-6321	Johns	Francine	10-07-1997	257
549-77-1001	Buckley	Bill	02-17-1979	632
098-40-1370	Fiske	Steven	01-05-1985	598

KEY FIELD

ATTRIBUTES (fields)

ENTITIES (records)

FIGURE **5.2**

Keys and attributes

The key field is the employee number. The attributes include last name, first name, hire date, and department number.

Most organizations use attributes and data items. Many governments use attributes and data items to help in criminal investigations. The U.S. Federal Bureau of Investigation (FBI) is building a huge database of peoples' physical characteristics or biometrics.[5] At a cost of $1 billion, the database management system named Next Generation Identification will catalog digital images of faces, fingerprints, and palm prints of U.S. citizens and visitors. Each person in the database is an entity, each biometric category is an attribute, and each image is a data item. The information will be used as a forensics tool and to increase homeland security. In another example, SecureAlerts uses location data to monitor about 15,000 ex-convicts using GPS signals sent from the cuffs attached to the former inmates.[6] The data is used to make sure that the ex-convicts under house arrest stay in their houses. Some people, however, fear that collecting data items such as a person's location using GPS features of today's smartphones and other mobile devices could be an invasion of privacy.[7]

key: A field or set of fields in a record that is used to identify the record.

primary key: A field or set of fields that uniquely identifies the record.

As discussed earlier, a collection of fields about a specific object is a record. A **key** is a field or set of fields in a record that identifies the record. A **primary key** is a field or set of fields that uniquely identifies the record. No other record can have the same primary key. For an employee record, such as the one shown in Figure 5.2, the employee number is an example of a primary key. The primary key is used to distinguish records so that they can be accessed, organized, and manipulated. Primary keys ensure that each record in a file is unique. For example, eBay assigns an "Item number" as its primary key for items to make sure that bids are associated with the correct item (see Figure 5.3).

Locating a particular record that meets a specific set of criteria might be easier and faster using a combination of secondary keys. For example, a customer might call a mail-order company to place an order for clothes.

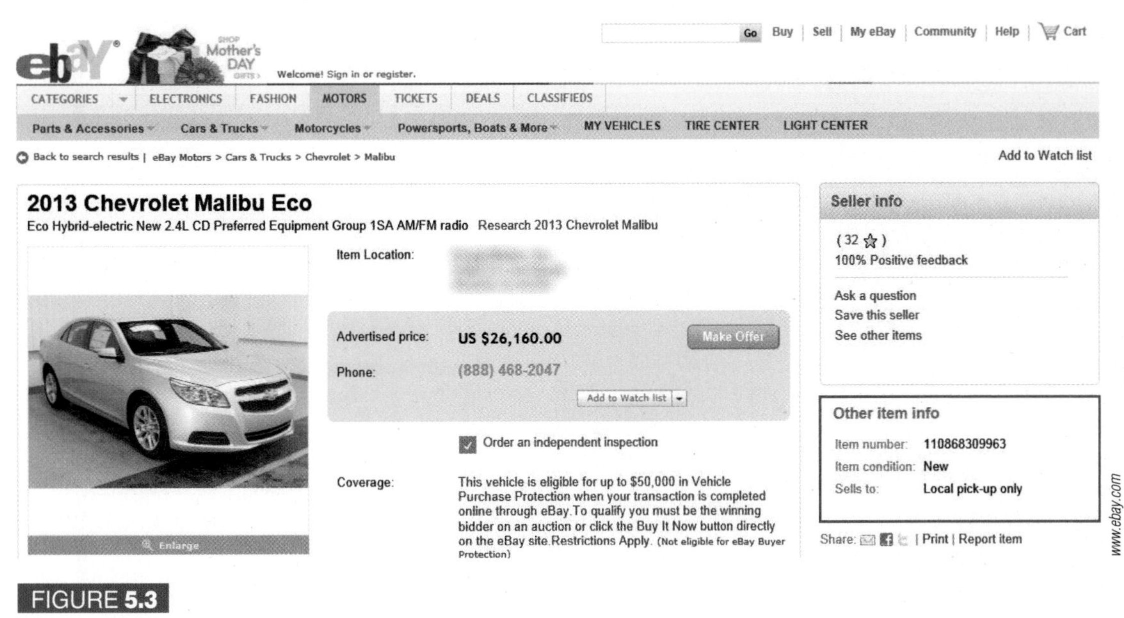

FIGURE 5.3

Primary key

eBay assigns an Item number as a primary key to keep track of each item in its database.

The order clerk can easily access the customer's mailing and billing information by entering the primary key—usually a customer number—but if the customer does not know the correct primary key, a secondary key such as last name can be used. In this case, the order clerk enters the last name, such as Adams. If several customers have a last name of Adams, the clerk can check other fields, such as address, first name, and so on, to find the correct customer record. After locating the correct record, the order can be completed and the clothing items shipped to the customer.

The Database Approach

At one time, information systems referenced specific files containing relevant data. For example, a payroll system would use a payroll file. Each distinct operational system used data files dedicated to that system. This approach to data management is called the **traditional approach to data management**.

traditional approach to data management: An approach to data management whereby each distinct operational system uses data files dedicated to that system.

database approach to data management: An approach to data management where multiple information systems share a pool of related data.

Today, most organizations use the **database approach to data management**, where multiple information systems share a pool of related data. A database offers the ability to share data and information resources. Federal databases, for example, often include the results of DNA tests as an attribute for convicted criminals. The information can be shared with law enforcement officials around the country. Often, distinct yet related databases are linked to provide enterprisewide databases. For example, many Walmart stores include in-store medical clinics for customers. Walmart uses a centralized electronic health records database that stores the information of all patients across all stores.[8] The database is interconnected with the main Walmart database to provide information about customer's interactions with the clinics and stores.

To use the database approach to data management, additional software—a database management system (DBMS)—is required. As previously discussed, a DBMS consists of a group of programs that can be used as an interface between a database and the user of the database. Typically, this software acts as a buffer between the application programs and the database itself. Figure 5.4 illustrates the database approach.

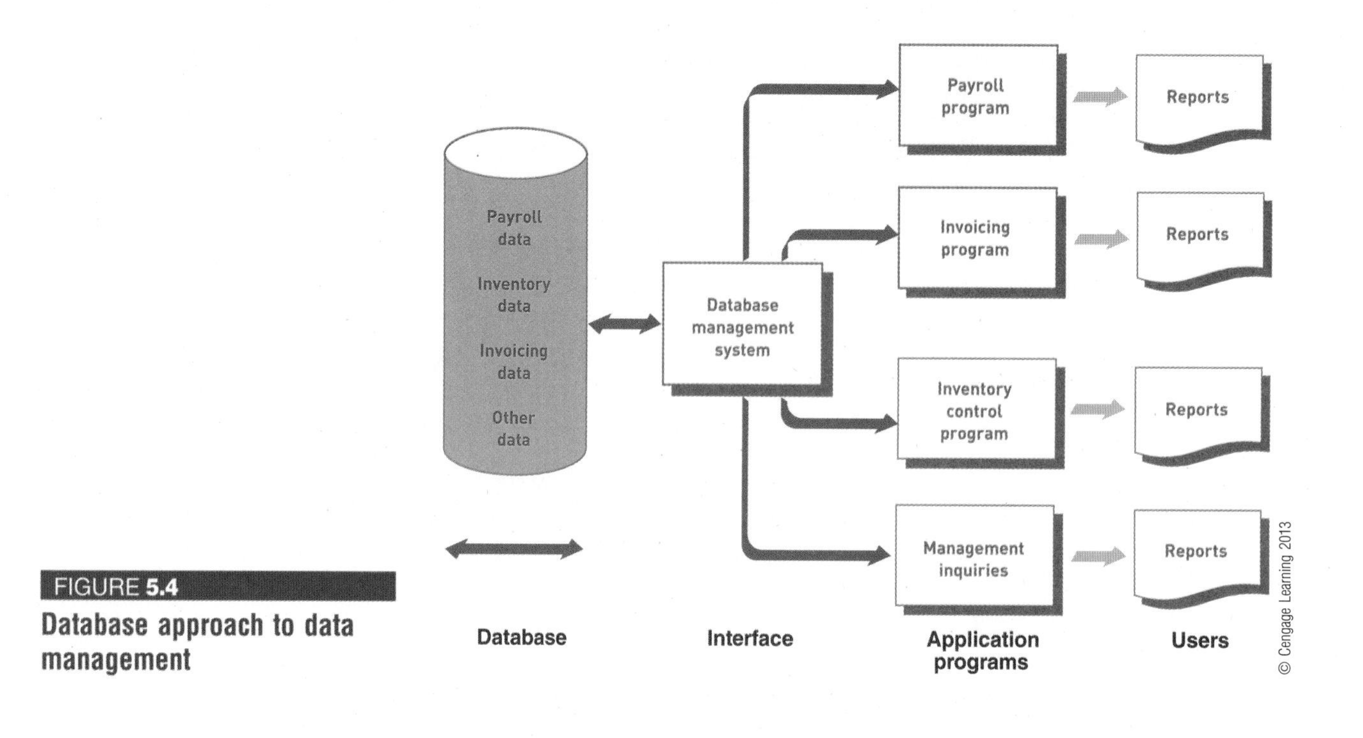

FIGURE 5.4

Database approach to data management

DATA CENTERS, DATA MODELING, AND DATABASE CHARACTERISTICS

Because today's businesses have so many elements, they must keep data organized so that it can be used effectively. A database should be designed to store all data relevant to the business and provide quick access and easy modification. Moreover, it must reflect the business processes of the organization. When building a database, an organization must carefully consider these questions:

- **Content.** What data should be collected and at what cost?
- **Access.** What data should be provided to which users and when?
- **Logical structure.** How should data be arranged so that it makes sense to a given user?
- **Physical organization.** Where should data be physically located?

Data Center

data center: A climate-controlled building or a set of buildings that houses database servers and the systems that deliver mission-critical information and services.

A **data center** is a climate-controlled building or a set of buildings that houses database servers and the systems that deliver mission-critical information and services.[9] Data centers of large organizations are often distributed among several locations.[10] IBM is helping to construct a huge data center complex in China involving at least seven separate data centers in multiple buildings totaling more than 6 million square feet.[11] FedEx opened a large data center in Colorado Springs, Colorado.[12] While a company's data sits in large super-cooled data centers, the people accessing that data are typically in offices spread across the country or around the world.

Traditional data centers consist of warehouses filled with row upon row of server racks and powerful cooling systems to compensate for the heat generated by the processors. Microsoft, Google, Dell, Hewlett-Packard, and others have adopted a modular data center approach, which uses large shipping containers like the ones that transport consumer goods around the world.[13] Google, for example, has built a $273-million modular data center in Finland.[14] The huge containers, such as the HP Ecopod, are packed with racks of servers prewired and cooled to easily connect and set up.[15] Microsoft constructed a 700,000-square-foot data center in Northlake, Illinois. It is considered to be one of the largest in the world, taking up 16 football fields of space. The mega facility is filled with 220 shipping containers packed with servers. Microsoft says that a new shipping container can be wheeled into place and connected to the Internet within hours.[16] According to the associate vice president of academic technologies for Purdue University discussing modular data centers, "From a business position, on keeping costs down and trying to get as efficient a solution as possible, this is a very, very viable solution."[17] See Figure 5.5.

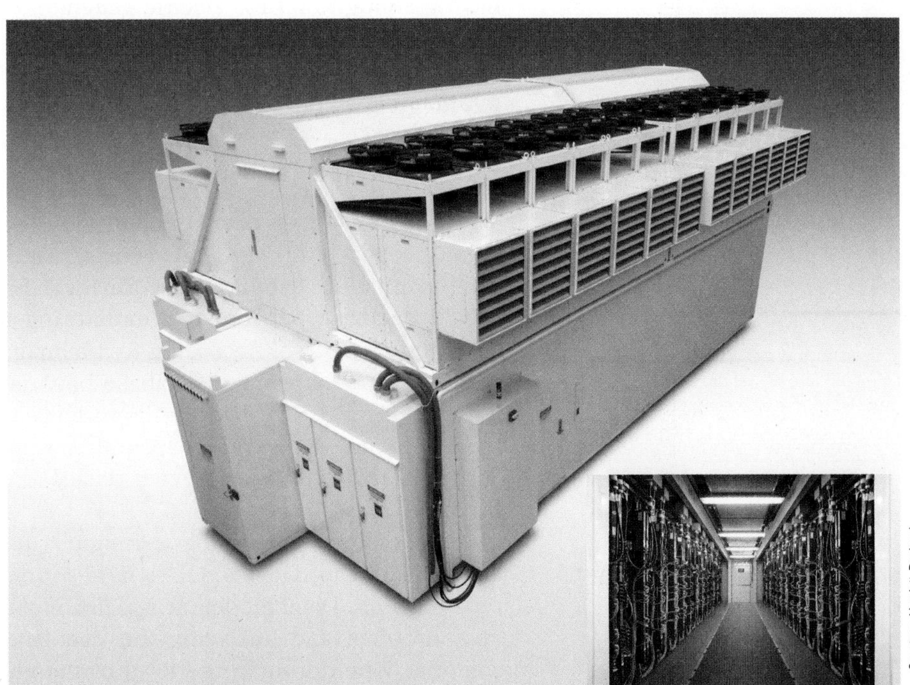

Courtesy Hewlett-Packard

FIGURE **5.5**

Modular data center

Modular data centers, such as the HP Ecopod, use large shipping containers to house servers.

As data centers continue to expand in terms of the quantity of data that they store and process, their energy demands are increasing. According to a Google spokesperson, "When building a data center, there are a whole bunch of cost items involved. But what has been the main focal point in recent years is the cost of cooling servers."[18] Businesses and technology vendors are working to develop green data centers that run more efficiently and require less energy for processing and cooling. Because of lower energy and land costs, rural North Carolina is becoming popular for large data centers.[19] Apple's $1-billion data center, Google's $600-million data center, and Facebook's $450-million data center are located in North Carolina.

Storage capacity is another concern.[20] The U.S. Library of Congress, for example, has about 450 billion objects stored in almost 20,000 disk drives attached to about 600 servers, and its storage requirements are growing.[21] According to one study, about one-third to one-half of all data centers will run out of space in the next several years.[22] Of those organizations needing more database capacity, about 40 percent indicated that they would build new data centers, about 30 percent said they would lease additional space, and the rest indicated that they would investigate other options, including the use of cloud computing. In addition, companies like IBM are investigating new database storage systems that will be much faster and store more data in a smaller amount of space.[23]

Backup and security procedures for data centers can be a concern.[24] After a hurricane disrupted its data center, the Situs Companies decided to back up its data center on the Internet using EVault, a subsidiary of Seagate.[25] When creating backup databases, some companies use software to eliminate any duplicated data to save storage capacity and reduce costs, an approach some call data de-duplication.[26] According to a Forrester report, "It's much more likely that a CIO or other executive will approve a budget for an upgrade if you can explain that in the next five years there is a 20% probability that a severe winter storm will knock out power to the data center and cost $500,000 in lost revenue and employee productivity."[27] Most Japanese data centers survived the devastating earthquake that shook Japan in 2011 by having enough fuel for electric generators and by building data centers using advanced construction techniques.[28] According to the IDC Digital Universe Study, however, only about a third of digital information has at least a minimum level of security.[29]

Data Modeling

When organizing a database, key considerations include determining what data to collect, who will have access to it, how they might want to use it, and how to monitor database performance. For example, James River Insurance uses a database performance monitoring tool from Confio to improve database performance.[30] According to a manager for James River, "Too often companies wait until a major database bottleneck has caused significant business issues before they realize the importance of deploying performance management software. James River Insurance Company had the foresight to take an approach that will help prevent database problems before they wreak havoc on the entire IT environment."[31]

data model: A diagram of data entities and their relationships.

enterprise data modeling: Data modeling done at the level of the entire enterprise.

entity-relationship (ER) diagrams: Data models that use basic graphical symbols to show the organization of and relationships between data.

One of the tools database designers use to show the logical relationships among data is a data model. A **data model** is a diagram of entities and their relationships. Data modeling usually involves understanding a specific business problem and analyzing the data and information needed to deliver a solution. When done at the level of the entire organization, this procedure is called enterprise data modeling. **Enterprise data modeling** is an approach that starts by investigating the general data and information needs of the organization at the strategic level and then examines more specific data and information needs for the various functional areas and departments within the organization.[32] Various models have been developed to help managers and database designers analyze data and information needs. An entity-relationship diagram is an example of such a data model.

Entity-relationship (ER) diagrams use basic graphical symbols to show the organization of and relationships between data. In most cases, boxes in ER diagrams indicate data items or entities contained in data tables, and diamonds show relationships between data items and entities. In other words, ER diagrams show data items in tables (entities) and the ways they are related.

ER diagrams help ensure that the relationships among the data entities in a database are correctly structured so that any application programs developed are consistent with business operations and user needs. In addition, ER diagrams can serve as reference documents after a database is in use. If changes are made to the database, ER diagrams help design them. Figure 5.6 shows an ER diagram for an order database. In this database design, one salesperson serves many customers. This is an example of a one-to-many relationship, as indicated by the one-to-many symbol (the "crow's-foot") shown in Figure 5.6. The ER diagram also shows that each customer can place one-to-many orders, that each order includes one-to-many line items, and that many line items can specify the same product (a many-to-one relationship). This database can also have one-to-one relationships. For example, one order generates one invoice.

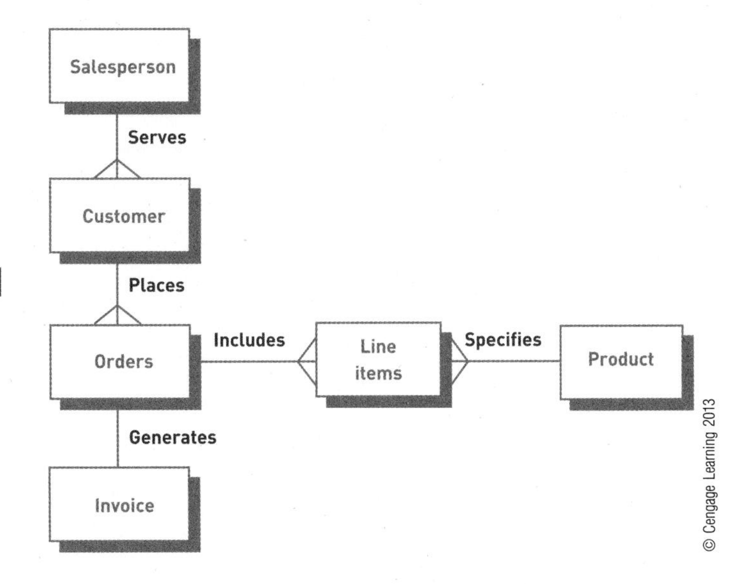

FIGURE 5.6
Entity-relationship (ER) diagram for a customer order database
Development of ER diagrams helps ensure that the logical structure of application programs is consistent with the data relationships in the database.

The Relational Database Model

relational model: A database model that describes data in which all data elements are placed in two-dimensional tables called *relations*, which are the logical equivalent of files.

Although there are a number of different database models, including flat files, hierarchical, and network models, the relational model has become the most popular, and use of this model will continue to increase. The **relational model** describes data using a standard tabular format; all data elements are placed in two-dimensional tables called *relations*, which are the logical equivalent of files. The tables in relational databases organize data in rows and columns, simplifying data access and manipulation. It is normally easier for managers to understand the relational model than other database models (see Figure 5.7).

Databases based on the relational model include IBM DB2, Oracle, Sybase, Microsoft SQL Server, Microsoft Access, and MySQL. Oracle is currently the market leader in general-purpose databases, with about half of the multibillion dollar database market.[33] Oracle's most recent edition of its relational database is highly sophisticated and uses database grids that allow a single database to run across a cluster of computers.[34] Oracle is also involved in maintaining database systems for its customers along with other companies specializing in database maintenance.[35]

domain: The allowable values for data attributes.

In the relational model, each row of a table represents a data entity—a record—and each column of the table represents an attribute—a field. Each attribute can accept only certain values. The allowable values for these attributes are called the **domain**. The domain for a particular attribute indicates what values can be placed in each column of the relational table. For instance, the domain for an attribute such as gender would be limited to male or female. A domain for pay rate would not include negative numbers. In this way, defining a domain can increase data accuracy.

Manipulating Data

selecting: Manipulating data to eliminate rows according to certain criteria.

After entering data into a relational database, users can make inquiries and analyze the data. Basic data manipulations include selecting, projecting, and joining. **Selecting** involves eliminating rows according to certain criteria. Suppose a project table contains the project number, description, and

Data Table 1: Project Table

Project	Description	Dept. number
155	Payroll	257
498	Widgets	632
226	Sales manual	598

Data Table 2: Department Table

Dept.	Dept. name	Manager SSN
257	Accounting	005-10-6321
632	Manufacturing	549-77-1001
598	Marketing	098-40-1370

Data Table 3: Manager Table

SSN	Last name	First name	Hire date	Dept. number
005-10-6321	Johns	Francine	10-07-1997	257
549-77-1001	Buckley	Bill	02-17-1979	632
098-40-1370	Fiske	Steven	01-05-1985	598

FIGURE 5.7

Relational database model

In the relational model, all data elements are placed in two-dimensional tables, or relations. As long as they share at least one common element, these relations can be linked to output useful information.

department number for all projects a company is performing. The president of the company might want to find the department number for Project 226, a sales manual project. Using selection, the president can eliminate all rows but the one for Project 226 and see that the department number for the department completing the sales manual project is 598.

projecting: Manipulating data to eliminate columns in a table.

Projecting involves eliminating columns in a table. For example, a department table might contain the department number, department name, and Social Security number (SSN) of the manager in charge of the project. A sales manager might want to create a new table with only the department number and the Social Security number of the manager in charge of the sales manual project. The sales manager can use projection to eliminate the department name column and create a new table containing only the department number and SSN.

joining: Manipulating data to combine two or more tables.

Joining involves combining two or more tables. For example, you can combine the project table and the department table to create a new table with the project number, project description, department number, department name, and Social Security number for the manager in charge of the project.

linking: The ability to combine two or more tables through common data attributes to form a new table with only the unique data attributes.

As long as the tables share at least one common data attribute, the tables in a relational database can be linked to provide useful information and reports. **Linking**, the ability to combine two or more tables through common data attributes to form a new table with only the unique data attributes, is one of the keys to the flexibility and power of relational databases. Suppose the president of a company wants to find out the name of the manager of the sales manual project and the length of time the manager has been with the company. Assume that the company has the manager, department, and project tables shown in Figure 5.7. A simplified ER diagram showing the relationship between these tables is shown in Figure 5.8. Note the crow's-foot by the project table. This symbol indicates that a department can have many projects. The president would make the inquiry to the database, perhaps via a personal computer. The DBMS would start with the project description and search the project table to find out the project's department number. It would then use the department number to search the department table for the manager's Social Security number. The department number is also in the department table and is the common element that links the project table to the department table. The DBMS uses the manager's Social Security number to search the manager table for the manager's hire date. The manager's Social Security number is the common element between the department table and the manager table. The final result is that the manager's name and hire date are presented to the president as a response to the inquiry (see Figure 5.9).

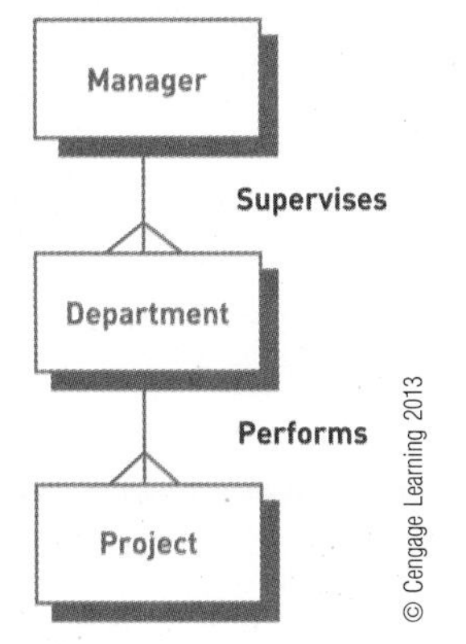

FIGURE 5.8

Simplified ER diagram showing the relationship between the Manager, Department, and Project tables

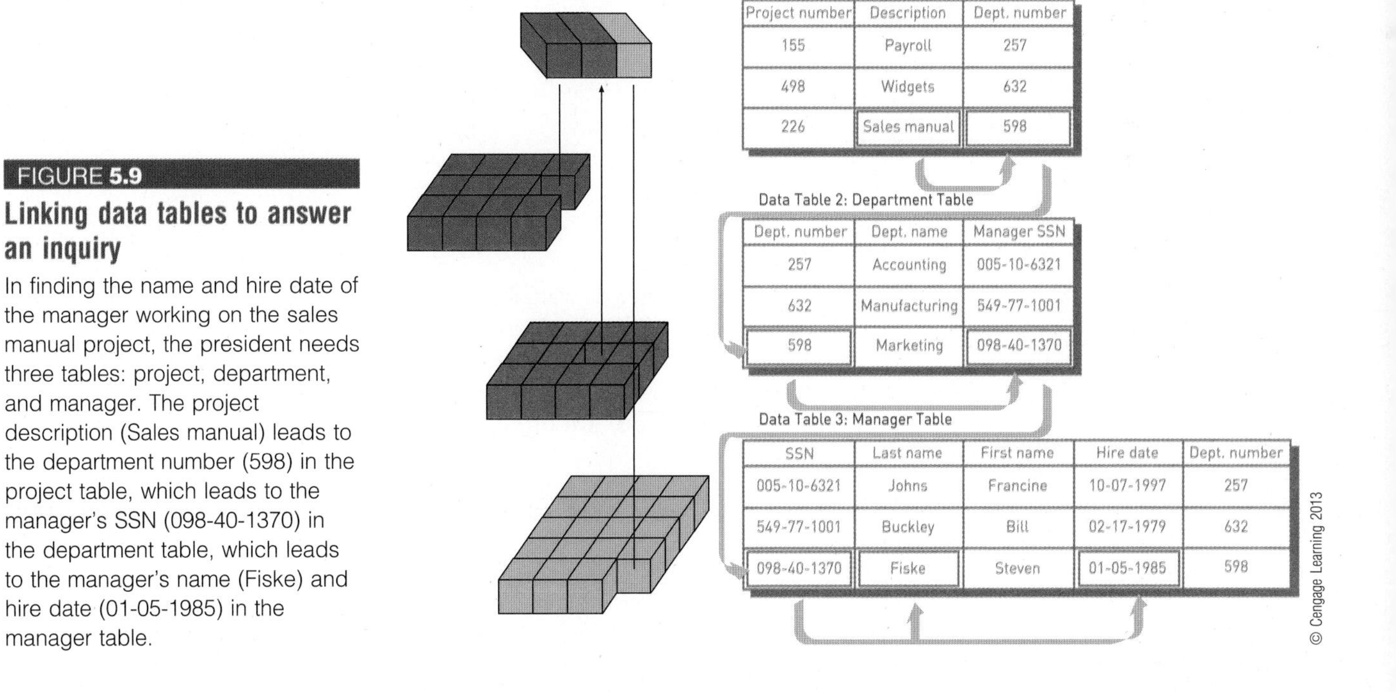

Data Table 1: Project Table

Project number	Description	Dept. number
155	Payroll	257
498	Widgets	632
226	Sales manual	598

Data Table 2: Department Table

Dept. number	Dept. name	Manager SSN
257	Accounting	005-10-6321
632	Manufacturing	549-77-1001
598	Marketing	098-40-1370

Data Table 3: Manager Table

SSN	Last name	First name	Hire date	Dept. number
005-10-6321	Johns	Francine	10-07-1997	257
549-77-1001	Buckley	Bill	02-17-1979	632
098-40-1370	Fiske	Steven	01-05-1985	598

FIGURE 5.9

Linking data tables to answer an inquiry

In finding the name and hire date of the manager working on the sales manual project, the president needs three tables: project, department, and manager. The project description (Sales manual) leads to the department number (598) in the project table, which leads to the manager's SSN (098-40-1370) in the department table, which leads to the manager's name (Fiske) and hire date (01-05-1985) in the manager table.

One of the primary advantages of a relational database is that it allows tables to be linked, as shown in Figure 5.9. This linkage reduces data redundancy and allows data to be organized more logically. The ability to link to the manager's SSN stored once in the manager table eliminates the need to store it multiple times in the project table.

The relational database model is by far the most widely used. It is easier to control, more flexible, and more intuitive than other approaches because it organizes data in tables. As shown in Figure 5.10, a relational database management system, such as Access, can be used to store data in rows and columns. In this figure, tabs at the top of the Access database can be used to create, edit, and manipulate the database. The ability to link relational tables also allows users to relate data in new ways without having to redefine complex relationships. Because of the advantages of the relational model, many companies use it for large corporate databases, such as those for marketing and accounting. The relational model can also be used with personal computers and mainframe systems. A travel reservation company, for example, can develop a fare-pricing system by using relational database technology that can handle millions of daily queries from online travel companies such as Expedia, Travelocity, and Orbitz.

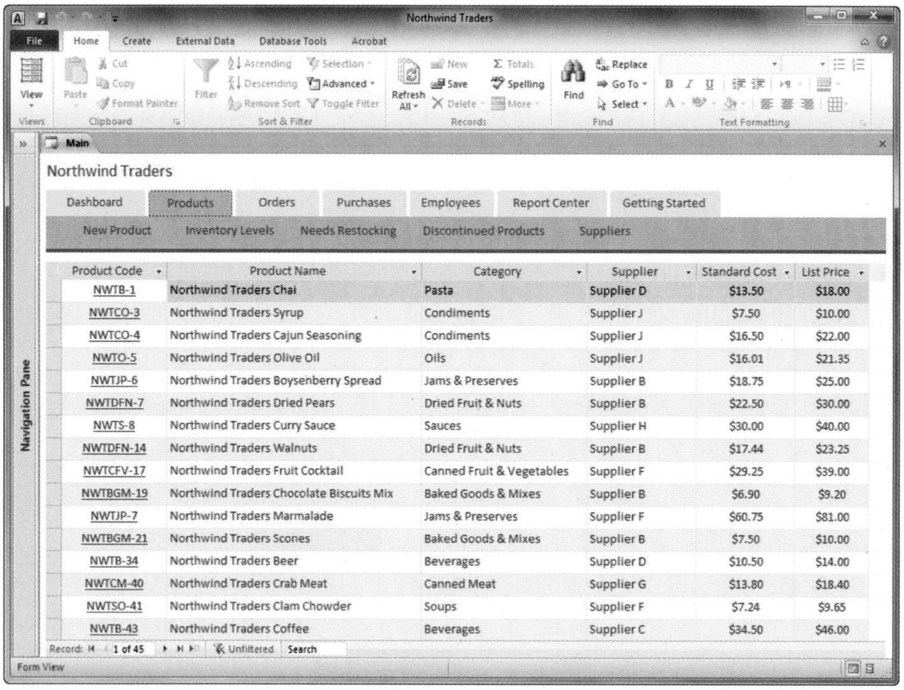
Northwind Traders
File
Home
Create
External Data
Database Tools
Acrobat
Main
Northwind Traders
Dashboard
Products
Orders
Purchases
Employees
Report Center
Getting Started
New Product
Inventory Levels
Needs Restocking
Discontinued Products
Suppliers
Product Code | Product Name | Category | Supplier | Standard Cost | List Price
NWTB-1 | Northwind Traders Chai | Pasta | Supplier D | $13.50 | $18.00
NWTCO-3 | Northwind Traders Syrup | Condiments | Supplier J | $7.50 | $10.00
NWTCO-4 | Northwind Traders Cajun Seasoning | Condiments | Supplier J | $16.50 | $22.00
NWTO-5 | Northwind Traders Olive Oil | Oils | Supplier J | $16.01 | $21.35
NWTJP-6 | Northwind Traders Boysenberry Spread | Jams & Preserves | Supplier B | $18.75 | $25.00
NWTDFN-7 | Northwind Traders Dried Pears | Dried Fruit & Nuts | Supplier B | $22.50 | $30.00
NWTS-8 | Northwind Traders Curry Sauce | Sauces | Supplier H | $30.00 | $40.00
NWTDFN-14 | Northwind Traders Walnuts | Dried Fruit & Nuts | Supplier B | $17.44 | $23.25
NWTCFV-17 | Northwind Traders Fruit Cocktail | Canned Fruit & Vegetables | Supplier F | $29.25 | $39.00
NWTBGM-19 | Northwind Traders Chocolate Biscuits Mix | Baked Goods & Mixes | Supplier B | $6.90 | $9.20
NWTJP-7 | Northwind Traders Marmalade | Jams & Preserves | Supplier F | $60.75 | $81.00
NWTBGM-21 | Northwind Traders Scones | Baked Goods & Mixes | Supplier B | $7.50 | $10.00
NWTB-34 | Northwind Traders Beer | Beverages | Supplier D | $10.50 | $14.00
NWTCM-40 | Northwind Traders Crab Meat | Canned Meat | Supplier G | $13.80 | $18.40
NWTSO-41 | Northwind Traders Clam Chowder | Soups | Supplier F | $7.24 | $9.65
NWTB-43 | Northwind Traders Coffee | Beverages | Supplier C | $34.50 | $46.00
Navigation Pane
Form View

FIGURE **5.10**

Building and modifying a relational database

Relational databases provide many tools, tips, and shortcuts to simplify the process of creating and modifying a database.

Data Cleanup

data cleanup: The process of looking for and fixing inconsistencies to ensure that data is accurate and complete.

As discussed in Chapter 1, valuable data is accurate, complete, economical, flexible, reliable, relevant, simple, timely, verifiable, accessible, and secure. The database must also be properly designed. The purpose of **data cleanup** is to develop data with these characteristics. Formalized approaches, such as *database normalization*, are often used to clean up problems with data.

DATABASE MANAGEMENT SYSTEMS

Creating and implementing the right database system ensures that the database will support both business activities and goals. But how do we actually create, implement, use, and update a database? The answer is found in the database management system. As discussed earlier, a DBMS is a group of programs used as an interface between a database and application programs or a database and the user. The capabilities and types of database systems, however, vary considerably.[36]

Overview of Database Types

Database management systems can range from small inexpensive software packages to sophisticated systems costing hundreds of thousands of dollars.

The following sections discuss a few popular alternatives (see Figure 5.11 for one example).

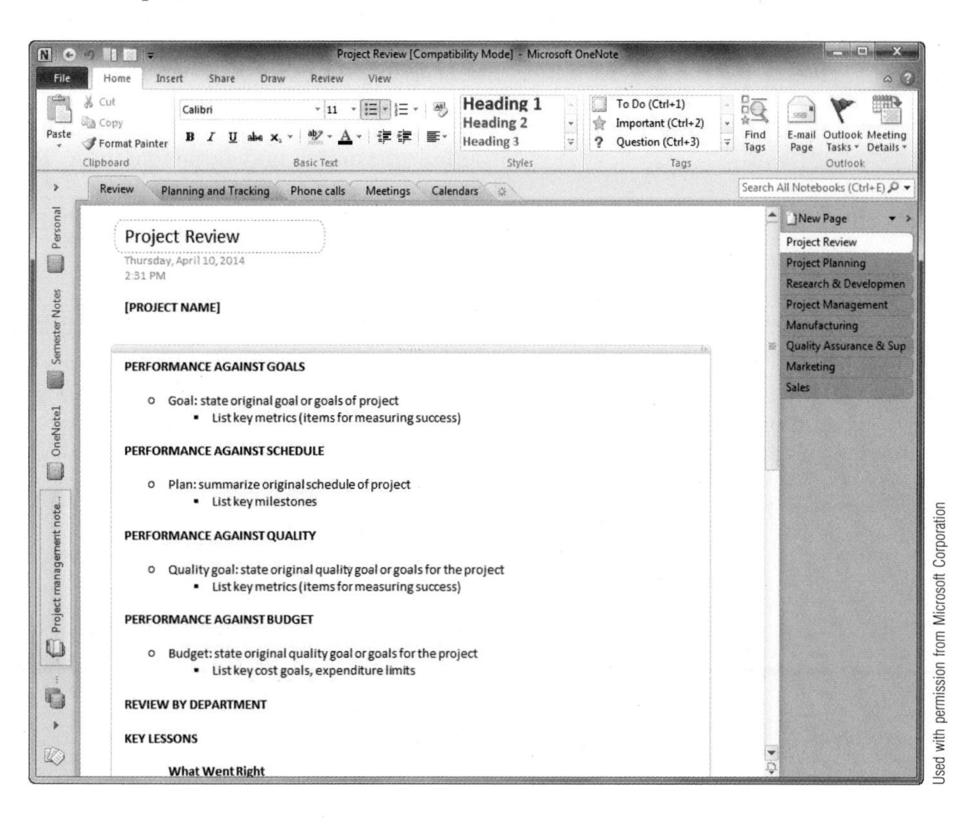

FIGURE **5.11**

Microsoft OneNote

Microsoft OneNote lets you gather any type of information and then retrieve, copy, and paste the information into other applications, such as word-processing and spreadsheet programs.

Flat File

A flat file is a simple database program whose records have no relationship to one another. Flat file databases are often used to store and manipulate a single table or file; they do not use any of the database models discussed previously, such as the relational model. Many spreadsheet and word-processing programs have flat file capabilities. These software packages can sort tables and make simple calculations and comparisons. Microsoft OneNote is designed to let people put ideas, thoughts, and notes into a flat file.[37] Similar to OneNote, EverNote is a free online database service that can store notes and other pieces of information, including photos, voice memos, or handwritten notes.[38] EverNote can be used on computers, smartphones, tablet computers, and other mobile devices.[39]

Single User

A database installed on a personal computer is typically meant for a single user. Microsoft Access and FileMaker Pro are designed to support single-user implementations. Microsoft InfoPath is another example of a database program that supports a single user.[40]

Multiple Users

Small, midsize, and large businesses need multiuser DBMSs to share information throughout the organization over a network. These more powerful, expensive systems allow dozens or hundreds of people to access the same database system at the same time. Popular vendors for multiuser database systems include Oracle, Microsoft, Sybase, and IBM. Many single-user databases, such as Microsoft Access, can be implemented for multiuser support over a network, though they often are limited in the number of users they can support.

Providing a User View

schema: A description of the entire database.

Because the DBMS is responsible for access to a database, one of the first steps in installing and using a large database involves "telling" the DBMS the logical and physical structure of the data and the relationships among the data for each user. This description is called a **schema** (as in schematic diagram). Large database systems, such as Oracle, typically use schemas to define the tables and other database features associated with a person or user. A schema can be part of the database or a separate schema file. The DBMS can reference a schema to find where to access the requested data in relation to another piece of data.

Creating and Modifying the Database

data definition language (DDL): A collection of instructions and commands used to define and describe data and relationships in a specific database.

Schemas are entered into the DBMS (usually by database personnel) via a data definition language. A **data definition language (DDL)** is a collection of instructions and commands used to define and describe data and relationships in a specific database. A DDL allows the database's creator to describe the data and relationships that are to be contained in the schema. In general, a DDL describes logical access paths and logical records in the database. Figure 5.12 shows a simplified example of a DDL used to develop a general schema. The use of the letter *X* in Figure 5.12 reveals where specific information concerning the database should be entered. File description, area description, record description, and set description are terms the DDL defines and uses in this example. Other terms and commands can be used, depending on the DBMS employed.

```
SCHEMA DESCRIPTION
SCHEMA NAME IS XXXX
AUTHOR      XXXX
DATE        XXXX
FILE DESCRIPTION
    FILE NAME IS XXXX
      ASSIGN XXXX
    FILE NAME IS XXXX
      ASSIGN XXXX
AREA DESCRIPTION
    AREA NAME IS XXXX
RECORD DESCRIPTION
    RECORD NAME IS XXXX
    RECORD ID IS XXXX
    LOCATION MODE IS XXXX
    WITHIN XXXX AREA FROM XXXX THRU XXXX
SET DESCRIPTION
    SET NAME IS XXXX
    ORDER IS XXXX
    MODE IS XXXX
    MEMBER IS XXXX
    .
    .
    .
```


FIGURE **5.12**

Using a data definition language to define a schema

data dictionary: A detailed description of all the data used in the database.

Another important step in creating a database is to establish a **data dictionary**, a detailed description of all data used in the database. The data dictionary contains the following information:

- Name of the data item
- Aliases or other names that may be used to describe the item
- Range of values that can be used

- Type of data (such as alphanumeric or numeric)
- Amount of storage needed for the item
- Notation of the person responsible for updating it and the various users who can access it
- List of reports that use the data item

A data dictionary can also include a description of data flows, the way records are organized, and the data-processing requirements. Figure 5.13 shows a typical data dictionary entry.

NORTHWESTERN MANUFACTURING

PREPARED BY:	D. BORDWELL
DATE:	04 AUGUST 2013
APPROVED BY:	J. EDWARDS
DATE:	13 OCTOBER 2013
VERSION:	3.1
PAGE:	1 OF 1
DATA ELEMENT NAME:	PARTNO
DESCRIPTION:	INVENTORY PART NUMBER
OTHER NAMES:	PTNO
VALUE RANGE:	100 TO 5000
DATA TYPE:	NUMERIC
POSITIONS:	4 POSITIONS OR COLUMNS

FIGURE **5.13**
Typical data dictionary entry

For example, the information in a data dictionary for the part number of an inventory item can include the following information:

- Name of the person who made the data dictionary entry (D. Bordwell)
- Date the entry was made (August 4, 2013)
- Name of the person who approved the entry (J. Edwards)
- Approval date (October 13, 2013)
- Version number (3.1)
- Number of pages used for the entry (1)
- Part name (PARTNO)
- Other part names that might be used (PTNO)
- Range of values (part numbers can range from 100 to 5,000)
- Type of data (numeric)
- Storage required (four positions are required for the part number)

A data dictionary is valuable in maintaining an efficient database that stores reliable information with no redundancy, and it makes it easy to modify the database when necessary. Data dictionaries also help computer and system programmers who require a detailed description of data elements stored in a database to create the code to access the data.

Storing and Retrieving Data

One function of a DBMS is to be an interface between an application program and the database. When an application program needs data, it requests the data through the DBMS. Suppose that to calculate the total price of a new car, a pricing program needs price data on the engine option—six cylinders instead of the standard four cylinders. The application program requests this data from the DBMS. In doing so, the application program follows a logical access path. Next, the DBMS, working with various system programs, accesses a storage device, such as disk drives and solid-state storage devices (SSDs), where the data is stored.[41] When the DBMS goes to this storage device to retrieve the data, it follows a path to the physical location (physical access path) where the price of this option is stored. In the pricing example, the DBMS might go to a disk drive to retrieve the price data for six-cylinder engines. This relationship is shown in Figure 5.14.

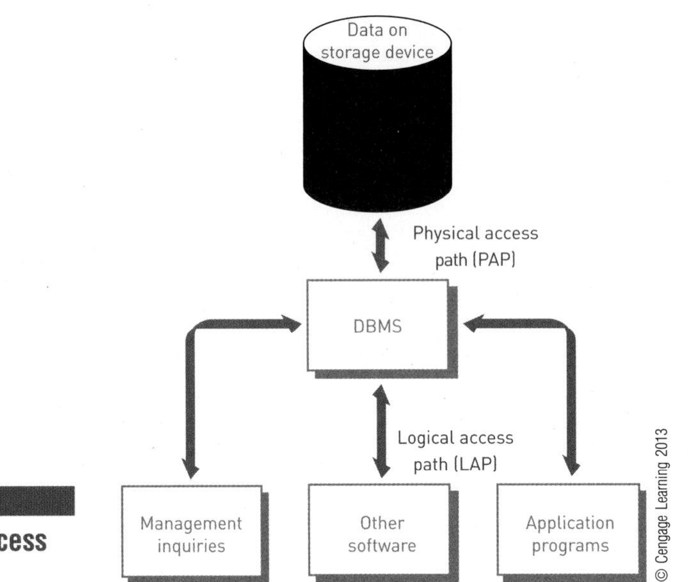

FIGURE 5.14
Logical and physical access paths

This same process is used if a user wants to get information from the database. First, the user requests the data from the DBMS. For example, a user might give a command, such as LIST ALL OPTIONS FOR WHICH PRICE IS GREATER THAN 200 DOLLARS. This is the logical access path (LAP). Then, the DBMS might go to the options price section of a disk to get the information for the user. This is the physical access path (PAP).

Two or more people or programs attempting to access the same record at the same time can cause a problem. For example, an inventory control program might attempt to reduce the inventory level for a product by ten units because ten units were just shipped to a customer. At the same time, a purchasing program might attempt to increase the inventory level for the same product by 200 units because inventory was just received. Without proper database control, one of the inventory updates might be incorrect, resulting in an inaccurate inventory level for the product. **Concurrency control** can be used to avoid this potential problem. One approach is to lock out all other application programs from access to a record if the record is being updated or used by another program.

concurrency control: A method of dealing with a situation in which two or more users or applications need to access the same record at the same time.

Manipulating Data and Generating Reports

After a DBMS has been installed, employees, managers, and consumers can use it to review reports and obtain important information. Using a DBMS, a company can manage this requirement. According to a professor at MIT, "CIOs can make sure that the information systems are available for aggregating reports from the field or news media."[42]

Some databases use *Query by Example (QBE)*, which is a visual approach to developing database queries or requests. Like Windows and other GUI operating systems, you can perform queries and other database tasks by opening windows and clicking the data or features you want (see Figure 5.15).

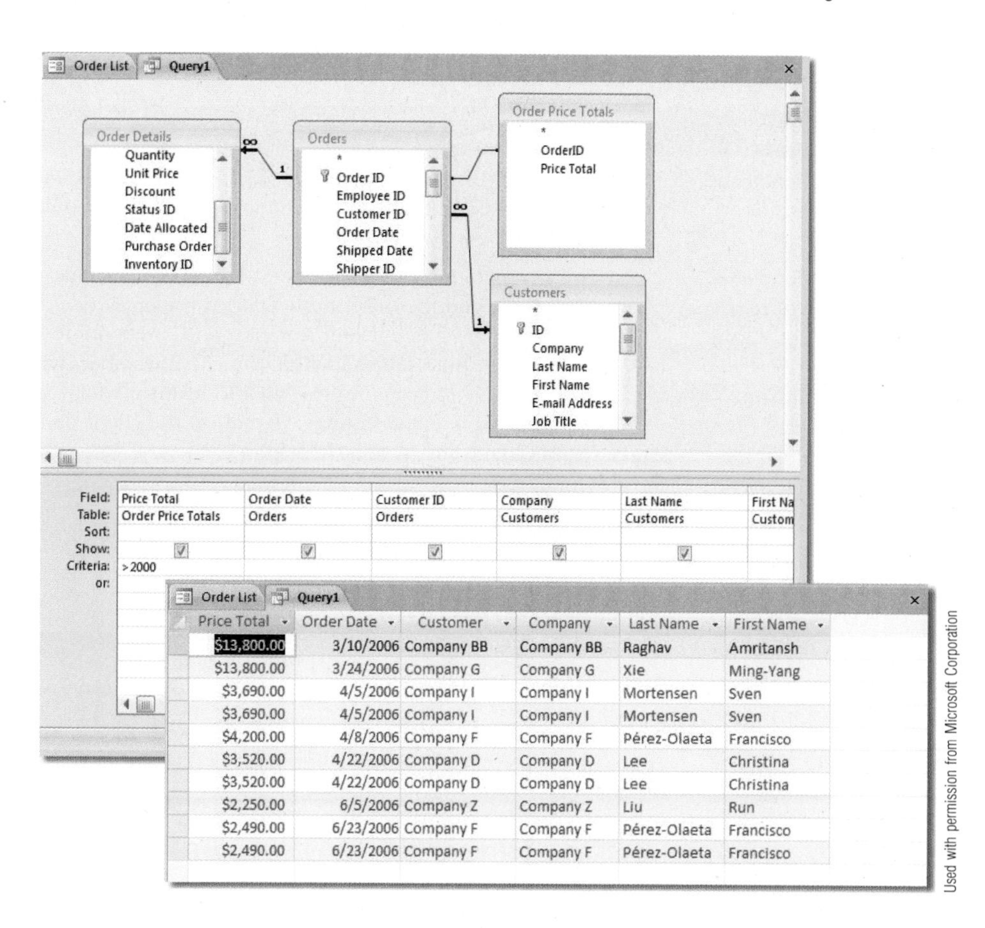

FIGURE **5.15**

Query by Example

Some databases use Query by Example (QBE) to generate reports and information.

In other cases, database commands can be used in a programming language. For example, C++ commands can be used in simple programs that will access or manipulate certain pieces of data in the database. Here's another example of a DBMS query: SELECT * FROM EMPLOYEE WHERE JOB_CLASSIFICATION="C2." The asterisk (*) tells the program to include all columns from the EMPLOYEE table. In general, the commands that are used to manipulate the database are part of the **data manipulation language (DML)**. This specific language, provided with the DBMS, allows managers and other database users to access and modify the data, make queries, and to generate reports. Again, the application programs go through schemas and the DBMS before getting to the data stored on a device such as a disk.

data manipulation language (DML): A specific language, provided with a DBMS, which allows users to access and modify the data, to make queries, and to generate reports.

In the 1970s, D. D. Chamberlain and others at the IBM Research Laboratory in San Jose, California, developed a standardized data manipulation language called *Structured Query Language (SQL)*, pronounced like the word *sequel* or spelled out as *SQL*. The EMPLOYEE query shown earlier is written in SQL. In 1986, the American National Standards Institute (ANSI) adopted SQL as the standard query language for relational databases. Since ANSI's acceptance of SQL, interest in making SQL an integral part of relational databases on both mainframe and personal computers has increased. SQL has many built-in functions, such as average (AVG), the largest value (MAX), the smallest value (MIN), and others. Table 5.1 contains examples of SQL commands.

SQL lets programmers learn one powerful query language and use it on systems ranging from PCs to the largest mainframe computers (see Figure 5.16). Programmers and database users also find SQL valuable because SQL statements can be embedded into many programming languages, such as the widely used C++, Java, and other languages. Because SQL uses standardized and simplified procedures for retrieving, storing, and manipulating data, the popular database query language can be easy to understand and use.

TABLE 5.1 Examples of SQL commands

SQL Command	Description
SELECT ClientName, Debt FROM Client WHERE Debt > 1000	This query displays all clients (ClientName) and the amount they owe the company (Debt) from a database table called Client for clients who owe the company more than $1,000 (WHERE Debt > 1000).
SELECT ClientName, ClientNum, OrderNum FROM Client, Order WHERE Client.ClientNum=Order.ClientNum	This command is an example of a join command that combines data from two tables: the client table and the order table (FROM Client, Order). The command creates a new table with the client name, client number, and order number (SELECT ClientName, ClientNum, OrderNum). Both tables include the client number, which allows them to be joined. This ability is indicated in the WHERE clause, which states that the client number in the client table is the same as (equal to) the client number in the order table (WHERE Client.ClientNum=Order.ClientNum).
GRANT INSERT ON Client to Guthrie	This command is an example of a security command. It allows Bob Guthrie to insert new values or rows into the Client table.

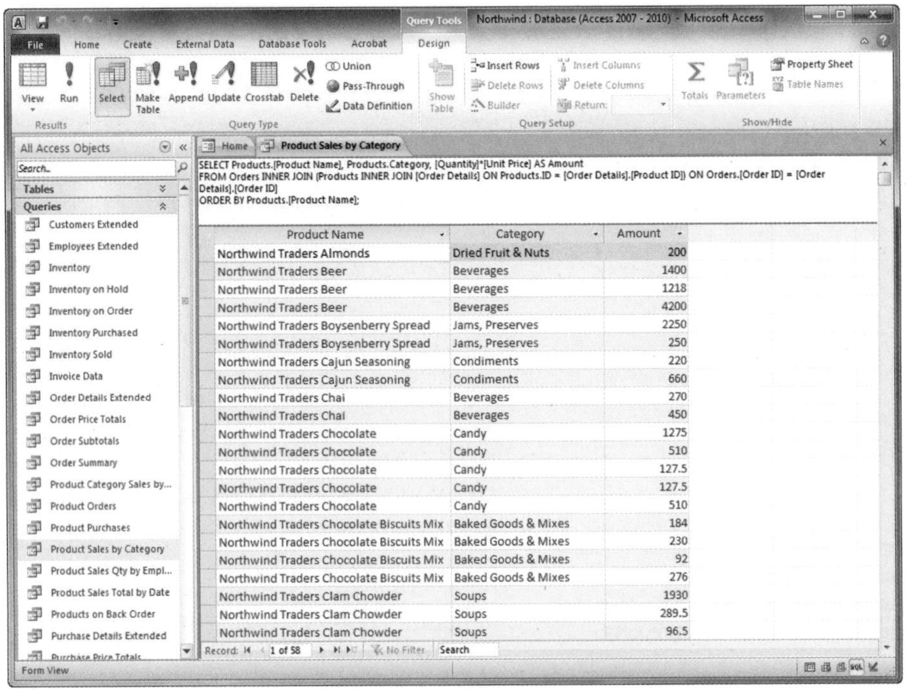

FIGURE 5.16

Structured Query Language

Structured Query Language (SQL) has become an integral part of most relational databases, as shown by this example from Microsoft Access 2010.

After a database has been set up and loaded with data, it can produce desired reports, documents, and other outputs (see Figure 5.17). These outputs usually appear in screen displays or hard copy printouts. The output-control features of a database program allow you to select the records and fields you want to appear in reports. You can also make calculations specifically for the report by manipulating database fields. Formatting controls and organization options (such as report headings) help you to customize reports and create flexible, convenient, and powerful information-handling tools.

A DBMS can produce a wide variety of documents, reports, and other output that can help organizations achieve their goals. The most common reports select and organize data to present summary information about some aspect of company operations. For example, accounting reports often summarize financial data such as current and past-due accounts. Many companies base

Monthly Sales Report

Monthly Sales Report

Sunday, July 29, 2013 5:36:07 PM

June, 2012

Product	Sales
Northwind Traders Boysenberry Spread	$2,250.00
Northwind Traders Dried Apples	$1,590.00
Northwind Traders Fruit Cocktail	$1,560.00
Northwind Traders Chocolate	$1,020.00
Northwind Traders Dried Pears	$900.00
Northwind Traders Cajun Seasoning	$660.00
Northwind Traders Coffee	$230.00
Northwind Traders Clam Chowder	$96.50
June Sales Total	$8,306.50

Page 1 of 1

FIGURE 5.17

Database output

A database application offers sophisticated formatting and organization options to produce the right information in the right format.

their routine operating decisions on regular status reports that show the progress of specific orders toward completion and delivery.

Database Administration

Database systems require a skilled database administrator (DBA), who is expected to have a clear understanding of the fundamental business of the organization, be proficient in the use of selected database management systems, and stay abreast of emerging technologies and new design approaches. The role of the DBA is to plan, design, create, operate, secure, monitor, and maintain databases. Typically, a DBA has a degree in computer science or management information systems and some on-the-job training with a particular database product or more extensive experience with a range of database products (see Figure 5.18).

FIGURE 5.18

Database administrator

The role of the database administrator (DBA) is to plan, design, create, operate, secure, monitor, and maintain databases.

The DBA works with users to decide the content of the database—to determine exactly what entities are of interest and what attributes are to be recorded about those entities. Thus, personnel outside of IS must have some idea of what the DBA does and why this function is important. The DBA can play a crucial role in the development of effective information systems to benefit the organization, employees, and managers.

The DBA also works with programmers as they build applications to ensure that their programs comply with database management system standards and conventions. After the database is built and operating, the DBA monitors operations logs for security violations. Database performance is also monitored to ensure that the system's response time meets users' needs and that it operates efficiently. If there is a problem, the DBA attempts to correct it before it becomes serious.

A database failure can cause huge financial losses for a business, a government, or nonprofit organization. For example, some fear that the older database system used by the Social Security Administration (SSA) might be vulnerable to a service interruption or failure.[43] According to SSA's inspector general, "Service interruption would severely affect the American public delaying the delivery of benefits to citizens who depend on these funds in their day-to-day lives and likely hinder people's ability to obtain employment, driver's licenses, and even loans and mortgages." A failure due to mechanical problems, controller failures, viruses or attacks, or human failure can cause productivity in an organization to grind to a halt. A large responsibility of a DBA is to protect the database from attack or other forms of failure. DBAs use security software, preventive measures, and redundant systems to keep data safe and accessible.

data administrator: A nontechnical position responsible for defining and implementing consistent principles for a variety of data issues.

Some organizations have also created a position called the **data administrator**, a nontechnical, but important position responsible for defining and implementing consistent principles for a variety of data issues, including setting data standards and data definitions that apply across all the databases in an organization. For example, the data administrator would ensure that a term such as "customer" is defined and treated consistently in all corporate databases. This person also works with business managers to identify who should have read or update access to certain databases and to selected attributes within those databases. This information is then communicated to the database administrator for implementation. The data administrator can be a high-level position reporting to top-level managers.

Popular Database Management Systems

Some popular DBMSs for single users include Microsoft Access and FileMaker Pro. The complete DBMS market encompasses software used by professional programmers and that runs on midrange servers, mainframes, and supercomputers. The entire market generates billions of dollars per year in revenue by companies including IBM, Oracle, and Microsoft.

Like other software products, a number of open-source database systems are available, including PostgreSQL and MySQL. CouchDB by Couchbase is another example of an open-source database system used by Zynga and others.[44] Zynga, the developer of the popular Internet game Farmville, uses CouchDB to process 250 million visitors a month.[45] Apache Hadoop is an open-source database that can be used to manage large unstructured databases in conjunction with traditional relational database management systems (RDBMSs).[46] JP Morgan Chase, for example, used Hadoop for fraud detection and IS risk management.[47] IBM, Google, the *New York Times*, and many other organizations also use Hadoop.[48] Open-source software is described in Chapter 4.

Some refer to a new form of database system as *Database as a Service (DaaS)*; others call it Database 2.0. DaaS is similar to Software as a Service (SaaS). Recall that a SaaS system is one in which the software is stored on a service provider's servers and is accessed by the client company over a network. In DaaS, the database is stored on a service provider's servers and accessed by the client over a network, typically the Internet.[49] In DaaS, database administration is provided by the service provider. SaaS and DaaS are both part of the larger cloud-computing trend. Recall from Chapter 3 that cloud computing uses a giant cluster of computers that run high-performance applications. In cloud computing, all information systems and data are maintained and managed by service providers and delivered over the Internet. Apple Computer, for example, has developed a service called iCloud to allow people to store their music and other documents on the Internet.[50] Box.net allows people to store, sync, and share information and documents on the Internet.[51] Oracle and other database companies have also embraced cloud computing.[52]

More than a dozen companies are moving in the DaaS direction. They include Google, Microsoft, Oracle, Amazon, Intuit, MyOwnDB, and Trackvia. Oracle's DaaS combines cloud computing with grid computing and virtualization, discussed next, to provide cost-effective, reliable, and scalable database solutions. Oracle provides both private clouds, accessible only to users on a private network, and public clouds, accessible to the public over the Internet.[53] Razorfish, a digital advertising and marketing firm, uses Amazon's Elastic Cloud service to collect and analyze data (not personally identifiable) from browsing sessions to develop effective marketing campaigns.[54] Procter and Gamble consolidated hundreds of projects into a single cloud database, which saved operational costs and reduced meeting time and data entry time for employees.[55]

Database Virtualization

The virtualization approach was first introduced in Chapter 3 and briefly discussed in Chapter 4. As mentioned in Chapter 3, a virtual server is a method of logically dividing the resources of a single physical server to create multiple logical servers, each acting as its own dedicated machine.[56] In Chapter 4, we mentioned that virtualization tools allow various operating systems to run on a single server with advanced security features and robust administrative support. *Database virtualization* uses virtual servers and operating systems to allow two or more database systems, including servers and databases management systems (DBMSs), to act like a single unified database system. A virtual database acts independently of the physical servers, disk drives, and other database components that can be located around the world. Virtualization allows organizations to more efficiently use computing resources, reduce costs, and provide better access to critical information. Database applications can be customized and targeted to individual users without having to conform to or be limited by the physical hardware.[57]

Database virtualization, however, does present some challenges.[58] Security is always a concern.[59] In one case, a fired employee was able to gain access to a virtual database and delete important applications, e-mail, and other documents, costing the company about $800,000 in losses, according to the FBI.[60] Organizations can use a number of approaches to prevent security problems. Every layer or component of a virtual database should be secured with firewalls, IDs, passwords, and similar approaches. Employee access to a virtual database can be limited to certain aspects or areas of the database. All physical components of the virtual database, including servers and networks, should be completely secured as well. Backup is also an issue with virtual databases.[61] Disk files and other physical hardware components should have adequate backup procedures. Some companies that use virtual databases have two or more data centers that back up each other.[62] How to recover from natural and human-made disasters is also an issue.[63] Instead of using traditional hardware and software disaster recovery procedures, specialized recovery software and procedures for virtual databases should be used. Companies such as Acronis provide this type of specialized backup and recovery tools for virtual databases.[64]

Special-Purpose Database Systems

In addition to the popular database management systems just discussed, some specialized database packages are used for specific purposes or in specific industries. NoSQL (Not only SQL) are database management systems that can handle or accommodate data that does not fit into tables required by traditional relational databases previously discussed. Many NoSQL databases are open source, including Hadoop, Cassandra, Hypertable, and others.[65] As another example, Apple's iTunes software uses a special-purpose database system that includes fields for song name, rating, file size, time, artist, album, and genre. When iTunes users go to the iTunes store and search for an artist, they are actually querying the central iTunes database (see Figure 5.19).

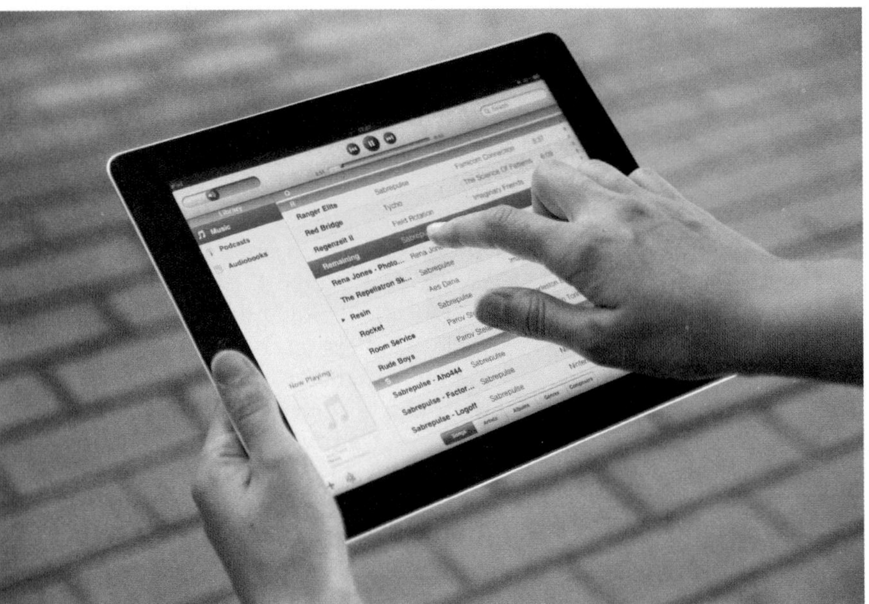

FIGURE 5.19
iTunes database
Apple's iTunes software uses a database to catalog and access music.

Using Databases with Other Software

Database management systems are often used with other software and with the Internet. A DBMS can act as a front-end application or a back-end application. A *front-end application* is one that people interact with directly. Marketing researchers often use a database as a front end to a statistical analysis program. The researchers enter the results of market questionnaires or surveys into a database. The data is then transferred to a statistical analysis program to determine the potential for a new product or the effectiveness of an advertising campaign. A *back-end application* interacts with other programs or applications; it only indirectly interacts with people or users. When people request information from a Web site, the site can interact with a database (the back end) that supplies the desired information. For example, you can connect to a university Web site to find out whether the university's library has a book you want to read. The site then interacts with a database that contains a catalog of library books and articles to determine whether the book you want is available (see Figure 5.20).

INFORMATION SYSTEMS @ WORK

Managing the Database: It Can't Stop

A database is like any other organizational asset: someone must watch over it for the organization to obtain the greatest value from its investment. As you read in this chapter, that person is the database administrator (DBA).

In a company like Vodafone, database administration is a big job. Vodafone Group is one of the world's largest mobile communications companies with operations in Europe, the Middle East, Africa, Asia Pacific, and the United States (where it owns 45 percent of Verizon Wireless). Vodafone provides communications services to 391 million registered mobile customers.

Vodafone has 3,650 Oracle databases in Europe alone. These databases support critical applications that include online Web services, online Vodafone shops, automatic teller machines (ATMs) for prepaid mobile top-ups (refills), and billing systems.

Vodafone needed to manage these databases effectively without increased staffing. It wanted to become proactive in database management to improve overall system stability and availability. Specific challenges included the following:

- Improve reporting to ensure that database configuration issues are flagged and resolved before they become major and affect customer experience.
- Improve system availability and stability to ensure that the business-critical applications (including billing, online Web services, and retail) are always available.
- Provide faster access to the knowledge of customer support specialists, thus speeding problem resolution.

Fortunately, DBAs need not go into this battle unarmed. The same technology that gave rise to these needs also helps companies deal with them. Oracle's solution, which Vodafone (already a user of Oracle database management software) adopted, is called Oracle Enterprise Manager, now in version 12c. Other database suppliers have corresponding products.

For example, Enterprise Manager 12c has a module called the Automatic Database Diagnostic Monitor (ADDM). Its purpose is to diagnose why a system is slow, a time-consuming task for DBAs. ADDM focuses on activities that are the most time consuming for databases. It drills down through a problem classification tree to find the causes of problems. ADDM's goal is to discover the cause behind performance problems rather than just reporting symptoms. Each ADDM finding has an associated impact and benefit measure to rank issues and respond first to the most critical ones. To better understand the effect of the findings over time, each finding has a descriptive name that DBAs can use to apply filters, search for data, and retrieve previous occurrences of the finding in the last 24 hours.

Vodafone began by carrying out a pilot project using Enterprise Manager 12c on 108 of its databases, just under 3 percent of the total, in the third quarter of 2011. Using data from this pilot project, Vodafone calculated that it would need 85 DBAs in Europe for all its databases, an average of 43 databases each—less than an hour a week per database. The reduction in time spent on technical tasks freed up time for business-generating activities, such as designing improved solutions to business needs.

Using Oracle's support services and Enterprise Manager 12c, "we have improved response times by more than 50 percent and are much more proactive, fixing issues before they become a problem for our customers," reports Peter O'Brien, manager of technology and infrastructure services for Oracle products at Vodafone Group. Informed use of technology support tools clearly pays off for this business.

Discussion Questions

1. Database support products are not interchangeable: each vendor's support software is designed to work with its databases. How would you approach the problem of selecting a database supplier if one of them had the best support software but was deficient in other areas? Or if that database was the most expensive?
2. Vodafone's customers make mobile telephone calls at any time. This usage means that at least some of its databases must be available around the clock. Downtime of more than a few seconds will cause Vodafone to lose customers. By contrast, your school could survive a database outage of several minutes to a few hours with little harm beyond annoyance. Should your school invest in a product like Enterprise Manager?

Critical Thinking Questions

1. Do you think you would enjoy being a database administrator? Do you think you would be good at this job? Explain why.

2. From the description of the difference between the jobs of database administrator and of data administrator, would a data administrator use a product like Enterprise Manager? Why or why not?

SOURCES: Oracle Enterprise Manager 12c for Database Management, *www.oracle.com/technetwork/oem/db-mgmt/db-mgmt-093445.html,* accessed May 5, 2012; Staff, "Vodafone Group plc Embraces Proactive Support, Improving Pan-European Database Performance to Ensure Reliable Mobile Communications for 391 Million Customers," Oracle, *www.oracle.com/us/corporate/customers/customersearch/vodafone-group-1-db-ss-1530452.html,* February 22, 2012; Vodafone Web site, *www.vodafone.com,* accessed May 5, 2012.

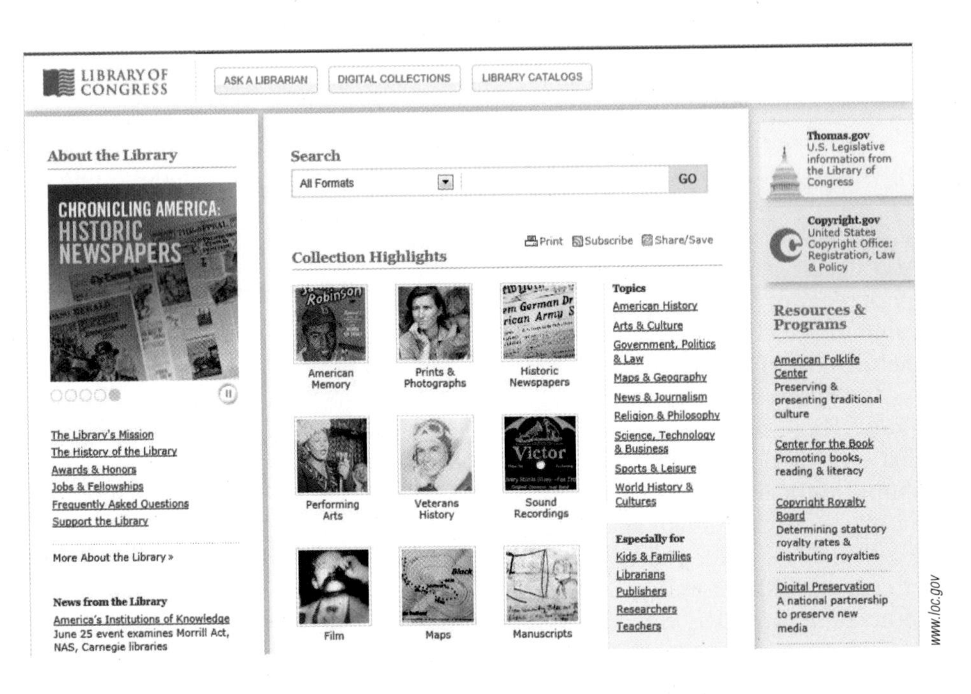

FIGURE 5.20

Library of Congress Web site

The Library of Congress (LOC) provides Web access to its databases, which include references to books and digital media in the LOC collection.

DATABASE APPLICATIONS

Today's database applications manipulate the content of a database to produce useful information. These applications allow organizations to store and use a large amount of unstructured data (big data). They can also help users to link the company databases to the Internet, set up data warehouses and marts, use databases for strategic business intelligence, place data at different locations, use online processing and open connectivity standards for increased productivity, develop databases with the object-oriented approach, and search for and use unstructured data, such as graphics, audio, and video.

Big Data Applications

In the past, organizations collected, stored, and processed data from their transaction processing system. This is still the case, but today, organizations are collecting huge amounts of unstructured data from other sources, such as the Internet, photos, video, audio, social networks, and sensors. This is often called *big data*—large amounts of unstructured data that is difficult or impossible to capture, store, and manipulate using traditional database management systems.

These large amounts of unstructured data can provide valuable information and insights to help organizations achieve their goals.[66] Big data, for example, can reveal the customers that are most likely to purchase products and services from the business. It can identify where and when specific customers shop. It can even determine the price a customer is willing to pay for a product or service. Because of the analysis done on big data, some companies are starting to offer different prices to different customers based on what they might be willing to pay, which can increase total revenues and profits. "This is an opportunity to walk into the CEO's office and say, I can change this business and provide knowledge at your fingertips in a matter of seconds for a price I couldn't touch five years ago," says the CIO of Catalina Marketing.[67]

While some companies have tried to use traditional relational database management systems discussed previously in this chapter, special big data hardware and software tools can be more effective and efficient to use. Hadoop, discussed previously, can be used to store, organize, and analyze big data. Yahoo, for example, used Hadoop to collect and analyze exabytes (millions of terabytes) of data.[68] Oracle has developed its Big Data Appliance, which is a combination of hardware and software specifically designed to capture, store, and analyze large amounts of unstructured data. SAS also supports big data analysis. IBM has developed InfoSphere BigInsights, based on the open-source Hadoop database program. IBM also offers BigSheets, which helps organizations analyze continuously created data.

Not everyone, however, is happy with big data applications. Some people have privacy concerns that corporations are harvesting huge amounts of personal data that can be shared with other organizations. With all this data, organizations can develop extensive profiles of people without their knowledge or consent. Big data also introduces security concerns. Can an organization keep big data secure from competitors and malicious hackers? Some experts believe companies that collect and store big data could be open to liability suits from individuals and organizations. Even with these potential disadvantages, many companies are rushing into big data with its potential treasure trove of information and new applications.

Linking the Company Database to the Internet and Mobile Devices

The ability to link databases to the Internet is one reason the Internet is so popular. A large percentage of corporate databases are accessed over the Internet through a standard Web browser. Being able to access bank account data, student transcripts, credit card bills, product catalogs, and a host of other data online is convenient for individual users and increases effectiveness and efficiency for businesses and organizations. Amazon.com, eHarmony.com, eBay, and many others have made billions of dollars by combining databases, the Internet, and smart business models. Of course, security is always a concern when linking a database to the Internet. Blue Cross and Blue Shield, for example, spent about $6 million to encrypt data stored in its databases to protect client privacy and secure its databases.[69]

Developing a seamless integration of databases with the Internet is sometimes called a *semantic Web.*[70] A semantic Web provides metadata with all Web content using technology called the Resource Description Framework (RDF).[71] The result is a better organized Web that acts like one large database system. The World Wide Web Consortium (W3C) has established standards, including an RDF, for a semantic Web in hopes of bringing content providers onboard. Other database tools, including data fusion products by TLO, allow organizations to connect different public and private databases together and make them easy to use and search.[72]

The use of smartphones and tablet computers to connect to corporate databases has increased. Data can be sent to smartphones or tablet computers when they are connected to larger computers or wirelessly.[73] The president of Edmunds believes that the ability to search corporate databases for information is a critical business tool.[74] He uses an Apple iPad to get instant access to information from any location or while traveling. The large clothing retailer Guess uses mobile business intelligence to deliver sales and marketing data

HanaTour Strengthens Database Security

HanaTour International Service is South Korea's largest provider of overseas travel services and air tickets. HanaTour employs nearly 2,000 people in Korea and travel agents outside Korea to provide clients with travel information about 26 regions worldwide.

HanaTour customers who book travel provide the company with personal details, including their addresses, contact phone numbers, dates of birth, passport numbers, and payment information. These details, along with their airline and tour bookings and travel itineraries, are stored in HanaTour's database. The confidential nature of this information means HanaTour must have security measures in place to protect the database from unauthorized access.

In addition to these marketplace requirements, HanaTour must comply with South Korea's Electronic Communication Privacy Act. That act requires industries to take measures to protect the privacy of personal information. Thus, protecting customer data is not only good business, but also a legal requirement.

Even without the legal mandate, data security is a good idea for the following reasons, among others:

- **Building customer trust**. Customers who trust a firm are more likely to return.
- **Ensuring privacy**. A firm cannot ensure privacy if it does not have security.
- **Avoiding unnecessary costs**. Security breaches can be expensive, both directly in terms of legal costs and customer compensation and indirectly in terms of lost profits due to customers going elsewhere.
- **Maintaining a positive public image**. With so much publicity about database breaches, a strong security effort will protect a firm's reputation.
- **Gaining a competitive edge**. A firm's reputation for data security helps to leverage the competitive forces of rivalry with existing competitors and erect barriers to new entrants, which you read about in Chapter 2.

Among the actions that HanaTour took to improve database security were:

- Added data encryption, both in the database and during transmission
- Implemented access control based on individual authorizations and assigned tasks
- Discouraged hacker attacks by blocking database access even if a hacker obtained top-level administrator privileges for the system
- Created an audit trail of database access to spot suspicious activities so that action could be taken immediately
- Implemented reports to show compliance with security requirements
- Used audit information to develop further security plans

Like most small and medium-sized firms, HanaTour does not need the skills that this security upgrade called for on a permanent full-time basis. Rather than hiring and training staff members to address short-term needs and then releasing or finding other work for these employees, HanaTour engaged specialists. The company worked with Korean database consulting firm Wizbase. HanaTour had worked with Wizbase previously, so they didn't have to spend time explaining basic information about how HanaTour's business works.

The net result of these actions was to make it much more difficult for unauthorized people to see any of the personal information that HanaTour customers supplied. Did this help HanaTour? According to Kim Jin-hwan, director of the HanaTour's IT department, "Our business is based on service. We do not want anything to go wrong on a customer's holiday that will inconvenience them. Lost data or any disruptions to our system would affect our ability to provide optimum service. We upgraded our database to improve performance and take advantage of new security features, which would minimize the risk of losing confidential customer data and strengthen our database and systems from unlawful access."

Discussion Questions

1. From the user side, Mr. Kim said that HanaTour upgraded to a new release of its database management software due to its improved security features. What are some potential disadvantages of upgrading to the latest release of a DBMS? When might an organization decide *not* to upgrade because of those disadvantages, even if the new release offers improved security features?
2. From the vendor side, given the tradeoffs that the previous question suggests, some users do not upgrade to the latest release of any database management system. What can a vendor do to motivate those users to upgrade? (You may want to think about the previous question before you answer this one.)

Critical Thinking Questions

1. HanaTour chose Wizbase as its implementation partner in part because of prior experience with that firm. Suppose Wizbase didn't have database security experts but promised to hire some if HanaTour selected Wizbase for this project. (Wizbase might feel that database security experts would be useful on other projects later.) Another firm might have data security experts on its staff, but HanaTour has no prior experience with that company. What is the relative importance of these two factors? What other factors would you consider in your choice of a partner on a database security project?
2. Think of the data that your university's database has about students as a large table, with a row for each student and a column for each data element. Group the data into major categories such as contact data, medical data, financial data, academic data and so on. Which groups of people, by job, should have access to each category? Within a group, who should have access to only one row, who should have access to more than one row but not all rows, and who should have access to all rows of the table? Should anyone be allowed to see data but not change it?

SOURCES: Forrester Research, "Formulate a Database Security Strategy To Ensure Investments Will Actually Prevent Data Breaches and Satisfy Regulatory Requirements," *www.oracle.com/us/corporate/analystreports/infrastructure/forrester-thlp-db-security-1445564.pdf*, July 13, 2011; HanaTour Web site (mostly in Korean), *www.hanatour.com*, accessed May 8, 2012; Staff, "HanaTour International Service Tightens Customer Data Security by Introducing Data Encryption, Access Control, and Audit Solutions," Oracle, *www.oracle.com/us/corporate/customers/customersearch/hanatour-intl-1-database-cs-1521219.html*, February 13, 2012; TD Merchant Services, *Contact*, *www.tdcanadatrust.com/document/PDF/merchantservices/tdct-merchantservices-contact-fall2008.pdf*, Fall 2008; Wizbase Web site (mostly in Korean), *www.wizbase.co.kr*, accessed May 7, 2012.

to its managers and workers with smart phones, tablet computers, and other mobile devices.[75]

Data Warehouses, Data Marts, and Data Mining

The raw data necessary to make sound business decisions is stored in a variety of locations and formats. This data is initially captured, stored, and managed by transaction processing systems that are designed to support the day-to-day operations of the organization. For decades, organizations have collected operational, sales, and financial data with their online transaction processing (OLTP) systems. The data can be used to support decision making through data warehouses, data marts, and data mining.

Data Warehouses

data warehouse: A large database that collects business information from many sources in the enterprise, covering all aspects of the company's processes, products, and customers, in support of management decision making.

A **data warehouse** is a database that holds business information from many sources in the enterprise, covering all aspects of the company's processes, products, and customers. Data warehouses allow managers to *drill down* to get more detail or *roll up* to take detailed data and generate aggregate or summary reports. The primary purpose is to relate information in innovative ways and help managers and executives make better decisions. A data warehouse stores historical data that has been extracted from operational systems and external data sources (see Figure 5.21).

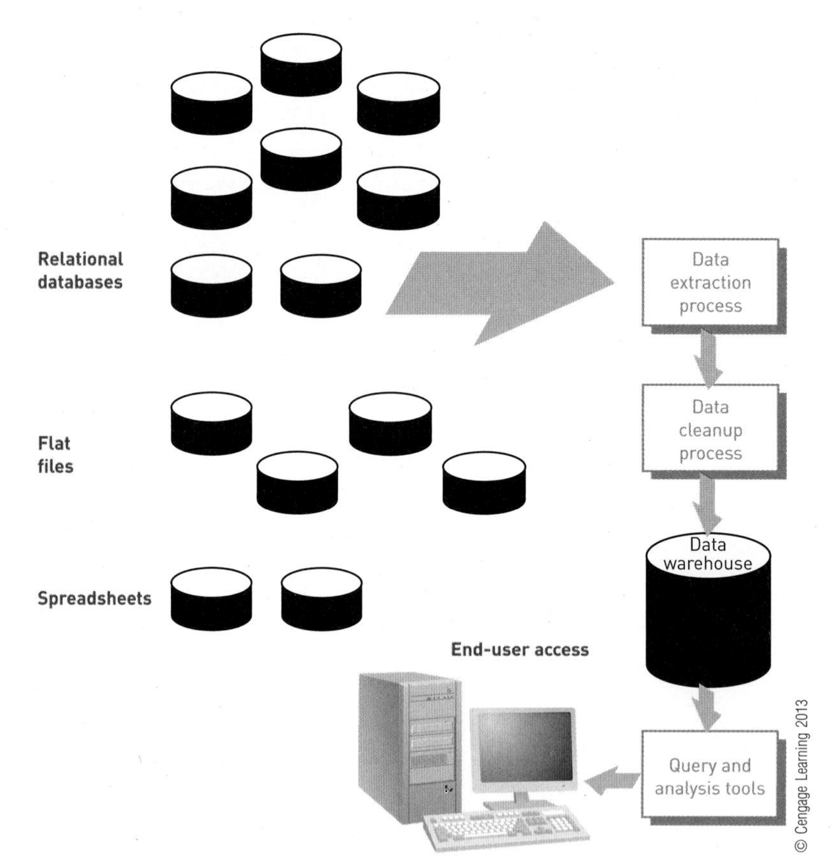

FIGURE 5.21
Elements of a data warehouse

Data warehouses typically start out as very large databases, containing millions and even hundreds of millions of data records. To keep it fresh and accurate, the data warehouse receives regular updates. Old data that is no longer needed is purged from the data warehouse. Data warehouses can also acquire data from unique sources. Oracle's Warehouse Management software, for example, can accept information from Radio Frequency Identification (RFID) technology, which is being used to tag products as they are shipped or moved from one location to another.[76] As another example, Hewlett-Packard's acquisition of Vertica Systems helps the company deliver data warehousing products.[77] Vertica Systems provides data-warehousing products to financial services companies, health care organizations, and companies in a variety of other industries.[78]

© iStockphoto/nullplus

Data warehouses can use tools such as Oracle's Warehouse Management software to acquire data from unique sources such as scans of RFID tags.

Data Marts

data mart: A subset of a data warehouse that is used by small and medium-sized businesses and departments within large companies to support decision making.

A **data mart** is a subset of a data warehouse. Data marts bring the data warehouse concept—online analysis of sales, inventory, and other vital business data that has been gathered from transaction processing systems—to small and medium-sized businesses and to departments within larger companies. Rather than store all enterprise data in one monolithic database, data marts contain a subset of the data for a single aspect of a company's business—for example, finance, inventory, or personnel. In fact, a specific area in the data mart might contain more detailed data than the data warehouse. In one survey, almost 50 percent of responding firms said they used data marts extensively.[79] Another 20 percent said they used them on a more limited basis.

Data Mining and Business Intelligence

data mining: An information-analysis tool that involves the automated discovery of patterns and relationships in a data warehouse.

Data mining is an information-analysis tool that involves the automated discovery of patterns and relationships in a data warehouse. According to the CEO of Blue Cross and Blue Shield, "Creating a warehouse with 3.5 billion pieces of data was the easy part. The hard part is turning that into actionable information and insights that will improve the quality and delivery of care, ultimately bending the cost curve in America."[80] Like gold mining, data mining sifts through mountains of data to find a few nuggets of valuable information. For example, Brooks Brothers, the oldest clothing retailer in the United States, used data mining to provide store managers with reports that help improve store performance and customer satisfaction.[81] Health care facilities use data mining to help doctors analyze vast medical databases to improve diagnosis and patient care.[82] Increasingly, companies are using data mining to gain a competitive advantage over their rivals.[83] The private equity firm Kohlberg, Kravis, and Roberts (KKR), for example, used data mining to analyze all of its existing businesses to find opportunities for cost savings or new revenues. American International Group (AIG) used data mining to analyze insurance risk. According to the Chief Executive of AIG, "There's an incredible amount of data we can leverage to make better decisions. We need to stretch our thinking."[84] Data mining is also being used by governments and nonprofit organizations. Data mining can be used to uncover Medicare fraud and help find which care providers are charging the government a lot of money to perform the same or similar procedures on the same patients.[85] Data mining can also be used to find people and organizations that attempt to launder money.[86] Data mining software from Verafin, for example, works by analyzing large numbers of financial transactions looking for suspicious transactions or patterns, including money laundering.[87]

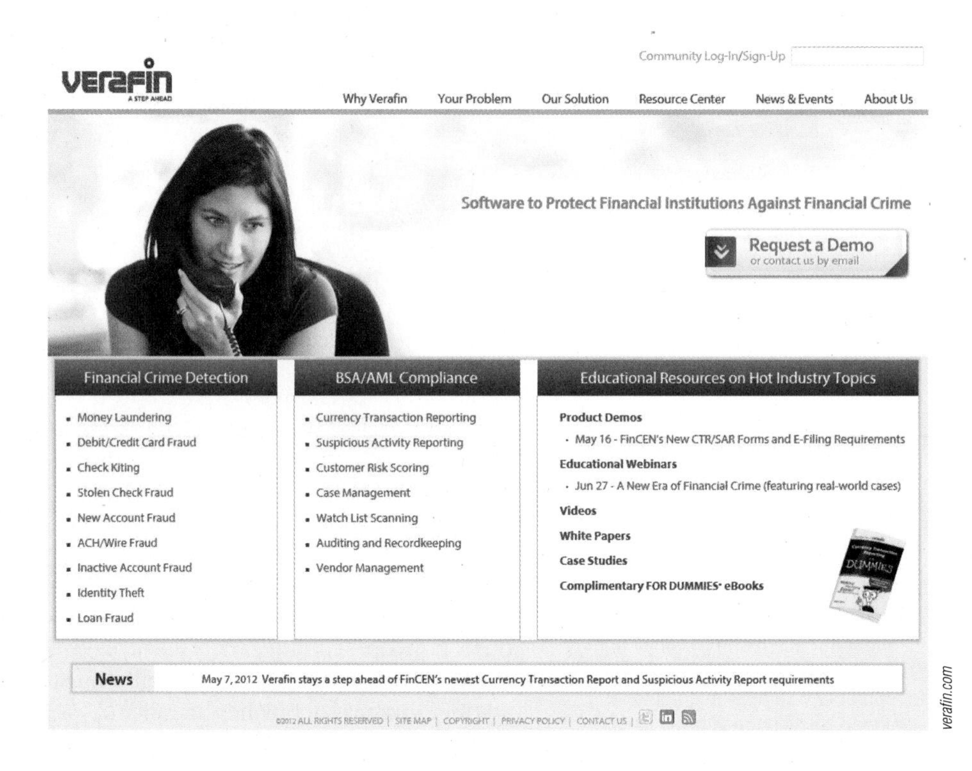

verafin.com

To meet its mission of helping financial institutions fight fraud, Verafin develops data mining software to analyze transactions and detect financial crimes.

predictive analysis: A form of data mining that combines historical data with assumptions about future conditions to predict outcomes of events, such as future product sales or the probability that a customer will default on a loan.

Predictive analysis is a form of data mining that combines historical data with assumptions about future conditions to predict outcomes of events, such as future product sales or the probability that a customer will default on a loan. Retailers use predictive analysis to upgrade occasional customers into frequent purchasers by predicting what products they will buy if offered an appropriate incentive. According to the CEO of Fair Isaac, "We know what you're going to do tomorrow."[88] Genalytics, Magnify, NCR Teradata, SAS Institute, Sightward, SPSS, Exalytics by Oracle, and Quadstone have developed predictive analysis tools.[89] Also called *Analytics* by some, predictive analysis often uses sophisticated mathematical techniques to analyze vast stores of data.[90] According to the chief technology officer of Bundle, a financial analysis company, "Just having a pile of data is not that useful. You have to ask the right questions and process the data in the right way where you can do meaningful analytics."[91] Proctor & Gamble has developed the Business Sufficiency program to help it predict future sales up to a year into the future.[92] The company is also hiring people with expertise in predictive analysis.[93] Traditional DBMS vendors are well aware of the great potential of data mining. Thus, companies such as Oracle, Sybase, Tandem, and Red Brick Systems are all incorporating data-mining functionality into their products.

business intelligence (BI): The process of gathering enough of the right information in a timely manner and usable form and analyzing it to have a positive impact on business strategy, tactics, or operations.

The use of databases for business-intelligence purposes is closely linked to the concept of data mining.[94] **Business intelligence (BI)** involves gathering enough of the right information in a timely manner and usable form and analyzing it so that it can have a positive effect on business strategy, tactics, or operations.[95] The Pirates baseball team used business intelligence to help increase attendance at games.[96] According to the director of business analytics, "We interact with fans in several ways, and the business intelligence tools help analyze data from multiple sources." BI turns data into useful information that is then distributed throughout an enterprise. It provides insight into the causes of problems, and when implemented, it can improve business operations. The Copper Mountain Ski Resort, for example, used Visual One software from Agilysys to provide business intelligence.[97] According the information technology director, "It's a highly flexible system that integrates easily with other solutions and enables us to view data in new and creative ways." Some university programs, including Fordham University's Graduate School of Business and Indiana University's School of Business, are starting to offer courses in business intelligence.[98] According to a senior vice president of IBM, "The more students that graduate knowledgeable in areas we care about, the better it is not just for our company but the companies we work with."

competitive intelligence: One aspect of business intelligence limited to information about competitors and the ways that knowledge affects strategy, tactics, and operations.

Competitive intelligence is one aspect of business intelligence and is limited to information about competitors and the ways that knowledge affects strategy, tactics, and operations. Competitive intelligence is a critical part of a company's ability to see and respond quickly and appropriately to the changing marketplace. Competitive intelligence is not espionage—the use of illegal means to gather information. In fact, almost all the information a competitive-intelligence professional needs can be collected by examining published information sources, conducting interviews, and using other legal and ethical methods. Using a variety of analytical tools, a skilled competitive-intelligence professional can by deduction fill the gaps in information already gathered.

counterintelligence: The steps an organization takes to protect information sought by "hostile" intelligence gatherers.

The term **counterintelligence** describes the steps an organization takes to protect information sought by "hostile" intelligence gatherers. One of the most effective counterintelligence measures is to define "trade secret" information relevant to the company and control its dissemination.

online analytical processing (OLAP): Software that allows users to explore data from a number of perspectives.

Online analytical processing (OLAP) allows users to explore data from a number of perspectives. OLAP databases support business intelligence discussed above and have been optimized to provide useful reports and analysis. The leading OLAP database vendors include Microsoft, Cognos, SAP, Business Objects, MicroStrategy, Applix, Infor, and Oracle.

Unlike data mining that provides bottom-up, discovery-driven analysis, OLAP provides top-down, query-driven data analysis. Whereas data mining requires no assumptions and instead identifies facts and conclusions based on patterns discovered, OLAP requires repetitive testing of user-originated theories. OLAP, or multidimensional analysis, requires a great deal of human ingenuity and interaction with the database to find information in the database. A user of a data-mining tool does not need to figure out what questions to ask; instead, the approach is "Here's the data, tell me what interesting patterns emerge." For example, a data-mining tool in a credit card company's customer database can construct a profile of fraudulent activity from historical information. Then, this profile can be applied to all incoming transaction data to identify and stop fraudulent behavior, which might otherwise go undetected. Table 5.2 compares OLAP and data mining.

TABLE 5.2 Comparison of OLAP and Data Mining

Characteristic	OLAP	Data Mining
Purpose	Supports data analysis and decision making	Supports data analysis and decision making
Type of analysis supported	Top-down, query-driven data analysis	Bottom-up, discovery-driven data analysis
Skills required of user	Must be very knowledgeable of the data and its business context	Must trust in data-mining tools to uncover valid and worthwhile hypotheses

data loss prevention (DLP): Systems designed to lock down—to identify, monitor, and protect—data within an organization.

Data loss prevention (DLP) refers to systems designed to lock down data within an organization. DLP software from RSA, Symantec, Code Green, Safend, Trend Micro, Sophos, and others are designed to identify, monitor, and protect data wherever it may exist on a system. That includes data stored on disk, passing over a network, and found in databases, in files, in e-mail, and elsewhere. DLP is a powerful tool for counterintelligence and is a necessity in complying with government regulations that require companies to safeguard private customer data.

Distributed Databases

distributed database: A database in which the data can be spread across several smaller databases connected through telecommunications devices.

Distributed processing involves placing processing units at different locations and linking them via telecommunications equipment. A **distributed database**, a database in which the data can be spread across several smaller databases connected through telecommunications devices, works on much the same principle. A user in the Milwaukee branch of a clothing manufacturer, for example, might make a request for data that is physically located at corporate headquarters in Milan, Italy. The user does not have to know where the data is physically stored (see Figure 5.22).

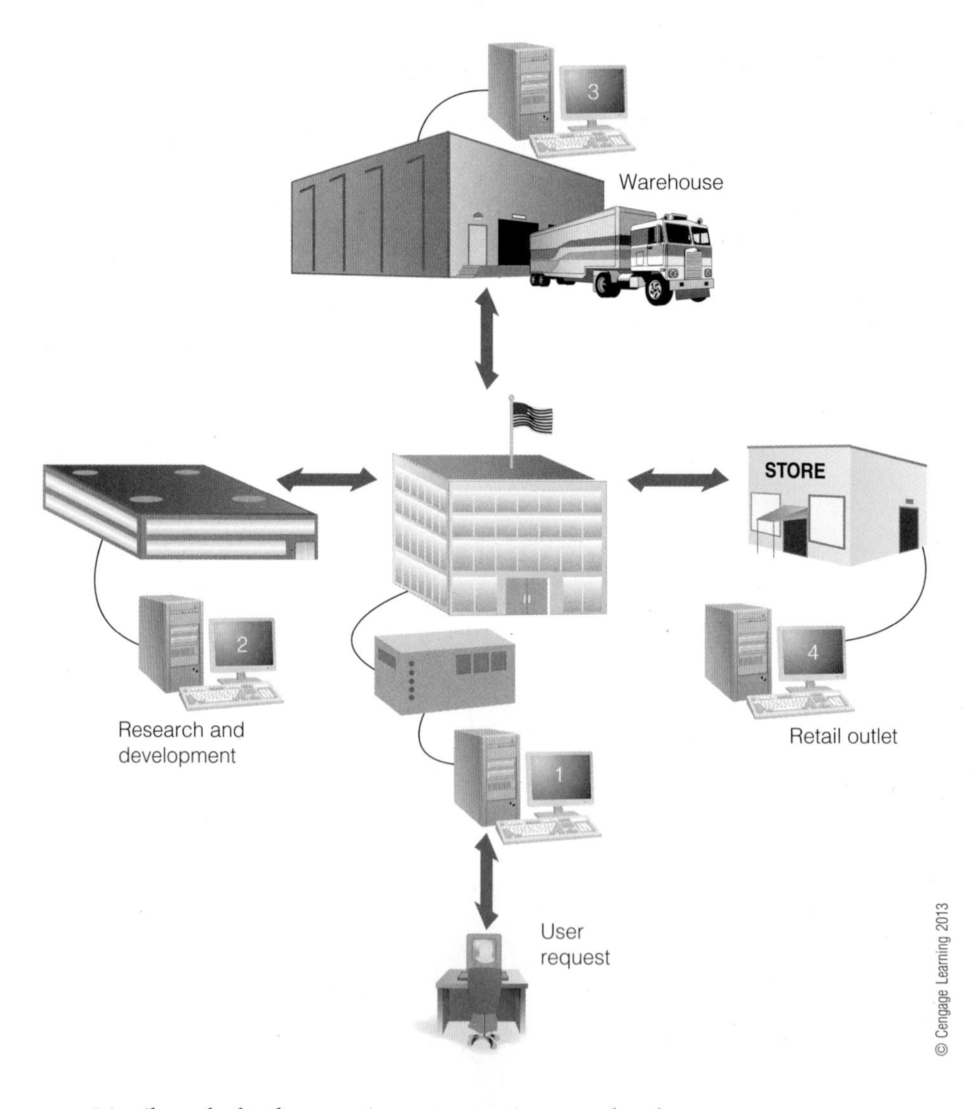

FIGURE 5.22
Use of a distributed database
For a clothing manufacturer, computers might be located at corporate headquarters, in the research and development center, in the warehouse, and in a company-owned retail store. Telecommunications systems link the computers so that users at all locations can access the same distributed database, no matter where the data is actually stored.

Distributed databases give corporations and other organizations more flexibility in how databases are organized and used. Local offices can create, manage, and use their own databases, and people at other offices can access and share the data in the local databases. Giving local sites more direct access to frequently used data can improve organizational effectiveness and efficiency significantly. The New York City Police Department, for example, has thousands of officers searching for information located on servers in offices around the city.

Despite its advantages, distributed processing creates additional challenges in integrating different databases (information integration) and maintaining data security, accuracy, timeliness, and conformance to standards. Distributed databases allow more users direct access at different sites; however, controlling who accesses and changes data is sometimes difficult. Also, because distributed databases rely on telecommunications lines to transport data, access to data can be slower.

replicated database: A database that holds a duplicate set of frequently used data.

To reduce telecommunications costs, some organizations build a replicated database. A **replicated database** holds a duplicate set of frequently used data. The company sends a copy of important data to each distributed processing location when needed or at predetermined times. Each site sends the changed data back to update the main database on an update cycle that meets the needs of the organization. This process, often called *data synchronization*, is used to make sure that replicated databases are accurate, up to date, and consistent with each other. A railroad, for example, can use a replicated database to increase punctuality, safety, and reliability. The primary database can hold data on fares, routings, and other essential information. The data can be continually replicated and downloaded on a read-only basis from the master database to hundreds of remote servers across the country. The remote locations can send back to the main database the latest figures on ticket sales and reservations.

Object-Relational Database Management Systems

object-oriented database: A database that stores both data and its processing instructions.

An **object-oriented database** uses the same overall approach of object-oriented programming that was discussed in Chapter 4. With this approach, both the data and the processing instructions are stored in the database. For example, an object-oriented database could store monthly expenses and the instructions needed to compute a monthly budget from those expenses. A traditional DBMS might only store the monthly expenses. Object-oriented databases are useful when a database contains complex data that needs to be processed quickly and efficiently.

In an object-oriented database, a *method* is a procedure or action. A sales tax method, for example, could be the procedure to compute the appropriate sales tax for an order or sale—for example, multiplying the total amount of an order by 7 percent, if that is the local sales tax. A *message* is a request to execute or run a method. For example, a sales clerk could issue a message to the object-oriented database to compute sales tax for a new order. Many object-oriented databases have their own query language, called *object query language* (*OQL*), which is similar to SQL, discussed previously.

object-oriented database management system (OODBMS): A group of programs that manipulate an object-oriented database and provide a user interface and connections to other application programs.

An object-oriented database uses an **object-oriented database management system (OODBMS)** to provide a user interface and connections to other programs. Computer vendors who sell or lease OODBMSs include Versant and Objectivity.[99] Many organizations are selecting object-oriented databases for their processing power. Versant's OODBMS, for example, is being used by companies in the telecommunications, defense, online gaming, and health care industries and by government agencies.[100] The *Object Data Standard* is a design standard created by the *Object Database Management Group* for developing object-oriented database systems.[101]

object-relational database management system (ORDBMS): A DBMS capable of manipulating audio, video, and graphical data.

An **object-relational database management system (ORDBMS)** provides a complete set of relational database capabilities plus the ability for third parties to add new data types and operations to the database. These new data types can be audio, images, unstructured text, spatial, or time series data that require new indexing, optimization, and retrieval features. Each of the vendors offering ORDBMS facilities provides a set of application programming interfaces to allow users to attach external data definitions and methods associated with those definitions to the database system. They are essentially offering a standard socket into which users can plug special instructions. DataBlades, Cartridges, and Extenders are the names applied by Oracle and IBM to describe the plug-ins to their respective products. Other plug-ins serve as interfaces to Web servers.

Visual, Audio, and Other Database Systems

In addition to raw data, organizations are finding a need to store large amounts of visual and audio signals in an organized fashion. Credit card companies, for example, enter pictures of charge slips into an image database using a scanner. The images can be stored in the database and later sorted by customer name, printed, and sent to customers along with their monthly statements. Image databases are also used by physicians to store X-rays and transmit them to clinics away from the main hospital. Financial services, insurance companies, and government branches are using image databases to store vital records and replace paper documents. Drug companies often need to analyze many visual images from laboratories. Visual databases can be stored in some object-relational databases or special-purpose database systems. Many relational databases can also store images.

In addition to visual, audio, and virtual databases, other special-purpose database systems meet particular business needs. *Spatial data technology* involves using a database to store and access data according to the locations it describes and to permit spatial queries and analysis. Builders and insurance companies use spatial data to make decisions related to natural hazards. Spatial data can even be used to improve financial risk management with information stored by investment type, currency type, interest rates, and time. MapInfo software from Pitney Bowes allows businesses such as Home Depot, Sonic Restaurants, CVS Corporation, and Chico's to choose the optimal location for new stores and restaurants based on geospatial demographics. The software provides information about local competition, populations, and traffic patterns to predict how a business will fare in a particular location. Spatial data technology also can be used to assist law enforcement agencies and emergency response teams to prepare for emergencies and provide community protection in an efficient manner.[102]

Spatial data technology is used by firefighters and other emergency responders to respond to incidents immediately.

SUMMARY

Principle:

Data management and modeling are key aspects of organizing data and information.

Data is one of the most valuable resources that a firm possesses. It is organized into a hierarchy that builds from the smallest element to the largest. The smallest element is the bit, a binary digit. A byte (a character such as a letter or numeric digit) is made up of eight bits. A group of characters, such as a name or number, is called a field (an object). A collection of related fields is a record; a collection of related records is called a file. The database, at the top of the hierarchy, is an integrated collection of records and files.

An entity is a generalized class of objects for which data is collected, stored, and maintained. An attribute is a characteristic of an entity. Specific values of attributes—called data items—can be found in the fields of the record describing an entity. A data key is a field within a record that is used to identify the record. A primary key uniquely identifies a record, while a secondary key is a field in a record that does not uniquely identify the record.

Traditional file-oriented applications are often characterized by program-data dependence, meaning that they have data organized in a manner that cannot be read by other programs. To address problems of traditional file-based data management, the database approach was developed. Benefits of this approach include reduced data redundancy, improved data consistency and integrity, easier modification and updating, data and program independence, standardization of data access, and more efficient program development.

When building a database, an organization must consider content, access, logical structure, and physical organization of the database. Many enterprises build a data center to house the servers that physically store databases and the systems that deliver mission-critical information and services. One of the tools that database designers use to show the logical structure and relationships among data is a data model. A data model is a map or diagram of entities and their relationships. Enterprise data modeling involves analyzing the data and information needs of an entire organization. Entity-relationship (ER) diagrams can be used to show the relationships among entities in the organization.

The relational model places data in two-dimensional tables. Tables can be linked by common data elements, which are used to access data when the database is queried. Each row represents a record, and each column represents an attribute (or field). Allowable values for these attributes are called the domain. Basic data manipulations include selecting, projecting, and joining. The relational model is easier to control, more flexible, and more intuitive than the other models because it organizes data in tables.

Principle:

A well-designed and well-managed database is an extremely valuable tool in supporting decision making.

A DBMS is a group of programs used as an interface between a database and its users and other application programs. When an application program requests data from the database, it follows a logical access path. The actual retrieval of the data follows a physical access path. Records can be considered in the same way: a logical record is what the record contains; a physical record is where the record is stored on storage devices. Schemas are used to describe the entire database, its record types, and their relationships to the DBMS.

A DBMS provides four basic functions: offering user views, creating and modifying the database, storing and retrieving data, and manipulating data and generating reports. Schemas are entered into the computer via a data definition language, which describes the data and relationships in a specific database. Another tool used in database management is the data dictionary, which contains detailed descriptions of all data in the database.

After a DBMS has been installed, the database can be accessed, modified, and queried via a data manipulation language. A more specialized data manipulation language is the query language, the most common being Structured Query Language (SQL). SQL is used in several popular database packages today and can be installed on PCs and mainframes.

Database as a Service (DaaS) is a new form of database service in which clients lease use of a database on a service provider's site. In DaaS, the database is stored on a service provider's servers and accessed by the client over a network, typically the Internet. In DaaS, database administration is provided by the service provider. *Database virtualization* uses virtual servers and operating systems to allow two or more database systems, including servers and databases management systems (DBMSs), to act like a single unified database system. A virtual database acts independently of the physical servers, disk drives, and other database components that can be located around the world. Virtualization allows organizations to more efficiently use computing resources, reduce costs, and provide better access to critical information.

A database administrator (DBA) plans, designs, creates, operates, secures, monitors, and maintains databases. Attacks on databases such as SQL injection attacks are an all-too-common threat that DBAs must guard against. Selecting a DBMS begins by analyzing the information needs of the organization. Important characteristics of databases include the size of the database, the number of concurrent users, performance of the database, the ability of the DBMS to be integrated with other systems, the features of the DBMS, the vendor considerations, and the cost of the database management system.

Principle:

The number and types of database applications will continue to evolve and yield real business benefits.

Traditional online transaction processing (OLTP) systems put data into databases very quickly, reliably, and efficiently, but they do not support the types of data analysis that today's businesses and organizations require. To address this need, organizations are building data warehouses, which are relational database management systems specifically designed to support management decision making. Data marts are subdivisions of data warehouses and are commonly devoted to specific purposes or functional business areas.

Data mining, which is the automated discovery of patterns and relationships in a data warehouse, is a practical approach to generating hypotheses about the data that can be used to predict future behavior.

Predictive analysis is a form of data mining that combines historical data with assumptions about future conditions to forecast outcomes of events such as future product sales or the probability that a customer will default on a loan.

Business intelligence is the process of getting enough of the right information in a timely manner and usable form and analyzing it so that it can have a positive effect on business strategy, tactics, or operations. Competitive intelligence is one aspect of business intelligence limited to information about competitors and the ways that information affects strategy, tactics, and operations. Competitive intelligence is not espionage—the use of illegal means to gather information. Counterintelligence describes the steps an organization takes to protect information sought by "hostile" intelligence gatherers. Data loss prevention (DLP) refers to systems designed to lock down data within an organization.

Multidimensional databases and online analytical processing (OLAP) programs are being used to store data and allow users to explore the data from a number of different perspectives.

An object-oriented database uses the same overall approach of object-oriented programming, first discussed in Chapter 4. With this approach, both the data and the processing instructions are stored in the database. An object-relational database management system (ORDBMS) provides a complete set of relational database capabilities, plus the ability for third parties to add new data types and operations to the database. These new data types can be audio, video, and graphical data that require new indexing, optimization, and retrieval features.

In addition to raw data, organizations are finding a need to store large amounts of visual and audio signals in an organized fashion. A number of special-purpose database systems are also being used.

CHAPTER 5: SELF-ASSESSMENT TEST

Data management and modeling are key aspects of organizing data and information.

1. A group of programs that manipulate the database and provide an interface between the database and the user of the database and other application programs is called a(n) ______________.
   a. GUI
   b. operating system
   c. DBMS
   d. productivity software
2. A(n) ______________ is a skilled and trained IS professional who directs all activities related to an organization's database.
3. A record is made up of multiple fields. True or False?
4. A(n) ______________ is a field or set of fields that uniquely identifies a database record.
   a. attribute
   b. data item
   c. key
   d. primary key
5. The ______________ approach provides a pool of related data shared by multiple information systems.
6. Many businesses store their database and related systems in climate-controlled facilities called ______________.
7. What database model places data in two-dimensional tables?
   a. relational
   b. network
   c. normalized
   d. hierarchical

A well-designed and well-managed database is an extremely valuable tool in supporting decision making.

8. ______________ involves combining two or more database tables.
   a. Projecting
   b. Joining
   c. Selecting
   d. Data cleanup
9. Because the DBMS is responsible for providing access to a database, one of the first steps in installing and using a database involves telling the DBMS the logical and physical structure of the data and relationships among the data in the database. This description of an entire database is called a(n) ______________.
10. A(n) ______________ database acts independently of the physical servers, disk drives, and other database components that can be located around the world.
    a. virtual
    b. personal
    c. relational
    d. object-oriented
11. Access is a popular DBMS for ______________.
    a. personal computers
    b. graphics workstations
    c. mainframe computers
    d. supercomputers
12. A trend in database management, known as Database as a Service, places the responsibility of storing and managing a database on a service provider. True or False?

The number and types of database applications will continue to evolve and yield real business benefits.

13. A(n) _____________ holds business information from many sources in the enterprise, covering all aspects of the company's processes, products, and customers.
14. An information-analysis tool that involves the automated discovery of patterns and relationships in a data warehouse is called _____________.
    a. a data mart
    b. data mining
    c. predictive analysis
    d. business intelligence
15. _____________ allows users to predict the future based on database information from the past and present.
16. The process of gathering information in a timely manner and in a usable form so that it positively affects business strategy, tactics, and operations is called _____________.

CHAPTER 5: SELF-ASSESSMENT TEST ANSWERS

1. c
2. database administrator
3. True
4. d
5. database
6. data centers
7. a
8. b
9. schema
10. a
11. a
12. True
13. data warehouse
14. b
15. Predictive analysis
16. business intelligence

REVIEW QUESTIONS

1. What is an attribute? How is it related to an entity?
2. Define the term *database*. How is it different from a database management system?
3. What is the hierarchy of data in a database?
4. What is a relation, and what is its importance to relational databases?
5. What is the purpose of a primary key? How is it useful in controlling data redundancy?
6. What is a virtual database? What are the advantages and disadvantages of using a virtual database?
7. What are the advantages of the database approach over the traditional approach to database management?
8. What is data modeling? What is its purpose? Briefly describe three commonly used data models.
9. What is a data center, and why are they becoming increasingly important?
10. What is Database as a Service (DaaS)? What are the advantages and disadvantages of using the DaaS approach?
11. How can a data dictionary be useful to database administrators and DBMS software engineers?
12. Identify important characteristics in selecting a database management system.
13. What is the difference between a data definition language (DDL) and a data manipulation language (DML)?
14. What is the difference between projecting and joining?
15. What is a distributed database system?
16. What is a data warehouse, and how is it different from a traditional database used to support OLTP?
17. What is meant by the "front end" and the "back end" of a DBMS?
18. What is the relationship between the Internet and databases?
19. What is data mining? What is OLAP? How are they different?
20. What is an ORDBMS? What kind of data can it handle?
21. What is business intelligence? How is it used?
22. What is predictive analysis, and how does it assist businesses in gaining a competitive advantage?
23. In what circumstances might a database administrator consider using an object-oriented database?

DISCUSSION QUESTIONS

1. You have been selected to represent the student body on a project to develop a new student database for your school. What is the first step in developing the database? What actions might you take to fulfill this responsibility to ensure that the project meets the needs of students and is successful?
2. Your company wants to increase revenues from its existing customers. How can data mining be used to accomplish this objective?
3. You are going to design a database for your school's outdoors club to track its activities. Would cloud computing be a useful strategy to set up the database? Describe your reasoning. If not, what would you recommend?
4. Make a list of the databases in which data about you exists. How is the data in each database captured? Who updates each database and how often? Is it possible for you to request a printout of the contents of your data record from each database? What data privacy concerns do you have?
5. If you were the database administrator for the iTunes store, how might you use predictive analysis to determine which artists and movies will sell the most next year?
6. You are the vice president of information technology for a large multinational consumer packaged goods company (such as Procter & Gamble or Unilever). You must make a presentation to persuade the board of directors to invest $5 million to establish a competitive-intelligence organization—including people, data-gathering services, and software tools. What key points do you need to make in favor of this investment? What arguments can you anticipate that the board might make?
7. Identity theft, where people steal personal information, continues to be a problem for consumers and businesses. Assume that you are the database administrator for a corporation with a large database that is accessible from the Web. What steps would you implement to prevent people from stealing personal information from the corporate database?
8. You have been hired to set up a database for a company similar to Netflix that rents movies over the Internet. Describe what type of database management system you would recommend for this application.

PROBLEM-SOLVING EXERCISES

1. Develop a simple data model for the music you have on your digital music player or in your CD collection, where each row is a song. For each row, what attributes should you capture? What will be the primary key for the records in your database? Describe how you might use the database to expand your music exposure and enjoyment.
2. A video movie rental store is using a relational database to store information on movie rentals to answer customer questions. Each entry in the database contains the following items: Movie Number (the primary key), Movie Title, Year Made, Movie Type, MPAA Rating, Number of Copies on Hand, and Quantity Owned. Movie Types are comedy, family, drama, horror, science fiction, and western. MPAA ratings are G, PG, PG-13, R, NC-17, and NR (not rated). Use a single-user database management system to build a data-entry screen to enter this data. Build a small database with at least ten entries.
3. To improve service to their customers, the salespeople at the video rental store have proposed a list of changes being considered for the database in the previous exercise. From this list, choose two database modifications and modify the data-entry screen to capture and store this new information. Proposed changes are
   a. To help store clerks locate the newest releases, add the date that the movie was first available.
   b. Add the director's name.
   c. Add the names of three primary actors in the movie.
   d. Add a rating of one, two, three, or four stars.
   e. Add the number of Academy Award nominations.
4. Using a graphics program, develop an entity-relationship diagram for a database application for an Internet bookstore where students buy textbooks from a salesperson and receive invoices for their purchases. Use Figure 5.6 as a guide.

TEAM ACTIVITIES

1. Each team member should separately list, describe, and diagram the reports he or she would like to see in a class registration application from the perspective of a university administrator. The team should then rank the reports from most important to least important. Write a report that lists, describes, and diagrams the top three reports. The report should also include an appendix that lists all reports from each team member.
2. As a team of three or four classmates, interview business managers from three different businesses that use databases. What data entities and data attributes are contained in each database? What database company did each company select to provide their database and why? How do they access the database to perform analysis? Have they received training in any query or reporting tools? What do they like about their databases, and what could be improved? Do any of them use data-mining or OLAP techniques? Weighing the information obtained, select one of these databases as being most strategic for the firm and briefly present your selection and the rationale for the selection to the class.
3. Imagine that you and your classmates are a research team developing an improved process for evaluating loan applicants for automobile purchases. The goal of the research is to predict which applicants will become delinquent or forfeit their loan. Those who score well on the application will be accepted, and those who score exceptionally well will be considered for lower-rate loans. Prepare a brief report for your instructor addressing these questions:
   a. What data do you need for each loan applicant?
   b. What data might you need that is not typically requested on a loan application form?
   c. Where might you get this data?
   d. Take a first cut at designing a database for this application. Using the material in this chapter on designing a database, draw the logical structure of the relational tables for this proposed database. In your design, include the data attributes you believe are necessary for this database and show the primary keys in your tables. Keep the size of the fields and tables as small as possible to minimize required disk drive storage space. Fill in the database tables with the sample data for demonstration purposes (ten records). After your design is complete, implement it using a relational DBMS.

WEB EXERCISES

1. Use a Web search engine to find information on database virtualization. Write a brief report describing what you found, including companies that have used this approach and what others have said about the advantages and disadvantages of this approach.
2. More information is being produced than can currently be stored in data centers, and yet, existing data centers are consuming huge amounts of energy and putting a strain on the environment and on budgets. Go online to research "Green Data Center" to learn what can be done to store more data using fewer resources. Students with the most unique and useful suggestions may be awarded extra credit.

CAREER EXERCISES

1. Describe what database privacy policies and approaches you would like to see at work in a career of your choice. Write a report on your privacy recommendations.
2. How could you use business intelligence (BI) to do a better job at work? Give some specific examples of how BI can give you a competitive advantage.

CASE STUDIES

Case One

Helping Marines and Sailors to Quit Smoking

The health impact of tobacco smoking is well known. Yet, the incidence of smoking remains high for three reasons: It is legal in most places, powerful organizations have an economic interest in promoting it, and many smokers become physically addicted to the nicotine in tobacco. However, many organizations whose performance depends on the physical fitness of their members make a great effort to reduce the use of tobacco.

The United States military is one such organization. In the field, medical officers such as Cdr. Alfredo Baker of the Navy often initiate tobacco cessation programs. Cdr. Baker, deployed as a psychiatrist with a regimental combat team in Afghanistan, wrote to Matt Greger, head of database consulting firm The Business Helper, Inc. (which closed in March, 2012), asking for help:

> I recently purchased FileMaker Pro 11 because I wanted to learn how to use a database for first time and this software was highly recommended.... I have a question. I recently started a Tobacco Smoking Cessation Program for my Marines and Sailors and I would like to gather not only their information (no personal ID) but their amount of tobacco used, type, years, quitting times, etc., so I follow their progress and probably gather enough data to publish my findings later on. Where do I start? Could you be kind in guiding me on the basics steps in building a database and relational tables?

Greger replied that he'd be honored to help. He began by suggesting a few online or downloadable tutorials to get Cdr. Baker started. Kevin Mallon, senior public relations manager at FileMaker, Inc., read of Baker's request on Greger's blog and offered additional help.

After Baker read the materials that Greger and Mallon had suggested, he and Greger exchanged additional e-mail correspondence about the database. They decided on two entities: People (Marines, sailors, and soldiers), and Appointments (to track each interaction between a person and Cdr. Baker).

The People table had only information that would not change: name, last four digits of Social Security number, sex, date of birth, and the age at which the person began to use tobacco.

The Appointment table, aside from information about the appointment itself, had information that could change over time such as current tobacco use, family tobacco use, and methods used in an attempt to quit. The table also contained the person's data for two standard tobacco dependency tests: the Fagerstrom Test for Nicotine Dependence and the Horn Psychological Test. Cdr. Baker hoped to establish connections between scores on these tests, other data about the tobacco user, and the success of the cessation program.

In November 2011, Cdr. Baker reported that his program seemed to be succeeding:

> "Some of [the Marines and sailors] have told that the desire of smoking again is almost gone, others reported decreased desire to smoke as it tasted bad and it reminded them of the carcinogenic substances in the cigarette or dip can. Most of them have decreased their smoking ... about 40-70% of their daily consumption. I never expected such dramatic drop but it appears that the way I prepared the program, it looks like it is working ... it is early in the game but these are positive results."

Baker's database will be crucial to documenting what works and what doesn't so this success can be replicated elsewhere in the U.S. armed forces.

Discussion Questions

1. In terms of the data modeling concepts you read about in this chapter, what type is the relationship between the People and Appointment entities?
2. Suppose Cdr. Baker wanted to identify associations among people in the same military unit regarding tobacco use or cessation behavior. That would require adding a new entity to the database. Draw the resulting data model as an entity-relationship diagram.

Critical Thinking Questions

1. Most people who need databases for professional use are not on a remote military base in Afghanistan. In what ways did Cdr. Baker's situation and location help him on this database project? In what ways were they a hindrance?
2. Cdr. Baker could have implemented this simple database using a spreadsheet program such as Microsoft Excel. He would have one worksheet for People and a second for Appointments, with a column for each data element. He could then use one of Excel's LOOKUP functions to associate people with their appointments. Discuss the pros and cons of using a spreadsheet program versus a full-fledged personal database management program, such as FileMaker Pro or Microsoft Access, for this application.

SOURCES: Baker, A., "Tobacco Cessation Program Database," *fmprodb.com/wp-content/uploads/2011/11/TCP-Outline.pdf*, October 25, 2011; FileMaker, Inc., Web site, *www.filemaker.com*, accessed May 11, 2012; FileMaker Pro 11 Tutorial, FileMaker, Inc., *www.filemaker.com/support/product/docs/fmp/fmp11_tutorial.pdf*, December 10, 2009; Greger, M., "Helping Our Soldiers Quit Smoking," *fmprodb.com/category/casestudies*, October 25, October 27, and November 15, 2011; Heatherton, T., et al, "The Fagerstrom Test for Nicotine Dependence: A Revision of the Fagerstrom Tolerance Questionnaire," *British Journal of Addiction*, *www.mendeley.com/research/the-fagerstrm-test-for-nicotine-dependence-a-revision-of-the-fagerstrom-tolerance-questionnaire* (requires free registration), September 1991.

Case Two

The Giants, the Pats, the Money, and the Tweets

Social networks are big in professional sports. About 2.8 million people follow the National Football League (NFL) on Twitter, and about 4.5 million "like" its Facebook page. The 2012 NFL championship game, Super Bowl 46 between the New York Giants and the New England Patriots, was watched by about 100 million television viewers. During the game, people posted 7,366,400 tweets about it. Of those, 324,221 mentioned one or more advertisers. With social media attention on such a large scale, it's no wonder that advertisers analyze social media as one indicator of whether their money was well spent.

Aside from sheer volume, the data from social media is unstructured. That puts this social media activity squarely into the "big data" category. What can advertisers learn from it?

If one measures advertising effectiveness by cost per tweet, Swedish high-fashion clothing chain H&M did best. According to Brandwatch, a firm that analyzes big data on the Web, H&M's 30-second commercial was mentioned in 17,190 tweets and cost $3.5 million for $204 per tweet. Pepsi, which had the second-highest number of tweets with 28,996, ran two such commercials for a second-lowest $253. Budweiser, the highest-spending brand in the 10 most tweeted about (estimated $35 million), had only 13,910 tweets for a cost of $2,265 each.

Do these findings mean that Budweiser wasted its money? Not necessarily. Perhaps beer drinkers tweet less than high-fashion customers or cola drinkers. It's more useful to compare within a category. Coca-Cola spent half again as much as Pepsi but had slightly more than half as many tweets (17,334). Similarly, Volkswagen had 20,818 tweets for $7 million worth of air time, $336 per tweet. Hyundai, with 4,325 tweets for the same cost, was far less effective in generating Twitter buzz.

One can also look for the sentiment of tweets. The way words are used can affect their meaning, but when taken over a large number of messages and compared across advertisers, the results of *sentiment analysis* are informative. Most tweets are neutral, and most advertisers show a small excess of positive over negative tweets. For example, Pepsi had 7 percent positive tweets, 2 percent negative, with the rest neutral. Twitter sentiment much below that suggests a problem. Skechers had 11 percent negative tweets and only 4 percent positive. The data don't say why, but they suggest that Skechers ought to look into the reason. (It may have been due to a protest based on their commercial having been filmed at a location where dogs were allegedly mistreated. If that's the reason, it's unlikely to affect shoe sales.)

The Brandwatch process begins by *gathering data*. Brandwatch monitors blogs, microblogs such as Twitter, social sites such as Facebook, image sites such as Flickr, video sites such as YouTube and Vimeo, discussion forums, and news sites.

Brandwatch then *cleans the data*. It removes duplicates, eliminates navigation text and ads, separates actual mentions of a brand from uses of the same word (an apple isn't necessarily an Apple), and analyzes the site to determine its date so that trends can be tracked.

The third step is *data analysis*, including sentiment analysis. Combined with date information, data analysis gives sentiment trends over time.

Finally, Brandwatch *presents* the data in several ways, including a digital dashboard. The net result is insight into the success of an ad campaign that could probably not be obtained any other way.

Discussion Questions

1. Brandwatch can create a graph of Twitter activity over time. The time at which a commercial airs can be indicated on that graph. What could an advertiser do with this information?
2. Suppose you're Hyundai's U.S. advertising manager. Your boss asks you to explain your ad's poor performance relative to competitor Volkswagen as measured by cost per tweet. How do you respond?

Critical Thinking Questions

1. Draw an entity-relationship diagram for the tables that a database would need to store information about tweets related to Super Bowl ads after data cleaning is complete. List the attributes that the database must store for each entity.
2. Web search strings are a source of big data. How can a company such as Google use them to improve its search results? For example, if you search for Robert Smith, how can a search engine improve the likelihood that its top hits will be the Robert Smith you want, not any of the thousands of other Robert Smiths in the world?

SOURCES: Brandwatch Web site, *www.brandwatch.com,* accessed May 16, 2012; Staff, "Visualizing Big Social Media Data," Brandwatch, *www.brandwatch.com/wp-content/uploads/brandwatch/The-Brandwatch-Super-Bowl-2012.pdf,* March 14, 2012; Staff, "Brandwatch Superbowl 2012," Brandwatch, *labs.brandwatch.com/superbowl,* accessed May 16, 2012; Horovitz, B., "Even without Kardashian, Skechers Ad Stirs Controversy," *USA Today, www.usatoday.com/money/advertising/story/2012-01-11/kim-kardashian-skechers-super-bowl-ad/52506236/1,* January 12, 2012; Tofel, K., "Super Bowl 46 Mobility by the Numbers," *gigaom.com/mobile/super-bowl-46-mobility-by-the-numbers,* February 5, 2012.

Questions for Web Case

See the Web site for this book to read about the Altitude Online case for this chapter. Following are questions concerning this Web case.

Altitude Online: Database Systems, Data Centers, and Business Intelligence

Discussion Questions

1. What work is involved in merging multiple databases into one central database, as Altitude Online is doing?
2. Why do you think Altitude Online found it necessary to hire a database administrator? How will the ERP affect the responsibilities of IS personnel across the organization?

Critical Thinking Questions

1. In a major move such as this, what opportunities can Altitude Online take advantage of as it totally revamps its database system that it perhaps wouldn't have considered before?
2. Why do you think Altitude Online is beginning work on its database prior to selecting an ERP vendor?

NOTES

Sources for the opening vignette: B. I. Practice Web site, *www.bipractice.co.za*, accessed May 4, 2012; Medihelp Web site, *www.medihelp.co.za,* accessed May 4, 2012; Medihelp, "Business Critical High-Performance Data Warehouse," submitted as *Computerworld* case study, *www.eiseverywhere.com/file_uploads/8afb2148c7782bb37e9ae1fa220b9824_Medihelp_-_Business_Critical_High-Performance_Data_Warehouse.pdf*, accessed January 16, 2012; Staff, "Medihelp Customer Case Study," *www.sybase.com/files/Success_Stories/Medihelp-CS.pdf*, October 14, 2011.

1. Horwitt, E., "BI Catches On," *Computerworld,* January 24, 2011, p. 30.
2. Chickowski, Ericka, "Database Firewall Brouhaha," *InformationWeek,* March 14, 2011, p. 38.
3. Lev-Ram, L., "The Hot New Gig in Tech," *Fortune,* September 5, 2011, p. 29.
4. EMC Web site, *www.emc.com/collateral/demos/microsites/emc-digital-universe-2011/index.htm*, accessed October 13, 2011.
5. FDIAI Web site, *www.fdiai.org/articles/fbi%20DNA%20Database.htm*, accessed October 12, 2011.
6. Vance, Ashlee, "The Data Knows," *Bloomberg Businessweek,* September 12, 2001, p. 70.
7. Cox, John, "Microsoft Accused of Collecting Location Data," *Network World,* September 12, 2011, p. 12.
8. Walmart Web site, *www.walmart.com/cp/Walmart-Clinics/1078904*, accessed October 18, 2011.
9. Duffy, Jim, "Cisco Fleshes Out Data Center Switch Fabric Plan," *Network World,* April 4, 2011, p. 10.
10. Thibodeau, Patrick, "Data Center Double Duty," *Computerworld*, May 9, 2011, p. 44.
11. Babcock, Charles, "IBM to Build Cloud Data Centers in China," *InformationWeek,* February 14, 2011, p. 16.
12. Murphy, Chris, "FedEx CIO Explains the Real Power of Cloud," *InformationWeek,* February 14, 2011, p. 18.
13. Morthen, Ben, "Data Centers Boom," *The Wall Street Journal,* April 9, 2011, p. B6.
14. Grunberg, S. and Rolander, N., "For Data Center, Google Goes for the Cold," *The Wall Street Journal,* September 12, 2011, p. B10.
15. HP Web site, *http://h18004.www1.hp.com/products/servers/solutions/datacentersolutions/index.html?jumpid=ex_R2849/us/en/ISS/cclicks/4AA2-9292ENW/discover/pod*, accessed October 18, 2011.
16. Cortera Web site, *http://start.cortera.com/company/research/k9s9mzm5s/microsoft-data-center*, accessed October 18, 2011.
17. Edwards, John, "Make Mine Modular," *Computerworld,* August 8, 2011, p. 26.
18. Grunberg, S. and Rolander, N., "For Data Center, Google Goes for the Cold," *The Wall Street Journal,* September 12, 2011, p. B10.
19. Thibodeau, P., "Rural N.C. Becomes Popular IT Location," *Computerworld,* June 20, 2011, p. 2.
20. Troianovski, A., "Storage Wars: Web Growth Sparks Data Center Boom," *The Wall Street Journal,* July 7, 2011, p. B1.
21. Brandon, John, "Extreme Storage," *Computerworld,* October 10, 2011, p. 25.
22. Staff, "Data Centers Expand at Furious Pace," *Network World,* May 23, 2011, p. 8.
23. Lawson, S., "IBM's Futuristic Storage Aims for Speed, Density," *Network World,* September 12, 2011, p. 16.
24. Barbee, T., "Disaster Recovery on a Budget," *CIO,* May 1, 2011, p. 48.
25. Stantosus, Megan, "Safe Storage for Stormy Weather," *CIO,* October 1, 2011, p. 22.
26. Harbaugh, Logan, "Recoup with Data Dedupe," *Network World,* September 12, 2011, p. 24.
27. Collett, Stacy, "Calculated Risk," *Computerworld,* February 7, 2011, p. 24.
28. Niccolai,, James, "Data Centers Survived Japan's Quake," *Computerworld,* July 18, 2011, p. 2.
29. EMC Web site, *www.emc.com/collateral/demos/microsites/emc-digital-universe-2011/index.htm*, accessed October 13, 2011.
30. Confio Web site, *www.confio.com*, accessed October 14, 2011.
31. Staff, "James River Insurance Selects Confio Software," *Business Wire,* August 2, 2011.
32. Falgout, Jim, "Dataflow's Big Data Edge," *InformationWeek,* September 5, 2011, p. 37.
33. Worthen, Ben, "Oracle Displays Resilience," *The Wall Street Journal,* September 21, 2011, p. B2.
34. Oracle Web site, *www.oracle.com/us/products/database,* accessed October 18, 2011.
35. Winkler, Rolfe, "Is the Maintenance Business a Software Target at Oracle," *The Wall Street Journal,* August 31, 2011, p. C14.
36. Google Web site, *www.google.com/streetview*, accessed October 18, 2011.
37. Microsoft Web iste, *http://office.microsoft.com/en-us/onenote*, accessed October 24, 2011.
38. Mangalindan, J.P., "Can This Software Make You Smarter?" *Fortune,* October 17, 2011, p. 70.
39. EverNote Web site, *www.evernote.com*, accessed October 20, 2011.
40. Microsoft Web site, *http://office.microsoft.com/en-us/infopath*, accessed October 24, 2011.

41. Mearian, Lucas, "IT Moving to SSD For Robust Applications," *Computerworld,* September 22, 2011, p. 14.
42. Cusumano, M., "Fight Uncertainty with Data," *CIO,* February 1, 2011, p. 18.
43. Staff, "Social Security Confronts IT Obsolescence," *InformationWeek,* March 14, 2011, p. 18.
44. Couchbase Web site, *www.couchbase.com*, accessed October 19, 2011.
45. Babcock, Charles, "Merged Startups Play Well With Big Data," *InformationWeek,* February 28, 2011, p. 12.
46. Vijayan, J., "Hadoop Works Alongside RDBMS," *Computerworld,* August 22, 2011, p. 5.
47. Vijayan, J. "Hadoop Is Ready for the Enterprise," *Computerworld,* November 21, 2011, p. 8.
48. Weiss, Todd, "Look Before You Leap Into Hadoop," *Computerworld,* February 27, 2012, p. 8.
49. Kane, Y. and Smith E., "Apple Readies iCloud Service," *The Wall Street Journal,* June 1, 2011, p. B1.
50. Apple Web site, *www.apple.com/icloud*, accessed October 18, 2011.
51. Levy, Ari, "Getting Business to Think inside the Box.net," *Bloomberg Businessweek,* September 26, 2011, p. 48.
52. Worthen, Ben, "Oracle Continues Its Push into Cloud Computing with RightNow Deal," *The Wall Street Journal,* October 25, 2011, p. B6.
53. Palazzolo, J. and Rubenfield, S., "U.S. Probes Oracle Dealings," *The Wall Street Journal,* August 31, 2011, p. B1.
54. Amazon Web Services Web site, *http://aws.amazon.com/ec2/*, accessed October 18, 2011.
55. CloudTweaks Web site, *www.cloudtweaks.com/2011/07/procter-and-gamble-collaborates-in-the-cloud-with-box-net*, accessed October 18, 2011.
56. Doyle, Jeff, "Open Flow," *InformationWeek,* October 17, 2011, p. 38.
57. Dernan, Andy, "Virtual vs. Physical," *InformationWeek,* August 15, 2011, p. 21.
58. Foskett, Stephen, "Virtualization's Storage Effect," *InformationWeek,* October 10, 2011, p. 31.
59. Marko, Kurt, "Virtualization vs. the Network," *InformationWeek,* October 17, 2011, p. 50.
60. Davis, Michael, "Virtualization Security Checklist," *InformationWeek,* October 17, 2011, p. 44.
61. McTigue, Jake, "Mix and Match For Safe VMs," *InformationWeek,* October 10, 2011, p. 44.
62. Babcock, Charles, "Global Data Centers Chains Promise Easier Backup," *InformationWeek,* October 17, 2011, p. 18.
63. Boles, Jeff, "Storage Administrators List Five Ways That Server Virtualization Complicates Their Lives," *Computerworld,* October 10, 2011, p. 16.
64. Acronis Web site, *www.acronis.com*, accessed October 20, 2011.
65. NoSQL Databases Web site, *http://nosql-database.org*, accessed October 20, 2011.
66. Friedenberg, Michael, "Big Data, Big Dividends," *CIO,* December 1, 2011, p. 12.
67. Stackpole, Beth, "Your Big Data To Do List," *Computerworld,* February 12, 2012, p. 22.
68. Bednarz, Ann, "Rise of Hadoop Challenging for IT," *Network World,* February 13, 2012, p. 1.
69. Mearian, Lucas, "Health Insurer Encrypts all Stored Data," *Computerworld,* August 22, 2011, p. 4.
70. W3C Web site, *www.w3.org/standards/semanticweb*, accessed October 24, 2011.
71. W3C Web site, *www.w3.org/TR/rdf-primer*, accessed October 18, 2011.
72. Woolner, Ann, "You Can Run, But It's Hard to Hide from TLO," *Bloomberg Businessweek,* September 19, 2011, p. 44.
73. Mossberg, Walter, "For iPad and Mobile Devices, a Port Out of the Norm," *The Wall Street Journal,* February 9, 2012, p. D4.
74. Konrad, Alex, "Tablets Storm the Corner Office," *Fortune,* October 17, 2011.
75. Nash, K., "Tailor-Made Mobile," *CIO,* September 15, 2011, p. 17.
76. Oracle Web site, *www.oracle.com/us/products/applications/ebusiness/scm/051318.html*, accessed October 24, 2011.
77. Deal, V., "HP's Murky Move Back into Data Warehousing," *Information Week,* February 28, 2011, p. 14.
78. Vertica Web site, *www.vertica.com/*, accessed August 31, 2011.
79. Sharpe, Michael, "The Data Mastery Imperative," *InformationWeek,* July 11, 2011, p. 29.
80. Frank, Diana, "Healthier Intelligence," *CIO,* March 1, 2012, p. 50.
81. Brooks Brothers Web site, *www.brooksbrothers.com*, accessed October 18, 2011.
82. Hobson, Katherine, "Getting Docs to Use PCs," *The Wall Street Journal*, March 15, 2011, p. B5.
83. Nash, K., "Do It Yourself," *CIO,* September 1, 2011, p. 28.
84. Holm, Erik, "New Hire Will Mine AIG's Data," *The Wall Street Journal,* January 9, 2012, p. C5.
85. Carreyrou, J. and McGinty, T., "Medicare Records Reveal Troubling Trail of Surgeries," *The Wall Street Journal,* March 29, 2011, p. A1.
86. Tozzi, J., "Bank Data Miner," *Bloomberg Businessweek,* July 3, 2011, p. 41.
87. Verafin Web site, *http://verafin.com*, accessed September 15, 2011.
88. Thrum, Scott, "Data Mining Your Mind," *The Wall Street Journal,* October 27, 2011, p. B1.
89. Henschen, Doug, "Ellison Touts In-Memory, Flirts With NoSQL," *InformationWeek,* October 17, 2011, p. 13.
90. Overby, S., "Analytics Uncorked," *CIO,* September 1, 2011, p. 20.
91. Jackson, Joab, "A Better Slice of Data," *CIO,* June 15, 2011, p. 11.
92. Henschen, Doug, "P&G Turns Analytics into Action," *InformationWeek,* September 19, 2011, p. 77.
93. Murphy, Chris, "P&G's Hiring Priority: Analytics Experts," *Information Week,* February 27, 2012, p. 8.
94. Bruni, M., "5 Steps to Agile BI," *InformationWeek,* June 13, 2011, p. 33.

95. Howson, C., "Predictions of What's in Store for the BI Market," *Information Week,* January 31, 2011, p. 22.
96. Vijayan, J., "Pirates Tap BI to Boost Attendance," *Computerworld,* October 10, 2011, p. 2.
97. Staff, "Copper Mountain Resort Selects Agilysys Visual One," *PR Newswire,* August 2011.
98. Korn, M. and Tibken, S., "Schools Plan Leap into Data," *The Wall Street Journal,* August 4, 2011, p. B12.
99. Objectivity Web site, *www.objectivity.com*, accessed October 26, 2011.
100. Versant Web site, *www.versant.com*, accessed October 26, 2011.
101. ODBMS.org Web site, *www.odbms.org/odmg*, accessed October 26, 2011.
102. Pitney-Bowes Web site, *www.pbinsight.com/resources/case-studies/details/cumberland-county*, accessed October 18, 2011.

6 Telecommunications and Networks

Principles	Learning Objectives
• A telecommunications system consists of several fundamental components.	• Identify and describe the fundamental components of a telecommunications system. • Discuss two broad categories of telecommunications media and their associated characteristics. • Briefly describe several options for short-range, medium-range, and long-range communications.
• Networks are an essential component of an organization's information technology infrastructure.	• Identify the benefits of using a network. • Describe three distributed processing alternatives and discuss their basic features. • Identify several telecommunications hardware devices and discuss their functions.
• Network applications are essential to organizational success.	• List and describe several network applications that organizations benefit from today.

Information Systems in the Global Economy

NETHOPE, UNITED STATES

NetHope Worldwide Disaster Relief

When disaster strikes, it has both an economic and human impact. People who are not affected by the disaster often feel a moral obligation to come to the aid of the disaster's victims. People feel this obligation even more strongly if they have specific skills that can help alleviate the effects of the disaster.

One skill needed in many natural disasters is the ability to set up networks so the people affected, their governments, and organizations that want to help can communicate with each other. That need for long-distance communication places a particular responsibility on the technology community. Many high-tech companies fulfill this responsibility under the auspices of NetHope, a group of large humanitarian relief organizations.

Consider, for example, the 2010 floods in Pakistan. They affected over 20 million people and had an economic impact estimated at $9.5 billion. If relief is not provided quickly in such situations, the effects of a disaster can worsen and spread.

NetHope and its partners have been on site in all recent natural disasters. In the first two months after the Haiti earthquake of 2010, NetHope and its partner Microsoft helped launch a Web site for interagency collaboration, set up cloud computing solutions for Haiti's government and for organizations working in the country, and had Bing and MSN each set up Web pages where people could donate to Haiti. Communication between Haitians and aid workers was hampered because few workers spoke Haitian Creole. To solve this problem, Microsoft added Haitian Creole to Microsoft Translator, a free automatic translation tool, and provided the tool to aid workers.

In the 2011 tsunami in Japan, NetHope partners provided a cloud-based community communication portal for Second Harvest Japan. The organization uses the portal to coordinate food donors, transportation providers, and distributors in the Japanese relief effort. Cloud services avoid problems such as damaged infrastructure and equipment, power shortages, and telecommunications service interruptions.

One of NetHope's five major missions is connectivity. According to the NetHope Web site, the connectivity objective is to "improve communications between organizations and field offices in remote parts of the world, where infrastructure is limited or absent." Until recently, NetHope tried to meet this objective by placing very small aperture terminals (VSATs) in remote areas with little to no terrestrial infrastructure. VSAT systems include an Earth station (usually less than 3 meters, or 10 feet, wide) placed outdoors in line of sight to the sky to link to a satellite in geosynchronous orbit. The satellite can relay messages to anywhere else on Earth, permitting communication with isolated areas.

However, as Gisli Olafsson, Emergency Response Director of NetHope (and a former Microsoft employee) learned, "using VSAT as the preferred way to connect is not always the most effective and economical method." Olafsson continues: "With most countries moving toward a 3G wireless broadband mobile network,...we have seen that mobile networks are becoming more resilient to large-scale disasters, with core services generally being available within two weeks of a major incident.... It is more economical and easier to stockpile and transfer 3G modems than VSAT kits."

Technology, of course, is never the entire answer. People are an important part of any system. After the 2010 Haiti earthquake, NetHope launched NetHope Academy to provide IT skills training and on-the-job work experience to unemployed Haitians so they could build in-country technical expertise. The first group of NetHope Academy interns spent three weeks in intensive boot camp-style classroom training. They were then placed with teams rebuilding devastated areas of Haiti, using their new skills to help team members keep in touch with people outside their immediate area.

In addition to Microsoft, NetHope partners include such well-known technology firms as Accenture, Cisco, Hewlett-Packard, and Intel. The technology community can be proud of its commitment to humanitarian aid and economic recovery.

As you read this chapter, consider the following:

- Major natural disasters, while not as rare as one might wish, are not common enough that companies develop network and communications products for that use alone. The technologies that NetHope partners use were originally developed for other reasons. What additional uses could these technologies have?
- Which of the telecommunications and networking technologies in this chapter would be useful in natural disasters?

WHY LEARN ABOUT TELECOMMUNICATIONS AND NETWORKS?

© Andrey Burmakin/Shutterstock

Effective communication is essential to the success of every major human undertaking, from building great cities to waging war to running a modern organization. Today, we use electronic messaging and networking to enable people everywhere to communicate and interact effectively without requiring face-to-face meetings. Regardless of your chosen major or career field, you will need the communications capabilities provided by telecommunications and networks, especially if your work involves the supply chain. Among all business functions, supply chain management probably uses telecommunications and networks the most because it requires cooperation and communications among workers in inbound logistics, warehouse and storage, production, finished product storage, outbound logistics, and most important, with customers, suppliers, and shippers. All members of the supply chain must work together effectively to increase the value perceived by the customer, so partners must communicate well. Other employees in human resources, finance, research and development, marketing, and sales positions must also use communications technology to communicate with people inside and outside the organization. To be a successful member of any organization, you must be able to take advantage of the capabilities that these technologies offer you. This chapter begins by discussing the importance of effective communications.

In today's high-speed global business world, organizations need always-on, always-connected computing for traveling employees and for network connections to their key business partners and customers. Forward-thinking organizations strive to increase revenue, reduce time to market, and enable collaboration with their suppliers, customers, and business partners by using telecommunications systems. Here are just a few examples of organizations using telecommunications and networks to move ahead:

- The West Yorkshire Fire and Rescue Service in the United Kingdom is fully operational 24 hours a day each day of the year. It encompasses 48 fire stations and 1,745 full-time and part-time firefighters. The service switched to a unified communications solution consisting of both hardware and software that enables firefighters and support people to use one device to receive calls, voice messages, audio and video conferencing, e-mail, and instant messages. Prior to this change, the service's communications channels were not integrated and the firefighters had to use multiple devices to perform these various functions. Precious seconds could be lost, or worse, an emergency call could be dropped as firefighters switched from one device to another.[1]

- Over three dozen hospitals in the United States are using telecommunications to enable remote doctors to monitor and care for patients. Recently, doctors in the intensive care unit of Banner Estrella Medical Center in Phoenix, Arizona, collaborated with a doctor in Tel Aviv, Israel, to treat and monitor a patient with a life-threatening diabetic reaction. While the doctors and nurses in Phoenix provided hands-on care, the doctor in Tel Aviv was able to monitor the patient through the night as critical patient data such as heart and breathing rates as well as live video was transmitted to the doctor. This level of monitoring would not have been possible using only the resources at Banner because of a shortage of critical care specialists.[2]
- According to a recent survey, 80 percent of physicians and 75 percent of nurses own a smartphone and are increasingly using them as a valuable tool to provide patient services. Physicians can download applications to their phones that enable them to check drug references, perform common medical calculations, consult normal lab value charts, use decision-support tools, and view electronic medical records.[3]
- Thousands of companies use Webcasts to inform and educate potential customers about their products and services. For example, Amgen, a biotechnology pioneer, broadcast its fourth quarter and full year 2011 financial results using a Webcast.[4]

AN OVERVIEW OF TELECOMMUNICATIONS

Telecommunications refers to the electronic transmission of signals for communications by means such as telephone, radio, and television. Telecommunications is creating profound changes in business because it lessens the barriers of time and distance. Advances in telecommunications technology allow us to communicate rapidly with business partners, clients, and coworkers almost anywhere in the world. Telecommunications also reduces the amount of time needed to transmit information that can drive and conclude business actions. It not only is changing the way organizations operate but the nature of commerce itself. As networks connect to one another and transmit information more freely, a competitive marketplace demands excellent quality and service from all organizations.

Figure 6.1 shows a general model of telecommunications. The model starts with a sending unit (1) such as a person, a computer system, a terminal, or another device that originates the message. The sending unit transmits a signal (2) to a telecommunications device (3). The telecommunications device—a hardware component that facilitates electronic communication—performs many tasks, which can include converting the signal into a different form or from one type to another. The telecommunications device then sends the signal through a medium (4). A **telecommunications medium** is any material substance that carries an electronic signal to support communications between a sending and receiving device. Another telecommunications device (5) connected to the receiving device (6) receives the signal. The process can be reversed, and the receiving unit (6) can send a message to the original sending unit (1). An

telecommunications medium: Any material substance that carries an electronic signal to support communications between a sending and receiving device.

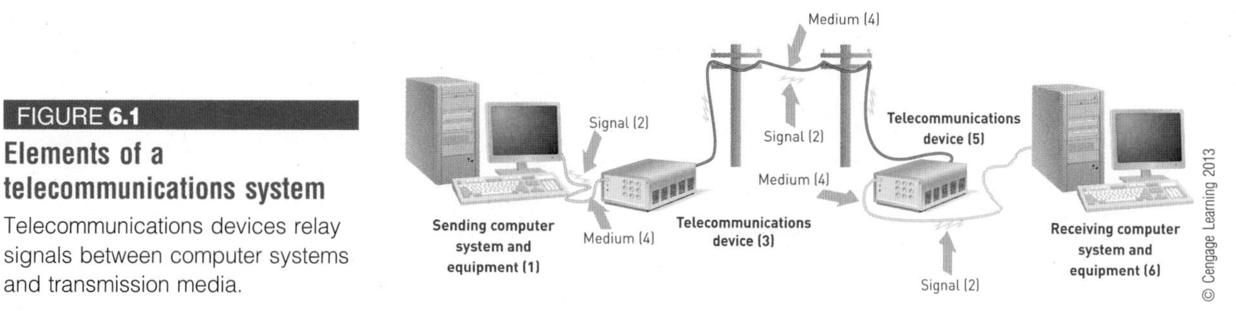

FIGURE 6.1
Elements of a telecommunications system
Telecommunications devices relay signals between computer systems and transmission media.

important characteristic of telecommunications is the speed at which information is transmitted, which is measured in bits per second (bps). Common speeds are in the range of thousands of bits per second (Kbps) to millions of bits per second (Mbps) and even billions of bits per second (Gbps).

networking protocol: A set of rules, algorithms, messages, and other mechanisms that enable software and hardware in networked devices to communicate effectively.

A **networking protocol** is a set of rules, algorithms, messages, and other mechanisms that enable software and hardware in networked devices to communicate effectively. The goal is to ensure fast, efficient, error-free communications and to enable hardware, software, and equipment manufacturers and service providers to build products that interoperate effectively. The Institute of Electrical and Electronics Engineers (IEEE) is a leading standards-setting organization whose IEEE 802 network standards are the basis for many telecommunications devices and services. The International Telecommunication Union (ITU) is a specialized agency of the United Nations with headquarters in Geneva, Switzerland. The international standards produced by the ITU are known as Recommendations and carry a high degree of formal international recognition.

Communications between two people can occur synchronously or asynchronously. With synchronous communications, the receiver gets the message as soon as it is sent. Voice and phone communications are examples of synchronous communications. With asynchronous communications, the receiver gets the message after some delay—a few seconds to minutes or hours or even days after the message is sent. Sending a letter through the post office or an e-mail over the Internet are examples of asynchronous communications. Both types of communications are important in business.

Most colleges and universities today offer distance learning courses using telecommunications hardware and software that enables instructors to connect to students anywhere in the United States. Students can elect to attend a synchronous class where they attend the online course with the instructor at the same time and where they can see and hear the instructor in real time and interact by asking questions. Other students can attend asynchronous classes in which they participate at any time that is convenient to them. They see a slide presentation, and if they have questions, they can correspond with the instructor via e-mail. Harvard University Extension School offers hundreds of distance learning courses each term. The courses employ either online video of faculty lecturing on campus or Web conferencing where students communicate in a live session with the instructor by chat or voice technology.[5]

Telecommunications technology enables businesspeople to communicate with coworkers and clients from remote locations.

Basic Telecommunications Channel Characteristics

The transmission medium carries messages from the source of the message to its receivers. A transmission medium can be divided into one or more telecommunications channels, each capable of carrying a message. Telecommunications channels can be classified as simplex, half-duplex, or full-duplex.

simplex channel: A communications channel that can transmit data in only one direction and is seldom used for business telecommunications.

half-duplex channel: A communications channel that can transmit data in either direction but not simultaneously.

full-duplex channel: A communications channel that permits data transmission in both directions at the same time; a full-duplex channel is like two simplex channels.

A **simplex channel** can transmit data in only one direction and is seldom used for business telecommunications. Doorbells and the radio operate using a simplex channel. A **half-duplex channel** can transmit data in either direction but not simultaneously. For example, A can begin transmitting to B over a half-duplex line, but B must wait until A is finished to transmit back to A. Personal computers are usually connected to a remote computer over a half-duplex channel. A **full-duplex channel** permits data transmission in both directions at the same time, so a full-duplex channel is like two simplex channels. Private leased lines or two standard phone lines are required for full-duplex transmission.

Channel Bandwidth

channel bandwidth: The rate at which data is exchanged, usually measured in bps.

broadband communications: A relative term but it generally means a telecommunications system that can transmit data very quickly.

In addition to the direction of data flow supported by a telecommunications channel, you must consider the speed at which data can be transmitted. Telecommunications **channel bandwidth** refers to the rate at which data is exchanged, usually measured in bps—the broader the bandwidth, the more information can be exchanged at one time. **Broadband communications** is a relative term, but it generally means a telecommunications system that can transmit data very quickly. For example, for wireless networks, broadband lets you send and receive data at a rate greater than 1.5 Mbps.

Telecommunications professionals consider the capacity of the channel when they recommend transmission media for a business. In general, today's organizations need more bandwidth than they did even a few years ago for increased transmission speed to carry out their daily functions.

Circuit Switching and Packet Switching

circuit-switching network: A network that sets up a circuit between the sender and receiver before any communications can occur; this circuit is maintained for the duration of the communication and cannot be used to support any other communications until the circuit is released and a new connection is set up.

packet-switching network: A network in which no fixed path is created between the communicating devices and the data is broken into packets, with each packet transmitted individually and capable of taking various paths from sender to recipient.

Circuit switching and packet switching are two different means for routing data from the communications sender to the receiver. In a **circuit-switching network**, a dedicated circuit is first established to create a path that connects the communicating devices. This path is committed and then used for the duration of the communication. Because it is dedicated, the circuit cannot be used to support communications from other users or sending devices until the circuit is released and a new connection is set up. The traditional telephone network is a circuit-switching network.

In a **packet-switching network**, no fixed path is created between the communicating devices, and the data is broken into packets for sending over the network. Each packet is transmitted individually and is capable of taking various paths from sender to receiver. Because the communication paths are not dedicated, packets from multiple users and communicating devices can travel over the same communication path. Once all the packets forming a message arrive at the destination, they are compiled into the original message. The Internet is an example of a packet-switching network that employs the packet-switching protocol called TCP/IP.

The advantage of a circuit-switching network is that it provides for the nonstop transfer of data without the overhead of assembling and disassembling data into packets and determining which routes packets should follow. Circuit-switching networks are best when data must arrive in exactly the same order in which it is sent. This is the case with most real-time data such as live audio and video.

Packet-switching networks are more efficient at using the channel bandwidth because packets from multiple conversations can share the same communications links. However, there can be delays in the delivery of messages due to network congestion and the loss of packets or the delivery of packets out of order. Thus, packet-switching networks are used to communicate data that can withstand some delays in transmission, such as e-mail messages and the sending of large data files. Another key consideration is the type of telecommunications media to use.

Telecommunications Media

Each telecommunications media type can be evaluated according to characteristics such as cost, capacity, and speed. In designing a telecommunications system, the transmission media selected depends on the amount of information to be exchanged, on the speed at which data must be exchanged, on the level of concern for data privacy, on whether the users are stationary or mobile, and on many other business requirements. The transmission media are selected to support the communications goals of the organization systems that are at the lowest cost but that still allow for possible modifications should business requirements change. Transmission media can be divided into two broad categories: *guided transmission media* in which telecommunications signals are guided along a solid medium and *wireless* in which the telecommunications signal is broadcast over airwaves as a form of electromagnetic radiation.

Guided Transmission Media Types

Guided transmission media are available in many types. Table 6.1 summarizes the guided media types by physical media type. These guided transmission media types are discussed in the sections following the table.

TABLE 6.1 Guided Transmission Media Types

Media Type	Description	Advantages	Disadvantages
Twisted-pair wire	Twisted pairs of copper wire, shielded or unshielded	Used for telephone service; widely available	Transmission speed and distance limitations
Coaxial cable	Inner conductor wire surrounded by insulation	Cleaner and faster data transmission than twisted-pair wire	More expensive than twisted-pair wire
Fiber-optic cable	Many extremely thin strands of glass bound together in a sheathing; uses light beams to transmit signals	Diameter of cable is much smaller than coaxial; less distortion of signal; capable of high transmission rates	Expensive to purchase and install

Twisted-Pair Wire

Twisted-pair wire contains two or more twisted pairs of wire, usually copper, as shown in Figure 6.2 (left). Proper twisting of the wire keeps the signal from “bleeding” into the next pair and creating electrical interference. Because the twisted-pair wires are insulated, they can be placed close together and packaged in one group, with hundreds of wire pairs being grouped into one large wire cable. Twisted-pair wires are classified by category (Category 2, 3, 5, 5E, and 6—there is no Category 1, and Category 4 is no longer used). The lower categories are used primarily in homes. Higher categories are used in networks and can carry data at higher speeds. For example, 10-Gigabit Ethernet is a standard for transmitting data in full-duplex mode at the speed of 10 billion bps for limited distances over Category 5 or 6 twisted-pair wire. The 10-Gigabit Ethernet cable can be used for the high-speed links that connect groups of computers or to move data stored in large databases on large computers to stand-alone storage devices.

The Charlotte County Public School District services 17,500 students in southwest Florida. The school district uses 10-Gigabit Ethernet cable to connect its two dozen schools and learning centers, enabling them to download high-resolution videos and to support live video feeds between classrooms separated by 20 miles or more.

Coaxial Cable

Figure 6.2 (middle) shows a typical coaxial cable, similar to that used in cable television installations. When used for data transmission, coaxial cable falls in the middle of the guided transmission media in terms of cost and performance. The cable itself is more expensive than twisted-pair wire but less than fiber-optic cable (discussed next). However, the cost of installation and other necessary communications equipment makes it difficult to compare the total costs of each medium. Coaxial cable offers cleaner and crisper data transmission (less noise) than twisted-pair wire and also a higher data transmission rate.

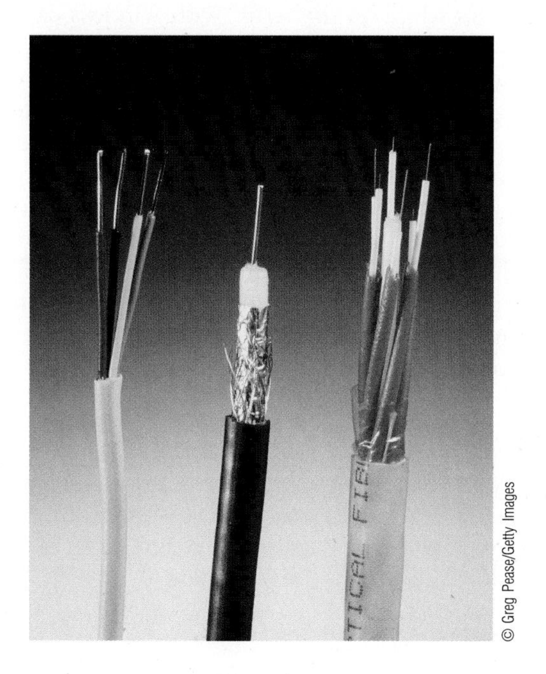

FIGURE **6.2**

Types of guided transmission media

Twisted-pair wire (left), coaxial cable (middle), fiber-optic cable (right).

Many cable companies, including Time Warner, Cox Communications, Cablevision, and Comcast, aggressively court customers for telephone service. They entice people away from phone companies by offering highly discounted rates for bundled high-speed Internet and phone services along with services phone companies do not offer, such as TV and new movies on demand the same day as the DVD is released.

Fiber-Optic Cable

Fiber-optic cable, consisting of many extremely thin strands of glass or plastic bound together in a sheathing (also known as a jacket), transmits signals with light beams (see Figure 6.2, right). These high-intensity light beams are generated by lasers and are conducted along the transparent fibers. These fibers have a thin coating, called *cladding*, which works like a mirror, preventing the light from leaking out of the fiber. The much smaller diameter of fiber-optic cable makes it ideal when there is no room for bulky copper wires—for example, in crowded conduits, which can be pipes or spaces carrying both electrical and communications wires. Fiber-optic cable and associated telecommunications devices are more expensive to purchase and install than their twisted-pair wire counterparts, although the cost is decreasing.

FiOS is a bundled set of communications services, including Internet, telephone, and high definition TV, that operates over a total fiber-optic communications network. With this service, fiber-optic cable is run from the carrier's local exchange all the way to the customer's premises. (Cable networks often use fiber optic in their network backbone that connects their local exchanges, but they do not run fiber optic to the customer's premises). FiOS is offered from Verizon in selected portions of the United States with Internet speeds of 150 Mbps, which requires just four minutes to download a two-hour movie (500 MB).[6] This same task would take more than a half hour over a 15-Mbps cable network. A shortcoming of this service is that a power outage at the premises means no FiOS service. A battery backup unit is advisable to avoid this potential problem.

Wireless Communications Options

Wireless communications coupled with the Internet are revolutionizing how and where we gather and share information, collaborate in teams, listen to music or watch videos, and stay in touch with our families, friends, and coworkers while on the road. With wireless capability, a coffee shop can become our living room, and the bleachers at a ball park can become our office. The many advantages and freedom provided by wireless communications are causing many organizations to consider moving to an all-wireless environment.

Wireless transmission involves the broadcast of communications in one of three frequency ranges: radio, microwave, or infrared, as shown in Table 6.2. In some cases, the use of wireless communications is regulated, and the signal must be broadcast within a specific frequency range to avoid interference with other wireless transmissions. For example, radio and TV stations must gain approval to use a certain frequency to broadcast their signals. Where wireless communications are not regulated, there is a high potential for interference between signals.

TABLE 6.2 Frequency ranges used for wireless communications

Technology	Description	Advantages	Disadvantages
Radio frequency range	Operates in the 3 kHz–300 MHz range	Supports mobile users; costs are dropping	Signal highly susceptible to interception
Microwave—terrestrial and satellite frequency range	High-frequency radio signal (300 MHz–300 GHz) sent through atmosphere and space (often involves communications satellites)	Avoids cost and effort to lay cable or wires; capable of high-speed transmission	Must have unobstructed line of sight between sender and receiver; signal highly susceptible to interception
Infrared frequency range	Signals in the 300 GHz–400 THz frequency range sent through air as light waves	Allows you to move, remove, and install devices without expensive wiring	Must have unobstructed line of sight between sender and receiver; transmission effective only for short distances

With the spread of wireless network technology to support devices such as smartphones, mobile computers, and cell phones, the telecommunications industry needed new protocols to define how these hardware devices and their associated software would interoperate on the networks provided by telecommunications carriers. More than 70 active groups are setting standards at the regional, national, and global levels, resulting in a dizzying array of communications standards and options. Some of the more widely used wireless communications options are discussed next.

Short-Range Wireless Options

Many wireless solutions provide communications over very short distances, including Near Field Communications, Bluetooth, ultra wideband, infrared transmission, and ZigBee.

Near Field Communication (NFC)

Near Field Communication (NFC): A very short-range wireless connectivity technology designed for consumer electronics, cell phones, and credit cards.

Near Field Communication (NFC) is a very short-range wireless connectivity technology designed for consumer electronics, cell phones, and credit cards. Once two NFC-enabled devices are in close proximity (touching or a few centimeters apart), they exchange the necessary communications parameters and passwords to enable Bluetooth, Wi-Fi, or other wireless communications between the devices. Because only two devices participate in the communications, NFC establishes a peer-to-peer network.

Speedpass is a contactless payment system based on NFC that enables ExxonMobil customers to pay for purchases easily and safely by waving their key tag across an area of a gasoline pump, car wash kiosk, or convenience store terminal. This action transmits a unique identification and security code to the Speedpass payment system. Payment is then instantly processed using the linked credit or debit card.[7]

Assa Abloy AB is a global leader in providing door-opening solutions with over 37,000 employees and a headquarters in Sweden. The firm is piloting the use of near field communications to enable BlackBerry smartphone users to swipe their phones past a door card reader to gain access to a building or their home or to open a garage door.[8]

Bluetooth: A wireless communications specification that describes how cell phones, computers, faxes, personal digital assistants, printers, and other electronic devices can be interconnected over distances of 10–30 feet at a rate of about 2 Mbps.

Bluetooth is a wireless communications specification that describes how cell phones, computers, printers, and other electronic devices can be interconnected over distances of 10–30 feet at a rate of about 2 Mbps and allows users of multifunctional devices to synchronize with information in a desktop computer, send or receive faxes, print, and in general, coordinate all mobile and fixed computer devices. The Bluetooth technology is named after the tenth-century Danish King Harald Blatand, or Harold Bluetooth in English. He had been instrumental in uniting warring factions in parts of what is now Norway, Sweden, and Denmark, just as the technology named after him is designed to allow collaboration among differing devices such as computers, phones, and other electronic devices.

As more states are banning the use of handheld cell phones in cars, one of the most important accessories for smartphone users is their hands-free Bluetooth headset.[9] With Bluetooth Office, BMW drivers can also use the hands-free system in their cars to have their e-mail messages, notes, calendar entries, and text messages read aloud over the car's speakers.[10]

Nokia Research is using Bluetooth technology to pilot the use of indoor location-based services. "We want to take what's been done in navigation outdoors and bring it inside," says Fabio Belloni, a Nokia researcher involved in the project. Once implemented, malls, exhibit halls, and other large buildings would be outfitted with Bluetooth low-energy antenna arrays that can track devices equipped with Bluetooth tags. Another idea is to use Bluetooth-equipped carts to track people as they move about large stores to capture data for studying shopper behavior.[11]

Ultra Wideband

ultra wideband (UWB): A form of short-range communications that employs extremely short electromagnetic pulses lasting just 50 to 1,000 picoseconds that are transmitted across a broad range of radio frequencies of several gigahertz.

Ultra wideband (UWB) communications involves the transmission of extremely short electromagnetic pulses lasting just 50 to 1,000 picoseconds. (One picosecond is one trillionth or one-millionth of one-millionth of a second.) The pulses are capable of supporting data transmission rates of 480 to 1,320 Mbps over a relatively short range of 10 to 50 meters.[12] UWB provides several advantages over other communications means: a high throughput rate, the ability to transmit virtually undetected and impervious to interception or jamming, and a lack of interference with current communications services.

Potential UWB applications include wirelessly connecting printers and other devices to desktop computers or enabling completely wireless home multimedia networks. Manufacturers of medical instruments are using UWB for video endoscopes, laryngoscopes, and ultrasound transducers.[13]

Infrared Transmission

infrared transmission: A form of communications that sends signals at a frequency of 300 GHz and above—higher than those of microwaves but lower than those of visible light.

Infrared transmission sends signals at a frequency of 300 GHz and above—higher than those of microwaves but lower than those of visible light. It is frequently used in wireless networks, intrusion detectors, home entertainment remote control, and fire sensors. Infrared transmission requires line-of-sight transmission and short distances—such as a few yards. It allows handheld computers to transmit data and information to larger computers within the same room and to connect a display screen, printer, and mouse to a computer.

Temperature and humidity monitoring sensors are frequently used in computer data centers, hospitals, laboratories, museums, warehouses, wine production, and storage facilities. TandD Corporation, a manufacturer of these sensors, uses infrared communications to transmit the data captured by the sensors to a data recorder device, thus eliminating the need to physically gather the sensors to log the data.[14]

ZigBee

ZigBee: A form of wireless communications frequently used in security systems and heating and cooling control systems.

ZigBee is a form of wireless communications frequently used in security systems and for sensing and controlling energy-consuming devices, such as lighting and heating, ventilation, and air-conditioning (HVAC) in both residential homes and commercial buildings. ZigBee is a relatively low-cost technology and requires little power, which allows longer life with smaller batteries.

Pacific Gas & Electric is conducting a 500-home pilot to allow customers to manage their home energy usage by smart meters that are linked to an in-house control dashboard via a ZigBee wireless home area network.[15] The Desmeules Chrysler Dodge dealership in Quebec implemented an access control system using ZigBee technology to transmit data from each door lock to a central computer that controls who can enter each door and at what time of day.[16]

Medium-Range Wireless Options

Wi-Fi: A medium-range wireless telecommunications technology brand owned by the Wi-Fi Alliance.

Wi-Fi is a wireless telecommunications technology brand owned by the Wi-Fi Alliance, which consists of about 300 technology companies, including AT&T, Dell, Microsoft, Nokia, and Qualcomm. The alliance exists to improve the interoperability of wireless local area network products based on the IEEE 802.11 series of telecommunications standards. IEEE stands for the Institute of Electrical and Electronic Engineers, a nonprofit organization and one of the leading standards-setting organizations. Table 6.3 summarizes several variations of this standard. Although 802.11n provides improvements in coverage and performance, it also brings with it increased security vulnerabilities. It can take twice as long to scan the frequencies used for 802.11n for malicious patterns (8 seconds versus 4 seconds) as it does for other 802.11 standards.

TABLE 6.3 IEEE 802.11 wireless local area networking standards

Wireless Networking Protocol	Theoretical Maximum Data Rate per Stream (Mbps)	Comments
IEEE 802.11a	54	Provides at least 23 non-overlapping channels rather than 802.11b and g where all channels overlap
IEEE 802.11b	11	First widely accepted wireless network standard; equipment using this protocol may occasionally suffer from interference from microwave ovens, cordless telephones, and Bluetooth devices.
IEEE 802.11g	54	Equipment using this protocol may occasionally suffer from interference from microwave ovens, cordless telephones, and Bluetooth devices
IEEE 802.11n	140	Employs multiple input, multiple output (MIMO) technology that allows multiple data streams to be transmitted over the same channel using the same bandwidth used for only a single data stream in 802.11a/b/g

With a Wi-Fi wireless network, the user's computer, smartphone, or cell phone has a wireless adapter that translates data into a radio signal and transmits it using an antenna. A wireless access point, which consists of a transmitter with an antenna, receives the signal and decodes it. The access point then sends the information to the Internet over a wired connection, as shown in Figure 6.3. When receiving data, the wireless access point takes the information from the Internet, translates it into a radio signal, and sends it to the device's wireless adapter. These devices typically come with built-in wireless transmitters and software that enable them to alert the user to the existence of a Wi-Fi network. The area covered by one or more interconnected wireless access points is called a "hot spot." Current Wi-Fi access points have a maximum range of about 300 feet outdoors and 100 feet within a dry-walled building. Wi-Fi has proven so popular that hot spots are established in many public places such as airports, coffee shops, college campuses, libraries, and restaurants.

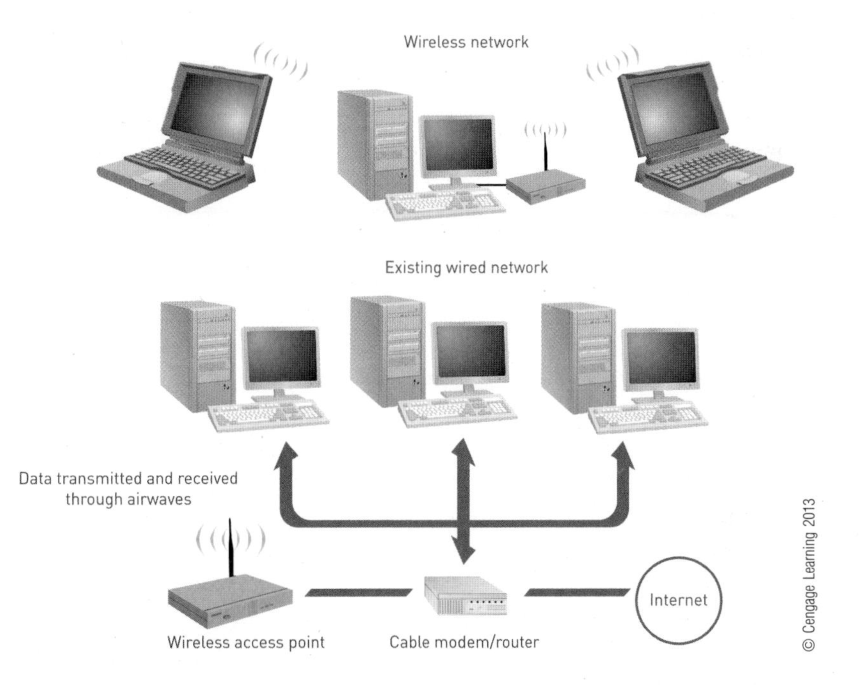

FIGURE 6.3
Wi-Fi network

The availability of free Wi-Fi within a hotel's premises has become very popular with business travelers. Many hotels offer or are considering offering free Wi-Fi including Best Western, Courtyard by Marriott, Fairmont, Hampton, Hilton, Holiday Inn, Hotel Indigo, Kimpton, Starwood, The Four Seasons Hotels and Resorts, and Wyndham hotels.[17]

Hundreds of cities in the United States have implemented municipal Wi-Fi networks for use by meter readers and other municipal workers and to partially subsidize Internet access to their citizens and visitors. Supporters of the networks believe that the presence of such networks stimulates economic development by attracting new businesses. Critics doubt the long-term viability of municipal Wi-Fi networks because the technology cannot easily handle rapidly increasing numbers of users. Also, because municipal Wi-Fi networks use an unlicensed bandwidth available to any user and they operate at up to 30 times the power of existing home and business Wi-Fi networks, critics claim interference is inevitable with these networks. As a result, cities are considering other options including WiMAX and fiber-optic networks, slowing the growth of city Wi-Fi networks.

One possible drawback to Wi-Fi network use is found in a study published in the medical journal *Fertility and Sterility* suggesting that "the use of a laptop computer wirelessly connected to the Internet and positioned near the male reproductive organs cause undesirable health problems. At present it is not known whether this effect is induced by all laptop computers connected by Wi-Fi to the Internet, or what use conditions heighten this effect."[18]

Wide Area Wireless Network Types

Many solutions provide wide area network options, including satellite and terrestrial microwave transmission, wireless mesh, 3G, 4G, and WiMAX.

Microwave Transmission

Microwave is a high-frequency (300 MHz–300 GHz) signal sent through the air, as shown in Figure 6.4. Terrestrial (Earthbound) microwaves are transmitted by line-of-sight devices, so the line of sight between the transmitter and receiver must be unobstructed. Typically, microwave stations are placed in a series—one station receives a signal, amplifies it, and retransmits it to the next microwave transmission tower. Such stations can be located roughly 30 miles apart before the curvature of the Earth makes it impossible for the towers to "see" one another. Microwave signals can carry thousands of channels at the same time.

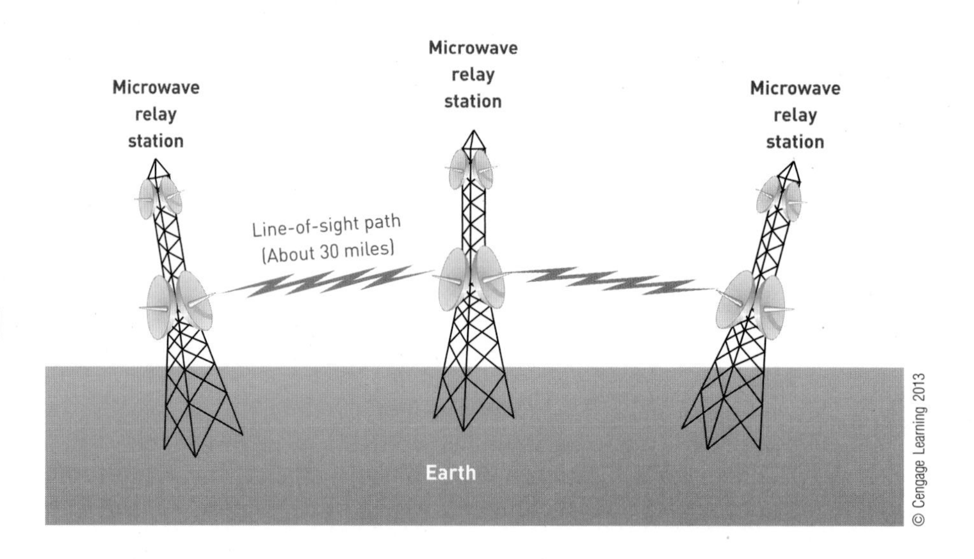

FIGURE **6.4**

Microwave communications

Because they are line-of-sight transmission devices, microwave dishes are frequently placed in relatively high locations, such as mountains, towers, or tall buildings.

A communications satellite also operates in the microwave frequency range (see Figure 6.5). The satellite receives the signal from the Earth station, amplifies the relatively weak signal, and then rebroadcasts it at a different frequency. The advantage of satellite communications is that satellites can receive and broadcast over large geographic regions. Such problems as the curvature of the Earth, mountains, and other structures that block the line-of-sight microwave transmission make satellites an attractive alternative. Geostationary, low earth orbit, and small mobile satellite stations are the most common forms of satellite communications.

A *geostationary satellite* orbits the Earth directly over the equator, approximately 22,300 miles above the Earth so that it appears stationary. The U.S. National Weather Service relies on the Geostationary Operational Environmental Satellite program for weather imagery and quantitative data to support weather forecasting, severe storm tracking, and meteorological research.

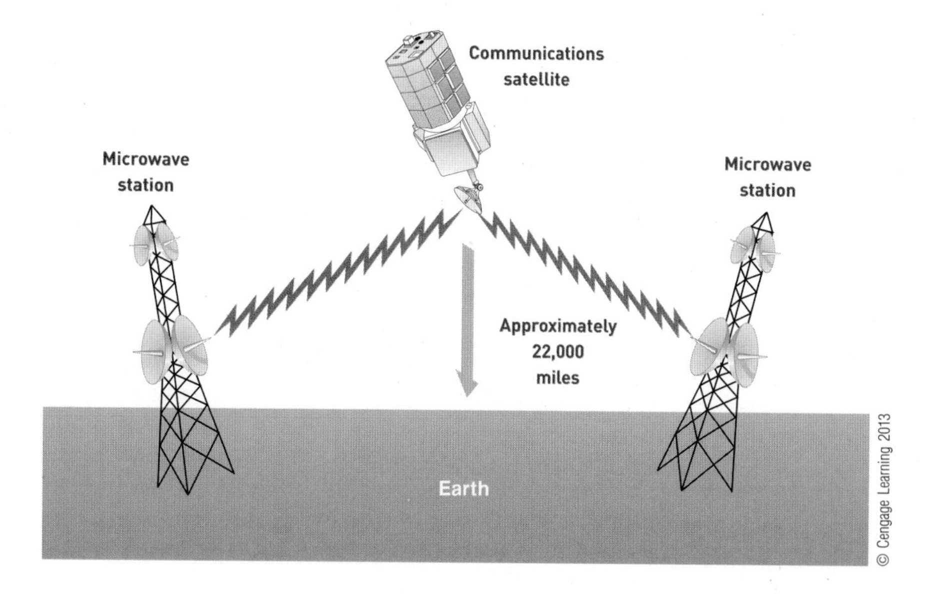

FIGURE 6.5
Satellite transmission
Communications satellites are relay stations that receive signals from one Earth station and rebroadcast them to another.

A *low earth orbit (LEO) satellite* system employs many satellites, each in an orbit at an altitude of less than 1,000 miles. The satellites are spaced so that, from any point on the Earth at any time, at least one satellite is on a line of sight. Iridium Communications, Inc., provides a global communications network that spans the entire Earth using 66 satellites in a near polar orbit at an altitude of 485 miles. Calls are routed among the satellites to create a reliable connection between call participants that cannot be disrupted by natural disasters such as earthquakes, tsunamis, or hurricanes that knock out ground-based wireless towers and wire- or cable-based networks.[19]

The Iridium satellite phone costs on the order of $1,400 or can be rented for $40 per week. Its AccessPoint Mail & Web software enables smartphone users to check their voice messages and e-mail provided they have access to an Iridium phone and its Wi-Fi hot spot device. However, users pay a $1-per-minute connect charge and Web browsing is at the slow rate of 12 kbps and 40 kbps for e-mail. Cubic Global Tracking Solutions works with customers such as the Port Authority of New York and New Jersey, the United States Army–Logistics Transformation Agency, and Horizon Shipping Lines to track their assets and verify their status throughout the supply chain. The firm employs the Iridium satellite network to rely information about its customers' assets.[20]

A *very small aperture terminal (VSAT)* is a satellite ground station with a dish antenna smaller than 3 meters in diameter. The Chinese National Petroleum Cooperation subsidiary BGP is implementing VSAT communications systems for its fleet of seismic vessels used to locate and pinpoint potential areas for oil drilling.[21] The VSAT systems provide Internet access and local area network communications.[22]

Wireless Mesh

wireless mesh: A form of communication that uses multiple Wi-Fi access points to link a series of interconnected local area networks to form a wide area network capable of serving a large campus or an entire city.

Wireless mesh uses multiple Wi-Fi access points to link a series of interconnected local area networks to form a wide area network capable of serving a large campus or an entire city. Communications are routed among network nodes by allowing for continuous connections and reconfiguration around blocked paths by "hopping" from node to node until a connection can be established. Mesh networks are very robust: if one node fails, all the other nodes can still communicate with each other, directly or through one or more intermediate nodes.

The city of Houston implemented a wireless mesh network to connect video cameras to a central location where first responders could monitor activity in its downtown area. The effort was funded by the Department of Homeland Security in an attempt to ensure public safety.[23]

3G Wireless Communications

Wireless communications has evolved through four generations of technology and services. The 1G (first generation) of wireless communications standards originated in the 1980s and was based on analog communications. The 2G (second generation) are fully digital networks that superseded 1G networks in the early 1990s. Phone conversations were encrypted, mobile phone usage was expanded, and short message services (SMS), or texting, was introduced.

The International Telecommunications Union (ITU) defines and adopts global standards to facilitate the communication among telecommunications network and equipment from manufacturers around the world. In 1999, the ITU established a single standard called IMT-2000 (now referred to as 3G) for cellular networks. The goal was to standardize future digital wireless communications and allow global roaming with a single handset with transmission speeds in the range of 2–4 Mbps. IMT-2000 was intended to become the single, unified, worldwide standard, but the 3G standard effort split into several different standards.

One variant of the 3G standard is the Universal Mobile Telephone System (UMTS), which is the preferred solution for European countries. Another 3G-based standard is Code-Division Multiple Access (CDMA), which is used in Australia, Canada, China, India, Israel, Mexico, South Korea, the United States, and Venezuela. China Mobil operates a 3G network based on a unique Time-Division Synchronous Code Division Multiple Access (TD-SCDMA) technology.

3G wireless communications supports wireless voice and broadband speed data communications in a mobile environment at speeds of 2–4 Mbps. It is called 3G for third generation of solutions for wireless voice and data communications. Additional capabilities include mobile video, mobile e-commerce, location-based services, mobile gaming, and the downloading and playing of songs.

The wide variety of 3G cellular communications protocols can support many business applications. The challenge is to enable these protocols to intercommunicate and support fast, reliable, global wireless communications. Existence of these standards enables network operators to select from a large number of competitive hardware suppliers.

4G Wireless Communications

The ITU has defined 4G as a network that has a meaningful improvement over 3G services. 4G broadband mobile wireless is expected to deliver more advanced versions of enhanced multimedia, smooth streaming video, universal access, portability across all types of devices, and, eventually, worldwide roaming. 4G will also deliver increased data transmission rates in the 5–20 Mbps range, 10 times the speed of 3G networks and faster than most home-based broadband services.[24]

Each of the four major U.S. network operators offers a different version of 4G wireless communications that provide significantly faster data speeds than its 3G networks. Verizon is rapidly expanding its 4G network based on the Long Term Evolution (LTE) standard with plans to cover 80 percent of the United States by 2013. The Sprint 4G network is based on the WiMAX standard, which is faster than 3G networks but not as fast as networks based on LTE. The current AT&T and T-Mobile 4G networks are based on the HSPA+ standard, with AT&T rapidly converting from HSPA+ to the LTE standard.

Long Term Evolution (LTE): A standard for wireless communications for mobile phones based on packet switching.

Long Term Evolution (LTE) is a standard for wireless communications for mobile phones based on packet switching, which is an entirely different approach from the circuit-switching approach employed in 3G telecommunications networks. Carriers must reengineer their voice call networks to convert to the LTE standard. Verizon Wireless began introducing its 4G LTE network in December 2010, and within a year, over 186 million Americans could access the network with 10 times the speed of Verizon's 3G network.[25] Carriers will eventually migrate to LTE Advanced that is expected to deliver download peak rates of 1 Gbps and upload peak rates of 0.5 Gbps and to become available in the 2014 time frame.[26]

Worldwide Interoperability for Microwave Access (WiMAX) is a 4G alternative based on a set of IEEE 802.16 wireless metropolitan area network standards that support various types of communications access. In many respects, WiMAX operates like Wi-Fi, only over greater distances and at faster transmission speeds. Fewer WiMAX base stations are required to cover the same geographical area than when Wi-Fi technology is used.

Clearwire operates a 4G mobile WiMAX network, and Sprint and Comcast sell airtime on the network. This network enables the companies to compete with the wireless data offerings of AT&T and Verizon.

Most telecommunications experts agree that WiMAX is an attractive option for developing countries with little or no wireless telephone infrastructure. However, it is not clear whether WiMAX will be as successful in developed countries such as the United States, where regular broadband is plentiful and cheap and where 3G wireless networks already cover most major metropolitan areas. In addition, AT&T and Verizon Wireless have chosen a different direction and plan to upgrade their wireless networks with the LTE communications technology.[27] An advanced version of WiMAX called WiMAX-2 is under development based on the 802.16m standard and is expected to be able to transmit at up to 300 Mbps.[28]

HSPA+ (Evolved High Speed Packet Access): A broadband telecommunications standard that provides data rates of up to 42 Mbps on the downlink and 22 Mbps on the uplink.

HSPA+ (Evolved High Speed Packet Access) is a broadband telecommunications standard that provides data rates of up to 42 Mbps on the downlink and 22 Mbps on the uplink.

Growth in Wireless Data Traffic

Over the next several years, the growth in the amount of wireless data traffic will create many opportunities for innovators to solve network capacity problems and avoid user service issues. Ericsson expects that by 2016 the amount of data traffic generated by mobile smartphones will be about equal to the volume of traffic generated by mobile computers.[29] Cisco (a major provider of networking technology and services) forecasts that mobile data traffic will grow at an average rate of 92 percent a year from 2011 to 2016—a 25-fold increase. Most of the growth will come from online video. Although 4G LTE networks have 20 times the data-carrying capacity of 3G networks, even 4G networks will not be able to keep pace with such rapid growth rates.[30]

Alcatel-Lucent (a communications company whose Bell Labs is world renown for research and innovation in communications technology) estimates that wireless carriers are currently spending billions of dollars each year to increase their data-carrying capacity. One approach to solve the capacity problem is to predict what users want to watch and store those videos nearer to the users rather than in a central location. This practice would reduce the amount of network capacity required. Another approach is to automatically offload 4G traffic onto nearby Wi-Fi networks.[31] Indeed, market research company Informa Telecoms and Media predicts that the number of public Wi-Fi hotspots will increase by 350 percent between 2012 and 2015 to 5.8 million worldwide to support this offloading.[32]

NETWORKS AND DISTRIBUTED PROCESSING

computer network: The communications media, devices, and software needed to connect two or more computer systems or devices.

A **computer network** consists of communications media, devices, and software needed to connect two or more computer systems or devices. The computers and devices on the networks are also called *network nodes*. After they are connected, the nodes can share hardware and software and data, information, and processing jobs. Increasingly, businesses are linking computers in networks to streamline work processes and enable employees to collaborate on projects. If a company uses networks effectively, it can grow into an agile, powerful, and creative organization, giving it a long-term competitive advantage. Organizations can use networks to share hardware, programs, and databases. Networks can transmit and receive information to improve organizational effectiveness and efficiency, and they enable geographically separated workgroups to share documents and opinions, which fosters teamwork, innovative ideas, and new business strategies.

Network Types

Depending on the physical distance between nodes on a network and the communications and services it provides, networks can be classified as personal area, local area, metropolitan area, or wide area.

Personal Area Networks

personal area network (PAN): A network that supports the interconnection of information technology close to one person.

A **personal area network (PAN)** is a wireless network that connects information technology devices close to one person. With a PAN, you can connect a laptop, digital camera, and portable printer without cables. You can download digital image data from the camera to the laptop and then print it on a high-quality printer—all wirelessly. Additionally, a PAN enables data captured by the sensors placed on your body to be transmitted to your smartphone as input to applications that can serve as calorie trackers, heart monitors, glucose monitors, and pedometers.[33] The IEEE 802.15 is a working group of the standards committee that specifies wireless personal area network standards.[34]

Local Area Networks

local area network (LAN): A network that connects computer systems and devices within a small area, such as an office, home, or several floors in a building.

A network that connects computer systems and devices within a small area, such as an office, home, or several floors in a building, is a **local area network (LAN)**. Typically, LANs are wired into office buildings and factories, as shown in Figure 6.6. Although LANs often use unshielded twisted-pair wire, other media—including fiber-optic cable—are also popular. Increasingly, LANs are using some form of wireless communications. You can build LANs to connect personal computers, laptop computers, or powerful mainframe computers.

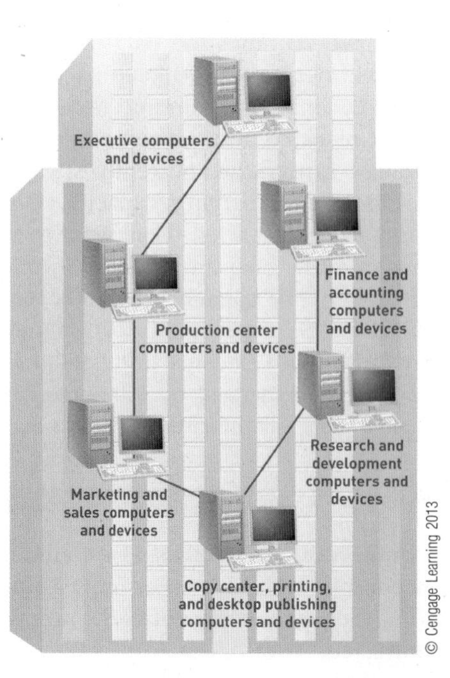

FIGURE 6.6

Typical LAN

All network users within an office building can connect to each other's devices for rapid communication. For instance, a user in research and development could send a document from her computer to be printed at a printer located in the desktop publishing center.

Deloitte Belgium, one of the largest accounting and consulting organizations in Belgium, implemented a wireless local area network to connect employees at its 10 offices and two data centers. Employees can use the LAN for both data and voice communications. "As workers become increasingly mobile, wireless has moved from simply being a convenience to the primary means of network access, and large organizations like Deloitte Belgium need to expand their infrastructure rapidly to provide employees and customers with ubiquitous, secure, and uninterrupted connectivity from multiple mobile devices," says Mike Marcellin, a Juniper Networks marketing vice president.[35]

A basic type of LAN is a simple peer-to-peer network that a small business might use to share files and hardware devices such as printers. In a peer-to-peer network, you set up each computer as an independent computer, but you let other computers access specific files on its hard drive or share its printer. These types of networks have no server. Instead, each computer is connected to the next machine. Examples of peer-to-peer networks include Windows for Workgroups, Windows NT, Windows 2000, AppleShare, and Windows 7 Homegroup. Performance of the computers on a peer-to-peer network is usually slower because one computer is actually sharing the resources of another computer.

With more people working at home, connecting home computing devices and equipment into a unified network is on the rise. Small businesses are also connecting their systems and equipment. A home or small business can connect network resources, computers, printers, scanners, and other devices. A person working on one computer, for example, can use data and programs stored on another computer. In addition, several computers on the network can share a single printer. To make home and small business networking a reality, many companies are offering networking standards, devices, and procedures.

Metropolitan Area Networks

metropolitan area network (MAN): A telecommunications network that connects users and their computers in a geographical area that spans a campus or city.

A **metropolitan area network (MAN)** is a telecommunications network that connects users and their computers in a geographical area that spans a campus or city. A MAN might redefine the many networks within a city into a single larger network or connect several LANs into a single campus LAN. Often the MAN is owned either by a consortium of users or by a single network provider who sells the service to users.[36]

The Dallas zoo, which includes 40 buildings spread out over 106 acres, implemented a MAN to provide voice and data services for 196 phones and 300 personal computer users. Future plans call for adding security and safety cameras and information kiosks for zoo visitors.[37]

Wide Area Networks

wide area network (WAN): A telecommunications network that connects large geographic regions.

A **wide area network (WAN)** is a telecommunications network that connects large geographic regions. A WAN might be privately owned or rented and includes public (shared-users) networks. When you make a long-distance phone call or access the Internet, you are using a WAN. WANs usually consist of computer equipment owned by the user, together with data communications equipment and telecommunications links provided by various carriers and service providers (see Figure 6.7).

WANs often provide communications across national borders, which involves national and international laws regulating the electronic flow of data across international boundaries or *transborder data flow*. Many countries, including those in the European Union, have strict laws placing limits on the transmission of personal data about customers and employees across national borders.

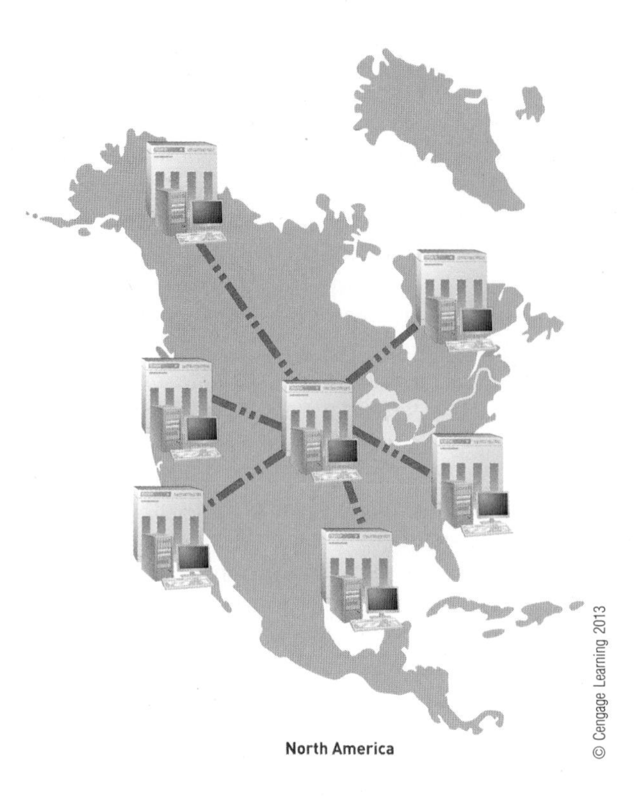

FIGURE 6.7
Wide area network

WANs are the basic long-distance networks used around the world. The actual connections between sites, or nodes (shown by dashed lines), might be any combination of guided and wireless media. When you make a long-distance telephone call or access the Internet, you are using a WAN.

CrescoAg LLC, an agriculture information management company, and Smartfield, Inc., a maker of in-crop analysis tools, have partnered to create a WAN that enables farmers to monitor and manage their crops. The system employs sensors that capture crop temperature, amount of moisture, and other environmental parameters. The data is transmitted to a computer for analysis to identify when a crop's high stress level has been reached. The system then sends an alert that irrigation or some type of topical application is necessary.[38]

Basic Processing Alternatives

centralized processing: An approach to processing wherein all processing occurs in a single location or facility.

decentralized processing: An approach to processing wherein processing devices are placed at various remote locations.

distributed processing: An approach to processing wherein processing devices are placed at remote locations but are connected to each other via a network.

When an organization needs to use two or more computer systems, it can implement one of three basic processing alternatives: centralized, decentralized, or distributed. With **centralized processing**, all processing occurs in a single location or facility. This approach offers the highest degree of control because a single centrally managed computer performs all data processing. The Ticketmaster reservation service is an example of a centralized system. One central computer with a database stores information about all events and records the purchases of seats. Ticket clerks at various ticket selling locations can enter order data and print the results, or customers can place orders directly over the Internet.

With **decentralized processing**, processing devices are placed at various remote locations. Each processing device is isolated and does not communicate with any other processing device. Decentralized systems are suitable for companies that have independent operating units, such as 7-Eleven, where each of its 6,500 U.S. stores is managed to meet local retail conditions. Each store has a computer that runs more than 50 business applications, such as cash register operations, gasoline pump monitoring, and merchandising.

With **distributed processing**, processing devices are placed at remote locations but are connected to each other via a network. One benefit of distributed processing is that managers can allocate data to the locations that can process it most efficiently. Kroger operates over 3,600 supermarkets, each with its own computer to support store operations such as customer checkout and inventory management. These computers are connected to a network so that sales data gathered by each store's computer can be sent to a huge data repository on a mainframe computer for efficient analysis by marketing analysts and product supply chain managers.

Ongoing terrorist attacks around the world and the heightened sensitivity to natural disasters (such as earthquakes in Argentina, Chile, and Japan; major flooding of the Mississippi; and dozens of tornados in the South and Midwest sections of the United States, all happening in 2011) have motivated many companies to distribute their workers, operations, and systems much more widely, a reversal of the previous trend toward centralization. The goal is to minimize the consequences of a catastrophic event at one location while ensuring uninterrupted systems availability.

File Server Systems

Users can share data through file server computing, which allows authorized users to download entire files from certain computers designated as file servers. After downloading data to a local computer, a user can analyze, manipulate, format, and display data from the file, as shown in Figure 6.8.

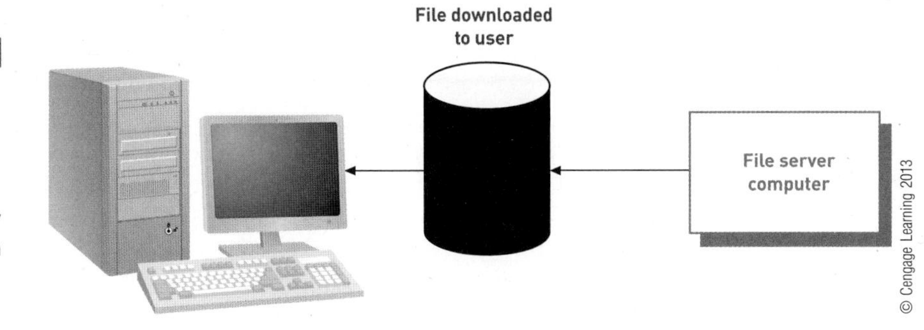

FIGURE 6.8

File server connection

The file server sends the user the entire file that contains the data requested. The user can then analyze, manipulate, format, and display the downloaded data with a program that runs on the user's personal computer.

Client/Server Systems

client/server architecture: An approach to computing wherein multiple computer platforms are dedicated to special functions, such as database management, printing, communications, and program execution.

In **client/server architecture**, multiple computer platforms are dedicated to special functions, such as database management, printing, communications, and program execution. These platforms are called *servers*. Each server is accessible by all computers on the network. Servers can be computers of all sizes; they store both application programs and data files and are equipped with operating system software to manage the activities of the network. The server distributes programs and data to the other computers (clients) on the network as they request them. An application server holds the programs and data files for a particular application, such as an inventory database. The client or the server can do the processing.

A client is any computer (often a user's personal computer) that sends messages requesting services from the servers on the network. A client can converse with many servers concurrently. For example, a user at a personal computer initiates a request to extract data that resides in a database somewhere on the network. A data request server intercepts the request and determines on which database server the data resides. The server then formats the user's request into a message that the database server will understand. When it receives the message, the database server extracts and formats the requested data and sends the results to the client. The database server sends only the data that satisfies a specific query—not the entire file (see Figure 6.9). As with the file server approach, when the downloaded data is on the user's machine, it can then be analyzed, manipulated, formatted, and displayed by a program that runs on the user's personal computer.

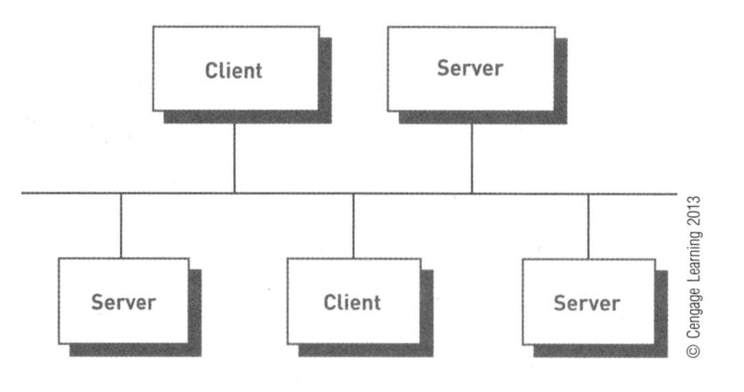

FIGURE 6.9
Client/server connection
Multiple computer platforms, called servers, are dedicated to special functions. Each server is accessible by all computers on the network. The client requests services from the servers, provides a user interface, and presents results to the user.

Table 6.4 lists the advantages and disadvantages of client/server architecture.

TABLE 6.4 Advantages and disadvantages of client/server architecture

Advantages	Disadvantages
Moving applications from mainframe computers and terminal-to-host architecture to client/server architecture can yield significant savings in hardware and software support costs.	Moving to client/server architecture is a major two- to five-year conversion process.
Minimizes traffic on the network because only the data needed to satisfy a user query is moved from the database to the client device.	Controlling the client/server environment to prevent unauthorized use, invasion of privacy, and viruses is difficult.
Security mechanisms can be implemented directly on the database server through the use of stored procedures.	Using client/server architecture leads to a multivendor environment with problems that are difficult to identify and isolate to the appropriate vendor.

An electronic health record (EHR) is a summary of health information generated by each patient encounter in any health care delivery setting. It includes patient demographics, medical history, immunization records, lab data, medications, vital signs, and other such data. The federal government earmarked $33 billion in incentives for health care providers that can demonstrate they are "meaningful users" of EHR technology. To earn this incentive, which can amount to millions of dollars, a hospital must show that it is achieving key desired policy outcomes in efficiency, patient safety, and care coordination. Mecosta County Medical Center is a 74-bed acute-care hospital in Michigan that implemented a client/server hospital information system to capture and display meaningful use data to ensure that the hospital is on track to earn its incentive.[39]

Telecommunications Hardware

Networks require various telecommunications hardware devices to operate, including smartphones, modems, multiplexers, private branch exchanges, switches, bridges, routers, and gateways.

Smartphones

As discussed in Chapter 3, a smartphone combines the functionality of a mobile phone, camera, Web browser, e-mail tool, MP3 player, and other devices into a single handheld device. For example, the Apple iPhone is a combination mobile phone, widescreen iPod, and Internet access device capable of supporting e-mail and Web browsing. An iPhone user can connect to the Internet either via Wi-Fi or AT&T's Edge data network. The iPhone 4S comes with a "virtual assistant," Siri, that recognizes voice commands and can answer basic questions such as "What is the weather today in Columbus?" It can also schedule appointments, dictate text messages, and conduct Web searches.

ETHICAL & SOCIETAL ISSUES

Using Networks to Share Medical Data

In today's mobile society, people often become ill or suffer an injury away from their regular physicians. With increasing medical specialization, the need for multiple physicians to access data for the same patient is increasing, while the best specialist to review a case may be far away. Electronic medical records (EMRs), accessed over data communications networks such as those discussed in this chapter, can alleviate these and similar problems in sharing medical data.

However, a health care facility can't just install a few wireless routers and an Internet gateway and then post EMRs online. The most difficult issues with EMRs involve ethics and privacy, not technology. Information systems specialists, however, are often focused on technology. They tend to concentrate on the technical issues rather than the human ones. Organizations must have a broader focus.

Lucile Packard Children's Hospital (LPCH) at Stanford in Palo Alto, California, decided that the best approach for patient care would be to combine an EMR, which is optimized for use inside a hospital, with an electronic health record (EHR) to be shared with medical providers outside it. That would enable all those responsible for the care of a given patient to stay informed on what the other members of the team were doing. The technology to link EMRs and EHRs was not complex, but the effort involved operational and cultural challenges.

The three major challenges LPCH faced in this project were finding the best party to administer, support, and maintain the system; negotiating expectations and cultural differences of the hospital and the software company providing the EHR; and implementing data flow between EHR and EMR. Here's what they did:

- The hospital's Health Information Management Services (HIMS) Department emerged as the most logical system "owner," the entity that could best administer, support, and maintain the EMR-EHR system. HIMS has expertise in medical information issues and a vision that wide adoption of a EHR system might facilitate a more efficient, timely, and patient-friendly way to release vital medical records.
- Open, ongoing, face-to-face communication and negotiation at multiple levels between LPCH and the software company helped to ensure that each organization understood the needs, concerns, and capacities of the other.
- By late 2011, the issue of data exchange wasn't entirely solved. One issue is motivating hospital physicians to check EHRs between patient visits. Failing to check reduces the value of the EHR as a communication vehicle with other caregivers.

Unauthorized people want to access other people's medical data for many reasons, ranging from curiosity to blackmail. Privacy considerations, therefore, are vital in a system such as this. Medical data shared with people outside the originating organization must be protected from malicious attempts to intercept it. Ponemon Institute reports that medical data breaches rose over 30 percent in 2011, with 96 percent of health care organizations reporting at least one privacy breach in 2010–2011. Hospitals are trying new technical approaches to ensure privacy via improved security and innovative system design. For example, one approach involves storing personal information in one file with a pseudonym and storing the real name that corresponds to that pseudonym in a separate table. The benefit is that, "if one part of data is stolen, the information is useless. Neither the mapping table nor the medical database with pseudonyms is really useful alone." This is not the only type of security precaution that people must take in setting up a medical information system, but it is one example. (The specific methods that a particular system uses are generally kept secret, to avoid giving potential attackers any information that would help them penetrate the system.)

Early feedback on the LPCH system is positive. A survey found that 89 percent of responders report that the system improved the management of children's health care at least somewhat, with only 11 percent saying that it improved "very little" or "not at all." This result is a credit to how well the team dealt with the human as well as the technical issues.

Discussion Questions

1. Consider a child who has surgery at LPCH, after which the surgeon enters data into the EMR. The child goes home, where a pediatrician or others outside the hospital enter data into the EHR. The EMR and EHR then update each other. Discuss the pros and cons of that approach, compared to a system where all of a patient's caregivers (both in and out of LPCH) share a single patient information database.
2. Suppose you are injured far from home and are treated at a hospital that has a system like the one in this case. When you go home, your family doctor says, "I don't have time for that new-fangled stuff, and besides, nobody pays me for my time while I'm using it." How would you respond in terms of getting better care?

Critical Thinking Questions

1. Suppose you are a physician in a hospital. You see a patient from a remote town once every three months to review his case and adjust his medications for the next three months. In between, he is under the care of a visiting nurse in his town. When this nurse has a question, he usually calls you. He would prefer to put a note in the patient's file so that you would have the entire file when you read the note. However, he knows that you are not likely to check the file for notes in between visits because you are not paid to check the files. E-mail would be slower than a phone call because you do not check it while you are seeing patients and the messages do not go to a person who can interrupt you if a question is urgent. What solution do you suggest?
2. Suppose you are the parent of a child who will have a series of operations a year apart over several years to correct a birth defect. (Once they are over, your child will have a normal life in every respect.) Those operations will be performed at a hospital located several hours travel from your home. Discuss how you, the hospital (which is not LPCH), and your local pediatrician could each use a system such as the one described in this case.

SOURCES: Anoshiravani, A., et al, "Implementing an Interoperable Personal Health Record in Pediatrics: Lessons Learned at an Academic Children's Hospital," *Journal of Participatory Medicine*, *www.jopm.org/columns/innovations/2011/07/11/implementing-an-interoperable-personal-health-record-in-pediatrics-lessons-learned-at-an-academic-children's-hospital*, July 11, 2011; Lucille Packard Children's Hospital Web site, *www.lpch.org*, accessed January 29, 2012; Maragioglio, J., "Hackers Target Patient Records on Mobile Devices," *Mobiledia*, *www.mobiledia.com/news/121550.html*, December 21, 2011; Benzschawel, S. and Da Silveira, M., "Protecting Patient Privacy When Sharing Medical Data," eTELEMED 2011: The Third International Conference on eHealth, Telemedicine, and Social Medicine, Gosier, Guadeloupe, France, *www.thinkmind.org/index.php?view=article&articleid=etelemed_2011_5_10_40072*, February 23, 2011.

Apple iPhone

The iPhone is a combination mobile phone, widescreen iPod, and Internet access device.

The phone uses a powerful processor known as the A5, the same microprocessor found inside the iPad. Table 6.5 lists some of the more popular smartphones.

Smartphones have their own software operating systems and are capable of running applications that have been created for their particular operating system. As a result, the capabilities of smartphones will continue to evolve as new applications become available. The Apple iPhone (iPhone OS), BlackBerry (RIM OS), and Palm (Palm OS) smartphones come with their own proprietary operating systems. The Android, Windows Phone 7, and Symbian operating systems are used on various manufacturers' smartphones.

TABLE 6.5 Partial list of 4G LTE smartphones

4G Smartphone	Manufacturer
Blackberry Torch	Research in Motion
Droid Charge	Samsung
Droid Razr	Motorola
Infuse 4G	Samsung
iPhone 4	Apple
Photon	Motorola
Rezound	HTC
Revolution	LG
Sensation	HTC

Smartphone applications are developed by the manufacturers of the handheld device, by the operators of the communications network on which they operate, and by third-party software developers. Table 6.6 lists the number of applications available for each of the most common smartphone operating systems.[40] Table 6.7 lists the top application categories.[41]

TABLE 6.6 Number of applications by operating system (December 2011)

Platform	Number of Applications Available
Apple	590,138
Android	320,315
Blackberry	43,544
Microsoft	35,479

TABLE **6.7** Top application categories

Category	Percent of Total Applications across All Operating Systems
Entertainment	17%
Games	14%
Books	10%
Lifestyle	8%
Utilities	7%
Education	7%
Travel	5%
Business	4%
References	4%
Music	3%
Health and fitness	3%
Productivity	3%
Sports	3%
News	3%
Social networking	3%

Courtesy of Verizon Wireless

Droid smartphone

The Droid smartphone is manufactured by Motorola and uses the Android operating system from Google.

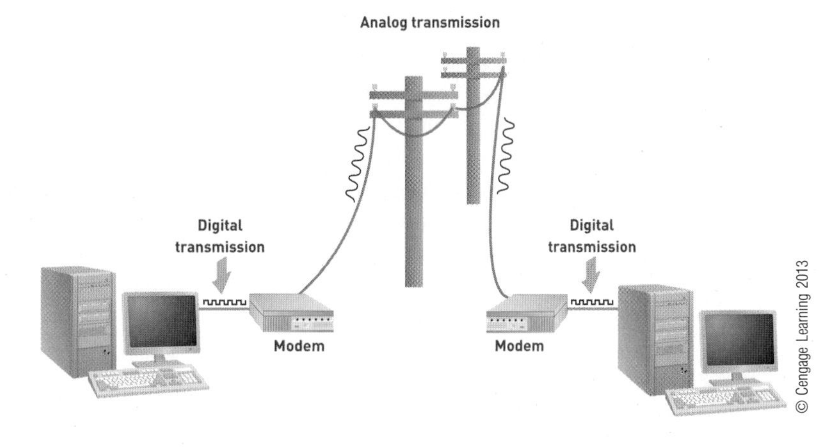

© Cengage Learning 2013

FIGURE **6.10**

How a modem works

Digital signals are modulated into analog signals, which can be carried over existing phone lines. The analog signals are then demodulated back into digital signals by the receiving modem.

Modems

At each stage of the communications process, transmission media of differing types and capacities may be used. If you use an analog telephone line to transfer data, it can accommodate only an analog signal. Because a computer generates a digital signal represented by bits, you need a special device to convert the digital signal to an analog signal and vice versa (see Figure 6.10). Translating data from digital to analog is called *modulation*, and translating data from analog to digital is called *demodulation*. Thus, these devices are modulation/demodulation devices, or **modems**. Penril/Bay Networks, Hayes, Microcom, Motorola, and U.S. Robotics are modem manufacturers.

modem: A telecommunications hardware device that converts (modulates and demodulates) communications signals so they can be transmitted over the communication media.

Modems can dial telephone numbers, originate message sending, and answer incoming calls and messages. Modems can also perform tests and checks on how well they are operating. Some modems can vary their transmission rates based on detected error rates and other conditions. Wireless modems in laptop personal computers allow people on the go to connect to wireless networks and communicate with other users and computers.[42]

With a wireless modem, you can connect to other computers while in your car, on a boat, or in any area that has wireless transmission service. You can use PC memory card expansion slots for standardized credit card-sized PC modem cards, which work like standard modems. PC modems are becoming increasingly popular with notebook and portable computer users.

Cable company network subscribers use a cable modem, which has a low initial cost and can transmit at speeds up to 10 Mbps. The cable modem is always on so that you can be connected to the Internet around the clock. Digital subscriber line (DSL) refers to a family of services that provides high-speed digital data communications service over the wires of the local telephone company. Subscribers use a DSL modem to connect their computers to this service.

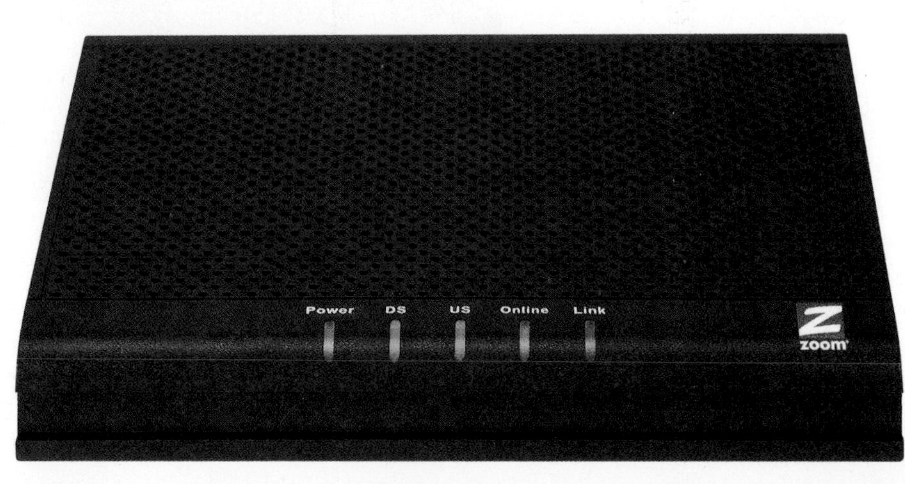

Courtesy of Zoom Telephonics, Inc.

A cable modem can deliver network and Internet access up to 500 times faster than a standard modem and phone line.

multiplexer: A device that combines data from multiple data sources into a single output signal that carries multiple channels, thus reducing the number of communications links needed and lowering telecommunications costs.

Multiplexers

A **multiplexer** is a device that combines data from multiple data sources into a single output signal that carries multiple channels, thus reducing the number of communications links needed and, therefore, lowering telecommunications costs (see Figure 6.11). Multiplexing is commonly used on long-distance phone lines, combining many individual phone calls onto a single long-distance line without affecting the speed or quality of an individual call. At the receiving end, a demultiplexer chooses the correct destination from the many possible destinations and routes each individual call to its correct destination.

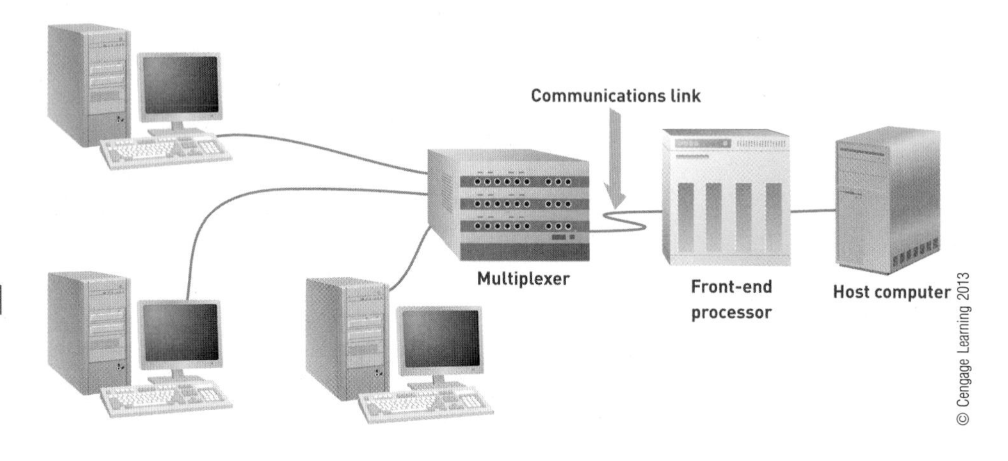

FIGURE 6.11
Use of a multiplexer to consolidate data communications onto a single communications link

Worldcall is a large telecommunications carrier that services Pakistan. It needed to extend its network to connect with the leading cellular operator in Pakistan (17 million customers). Worldcall made the connection using wireless communications employing multiplexers to minimize the cost.[43]

Telecommunications networks require state-of-the-art computer software technology to continuously monitor the flow of voice, data, and image transmission over billions of circuit miles worldwide.

private branch exchange (PBX): A telephone switching exchange that serves a single organization.

Private Branch Exchange (PBX)

A **private branch exchange (PBX)** is a telephone switching exchange that serves a single organization. It enables users to share a certain number of outside lines (trunk lines) to make telephone calls to people outside the organization. This sharing reduces the number of trunk lines required, which in turn reduces the organization's telephone expense. A PBX also enables the routing of calls from department to department or individual to individual with the organization. The PBX can also provide many other functions, such as voice mail, voice paging, three-way calling, call transfer, and call waiting. A VoIP-PBX can accept Voice over IP (VoIP) calls as well as traditional analog phone calls. With Voice over IP calls, the callers' voice communications are converted into packets of data for routing over the Internet.

FiftyFlowers maintains a Web site to help customers select from over 2,000 flower types for weddings or other special events.[44] It also employs a VoIP-PBX system to route calls from customers around the world to an available member of its customer service team. This minimizes customer wait time, reduces operational costs for both the customers and FiftyFlowers, and improves customer service.[45]

Switches, Bridges, Routers, and Gateways

Telecommunications hardware devices switch messages from one network to another at high speeds. A **switch** uses the physical device address in each incoming message on the network to determine to which output port it should forward the message to reach another device on the same network. A **bridge** connects two LANs together using the same telecommunications protocol. A **router** forwards data packets across two or more distinct networks toward their destinations through a process known as routing. Often, an Internet service provider (ISP) installs a router in a subscriber's home that connects the ISP's network to the network within the home. A **gateway** is a network device that serves as an entrance to another network. A failure in any one of these hardware devices can cause significant service disruptions for users. For example, service for millions of Time Warner Cable Internet and phone customers was temporarily disrupted when a router crashed while a software update was being performed.[46]

switch: A telecommunications device that uses the physical device address in each incoming message on the network to determine to which output port it should forward the message to reach another device on the same network.

bridge: A telecommunications device that connects two LANs together using the same telecommunications protocol.

router: A telecommunications device that forwards data packets across two or more distinct networks toward their destinations through a process known as routing.

gateway: A telecommunications device that serves as an entrance to another network.

Telecommunications Software

A **network operating system (NOS)** is systems software that controls the computer systems and devices on a network and allows them to communicate with each other. The NOS performs similar functions for the network as operating system software does for a computer, such as memory and task management and coordination of hardware. When network equipment (such as printers, plotters, and disk drives) is required, the NOS makes sure that these resources are used correctly. Novell NetWare, Windows 2000, Windows 2003, and Windows 2008 are common network operating systems.

network operating system (NOS): Systems software that controls the computer systems and devices on a network and allows them to communicate with each other.

Standard Chartered Bank operates with 85,000 employees in over 1,700 branches and outlets in more than 70 countries around the world. It derives 90 percent of its income and profits from operations in Asia, Africa, and the Middle East.[47] For any bank, processing and telecommunications system availability is essential. Standard Chartered implemented a telecommunications infrastructure consisting of switches and gateways running under the Junos network operating system to lower the cost and complexity of managing its networks and to provide full redundancy.[48]

Because companies use networks to communicate with customers, business partners, and employees, network outages or slow performance can mean a loss of business. Network management includes a wide range of technologies and processes that monitor the network and help identify and address problems before they can create a serious impact.

Software tools and utilities are available for managing networks. With **network-management software**, a manager on a networked personal computer can monitor the use of individual computers and shared hardware (such as printers), scan for viruses, and ensure compliance with software licenses. Network-management software also simplifies the process of updating files and

network-management software: Software that enables a manager on a networked desktop to monitor the use of individual computers and shared hardware (such as printers), scan for viruses, and ensure compliance with software licenses.

programs on computers on the network—a manager can make changes through a communications server instead of having to visit each individual computer. In addition, network-management software protects software from being copied, modified, or downloaded illegally. It can also locate telecommunications errors and potential network problems. Some of the many benefits of network-management software include fewer hours spent on routine tasks (such as installing new software), faster response to problems, and greater overall network control.

Today, most IS organizations use network-management software to ensure that their network remains up and running and that every network component and application is performing acceptably. The software enables IS staff to identify and resolve fault and performance issues before they affect customers and service. The latest network-management technology even incorporates automatic fixes: The network-management system identifies a problem, notifies the IS manager, and automatically corrects the problem before anyone outside the IS department notices it.

Weill Cornell Medical College located in New York City is a top-ranked clinical and medical research center. Its complex telecommunications network (including Ethernet, 802.11a/b/g/n, VoIP, video, and WAN links to remote offices) requires daily monitoring and analysis. The college employs network-management software to identify faults and fix problems quickly to maximize network uptime and user satisfaction.[49]

Banks use a special form of network-management software to monitor the performance of their automatic teller machines (ATMs). Status messages can be sent over the network to a central monitoring location to inform support people about situations such as low cash or receipt paper levels, card reader problems, and printer paper jams. Once a status message is received, a service provider or branch location employee can be dispatched to fix the ATM problem. First Charter Bank of Charlotte, North Carolina, deployed network-management software to manage its network of 138 ATMs from a central location to reduce costs, downtime, and service visits.[50]

Securing Data Transmission

The interception of confidential information by unauthorized individuals can compromise private information about employees or customers, reveal marketing or new product development plans, or cause organizational embarrassment. Organizations with widespread operations need a way to maintain the security of communications with employees and business partners, wherever their facilities are located.

Guided media networks have an inherently secure feature: Only devices physically attached to the network can access the data. Wireless networks, on the other hand, are surprisingly often configured by default to allow access to any device that attempts to "listen to" broadcast communications. Action must be taken to override the defaults.

encryption: The process of converting an original message into a form that can be understood only by the intended receiver.

encryption key: A variable value that is applied (using an algorithm) to a set of unencrypted text to produce encrypted text or to decrypt encrypted text.

Encryption of data is one approach taken to protect the security of communications over both wired and wireless networks. **Encryption** is the process of converting an original message into a form that can be understood only by the intended receiver. An **encryption key** is a variable value that is applied (using an algorithm) to a set of unencrypted text to produce encrypted text or to decrypt encrypted text (see Figure 6.12). The key is chosen from one of a large number of possible encryption keys. The longer the key, the greater the number of possible encryption keys. An encryption protocol based on a 56-bit key, for example, has 256 different possible keys, while one based on a 128-bit key has 2128 different possible keys. Of course, it is essential that the key be kept secret from possible interceptors. A hacker who obtains the key by whatever means can recover the original message from the encrypted data.

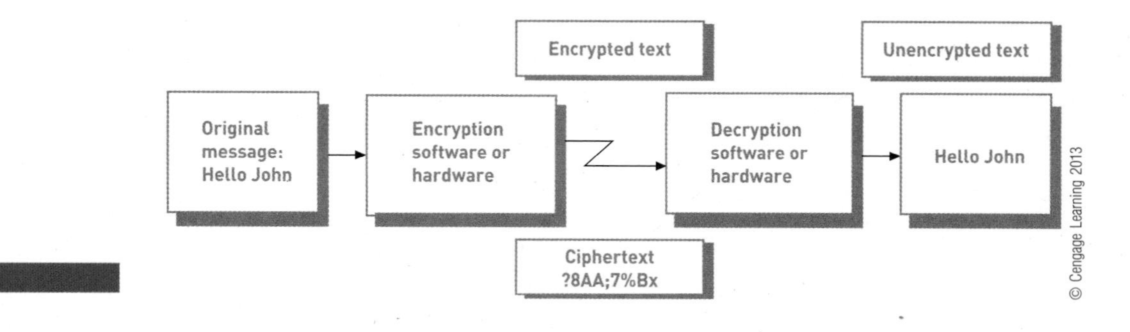

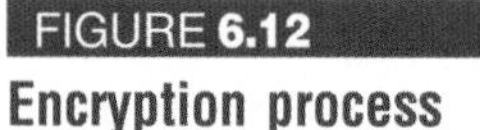
FIGURE 6.12
Encryption process

Encryption methods rely on the limitations of computing power for their security. If breaking a code requires too much computing power, even the most determined hacker cannot be successful.

With headquarters in Dallas, Texas, 7-Eleven operates, franchises, or licenses over 8,800 stores in North America and another 33,900 in 16 countries. The company uses encryption to secure its e-mail. Todd Cohen, leader of 7-Eleven's Information Security, Risk and Compliance practice, states that "The protection of sensitive partner information is essential to our leadership as a trusted retailer. E-mail is a critical communication tool in everyday business with our partners, and [encryption services] enable us to use email securely and confidently."[51]

Securing Wireless Networks

WEP and WPA are the two main approaches to securing wireless networks such as Wi-Fi and WiMAX. Wired equivalent privacy (WEP) used to use encryption based on 64-bit key, which has been upgraded to a 128-bit key. WEP represents an early attempt at securing wireless communications and is not difficult for hackers to crack. Most wireless networks now use the Wi-Fi Protected Access (WPA) security protocol that offers significantly improved protection over WEP.

The following steps, while not foolproof, help safeguard a wireless network:

- Connect to the router and change the default logon (admin) and password (password) for the router. These defaults are widely known by hackers.
- Create a service set identifier (SSID). This is a 32-character unique identifier attached to the header portion of packets sent over a wireless network that differentiates one network from another. All access points and devices attempting to connect to the network must use the same SSID.
- Configure the security to WPA. Surprisingly, many routers are shipped with encryption turned off.
- Disable SSID broadcasting. By default, wireless routers broadcast a message communicating the SSID so wireless devices within range (such as a laptop) can identify and connect to the wireless network. If a device doesn't know the wireless network's SSID, it cannot connect. Disabling the broadcasting of the SSID will discourage all but the most determined and knowledgeable hackers.
- Configure each wireless computer on the network to access the network by setting the security to WPA and entering the same password entered to the router.

War driving involves hackers driving around with a laptop and antenna trying to detect insecure wireless access points. Once connected to such a network, the hacker can gather enough traffic to analyze and crack the encryption.

INFORMATION SYSTEMS @ WORK

Is E-mail History?

It's hard to imagine business without e-mail, which is as common as the water cooler—and probably more useful. Is it, however, time to move on?

That's what IT services firm Atos, based in Bezons just outside Paris, France, thinks. It aims to become a "zero email company" within three years to help tackle what it calls "information pollution," which Atos sees as bogging down company progress because employees overuse e-mail rather than turning to more effective forms of interaction.

As one example of information pollution, consider a six-person project group. One member sends the others an e-mail with a suggestion. The other five respond, with copies to all group members (as is standard business e-mail practice). Three of the members reply to the other comments. Soon each project member receives two or three dozen e-mail messages. Most of those messages repeat the same content with a new sentence or two at the top.

Atos chairman and CEO Thierry Breton, speaking at a conference in February 2011, said he plans to end all internal e-mails among Atos employees. "We are producing data on a massive scale that is fast polluting our working environments and also encroaching into our personal lives," he said. "We are taking action now to reverse this trend, just as organizations took measures to reduce environmental pollution after the industrial revolution."

Though its name may not be a household word in the United States, Atos is not a small firm. It has 74,000 employees in 42 countries and annual revenues in excess of U.S. $10 billion. If it can eliminate e-mail, size should not prevent other companies from doing the same.

As in most companies, Atos employees regularly use e-mail to communicate with one another and to share documents and other files. What will they use instead? Collaboration and social media tools. A collaboration tool lets employees use an online discussion group instead of sending e-mails to each other. (An online discussion group would have helped the six-person project team in the earlier example.) Social media tools include using familiar sites such as Facebook to keep others informed about their business activities. Groups can also use wikis to create shared repositories of information on topics of interest. Atos has found that making such tools available reduces e-mail volume by 10 to 20 percent immediately.

The effort to reduce e-mail volume concerns internal e-mail messages only. Breton and other Atos managers do not expect customers to stop sending e-mail messages to company employees, nor do they expect employees who receive e-mail inquiries from customers and suppliers to respond in any other way. They just hope to eliminate e-mail as an *internal* communication medium.

Not everyone thinks that eliminating internal e-mail is practical. Industry analyst Brian Prentice of Gartner Group faults the bureaucracy that leads to overuse of e-mail, not the medium itself. In his view, "The only solution is to tackle the ballooning administration and bureaucracy overhead in organizations that is fuelling the number of emails being generated. Specifically, our criticism of email as a collaboration tool needs to shift towards the unchecked growth of bureaucracy it enables." In other words, e-mail is a symptom, not the problem.

Not surprisingly, Hubert Tardieu, an advisor to Atos CEO Breton, disagrees. He believes that e-mail does not lend itself to creating communities of shared interests within an organization but that other forms of electronic communication do. "Zero mail is not an objective in itself," he writes, "but the recognition that companies are suffering from email overload." While conceding that the root problem is modern bureaucracy, he feels that e-mail encourages the growth of that bureaucracy while other media don't. Tardieu concludes, "The bet of the Social Organization is that cooperation within communities associated with appropriate processes will create a new style of organization and a new style of management more suitable to the digital generation. In the Social Organization, we shall communicate through the enterprise social network across the various communities we belong to, reducing the usage of mail to formal communication."

In early 2012, a year into Atos's three-year effort, the jury is still out. Stay tuned!

Discussion Questions

1. Do you agree with Brian Prentice of the Gartner Group that e-mail is a symptom, not the problem itself? Will information overload decline if people start using other tools instead of e-mail?
2. Think about your use of social media such as Facebook. If you could count on all your friends to see updates to your Facebook page and if they

knew that was the only way they would hear from you, what fraction of your e-mails do you think you could avoid sending? Would it be more or less work to update your Facebook page than to send e-mail messages?

Critical Thinking Questions

1. Recall a group project you completed in school recently, and then answer one of the following questions:
   a. If you communicated with group members primarily by e-mail, would the project have gone more smoothly if you had a different way to communicate? Why or why not? What would be the best way to communicate with each other?
   b. If you communicated using a tool other than e-mail, was the tool better than e-mail? Why or why not? Could a third method, neither e-mail nor what you used, be better?
2. Atos is a large firm and can afford to set up several social media and collaboration sites, each of which eliminates some need for e-mail. Suppose you work for a small firm with fewer technology resources and people to support them. Can you use this approach? If not all of it, then can you use any of it? Should you?

SOURCES: Savvas, A., "Atos Origin Abandoning Email," *Computerworld UK, www.computerworlduk.com/news/it-business/3260053/atos-origin-abandoning-email*, February 9, 2011; Savvas, A., "Defiant Atos Sticks with Company-wide Email Ban," *Computerworld UK, www.computerworlduk.com/news/it-business/3323504/defiant-atos-sticks-with-company-wide-email-ban*, December 7, 2011; Prentice, B., "Why Will 'Zero Email' Policies Fail? Bureaucracy!," Gartner blog *blogs.gartner.com/brian_prentice/2011/12/11/why-will-zero-email-policies-fail-bureaucracy*, December 11, 2011; Tardieu, H., "Achieving a Zero Email Culture: Is Bureaucracy a Showstopper?," Atos blog, *blog.atos.net/sc/2011/12/21/achieving-a-zero-email-culture-is-bureaucracy-a-showstopper*, December 21, 2011; Atos S.A. Web site, *www.atos.net*, accessed January 27, 2012.

Virtual Private Network (VPN)

virtual private network (VPN): A private network that uses a public network (usually the Internet) to connect multiple remote locations.

The use of a virtual private network is another means used to secure the transmission of communications. A **virtual private network (VPN)** is a private network that uses a public network (usually the Internet) to connect multiple remote locations. A VPN provides network connectivity over a potentially long physical distance and thus can be considered a form of wide area network. VPNs support secure, encrypted connections between a company's employees and remote users through a third-party service provider. Telecommuters, salespeople, and frequent travelers find the use of a VPN to be a safe, reliable, low-cost way to connect to their corporate intranets. Often, users are provided with a security token that displays a constantly changing password to log onto the VPN. This solution avoids the problem of users forgetting their password while providing added security through use of a password constantly changing every 30 to 60 seconds.

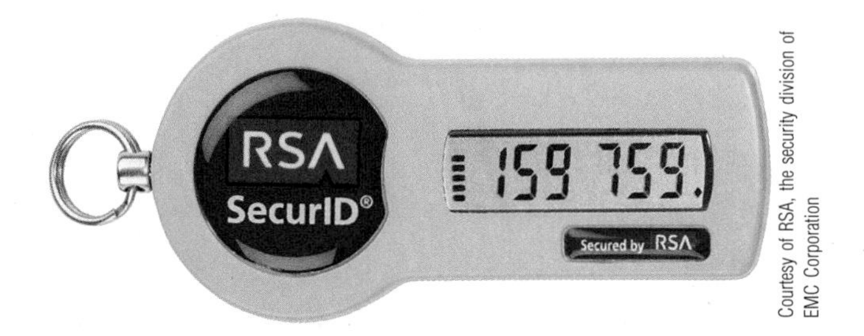

Courtesy of RSA, the security division of EMC Corporation

RSA SecurID security token

The six digits displayed on the token are used as an access code to gain access to a VPN network. The digits change every 60 seconds.

Werner Enterprises is a transportation and logistics company with a fleet of 7,250 trucks, nearly 25,000 tractors, and more than 13,000 employees and independent contractors.[52] Most of its business is in North America, but it is expanding globally with customers in Africa, China, Europe, and Latin America. Werner employs a VPN solution to link its Shanghai operations center to its headquarters and data center in Omaha, Nebraska, so that employees can access load management systems to support global operations.[53]

TELECOMMUNICATIONS SERVICES AND NETWORK APPLICATIONS

Telecommunications and networks are vital parts of today's information systems. In fact, it is hard to imagine how organizations could function without them. For example, when a business needs to develop an accurate monthly production forecast, a manager simply downloads sales forecast data gathered directly from customer databases. Telecommunications provides the network link, allowing the manager to access the data quickly and generate the production report, which supports the company's objective of better financial planning. This section looks at some of the more significant telecommunications services and network applications.

Cellular Phone Services

The cell phone has become ubiquitous and is an essential part of life in the twenty-first century. The ITU estimates that at the end of 2008, there were 2.2 billion mobile phone users in developing countries. Although many of the people who live in these countries do not have ready access to clean drinking water, electricity, or the Internet, more than half of these individuals are expected to have mobile phones by 2012.

Cellular phones operate using radio waves to provide two-way communications. With cellular transmission, a local area such as a city is divided into cells. As a person with a cellular device such as a mobile phone moves from one cell to another, the cellular system passes the phone connection from one cell to another (see Figure 6.13). The signals from the cells are transmitted to a receiver and integrated into the regular phone system. Cellular phone users can thus connect to anyone who has access to either a cell phone or regular phone service, from a child at home to your friend on the road to a business associate in another country. Because cellular transmission uses radio waves, people with special receivers can listen to cellular phone conversations, so such conversations are not secure.

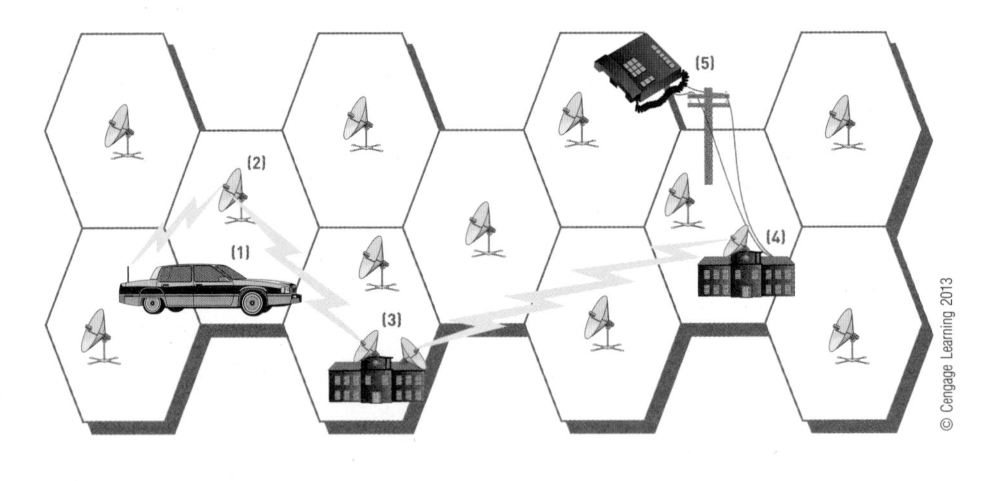

FIGURE 6.13

Typical cellular transmission scenario

Using a cellular car phone, the caller dials the number (1). The signal is sent from the car's antenna to the low-powered cellular antenna located in that cell (2). The signal is sent to the regional cellular phone switching office, also called the mobile telephone subscriber office (MTSO) (3). The signal is switched to the local telephone company switching station located nearest to the call destination (4). Now integrated into the regular phone system, the call is switched to the number originally dialed (5), all without the need for operator assistance.

More and more workers rely on their mobile phones as their primary business phones. However, they frequently encounter problems with poor coverage and can find it difficult to place calls or conduct an extended conversation. A femtocell is a miniature cellular base station designed to serve a very small area such as inside homes, small offices, and outdoor public spaces. Many communications companies now offer femtocell solutions to boost cell phone signals or enable the cell phone to operate over other wireless networks, thus guaranteeing a strong, reliable cell signal. Cell phones may also be linked to a cordless phone via a Bluetooth connection so that if someone calls you on your cell, you can answer the call on the cordless phone. This improves the range of cell phone coverage and eliminates dropped calls, especially for those times you cannot find your cell in time.[54]

The number of femtocells in the United States is now greater than the number of cell towers. At the end of 2010, there were around 256,000 cell towers in the United States and nearly 350,000 femtocells.[55]

Digital Subscriber Line (DSL) Service

digital subscriber line (DSL): A telecommunications service that delivers high-speed Internet access to homes and small businesses over the existing phone lines of the local telephone network

A **digital subscriber line (DSL)** is a telecommunications service that delivers high-speed Internet access to homes and small businesses over the existing phone lines of the local telephone network (see Figure 6.14). Most home and small business users are connected to an *asymmetric DSL (ADSL)* line designed to provide a connection speed from the Internet to the user (download speed) that is three to four times faster than the connection from the user back to the Internet (upload speed). ADSL does not require an additional phone line and yet provides "always-on" Internet access. A drawback of ADSL is that the farther the subscriber is from the local telephone office, the poorer the signal quality and the slower the transmission speed. ADSL provides a dedicated connection from each user to the phone company's local office, so the performance does not decrease as new users are added. Cable modem users generally share a network loop that runs through a neighborhood so that adding users means lowering the transmission speeds. *Symmetric DSL (SDSL)* is used mainly by small businesses; it does not allow you to use the phone at the same time, but the speed of receiving and sending data is the same.

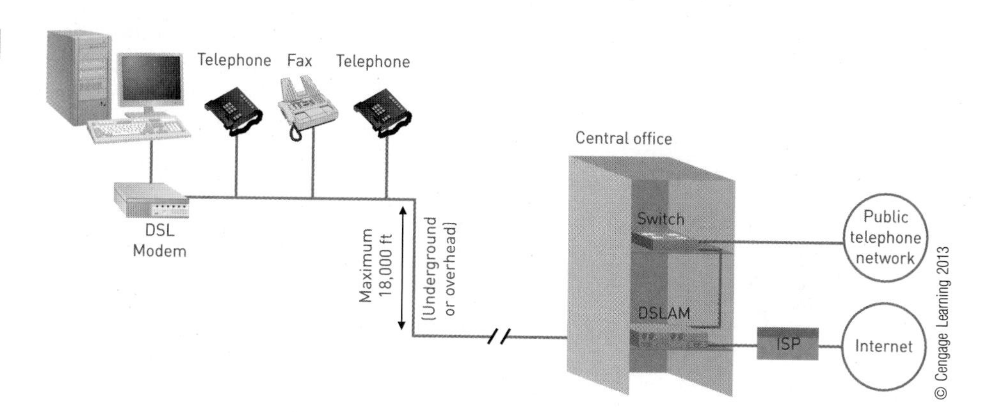

FIGURE 6.14
Digital subscriber line (DSL)
At the local telephone company's central office, a DSL Access Multiplexer (DSLAM) takes connections from many customers and multiplexes them onto a single, high-capacity connection to the Internet. Subscriber phone calls can be routed through a switch at the local telephone central office to the public telephone network.

Linking Personal Computers to Mainframes and Networks

One of the most basic ways that telecommunications connect users to information systems is by connecting personal computers to mainframe computers so that data can be downloaded or uploaded. For example, a user can download a data file or document file from a database to a personal computer. Some telecommunications software programs instruct the computer to connect to another computer on the network, download or send information, and then disconnect from the telecommunications line. These programs are called *unattended systems* because they perform the functions automatically without user intervention.

Voice Mail

voice mail: Technology that enables users to send, receive, and store verbal messages to and from other people around the world.

With **voice mail**, users can send, receive, and store verbal messages to and from other people around the world. Some voice mail systems assign a code to a group of people. Suppose the code 100 stands for all the sales representatives in a company. If anyone calls the voice mail system, enters the number 100, and leaves a message, all the sales representatives receive the same message. Call management systems can be linked to corporate e-mail and instant messaging systems. Calls to employees can be generated from instant messages or converted into e-mail messages to ensure quicker access and response.

The city of Vancouver, British Columbia, established a community voice mail service that provides people who are temporarily homeless or without phone service a means to receive incoming calls. The Vancouver Foundation spent $50,000 over two years to set up the program and purchase 500 phone numbers. Case workers in the community assign the numbers to clients, who can then give out a phone number and retrieve their messages from anywhere to obtain health care, seek housing, or apply for a job.[56]

Voice Mail-to-Text Services

voice mail-to-text service: A service that captures voice mail messages, converts them to text, and sends them to an e-mail account.

Voice mail is more difficult to manage than e-mail because you must deal with messages one by one without knowing who has called you and without being able to prioritize the messages. In recognition of these shortcomings of voice mail, several services (e.g., Jott, SpinVox, GotVoice, and SimulScribe) are now available to convert speech to text so that you can manage voice mails more effectively. If you subscribe to a **voice mail-to-text service**, your voice mail no longer reaches your phone service provider's voice mail service. Instead, it is rerouted to the voice-to-text service, translated into text, and then sent to your regular e-mail account or a special account for converted e-mail messages. You can also temporarily disable the voice-to-text service and receive voice messages.

Reverse 911 Service

reverse 911 service: A communications solution that delivers emergency notifications to users in a selected geographical area.

Reverse 911 service is a communications solution that delivers recorded emergency notifications to users in a selected geographical area. The technology employs databases of phone numbers and contact information. Some systems can send more than 250,000 voice or text messages per hour via phone, pager, cell phone, and e-mail.

Bladen County in North Carolina purchased a reverse 911 service integrated with a geographic information system to enable targeted areas simply by choosing a street or drawing a circle around a portion of a county map. Initially the database is filled only with the landline numbers of residents. But residents can enter unlisted phone numbers, cell phone numbers, and secondary numbers into the system for future contact.[57]

Police in the small town of Beverly, Massachusetts (population 39,000), put their reverse 911 service to innovative use to identify the victim of a theft. Police caught two people suspected of stealing a TV; however, nobody called and reported the missing item. The police sent a reverse 911 call to the neighborhood where the suspects were apprehended. Within 45 minutes of the call going out, the rightful owner called to claim his stolen property.[58]

Home and Small Business Networks

Some small businesses and many families own more than one computer and want to set up a simple network to share printers or an Internet connection; access shared files, such as photos, MP3 audio files, spreadsheets, and documents, on different machines; play games that support multiple concurrent players; and send the output of network-connected devices, such as a security camera or DVD player, to a computer.

One simple solution is to establish a wireless network that covers your home or small business. To do so, you can buy an 802.11n access point, connect it to your cable modem or DSL modem, and then use it to communicate with all your devices. For less than $100, you can purchase a combined router, firewall, Ethernet hub, and wireless hub in one small device. Computers in your network connect to this box with a wireless card, which is connected by cable or DSL modem to the Internet. This box enables each computer in the network to access the Internet. The firewall filters the information coming from the Internet into your network. You can configure it to reject information from offensive Web sites or potential hackers. The router can also encrypt all wireless communications to keep your network secure.

In addition, you can configure your computers to share printers and files. Windows 7 and Windows Vista include a Network and Sharing Center that helps with network configuration. Some of the basic configuration steps include assigning each computer to a workgroup and giving it a name, identifying the files you want to share (placing an optional password on some files), and identifying the printers you want to share.

Electronic Document Distribution

electronic document distribution: A process that enables the sending and receiving of documents in a digital form without being printed (although printing is possible).

Electronic document distribution lets you send and receive documents in a digital form without printing them (although printing is possible). It is much faster to distribute electronic documents via networks than to mail printed forms. While creating the new edition of this textbook, the authors distributed drafts of the chapters to a developmental editor, copyeditor, proofreader, and reviewers to obtain quick feedback and suggestions for improvements to be incorporated into a second and then final draft. Viewing documents on screen instead of printing them also saves paper and document storage space. Accessing and retrieving electronic documents is also much faster.

The YMCA of metropolitan Los Angeles serves over 200,000 members in 25 branch locations across the city. It linked its procurement system with an electronic document distribution system to produce documents that can be delivered quickly and cost effectively to its suppliers via print, fax, e-mail, or the Web. The documents provide a full audit trail of all purchasing transactions.[59]

Call Centers

A call center is a location where an organization handles customer and other telephone calls, usually with some amount of computer automation. Call centers are used by customer service organizations, telemarketing companies, computer product help desks, charitable and political campaign organizations, and any organization that uses the telephone to sell or support its products and services. An automatic call distributor (ACD) is a telephone facility that manages incoming calls, handling them based on the number called and an associated database of instructions. Call centers frequently employ an ACD to validate callers, place outgoing calls, forward calls to the right party, allow callers to record messages, gather usage statistics, balance the workload of support personnel, and provide other services.

The Philippines now provides more call center agents (some 400,000) than India (350,000) despite having one-tenth the population. Filipino call center agents also earn more than their Indian counterparts ($300 per month on average versus $250 per month). U.S. companies prefer Filipinos because their customers can more easily understand them than they can Indian agents. In addition, the Philippines has a more reliable power grid, and its cities have better public transportation so employees can get to work safely and on time.[60]

Some U.S. companies are moving their call centers back to the United States. For example, Delta Airlines and U.S. Airways moved their call centers back to the United States in response to angry customers who wanted better English. The companies made this move in spite of the fact that the U.S. employees are paid more than five times what the Indian or Filipino employees make.[61]

The National Do Not Call Registry was set up in 2003 by the U.S. Federal Trade Commission. Telemarketers who call numbers on the list face penalties of up to $11,000 per call, as well as possible consumer lawsuits. More than 209 million active registrants have been created by individuals logging on to the site *www.donotcall.gov* or calling 888-382-1222. The Do Not Call Rules require that, at least every 31 days, sellers and telemarketers remove from their call lists the numbers found in the registry. The number of complaints received during 2011 was 2.2 million.[62]

Even if you have registered, you can still receive calls from political organizations, charities, educational organizations, and telephone surveyors. You can also receive calls from companies with which you have an existing business relationship, such as your bank or credit card companies. Although the registry has greatly reduced the number of unwanted calls to consumers, it has created several compliance-related issues for direct marketing companies.

The Federal Trade Commission closed down two debt relief operations for making deceptive telemarketing calls, calling consumers on the Do Not Call Registry, and using illegal robocalls (a call placed by machines that automatically dial consumers and play a recorded message).[63]

Offshore call centers provide technical support services for many technology vendors and their customers.

Telecommuting and Virtual Workers and Workgroups

telecommuting: The use of computing devices and networks so that employees can work effectively away from the office.

Employees are performing more and more work away from the traditional office setting. Many enterprises have adopted policies for **telecommuting** so that employees can work away from the office using computing devices and networks. This practice means workers can be more effective and companies can save money on office and parking space and office equipment. According to a recent study by Forrester Research, 17 percent of North American enterprises and 14 percent of European enterprises report having employees who spend at least 20 percent of their time away from their normal work desk or who work from home.

Telecommuting is popular among workers for several reasons. Parents find that eliminating the daily commute helps balance family and work responsibilities. Qualified workers who otherwise might be unable to participate in the normal workforce (e.g., those who are physically challenged or who live in rural areas too far from the city office to commute regularly) can use telecommuting to become productive workers. When gas prices soar, telecommuting can help workers reduce significant expenses. Extensive use of telecommuting can lead to decreased need for office space, potentially saving a large company millions of dollars. Corporations are also being encouraged by public policy to try telecommuting as a means of reducing their carbon footprint and traffic congestion. Large companies also view telecommuting as a means to distribute their workforce and reduce the impact of a disaster at a central facility.

Some types of jobs are well suited to telecommuting, including jobs held by salespeople, secretaries, real estate agents, computer programmers, and legal assistants, to name a few. Telecommuting also requires a certain personality type to be effective. Telecommuters need to be strongly self-motivated, be organized, be focused on their tasks with minimal supervision, and have a low need for social interaction. Jobs unsuitable for telecommuting include those that require frequent face-to-face interaction, need much supervision, and have many short-term deadlines. Employees who choose to work at home must be able to work independently, manage their time well, and balance work and home life.

Blue Cross Blue Shield of North Dakota implemented a telecommuting program as a means to deal with the treacherous winter weather. In addition, the program saved office space and enabled the organization to increase its employee pool by reaching out further than a reasonable commuting area. More than two-thirds of the employees involved in the program reported increased productivity and job satisfaction.[64]

Electronic Meetings

videoconferencing: A set of interactive telecommunications technologies that enable people at multiple locations to communicate using simultaneous two-way video and audio transmissions.

Videoconferencing comprises a set of interactive telecommunications technologies that enable people at multiple locations to communicate using simultaneous two-way video and audio transmissions. Videoconferencing can reduce travel expenses and time, and it can increase managerial effectiveness through faster response to problems, access to more people, and less duplication of effort by geographically dispersed sites. Almost all videoconferencing systems combine video and phone call capabilities with data or document conferencing, as shown in Figure 6.15. You can see the other person's face, view the same documents, and swap notes and drawings. With some systems, callers can change live documents in real time. Many businesses find that the document- and application-sharing feature of the videoconference enhances group productivity and efficiency. It also fosters teamwork and can save corporate travel time and expense.

Jon Feingersh/Getty Images

FIGURE 6.15

Videoconferencing

Videoconferencing allows participants to conduct long-distance meetings "face to face" while eliminating the need for costly travel.

Group videoconferencing is used daily in a variety of businesses as an easy way to connect work teams. Members of a team meet in a specially prepared videoconference room equipped with sound-sensitive cameras that automatically focus on the person speaking, large TV-like monitors for viewing the participants at the remote location, and high-quality speakers and microphones. Videoconferencing costs have declined steadily, while video quality and synchronization of audio to video—once weak points for the technology—have improved.

Facebook has rapidly expanded to hundreds of millions of active users around the globe. It is a challenge to get members of a global operation to collaborate effectively and perform at a high level, especially when the members are separated by thousands of miles and multiple time zones as well as vast differences in culture and experiences. Videoconferencing has been a key strategy at Facebook to meet these challenges. Early on the firm implemented large room videoconferencing at its headquarters and larger office locations. More recently it has set up the Blue Jeans Network, a cloud-based videoconferencing service, to enable videoconferencing for employees who do not have easy access to these sites but can use desktop and mobile devices.[65]

Electronic Data Interchange

Electronic data interchange (EDI) is a way to communicate data from one company to another and from one application to another in a standard format, permitting the recipient to perform a standard business transaction, such as processing purchase orders. Connecting corporate computers among organizations is the idea behind EDI, which uses network systems and follows standards and procedures that can process output from one system directly as input to other systems—without human intervention. EDI can link the computers of customers, manufacturers, and suppliers, as shown in Figure 6.16. This technology eliminates the need for paper documents and substantially cuts down on costly errors. Customer orders and inquiries are transmitted from the customer's computer to the manufacturer's computer. The manufacturer's computer determines when new supplies are needed and can place orders by connecting with the supplier's computer.

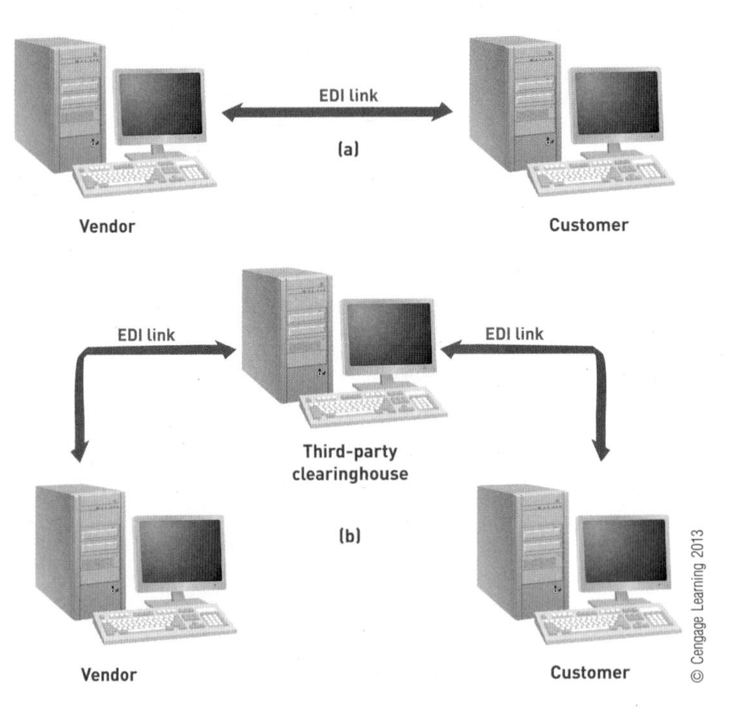

FIGURE 6.16

Two approaches to electronic data interchange

Many organizations now insist that their suppliers use EDI systems. Often, the vendor and customer (a) have a direct EDI connection or (b) the link is provided by a third-party clearinghouse that converts data and performs other services for the participants.

Audubon Metals recovers metal from automobile scrap including brass, copper, magnesium, and zinc. The company produces on the order of 120 million pounds of aluminum alloy per year. One of the purchasers of its recycled metal, Nissan, demands accurate and time EDI processing from all of its suppliers so that it can streamline the processing of purchase orders, invoices, and other documents.[66]

Electronic Funds Transfer

Electronic funds transfer (EFT) is a system of transferring money from one bank account directly to another without any paper money changing hands. It is used for both credit transfers, such as payroll payments, and for debit transfers, such as mortgage payments. The benefits of EFT include reduced administrative costs, increased efficiency, simplified bookkeeping, and greater security. One of the most widely used EFT programs is direct deposit, which deposits employee payroll checks directly into the designated bank accounts. The two primary components of EFT, wire transfer and automated clearing house, are summarized in Table 6.8.

TABLE **6.8** Comparison of ACH payments and wire transfers

	ACH Payments	Wire Transfers
When does payment clear?	Overnight	Immediately
Can payment be canceled?	Yes	No
Are sufficient funds guaranteed?	No	Yes
What is the approximate cost per transaction?	$0.25	$10–$40

The STAR Network is part of the payment services offered by First Data Corporation and provides EFT services to some 2 million retailers, their financial institutions, and customers nationwide.[67]

Unified Communications

Unified communications provides a simple and consistent user experience across all types of communications, such as instant messaging, fixed and mobile phone, e-mail, voice mail, and Web conferencing. The concept of *presence* (knowing where one's desired communication participants are and if they are available at this instant) is a key component of unified communications. The goal is to reduce the time required to make decisions and communicate results, thus greatly improving productivity.

All of the ways that unified communications can be implemented rely on fast, reliable communications networks. Typically, users have a device capable of supporting the various forms of communications (e.g., laptop with microphone and video camera or a smartphone) that is loaded with software supporting unified communications. The users' devices also connect to a server that keeps track of the presence of each user.

Quick Response Codes

Quick Response (QR) codes are a type of two-dimensional barcode that can be scanned by users with a smartphone camera. The camera must be equipped with the appropriate software to display text or connect to a wireless network and open a Web page in the smartphone's browser. The code consists of black modules arranged in a square pattern on a white background (see Figure 6.17). Air New Zealand was one of the first airlines to use QR codes to read boarding passes.[68] Sign posts with QR codes can be found in some parks and nature areas. When you scan the code, you are provided with information about the trail, flora and fauna of the location, and interesting spots along the trail.[69] At Findlay Market in Cincinnati, users can scan a QR code on a sign for squash at Daisy Mae's Market using their smartphone to pull up a short YouTube video on "How to Make Spaghetti Squash." In addition, the Cincinnati zoo uses the QR codes on exhibits, zoo maps, and membership booklets.[70]

FIGURE 6.17
Quick Response code

Global Positioning System Applications

The Global Positioning System (GPS) is a global navigation satellite system that uses two dozen satellites orbiting roughly 11,000 miles above the Earth. These satellites are used as reference points to calculate positions on Earth to an accuracy of a few yards or even less. GPS receivers have become as small as a cell phone and relatively inexpensive, making the technology readily accessible. The technology, originally developed for national defense and military applications, has migrated to consumer devices and is used in navigational and location tracking devices. GPS receivers are commonly found in automobiles, boats, planes, laptop computers, and cell phones.

To determine its position, a GPS receiver receives the signals from four GPS satellites and determines its exact distance from each satellite. (While the position of the receiver could be determined with just three measurements, a fourth distance measurement helps to adjust for any impreciseness in the other three.) This determination is done by very accurately measuring the time it takes for a signal to travel at the speed of light from the satellite to the receiver (distance = time × 186,000 miles/second). The GPS receiver then uses these distances to triangulate its precise location in terms of latitude, longitude, and altitude.

GPS tracking technology has become the standard by which fleet managers monitor the movement of their cars, trucks, and vehicles. GPS tracking quickly exposes inefficient routing practices, wasted time on the job, and speeding. Even small fleet operators can achieve significant benefits from the use of GPS tracking.

Computer-based navigation systems are also based on GPS technology. These systems come in all shapes and sizes and with varying capabilities—from PC-based systems installed in automobiles for guiding you across the country to handheld units you carry while hiking. All systems need a GPS antenna to receive satellite signals to pinpoint your location. On most of these systems, your location is superimposed on a map stored on CDs or a DVD. Portable systems can be moved from one car to another or carried in your backpack. Some systems come with dynamic rerouting capability, where the path recommended depends on weather and road conditions, which are continually transmitted to your car via a receiver connected to a satellite radio system.

All of the major U.S. wireless carriers offer cell phones that include an additional GPS antenna and internal GPS chip that can pinpoint a given location within a hundred yards or so. (The accuracy really depends on the accuracy of the timing mechanism of the cell phone. A tiny timing error can result in an error of 1,000 feet). Verizon customers can use Verizon VZ Navigator software, while AT&T, Sprint, T-Mobile, and Verizon Wireless customers can use TeleNav's GPS Navigator software.

Some employers use GPS-enabled phones to track their employees' locations. The locator phone provides GPS-enabled tracking devices that can provide its wearer's coordinates and dial emergency phone numbers. In

A Global Positioning System (GPS) can guide you along a route whether you are mobile or on foot.

addition to employers, parents and caregivers can track the phone's location by phone or online and can receive notification if it leaves a designated safe area. Many people suffering from dementia have wandered off but were quickly found because they were wearing such a device.[71]

The snow removal vehicles of Wayne County, Michigan (Detroit and surrounding area), are equipped with GPS systems to monitor and improve vehicle routing. The vehicle location data is also posted to a Web site that allows residents to view the location of vehicles superimposed on a map along with weather radar information from the National Weather Forecast Office.[72]

SUMMARY

Principle:

A telecommunications system consists of several fundamental components.

In a telecommunications system, the sending unit transmits a signal to a telecommunications device, which performs a number of functions such as converting the signal into a different form or from one type to another. The telecommunications device then sends the signal through a medium that carries the electronic signal. The signal is received by another telecommunications device that is connected to the receiving computer.

A networking protocol defines the set of rules that governs the exchange of information over a telecommunications channel to ensure fast, efficient, error-free communications and to enable hardware, software, and equipment manufacturers and service providers to build products that interoperate effectively. There is a myriad of telecommunications protocols, including international, national, and regional standards.

Communications among people can occur synchronously or asynchronously.

A transmission medium can be divided into one or more communications channels, each capable of carrying a message. Telecommunications channels can be classified as simplex, half-duplex, or full-duplex.

Channel bandwidth refers to the rate at which data is exchanged, usually expressed in bits per second.

A circuit-switching network uses a dedicated path for the duration of the communications. A packet-switching network does not employ a dedicated path for communications and breaks data into packets for transmission over the network.

The telecommunications media that physically connect data communications devices can be divided into two broad categories: guided transmission media and wireless media. Guided transmission media include twisted-pair wire, coaxial cable, and fiber-optic cable. Wireless transmission involves the broadcast of communications in one of three frequency ranges: radio, microwave, or infrared.

Wireless communications solutions for very short distances include Near Field Communications, Bluetooth, ultra wideband, infrared transmission, and ZigBee. Wi-Fi is a popular wireless communications solution for medium-range distances. Wireless communications solutions for long distances include satellite and terrestrial microwave transmission, wireless mesh, 3G and 4G cellular communications service, and WiMAX.

Principle:

Networks are an essential component of an organization's information technology infrastructure.

The geographic area covered by a network determines whether it is called a personal area network (PAN), local area network (LAN), metropolitan area network (MAN), or wide area network (WAN).

The electronic flow of data across international and global boundaries is often called transborder data flow.

When an organization needs to use two or more computer systems, it can follow one of three basic data-processing strategies: centralized (all processing at a single location, high degree of control), decentralized (multiple processors that do not communicate with one another), or distributed (multiple processors that communicate with each other). Distributed processing minimizes the consequences of a catastrophic event at one location while ensuring uninterrupted systems availability.

A client/server system is a network that connects a user's computer (a client) to one or more host computers (servers). A client is often a PC that requests services from the server, shares processing tasks with the server, and displays the results.

Numerous popular telecommunications devices include smartphones, modems, multiplexers, PBX systems, switches, bridges, routers, and gateways.

Telecommunications software performs important functions, such as error checking and message formatting. A network operating system controls the computer systems and devices on a network, allowing them to communicate with one another. Network-management software enables a manager to monitor the use of individual computers and shared hardware, scan for viruses, and ensure compliance with software licenses.

The interception of confidential information by unauthorized parties is a major concern for organizations. Encryption of data and the use of virtual private networks are two common solutions to this problem. Special measures must be taken to secure wireless networks.

Principle:

Network applications are essential to organizational success.

Telecommunications and networks are creating profound changes in business because they remove the barriers of time and distance.

The effective use of networks can turn a company into an agile, powerful, and creative organization, giving it a long-term competitive advantage. Networks let users share hardware, programs, and databases across the organization. They can transmit and receive information to improve organizational effectiveness and efficiency. They enable geographically separated workgroups to share documents and opinions, thus fostering teamwork, innovative ideas, and new business strategies.

The wide range of telecommunications and network applications includes cellular phone services, digital subscriber line (DSL), linking personal computer to mainframes, voice mail, voice-to-text services, reverse 911 service, home and small business networks, electronic document distribution, call centers, telecommuting, videoconferencing, electronic data interchange, electronic funds transfer, unified communications, Quick Response codes, and global positioning system applications.

CHAPTER 6: SELF-ASSESSMENT TEST

A telecommunications system consists of several fundamental components.

1. The Institute of Electrical and Electronics Engineers is a leading standards-setting organization whose standards are the basis for many telecommunications devices and services. True or False?
2. Telecommunications channel ________________ refers to the rate at which data is exchanged.
3. A ________________ defines a set of rules, algorithms, messages, and other mechanisms that enable software and hardware in networked devices to communicate effectively.
   a. network channel
   b. network bandwidth
   c. network protocol
   d. circuit switching
4. Two different means for routing data from the communications sender to the receiver are ________________.
   a. circuit switching and packet switching
   b. shielded and unshielded
   c. twisted and untwisted
   d. infrared and microwave
5. ________________ is a bundled set of communications services, including Internet, telephone, and high-definition TV, that operates over a total fiber-optic communications network.
6. ________________ is a very short-range wireless connectivity technology designed for consumer electronics, cell phones, and credit card that is used in the Speedpass contactless payment system.

Networks are an essential component of an organization's information technology infrastructure.

7. A(n) ________________ is a network that connects users and their computers in a geographic area that spans a campus or city.
8. The Ticketmaster reservation service is an example of an organization that uses decentralized processing to store information about events and record the purchases of seats. True or False?
9. A ________________ is a network device that serves as an entrance to another network.
   a. modem
   b. gateway
   c. bridge
   d. PBX

Network applications are essential to organizational success.

10. A(n) ________________ is a type of two-dimensional barcode that can be scanned by users with a smartphone camera to display text or open a Web page.
11. ________________ enables a manager on a networked personal computer to monitor the use of individual computers and shared hardware, scan for viruses, and ensure compliance with software licenses.
12. ________________ is the process of converting an original message into a form that can only be understood by the original receiver.
13. The use of telecommuting to enable employees to work away from the office is declining. True or False?
14. Connecting corporate computers among organizations is the idea behind ________________, which uses network systems and can process output from one system directly as input to other systems without human intervention.

CHAPTER 6: SELF-ASSESSMENT TEST ANSWERS

1. True
2. bandwidth
3. c
4. a
5. FiOS
6. Near Field Communications
7. metropolitan area network
8. False

9. b
10. Quick Response code
11. Network management software
12. Encryption
13. False
14. electronic data interchange or EDI

REVIEW QUESTIONS

1. Define the term telecommunications medium. Name three media types.
2. What is meant by Mbps and Gbps?
3. What is a telecommunications protocol? Give the names of two specific telecommunications protocols.
4. What are the names of the three primary frequency ranges used in wireless communications?
5. Briefly describe the differences between a circuit-switching and packet-switching network.
6. What is FiOS?
7. What is the difference between guided and unguided transmission media?
8. Is telecommuting aimed solely at enabling employees to work from home? Explain your answer.
9. What is the difference between Wi-Fi and WiMAX communications?
10. What is a wireless mesh network? What is one of the key advantages of such a network?
11. What distinguishes 3G from 4G communications? Identify three 4G communications standards.
12. Define the term smartphone. Identify four common smartphone platforms.
13. What are some of the advantages of videoconferencing? Describe a recent meeting you attended that could have been conducted using videoconferencing.
14. What is the difference between a network operating system and network-management software?
15. Identify two approaches to securing the transmission of confidential data.
16. What is a reverse 911 service? Give two examples of how such a system might be used.
17. What is electronic data interchange? List three types of data that a manufacturer and supplier might exchange electronically.

DISCUSSION QUESTIONS

1. What are the risks of transmitting data over an unsecured Wi-Fi network? What steps are necessary to secure a Wi-Fi Protected Access network?
2. Briefly discuss the expected growth in wireless data traffic. What are the implications of this growth?
3. Briefly discuss the terms LAN, PAN, MAN, and WAN and identify what distinguishes these types of networks.
4. What is the issue associated with transborder data flow? How might this issue limit the use of an organization's WAN?
5. Distinguish between centralized and distributed data processing.
6. Briefly describe how a cellular phone service works.
7. Briefly explain how the GPS global navigation satellite system determines the position of a transmitter.
8. Briefly discuss some of the changes that are affecting the operations of a telephone call center.
9. Imagine that you are responsible for signage on your campus to help visitors learn about your school and more easily find their way around. Discuss how you might employ quick response code technology in this effort.

PROBLEM-SOLVING EXERCISES

1. You have been hired as a telecommunications consultant to help an organization select and purchase 500 smartphones for its sales force to use. Develop a list of key business requirements that must be met so that salespeople can use a smartphone effectively. Identify three different possible candidate smartphones. Which one would you recommend and why? Summarize your selection process and support your recommendation in a one-page memo that includes a table comparing the candidate smartphones.
2. As a member of the Information Systems organization of a mid-sized firm, you are convinced that telecommuting represents an

excellent opportunity for the firm to both reduce expenses and improve employee morale. Use PowerPoint or similar software to make a convincing presentation to management for adopting such a program. Your presentation must identify benefits and potential issues that must be overcome to make such a program a success.

TEAM ACTIVITIES

1. Form a team to interview a manager employed in a telecommunications service provider. Identify the greatest challenges and opportunities that the company faces in the next two to three years.
2. Form a team to identify the public locations (such as an airport, public library, or café) in your area where wireless LAN connections are available. Visit two locations and write a brief paragraph discussing your experience at each location trying to connect to the Internet.

WEB EXERCISES

1. Do research on the Web to identify the latest 4G communications developments. In your opinion, which carrier's 4G network (AT&T, Sprint Nextel, T-Mobil, or Verizon) is the most widely deployed? Write a short report on what you found.
2. Go online to find out more about the rapid growth of wireless data traffic. Document the current growth trend. Identify actions taken by two different wireless carriers to meet this increase in demand. Briefly summarize your findings in a written report.

CAREER EXERCISES

1. Identify three recent college graduates who are working as business managers. Discuss with them their use of telecommunications technology and its impact on their effectiveness. Document your findings.
2. Do research to assess potential career opportunities in the telecommunications or networking industry. Consider resources such as the Bureau of Labor Statistics list of fastest growing positions, *Network World*, and *Computerworld*. Are there particular positions within these industries that offer good opportunities? What sort of background and education is required for candidates for these positions? You might be asked to summarize your findings for your class in a written or oral report.

CASE STUDIES

Case One

No More Wires in the Dorms at the University of Massachusetts Amherst

In the summer of 2011, the University of Massachusetts (UMass) Amherst unplugged its dorm rooms. No, the university didn't take out the electric wiring. Resident students don't have to study by candlelight, and they can still charge MP3 players and electric toothbrushes. However, the school removed the Ethernet cables that previously provided 10 Mbit/second service to dorm rooms. Instead, students now use an 802.11n Wi-Fi network that is several times as fast.

UMass Amherst is the largest of the five University of Massachusetts campuses. Established as an agricultural school in 1863, it occupies a 1,450-acre site about 90 miles west of Boston. It has about 22,000 undergraduate students, over half of whom live on campus, and about 6,000 graduate students.

In the 1980s, UMass Amherst installed the latest telecommunications technology for resident students: Ethernet, operating at 10 Mbps (10 million bits per second). This required running high-speed wires through all dormitory buildings and "tapping" off these wires to run wires to a wall jack for every student. Students could then connect their computers to those wall jacks to obtain faster data communication than the only other technology then in use: modems over telephone lines. The Ethernet system had the added advantage of not interfering with the phone line. Cell (mobile) phones were still rare at the time.

By the twenty-first century, this network was outmoded. The university, therefore, introduced wireless access in areas where students gathered to study, then in classrooms, libraries, and administrative buildings. A pilot program, testing wireless-only access in a single residence hall with 139 students, was well received. The university decided to replicate that system for all resident students, completely replacing the existing wired network.

A vendor selection process, conducted under strict state procurement guidelines, selected AP-125 access points from Aruba Networks as the wireless connection device of choice. These devices can operate in any of several frequency bands, as permitted by the 802.11n version of the Wi-Fi standard. UMass Amherst's IT staff chose to use the 5 MHz frequency band, which provides faster data transfer than the 2.5 MHz band but which doesn't radiate as far, thus requiring more access points to cover a given area. Dan Blanchard, senior advisor to the university's CIO, explains the selection, saying that the higher speed is "simply what our customers want."

The conversion to the new system cost $6 million in total: devices, wiring, and labor. However, that was less than it would have cost to upgrade the wired network to current wired technology.

The issues that UMass Amherst had to deal with weren't only technical. During the era of wired service, students had installed their own "rogue" access points to give them more flexibility in moving around their rooms and in providing computer access for visitors. These had to be identified and removed, while explaining to students that they were no longer necessary.

The system replaced only the wired access in dorms, not throughout the campus, because today's high-speed wired networks can provide higher speeds than wireless. Research labs and data centers use this type of high-speed service. However, the Wi-Fi network now covers 80 percent of the campus.

Discussion Questions

1. UMass Amherst says that upgrading the dorms from the older Ethernet network to 802.11n Wi-Fi cost less than upgrading to a newer wired network, but the system also gives students slower communications speeds than a new wired network (such as the one used in research labs and data centers) could give them. Do you agree with this decision? Why or why not?
2. Suppose you're planning to wire a fraternity house for computers. You don't have 12,000 students; you have about 15 who live there and another 15 or so who come for meals and to study. Discuss the pros and cons of wired versus wireless access. Which would you choose? (Assume the connection from the house to its ISP is the same in either case.)

Critical Thinking Questions

1. The $6 million cost of the network works out to $500 per student. Assume the new network will have a useful life of five years, making its cost $100 per student per year and that students will bear this cost. If you were a UMass Amherst student, would you prefer to have this new network or to use wired access and save $100 per year? Explain why.
2. A UMass Amherst official says that maintaining the earlier wired network was easier and that "We have to send [technicians] out more frequently than before, because there's no way of testing links remotely." Aruba Networks says that "trouble tickets have been ... reduced." Can they both be right? How?

SOURCES: Aruba Networks, "University of Massachusetts Amherst Gives 12,000 Residential Students Wireless Access with Aruba MOVE Architecture," *Yahoo! Finance*, *finance.yahoo.com/news/University-of-Massachusetts-bw-3062767980.html*, October 18, 2011; Aruba Networks, product information page for AP-124 and AP-125 access points, *www.arubanetworks.com/product/aruba-ap-124-ap-125-access-points*, accessed October 21, 2011; Hamblen, M., "One Big Wi-Fi Rollout: 12,000 College Dorm Residents," *Computerworld* *www.computerworld.com/s/article/9221083/One_big_Wi_Fi_rollout_12_000_college_dorm_residents*, October 21, 2011.

Case Two

JW Marriott Marquis Miami's Luxurious Communications Technology

When the JW Marriott Marquis Miami opened in November 2010, its developers, MDM Development Group and MetLife, wanted to differentiate it from other hotels through advanced technology. "Our goal for the JW Marriott Marquis Miami and Hotel Beaux Arts Miami [a separate section on the 39th floor] is to create hotels for the 21st century, giving guests the benefit of advanced technology in a luxury setting, in an urban environment," says hotel general manager Florencia Tabeni. "We wanted to provide every technology amenity available to travelers, whether they joined us for business or pleasure."

To accomplish that goal, the developers selected communications technology from Cisco Systems Inc. Among the technology capabilities they installed in their new hotel were the following features:

- An NBA-approved basketball arena, convertible to a tennis court or ballroom, has a giant video wall built from a 5 x 5 array of 52-inch TV screens. At a wedding reception, for example, the video wall can show videos of the ceremony earlier in the day or video streamed live from camcorders that support wireless transmission.
- In the lobby, a concierge greets guests from a Virtual Concierge unit with audio and video. Units can also be installed in conference rooms, providing face-to-face access to concierge services without making meeting participants visit the lobby.
- Touch-screen telephones in guest rooms can do more than just make phone calls. For example, they can display the breakfast menu for in-room dining.
- Telepresence video conferencing systems are installed in conference rooms, enabling people to interact face-to-face with others in similarly equipped rooms elsewhere, including other participating Marriott locations. Future plans include providing similar systems in certain suites so that business travelers can meet with others without renting a conference room or even leaving their own rooms.
- Seventy video surveillance cameras are monitored and controlled from a central location.
- Calls to the 911 emergency number are simultaneously routed to first responders and to appropriate hotel personnel.

All of this gear is connected via Gigabit Ethernet (one billion bits per second) links to a switch on every three floors. Each guest room has six connections: three phones, the TV set, a computer link, and the minibar (which senses when guests remove food and beverage items). The switches, in turn, are connected to each other and to the hotel data center via multiple 10 Gigabit Ethernet (ten billion bits per second) links. Wi-Fi access points are also located throughout the hotel.

The property developers engaged Modcomp, a system developer based in Deerfield Beach, Florida, to plan, design, and implement this system. Modcomp combined a dozen Cisco product lines into the hotel system. "In Cisco we found a single, trusted vendor for all of our video and collaboration needs: for business, entertainment, and physical safety," says Nicholas Corrochano, MDM's vice president of information technology. "The result is a stay that's memorable for all guests, and highly productive for business guests."

Discussion Questions

1. You are traveling to the Miami area with friends for spring break. You pool your funds to stay at the JW Marriott Marquis Miami. Which of the capabilities mentioned in this case would you find useful? Which would not be useful?
2. You are on a business trip to the Miami area with your boss to meet with potential customers. Your company books you into the JW Marriott Marquis Miami. Which of the capabilities mentioned in this case would you find useful? Which would not be useful?

Critical Thinking Questions

1. Modcomp acquired all the system components from one supplier. Using a single vendor can make it easier to connect system elements but may mean passing up superior technologies from other vendors in specific areas. (For example, a company that specializes in surveillance cameras might have better ones than Cisco does.) Discuss the pros and cons of a single-vendor approach versus a "best of breed" approach, where each component is selected individually as being the best of its kind.
2. At the end of the case, Nicholas Corrochano praises Cisco, which offers products and services to support all these capabilities. However, so do other vendors. Given that, do you feel that Corrochano's comments were appropriate?

SOURCES: JW Marriott Marquis Miami Web site, *www.marriott.com/hotels/travel/miamj-jw-marriott-marquis-miami*, accessed January 31, 2012; Cisco Systems, Inc., "Connected Hotels," *www.cisco.com/web/strategy/trec/trec.html*, accessed January 31, 2012; Cisco Systems, Inc., "JW Marriott Marquis Miami" (video), *www.cisco.com/en/US/prod/collateral/voicesw/ps6788/vcallcon/ps556/jw_marriott_marquis_miami_ps430_Products_Case_Study.html*, accessed January 31, 2012; Cisco Systems, Inc., "Hotel Transforms Guest Experience through Technology," *www.cisco.com/en/US/prod/collateral/voicesw/ps6788/vcallcon/ps556/case_study_c36-648251_ps430_Products_Case_Study.html*, downloaded January 31, 2012.

Questions for Web Case

See the Web site for this book to read about the Altitude Online case for this chapter. The following are questions concerning this Web case.

Altitude Online: Telecommunications and Networks

Discussion Questions

1. What telecommunications equipment is needed to fulfill Altitude Online's vision?
2. Why is it necessary to lease a line from a telecommunication company?

Critical Thinking Questions

1. What types of services will be provided over Altitude Online's network?
2. What considerations should Jon and his team take into account as they select telecommunications equipment?

NOTES

Sources for the opening vignette: Microsoft case study, "Microsoft Disaster Response," Computerworld Honors Awards, 2011, *www.eiseverywhere.com/file_uploads/1731e3ed9282e5b4b81db8572c9d5e4f_Microsoft_Corporation_-_Microsoft_Disaster_Response.pdf*, downloaded January 28, 2012; Microsoft Web site, "Microsoft Supports Relief Efforts in Haiti," (video), *www.microsoft.com/en-us/showcase/details.aspx?uuid=ed1d948f-5dfb-45f1-9c39-20050b7d752c*, August 19, 2010; Microsoft Citizenship Team, "How Technology Is Helping Distribute Food in Japan," *Microsoft Citizenship*, blog, *blogs.technet.com/b/microsoftupblog/archive/2011/03/18/how-technology-is-helping-distribute-food-in-japan.aspx*, March 18, 2011; Microsoft Web site, "Serving Communities: Disaster and Humanitarian Response," *www.microsoft.com/about/corporatecitizenship/en-us/serving-communities/disaster-and-humanitarian-response*, accessed January 30, 2012; NetHope Web site, *www.nethope.org*, accessed January 31, 2012; Olafsson, G., "Information and Communication Technology Usage in the 2010 Pakistan Floods," *NetHope*, blog, *blog.disasterexpert.org/2011/09/pakistan-floods-use-of-information-and.html*, September 9, 2011.

1. "West Yorkshire Fire & Rescue Service Adopts Unified Communications Solution, Increases Productivity 35 Percent," Microsoft Case Studies, *www.microsoft.com/sv-se/kundreferenser/Microsoft-Lync-Server-2010/West-Yorkshire-Fire-Rescue-Service/Fire-Service-Adopts-Unified-Communications-Solution-Increases-Productivity-35-Percent/4000011194*, October 6, 2011.
2. Alltucker, Ken, "Hospitals Turn to Telemedicine for Remote Care of Patients," *USA Today*, July 3, 2011.
3. Gullo, Chris, "75 Percent of Nurses Own a Smartphone or Tablet," *Mobilehealthnews.com*, *http://webcache.googleusercontent.com/search?hl=en&gbv=2&q=*

cache: 4pxeGA8PtL8J:http://mobihealthnews.com/14361/75-percent-of-nurses-own-smartphones-or-tablets/+what+percent+of+physicians+own+a+smartphone+%26+2011&ct=clnk, November 1, 2011.

4. "Amgen Announces Webcast of 2011 Fourth Quarter and Full Year Financial Results," *Fierce Biotech*, *www.fiercebiotech.com/press-releases/amgen-announces-webcast-2011-fourth-quarter-and-full-year-financial-results*, January 24, 2012.
5. "How Distance Education Works," Harvard Extension School, *www.extension.harvard.edu/distance-education/how-distance-education-works*, accessed February 3, 2012.
6. Verizon Web site, "FiOS: The Network Built to Power Your Family's Devices," *www22.verizon.com/home/aboutfios*, accessed December 5, 2011.
7. CSD Staff, "ExxonMobile Speedpass Customers Awarded Gas Savings," *Convenience Store Decisions*, *www.csdecisions.com/2011/05/02/exxonmobil-speedpass-customers-awarded-gas-savings*, May 2, 2011.
8. Rising, Malin, "RIM Wants into the Business," *USA Today*, November 8, 2011.
9. Pierce, David, "The 10 Best Bluetooth Headsets," *PC Magazine*, *www.pcmag.com/article2/0,2817,2369586,00.asp*, September 13, 2011.
10. Bluetooth Web site, "Moving Beyond Hands Free Audio in the Car with BMW," *www.bluetooth.com/Pages/Bluetooth-Home.aspx*, accessed December 5, 2011.
11. Merritt, Rick, "Nokia Tweaks Bluetooth for Indoor Navigation," *EETimes*, *www.eetimes.com/electronics-news/4230993/Nokia-tweaks-Bluetooth-for-indoor-navigation*, November 29, 2011.
12. Gelke, Hans, "Harnessing Ultra-Wideband for Medical Applications," *Medical Electronic Design www.medical-electronicsdesign.com/article/harnessing-ultra-wide-band-medical-applications*, accessed November 14, 2011.
13. Ibid.
14. "New Temp & Humidity Data Loggers with Infrared Communications," *Elsevier's Analytical Chemistry Journals*, blog, *http://elsevieranalyticalchemistry.blogspot.com/2011/05/new-temp-humidity-data-loggers-with.html*, May 11, 2011.
15. St. John, Jeff, "PG&E's Home Area Network Pilot: Silver Spring and Control4 on Board," *Green Tech Grid*, *http://cleantechvc.greentechmedia.com/articles/read/pges-home-area-network-pilot-silver-spring-and-control4-on-board*, November 29, 2011.
16. Security Technology Executive staff, "Access Control Case in Point: Wireless System Helps Eliminate Break-Ins," *www.securityinfowatch.com/access-control-case-point-wireless-system-helps-eliminate-break-ins*, July 14, 2011.
17. Hotel Chatter Annual Wi-Fi Report 2011, *Hotel Chatter*, at *www.hotelchatter.com/special/Best-WiFi-Hotels-2011*, accessed December 8, 2011.
18. Murphy, Samantha, "Laptop Wi-Fi Might Cause Male Fertility Problems," *Mashable Tech*, *http://mashable.com/2011/11/30/laptop-wi-fi-fertility*, November 20, 2011.
19. Iridium Web site, "Iridium Global Network, The Satellite Constellation," *www.iridium.com/About/IridiumGlobalNetwork/SatelliteConstellation.aspx*, accessed November 21. 2011.
20. Cubic Web site, "Industries Served, Cubic Global Tracking Solutions," *www.cubic.com/Solutions/Cubic-Global-Tracking-Solutions/Industries-Served*, accessed December 3, 2011.
21. Satellite press releases, "Chinese Petroleum Subsidiary Continues VSAT Rollout," *Digital Ship*, *http://webcache.googleusercontent.com/search?hl=en&gbv=2&q=cache:WqZAceyKNk4J:http://www.satprnews.com/2011/11/29/chinese-petroleum-subsidiary-continues-vsat-rollout/+%E2%80%9CChinese+Petroleum+Subsidiary+Continues+VSAT+Rollout%E2%80%9D&ct=clnk*, November 29, 2011.
22. Marlink Web site, "Sealink," *www.marlink.com/Product.aspx?m=13*, accessed December 8, 2011.
23. Vos, Esme, "Houston, Texas Deploys Wireless Mesh Network to Improve Public Safety," *MuniWireless*, *www.muniwireless.com/2011/09/20/houston-texas-deploys-wireless-mesh-network-for-public-safety*, September 20, 2011.
24. Goldman, David, "4G Won't Solve 3G's Problems," *CNN Money*, March 29, 2011.
25. Verizon Web site, "Verizon LTE Information Center," *www.verizonwireless.com/lte*, accessed December 15, 2011.
26. 4G Americas Web site, "LTE Advanced," *www.4gamericas.org/index.cfm?fuseaction=page§ionid=352*, accessed December 15, 2011.
27. Goldman, David, "4G Won't Solve 3G's Problems," *CNN Money*, March 29, 2011.
28. Murph, Darren, "IEEE Approves Next Generation WiMAX Standard, Invites You to Meet 802.16m," *engadget*, *www.engadget.com/2011/04/01/ieee-approves-next-generation-wimax-standard-invites-you-to-mee*, April 1, 2011.
29. Hettick, Larry, "Ericsson Predicts Tenfold Mobile Data Traffic Growth by 2016," *Network World*, *www.networkworld.com/newsletters/converg/2011/111411convergence1.html*, November 11, 2011.
30. Goldman, David, "4G Won't Solve 3G's Problems, *CNN Money*, March 29, 2011.
31. Ibid.
32. Ricknäs, Mikael, "Number of Wi-Fi Hotspots to Quadruple by 2015," *Computerworld*, November 9, 2011.
33. Higginbotham, Stacey, "Bluetooth to Battle for Personal Area Network Crown," *GigaOM*, *http://gigaom.com/2011/01/26/bluetooth-to-battle-for-personal-area-network-crown*, January 26, 2011.
34. Javvin Web site, "IEEE 802.14 and Bluetooth: WPAN Communications," *www.javvin.com/protocolBluetooth.html*, December 4, 2011.
35. "Deloitte Belgium Implements Juniper Networks Wireless LAN to Deliver Fast, Reliable and Secure Network," Juniper Networks press release, *www.juniper.net/us/en/company/press-center/press-releases/2011/pr_2011_09_22-03_01.html*, September 22, 2011.
36. Arsip blog, "Metropolitan Area Network," *http://localare.blogspot.com/2011/08/metropolitan-area-network-man.html*, accessed December 9, 2011.
37. D-Link Success Story, "Dallas Zoo," *www.dlink.com/tools/FrameContent.aspx?type=1&rid=104*, accessed December 28, 2011.

38. SmartGrid Web site, "CrescoAg and Smartfield Partner to Deliver In-Season Crop Monitoring Technologies to Growers," *http://smart-grid.tmcnet.com/news/2011/10/18/5864599.htm*, October 18, 2011.
39. Iatric Systems Success Story, "Mecosta County Medical Center," *www.iatric.com/download/public/Iatric_Meaningful_Use_Manager_Success_Mecosta_12-2011.pdf*, accessed December 28, 2011.
40. Epstein, Zach, "Available Apps across Major Mobile Platforms Approaching Million-App Milestone," *BGR*, *www.bgr.com/2011/12/05/available-apps-across-major-mobile-platforms-approach-million-app-milestone*, December 5, 2011.
41. Ibid.
42. Sheikh, Musarrat, "An Introduction to Wireless Modems," *http://goarticles.com/article/An-Introduction-to-Wireless-Modems/4133760*, February 6, 2011.
43. Smart Bridges Customer Success Stories, "Telco Provides Wireless E1 for Instant Connectivity," *www.smartbridges.com/customer-success-stories/telco-provides-instant-wireless-e1-service*, accessed January 5, 2012.
44. FiftyFlowers Web site, "Our Team," *www.fiftyflowers.com/custom/OurTeam.htm*, accessed December 12, 2011.
45. "FiftyFlowers.com Chooses Virtual PBX to Save Time, Money and Resources," FiftyFlowers press release, *http://finance.yahoo.com/news/FiftyFlowerscom-Chooses-iw-2733942885.html*, June 14, 2011.
46. FierceIPTV Web site, "Time Warner Cable Outage Traced to Level 3 Router Glitch," *www.fierceiptv.com/story/time-warner-cable-outage-traced-level-3-router-glitch/2011-11-07*, November 7, 2011.
47. Standard Chartered Web site, "Who We Are," *www.standardchartered.com/en/about-us/who-we-are/index.html*, accessed December 13, 2011.
48. Marketwire Web site, "Standard Chartered Bank Adopts Juniper Networks Infrastructure," *www.marketwire.com/printer_friendly?id=1575550*, October 19, 2011.
49. Wild Packets Web site, "Cornell University Weill Medical College," *www.wildpackets.com/about_us/cornell*, accessed January 6, 2012.
50. Diebold Web site, "Case Studies & Testimonials," *www.diebold.com/gssssps/case.htm*, accessed December 13, 2011.
51. 4-Traders Web site, "ZIX: 08/23/2011 7-Eleven Expands Secure Email Services with ZIX Corporation," *4-Traders*, *www.4-traders.com/ZIX-CORPORATION-11477/news/ZIX-08-23-2011-7-Eleven-Expands-Secure-Email-Services-with-Zix-Corporation-13767214*, August 24, 2011.
52. Werner Enterprises Web site, "About Werner," *www.werner.com/content/about*, accessed December 15, 2011.
53. AT&T Case Study, "Werner Enterprises Sees Green Light for Growth in China," *www.business.att.com/enterprise/resource_item/Insights/Case_Study/werner_enterprises*, accessed December 15, 2011.
54. Tatara Systems Web site, "What is a Femtocell?" *www.tatarasystems.com/contentmgr/showdetails.php/id/444*, accessed November 21, 2011.
55. FierceWireless Web site, "UPDATED: Sprint: We've Got 250,000 Femtocells on our Network," *www.fiercewireless.com/ctialive/story/sprint-weve-got-100000-femtocells-our- network/2011-03-23#ixzz1fIKlCYWp*, March 23, 2011.
56. "Lifelines," Vancouver Foundation, *www.vancouver-foundationstories.ca/story.php?recordID=284*, accessed February 3, 2012.
57. Futch, Michael, "Reverse 911 Systems Has Bladen County Wired for Safety," *Fay Observer*, *http://fayobserver.com/articles/2011/12/18/1142620*, December 18, 2011.
58. Roman, Jesse, "Couple Arrested after Reverse 911 Locates Theft Victim," *The Salem News*, November 11, 2011.
59. Bottomline Technologies Web site, "Success Stories: YMCA of Metropolitan Los Angeles," *www.bottomline.com/customer_success/industry.html*, accessed December 19, 2011.
60. Bajaj, Vikas, "A New Capital of Call Centers," *New York Times*, November 25, 2011.
61. Ibid.
62. "National Do Not Call Registry Data Book 2011," Federal Trade Commission, November 2011.
63. "FTC Settlements Put Debt Relief Operations Out of Business," Federal Trade Commission, *www.ftc.gov/opa/2011/05/amsdynamic.shtm*, May 26, 2011.
64. "The Growing WAH Movement," *Work Online Journal*, *http://workonlinejournal.com/the-growing-wah-movement*, July 2011.
65. Blue Jeans Network Web site, "Case Study: Facebook," *http://bluejeans.com/wp-content/uploads/2011/06/Facebook_CaseStudy_6_23_11.pdf*, accessed December 19, 2011.
66. Rivers, Jon, "Microsoft Dynamics GP Customer Overcomes Stall and Revs Engine with Dynamics EDI," *ERP Software Blog*, *www.erpsoftwareblog.com*, May 24, 2011.
67. "The Bankers Bank Names STAR Its Preferred Debit Network," First Data press release, *www.firstdata.com/en_us/about-first-data/media/press-releases/10_06_11_2.html*, October 6, 2011.
68. "Cemix First in Hardware Industry to Use QR Codes," *Tool Magazine*, *www.cemix.co.nz/cemix-use-qr-codes*, Oct/Nov Issues 2011.
69. Odell, Jolie, "Who's Really Scanning All Those QR Codes," *Mashable Tech*, *http://mashable.com/2011/03/04/qr-codes-infographic*, November 3, 2011.
70. Baverman, Laura, "Smarty Tags Takes Next Step", *Cincinnati Enquirer*, *www.ongo.com/v/880972/-1/8B56136933D2D2BE/smartytags-takes-next-step*, accessed November 21, 2011.
71. Meitrack Web site, "A Success Story—Meitrack's Trackers in Real Life Application," *www.meitrack.net/about-meitrack/news/283-a-success-story-meitracks-trackers-in-real-life-application*, November 9, 2011.
72. "Michigan Municipalities Allowing Residents to Track City Snow Plows Online," *Government Fleet*, *www.government-fleet.com/Channel/GPS-Telematics/News/Story/2011/12/Ann-Arbor-Allowing-Residents-to-Track-City-Snow-Plows-Online.aspx*, December 29, 2011.

7 The Internet, Web, Intranets, and Extranets

Principles	Learning Objectives
• The Internet provides a critical infrastructure for delivering and accessing information and services.	• Briefly describe how the Internet works, including methods for connecting to it and the role of Internet service providers.
• Originally developed as a document-management system, the World Wide Web has grown to become a primary source of news and information, an indispensable conduit for commerce, and a popular hub for social interaction, entertainment, and communication.	• Describe the World Wide Web and how it works. • Explain the use of markup languages, Web browsers, and Web servers. • Identify and briefly describe the process of creating software applications for the Web.
• The Internet and Web provide numerous resources for finding information, communicating and collaborating, socializing, conducting business and shopping, and being entertained.	• List and describe several sources of information on the Web. • Describe methods of finding information on the Web. • List and describe several forms of online communication, along with the benefits and drawbacks of each, in terms of convenience and effectiveness. • Explain Web 2.0 and provide examples of Web 2.0 sites. • List and describe sources of online media and entertainment. • Explain how Web resources are used to support shopping and travel. • Briefly name and describe two useful Internet utilities.
• Popular Internet and Web technologies have been applied to business networks in the form of intranets and extranets.	• Explain how intranets and extranets use Internet and Web technologies and describe how the two differ.

Information Systems in the Global Economy

PEPSICO, RUSSIA

Social Networking inside a Business

PepsiCo Russia, an organizational element of PepsiCo Europe, is the largest food and beverage business in Russia and the countries of the former Soviet Union. Much of PepsiCo Russia's growth has been through acquisitions since PepsiCo management recognizes the business potential of Russia and neighboring countries. For example, its $3.8 billion acquisition of 18,000-employee Will-Bimm-Dann, completed in 2011, is the largest foreign investment to date in the Russian food industry. PepsiCo had previously acquired a majority stake in JSC Lebedyanski, Russia's leading juice producer and a major baby food company, for $1.4 billion. Because of these and other acquisitions, PepsiCo Russia now consists of dozens of employee groups that have no shared history. Making them work as a unit creates a management challenge.

To deal with this challenge, PepsiCo Russia decided to create an employee intranet portal with a social focus. To learn what users expect of social sites, the project team looked at Facebook, VK (previously VKontakte, a Russian social site comparable to Facebook), LinkedIn, and Google+. "We studied the [user] experience of the world's best social networking sites, and combined it with the concepts of enterprise portals," says project leader Eugene Karpov.

PepsiCo Russia used Microsoft SharePoint collaboration software to integrate social networking concepts with its portal. The current SharePoint release has social networking features: users can find information from others with matching interests; they can bookmark, tag, and rate content, making it accessible to those on a team; and they get consolidated views of what other users are tracking or have written. Wikis also made their debut in SharePoint 2010. SharePoint 15, anticipated for 2013, is expected to extend the package's social networking capabilities.

The intranet uses a Quick Poll feature to gauge employee sentiment on topics of interest and to provide feedback on the portal itself. For example, the Quick Poll of March 15, 2012, asked, "What news does the portal lack?" It offered five choices: PepsiCo Russia business achievements, PepsiCo global news, news from plants and regions, corporate citizenship, and life at PepsiCo (sports, contests, and so on). Like most other content, this poll can be displayed in either Russian or English as a user prefers.

Since PepsiCo Russia's programmers had little experience with SharePoint, the company partnered with an experienced SharePoint developer to build its intranet site. This firm, WSS Consulting, carried out the project for a total cost of $100,000, not including SharePoint license fees. One reason WSS could do it for this relatively small sum is that it had already developed a general-purpose portal for SharePoint, WSS Portal. WSS Portal provided a ready-made basis for PepsiCo Russia's portal, thus reducing its cost, shortening its development time, and ensuring good performance: the server typically returns pages in less than 0.2 seconds.

Features of the new portal include social profiles, community membership, quick polls, document management, and location-specific information such as weather. In March 2012, 5,000 office employees had portal access, but access will be extended to thousands of factory-floor workers in the future. Employee reception of the portal has been positive, and there are many plans for future enhancements.

As you read this chapter, consider the following:

- How can the Internet help companies communicate with their customers, their suppliers, and internally?
- How can social networks help companies achieve their business objectives?

WHY LEARN ABOUT THE INTERNET, WEB, INTRANETS, AND EXTRANETS?

To say that the Internet has had a big impact on organizations of all types and sizes would be an understatement. Since the early 1990s, when the Internet was first used for commercial purposes, it has affected all aspects of business. Businesses use the Internet to sell and advertise their products and services, reaching out to new and existing customers. If you are undecided about a career, you can use the Internet to investigate career opportunities and salaries using sites, such as *www.monster.com* and *www.linkedin.com*. Most companies have Internet sites that list job opportunities, descriptions, qualifications, salaries, and benefits. If you have a job, you probably use the Internet daily to communicate with coworkers and your boss. People working in every field and at every level use the Internet in their work. Purchasing agents use the Internet to save millions of dollars in supplies every year. Travel and events-management agents use the Internet to find the best deals on travel and accommodations. Automotive engineers use the Internet to work with other engineers around the world developing designs and specifications for new automobiles and trucks. Property managers use the Internet to find the best prices and opportunities for commercial and residential real estate. Whatever your career, you will probably use the Internet daily. This chapter starts by exploring how the Internet works and then investigates the many exciting opportunities for using the Internet to help you achieve your goals.

The Internet is the world's largest computer network. Recall from Chapter 1 that the Internet is actually a collection of interconnected networks, all freely exchanging information using a common set of protocols. More than 800 million computers, or hosts, make up today's Internet, supporting nearly 2 billion users.[1] Those numbers are expected to continue growing.[2] Figure 7.1 shows the staggering growth of the Internet, as measured by the number of Internet host sites, or domain names. Domain names are discussed later in the chapter.

USE AND FUNCTIONING OF THE INTERNET

The Internet is truly international in scope, with users on every continent—including Antarctica. Although the United States has high Internet penetration among its population, it does not constitute the majority of people online. Of all the people using the Internet, citizens of Asian countries make up about 40 percent, Europeans about 20 percent, and North Americans about 15 percent.[3] China has by far the most Internet users at almost 400 million, which is more people than the total U.S. population but that is only about 30 percent of China's total population. Being connected to the Internet provides global economic opportunity to individuals, businesses, and countries.

Internet sites can have a profound impact on world politics.[4] These sites allow people to express their feelings about their governments. The beating of one person by the police force and his subsequent death may have caused a revolution in Egypt.[5] An executive at a large Internet company may have also been instrumental in the street demonstrations when he posted "Welcome back Egypt" using the Internet.[6]

Some countries try to control the content and services provided by search engines and social networking sites.[7] In one case, an Asian country tried to force a popular search engine company to locate its computer servers inside the country.[8] The search engine company decided to divert Internet traffic from the Asian country to other countries. According to the chairman of the company, "I'm very concerned that we will end up with an Internet per country."

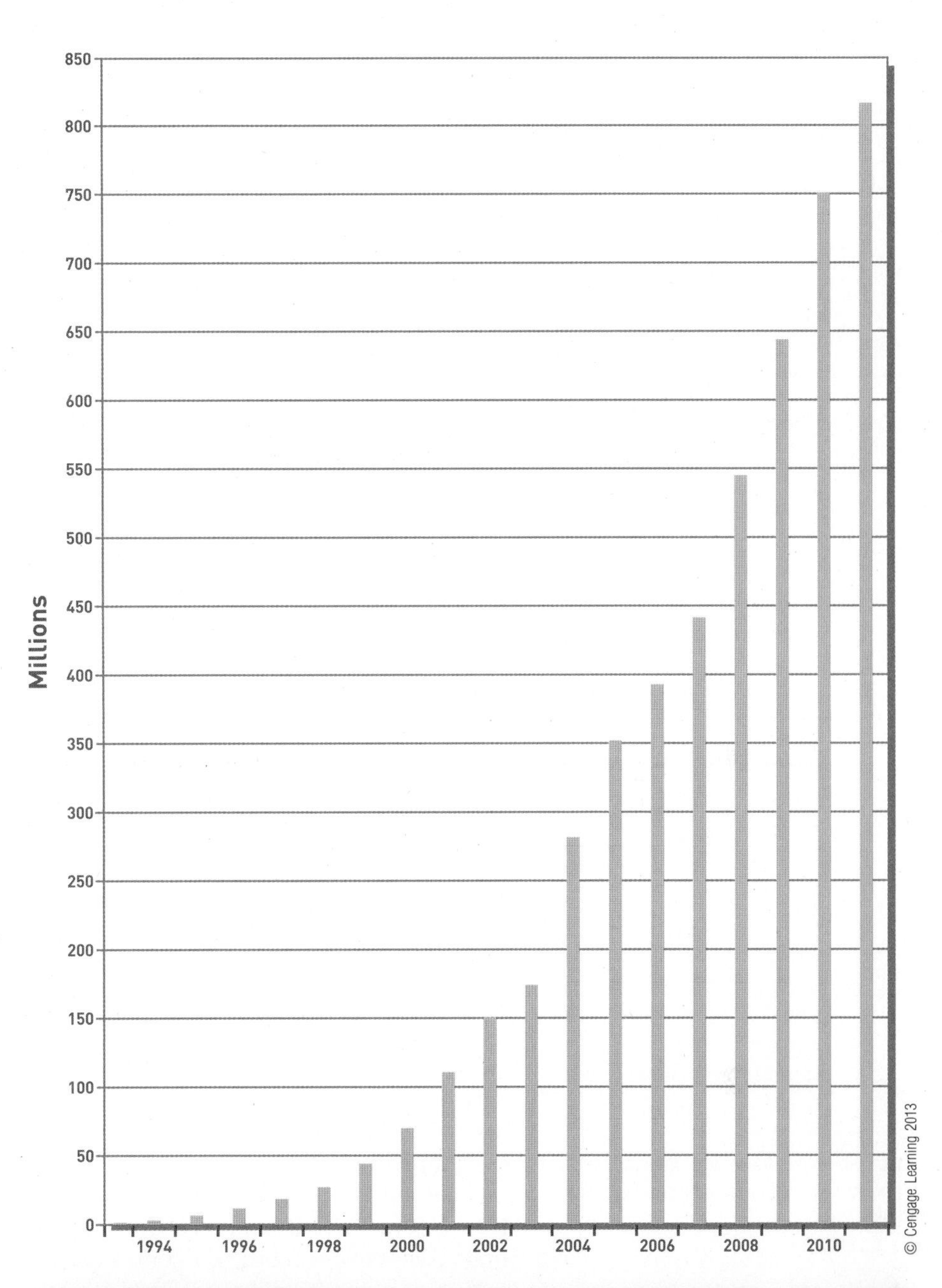

FIGURE **7.1**

Internet growth: Number of Internet domain names

(Source: Data from "The Internet Domain Survey," *www.isc.org*.)

More than 2 billion people use the Internet worldwide.

ARPANET: A project started by the U.S. Department of Defense (DoD) in 1969 as both an experiment in reliable networking and a means to link the DoD and military research contractors, including many universities doing military-funded research.

The ancestor of the Internet was the **ARPANET**, a project started by the U.S. Department of Defense (DoD) in 1969. The ARPANET was both an experiment in reliable networking and a means to link the DoD and military research contractors, including many universities doing military-funded research. (*ARPA* stands for the Advanced Research Projects Agency, the branch of the DoD in charge of awarding grant money. The agency is now known as DARPA—the added "D" is for *Defense*.[9]) The ARPANET was eventually broken into two networks: MILNET, which included all military sites, and a new, smaller ARPANET, which included all the nonmilitary sites. The two networks remained connected, however, through use of the **Internet Protocol (IP)**, which enables computers to route communications traffic from one network to another as needed. All the networks connected to the Internet use IP so that they can communicate.

Internet Protocol (IP): A communication standard that enables computers to route communications traffic from one network to another as needed.

To speed Internet access, a group of corporations and universities called the University Corporation for Advanced Internet Development (UCAID) is working on a faster alternative Internet. Called Internet2 (I2), Next Generation Internet (NGI), or Abilene, depending on the universities or corporations involved, the new Internet offers the potential of faster Internet speeds—up to 2 Gb per second or more.[10] The *National LambdaRail* (NLR) is a cross-country, high-speed (10 Gbps) fiber-optic network dedicated to research in high-speed networking applications.[11] The NLR provides a "unique national networking infrastructure" to advance networking research and next-generation network-based applications in science, engineering, and medicine. This new high-speed fiber-optic network will support the ever-increasing need of scientists to gather, transfer, and analyze massive amounts of scientific data.

How the Internet Works

In the early days of the Internet, the major telecommunications (telecom) companies around the world agreed to connect their networks so that users on all the networks could share information over the Internet. These large telecom companies are called *network service providers* (*NSPs*). Examples include Verizon, Sprint, British Telecom, and AT&T. The cables, routers, switching stations, communication towers, and satellites that make up these networks are the hardware over which Internet traffic flows. The combined hardware of these and other NSPs—the fiber-optic cables that span the globe over land and under sea—make up the Internet **backbone**.

backbone: One of the Internet's high-speed, long-distance communications links.

The Internet transmits data from one computer, called a *host*, to another (see Figure 7.2). If the receiving computer is on a network to which the first computer is directly connected, it can send the message directly.

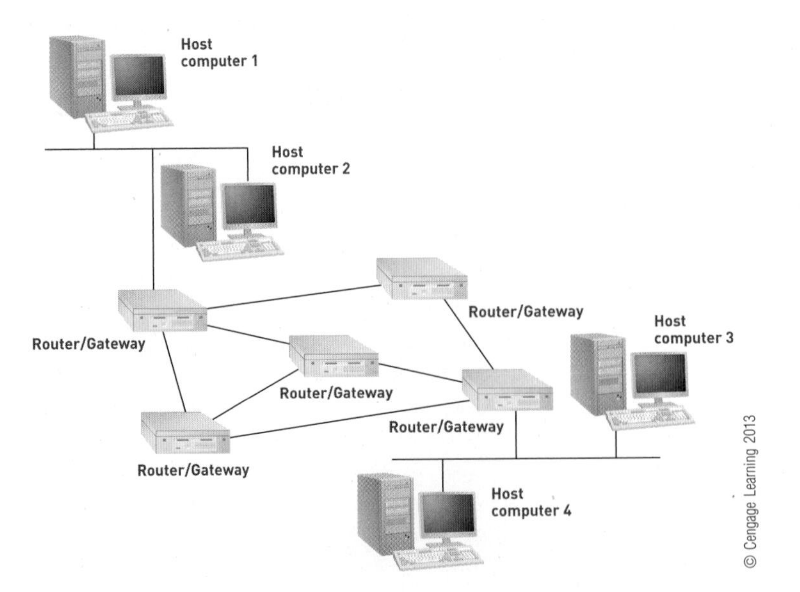

FIGURE 7.2
Routing messages over the Internet

If the receiving and sending computers are not directly connected to the same network, the sending computer relays the message to another computer that can forward it. The message is typically sent through one or more routers to reach its destination. It is not unusual for a message to pass through a dozen or more routers on its way from one part of the Internet to another. Thus, the Internet routes data packets over the network backbone from router to router to reach their destinations.

The various telecommunications networks that are linked to form the Internet work much the same way—they pass data around in chunks called *packets*, each of which carries the addresses of its sender and its receiver along with other technical information. The set of conventions used to pass packets from one host to another is the IP. Many other protocols are used in connection with IP. The best known is the **Transmission Control Protocol (TCP)**. Many people use "TCP/IP" as an abbreviation for the combination of TCP and IP used by most Internet applications. After a network following these standards links to the Internet's backbone, it becomes part of the worldwide Internet community.

Transmission Control Protocol (TCP): The widely used Transport-layer protocol that most Internet applications use with IP.

Each computer on the Internet has an assigned address, called its IP address, that identifies it on the Internet. An **IP address** is a 64-bit number that identifies a computer on the Internet. The 64-bit number is typically divided into four bytes and translated to decimal; for example, 69.32.133.79. The Internet will be upgraded to Internet Protocol version 6 (IPv6), which uses 128-bit addresses to provide for many more devices.[12] Because people prefer to work with words rather than numbers, a system called the Domain Name System (DNS) was created. Domain names such as *www.cengage.com* are mapped to IP addresses such as 69.32.133.79 using the DNS. If you type either *www.cengage.com* or 69.32.133.79 into your Web browser, you will access the same Web site. To make room for more Web addresses, efforts are underway to increase the number of available domain names.[13] Today, the .com domain has more than 90 million Web addresses, .net has more than 10 million addresses, and .org has about 9 million Web addresses.

IP address: A 64-bit number that identifies a computer on the Internet.

A **Uniform Resource Locator (URL)** is a Web address that specifies the exact location of a Web page using letters and words that map to an IP address and a location on the host. The URL gives those who provide information over the Internet a standard way to designate where Internet resources such as servers and documents are located. Consider the URL for Course Technology, *http://www.cengage.com/coursetechnology*.

Uniform Resource Locator (URL): A Web address that specifies the exact location of a Web page using letters and words that map to an IP address and a location on the host.

The "http" specifies the access method and tells your software to access a file using the Hypertext Transport Protocol. This is the primary method for interacting with the Internet. In many cases, you don't need to include http:// in a URL because it is the default protocol. The "www" part of the address sometimes, but not always, signifies that the address is associated with the World Wide Web service. The URL *www.cengage.com* is the domain name that identifies the Internet host site. The part of the address following the domain name—/coursetechnology—specifies an exact location on the host site.

Domain names must adhere to strict rules. They always have at least two parts, with each part separated by a dot (period). For some Internet addresses, the far right part of the domain name is the country code, such as au for Australia, ca for Canada, dk for Denmark, fr for France, de (Deutschland) for Germany, and jp for Japan. Many Internet addresses have a code denoting affiliation categories, such as com for business sites and edu for education sites. Table 7.1 contains a few popular categories. The far left part of the domain name identifies the host network or host provider, which might be the name of a university or business. Countries outside the United States use different top-level domain affiliations from the ones described in the table.

TABLE 7.1 U.S. top-level domain affiliations

Affiliation ID	Affiliation
com	Business sites
edu	Educational sites
gov	Government sites
net	Networking sites
org	Nonprofit organization sites

The Internet Corporation for Assigned Names and Numbers (ICANN) is responsible for managing IP addresses and Internet domain names.[14] One of its primary concerns is to make sure that each domain name represents only one individual or entity—the one that legally registers it. For example, if your teacher wanted to use *www.course.com* for a course Web site, he or she would discover that domain name has already been registered by Course Technology and is not available. ICANN uses companies called *accredited domain name registrars* to handle the business of registering domain names. For example, you can visit *www.namecheap.com*, an accredited registrar, to find out if a particular name has already been registered; If not, you can register the name for around $9 per year. Once you do so, ICANN will not allow anyone else to use that domain name as long as you pay the yearly fee.

Accessing the Internet

Although you can connect to the Internet in numerous ways, Internet access is not distributed evenly throughout the world or even throughout a city. Which access method you choose is determined by the size and capability of your organization or system, your budget, and the services available to you (see Figure 7.3).

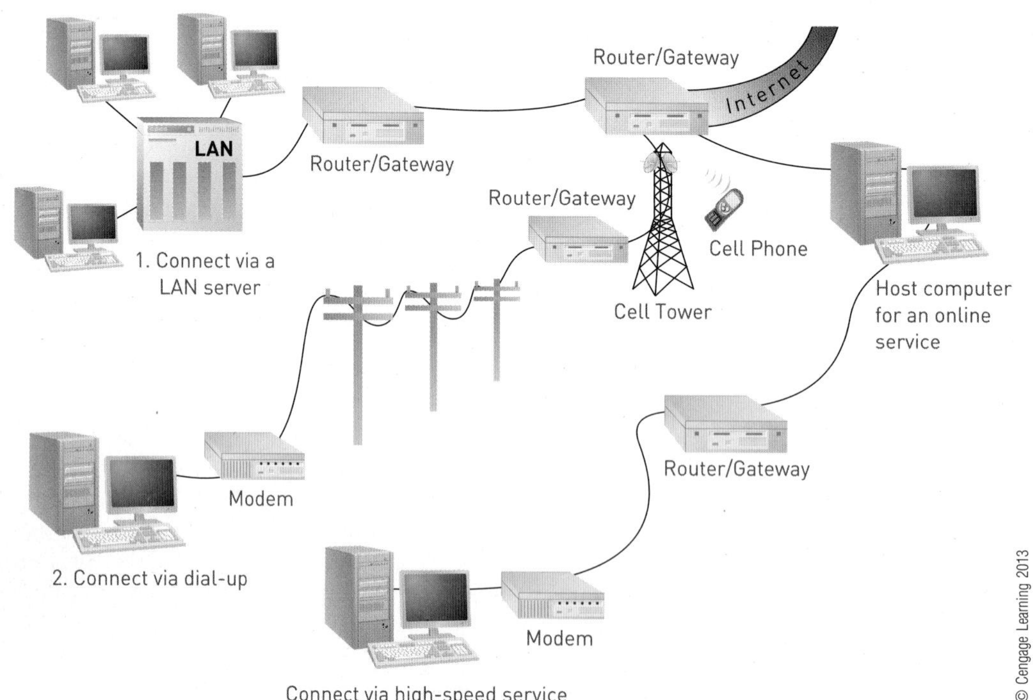

FIGURE 7.3

Several ways to access the Internet

Users can access the Internet in several ways, including using a LAN server, telephone lines, a high-speed service, or a wireless network.

Connecting via LAN Server

Businesses and organizations that manage a local area network (LAN) connect to the Internet via server. By connecting a server on the LAN to the Internet using a router, all users on the LAN are provided access to the Internet. Business LAN servers are typically connected to the Internet at very fast data rates, sometimes in the hundreds of Mbps. In addition, you can share the higher cost of this service among several dozen LAN users to allow a reasonable cost per user.

Connecting via Internet Service Providers

Internet service provider (ISP): Any organization that provides Internet access to people.

Companies and residences unable to connect directly to the Internet through a LAN server must access the Internet through an Internet service provider. An **Internet service provider (ISP)** is any organization that provides Internet access to people. Thousands of organizations serve as ISPs, ranging from universities that make the Internet available to students and faculty to small Internet businesses to major telecommunications giants such as AT&T and Comcast. To connect to the Internet through an ISP, you must have an account with the service provider (for which you usually pay) along with software (such as a browser) and devices (such as a computer or smartphone) that support a connection via TCP/IP.

Perhaps the least expensive but slowest connection provided by ISPs is a dial-up connection. A *dial-up Internet connection* uses a modem and standard phone line to "dial-up" and connect to the ISP server.

Several high-speed Internet services are available for home and business. They include cable modem connections from cable television companies, DSL connections from phone companies, and satellite connections from satellite television companies. These technologies were discussed in Chapter 6. High-speed services provide data transfer rates between 1 and 15 Mbps. Some businesses and universities use the very fast T1 or T3 lines to connect to the Internet.

In addition to connecting to the Internet through wired systems such as phone lines and television cables, wireless Internet over cellular and Wi-Fi networks has become common. Thousands of public Wi-Fi services are available in coffee shops, airports, hotels, and elsewhere, where Internet access is provided free, for an hourly rate, or for a monthly subscription fee.

Cell phone carriers also provide Internet access for handsets, notebooks, and tablets. New 4G mobile phone services rival wired high-speed connections enjoyed at home and work.[15] Sprint, Verizon, AT&T, and other popular carriers are working to bring 4G service to subscribers, beginning in large metropolitan areas.

When Apple introduced the iPhone, one of its slogans was the "Internet in your pocket." The iPhone proves the popularity of and the potential for Internet services over a handset. Many other smartphones followed hot on the heels of the iPhone, offering similar services on all of the cellular

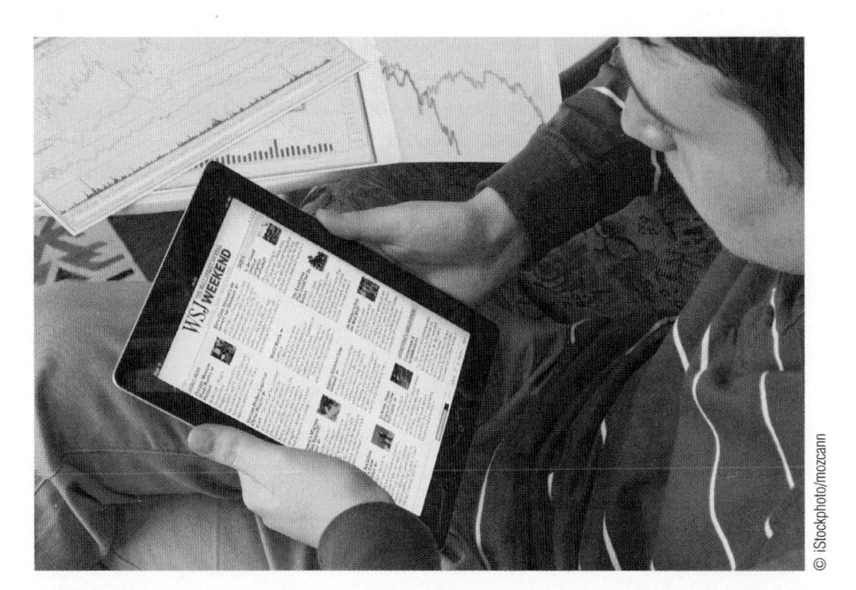

© iStockphoto/mozcann

Connecting wirelessly

The iPad connects to the Internet over cellular or Wi-Fi networks.

Bringing High-Speed Internet to Poland

Full business or personal participation in today's society is impossible without good Internet access. A country that wants to progress economically and provide opportunities for its citizens must ensure that Internet access is widely available.

The government of Poland understands this principle. The Polish government has taken several steps to make high-speed Internet connections widely available.

In 2009, Poland passed legislation to support and encourage the development of telecommunications networks, reducing regulatory barriers to new infrastructure and increasing competition. Many new projects were initiated after this legislation was adopted. For example, on January 14, 2011, the Łódzkie *voivodship* (province) in central Poland opened the bidding for operating a network that will give all of its nearly 3 million residents Internet access in their homes.

Consistent with this philosophy, in October 2009, the Polish government reached an agreement with the largest telecommunications carrier, Telekomunikacja Polska (TP), to deploy at least 1.2 million broadband lines by the end of 2012. By the end of 2010, TP had built more than 454,000 such lines, including more than 420,000 over 6 Mbps. The company also increased the percentage that will go into unprofitable rural areas from the initially planned 23 percent to 30 percent. In April 2011, TP started regulatory discussions about deploying 3 million Fiber to the Home (FTTH) lines, beginning in 2012.

Other legislation supports this aim. For example, new apartment buildings must have high-speed data connections from the building access point to each unit. Knowing that the most expensive part of broadband installation is already done for them, telecommunications companies are more likely to bring high-speed Internet connections to the building itself and to compete in offering that service to the building's residents.

The European Union (EU) is also contributing to Poland's Internet infrastructure. The EU gets a portion of each member country's value-added tax (VAT revenue) collections and allocates those funds to development projects throughout the EU. The Broadband Network of Eastern Poland will provide broadband Internet access to most residents of the five low-income *voivodships* in that part of the country. This endeavor is the largest EU-funded information technology project anywhere, with a total budget of PLN 1.4 billion (about U.S. $400 million) through 2015. The EU will supply about 85 percent of that budget, with Poland providing the remaining 15 percent.

As of the end of 2011, 62 percent of Poland's residents had high-speed Internet access. This percentage is above the worldwide average of 32.7 percent but below the EU average of 71.5 percent. Poland's current efforts should increase the percentage to the European average—which is, of course, a moving target as all EU countries are also moving forward. The International Telecommunications Union estimates that 90 percent of all Poles will have broadband Internet access at a fixed location by the end of 2015. Since many of the remaining 10 percent will have mobile access and since many of those who have no personal access will have convenient Internet access through their public libraries, high-speed Internet access in some form will be nearly universal. This access will be a key factor in Poland's future economic success.

Discussion Questions

1. The case states that "A country that wants to progress economically and provide opportunities for its citizens must ensure that Internet access is widely available." Do you agree? Why or why not?

2. The case states that the EU takes a fraction of each member's VAT revenue and returns that money to member countries for economic development projects. Less prosperous countries get back more than they put in; more prosperous ones get back less. The intent of this European Funds program is to reduce developmental differences among regions. Poland gets back considerably more than it contributes to this fund. About 30 percent of the funds Poland gets go into telecommunications infrastructure, with most of the rest going to transportation infrastructure. Do you think this is an appropriate split? How would you divide these funds? Consider other uses besides these two as well.

Critical Thinking Questions

1. As a university student, you almost certainly have high-speed Internet access on campus and most likely where you live as well. If you did not have such access, what difficulties would you encounter? How hard would it be to do the job you hope to have after graduation without good Internet access?
2. As of March 31, 2012, 7.5 million Poles out of a population of 38.5 million were Facebook members—slightly less than 20 percent. The corresponding fraction for the United States was just over 50 percent. How much of the difference do you think is due to lack of high-speed Internet access and how much to other factors? What do you think will be the effect of better high-speed Internet availability on Facebook membership in Poland?

SOURCES: Staff, "Poland—Impact of the Regulations on the Stimulation of the Infrastructural Investments and Actions Concerning Development of the Information Society," International Telecommunications Union, *www.itu.int/ITU-D/eur/NLP-BBI/CaseStudy/CaseStudy_POL_Impact_of_Regulation.html*, June 29, 2011 Staff, "Poland Broadband Overview," Point-Topic, *point-topic.com/content/operatorSource/profiles2/poland-broadband-overview.htm*, August 26, 2011 Staff, "Broadband Network in Eastern Poland," Polish Information and Foreign Investment Agency, *www.paiz.gov.pl/20111114/broadband_network_in_eastern_poland*, November 14, 2011; Internet World Stats, "Internet Usage in Europe," *www.internetworldstats.com/stats4.htm*, April 11, 2012.

networks. More recently, the iPhone brought video calling into vogue, while the iPad and other tablets provide anywhere, anytime access to all types of Internet services on a larger display.

Cloud Computing

cloud computing: A computing environment where software and storage are provided as an Internet service and are accessed with a Web browser.

Cloud computing refers to a computing environment where software and storage are provided as an Internet service and accessed with a Web browser (see Figure 7.4). Google and Yahoo!, for example, store the e-mail of many users, along with calendars, contacts, and to-do lists. Apple Computer has developed a service called iCloud to allow people to store their music and other documents on its Internet site.[16] Facebook provides social interaction and can store personal photos, as can Flickr and a dozen other photo sites. Pandora delivers music and Hulu and YouTube deliver movies. Google Docs, Zoho, 37signals, Flypaper, Adobe Buzzword, and others provide Web-delivered productivity and information management software. With its Office 365 software product, Microsoft is emphasizing cloud computing to a greater extent.[17] Office 365 competes with other online software suites.[18] Communications, contacts, photos, documents, music, and media will be available to you from any Internet-connected device with cloud computing.

FIGURE 7.4

Cloud computing

Cloud computing uses applications and resources delivered via the Web.

Cloud computing offers many advantages to businesses. By outsourcing business information systems to the cloud, a business saves on system design, installation, and maintenance. The New York Stock Exchange (NYSE), for example, is starting to offer cloud-computing applications that let customers pay for the services and data they use on Euronext, a European market for stocks, bonds, and other investments.[19]

Cloud computing can have several methods of deployment. Those that have been discussed thus far are considered public cloud services. *Public cloud* refers to service providers that offer their cloud-based services to the general public, whether that is an individual using Google calendar or a corporation using Salesforce.com. There is also a *private cloud* deployment where cloud technology is used within the confines of a private network. Corus Automotive Engineering Group found that private cloud technology from Univa UD was ideal for managing its massive parallel processing technical applications.[20] Some businesses like private cloud computing because it provides more control over infrastructure and security than public cloud computing. The chief executive officer of Salesforce.com helped Burberry, a popular maker of men's and women's clothing and accessories, with its online marketing and sales by tapping into its private cloud during a meeting at California's Half Moon Bay.[21] Within 10 minutes of sending questions to Salesforce staff using its private cloud called Chatter, responses and ideas flowed into the meeting at Half Moon Bay.[22] According to the CEO of Burberry, "We want to get [Burberry] globally connected after 155 years." Businesses may also elect to combine public cloud and private cloud services to create a *hybrid cloud*. In another version of cloud computing, known as a *community cloud*, several businesses share cloud-computing resources.

Today, no universally accepted standards allow data and programs to be shared between different cloud-computing applications.[23] Managing cloud-computing applications can be difficult when 10 or more cloud-computing providers are involved.[24] Increasingly, companies are developing formal cloud-computing strategies.[25] According to the director of solutions for NetApp, "Our customers are interested in a comprehensive cloud computing strategy that includes both private and public cloud models depending on their application needs." Today, a number of tools are available for managing cloud-computing applications, such as Hewlett-Packard's CloudSystem Matrix, TurnKey Linux Backup, and others.[26]

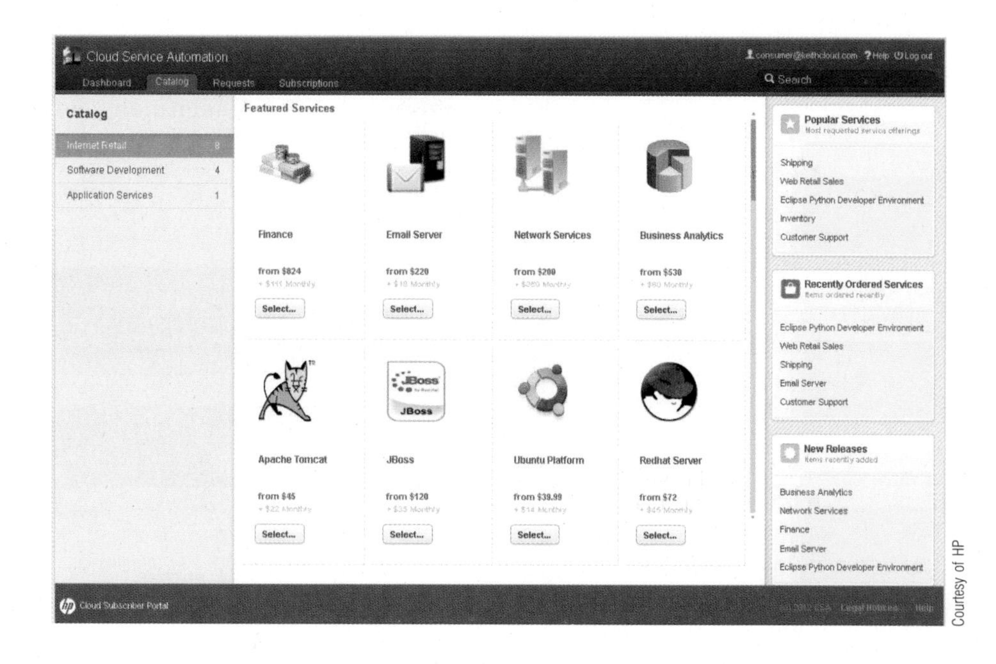

Hewlett-Packard CloudSystem Matrix is designed for managing cloud-computing applications.

Courtesy of HP

THE WORLD WIDE WEB

The World Wide Web was developed by Tim Berners-Lee at CERN, the European Organization for Nuclear Research in Geneva, Switzerland. He originally conceived of it as an internal document-management system. From this modest beginning, the Web has grown to become a primary source of news and information, an indispensable conduit for commerce, and a popular hub for social interaction, entertainment, and communication.

How the Web Works

While the terms Internet and Web are often used interchangeably, technically, the two are different technologies. The Internet is the infrastructure on which the Web exists. The Internet is made up of computers, network hardware such as routers and fiber optic cables, software, and the TCP/IP protocols. The **Web**, on the other hand, consists of server and client software, the Hypertext Transfer Protocol (HTTP), standards, and markup languages that combine to deliver information and services over the Internet.

Web: Server and client software, the Hypertext Transfer Protocol (HTTP), standards, and markup languages that combine to deliver information and services over the Internet.

The Web was designed to make information easy to find and organize. It connects billions of documents, which are now called Web pages, stored on millions of servers around the world. These are connected to each other using **hyperlinks**, specially denoted text or graphics on a Web page that, when clicked, open a new Web page containing related content. Using hyperlinks, users can jump between Web pages stored on various Web servers, creating the illusion of interacting with one big computer. Because of the vast amount of information available on the Web and the wide variety of media, the Web has become the most popular means of information access in the world today.

hyperlink: Highlighted text or graphics in a Web document that, when clicked, opens a new Web page containing related content.

In short, the Web is a hyperlink-based system that uses the client/server model. It organizes Internet resources throughout the world into a series of linked files, called pages, accessed and viewed using Web client software called a **Web browser**. Internet Explorer, Firefox, Chrome, and Safari are popular Web browsers (see Figure 7.5). A collection of pages on one particular topic, accessed under one Web domain, is called a Web site. The Web was

Web browser: Web client software, such as Internet Explorer, Firefox, Chrome, and Safari, are used to view Web pages.

originally designed to support formatted text and pictures on a page. It has evolved to support many more types of information and communication, including user interactivity, animation, and video. Web *plug-ins* help provide additional features to standard Web sites. Adobe Flash and Real Player are examples of Web plug-ins.

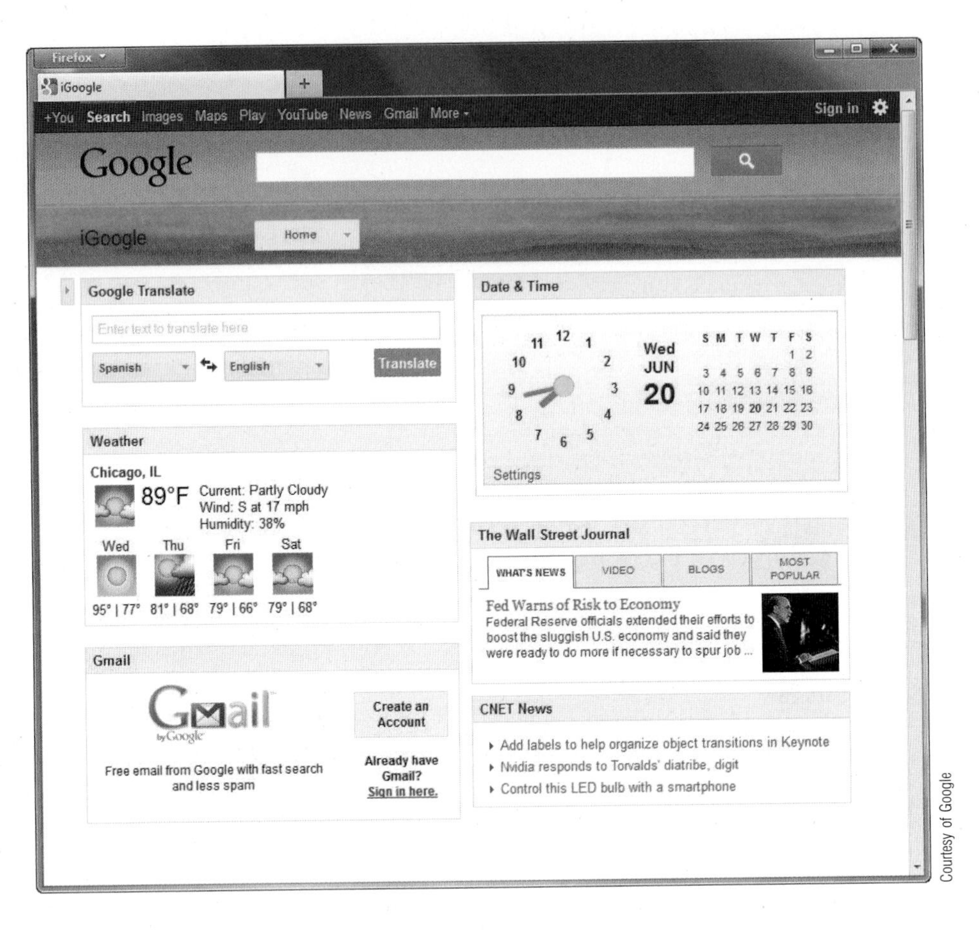

Courtesy of Google

FIGURE 7.5
Mozilla Firefox
Web browsers such as Firefox let you access Internet resources such as this customizable Web portal from Google.

Hypertext Markup Language (HTML): The standard page description language for Web pages.

HTML tags: Codes that tell the Web browser how to format text—as a heading, as a list, or as body text, for example—and whether images, sound, and other elements should be inserted.

Hypertext Markup Language (HTML) is the standard page description language for Web pages. HTML is defined by the World Wide Web Consortium (referred to as "W3C") and has developed through numerous revisions. It is currently in its fifth revision—HTML5.[27] HTML tells the browser how to display font characteristics, paragraph formatting, page layout, image placement, hyperlinks, and the content of a Web page. HTML uses **tags**, which are codes that tell the browser how to format the text or graphics, as a heading, list, or body text, for example. Web site creators mark up a page by placing HTML tags before and after one or more words. For example, to have the browser display a sentence as a heading, you place the <h1> tag at the start of the sentence and an </h1> tag at the end of the sentence. When you view this page in your browser, the sentence is displayed as a heading. HTML also provides tags to import objects stored in files, such as photos, pictures, audio, and movies, into a Web page. In short, a Web page is made up of three components: text, tags, and references to files. The text is your Web page content, the tags are codes that mark the way words will be displayed, and the references to files insert photos and media into the Web page at specific locations. All HTML tags are enclosed in a set of angle brackets (< and >), such as

<h2>. The closing tag has a forward slash in it, such as </b> for closing bold. Consider the following text and tags:

```
<html>
<head>
<title>Table of Contents</title>
<link href="style.css" rel="stylesheet" type="text/css" />
</head>
<body style="background-color:#333333">
<div id="container">
<p><img src="header.png" width="602" height="78" /></p>
<h1 align=center>Principles of Information Systems</h1>
<ol>
<li>An Overview</li>
<li>Information Technology Concepts</li>
<li>Business Information Systems</li>
<li>Systems Development</li>
<li>Information Systems in Business and Society</li>
</ol>
</div>
</body>
</html>
```

The <html> tag identifies this as an HTML document. HTML documents are divided into two parts: the <head> and the <body>. The <body> contains everything that is viewable in the Web browser window, and the <head> contains related information such as a <title> to place on the browser's title bar. The background color of the page is specified in the <body> tag using a hexadecimal code. The heading "Principles of Information Systems" is identified as the largest level 1 heading with the <h1> tag, typically a 16-18 point font, centered on the page. The <ol> tag indicates an ordered list, and the <li> tags indicate list items. The resulting Web page is shown in Figure 7.6.

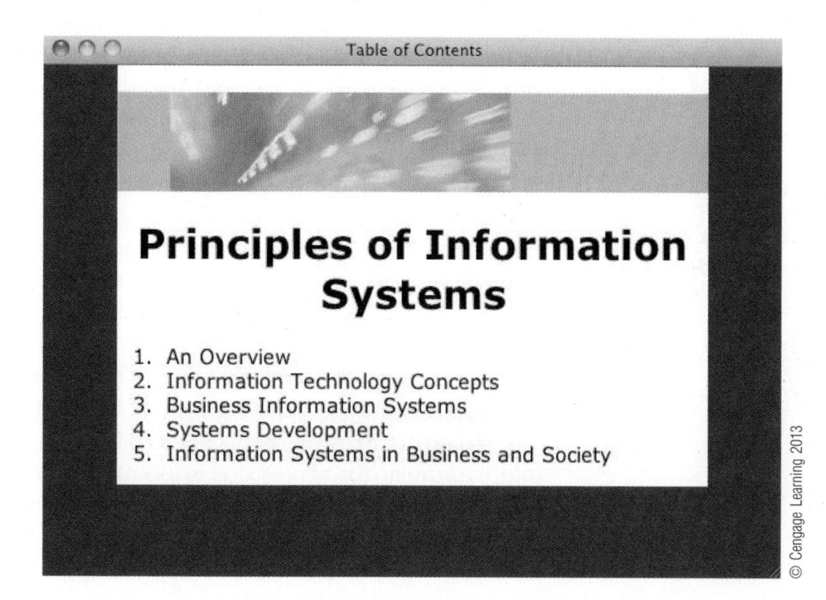

FIGURE **7.6**

HTML code interpreted by a browser

The example HTML code as interpreted by the Firefox Web browser on a Mac.

Cascading Style Sheets (CSS): A markup language for defining the visual design of a Web page or group of pages.

HTML works hand in hand with another markup language called CSS. CSS, which stands for **Cascading Style Sheets**, has become a popular tool for designing groups of Web pages.[28] CSS uses special HTML tags to globally define font characteristics for a variety of page elements as well as how those elements are laid out on the Web page. Rather than having to specify a font for each occurrence of an element throughout a document, formatting can be specified once and applied to all occurrences. CSS styles are often defined in a separate file and then can be applied to many pages on a Web site. In the previous example code, you may have noticed the <link> tag that refers to an external style sheet file, style.css.

Extensible Markup Language (XML): The markup language designed to transport and store data on the Web.

Extensible Markup Language (XML) is a markup language designed to transport and store data on the Web. Rather than using predefined tags like HTML, XML allows the coder to create custom tags that define data. For example, the following XML code identifies the components of a book:

```
<book>
<chapter>Hardware</chapter>
<topic>Input Devices</topic>
<topic>Processing and Storage Devices</topic>
<topic>Output Devices</topic>
</book>
```

XML is extremely useful for organizing Web content and making data easy to find. Many Web sites use CSS to define the design and layout of Web pages, XML to define the content, and HTML to join the content (XML) with the design (CSS) (see Figure 7.7). This modular approach to Web design allows you to change the visual design without affecting the content and to change the content without affecting the visual design.

Web Programming Languages

Many of the services offered on the Web are delivered through the use of programs and scripts. A Web program may be something as simple as a menu that expands when you click it or as complicated as a full-blown spreadsheet application. Web applications may run on the Web server, delivering the results of the processing to the user, or they may run directly on the client, the user's PC. These two categories are commonly referred to as server-side and client-side software.

Java: An object-oriented programming language from Sun Microsystems based on the C++ programming language, which allows applets to be embedded within an HTML document.

JavaScript is a popular programming language for client-side applications. Using JavaScript, you can create interactive Web pages that respond to user actions. JavaScript can be used to validate data entry in a Web form, to display photos in a slideshow style, to embed simple computer games in a Web page, and to provide a currency conversion calculator. **Java** is an object-oriented programming language from Sun Microsystems based on the C++ programming language, which allows small programs, called *applets*, to be embedded within an HTML document. When the user clicks the appropriate part of an HTML page to retrieve an applet from a Web server, the applet is downloaded onto the client workstation where it begins executing. Unlike other programs, Java software can run on any type of computer. It can be used to develop client-side or server-side applications. Programmers use Java to make Web pages come alive, adding splashy graphics, animation, and real-time updates.

Hypertext Preprocessor, or *PHP*, is an open-source programming language that is popular for server-side application development. Unlike some other Web programming languages, PHP is easy to use because its code, or instructions, can be embedded directly into HTML code. PHP can be used with a variety of database management systems, such as MySQL, DB2, Oracle, Informix, and many others. PHP's flexibility, power, and ease of use make it popular with many Web developers. Perl is another popular server-side programming language.

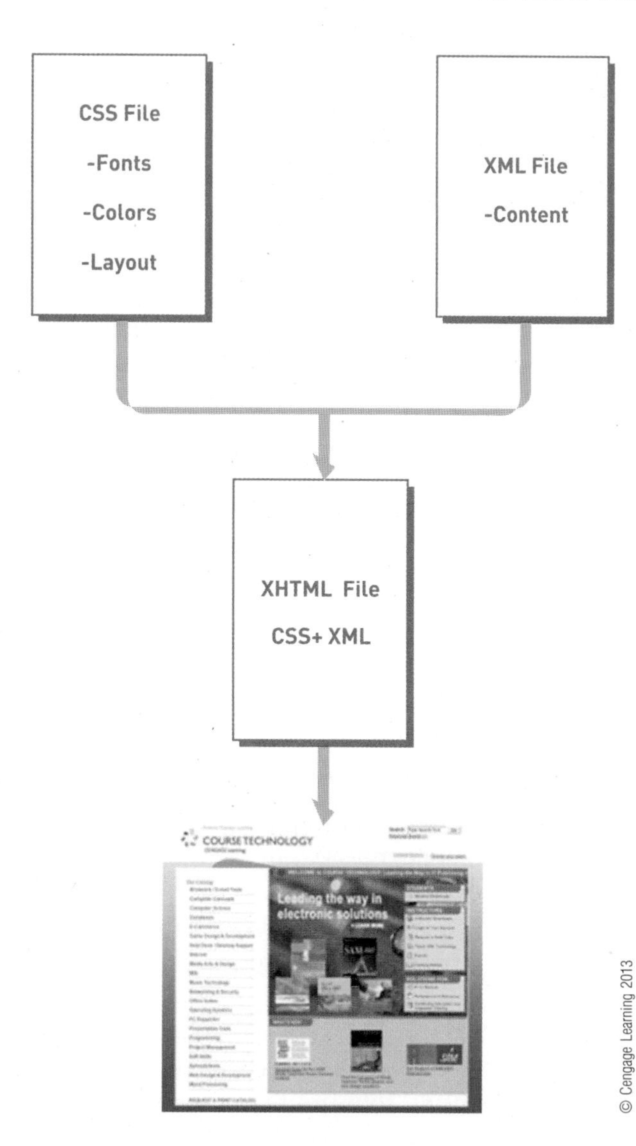

FIGURE 7.7

XML, CSS, and HTML

Today's Web sites are created using XML to define content, CSS to define the visual style, and HTML to put it all together.

Adobe Flash and *Microsoft Silverlight* provide development environments for creating rich Web animation and interactive media. Both Flash and Silverlight require the user to install a browser plug-in to run. Flash became so common that popular browsers included it as a standard feature. The introduction of HTML5 in 2010 provided Web developers the ability to create interactive Web content and media natively in HTML without the need for Flash or Silverlight. A number of technology companies, led by Apple, are moving away from Flash to HTML5 (see Figure 7.8).

HTML5 Video Events and API

This page demonstrates the new HTML5 video element, its media API, and the media events. Play, pause, and seek in the entire video, change the volume, mute, change the playback rate (including going into negative values). See the effect on the video and on the underlying events and properties.

FIGURE 7.8

HTML5 video

HTML5 supports interactive media including video and audio.

Web Services

Web services consist of standards and tools that streamline and simplify communication among Web sites and that promise to revolutionize the way we develop and use the Web for business and personal purposes. Internet companies, including Amazon, eBay, and Google, are now using Web services. Amazon, for example, has developed Amazon Web Services (AWS) to make the contents of its huge online catalog available to other Web sites or software applications.[29]

The key to Web services is XML. Just as HTML was developed as a standard for formatting Web content into Web pages, XML is used within a Web page to describe and transfer data between Web service applications. It is easy to read and has wide industry support. In addition to XML, three other components are used in Web service applications:

1. SOAP (Simple Object Access Protocol) is a specification that defines the XML format for messages. It allows businesses, their suppliers, and their customers to communicate with each other. It provides a set of rules that makes information and data easier to move over the Internet.
2. WSDL (Web Services Description Language) provides a way for a Web service application to describe its interfaces in enough detail to allow a user to build a client application to talk to it. In other words, WSDL allows one software component to connect to and work with another software component on the Internet.
3. UDDI (Universal Discovery Description and Integration) is used to register Web service applications with an Internet directory so that potential users can easily find them and carry out transactions over the Web.

Developing Web Content and Applications

Popular tools for creating Web pages and managing Web sites include Adobe Dreamweaver, Microsoft Expression Web, and Nvu (see Figure 7.9). Today's Web development applications allow developers to create Web sites using software that resembles a word processor. The software includes features that allow the developer to work directly with the HTML code or to use auto-generated code. Web development software also helps the designer keep track of all files in a Web site and the hyperlinks that connect them.

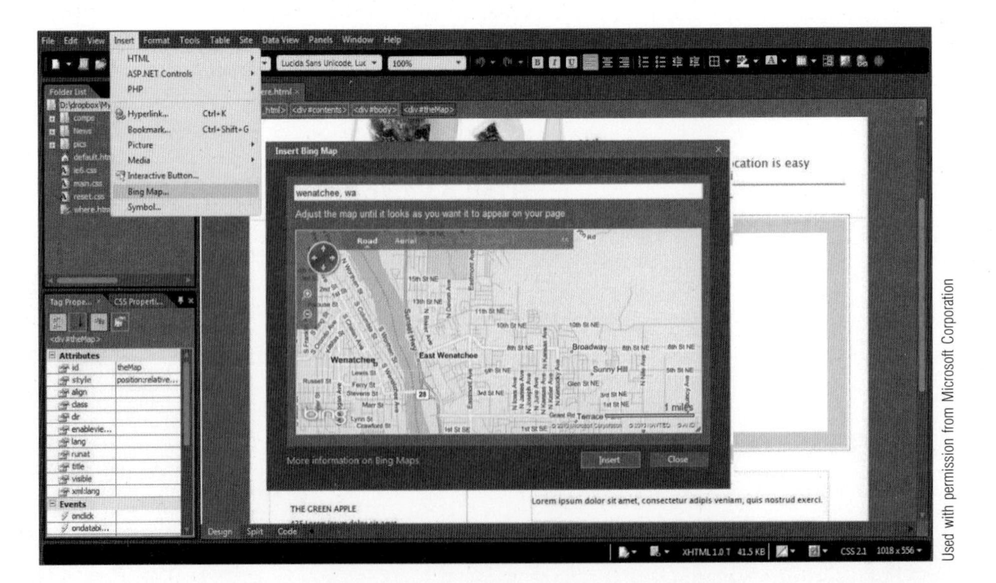

FIGURE 7.9
Creating Web pages

Microsoft Expression Web makes Web design nearly as easy as using a word processor.

Web application framework: Web development software that provides the foundational code—or framework—for a professional interactive Web site, allowing developers to customize the code to specific needs.

Web application frameworks have arisen to simplify Web development by providing the foundational code—or framework—for a professional interactive Web site, allowing developers to customize the code to specific needs. They include popular development software, such as Drupal and Joomla!, and range in complexity from WordPress, which allows nonprogrammers to create Web sites, to Ruby on Rails, which requires significant experience with programming. Web application frameworks that support full enterprise-level needs are referred to as online content management systems or frameworks. Most frameworks use a database to store and deliver Web content. The federal government used the Drupal framework to develop the *www.whitehouse.gov* Web site (see Figure 7.10).[30]

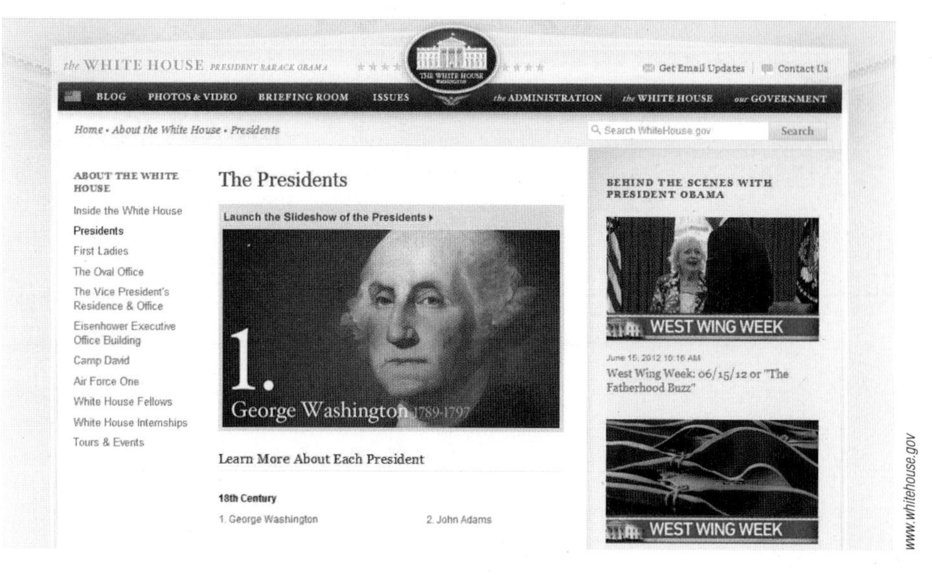

FIGURE **7.10**

Web application framework

Whitehouse.gov was created with the Drupal Web application framework.

www.whitehouse.gov

Web sites are typically developed on personal computers and then uploaded to a Web server. Although a business may manage its own Web server, the job is often outsourced to a Web-hosting company. Web hosts maintain Web servers, storage systems, and backup systems, and they provide Web development software and frameworks, Web analytics tools, and e-commerce software when required. A Web host can charge $15 or more per month, depending on the services delivered. Some Web-hosting sites also include domain name registration and Web site design services.

Many products make it easy to develop Web content and interconnect Web services, as discussed in the next section. Microsoft, for example, provides a development and Web services platform called .NET, which allows developers to use various programming languages to create and run programs, including those for the Web. The .NET platform also includes a rich library of programming code to help build XML Web applications. Other popular Web development platforms include JavaServer Pages, Microsoft ASP, and Adobe ColdFusion.

INTERNET AND WEB APPLICATIONS

The types of Internet and Web applications available are vast and ever expanding. Individuals and organizations around the world rely on Internet and Web applications. Using the Internet, entrepreneurs can start online companies and thrive. Graduates of the University of Pennsylvania's Wharton School, for example, started an Internet prescription eyeglass company.[31] The prescription eyeglasses can sell for less than $100. Internet companies such as *www.frelancer.com* and *www.livework.com* can help entrepreneurs prosper on the Internet. A former Apple Computer lawyer quit his job and started LawPivot to help entrepreneurs and startup Internet companies get legal advice and find venture capital.[32] The Internet also allows customers to determine the best delivery of products and services, depending on their speed, cost, and convenience requirements.[33]

Businesses could not survive in today's competitive environment without the use of the Internet. Ratiophram Canada, a pharmaceutical company, used the Internet to help solve a drug distribution problem in which demand for its generic drugs varied considerably.[34] Ratiophram employees used the Internet to share information and collaborate about the varying demand. As a result, the percentage of orders filled on time went from below 90 percent to above 95 percent. The popular enterprise resource planning (ERP) company SAP is teaming up with Google to mashup, or integrate, enterprise data from SAP with Internet geographic data from Google.[35] The result will be creative graphical reports, such as sales by region, loan defaults by neighborhood, and similar reports placed over Google maps.

For businesses of all sizes, lackluster Internet sites mean fewer people visiting the sites, which usually translates into lower sales and profits.[36] Internet consultants are often hired to make Internet sites more interesting and fun to visit. The value of some Internet companies that go public by offering their stock to investors through initial public offerings (IPOs) has soared in some cases. In one case, an IPO for a popular Internet company gained about 79 percent in one day.[37] Of course, this one-day return on investment is not guaranteed or usual.

Today, many organizations are concerned about the profitability of their Internet business model.[38] In the early days of the Internet, many media companies placed free content on their Web sites. Now, media companies and others are investigating ways to generate revenues from their Web sites or removing content from their Web sites that could compete with traditional newspapers, magazines, or TV content and hurt profitability in their traditional businesses.[39] *The New York Times*, for example, is considering charging for some of its online content while offering other content free as usual.[40] Not all media companies, however, have been profitable on the Internet. *Slate*, an online magazine owned by the *Washington Post*, had to lay off a number of key employees because of a lack of revenues and profits.[41]

Internet advertising has been an important revenue source for many organizations. Internet companies, however, have to be careful about what business or advertising they accept. A popular Internet search company, for example, paid a $500 million fine to the U.S. Justice Department for accepting advertising that illegally promoted drugs from Canadian Internet drug companies.[42] Social Web sites, such as Facebook, represent a large percentage of all Internet advertising.[43] Without question, social Web sites like Facebook and newer Web approaches have exploded in popularity and importance.

Web 2.0 and the Social Web

Over the years, the Web has evolved from a one-directional resource where users only obtain information to a two-directional resource where users obtain and contribute information. Consider Web sites such as YouTube, Wikipedia, and Facebook as examples. The Web has also grown in power to support full-blown software applications such as Google Docs and is becoming a computing platform itself. These two major trends in how the Web is used and perceived have created dramatic changes in how people, businesses, and organizations use the Web, creating a paradigm shift to **Web 2.0**.

Web 2.0: The Web as a computing platform that supports software applications and the sharing of information among users.

The Social Web

The original Web—Web 1.0—provided a platform for technology-savvy developers and the businesses and organizations that hired them to publish information for the general public to view. Web sites such as YouTube and Flickr allow users to share video and photos with other people, groups, and the world. Microblogging sites such as Twitter allow people to post thoughts and ideas throughout the day for friends to read (see Figure 7.11).

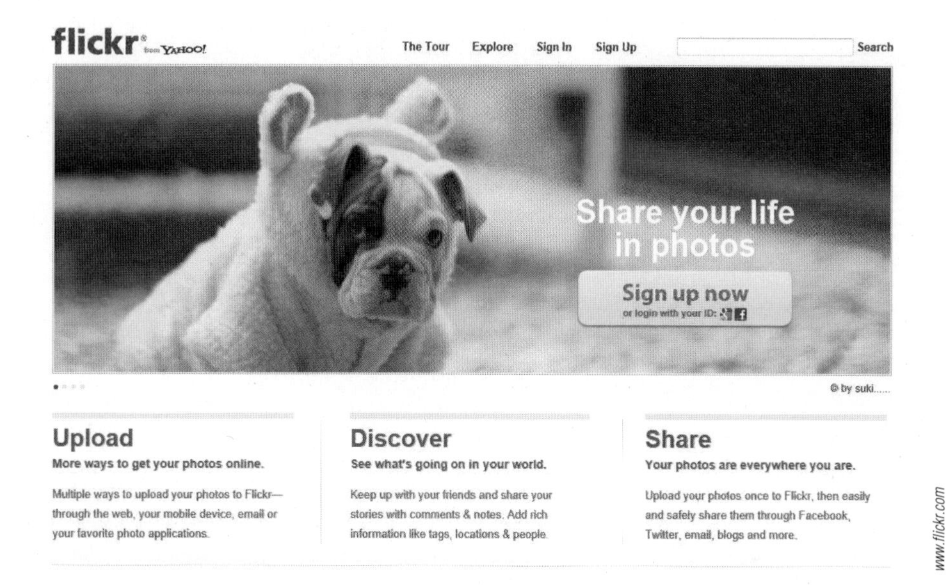

FIGURE **7.11**

Flickr

Flickr allows users to share photos with other people around the world.

Social networking Web sites provide Web-based tools for users to share information about themselves with people on the Web and to find, meet, and converse with other members. Some of these characteristics can be seen in *The Social Network*, a popular movie about the start and growth of Facebook. Google is experimenting with Google+, a social networking site that could compete with social networks like Facebook.[44] Some social networking sites are using celebrities to promote and generate interest in their sites.[45] LinkedIn is unique in that it is designed for professional use to assist its members with creating and maintaining valuable professional connections.[46] Ning provides tools for Web users to create their own social networks dedicated to a topic or interest.[47]

Social networking sites provide members with a personal Web page and allow them to post photos and information about themselves. Special interest groups can be created and joined as well. Social networking Web sites have also been used to help encourage medical research on unusual diseases.[48] In one case, a young woman was diagnosed with spontaneous coronary artery disease (SCAD) after several doctor visits. When she posted her situation on the Internet and connected with other people with the same rare disease, medical researchers at the Mayo Clinic got interested in the disease and started conducting a pilot study of the disease by investigating message boards on the Internet from people around the world. Social networking sites have also been used to help people find important health information and even locate organ donors. Facebook, for example, allows people to sign up as organ donors.[49] More than 100,000 people in the United States may be waiting for organs at any given time.

Social networks have become very popular for finding old friends, staying in touch with current friends, and making new friends. Besides their personal value, these networks provide a wealth of consumer information and opportunity to businesses as well.[50] Some businesses are including social networking features in their workplaces. The use of social media in business is called Enterprise 2.0. Enterprise 2.0 applications, such as Salesforce's Chatter, bring Facebook-like interaction to the workplace. Employees post profiles, making it easy for their colleagues to find those with knowledge that is useful to the work environment.

Social networking sites have helped some of their founders and become wealthy. An acquisition of a photo-sharing company made its founders multimillionaires.[51] Facebook's initial public offering (IPO) made its founder a billionaire. Not everyone is happy with social networking sites, however. Employers might use social networking sites to get personal information about you. Some people worry that their privacy will be invaded or their personal information used without their knowledge or consent.[52]

Rich Internet Applications

The introduction of powerful Web-delivered applications, such as Google Docs, Adobe Photoshop Express, Xcerion Web-based OS, and Microsoft Office Web Apps, have elevated the Web from an online library to a platform for computing.[53] Many of the computer activities traditionally provided through software installed on a PC can now be carried out using rich Internet applications (RIAs) in a Web browser without installing any software. A **rich Internet application** is software that has the functionality and complexity of traditional application software but that runs in a Web browser and does not require local installation (see Figure 7.12). RIAs are the result of continuously improving programming languages and platforms designed for the Web.

rich Internet application (RIA): Software that has the functionality and complexity of traditional application software but that does not require local installation and runs in a Web browser.

FIGURE **7.12**

Rich Internet application

SlideRocket is a rich internet application for creating vibrant online presentations.

Most RIAs take advantage of being online by emphasizing their collaborative benefits. Microsoft and Google both support online document sharing and collaborative editing. *37signals* provides online project management, contact management, calendar, and group chat applications.[54] Microsoft SharePoint provides businesses with collaborative workspaces and social computing tools to allow people at different locations to work on projects together.[55]

Online Information Sources

The Web has become the most popular source for daily news, surpassing newspapers and television. It has become the first place people look when they want news or are faced with a challenge or question.

News

The Web is a powerful tool for keeping informed about local, state, national, and global news. It has an abundance of special-interest coverage and provides the capacity to deliver deeper analysis of the subject matter. Text and photos are supported by the HTML standard. Video (sometimes called a Webcast) and audio are provided in the browser through plug-in technology and in podcasts (see Figure 7.13).

© iStockphoto/Thomas Dickson

FIGURE **7.13**

Online news

Online news is available in text, audio, and video formats providing the ability to drill down into stories.

As traditional news sources migrate to the Web, new sources are emerging from online companies. News Web sites from Google, Yahoo!, Digg, and Newsvine provide popular or interesting stories from a variety of news sources. In a trend some refer to as social journalism or citizen journalism, ordinary citizens are more involved in reporting the news than ever before. The online community is taking journalism into its hands and reporting the news from each person's perspective using an abundance of online tools. Although social journalism provides important news not available elsewhere, its sources may not be as reliable as mainstream media sources. It is sometimes difficult to discern news from opinion.

Education and Training

As a tool for sharing information and a primary repository of information on all subjects, the Web is ideally suited for education and training. Advances in interactive Web technologies further support important educational relationships between teacher and student and among students (see Figure 7.14).

CENGAGE brain .com

New Perspectives on Computer Concepts 2012: Comprehensive, 14th Edition

Select a chapter...

Book Overview
Chapter 1
Concept Quiz
Flashcards
Glossary

Book Resources
Orientation Information

Web Quiz

Assignment Name: popn14q/popn14q_chp01

Start over | Review | Enter Answer

1. All of the following are examples of output EXCEPT ________.
- reports
- documents
- keyboard
- music

2. When the microprocessor completes an instruction, ________.
- the instruction pointer zeros out
- the ALU registers 999
- the control unit increments the instruction pointer to the memory address of the next instruction, and the instruction cycle begins again
- the data from memory is moved to the accumulator

3. Suppose a microprocessor is adding two numbers together. Where are the two numbers placed so that the addition operation can proceed?
- In the ALU registers.
- In the instruction pointer.
- In the accumulator.
- In the control unit.

© Cengage Learning 2013

FIGURE **7.14**

Cengage Brain instruction resources

The Internet supports education from pre-K to lifelong learning.

Today, schools at all levels provide online education and training. Kahn Academy, for example, provides free online training and learning in economics, math, banking and money, biology, chemistry, history, and many other subjects.[56] NPower trains young adults in information systems skills.[57] The nonprofit organization gives training and hope to hundreds of disadvantaged young adults through a 22-week training program that can result in certification from companies such as Microsoft and Cisco.[58] One study conducted by a Nobel Prize winner suggested that using technology to teach concepts can be as effective as or more effective than traditional learning approaches without the use of technology.[59] The director of technology at one Colorado school used online learning to help traveling students keep up with their studies.[60] Many of the students were skiers and often traveled to skiing competitions. New training programs are now available on PCs, tablet computers, and smartphones. College students are also starting to use these devices to read electronic textbooks instead of carrying heavy printed textbooks to class.[61]

Educational support products, such as Blackboard, provide an integrated Web environment that includes virtual chat for class members; a discussion group for posting questions and comments; access to the class syllabus and agenda, student grades, and class announcements; and links to class-related material. Conducting classes over the Web with no physical class meetings is called *distance education*.

In a program it calls Open Courseware, the Massachusetts Institute of Technology (MIT) offers all of its courses free online (see Figure 7.15). Organizations such as the Open Courseware Consortium and the Center for Open Sustainable Learning have been established to support open education around the world.

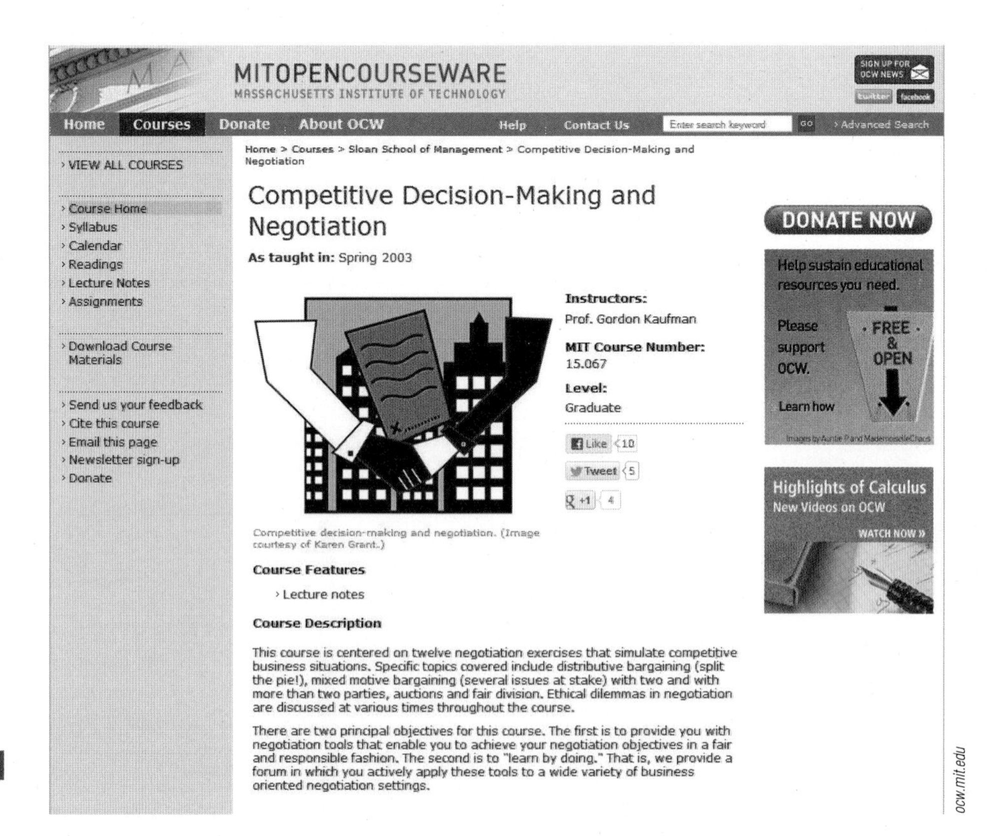

FIGURE **7.15**
MIT's Open Courseware

Beyond traditional education, corporations such as Skillsoft offer professional job skills training over the Web. Job seekers often use these services to acquire specialized business or technical training.[62] Museums, libraries, private businesses, government agencies, and many other types of organizations and individuals offer educational materials online for free or a fee. Consider eHow, the Web site that claims to teach you "How to do just about everything!"[63] Certiport offers training and testing for technology certification, such as for Microsoft and Adobe products.[64]

Business and Job Information

Providing news and information about a business and its products through the company's Web site and online social media can assist in increasing a company's exposure to the general public and improving its reputation. Providing answers to common product questions and customer support online can help keep customers coming back for more. For example, natural food company Kashi used its Web site to promote healthy living, with a blog about leading a natural lifestyle, recipes, and personal stories from Kashi employees. The Web site helps build a community around the Kashi brand and promotes awareness of Kashi's philosophy and products.[65]

The Web is also an excellent source of job-related information. People looking for their first jobs or seeking information about new job opportunities can find a wealth of information on the Web.[66] Search engines, such as Google or Bing (discussed next), can be a good starting point for searching for specific companies or industries.[67] You can use a directory on Yahoo's home page, for example, to explore industries and careers. Most medium and large companies have Web sites that list open positions, salaries, benefits, and people to contact for further information. The IBM Web site, *www.ibm.com*, has a link to "Jobs." When you click this link, you can find information on jobs with IBM around the world. In addition, several Internet sites specialize in helping you find job information and even apply for jobs online, including *www.linkedin.com*, *www.monster.com*, *www.hotjobs.com*, and *www.careerbuilder.com*.

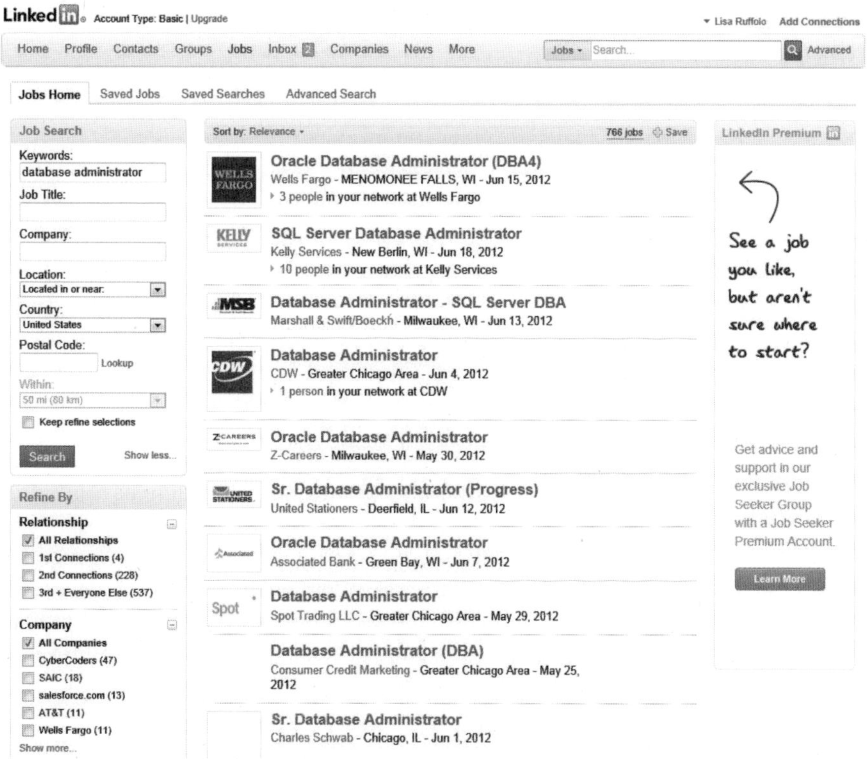

Several Internet sites specialize in helping people get job information and even apply for jobs online.

Search Engines and Web Research

search engine: A valuable tool that enables you to find information on the Web by specifying words that are key to a topic of interest, known as keywords.

A **search engine** is a valuable tool that enables you to find information on the Web by specifying words or phrases known as keywords, which are related to a topic of interest. You can also use operators such as OR and NOT for more precise search results. Table 7.2 provides examples of the use of operators in Google searches as listed on Google's help page (*www.google.com/help/cheatsheet.html*).

TABLE **7.2** Using operators in Google Web searches

Keywords and Operator Entered	Search Engine Interpretation
vacation Hawaii	The words "vacation" and "Hawaii"
Maui OR Hawaii	Either the word "Maui" or the word "Hawaii"
"To each his own"	The exact phrase "To each his own"
virus—computer	The word virus, but not the word computer
Star Wars Episode +I	The movie title "Star Wars Episode", including the Roman numeral I
~auto loan	Loan information for both the word "auto" and its synonyms, such as "truck" and "car"
define:computer	Definitions of the word "computer" from around the Web
red * blue	The words "red" and "blue" separated by one or more words

The search engine market is dominated by Google. Other search engines include Yahoo!, Microsoft Bing, and China's Baidu.[68] The rest of the market is divided among other companies such as Ask.com, AOL, and Mahalo. Google has taken advantage of its market dominance to expand into other Web-based services, most notably e-mail, scheduling, maps, Web-based applications, and cell phone software. Search engines like Google often have to modify how they display search results, depending on pending litigation from other Internet companies and government scrutiny, such as antitrust investigations.[69]

To help users get the information they want from the Web, most search engines use an automated approach that scours the Web with automated programs called spiders. These spiders follow all Web links in an attempt to catalog every Web page by topic; each Web page is analyzed and ranked using unique algorithms, and the resulting information is stored in a database. A keyword search at Yahoo!, Bing, or Google isn't a search of the Web but rather a search of a database that stores information about Web pages. The database is continuously checked and refreshed so that it is an accurate reflection of the current status of the Web.

Some search companies have experimented with human-powered and human-assisted search, such as Mahalo.[70] Human-powered search provides search results created by human researchers. Because the system is human powered, the search results are typically more accurate, definitive, and complete. The Web site *www.liveperson.com* takes human power one step further and allows visitors to chat and seek advice from human experts. The Web service contracts thousands of experts in a wide variety of fields to answer questions from users for a fee.[71]

The Bing search engine has attempted to innovate with its design. Bing refers to itself as a decision engine, providing more than just a long list of

INFORMATION SYSTEMS @ WORK

Improved Insight via Clickstream Analysis

When you visit a site, your *clickstream* is the sequence of pages you click as you spend time on the site. *Clickstream analysis* is the process of analyzing many clickstreams to understand visitors' collective behavior. The goal of clickstream analysis usually is to optimize a site for its users.

For example, clickstream analysis may find that many users want to see a list of a company's sales offices. Rather than making them reach that page by clicking "About Our Company," then "International Regions," then their local region, and finally, a list of its locations, the company might put a "Sales Offices" link on its home page. That would take the user directly to a page with a list of regions. Click a region, and it expands to show a list of its sales offices on the same page.

Besides improving visitor satisfaction with the site, such a change to the site design also reduces the page-serving load on the site owner's Web servers. That, in turn, improves performance and may defer the need for an expensive upgrade.

Clickstream analysis is vital to organizations that depend on the Web for their existence. Greg Linden explains that "Google [search] and Microsoft [Bing] learn from people using Web search. When people find what they want, Google notices. When other people do that same search later, Google has learned from earlier searchers, and makes it easier for the new searchers to get where they want to go."

Learning from clickstreams could be useful in online education. Discussing algebra, Linden notes, "As millions of students try different exercises, we [that is, our computer] forget the paths that consistently led to continued struggles, remember the ones that lead to rapid mastery, and, as new students come in, we put them on the successful paths we have seen before." Therefore, students learn algebra more quickly and more easily. The improved experience may affect their overall attitude toward learning mathematics.

The benefits of clickstream analysis aren't just for online companies. In an interview with M.I.T.'s *Sloan Management Review,* David Kreutter, Pfizer's vice-president of U.S. commercial operations, described its value to Pfizer: "When physicians visit our Web site, we know what they're clicking on, we know what they're clicking through to.... We've got more data from which to try to discern patterns, which we can use in a predictive way. That's really what we're trying to focus on now: can we detect patterns early-on, or at least much earlier than prescription writing, that will allow us to adapt more quickly to our customers' needs as well as to the competitive environment?"

Clickstream analysis helps Pfizer monitor what happens when its representatives visit physicians. Kreutter continues, "If our strategy is to deliver certain messages in a certain order, we can see if the message was delivered that way. For example, if we know that a certain segment of doctors in South Florida have a heavy proportion of elderly patients, they will often want to hear about drug-drug interactions first (since their patients are on many medications). We can track if we executed against that strategy, and we can track if that strategy had the impact, the literal prescribing behavior, that we anticipated. It ... helps us to figure out, if we don't have the impact we hoped for, if our strategy was right but the execution was flawed, or if the strategy fundamentally needs to be rethought."

Discussion Questions

1. Consider Greg Linden's example of search engines learning from watching which search results users chose to click. What are the benefits to the search engine company (e.g., Google or Microsoft in Linden's examples) of having this information?
2. How is clickstream analysis related to the competitive forces you learned about in Chapter 2?

Critical Thinking Questions

1. How could your college or university benefit from clickstream analysis? Give at least two areas. For each area, say what the institution would hope to learn, how it would learn it, and what the benefits of knowing it would be.
2. Consider Pfizer's use of clickstream analysis to track the behavior of its sales representatives during presentations to physicians. Is this monitoring ethical? Why or why not?

SOURCES: Kiron, D. and Shockley, R., "How Pfizer Uses Tablet PCs and Click-Stream Data to Track Its Strategy," *Sloan Management Review, sloanreview.mit.edu/the-magazine/2011-fall/53118/how-pfizer-uses-tablet-pcs-and-click-stream-data-to-track-its-strategy*, August 25, 2011; Linden, G., "Massive-Scale Data Mining for Education," *Communications of the ACM*, vol. 54, no. 11, November 2011, p. 13; Pfizer Web site, *www.pfizer.com*, accessed June 8, 2012.

links in its search results. Bing also includes media—music, videos, and games—in its search results (see Figure 7.16).

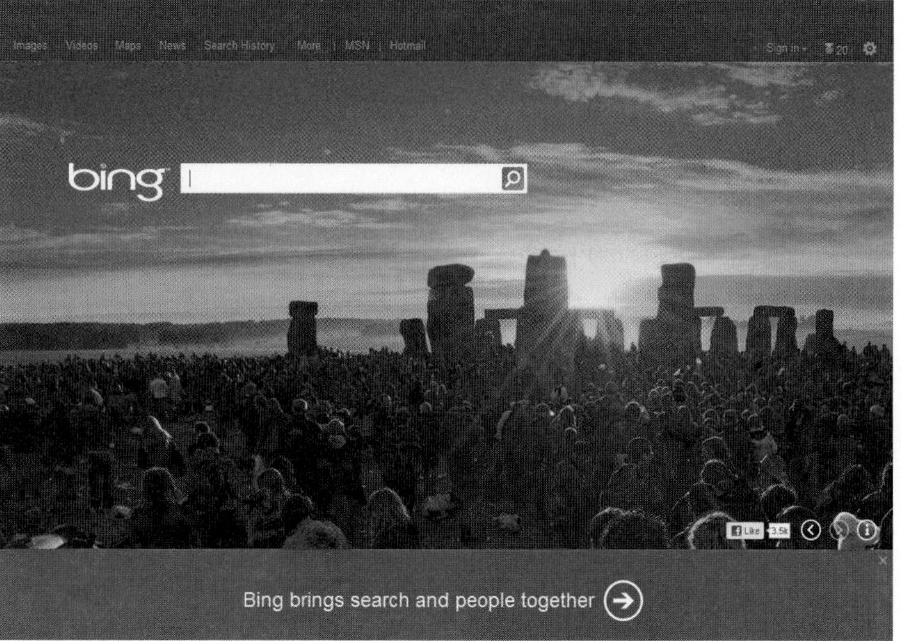

FIGURE **7.16**

Microsoft Bing decision engine

Microsoft calls its search engine a decision engine to distinguish it from other search software.

A meta search engine allows you to run keyword searches on several search engines at once. For example, a search run from *www.dogpile.com* returns results from Google, Yahoo!, MSN, Ask, and other search engines.

Savvy business owners know that the results gained from search engines are tools that draw visitors to the certain Web sites. Many businesses invest in search engine optimization (SEO)—a process for driving traffic to a Web site by using techniques that improve the site's ranking in search results. Normally, when a user gets a list of results from a Web search, the links listed highest on the first page of search results have a far greater chance of being clicked. SEO professionals, therefore, try to get the Web sites of their businesses to be listed with as many appropriate keywords as possible. They study the algorithms that search engines use, and then, they alter the contents of their Web pages to improve the page's chance of being ranked number one. SEO professionals use *Web analytics software* to study detailed statistics about visitors to their sites.

In addition to search engines, you can use other Internet sites to research information. Wikipedia, an online encyclopedia with over 3 million English-language entries created and edited by millions of users, is another example of a Web site that can be used to research information (see Figure 7.17). In Hawaiian, *wiki* means quick, so a "wikipedia" provides quick access to information. The Web site is both open source and open editing, which means that people can add or edit entries in the encyclopedia at any time. Besides being self-regulating, Wikipedia articles are vetted by around 1,700 administrators. However, even with so many administrators, it is possible that some entries are inaccurate and biased.

The wiki approach to content development is referred to as *crowd sourcing*, which uses the combined effort of many individuals to accomplish some task. Another example of crowd sourcing is the OpenStreetMap.org project. OpenStreetMap uses a wiki and the power of the crowd to develop a detailed map of the world.

Besides online catalogs, libraries typically provide links to public and sometimes private research databases on the Web. Online research databases

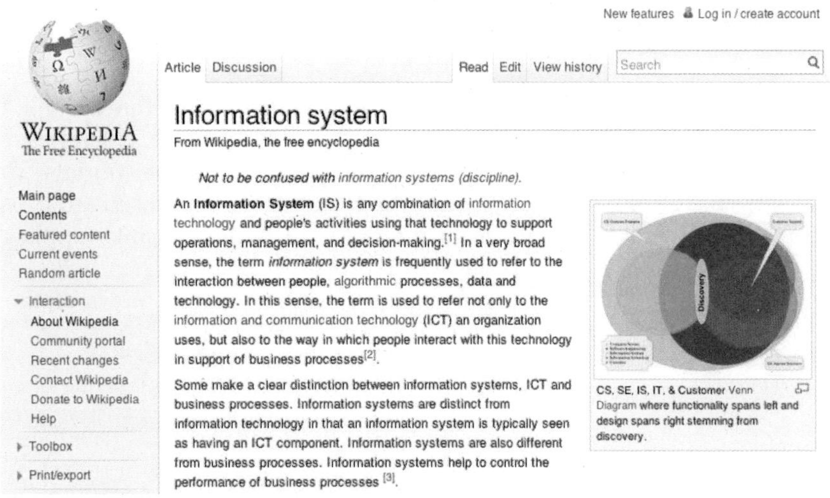

New features Log in / create account

WIKIPEDIA
The Free Encyclopedia

Main page
Contents
Featured content
Current events
Random article
Interaction
About Wikipedia
Community portal
Recent changes
Contact Wikipedia
Donate to Wikipedia
Help
Toolbox
Print/export

Article Discussion Read Edit View history Search

Information system

From Wikipedia, the free encyclopedia

Not to be confused with information systems (discipline).

An **Information System** (IS) is any combination of information technology and people's activities using that technology to support operations, management, and decision-making.[1] In a very broad sense, the term *information system* is frequently used to refer to the interaction between people, algorithmic processes, data and technology. In this sense, the term is used to refer not only to the information and communication technology (ICT) an organization uses, but also to the way in which people interact with this technology in support of business processes[2].

Some make a clear distinction between information systems, ICT and business processes. Information systems are distinct from information technology in that an information system is typically seen as having an ICT component. Information systems are also different from business processes. Information systems help to control the performance of business processes [3].

CS, SE, IS, IT, & Customer Venn Diagram where functionality spans left and design spans right stemming from discovery.

en.wikipedia.org

FIGURE 7.17
Wikipedia
Wikipedia captures the knowledge of tens of thousands of experts.

allow visitors to search for information in thousands of journal, magazine, and newspaper articles. Information database services are valuable because they offer the best in quality and convenience. They conveniently provide full-text articles from reputable sources over the Web. College and public libraries typically subscribe to many databases to support research. One of the most popular private databases is LexisNexis Academic Universe (see Figure 7.18).

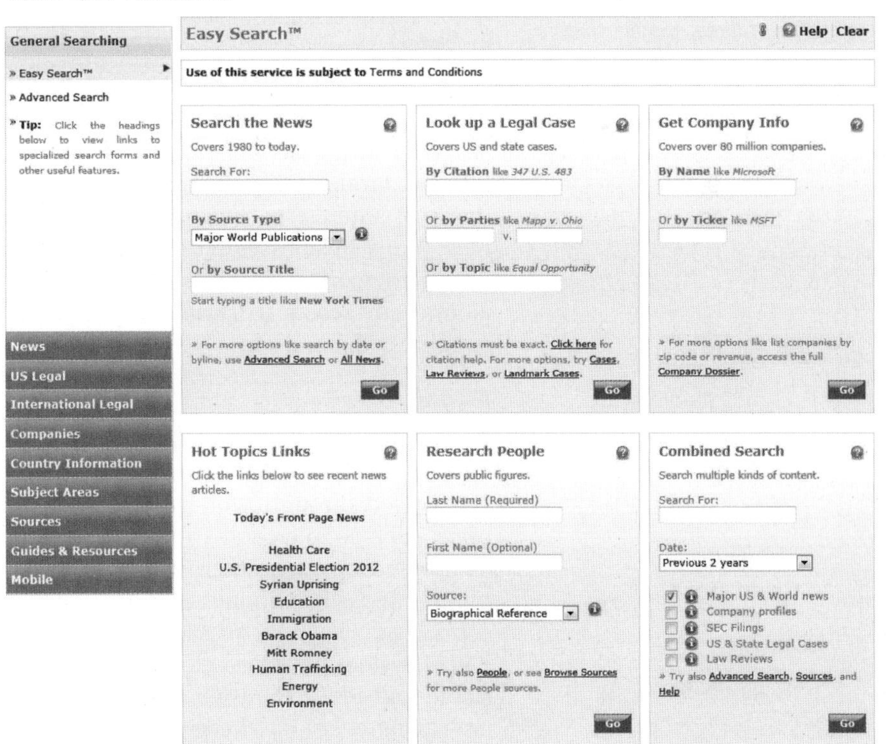

FIGURE 7.18
LexisNexis
At LexisNexis Academic Universe, you can search the news, legal cases, company information, people, or a combination of categories.

Web Portals

A **Web portal** is a Web page that combines useful information and links and acts as an entry point to the Web; portals typically include a search engine, a subject directory, daily headlines, and other items of interest. Because many

Web portal: A Web page that combines useful information and links and acts as an entry point to the Web; portals typically include a search engine, a subject directory, daily headlines, and other items of interest. Many people choose a Web portal as their browser's home page (the first page you open when you begin browsing the Web).

people choose a Web portal as their browser's home page (the first page you open when you begin browsing the Web), the two terms are used interchangeably.

Many Web pages have been designed to serve as Web portals. iGoogle, Yahoo!, AOL, and MSN are examples of horizontal portals; "horizontal" refers to the fact that these portals cover a wide range of topics. My Yahoo! and iGoogle allow users to custom design their pages, selecting from hundreds of widgets—small applications that deliver information and services. Yahoo also integrates with Facebook so that Facebook users can access their friends and news streams from the My Yahoo portal (see Figure 7.19).

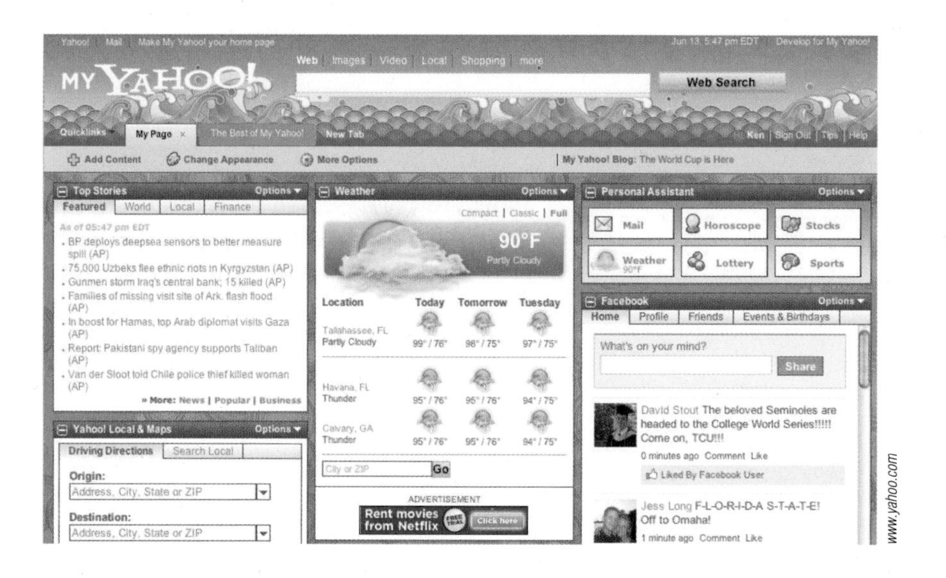

FIGURE **7.19**

MyYahoo! personalized portal

Personalized portals contain custom designs and widgets.

Vertical portals are pages that provide information and links for special-interest groups. For example, the portal at *www.iVillage.com* focuses on items of interest to women, and *www.AskMen.com* is a vertical portal for men. Many businesses set up corporate portals for their employees to provide access to work-related resources, such as corporate news and information, along with access to business tools, databases, and communication tools to support collaboration.

E-mail

E-mail is a useful form of Internet communication that supports text communication, HTML content, and sharing documents as e-mail attachments. E-mail is accessed through Web-based systems or through dedicated e-mail applications, such as Microsoft Outlook and Mozilla Thunderbird. E-mail can also be distributed through enterprise systems to desktop computers, notebook computers, and smartphones. Grady Health Systems, for example, upgraded its e–mail service from an older e-mail system to Microsoft's Exchange Online e-mail system based on cloud computing.[72] The new e-mail system was more stable and less expensive. According to the CIO of the hospital, "We clearly are saving every day because we don't have the expenses associated with our old instability."

Many people use online e-mail services, such as Hotmail, MSN, and Gmail (see Figure 7.20). Online e-mail services store messages on the server, not the user's computer, so that users need to be connected to the Internet to view, send, and manage e-mail. Other people prefer to use software such as Outlook, Apple Mail, or Thunderbird, which retrieve e-mail from the server and deliver it to the user's PC.

FIGURE 7.20

Gmail

Gmail is one of several free online e-mail services.

Courtesy of Google

Business users who access e-mail from smartphones, such as the BlackBerry, take advantage of a technology called push e-mail. Push e-mail uses corporate server software that transfers, or pushes, e-mail to the handset as soon as it arrives at the corporate e-mail server. To the BlackBerry user, it appears as though e-mail is delivered directly to the handset. Push e-mail allows the user to view e-mail from any mobile or desktop device connected to the corporate server. This arrangement allows users flexibility in where, when, and how they access and manage e-mail.

© Marvin Woodyatt/Photoshot

BlackBerry users have instant access to e-mail sent to their business accounts.

Some e-mail services scan for possible junk or bulk mail, called *spam*, and the service deletes it or places it in a separate folder. More than half of all e-mail can be considered spam. While spam-filtering software can prevent or discard unwanted messages, other software products can help users sort and answer large amounts of legitimate e-mail. For example, software from ClearContext, Seriosity, and Xobni rank and sort messages based on sender, content, and context, allowing individuals to focus on the most urgent and important messages first.

Instant Messaging

instant messaging: A method that allows two or more people to communicate online in real time using the Internet.

Instant messaging is online, real-time communication between two or more people who are connected to the Internet (see Figure 7.21). With instant messaging, participants build buddy lists, or contact lists, that let them see which contacts are currently logged on to the Internet and available to chat. If you send messages to one of your online buddies, a small dialog box opens on your buddy's computer and allows the two of you to chat via the keyboard. Although chat typically involves exchanging text messages with one other person, more advanced forms of chat exist. Today's instant messaging software supports not only text messages but also the sharing of images, sounds, files, and voice communications. Popular instant messaging services include America Online Instant Messenger (AIM), MSN Messenger, Google Talk, and Yahoo!.

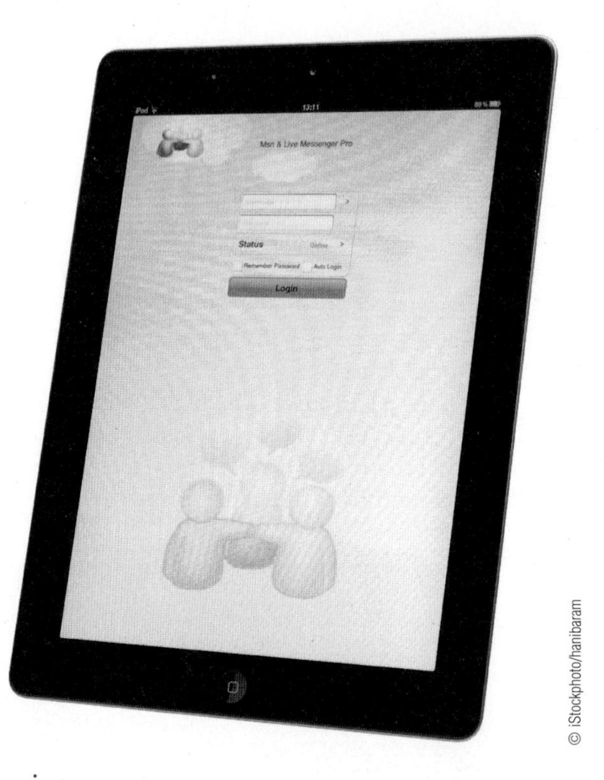

FIGURE **7.21**

Instant messaging

Instant messaging lets you converse with another Internet user by exchanging messages instantaneously.

Microblogging, Status Updates, and News Feeds

Twitter is a Web application that allows members to report on what they are doing throughout the day. Referred to as a microblogging service, Twitter allows users to send short text updates (up to 140 characters) from a cell phone or a Web account to their Twitter followers. While Twitter has been hugely successful for personal use, businesses are finding value in the service as well. Business people use Twitter to stay in close touch with associates by sharing their location and activities throughout the day. Businesses also find Twitter to be a rich source of consumer sentiment that can be tapped to improve marketing, customer relations, and product development. Many businesses have a presence on Twitter, dedicating personnel to communicate with customers by posting announcements and reaching out to individual users. Village Books, an independent bookstore, uses Twitter to build relationships with its customers and to make them feel part of their community.

The popularity of Twitter has caused social networks, such as Facebook, LinkedIn, and MySpace, to include Twitter-like news feeds. Previously referred to as Status Updates, Facebook users share their thoughts and activities with their friends by posting messages to Facebook's News Feed.

Conferencing

Some Internet technologies support real-time online conferencing. Participants dial into a common phone number to share a multiparty phone conversation. The Internet has made it possible for those involved in teleconferences to share computer desktops. Using services such as WebEx or GoToMeeting, conference participants log on to common software that allows them to broadcast their computer display to the group. This ability is quite useful for presenting with PowerPoint, demonstrating software, training, or collaborating on documents. Participants verbally communicate by phone or PC microphone. Some conferencing software uses Web cams to broadcast video of the presenter and group participants. For example, Papa Johns used GoToMeeting to conduct training sessions for managers around the world. Five online training sessions a year saved the company $50,000 in travel expenses.[73]

Telepresence takes video conferencing to the ultimate level. Telepresence systems such as those from Cisco and Polycom use high-resolution video and audio with high-definition displays to make it appear that conference participants are actually sitting around a table (see Figure 7.22). Participants enter a telepresence studio where they sit at a table facing display screens that show other participants in other locations. Cameras and microphones collect high-quality video and audio at all locations and transmit them over high-speed network connections to provide an environment that replicates actual physical presence. Document cameras and computer software are used to share views of computer screens and documents with all participants.

Courtesy of Polycom

FIGURE **7.22**

Halo Collaboration Meeting Room

The Halo telepresence system allows people at various locations to meet as though they were gathered around a table.

You don't need to be a big business to enjoy the benefits of video conversations. Free software is available to make video chat easy to use for anyone with a computer, a Webcam, and a high-speed Internet connection. Online applications such as Google Chat and Microsoft Messenger support video connections between Web users. For spontaneous, random video chat with strangers, you can use *www.Chatroulette.com* and Internet Conga Line. Software, such as Apple iChat and Skype, provide computer-to-computer video chat so users can speak to each other face-to-face. In addition to offering text, audio, and video chat on computers, Skype offers its video phone service over Internet-connected TVs. Recent Internet-connected sets from Panasonic and Samsung ship with the Skype software preloaded. You attach a Web cam to your TV to have a video chat from your sofa.

Blogging and Podcasting

Web log (blog): A Web site that people can create and use to write about their observations, experiences, and opinions on a wide range of topics.

A **Web log**, typically called a **blog**, is a Web site that people can create and use to write about their observations, experiences, and opinions on a wide range of topics. The community of blogs and bloggers is often called the *blogosphere.* A *blogger* is a person who creates a blog, while *blogging* refers to the process of placing entries on a blog site. A blog is like a journal. When people post information to a blog, it is placed at the top of the blog page. Blogs can include links to external information and an area for comments submitted by visitors. Video content can also be placed on the Internet using the same approach as a blog. This is often called a *video log* or *vlog.*

Internet users may subscribe to blogs using a technology called Really Simple Syndication (RSS). RSS is a collection of Web technologies that allow users to subscribe to Web content that is frequently updated, such as news sites and blogs. With RSS, you can receive a blog update and the latest headlines without actually visiting the blog or news Web site. Software used to subscribe to RSS feeds is called *aggregator software.* Google Reader is a popular aggregator for subscribing to blogs.

To set up a blog, you can go to the Web site of a blog service provider, such as *www.blogger.com* or *www.wordpress.com,* create a user name and password, select a theme, choose a URL, follow any other instructions, and start making your first entry. People who want to find a blog on a certain topic can use blog search engines, such as Technorati, Feedster, and Blogdigger. You can also use Google to locate a blog.

A *podcast* is an audio broadcast over the Internet. The name podcast originated from Apple's *iPod* combined with the word *broadcast.* A podcast is like an audio blog. Using PCs, recording software, and microphones, you can record podcast programs and place them on the Internet. Apple's iTunes provides free access to tens of thousands of podcasts, which are sorted by topic and searchable by key word (see Figure 7.23). After you find a podcast, you can download it to your PC (Windows or Mac), to an MP3 player such as the iPod, or any smartphone or tablet. You can also subscribe to podcasts using RSS software included in iTunes and other digital audio software.

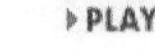

1.
▸ PLAY
NPR: Science Friday Podcast
by Ira Flatow
Science Friday, as heard on NPR, is a weekly discussion of the latest news in science, technology, health, and the environment hosted by Ira Flatow.
2.
TEDTALKS
▸ PLAY
TEDTalks Podcast
by Anthony Robbins
Each year, TED hosts some of the world's most fascinating people: Trusted voices and convention-breaking mavericks, icons and geniuses.
3.
Entrepreneurial Thought Leaders
▸ PLAY
Entrepreneurial Thought Leaders Podcast
by Forrest Glick
The DFJ Entrepreneurial Thought Leaders Seminar (ETL) is a weekly seminar series on entrepreneurship, co-sponsored by BASES (a student entrepreneurship group), Stanford Technology Ventures Program, and the Department of Management Science and Engineering.
4.
MIXERGY
▸ PLAY
Mixergy Video Podcast
by Andrew Warner
Interviews with a mix of successful online businesspeople. Andrew Warner asks them to teach ambitious startups how to build companies that leave a legacy...
5.
RAILS
Ruby on Rails Podcast
by Scott Barron
The Rails podcast is a super-agile way for you to get the inside scoop on the Rails community.
www.learnoutloud.com

FIGURE 7.23
Podcasts
iTunes and other sites provide free access to tens of thousands of podcasts.

Online Media and Entertainment

Like news and information, all forms of media and entertainment have followed their audiences online. Music, movies, television program episodes, user-generated videos, e-books, and audio books are all available online to download and purchase or stream.

content streaming: A method for transferring large media files over the Internet so that the data stream of voice and pictures plays more or less continuously as the file is being downloaded.

Content streaming is a method of transferring large media files over the Internet so that the data stream of voice and pictures plays more or less continuously as the file is being downloaded. For example, rather than wait for an entire 5 MB video clip to download before they can play it, users can begin viewing a streamed video as it is being received. Content streaming works best when the transmission of a file can keep up with the playback of the file.

Music

The Internet and the Web have made music more accessible than ever, with artists distributing their songs through online radio, subscription services, and download services. Pandora, Napster, and Grooveshark are just a few examples of free Internet music sites. Other Internet music sites charge a fee for music. Rhapsody has about 800,000 paid listeners, Slacker Radio has about 300,000 paid listeners, and Spotify has about 1.5 million paid listeners (see Figure 7.24). Internet music has even helped sales of classical music by Mozart, Beethoven, and others.[74] Internet companies, including Facebook, are starting to make music, movies, and other digital content available on their Web sites.[75] Facebook, for example, will allow online music companies, such as Spotify and Rdio, to post music activity on its Web site.

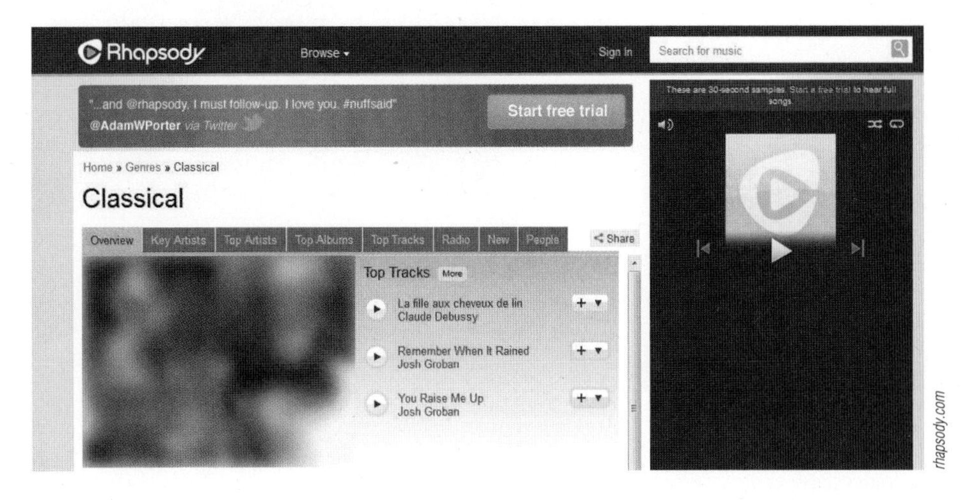

FIGURE 7.24

Rhapsody

Rhapsody provides streaming music by subscription.

Apple's iTunes was one of the first online music services to find success. Microsoft, Amazon, Walmart, and other retailers also sell music online. The going rate for music downloads is $0.89 to $0.99 per song. Downloaded music may include digital rights management (DRM) technology that prevents or limits the user's ability to make copies or to play the music on multiple players.

Podcasts are yet another way to access music on the Web. Many independent artists provide samples of their music through podcasts. Podcast Alley includes podcasts from unsigned artists.[76]

Movies, Video, and Television

Television and movies are expanding to the Web in leaps and bounds. Web sites such as Hulu and Internet-based television platforms such as Netflix and

Joost provide television programming from hundreds of providers, including most mainstream television networks.[77] See Figure 7.25. Walmart's acquisition of Vudu has allowed the big discount retailer to successfully get into the Internet movie business.[78] According to the general manager of Vudu, "The business we're in today, offering first-run movies a la carte, is doing very well right now and has tripled so far this year." Some TV networks, such as CNN and HLN, are streaming more programming over the Internet.[79] Increasingly, TV networks have iPad and other mobile applications (apps) that stream TV content to tablet computers and other mobile devices. Other TV networks are starting to charge viewers to watch their episodes on the Internet. Some only allow free viewing of an episode no more than once a week or longer after the episode first appears.[80] People that watch movies and TV programs over the Internet from companies like Netflix require large amounts of Internet capacity or bandwidth to download and watch the programs they want.[81] Some cable and wireless providers, such as Comcast, Verizon, and AT&T, are starting to charge customers more if they use more Internet capacity or bandwidth, often measured in megabytes (MB) or gigabytes (GB).

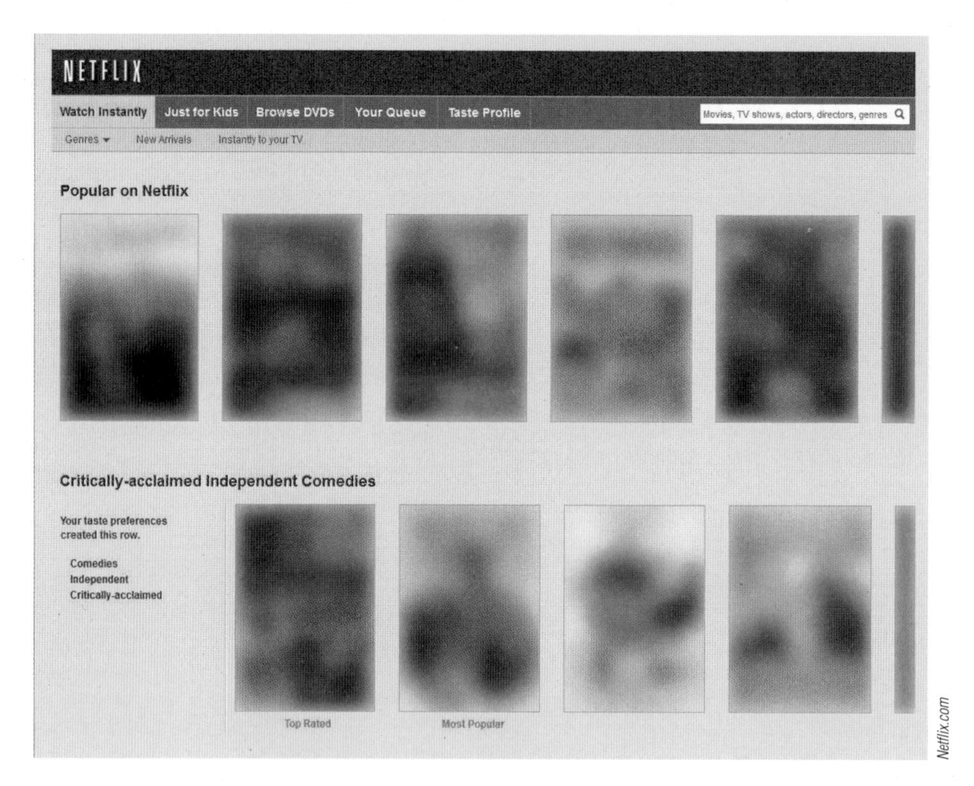

FIGURE **7.25**

Netflix

Netflix provides online access to thousands of movies and television shows.

No discussion of Internet video would be complete without mentioning YouTube. YouTube supports the online sharing of user-created videos. Every day, people upload hundreds of thousands of videos to YouTube and view hundreds of millions of these videos. YouTube videos are relatively short and cover a wide range of categories from the nonsensical to college lectures (see Figure 7.26). Other video-streaming sites include Google Video, Yahoo! Video, Metacafe, and AOL Video. As more companies create and post videos to Web sites like YouTube, some IS departments are creating a new position—video content manager.[82]

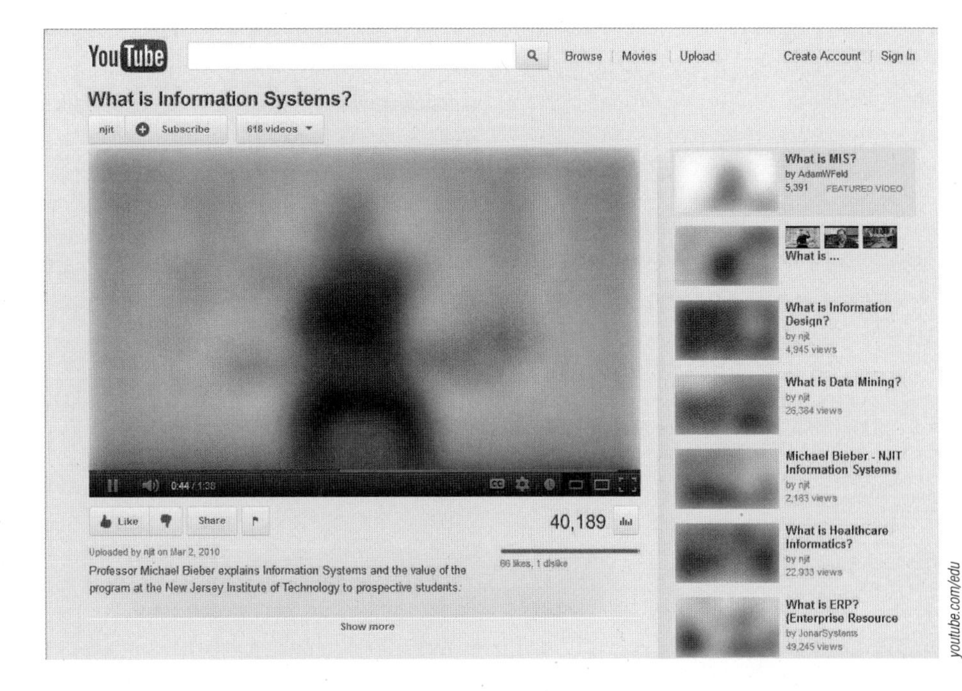

FIGURE **7.26**

YouTube EDU

YouTube EDU provides thousands of educational videos from hundreds of universities.

E-Books and Audio Books

An e-book is a book stored digitally rather than on paper and read on a display using e-book reader software. E-books have been available for quite a while, nearly as long as computers. However, it wasn't until the introduction of Amazon's eBook reading device, the Kindle, in 2007 that they became more widely accepted. Several features of the Kindle appeal to the general public. First, it features ePaper, a display that does not include backlighting like traditional displays. Some feel that ePaper is less harsh on your eyes than using a backlit display. Second, the Kindle is light and compact, similar in size and weight to a paperback book, although thinner than most books. Finally, Amazon created a vast library of eBooks that could be purchased and downloaded to the Kindle over whispernet—a wireless network provided free of charge by Sprint. Today, dozens of electronics manufacturers are offering eBook readers.

Apple's iPad changed the eBook industry by providing a form factor that is similar to but larger than the Kindle.[83] The iPad also includes a color backlit display. As an eBook reader, the iPad functions much like the Kindle; however, the iPad provides thousands of applications in addition to e-books. Besides using the Kindle, iPad, and other slate devices, you can access eBooks on the Web, download them as PDF files to view on your computer, or you can read them on your smartphone. While e-books are convenient, some have accused e-book publishers and distributors with conspiring to raise e-book prices.[84]

There are dozens of eBook formats. Some are proprietary, such as Kindle's .azw format, which can be viewed only on a Kindle. Others formats are open, such as Open eBook's .opf format and the .epub format, both of which can be read on many different devices and software packages, including Apple's iPad (see Figure 7.27).

Audio books have become more popular due to the popularity of the iPod, the iPhone, and other mobile devices along with services such as Audible.[85] Audio books are either read by a narrator without much inflection or varying voices, or they can be performed by actors who add dramatic interpretations of the book to the reading. Audio books may be abridged

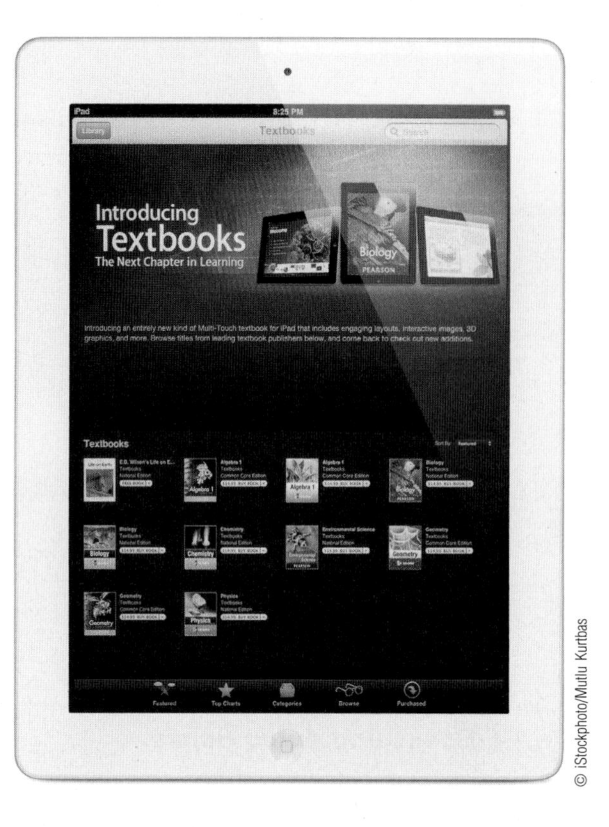

FIGURE 7.27
iPad publishing
The iPad provides a new interactive platform for magazines and books.

(consolidated and edited for audio format) or unabridged (read word for word from the book). Audio book services may allow you to purchase books individually or sign up for a membership and receive a new book each month.

Online Games and Entertainment

Video games have become a huge industry. It can generate over $20 billion annually, more than Hollywood movies.[86] A chewing gum maker, for example, has developed an alternate-reality game to attract potential game players and to advertise its chewing-gum brands.[87] According to the company's chief marketing manager, "This is the true nature of the brand, which is exploratory. You go down the rabbit hole, and you don't know how deep it is or where it goes." Zynga, a fast-growing Internet company, sells virtual horses and other virtual items for games, such as FarmVille.[88] The company, for example, sells a clown pony with colorful clothes for about $5. Zynga has a VIP club for people that spend a lot on virtual items it offers for sale. Some Internet companies also sell food for virtual animals.[89] People can feed and breed virtual animals and sell their offspring. With all the money being made with virtual animals and their pet food, lawsuits are likely. The market for Internet gaming is very competitive and constantly changing.[90] After Google included online games on its Web site, Facebook updated its online gaming offerings. Many video games are available online. They include single-user, multiuser, and massively multiuser games. The Web offers a multitude of games for all ages.

Game consoles such as the Wii, Xbox, and PlayStation provide multiplayer options for online gaming over the Internet. Subscribers can play with or against other subscribers in 3D virtual environments. They can even talk to each other using a microphone headset. Microsoft's Xbox LIVE provides features that allow users to keep track of their buddies online and match up with other players who are of the same skill level.[91]

Shopping Online

Shopping on the Web can be convenient, easy, and cost effective. You can buy almost anything via the Web, from books and clothing to cars and sports equipment. Groupon, for example, offers discounts at restaurants, spas, auto repair shops, music performances, and almost any other product or service offered in your area or city.[92] See Figure 7.28. Revenues for Groupon are expected to be hundreds of millions of dollars annually.[93] Groupon, however, has closed offices and fired employees in China.[94] It has also come under fire from states such as Connecticut that have consumer protection laws prohibiting gift card expiration dates.[95] According to one state official, "If Groupons meet the definition of a gift certificate, we feel there may be a violation of the state law."

© iStockphoto/Lee Rogers

FIGURE 7.28

Groupon

Groupon offers discounts at restaurants, spas, auto repair shops, music performances, and almost any other product or service offered in your area or city.

Other online companies offer different services. Dell.com and many other computer retailers provide tools that allow shoppers to specify every aspect and component of a computer system to purchase. ResumePlanet.com would be happy to create your professional résumé. Peapod or Amazon Grocery would be happy to deliver groceries to your doorstep. Products and services abound online.

Many online shopping options are available to Web users. E-tail stores—online versions of retail stores—provide access to many products that may be unavailable in local stores. JCPenney, Target, Walmart, and many others carry only a percentage of their inventory in their retail stores; the other inventory is available online. To add to their other conveniences, many Web sites offer free shipping and pickup for returned items that don't fit or otherwise meet a customer's needs.

Like your local shopping mall, cybermalls provide access to a collection of stores that aim to meet your every need. Cybermalls are typically aligned with popular Web portals such as Yahoo!, AOL, and MSN.

Web sites such as *www.mySimon.com*, *www.DealTime.com*, *www.PriceSCAN.com*, *www.PriceGrabber.com*, and *www.NexTag.com* provide product price quotations from numerous e-tailers to help you to find the best deal. An application for Android smartphones called Compare Everywhere allows users to compare the price of an item offered by many retailers. Even if the best price is offered at your local warehouse store, shopping online provides the assurance that you are getting the best deal.

Online clearinghouses, Web auctions, and marketplaces offer a platform for businesses and individuals to sell their products and belongings. Online clearinghouses, such as *www.uBid.com*, provide a method for manufacturers to liquidate stock and for consumers to find a good deal. Outdated or overstocked items are put on the virtual auction block and users bid on the items. The highest bidder when the auction closes gets the merchandise—often for less than 50 percent of the advertised retail price. Credit card numbers are collected at the time that bids are placed. A good rule to keep in mind is not to place a bid on an item unless you are prepared to buy it at that price.

The most popular online auction or marketplace is *www.eBay.com* (see Figure 7.29). eBay provides a public platform for global trading where anyone can buy, sell, or trade practically anything. It offers a wide variety of features and services that enable members to buy and sell on the site quickly and conveniently. Buyers have the option to purchase items at a fixed price or in auction-style format, where the highest bid wins the product.

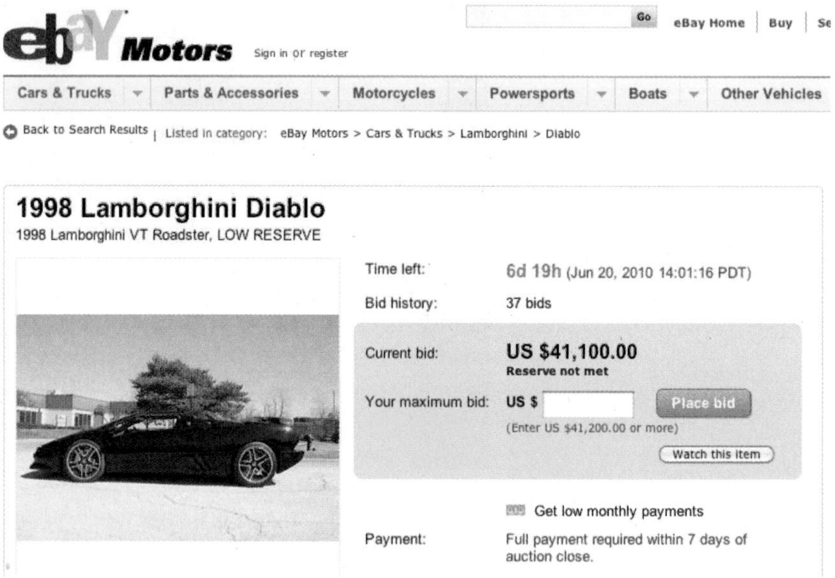

FIGURE **7.29**

eBay

eBay.com provides an online marketplace where anyone can buy, sell, or trade practically anything.

www.ebay.com

Internet auction sites have even been used by attorneys to market and advertise their services.[96] One Internet auction site, for example, allows lawyers to bid for clients and legal cases online.

Auction houses such as eBay accept limited liability for problems that buyers or sellers may experience in their transactions. Transactions that make use of eBay's PayPal service are protected. Others, however, may be risky. Participants should be aware that auction fraud is the most prevalent type of fraud on the Internet.

Craigslist is a network of online communities that provides free online classified advertisements. It is a popular online marketplace for purchasing items from local individuals.[97] Many shoppers turn to Craigslist rather than going to the classifieds in the local paper.

Businesses benefit from shopping online as well. *Global supply management (GSM)* online services provide methods for businesses to find the best deals on the global market for raw materials and supplies needed to manufacture their products. *Electronic exchanges* provide an industry-specific Web resource created to deliver a convenient centralized platform for B2B e-commerce among manufacturers, suppliers, and customers. You can read more about this topic in Chapter 8.

Travel, Geolocation, and Navigation

The Web has had a profound effect on the travel industry and the way people plan and prepare for trips. From getting assistance with short trips across town to planning long holidays abroad, travelers are turning to the Web to save time and money and overcome much of the risk involved in visiting unknown places.

Travel Web sites, such as *www.travelocity.com*, *www.expedia.com*, *www.kayak.com*, and *www.priceline.com*, help travelers find the best deals on flights, hotels, car rentals, vacation packages, and cruises. Priceline offers a slightly different approach from the other Web sites. It allows shoppers to name a price they're willing to pay for a ticket and then works to find an airline that can meet that price. After flights have been reserved, travelers can use these Web sites to book hotels and rental cars, often at discounted prices.

Mapping and geolocation tools are among the most popular and successful Web applications. MapQuest, Google Maps, and Bing Maps are examples (see Figure 7.30). By offering free street maps for cities around the world, these tools help travelers find their way. Provide your departure location and destination, and these online applications produce a map that displays the fastest route. Now with GPS technologies, these tools can detect your current location and provide directions from where you are.

Courtesy of Google

FIGURE **7.30**

Google Maps

Mapping software, such as Google Maps, provide streetside views of Times Square.

Google Maps also provides extensive location-specific business information, satellite imagery, up-to-the-minute traffic reports, and Street View. The latter is the result of Google employees driving the streets of the world's cities in vehicles with high-tech camera gear, taking 360-degree images. These images are integrated into Google Maps to allow users to get a "street view" of an area that can be manipulated as if they were actually walking down the street looking around. Bing Maps takes it a step further with high-resolution aerial photos and street-level 3D photographs.

Map applications like Google Maps provide tool kits that allow them to be combined with other Web applications. For example, Google Maps can be used in conjunction with Twitter to display the location where various tweets were posted. Likewise, Google Maps combined with Flickr can overlay photos of specific geographic locations. Combined Web applications are commonly referred to as a *mashup*.

Geographic information systems (GIS) provide geographic information layered over a map. For example, Google Earth provides options for viewing traffic, weather, local photos and videos, underwater features such as shipwrecks and marine life, local attractions, businesses, and places of interest. Software such as Google Latitude and Loopt allow you to find your friends on a map—with their permission—and will automatically notify you if a friend is near.

Geo-tagging is technology that allows for tagging information with an associated location. For example, Flickr and other photo software and services allow photos to be tagged with the location they were taken. Once tagged, it becomes easy to search for photos taken, for example, in Florida. Geo-tagging also makes it easy to overlay photos on a map, as Google Maps and Bing Maps have done. Twitter, Facebook, and other social networks have made it possible for users to geo-tag photos, comments, tweets, and posts.

Geolocation information does pose a risk to privacy and security. Many people prefer that their location remain unknown, at least to strangers and often to acquaintances and even friends. Recently, criminals have made use of location information to determine when people are away from their residences so that they can burglarize without fear of interruption.

Internet Utilities

Just as the Web is an application that runs on the Internet to provide a framework for delivering information and services, other applications have been designed to run on the Internet for other purposes. Many of these applications serve as utilities for accessing and maintaining resources on the Internet. A few such utilities that predate the Web and http, and still remain useful, are telnet, SSH, and FTP.

Telnet is a network protocol that enables users to log on to networks remotely over the Internet. Telnet software uses a command-line interface that allows the user to work on a remote server directly. Because Telnet is not secured with encryption, most users are switching to *secure shell (SSH)*, which provides Telnet functionality through a more secure connection.

File Transfer Protocol (FTP): A protocol that provides a file transfer process between a host and a remote computer and allows users to copy files from one computer to another.

File Transfer Protocol (FTP) is a protocol that supports file transfers between a host and a remote computer (see Figure 7.31). Using FTP, users can copy files from one computer to another. For example, the authors and editors of this book used an FTP site provided by the publisher, Cengage Learning, to share and transfer important files during the publication process. Chapter files and artwork, for example, were uploaded to a Cengage Learning FTP site and downloaded by authors and editors to review. Like Telnet, FTP connections are not encrypted, and are, therefore, not secure. Many users are switching to secure FTP (SFTP) for more protected file transfers.

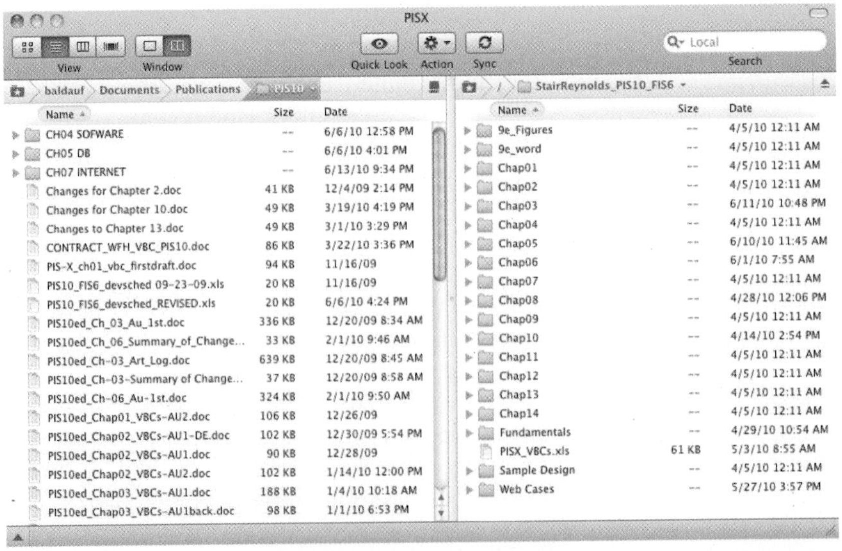

FIGURE 7.31

FTP applications

FTP applications allow you to transfer files between computers by clicking and dragging them from one window to another.

Courtesy of PISX

INTRANETS AND EXTRANETS

Recall from Chapter 1 that an intranet is an internal corporate network built using Internet and World Wide Web standards and technologies. Employees of an organization use it to gain access to corporate information. After getting their feet wet with public Web sites that promote company products and services, corporations are seizing the Web as a swift way to streamline—even transform—their organizations. These private networks use the infrastructure and standards of the Internet and the World Wide Web. Using an intranet offers one considerable advantage: many people are already familiar with Internet technology so that they need little training to make effective use of their corporate intranet.

An intranet is an inexpensive yet powerful alternative to other forms of internal communication, including conventional computer networks. One of an intranet's most obvious virtues is its ability to reduce the need for paper. Because Web browsers run on any type of computer, the same electronic information can be viewed by any employee. That means that all sorts of documents (such as internal phone books, procedure manuals, training manuals, and requisition forms) can be inexpensively converted to electronic form on the Web, easily distributed, and constantly updated. An intranet provides employees with an easy and intuitive approach to accessing information that was previously difficult to obtain. For example, it is an ideal solution for providing information to a mobile sales force that needs access to rapidly changing information.

A growing number of companies offer limited access to their private corporate network for selected customers and suppliers. Such networks are referred to as extranets; they connect people who are external to the company. An **extranet** is a network that links selected resources of the intranet of a company with its customers, suppliers, or other business partners. Like intranets, an extranet is built around Web technologies.

extranet: A network based on Web technologies that links selected resources of a company's intranet with its customers, suppliers, or other business partners.

Security and performance concerns are different for an extranet than for a Web site or network-based intranet. User authentication and privacy are critical on an extranet so that information is protected. Obviously, the network must perform well to provide quick response to customers and suppliers. Table 7.3 summarizes the differences between users of the Internet, intranets, and extranets.

TABLE 7.3 Summary of Internet, intranet, and extranet users

Type	Users	Need User ID and Password?
Internet	Anyone	No
Intranet	Employees and managers	Yes
Extranet	Business partners	Yes

Secure intranet and extranet access applications usually require the use of a *virtual private network (VNP)*, a secure connection between two points on the Internet. VPNs transfer information by encapsulating traffic in IP packets and sending the packets over the Internet, a practice called **tunneling**. Most VPNs are built and run by ISPs. Companies that use a VPN from an ISP have essentially outsourced their networks to save money on wide area network equipment and personnel.

tunneling: The process by which VPNs transfer information by encapsulating traffic in IP packets over the Internet.

INTERNET ISSUES

The Internet has greatly benefited individuals and organizations, but some Internet issues can have negative consequences.[98] Privacy invasion can be a potential problem with Internet and social networking sites.[99] A number of

Internet sites, for example, collect personal and financial information about people who visit their sites without the user's knowledge or consent.[100] Some Internet companies, however, are now starting to allow people to select a "do-not-track" feature that prevents personal and financial information from being gathered and stored.[101] Some people fear that new facial recognition software used by some Internet companies could also be an invasion of privacy.[102] Facial recognition software, for example, could be used to identify people in photos on social networking sites and other Internet sites. Additionally, some workers have been fired by their employers when they criticized them or their companies using Facebook, Twitter, and other social networking sites.[103] Some fired employees are fighting back by suing their employers.

Many states and local governments are trying to collect sales tax on Internet sales.[104] For one large Internet retailer, sales tax could be hundreds of millions of dollars annually. Sales tax for all Internet businesses could total more than $100 billion.

Internet attacks and hacks are also important Internet issues. It has been reported that a foreign TV program described software designed to hack into Web sites in the United States.[105] In another case, a foreign country may have hacked into e-mail accounts of officials and other important people in the United States.[106] The country involved has denied any involvement in the hacking case.[107] Increasingly, countries are charging and arresting people involved in Internet attacks by groups like *Anonymous*.[108]

SUMMARY

Principle:

The Internet provides a critical infrastructure for delivering and accessing information and services.

The Internet is truly international in scope, with users on every continent. It started with ARPANET, a project sponsored by the U.S. Department of Defense (DoD). Today, the Internet is the world's largest computer network. Actually, it is a collection of interconnected networks, all freely exchanging information. The Internet transmits data from one computer (called a host) to another. The set of conventions used to pass packets from one host to another is known as the Internet Protocol (IP). Many other protocols are used with IP. The best known is the Transmission Control Protocol (TCP). TCP is so widely used that many people refer to the Internet protocol as TCP/IP, the combination of TCP and IP used by most Internet applications. Each computer on the Internet has an assigned IP address for easy identification. A Uniform Resource Locator (URL) is a Web address that specifies the exact location of a Web page (using letters and words that map to an IP address) and the location on the host.

Cloud computing refers to a computing environment where software and storage are provided as an Internet service and accessed with a Web browser rather than installed and stored on PCs. As Internet connection speeds improve and wireless Internet access becomes pervasive, computing activities are increasing. Cloud computing offers many advantages. By outsourcing business information systems to the cloud, a business saves on system design, installation, and maintenance. Employees can also access corporate systems from any Internet-connected computer using a standard Web browser.

People can connect to the Internet backbone in several ways: via a LAN, whose server is an Internet host; a dial-up connection; high-speed service; or wireless service. An Internet service provider (ISP) is any organization that provides access to the Internet. To use this type of connection, you must have an account with the service provider and software that allows a direct link via TCP/IP.

Principle:

Originally developed as a document-management system, the World Wide Web has grown to become a primary source of news and information, an indispensable conduit for commerce, and a popular hub for social interaction, entertainment, and communication.

The Web is a collection of tens of millions of servers providing information via hyperlink technology to billions of users worldwide. Thanks to the high-speed Internet circuits connecting them and to hyperlink technology, users can jump between Web pages and servers effortlessly, creating the illusion of using one big computer. Because of its ability to handle multimedia objects and hypertext links between distributed objects, the Web is emerging as the most popular means of information access on the Internet today.

As a hyperlink-based system that uses the client/server model, the Web organizes Internet resources throughout the world into a series of linked files, called pages, accessed and viewed using Web client software, called a Web browser. Internet Explorer, Firefox, Chrome, and Safari are popular Web browsers. A collection of pages on one particular topic, accessed under one Web domain, is called a Web site.

Hypertext Markup Language (HTML) is the standard page description language for Web pages. The HTML tags tell the browser how to format the text: as a heading, as a list, or as body text, for example. HTML also indicates where images, sound, and other elements should be inserted. Some other Web standards have become nearly equal to HTML in importance, including Extensible Markup Language (XML), and Cascading Style Sheets (CSS).

Web 2.0 refers to the Web as a computing platform that supports software applications and the sharing of information among users with Web sites such as Facebook and Twitter. Over the past few years, the Web has been changing from a one-directional resource where users find information to a two-directional resource where users find and share information. The Web has also grown in power to support complete software applications and is becoming a computing platform itself. A rich Internet application (RIA) is software that has the functionality and complexity of traditional application software, but that runs in a Web browser and does not require local installation. Java, PHP, AJAX, MySQL, .NET, and Web application frameworks are all used to create interactive Web pages.

Principle:

The Internet and Web provide numerous resources for finding information, communicating and collaborating, socializing, conducting business and shopping, and being entertained.

The Web has become the most popular medium for distributing and accessing information. It is a powerful tool for keeping informed about local, state, national, and global news. As a tool for sharing information and a primary repository of information on all subjects, the Web is ideally suited for education and training. Museums, libraries, private businesses, government agencies, and many other types of organizations and individuals offer educational materials online for free or a fee. Many businesses use the Web browser as an interface to corporate information systems. Web sites have sprung up to support every subject and activity of importance.

A search engine is a valuable tool that enables you to find information on the Web by specifying words that are key to a topic of interest—known as keywords. Some search companies have experimented with human-powered and human-assisted searches. In addition to search engines, you can use other Internet sites to research information. *Wikipedia*, an online encyclopedia created and edited by millions of users, is an example of a Web site that can be used to research information. While *Wikipedia* is the best-known, general-purpose wiki, other wikis are designed for special purposes. Online research is also greatly assisted by traditional resources that have migrated from libraries to Web sites such as online databases.

A Web portal is a Web page that combines useful information and links onto one page and that often acts as an entry point to the Web—the first page you open when you begin browsing the Web. A Web portal typically includes a search engine, a subject directory, daily headlines, and other items of interest. It can be general or specific in nature.

The Internet and Web provide many applications for communication and collaboration. E-mail is an incredibly useful form of Internet communication that not only supports text communication but also supports HTML content and file sharing as e-mail attachments. Instant messaging is online, real-time communication between two or more people who are connected to the Internet. Referred to as a microblogging service, Twitter allows users to send short text updates (up to 140 characters long) from cell phone or Web to their Twitter followers. A number of Internet technologies support real-time online conferencing. The Internet has made it possible for those involved in teleconferences to share computer desktops. Using services such as WebEx or GoToMeeting, conference participants log on to common software that allows them to broadcast their computer displays to the group. Telepresence systems such as those from Cisco and HP use high-resolution video and audio with high-definition displays to make it appear that conference participants are actually sitting around a table.

Web sites such as YouTube and Flickr allow users to share video and photos with other people, groups, and the world. Microblogging sites like Twitter allow people to post thoughts and ideas throughout the day for friends to read. Social networking Web sites provide Web-based tools for users to share information about themselves with people on the Web and to find, meet, and converse with other members.

A Web log, typically called a blog, is a Web site that people can create and use to write about their observations, experiences, and opinions on a wide range of topics. Internet users may subscribe to blogs using a technology called Really Simple Syndication (RSS). RSS is a collection of Web technologies that allow users to subscribe to Web content that is frequently updated. A *podcast* is an audio broadcast over the Internet.

Like news and information, all forms of media and entertainment have followed their audiences online. The Internet and the Web have made music more accessible than ever, with artists distributing their songs through online radio, subscription services, and download services. With increasing amounts of Internet bandwidth available, streaming video and television are becoming commonplace. E-books have been available for quite a while, nearly as long as computers. However, it wasn't until the birth of Amazon's eBook reading device, the Kindle, in 2007 that they became more widely accepted. Online games include the many different types of single-user, multiuser, and massively multiuser games played on the Internet and the Web.

The Web has had a profound effect on the travel industry and the way people plan and prepare for trips. From getting assistance with short trips across town to planning long holidays abroad, travelers are turning to the Web to save time and money and overcome much of the risk involved in visiting unknown places. Mapping and geolocation tools are among the most popular and successful Web applications. MapQuest, Google Maps, and Bing Maps are examples. Geo-tagging is technology that allows for tagging information with an associated location.

Just as the Web is an application that runs on the Internet to provide a framework for delivering information and services, other applications have been designed to run on the Internet for other purposes. Telnet is a network protocol that enables users to log on to networks remotely over the Internet.

File Transfer Protocol (FTP) is a protocol that supports file transfers between a host and a remote computer. Like Telnet, FTP connections are not encrypted and are, therefore, not secure. Many users are switching to secure FTP (SFTP) for more secure file transfers.

Principle:

Popular Internet and Web technologies have been applied to business networks in the form of intranets and extranets.

An intranet is an internal corporate network built using Internet and World Wide Web standards and products. Because Web browsers run on any type of computer, the same electronic information can be viewed by any employee. That means that all sorts of documents can be converted to electronic form on the Web and constantly be updated.

An extranet is a network that links selected resources of the intranet of a company with its customers, suppliers, or other business partners. It is also built around Web technologies. Security and performance concerns are different for an extranet than for a Web site or network-based intranet. User authentication and privacy are critical on an extranet. Obviously, the network must perform well to provide quick response to customers and suppliers.

CHAPTER 7: SELF-ASSESSMENT TEST

The Internet provides a critical infrastructure for delivering and accessing information and services.

1. The ______________ was the ancestor of the Internet and was developed by the U.S. Department of Defense.
2. An IP address is a number that identifies a computer on the Internet. True or False?
3. On the Internet, what enables traffic to flow from one network to another?
   a. Internet Protocol
   b. ARPANET
   c. Uniform Resource Locator
   d. LAN server
4. Each computer on the Internet has an address called the *Transmission Control Protocol*. True or False?
5. ______________ is a computing environment where software and storage are provided as an Internet service and accessed with a Web browser.
   a. Cloud computing
   b. Internet Society (ISOC)
   c. The Web
   d. America Online (AOL)
6. A(n) ______________ is an organization that provides people with access to the Internet.

Originally developed as a document-management system, the World Wide Web has grown to become a primary source of news and information, an indispensable conduit for commerce, and a popular hub for social interaction, entertainment, and communication.

7. The World Wide Web was developed by Tim Berners-Lee at CERN. True or False?
8. Which technology was developed to assist in easily specifying the visual appearance of Web pages in a Web site?
   a. HTML
   b. XHTML
   c. XML
   d. CSS
9. Many of today's most popular online applications, including Gmail, Google Docs, Flickr, and Facebook, were developed with ______________.
10. What is the standard page description language for Web pages?
    a. Home Page Language
    b. Hypermedia Language
    c. Java
    d. Hypertext Markup Language (HTML)
11. Web development software that provides the foundational code—or framework—for a professional interactive Web site, allowing developers to customize the code to specific needs, is called a(n) ______________.

The Internet and Web provide numerous resources for finding information, communicating and collaborating, socializing, conducting business and shopping, and being entertained.

12. Web sites such as Facebook and LinkedIn are examples of ______________ Web sites.
    a. media sharing
    b. social network
    c. social bookmarking
    d. content streaming

13. A(n) ______________ is a valuable tool that enables you to find information on the Web by specifying words or phrases related to a topic of interest, known as keywords.
14. ______________ is an example of a microblogging service.
    a. Facebook
    b. WordPress
    c. Twitter
    d. YouTube
15. ______________ uses high-resolution video and audio with high-definition displays to make it appear that conference participants are actually sitting around a table.

Popular Internet and Web technologies have been applied to business networks in the form of intranets and extranets.

16. A(n) ______________ is a network based on Web technology that links customers, suppliers, and others to the company.
17. An intranet is an internal corporate network built using Internet and World Wide Web standards and products. True or False?

CHAPTER 7: SELF-ASSESSMENT TEST ANSWERS

1. ARPANET
2. True
3. a
4. False
5. a
6. Internet service provider (ISP)
7. True
8. d
9. AJAX
10. d
11. Web application framework
12. b
13. search engine
14. c
15. Telepresence
16. extranet
17. True

REVIEW QUESTIONS

1. What is the Internet? Who uses it and why?
2. What is ARPANET?
3. What is TCP/IP? How does it work?
4. Explain the naming conventions used to identify Internet host computers.
5. What is a Web browser? Provide four examples.
6. What is cloud computing?
7. Briefly describe three ways to connect to the Internet. What are the advantages and disadvantages of each approach?
8. What is an Internet service provider? What services do they provide?
9. How do Web application frameworks assist Web developers?
10. What are the advantages and disadvantages of Groupon?
11. What is a podcast?
12. How do human-powered search engines work?
13. For what are Telnet and FTP used, respectively?
14. What is content streaming?
15. What is instant messaging?
16. What is the Web? Is it another network like the Internet or a service that runs on the Internet?
17. What is a URL, and how is it used?
18. What are the advantages and disadvantages of streaming movies and TV programs over the Internet?
19. What is an intranet? Provide three examples of the use of an intranet.
20. What is an extranet? How is it different from an intranet?

DISCUSSION QUESTIONS

1. Social networks are widely used. Describe how this technology could be used in a business setting. Are there any drawbacks or limitations to using social networks in a business setting?
2. Your company is about to develop a new Web site. Describe how you could use Web services for your site.
3. Why is it important to have an organization that manages IP addresses and domain names?
4. What are the benefits and risks involved in using cloud computing?
5. You are the owner of a small business with five employees. Describe what approach you would use to connect to the Internet.

6. Describe how a company could use a blog and podcasting.
7. Briefly describe how the Internet phone service operates. Discuss the potential impact that this service could have on traditional telephone services and carriers.
8. Why is XML an important technology?
9. Discuss the advantages and disadvantages of a virtual private network.
10. Briefly describe the importance of Web services. What is involved?
11. Identify three companies with which you are familiar that are using the Web to conduct business. Describe their use of the Web.
12. What are the defining characteristics of a Web 2.0 site?
13. Name four forms of Internet communication and describe the benefits and drawbacks of each.
14. What social concerns surround geolocation technologies?
15. One of the key issues associated with the development of a Web site is getting people to visit it. If you were developing a Web site, how would you inform others about it and make it interesting enough that they would return and tell others about it?
16. Downloading music, radio, and video programs from the Internet is easier and more regulated than in the past, but some companies are still worried that people will illegally obtain copies of this programming without paying the artists and producers royalties. If you were an artist or producer, what would you do?
17. How could you use the Internet if you were a traveling salesperson?
18. Briefly summarize the differences in how the Internet, a company intranet, and an extranet are accessed and used.

PROBLEM-SOLVING EXERCISES

1. Do research on the Web to find several social networking sites. After researching these sites, use a word processor to write a report comparing and contrasting the services. Also discuss the advantages and potential problems of sharing personal information online. What information collected by social networking sites do you think should be kept private from the general public?
2. Develop a brief proposal for creating a business Web site. How could you use Web services to make creating and maintaining the Web site easier and less expensive? Develop a simple spreadsheet to analyze the income you need to cover your Web site and other business expenses.
3. Think of a business that you might like to establish. Use a word processor to define the business in terms of what product(s) or service(s) it provides, where it is located, and its name. Go to *www.godaddy.com* and find an appropriate domain name for your business that is not yet taken. Shop around online for the best deal on Web site hosting. Write a paragraph about your experience finding a name, why you chose the name that you did, and how much it will cost you to register the name and host a site.
4. You have been hired to develop a business model for an Internet site for a small local newspaper. Develop a spreadsheet that shows the revenues from the Web site and the costs involved in setting it up and running the Web site.
5. Develop a slide show using a graphics program, such as PowerPoint, to show how a new Internet site to sell used bicycles on campus can be started and profitably run.

TEAM ACTIVITIES

1. With your teammates, identify a company that is making effective use of Web 2.0 technologies on its Web site. Write a review of the site and why you believe it is effective.
2. Use Flickr.com to have a photo contest. Each group member should post four favorite photos taken by that member. Share account information among your group members and then use photo comment boxes to vote on your favorite photos. The photo with the most favorable comments wins.
3. Have your team describe a new and exciting Internet game. The game should include students, professors, and university administrators as players in the game. Write a report using Google docs or a similar word-processing program describing how your game will work.
4. Each team member should use a different search engine to find information about podcasting. Meet as a team and decide which search engine was the best for this task. Write a brief report to your instructor summarizing your findings.

WEB EXERCISES

1. This chapter covers a number of powerful Internet tools, including Internet phones, search engines, browsers, e-mail, extranets, and intranets. Pick one of these tools and find more information about it on the Internet. You might be asked to develop a report or send an e-mail message to your instructor about what you found.
2. Using the Internet, research three universities that extensively use online or distance learning. Write a report of what you found.
3. Research some of the potential disadvantages of using the Internet, such as privacy issues, fraud, or unauthorized Web sites. Write a brief report on what you found.

CAREER EXERCISES

1. Use three job-related Internet sites to explore starting salaries, benefits, and job descriptions for a career in an area that interests you. Describe the characteristics, advantages, and disadvantages of each job-related Internet site.
2. Consider how the Internet and Web can be useful to businesses in fields that interest you. Select two such businesses and research how they use the Web. Write up the results of your research, including the benefits of the Web to these businesses and recommendations for how they might extend their use of the Web to increase profits.

CASE STUDIES

Case One

University of Sydney Redesigns Its Web Site

Since its founding in 1850, the University of Sydney has grown to dozens of faculties, schools, and centers. Naturally, each one needed a Web site. The university eventually had over 600 distinct sites with millions of individual pages. These sites included intranets, online learning systems, and information for staff, students, and external stakeholders. Each unit chose its own tools to build its site, which was then hosted on one of more than 200 servers.

As often happens when a Web site "just grows," the result was a hodgepodge. The university had no site standards so that its sites did not have a consistent look and feel. Sites often contained duplicated, out-of-date, or inaccurate information.

Web publishing was also cumbersome. Many faculty members wanted to publish content themselves, but had to go through their unit's IT department. This requirement created an annoying bottleneck.

To address these challenges, the university needed publishing standards. University managers also wanted to remove IT bottlenecks by enabling faculty staff to publish web content themselves. "As well as a technology solution, we needed to change the way people thought about how information could be structured to meet users' needs," said Web services manager Charlie Forsyth.

The University of Sydney decided to implement a content management system. They hired advisory firm Gartner to draw up a short list of vendors before asking for bids. Gartner provides industry analysis, evaluating products for a range of users, and offers consulting services to customize the general analysis to the needs of a specific organization. With its Asia/Pacific headquarters in North Sydney and a research service solely for educational institutions, Gartner was a reasonable choice.

The content management bid was won by Hewlett-Packard subsidiary Autonomy. "Autonomy TeamSite had the scalability and power to tackle our mountain of content, the size of our Web presence, and the number of users we needed to service," said Forsyth. "Because TeamSite is a file-based, rather than database-driven, system, it requires a lower infrastructure cost than other enterprise-grade products." Marian Theobald, director of community engagement, adds that this "solution has given our staff a level of control and professionalism in presentation that was not available previously."

Initial implementation took four months. After that, the solution was rolled out to the group responsible for the central university site. This procedure allowed the university to develop processes and guidelines that it would later apply across its entire Web presence. In the next phase, the university created 130 Web sites with 25,000 pages of content. Today, content owners throughout the university can contribute directly to their sites while maintaining consistent presentation.

The university has leveraged this investment in other ways. Every year, its faculties publish 16 student handbooks with information on courses and units of study as well as academic regulations. Most run to hundreds of pages. These handbooks now use templates that each unit can populate, speeding their creation while ensuring presentation consistency. Students can get handbooks in the format they prefer: they can read a handbook online, view it as an e-book, receive it on a CD-ROM, or print it as a hard copy. The small number of students who choose printing has enabled the university to reduce handbook printing costs by 21 percent.

The new designs have cut the time it takes to find information. Before redeveloping the Web sites, a student survey showed 69 percent found it "easy to very easy" to find information on the university's main site. After redevelopment, this figure rose to 79 percent. Another success indicator is that the number of telephone inquiries from prospective students halved within two years, from 5,886 to 3,014, during a period of growth. The university attributes this decrease to improved information available on the Web.

Discussion Questions

1. Examine different parts of your college or university Web site. Does the site have a consistent look and feel? If you look at the Athletics section, the Admissions section, and the description of your academic program, can you tell at a glance that they belong to the same institution? If you can't, how would your school benefit from a more consistent appearance? If they are consistent, what would your school lose if they were not?
2. Suppose it costs twice as much to print a single copy of a handbook on demand as it costs to print each copy when they are printed in volume. What fraction of the students must use alternative access methods in order for the university to save 21 percent of its printing cost for handbooks overall? What if the cost ratio of single-copy printing to the cost of each copy when printed in volume is 5:1? If 85 percent of the students access handbooks online, how much more expensive can single-copy printing be for the university to break even on printing handbooks?

Critical Thinking Questions

1. Instead of implementing content management software, the University of Sydney could have issued Web site guidelines to its units and motivated them to use the same Web site development software by negotiating a site-wide license and offering training. How effective would that approach have been? Which of the benefits in this case study would the university probably have achieved in full, which would it probably have achieved in part, and which would it not have achieved at all? Should the university have considered this less expensive approach at all? Why or why not?
2. The case study gives one of the benefits for the new approach as the ability of individual faculty members to publish their own Web content. What are some disadvantages of this ability? On balance, do you think it is a good idea? Why or why not? Does a university differ from a profit-making corporation or a religious organization in this regard? If you think it does, what are the differences, and how do they affect this issue? If you think it doesn't differ, why don't other differences affect this issue?

SOURCES: Staff, "University of Sydney Makes the Grade with Autonomy," Autonomy, *publications.autonomy.com/pdfs/Promote/Case%20Studies/Education/20111110_CI_CS_University_of_Sydney_web.pdf* (requires free registration), November 10, 2011; Autonomy Web site, *www.autonomy.com*, accessed June 7, 2012; Gartner Group Web site, *www.gartner.com*, accessed June 8, 2012; University of Sydney Web site, *sydney.edu.au*, accessed June 7, 2012.

Case Two

Amanda Palmer Kickstarts

Kickstarter describes itself as "a funding platform for creative projects—everything from traditional forms of art (like theater and music) to contemporary forms (like design and games)." It lets people publicize their projects so that other interested people can fund them. Funders might get something in exchange, such as a copy of a CD that their funding helped produce or just the satisfaction of helping an idea get off the ground.

"Punk cabaret" singer Amanda Palmer ran one of the most successful campaigns in Kickstarter history to promote her new album, raising nearly $1.2 million. The average successful Kickstarter campaign, by contrast, brings in about $5,000—and not all Kickstarter campaigns are successful.

In return for funding, Palmer's fans can choose from a range of music-related items. For a pledge of $1, they can download music from her new album. For $25, they get a limited-edition version of the CD. For $300, they can attend a show with a special preshow backer party. For $5,000, a backer gets a private house concert. (When the campaign ended on May 31, 2012, 34 of the 35 available house concerts had been sold.) That is far from the complete list, which contains 24 items for pledges up to $10,000, but it suggests the range.

Ben Sisario, writing in *The New York Times,* describes Palmer as "one of music's most productive users of social media." Sisario continues, "She posts just-written songs to YouTube and is a prolific correspondent on Twitter, soliciting creative feedback from her 562,000 followers and selling tens of thousands of dollars of merchandise.... That engagement has brought her rare loyalty." In other words, Palmer is in a perfect position to benefit from a Kickstarter campaign—and she knows it. As music industry observer Jay Frank points out, she used Kickstarter effectively in several ways:

- **She set her goal low**. Palmer expected to exceed her $100,000 target in house concert revenue alone. Exceeding the goal becomes a story itself. That leads to more publicity and more pledges.
- **She bundled three campaigns in one**. She promoted a record, an art book, and a tour. Raising $1.2 million in three campaigns isn't as exciting as raising that much in one campaign, but it's still an impressive total.
- **She used her album as an inexpensive entry point**. The album seems to be what she's promoting, but most of her revenue will come from her live tour.

Even the title of Palmer's new album, *Theatre Is Evil,* reflects use of social media. She first announced the title on Saturday, June 2, 2012, via Twitter as *Theater Is Evil.* As she posted in her blog the following day, "i was barraged by a tweetstorm of brits, canadians, australians, south africans, and AMERICANS informing me that my title was WRONG WRONG WRONG and should be spelled THEATRE IS EVIL." She took an online poll in her discussion forum. The votes were 82 percent for the spelling "theatre," so as she posts, "i folded." The spelling of this word in the album title probably makes little difference, but the fan connections that she created and strengthened through this dialogue make a big difference.

Raising $1.2 million doesn't mean pocketing that much. Kickstarter takes 5 percent of all pledges. Palmer had about $250,000 in recording costs. She also had touring expenses, promotional expenses, and the costs of designing, producing, and shipping the packages fans get. What remains is taxable. One must also consider that time she spends working social media is time she can't spend on her music. Still, breaking even before an album's release is rare. Palmer did a lot better than that!

Discussion Questions

1. Amanda Palmer tweets, blogs, has an active presence on Facebook, and is on the lookout for new ways to use social media. How do you think that affected the success of her Kickstarter campaign?
2. You decide to launch a Kickstarter campaign to fund the development of a new type of athletic shoe that reduces fatigue in very long runs. Your goal is to raise $100,000 to put the shoe into production. Propose a range of offers to your backers for pledges of $5 to $5,000. Include at least six different price points: $5, $5,000, and four other amounts between these two.

Critical Thinking Questions

1. Suppose you and four friends have a band. You think you're pretty good, but you don't have even a college-wide reputation. Would Kickstarter be an effective way to promote your band? Why or why not?
2. An engineer at a large computer company proposed a new product but was rejected for internal R&D funding. With company permission, this engineer plans to look for funding on Kickstarter. Do you think this approach would be effective?

SOURCES: Amanda Palmer Web site, *www.amandapalmer.net*, accessed June 6, 2012; Frank, J., "Amanda Palmer Has a Huge Hack," *www.hypebot.com/hypebot/2012/05/amanda-palmer-has-a-huge-hack.html*, May 15, 2012; Houghton, B., "Amanda Palmer Blows Past $1 Million On Kickstarter," *www.hypebot.com/hypebot/2012/05/amanda-palmer-passes-1-million-on-kickstarter.html*, May 30, 2012; Kickstarter Web site, *www.kickstarter.com*, accessed June 6, 2012; Palmer, A., "Countdown Party and Album Title," blog, *blog.amandapalmer.net/post/24356061454/countdown-party-album-title*, June 3, 2012; Sisario, B., "Giving Love, Lots of It, to Her Fans," *The New York Times*, *www.nytimes.com/2012/06/06/arts/music/amanda-palmer-takes-connecting-with-her-fans-to-a-new-level.html*, June 5, 2012.

Questions for Web Case

See the Web site for this book to read about the Altitude Online case for this chapter. Following are questions concerning this Web case.

Altitude Online: The Internet, Web, Intranets, and Extranets

Discussion Questions

1. What impact will the new ERP system have on Altitude Online's public-facing Web site? How will it affect its intranet?
2. What types of applications will be available from the employee dashboard?

Critical Thinking Questions

1. Altitude Online employees have various needs, depending on their position within the enterprise. How might the dashboard and intranet provide custom support for individual employee needs?
2. What Web 2.0 applications should Altitude Online consider for its dashboard? Remember that the applications must be available only on the secure intranet.

NOTES

Sources for the opening vignette: Microsoft SharePoint Web site, *sharepoint.microsoft.com,* accessed June 6, 2012; PepsiCo Europe Web page, *www.pepsico.com/Company/The-Pepsico-Family/PepsiCo-Europe.html*, accessed June 6, 2012; Staff, "PepsiCo Announces Completion of Wimm-Bill-Dann Acquisition," PepsiCo, *www.pepsico.com/PressRelease/PepsiCo-Announces-Completion-of-Wimm-Bill-Dann-Acquisition09092011.html*, September 9, 2011; Schwartz, J., "What To Expect in SharePoint 15," *Redmond, redmondmag.com/articles/2012/04/01/whats-next-for-sharepoint.aspx,* April 4, 2012; Ward, T., "Social Intranet Case Study: PepsiCo Russia," *www.intranetblog.com/social-intranet-case-study-pepsico-russia/2012/03/28*, March 28, 2012; Weis, R. T., "How Pepsi Won the Cola Wars in Russia," *www.frumforum.com/how-pepsi-won-the-cola-wars-in-russia*, October 28, 2011; WSS Consulting Web site (in Russian/по-русски), *www.wss-consulting.ru*, accessed June 7, 2012.

1. The ISC Domain Survey, *www.isc.org/solutions/survey*, accessed September 7, 2011.
2. Internet Usage Statistics, *www.internetworldstats.com/stats.htm*, accessed September 7, 2011.
3. Internet Usage World Stats Web site, *www.internetworldstats.com*, accessed September 7, 2011.
4. Rhoads, Christopher, "Technology Poses Big Test for Regimes," *The Wall Street Journal*, February 12, p. A11.
5. Morozov, Evgeny, "Smart Dictators Don't Quash the Internet," *The Wall Street Journal*, February 19, 2011, p. C3.
6. Rhoads, Christopher, "Technology Poses Big Test for Regimes," *The Wall Street Journal*, February 12, p. A11.
7. Page, Jeremy, "China Co-Ops Social Media to Head Off Unrest," *The Wall Street Journal*, February 22, 2011, p. A8.
8. Cullison, A., "Google Sidesteps Edict," *The Wall Street Journal*, June 9, 2011, p. B7.

9. DARPA Web site, *www.darpa.mil*, accessed September 8, 2011.
10. Internet2 Web site, *www.internet2.edu*, accessed September 7, 2011.
11. National LambdaRail Web site, *www.nlr.net*, accessed June 20, 2012.
12. IPv6 Web site, *http://ipv6.com/*, accessed September 9, 2011.
13. Holmes, S. and Rhoads, C., "Web Addresses Enter New Era," *The Wall Street Journal*, June 21, 2011, p. B1.
14. ICANN Web site, *www.icann.org*, accessed September 9, 2011.
15. Sheridan, B., "The Apps Class of 2010," *Bloomberg Businessweek*, January 3, 2011, p. 80.
16. Kane, Y., and Smith E., "Apple Readies iCloud Service," *The Wall Street Journal*, June 1, 2011, p. B1.
17. Henschen, Doug, "Microsoft Places Bigger Bet on Cloud Apps," *InformationWeek*, July 11, 2011, p. 10.
18. Staff, "Office 365 vs. Google: Advantage Microsoft," *InformationWeek*, July 11, 2011, p. 10.
19. Babcock, B., "Specialized Clouds," *InformationWeek*, June 27, 2011, p. 34.
20. Univa Web site, *www.univa.com*, accessed September 7, 2011.
21. Burberry Web site, *http://us.burberry.com/store*, accessed September 6, 2011.
22. Gilette, Felix, "How Salesforce Tames Twitter for Big Business," *Bloomberg Businessweek*, September 4, 2011, p. 35.
23. Thibodeau, P., "The Race to Cloud Standards Gets Crowded," *Computerworld*, August 22, 2011, p. 24.
24. Schultz, B., "Managing Complexity," *Network World*, August 8, 2011, p. 16.
25. NetApp cloud Web site, *www.netapp.com/cloud*, accessed September 3, 2011.
26. Henderson, T., and Allen, B., "Five Cool Tools for Cloud Management," *Network World*, August 8, 2011, p. 20.
27. HTML5 Rocks Web site, *www.html5rocks.com/en*, accessed September 9, 2011.
28. W3C Web site, *www.w3.org/Style/CSS*, accessed September 9, 2011.
29. Amazon AWS Web site, *http://aws.amazon.com*, accessed September 9, 2011.
30. Drupal Web site, *http://drupal.org/project/framework*, accessed September 7, 2011.
31. Leiber, N., "A Startup's New Prescription for Eyewear," *Bloomberg Businessweek*, July 4, 2011, p. 49.
32. Kharif, Olga, "Startup Counsel," *Bloomberg Businessweek*, July 11, 2011, p. 41.
33. Vascellaro, J., "Building Loyalty on the Web," *The Wall Street Journal*, March 28, 2011, p. B6.
34. Gaudin, Sharon, "Drug Maker Fixes Supply Chain with Social Tools," *Computerworld*, August 8, 2011, p. 12.
35. Henschen, D., "SAP to Offer More Google Map Mashups," *InformationWeek*, April 15, 2011, p. 18.
36. MacMillan, D., "Creating Web Addicts for $10,000 a Month," *Bloomberg Businessweek*, January 24, 2011, p. 35.
37. Fowler, G. and Cowan, L., "Zillow IPO Zooms 79%," *The Wall Street Journal*, July 21, 2011, p. B1.
38. Schechner, S. and Vascellaro, J., "Hulu Reworks Its Script," *The Wall Street Journal*, January 27, 2011, p. A1.
39. Ibid.
40. Adams, Russell, "Times Prepares Pay Wall," *The Wall Street Journal*, January 24, 2011, p. B5.
41. Adams, Russell, "Slate's Layoffs Signal Flaws in Web Model," *The Wall Street Journal*, August 26, 2011, p. B7.
42. Catan, Thomas, "Repentant Google Settles on Drug Ads," *The Wall Street Journal*, August 25, 2011, p. B1.
43. Steel, E. and Fowler, S., "Facebook Gets New Friends," *The Wall Street Journal*, April 4, 2011, p. B4.
44. Efrati, A., "Google Takes on Friend Sprawl," *The Wall Street Journal*, June 29, 2011, p. B6.
45. Steel, E., "Websites Reach for the Stars," *The Wall Street Journal*, August 2, 2012, p. B4.
46. LinkedIn Web site, *www.linkedin.com*, accessed September 11, 2011.
47. Ning Web site, *http://get.ning.com*, accessed September 11, 2011.
48. Winslow, Ron, "When Patients Band Together," *The Wall Street Journal*, August 30, 2011, p. D1.
49. Staff, "Facebook Facilitates Organ Donation," *The Tampa Tribune*, May 2, 2012, p. 3.
50. Seth Godin Web site, *www.sethgodin.com/sg*, accessed September 7, 2011.
51. Raice, S. and Ante, S., "Insta-Rich: $1 Billion for Instagram," *The Wall Street Journal*, April 10, 2012, p. B1.
52. Angwin, J. and Singer-Vine, J., "The Selling of You," *The Wall Street Journal*, April 7–8, 2012, p. C1.
53. Efrati, A., "Google Opens Online App Store," *The Wall Street Journal*, February 3, 2011, p. B2.
54. 37Signals Web site, *http://37signals.com*, accessed September 11, 2011.
55. Microsoft Web site, *http://sharepoint.microsoft.com*, accessed September 11, 2011.
56. Khan Academy Web site, *www.khanacademy.org*, accessed May 5, 2011.
57. NPower Web site, *www.npower.org*, accessed September 6, 2011.
58. Evans, Bob, "Saving Lives and Changing the World via NPower," *InformationWeek*, March 28, 2011, p. 9.
59. Staff, "Method is More Vital Than Teacher," *The Tampa Tribune*, May 13, 2011, p. 1.
60. Pratt, A., "The Grill," *Computerworld*, May 23, 2011, p. 10.
61. MacMillan, D., "Textbook 2.0," *Bloomberg Businessweek*, June 13, 2011, p. 43.
62. Skillsoft Web site, *www.skillsoft.com*, accessed September 10, 2011.
63. eHow Web site, *www.ehow.com*, accessed September 10, 2011.
64. Certiport Web site, *www.certiport.com*, accessed September 10, 2011.
65. Kashi Web site, *www.kashi.com*, accessed September 7, 2011.
66. Henschen, D., "They're Ready," *InformationWeek*, February 14, 2011, p. 21.
67. Nash, Kim, "The Talent Advantage," *CIO*, March 1, 2011, p. 32.
68. Chao, L., "Baidu, Record Labels in Deal," *The Wall Street Journal*, July 19, 2011, p. B2.
69. Efrati, A., "Google Bows to Web Rivals," *The Wall Street Journal*, July 23, 2011, p. B1.
70. Mahalo Web site, *www.mahalo.com*, accessed September 7, 2011.

71. Live Person Web site, *www.liveperson.com*, accessed September 7, 2011.
72. Gaudin, Sharon, "Cloud Cures Hospital's Ailing Email System," *Computerworld*, July 18, 2011, p. 6.
73. Staff, "Case Study: Papa John's International," GoToMeeting, *www.gotomeeting.com/fec/images/pdf/caseStudies/GoToMeeting_Papa_Johns_Case_Study.pdf*, accessed September 7, 2011.
74. Smith, Ethan, "The Internet's $10 Million Mix Tapes," *The Wall Street Journal*, August 31, 2011, p. B1.
75. Smith, Ethan and Raice, Shayndi, "Facebook Adding Music Service," *The Wall Street Journal*, September 1, 2011, p. B8.
76. Podcast Alley Web site, *www.podcastalley.com*, accessed September 11, 2011.
77. Vascellaro, J. and Efrati, A., "Hulu Ponders Its Next Move," *The Wall Street Journal*, August 22, 2011, p. B5.
78. Bustillo, M. and Talley, K., "For Wal-Mart, a Rare Online Success," *The Wall Street Journal*, August 29, 2011, p. B1.
79. Schechner, S., "CNN, HLN to Stream on Web," *The Wall Street Journal*, July 18, 2011, p. B6.
80. Schechner, S., "Fox TV Shows Get Pay Wall," *The Wall Street Journal*, July 27, 2011, p. B6.
81. Peers, M., "The Time Bomb in Netflix's Strategy," *The Wall Street Journal*, July 14, 2011, p. C12.
82. Collett, S., "YouTube for the Enterprise," *Computerworld*, July 18, 2011, p. 26.
83. Trachtenberg, J., "Booksellers Alter Apple Sales," *The Wall Street Journal*, July 26, 2011, p. B6.
84. Catan et al., "U.S. Alleges E-Book Scheme," *The Wall Street Journal*, April 12, 2012, p. A1.
85. Audible Web site, *www.audible.com*, accessed September 11, 2011.
86. Leonard, D., "Master of the Game," *Bloomberg Businessweek*, July 4, 2011, p. 71.
87. Jargon, J., "Wrigley Targets Web Gamers," *The Wall Street Journal*, August 23, 2011, p. B6.
88. MacMillan, D. and Stone, B., "Zynga's Little Known Addiction: Whales," *Bloomberg Businessweek*, July 11, 2011, p. 37.
89. Scheck, J., "You Can Lead a Virtual Horse to Water, But You Might Get Sued along the Way," *The Wall Street Journal*, July 30, 2011, p. A1.
90. Raice, S., "Facebook Reacts to Google Games," *The Wall Street Journal*, August 13, 2011, p. B3.
91. Xbox Web site, *www.xbox.com/en-US/live*, accessed September 11, 2011.
92. Groupon Web site, *www.groupon.com*, accessed September 7, 2011.
93. Hickins, M., "Groupon Revenue Hit $760 Million, CEO Memo Shows," *The Wall Street Journal*, February 26, 2011, p. B3.
94. Chao, Loretta, "Groupon Stumbles in China," *The Wall Street Journal*, August 24, 2011, p. B1.
95. Hickins, Michael, "Groupon Deals Attract Connecticut's Scrutiny," *The Wall Street Journal*, July 15, 2011, p. B8.
96. O'Connell, V., "Building eBay for Lawyers," *The Wall Street Journal*, March 23, 2011, p. B7.
97. Craigslist Web site, *http:/craigslist.org*, accessed September 11, 2011.
98. Gorman, S. and Tibken, S., "Security Tokens Take Hit," *The Wall Street Journal*, June 7, 2011, p. B1.
99. Angwin, J. and Steel, E., "Web's Hot New Commodity: Privacy," *The Wall Street Journal*, February 28, 2011, p. A1.
100. Baker, J., "EU Law Tracking Cookies Ignored," *Computerworld*, June 6, 2011, p. 4.
101. Miller, John, "Yahoo Cookie Plan in Place," *The Wall Street Journal*, March 19, 2011, p. B3.
102. Fowler, G. and Lawton, C., "Facebook Again in Spotlight on Privacy," *The Wall Street Journal*, June 9, 2011, p. B1.
103. Borzo, Jeanette, "Employers Tread a Minefield," *The Wall Street Journal*, January 21, 2011, p. B6.
104. Woo, S., "Amazon Battles States over Sales Tax," *The Wall Street Journal*, August 3, 2011, p. A1.
105. Page, Jeremy, "Chinese State TV Alludes to U.S. Website Attacks," *The Wall Street Journal*, August 25, 2011, p. A8.
106. Efrati, A. and Gorman, S., "Google Mail Hack Is Blamed on China," *The Wall Street Journal*, June 2, 2011, p. A1.
107. Areddy, J., "Beijing Fires Back in Google Hack Row," *The Wall Street Journal*, June 3, 2011, p. B1.
108. Bryan-Low, Cassell, "Internet Attacks Yield Arrests," *The Wall Street Journal*, September 2, 2011, p. B5.

PART 1 | PART 2 | PART 3 | PART 4 | PART 5

Business Information Systems

CHAPTERS

8 Electron[illegible] Commerce

[illegible]	Learning Objectives
• Electronic an[illegible] providing ne[illegible] that present [illegible] problems.	• [illegible]e the current status of various forms [illegible]mmerce, including B2B, B2C, C2C, [illegible] Government. • [illegible]e a multistage purchasing model that [illegible]ribes how e-commerce works. • [illegible]ne m-commerce and identify some of [illegible] unique challenges.
• E-commerce and m-commerce can be used in many innovative ways to improve the operations of an organization.	• Identify several e-commerce and m-commerce applications. • Identify several advantages associated with the use of e-commerce and m-commerce.
• E-commerce and m-commerce offer many advantages yet raise many challenges.	• Identify the many benefits and challenges associated with the continued growth of e-commerce and m-commerce.
• Organizations must define and execute a strategy to be successful in e-commerce and m-commerce.	• Outline the key components of a successful e-commerce and m-commerce strategy.
• E-commerce and m-commerce require the careful planning and integration of a number of technology infrastructure components.	• Identify the key components of technology infrastructure that must be in place for e-commerce and m-commerce to work. • Discuss the key features of the electronic payment systems needed to support e-commerce and m-commerce.

Information Systems in the Global Economy
KRAMP GROUP, HOLLAND

A Million Spare Parts—and Counting

Kramp Group is Europe's largest distributor of accessories and parts for motorized equipment, agriculture, and construction machines. That may not sound glamorous, but "Modern agriculture is highly mechanized: it is impossible to run a farm successfully without tractors, harvesters and other machinery," explains IT manager Robert Varga. "If a critical component fails and puts one of our customers' machines out of action, the loss of productivity can cost them serious amounts of money. We have a catalogue of more than 300,000 spare parts which can be delivered within a single working day from any of our European warehouses to their nearest dealership, helping them get back up and running as quickly as possible."

Kramp Group CEO Eddie Perdok says, "We believe in the future and the power of e-commerce. Compared to other sales channels, the Internet gives us significant cost advantages."

Yet, to Kramp's customers, using the Internet isn't automatic. Kramp takes more than 50,000 customer orders every day from various channels. Prior to 2010, "nearly 40 percent of our customers still placed their orders by phone, which meant that our call center staff had to spend a lot of time on basic order-processing," says Varga. To reduce that figure, Kramp had to make its online store easier to use—but their existing store, which had been developed in-house, did not have the flexibility to achieve this goal.

Kramp turned to packages from IBM and German software firm Heiler AG to modernize its e-commerce systems. Hans Scholten, a member of Kramp Group's executive board, says, "We deliberately opt for the 'best of breed' solution for all packages. That means we choose the best available software for different applications." That philosophy helped determine the packages the company chose.

From IBM, the firm obtained WebSphere Commerce for its customer-facing side of its system. This software's multilanguage capability was important: Operating throughout Europe as Kramp does, being able to have one site that can operate in any of 10 languages was crucial. Kramp must translate the content into all the languages, since in 2012 even the best automatic translation software can't replace a skilled person, but the company doesn't have to develop and support different sites.

Kramp also uses Heiler Software's Product Information Management (PIM) solution. That software manages product data in the catalog behind WebSphere Commerce. Kramp wants to expand its deliverable stock to over one million items and could not do so without PIM. Expanding to more than one million stocked items is crucial to Kramp's *long tail* strategy: the concept that each of the slow-selling items may not account for much revenue but that the total of all slow-selling items is large enough to make a difference to Kramp's success.

Finally, though Kramp has the in-house capability to manage its e-commerce system, it turned to CDC Software to help integrate the pieces. Doing so itself would have required the company to hire additional staff, which it wouldn't need when the project was done.

The result was that after Kramp's new system had been online only a short while, Varga reported that "with 70 percent of our customers choosing to order online via the WebSphere Commerce solution, we are seeing a significant reduction in the average cost per transaction." He continued, "Our call center staff now has more time to help customers solve complex problems, which improves customer service. Better service and lower operational costs are helping Kramp Group achieve 10 to 12 percent annual growth, so the solution is making a real contribution to the success of our business."

As you read this chapter, consider the following:

- How do other types of e-commerce, such as business-to-consumer (B2C), differ from Kramp Group's business-to-business (B2B) e-commerce?
- Using the philosophy that Scholten expressed, Kramp got an excellent e-commerce package that might not work smoothly with its other software. Instead, it could have selected a single integrated package, all the pieces of which work well with each other but which might not be perfect for their e-commerce. As you read and think about different situations, consider which approach is better and why.

WHY LEARN ABOUT ELECTRONIC AND MOBILE COMMERCE?

© Andrey Burmakin/Shutterstock

Electronic and mobile commerce have transformed many areas of our lives and careers. One fundamental change has been the manner in which companies interact with their suppliers, customers, government agencies, and other business partners. As a result, most organizations today have set up business on the Internet or are considering doing so. To be successful, all members of the organization need to plan and participate in that effort. As a sales or marketing manager, you will be expected to help define your firm's e-commerce business model. As a customer service employee, you can expect to participate in the development and operation of your firm's Web site. As a human resource or public relations manager, you will likely be asked to provide Web site content for use by potential employees and shareholders. As an analyst in finance, you will need to know how to measure the business impact of your firm's Web operations and how to compare that to competitors' efforts. Clearly, as an employee in today's organization, you must understand what the potential role of e-commerce is, how to capitalize on its many opportunities, and how to avoid its pitfalls. The emergence of m-commerce adds an exciting new dimension to these opportunities and challenges. Future customers, potential employees, and shareholders will be accessing your firm's Web site via smartphones and tablet computers from places other than their homes or places of business. This chapter begins by providing a brief overview of the dynamic world of e-commerce.

AN INTRODUCTION TO ELECTRONIC COMMERCE

electronic commerce: Conducting business activities (e.g., distribution, buying, selling, marketing, and servicing of products or services) electronically over computer networks.

Electronic commerce is the conducting of business activities (e.g., distribution, buying, selling, marketing, and servicing of products or services) electronically over computer networks. It includes any business transaction executed electronically between companies (business-to-business), companies and consumers (business-to-consumer), consumers and other consumers (consumer-to-consumer), public sector and business (government-to-business), and the public sector to citizens (government-to-citizen). Business activities that are strong candidates for conversion to e-commerce are ones that are paper based, time consuming, and inconvenient for customers.

Business-to-Business E-Commerce

business-to-business (B2B) e-commerce: A subset of e-commerce in which all the participants are organizations.

Business-to-business (B2B) e-commerce is a subset of e-commerce in which all the participants are organizations. B2B e-commerce is a useful tool for connecting business partners in a virtual supply chain to cut resupply times and reduce costs. Although the business-to-consumer market grabs more of the news headlines, the B2B market is considerably larger and is growing more rapidly. For 2009, the most recent year for which U.S. Census Bureau numbers are available, B2B revenue was $3.8 trillion compared to business-to-consumer (B2C) revenue of $292 billion. In other words, B2B revenue was about 13 times the B2C revenue.[1]

An organization will use both *buy-side e-commerce* to purchase goods and services from its suppliers and *sell-side e-commerce* to sell products to its customers. Buy-side e-commerce activities may include identifying and comparing competitive suppliers and products, negotiating and establishing prices and terms, ordering and tracking shipments, and steering organizational buyers to preferred suppliers and products. Sell-side e-commerce activities may include enabling the purchase of products online, providing information for customers to evaluate the organization's goods and services, encouraging sales and generating leads from potential customers, providing a portal of information of interest to the customer, and enabling interactions among a community of consumers. Thus, buy-side and sell-side e-commerce activities support the organization's value chain and help the organization provide lower prices, better service, higher quality, or uniqueness of product and service, as first mentioned in Chapter 2.

Grainger is a B2B distributor of products for facilities maintenance, repair, and operations (a category called MRO) and sells nearly 1 million different items online. E-commerce accounts for 25 percent of Grainger's total sales but is expected to become 40 percent of its total business. The firm is adding click-to-call, or chat, capability to its Web site. According to Paul Miller, Grainger vice-president of e-commerce, "Better and faster access to information helps people make better decisions."[2]

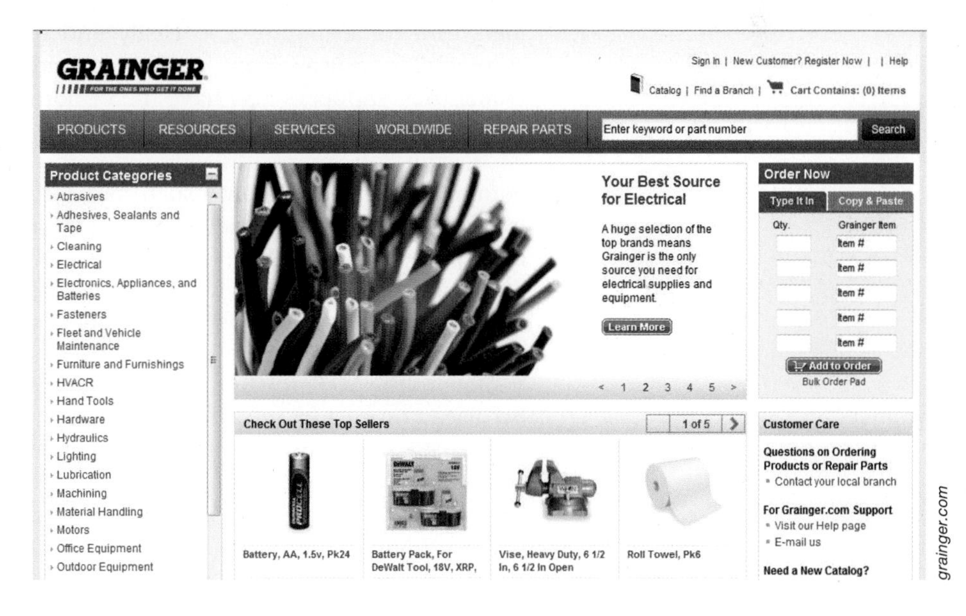

Grainger sells nearly 1 million different items online.

Business-to-Consumer E-Commerce

business-to-consumer (B2C) e-commerce: A form of e-commerce in which customers deal directly with an organization and avoid intermediaries.

Business-to-consumer (B2C) e-commerce is a form of e-commerce in which customers deal directly with an organization and avoid intermediaries. Early B2C pioneers competed with the traditional "brick-and-mortar" retailers in an industry selling their products directly to consumers. For example, in 1995, upstart Amazon.com challenged well-established booksellers Waldenbooks and Barnes and Noble. Amazon did not become profitable until 2003; the firm has grown from selling only books on a United States-based Web site to selling a wide variety of products through international Web sites in Canada, China, France, Germany, Japan, and the United Kingdom.

Global B2C e-commerce revenue in 2010 was estimated to be between $400 and $600 billion U.S. dollars with further growth estimated to increase to $700 to $950 billion by 2015.[3] In the United States, B2C e-commerce sales in the third quarter of 2011 amounted to $48.2 billion, an increase of 13.7 percent over the third quarter of 2010.[4] B2C e-commerce sales now amount to about 4.6 percent of total U.S. retail sales.[5] Figure 8.1 shows the forecasted growth by world region.[6]

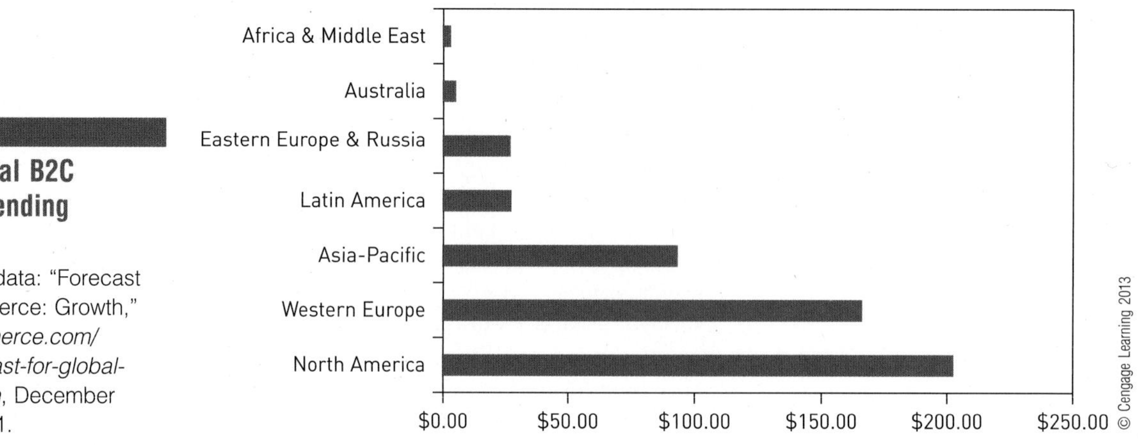

FIGURE **8.1**

Forecasted global B2C e-commerce spending for 2014

Source of the raw data: "Forecast for Global e-Commerce: Growth," *www.capturecommerce.com/blog/general/forecast-for-global-ecommerce-growth*, December 2010/ January 2011.

One reason for the steady growth in B2C e-commerce is shoppers find that many goods and services are cheaper when purchased via the Web, including stocks, books, newspapers, airline tickets, and hotel rooms. Consumers can, for example, also easily and quickly compare information about automobiles, cruises, loans, insurance, and home prices to find better values. Online B2C shoppers also have the opportunity to design their own personalized product or customize the packaging for some consumer goods products.

Easton Bell Sports is a leading designer, developer, and marketer of sports equipment for team sports, such as baseball, softball, hockey, football, and lacrosse, and for individual sports, such as cycling and skateboarding. With headquarters in Van Nuys, California, it has 31 facilities worldwide.[7] While the majority of its sales is to teams, the firm started a B2C Web site to enable individual consumers to personalize its standard products.[8]

Wrigley offers custom packaging for its Extra sugar-free brand of gum. Visitors to the MyExtraGum.com Web site can upload photos and messages to be placed on custom packs of gum. They can also specify fonts and colors for the packs and choose from among many package designs. The customer packs are pricey at $25 for five packs of gum, with each pack containing 15 sticks of gum. However, many consumers order the custom packs to celebrate important events in their lives or as personalized gifts.[9]

By using B2C e-commerce to sell directly to consumers, producers or providers of consumer products can eliminate the middlemen, or intermediaries, between them and the consumer. In many cases, this squeezes costs and inefficiencies out of the supply chain and can lead to higher profits for businesses and lower prices for consumers. The elimination of intermediate organizations between the producer and the consumer is called *disintermediation*.

More than just a tool for placing orders, the Internet is an extremely useful way to compare prices, features, and value, and to check other customers' opinions. Internet shoppers can, for example, unleash shopping bots or access sites such as eBay Shopping.com, Google Froogle, Shopzilla, PriceGrabber, Yahoo! Shopping, or Excite to browse the Internet and obtain lists of items, prices, and merchants. Many B2C merchants have added what is called "social commerce" to their Web sites by creating a section where shoppers can go to see only those products that have been reviewed and listed by other shoppers. Walmart implemented its Shopycat application that refers to information from a shopper's friends on Facebook to make gift recommendations for the shopper. He or she can then purchase these items from Walmart or other stores online or via a personal visit.[10]

One growing trend is consumers researching products online but then purchasing those products at their local brick-and-mortar stores. Sales in local stores that are stimulated through online marketing and research are called Web-influenced sales. Such sales are estimated to already exceed $1 trillion and are growing at the rate of billions of dollars per year in the United States.[11] Table 8.1 lists the 10 largest B2C retailers in the United States.

TABLE 8.1 Largest business-to-consumer retailers in the United States 2010

Rank	Company	Total Web Sales Volume 2011 (Billions of Dollars)
1	Amazon	$34.2
2	Staples	$10.2
3	Apple	$5.2
4	Dell	$4.8
5	Office Depot	$4.1
6	Walmart	$4.1
7	Sears	$3.1
8	Liberty Media (QVC, Liberty e-Commerce)	$3.0
9	OfficeMax	$2.9
10	CDW Corp.	$2.7

Source: "Top 500 Guide," *Internet Retailer*, *www.internetretailer.com/top500/list*, accessed January 16, 2012.

As a result of a 1992 Supreme Court ruling that says online retailers don't have to collect sales taxes in states where they lack a physical presence, millions of online shoppers do not pay state or local tax on their online purchases. Consumers who live in states with sales tax are supposed to keep track of their out-of-state purchases and report those "use taxes" on their state income tax returns. However, few tax filers report such purchases. Thus, despite having a legal basis to do so, the states find it very difficult to collect sales taxes on Internet purchases. This avoidance of sales tax creates a price advantage for online retailers over brick-and-mortar stores where sales taxes must be collected. It also results in the loss of about $23 billion in tax revenue that could go to state and local governments to provide services for their citizens.[12] (Illinois has specific lines on its income tax form to report Internet purchases. If you have not kept records, you are assessed a certain amount per $100,000 of income for Internet sales).

Consumer-to-Consumer E-Commerce

consumer-to-consumer (C2C) e-commerce: A subset of e-commerce that involves electronic transactions between consumers using a third party to facilitate the process.

Consumer-to-consumer (C2C) e-commerce is a subset of e-commerce that involves electronic transactions between consumers using a third party to facilitate the process. eBay is an example of a C2C e-commerce site; customers buy and sell items to each other through the site. Founded in 1995, eBay has become one of the most popular Web sites in the world.

Other popular C2C sites include Bidzcom, Craigslist, eBid, ePier, Ibidfree, Kijiji, Ubid, and Tradus. The growth of C2C is responsible for reducing the use of the classified pages of newspapers to advertise and sell personal items, so it has a negative impact on that industry. On the other hand, C2C has created an opportunity for many people to make a living out of selling items on auction Web sites.

Companies and individuals engaging in e-commerce must be careful that their sales do not violate the rules of various county, state, or country legal jurisdictions. More than 4,000 Web sites offer guns for sale including Craigslist where thousands of guns were offered despite the Web site's policy against such sales. Many sellers agree to sell guns to people who they think could not pass a background check—a felony offense.[13] New York City charges a $1.50 tax a pack on cigarettes, but residents can purchase thousands of cartons online to avoid the city tax and then resell the cigarettes illegally.[14] The resale of untaxed cigarettes is a problem in many states with high tax rates.

Table 8.2 summarizes the key factors that differentiate among B2B, B2C, and C2C e-commerce.

TABLE **8.2** Differences among B2B, B2C, and C2C

Factors	B2B	B2C	C2C
Value of sale	Thousands or millions of dollars	Tens or hundreds of dollars	Tens of dollars
Length of sales process	Days to months	Days to weeks	Hours to days
Number of decision makers involved	Several people to a dozen or more	One or two	One or two
Uniformity of offer	Typically a uniform product offering	More customized product offering	Single product offering, one of a kind
Complexity of buying process	Extremely complex, much room for negotiation on price, payment and delivery options, quantity, quality, and options and features	Relatively simple, limited discussion over price and payment and delivery options	Relatively simple, limited discussion over payment and delivery options; negotiation over price
Motivation for sale	Driven by a business decision or need	Driven by an individual consumer's need or emotion	Driven by an individual consumer's need or emotion

e-Government

e-Government: The use of information and communications technology to simplify the sharing of information, speed formerly paper-based processes, and improve the relationship between citizens and government.

e-Government is the use of information and communications technology to simplify the sharing of information, speed formerly paper-based processes, and improve the relationship between citizens and government. Government-to-citizen (G2C), government-to-business (G2B), and government-to-government (G2G) are all forms of e-Government, each with different applications.

Citizens can use G2C applications to submit their state and federal tax returns online, renew auto licenses, purchase postage, apply for student loans, and make campaign contributions. Citizens can purchase items from the U.S. government through its GSA Auctions Web site, which offers the general public the opportunity to bid electronically on a wide range of government assets. In addition, more than 11,000 federal government Web sites and many more state and local government Web sites publish useful information for citizens.[15]

G2B applications support the purchase of materials and services from private industry by government procurement offices, enable firms to bid on government contracts, and help businesses identify government contracts on which they may bid. Business.gov allows businesses to access information about laws and regulations and relevant forms needed to comply with federal requirements for their business. The Buyers.gov Web

INFORMATION SYSTEMS @ WORK

E-Commerce Web Sites: They Only Work if They Work

As consumers, when we go to a Web site we assume it will "just work." The retailer that publishes the site has to ensure that we have a smooth shopping experience and want to return.

Consider, for example, the cautionary tale of Target. Its site was unexpectedly offline on September 13, 2011, and again on October 25. Those were both before the year-end holiday shopping crush. If the service outages had occurred during that crush, Target's sales and reputation could have been seriously damaged.

"Target obviously needs to figure out what's causing the problems with their Web presence," said Dan Olds, an analyst with the Gabriel Consulting Group. "They need to put the same, or more, effort into maintaining their website as a physical store manager would to ensure his store is operating in an efficient and safe manner. Target needs to redouble these efforts as we get closer to the holiday sales season."

It's easy to observe Target's problems and say "it needs to do more to solve them." What, as a practical matter, should the company do? One answer is test, then test again.

Back in 2005, Best Buy's Web site experienced a catastrophic holiday failure and customers could not make online purchases. That same year competitors saw huge spikes in traffic, says Dave Karow, senior product manager of Web performance and testing at Keynote, a firm that monitors and tests mobile and Internet performance. "There's nothing like falling flat on your face to give you the conviction to do [the] right thing going forward. That was an extremely effective wakeup call for Best Buy," he continues, adding that the retailer now conducts several load tests throughout the year.

Online shoe retailer Zappos focuses on estimating the load on its site—the number of people who will try to access it at the same time and the complexity of the tasks those people will try to accomplish. The company then tests to make sure it can handle more than those projections. Kris Ongbongan, senior manager of technical operations and systems engineering at Zappos, explains "We have our finance and planning departments give us sales predictions and we take a multiple of that to see what traffic we can absorb and test to that," typically beginning in September. That gives Zappos enough time to make changes if necessary. "We have instrumentation around every transaction point on the website, from search pages to product detail pages to checkout," he says, "so we can look at each individually to see if there's any slowness or problems in any of those areas."

Alternative street fashion retailer Karmaloop learned its load-testing lesson the hard way. During the 2010 holiday season, it found that its content delivery network (CDN, the set of computers through which it delivers Web content to users) was not optimized for Cyber Monday traffic, says chief technical officer Joseph Finsterwald. (Cyber Monday is the Monday after the U.S. Thanksgiving Day holiday, typically the busiest day of the year for retail e-shopping.) "We worked with our CDN vendor Akamai to come up with a configuration that was a better fit for us," he says.

Michael Ebert is a partner in KPMG's IT Advisory Services practice. He recommends that companies use multiple networks and online points of presence to spread the load. That way, if one data center goes down, others can step in. The site "may be slow to respond, but at least I'm up and running. There's always a percentage of business you never regain if someone leaves the site."

Discussion Questions

1. Is load testing an e-commerce site only important for very large retailers such as Target and Best Buy or should smaller ones be concerned with it as well?
2. List three Web sites that your college or university operates. When does the load peak on those sites? Should your school be concerned with load testing?

Critical Thinking Questions

1. Suppose you are the CIO of an e-commerce firm, and your site goes down during a critical period. After the site is operating again, what business steps might you take to regain customer confidence and keep your customers from switching to your competitors?
2. Suppose you are the CIO of an e-commerce firm, and a competitor's site goes down during a critical period. What business steps might you take to capture some of that firm's customers? Keep ethical behavior in mind, and remember that your firm's site could go down at an equally critical time in the future.

SOURCES: Gaudin, S., "Target.com's Second Site Crash Could Become e-Comm Nightmare," *Computerworld*, *www.computerworld.com/s/article/9221221/Target.com_s_second_site_crash_could_become_e_comm_nightmare*, October 26, 2011; "Web Load Testing," Keynote Systems, Inc., *www.keynote.com/products/web_load_testing/load-testing-tools.html*, accessed March 2, 2012; Shein, E., "How to Bulletproof Your Website," *Computerworld*, *www.computerworld.com/s/article/9222177/How_to_bulletproof_your_website*, November 11, 2011.

site is a business and auction exchange that helps federal government agencies purchase information system products by using reverse auctions and by aggregating demand for commonly purchased products. FedBizOpps is a Web site where government agencies post procurement notices to provide an easy point of contact for businesses that want to bid on government contracts.

G2G applications support transactions between governments such as between the federal government and state or local governments. Government to Government Services online is a Web site that enables government organizations to report information, such as birth and death data, arrest warrant information, and information about the amount of state aid being received, to the Social Security Administration. This information can affect the payment of benefits to individuals. The State and Local Government on the Net Web site at *www.statelocalgov.net* provides a directory of thousands of state agencies and city and county governments.

AN INTRODUCTION TO MOBILE COMMERCE

A rapidly growing segment of e-commerce is mobile commerce. As discussed briefly in Chapter 1, mobile commerce (m-commerce) relies on the use of mobile, wireless devices, such as cell phones and smartphones, to place orders and conduct business. Handset manufacturers such as Ericsson, Motorola, Nokia, and Qualcomm are working with communications carriers such as AT&T, Cingular, Sprint/Nextel, and Verizon to develop such wireless devices, related technology, and services. The Internet Corporation for Assigned Names and Numbers (ICANN) created a .mobi domain to help attract mobile users to the Web. mTLD Top Level Domain Ltd of Dublin, Ireland, administers this domain and helps to ensure that the .mobi destinations work quickly, efficiently, and effectively with user handsets.

Mobile Commerce in Perspective

The market for m-commerce in North America is maturing much later than in Western Europe and Japan for several reasons. In North America, responsibility for network infrastructure is fragmented among many providers, consumer payments are usually made by credit card, and many Americans are unfamiliar with mobile data services. In most Western European countries, communicating via wireless devices is common, and consumers are much more willing to use m-commerce. Japanese consumers are generally enthusiastic about new technology and are therefore much more likely to use mobile technologies for making purchases.

In the United States, Forrester Research forecasts that the volume of mobile commerce (not counting sales via tablet-based computers) will reach $31 billion by 2016. This sum would represent only about 7 percent of total B2C e-commerce sales. Still, Forrester believes that consumers will use their smartphones to research purchases, look up information on products in retail stores, and in general, improve their in-store shopping experience.[16]

The number of mobile Web sites worldwide is expected to grow rapidly because of advances in wireless broadband technologies, the development of new and useful applications, and the availability of less costly but more powerful handsets. Experts point out that the relative clumsiness of mobile browsers and security concerns must be overcome to ensure rapid m-commerce growth.

M-Commerce Web Sites

A number of retailers have established special Web sites for users of mobile devices. Table 8.3 provides an alphabetical list of some of the best mobile Web sites that were launched in 2011 based on depth of content, personalization, social media integration, mobile coupons, and use of video.

TABLE 8.3 Highly rated m-commerce retail Web sites

Retailer	Products Offered	Web Site
Bath & Body Works	Personal care products	*www.bathandbodyworks.com*
BMW's Mini Financial Services	Financial services	*www.miniusa.com*
Brooks Brothers	Apparel	*www.brooksbrothers.com*
Cabela's Inc.	Outdoor gear	*www.cabelas.com*
Garnet Hill	Apparel and home decor	*www.garnethill.com*
L.L. Bean	Outdoor gear and apparel	*www.llbean.com*
New York & Company	Women's apparel	*www.nyandcompany.com*
Paul Fredrick	Men's apparel	*www.paulfredrick.com*
Staples	Office products	*www.staples.com*
Zales	Jewelry	*www.zales.com*

Source: Tsirulnik, Giselle, "Top 10 Mobile Commerce-Enabled Web Sites of 2011," *Mobile Commerce Daily*, *www.mobilecommercedaily.com/2011/11/17/top-10-mobile-commerce-enabled-web-sites-of-2011/print*, November 17, 2011.

An interesting service can be accessed from Twitter by sending a shopping-related question to @IMshopping. Your question is routed to an appropriate expert who can provide unbiased opinions and links to products within about 15 minutes. (Experts are rated over time based on their answers.) The expert's response is a URL link to a page at IMshopping.com that will provide a longer answer than the 140 character limit imposed by Twitter.

Advantages of Electronic and Mobile Commerce

Conversion to an e-commerce or m-commerce system enables organizations to reduce the cost of doing business, speed the flow of goods and information, increase the accuracy of order processing and order fulfillment, and improve the level of customer service.

Reduce Costs

By eliminating or reducing time-consuming and labor-intensive steps throughout the order and delivery process, more sales can be completed in the same period and with increased accuracy. With increased speed and accuracy of customer order information, companies can reduce the need for inventory—from raw materials to safety stocks and finished goods—at all the intermediate manufacturing, storage, and transportation points.

Equa-Ship is an e-commerce company that consolidates the goods from many small shippers into the same trailers so that the latter can earn the same high-volume shipping discounts that large companies like Amazon.com do. The downside is that the delivery time is a little longer. However, the cost savings enables small companies to offer free shipping at essentially the same cost to themselves as large companies. Shippers communicate their pick-up and delivery needs as well as print labels using online software provided by Equa-Ship.[17]

Speed the Flow of Goods and Information

When organizations are connected via e-commerce, the flow of information is accelerated because electronic connections and communications are already established. As a result, information can flow from buyer to seller easily, directly, and rapidly.

Bypass Lane is a mobile app available through the App Store, Android Market, and BlackBerry App World that enables consumers at a live sporting event to order food, drinks, and merchandise using their smartphones. The items can be delivered to their seat or picked up at a special window at the concession stand. The service is licensed by the sporting teams and is available in over 40 venues in the United States including the American Airlines Center, home of the Dallas Mavericks (NBA basketball) and Dallas Stars (NHL Hockey), as well as the homes of the Texas Longhorns (college football), Texas Rangers (MLB baseball), and St. Louis Cardinals and Philadelphia Eagles (NFL football). Participating teams license the software and pay Bypass Lane a per transaction fee. This mobile app not only increases the convenience of ordering food and enjoyment of the sporting event, it also increases concession stand sales by 30 to 40 percent.[18]

Increase Accuracy

By enabling buyers to enter their own product specifications and order information directly, human data-entry error on the part of the supplier is eliminated. R.O. Writer is shop management software used by thousands of auto repair, quick lube, and tire repair shops across the United States. It integrates with the electronic parts catalogs of various parts distributors to allow repair people to correctly order the specific parts needed for each repair job. The data is then transferred onto customer repair orders to produce timely and accurate customer invoices.[19]

Improve Customer Service

Increased and more detailed information about delivery dates and current status can increase customer loyalty. In addition, the ability to consistently meet customers' desired delivery dates with high-quality goods and services eliminates any incentive for customers to seek other sources of supply.

Ryder has been a provider of transportation, logistics, and supply chain management solutions for nearly 80 years. Its Ryder Location is a free smartphone app that provides improved customer service. This app enables customers to identify the closest Ryder facility either by specifying location (city, state, and/or zip code) or service needed (rental, maintenance, or fuel). Once the Ryder facility is selected, users can get turn-by-turn directions sent to their smartphone.[20]

Multistage Model for E-Commerce

A successful e-commerce system must address the many stages that consumers experience in the sales life cycle. At the heart of any e-commerce system is the user's ability to search for and identify items for sale; select those items and negotiate prices, terms of payment, and delivery date; send an order to the vendor to purchase the items; pay for the product or service; obtain product delivery; and receive after-sales support. Figure 8.2 shows how e-commerce can support each of these stages. Product delivery can involve tangible goods delivered in a traditional form (e.g., clothing delivered via a package service) or goods and services delivered electronically (e.g., software downloaded over the Internet).

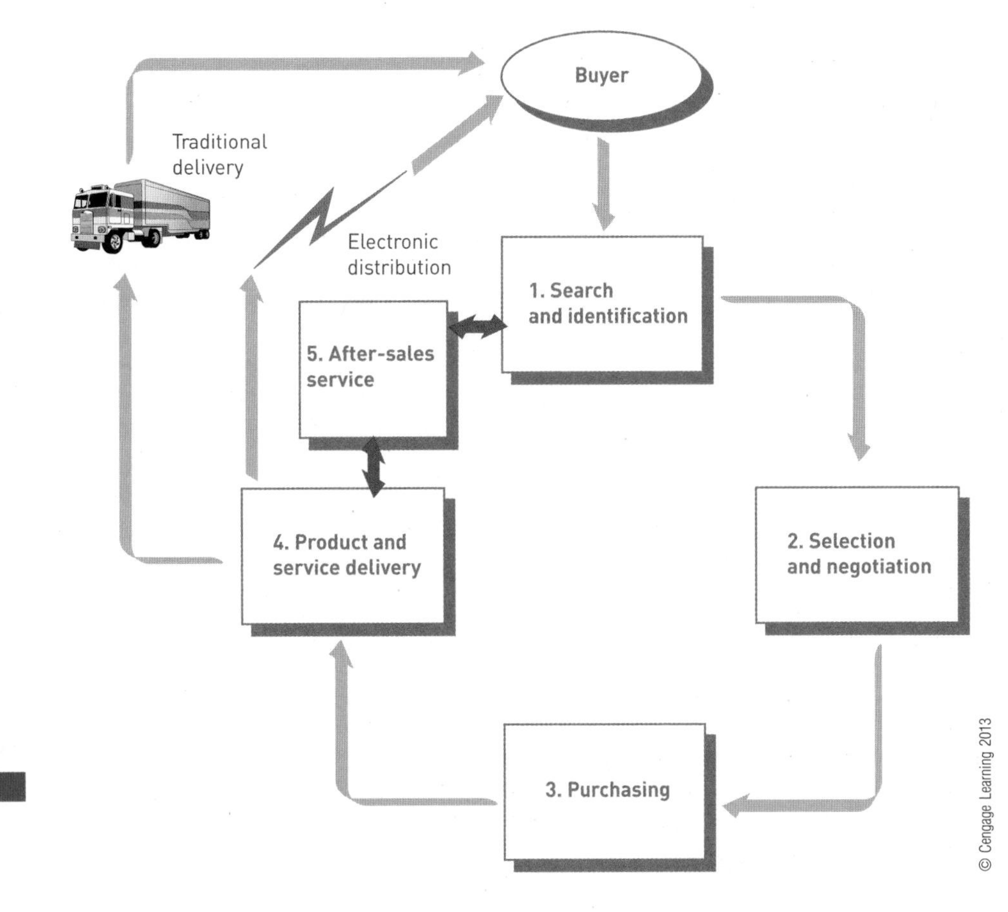

FIGURE 8.2
Multistage model for e-commerce (B2B and B2C)

Search and Identification

An employee ordering parts for a storeroom at a manufacturing plant would follow the steps shown in Figure 8.2. Such a storeroom stocks a wide range of office supplies, spare parts, and maintenance supplies. The employee prepares a list of needed items—for example, fasteners, piping, and plastic tubing. Typically, for each item carried in the storeroom, a corporate buyer has already identified a preferred supplier based on the vendor's price competitiveness, level of service, quality of products, and speed of delivery. The employee then logs on to the Internet and goes to the Web site of the preferred supplier.

From the supplier's home page, the employee can access a product catalog and browse until he or she finds the items that meet the storeroom's specifications. The employee fills out a request-for-quotation form by entering the item codes and quantities needed. When the employee completes the quotation form, the supplier's Web application calculates the total charge of the order with the most current prices and shows the additional cost for various forms of delivery—overnight, within two working days, or the next week. The employee might elect to visit other suppliers' Web home pages and repeat this process to search for additional items or obtain competing prices for the same items.

Selection and Negotiation

After the price quotations have been received from each supplier, the employee examines them and indicates by clicking the request-for-quotation form which items to order from a given supplier. The employee also specifies the desired delivery date. This data is used as input into the supplier's order-processing TPS. In addition to price, an item's quality and the supplier's service and speed of delivery can be important in the selection and negotiation process.

B2B e-commerce systems need to support negotiation between a buyer and the selected seller over the final price, delivery date, delivery costs, and any extra charges. However, these features are not fundamental requirements of most B2C systems, which offer their products for sale on a "take-it-or-leave-it" basis.

Purchasing Products and Services Electronically

The employee completes the purchase order specifying the final agreed-to terms and prices by sending a completed electronic form to the supplier. Complications can arise in paying for the products. Typically, a corporate buyer who makes several purchases from the supplier each year has established credit with the supplier in advance, and all purchases are billed to a corporate account. But when individual consumers make their first, and perhaps only, purchase from the supplier, additional safeguards and measures are required. Part of the purchase transaction can involve the customer providing a credit card number. Another approach to paying for goods and services purchased over the Internet is using electronic money, which can be exchanged for hard cash, as discussed later in the chapter.

Product and Service Delivery

Electronic distribution can be used to download software, music, pictures, videos, and written material through the Internet faster and for less expense than shipping the items via a package delivery service. Most products cannot be delivered over the Internet, so they are delivered in a variety of other ways: overnight carrier, regular mail service, truck, or rail. In some cases, the customer might elect to drive to the supplier and pick up the product.

Many manufacturers and retailers have outsourced the physical logistics of delivering merchandise to cybershoppers—those who take care of the storing, packing, shipping, and tracking of products. To provide this service, DHL, Federal Express, United Parcel Service, and other delivery firms have developed software tools and interfaces that directly link customer ordering, manufacturing, and inventory systems with their own system of highly automated warehouses, call centers, and worldwide shipping networks. The goal is to make the transfer of all information and inventory, from the manufacturer to the delivery firm to the consumer, fast and simple.

For example, when a customer orders a printer at the Hewlett-Packard (HP) Web site, that order actually goes to FedEx, which stocks all the products that HP sells online at a dedicated e-distribution facility in Memphis, Tennessee, a major FedEx shipping hub. FedEx ships the order, which triggers an e-mail notification to the customer that the printer is on its way and an inventory notice is sent to HP that the FedEx warehouse now has one fewer printers in stock (see Figure 8.3).

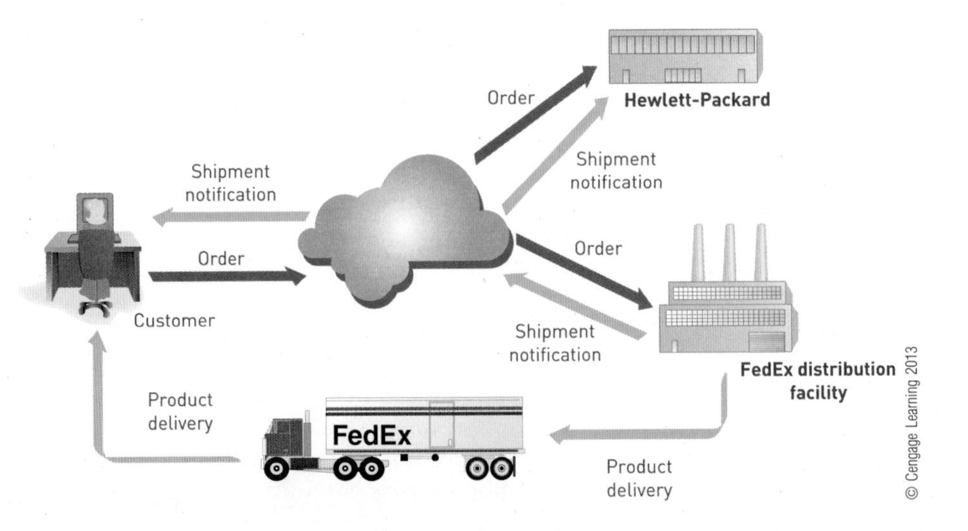

FIGURE 8.3

Product and information flow for HP printers ordered over the Web

For product returns, HP enters return information into its own system, which is linked to FedEx. This information signals a FedEx courier to pick up the unwanted item at the customer's house or business. Customers don't need to fill out shipping labels or package the item. Instead, the FedEx courier uses information transmitted over the Internet to a computer in his truck to print a label from a portable printer attached to his belt. FedEx has control of the return, and HP can monitor its progress from start to finish.

After-Sales Service

In addition to the information to complete the order, comprehensive customer information is also captured from the order and stored in the supplier's customer database. This information can include customer name, address, telephone numbers, contact person, credit history, and other details. For example, if the customer later contacts the supplier to complain that not all items were received, that some arrived damaged, or even that the product provides unclear instructions, any customer service representative can retrieve the order information from the database via a computing/communications device. Companies are adding to their Web sites the capability to answer many after-sales questions, such as how to maintain a piece of equipment, how to effectively use the product, and how to receive repairs under warranty.

E-Commerce Challenges

A company must overcome many challenges to convert its business processes from the traditional form to e-commerce processes, especially for B2C e-commerce. As a result, not all e-commerce ventures are successful. For example, Borders began an online Web site in the late 1990s, but after three years of operating in the red, the bookseller outsourced its e-commerce operations to Amazon in 2001. In 2006, Borders reversed course and decided to relaunch its own Borders.com Web site in May 2008. Since then, Borders generated disappointing sales figures. As a result of the substandard results, many decision makers were replaced, including the CIO and senior vice president of sales. Finally in early 2011, Borders applied for bankruptcy protection and began closing its stores.[21]

Dealing with Consumer Privacy Concerns

The following are three key challenges to e-commerce: (1) dealing with consumer privacy concerns, (2) overcoming consumers' lack of trust, and (3) overcoming global issues. While two-thirds of U.S. Internet users have purchased an item online and most Internet users say online shopping saves them time, about one-third of all adult Internet users will not buy anything online primarily because they have privacy concerns or lack trust in online merchants. In addition to having an effective e-commerce model and strategy, companies must carefully address consumer privacy concerns and overcome consumers' lack of trust.

According to the Privacy Rights Clearinghouse, the approximate number of computer records containing sensitive personal information involved in security breaches in the United States from 2005 to 2011 was in excess of 543 million![22] This figure represents the approximate number of records, not people affected, as some people were the victim of more than one breach. Following are a few examples of recent security beaches in which personal data was compromised:

- A massive security breach on the Sony PlayStation network compromised the personal information of over 100 million users.[23]
- Some 24 million customers of Zappos.com, an online shoes and clothing Web site, were informed that data for their accounts may have been illegally accessed by unauthorized parties.[24]

- Hackers managed to access the customer names, account numbers, and contact information for 2 million Citigroup customers.[25]
- The SEGA Pass Web site, which provides information about new products to registered members, was hit by hackers, and names, dates of birth, e-mail addresses, and encrypted passwords for 1.3 million users were compromised.[26]

identify theft: Someone using your personal identification information without your permission to commit fraud or other crimes.

In some cases, the compromise of personal data can lead to identity theft. According to the Federal Trade Commission (FTC), "**Identity theft** occurs when someone uses your personally identifying information, like your name, Social Security number, or credit card number, without your permission, to commit fraud or other crimes."[27] Thieves may use a consumer's credit card numbers to charge items to that person's accounts, use identification information to apply for a new credit card or a loan in a consumer's name, or use a consumer's name and Social Security number to receive government benefits.

Companies must be prepared to make a substantial investment to safeguard their customers' privacy or run the risk of losing customers and generating potential class action law suits should the data be compromised. It is not uncommon for customers to initiate a class action lawsuit for millions of dollars in damages for emotional distress and loss of privacy. In additional to potential damages, companies must frequently pay for customer credit monitoring and identity theft insurance to ensure that their customers' data is secure. Indeed a customer commenced a class action lawsuit against Zappos and parent company Amazon.com, alleging that she and millions of customers were harmed by the theft of personal account information.[28]

Most Web sites invest in the latest security technology and employ highly trained security experts to protect their consumers' data. The presence of the McAfee Secure security logon on the pages of an e-commerce Web site indicates that the site meets all guidelines set by the payment card industry. It also signifies that the site is secure from hackers and malicious software that could access customer identity information, passwords, or account numbers. AvidMax Outfitters is a member of the McAfee Secure program and has seen its visitor-to-sales conversion rate increase by 9 percent because visitors have increased confidence in the firm's payment security.[29]

Overcoming Consumers' Lack of Trust

Lack of trust in online sellers is one of the most frequently cited reasons that some consumers are not willing to purchase online. Can they be sure that the company or person with which they are dealing is legitimate and will send the item(s) they purchase? What if there is a problem with the product or service when it is received: for example, if it does not match the description on the Web site, is the wrong size or wrong color, is damaged during the delivery process, or does not work as advertised?

Online marketers must create specific trust-building strategies for their Web sites by analyzing their customers, products, and services. A perception of trustworthiness can be created by implementing one or more of the following strategies:

- Demonstrate a strong desire to build an ongoing relationship with customers by giving first-time price incentives, offering loyalty programs, or eliciting and sharing customer feedback.
- Demonstrate that the company has been in business for a long time.
- Make it clear that considerable investment has been made in the Web site.
- Provide brand endorsements from well-known experts or well-respected individuals.
- Demonstrate participation in appropriate regulatory programs or industry associations.

- Display Web site accreditation by the Better Business Bureau Online or TRUSTe programs.

Here are some tips to help online shoppers avoid problems:

- Only buy from a well-known Web site you can trust—one that advertises on national media, is recommended by a friend, or receives strong ratings in the media.
- Look for a seal of approval from organizations such as the Better Business Bureau Online or TRUSTe (see Figure 8.4).
- Review the Web site's privacy policy to be sure that you are comfortable with its conditions before you provide personal information.
- Determine what the Web site policy is for return of products purchased.
- Be wary if you must enter any personal information other than what's required to complete the purchase (name, credit card number, address, and telephone number).
- Do not, under any conditions, ever provide information such as your Social Security number, bank account numbers, or your mother's maiden name.
- When you open the Web page where you enter credit card information or other personal data, make sure that the Web address begins with "https," and check to see if a locked padlock icon appears in the Address bar or status bar, as shown in Figure 8.5.
- Consider using virtual credit cards, which expire after one use, when doing business.
- Before downloading music, change your browser's advanced settings to disable access to all computer areas that contain personal information.

FIGURE 8.4
Better Business Bureau Online and TRUSTe seals of approval

Courtesy Better Business Bureau and TRUSTe

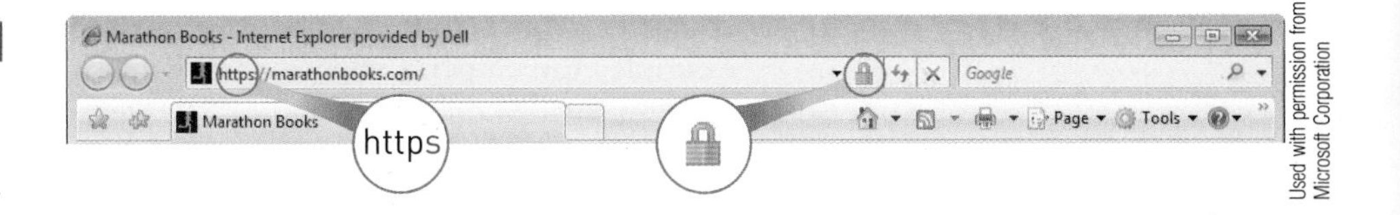

FIGURE 8.5
Web site security
Web site that uses "https" in the address and a secure site lock icon.

Used with permission from Microsoft Corporation

Overcoming Global Issues

E-commerce and m-commerce offer enormous opportunities by allowing manufacturers to buy supplies at a low cost worldwide. They also offer enterprises the chance to sell to a global market right from the start. Moreover, they offer great promise for developing countries, helping them to enter the prosperous global marketplace, which helps to reduce the gap between rich and poor countries. People and companies can get products and services from around the world, instead of around the corner or across town. These opportunities, however, come with numerous obstacles and issues, first identified in Chapter 1 as challenges associated with all global systems:

- **Cultural challenges**. Great care must be taken to ensure that a Web site is appealing, easy to use, and inoffensive to people around the world.
- **Language challenges**. Language differences can make it difficult to understand the information and directions posted on a Web site.

- **Time and distance challenges**. Significant time differences make it difficult for some people to be able to speak to customer services representatives or to get technical support during regular waking hours.
- **Infrastructure challenges**. The Web site must support access by customers using a wide variety of hardware and software devices.
- **Currency challenges**. The Web site must be able to state prices and accept payment in a variety of currencies.
- **State, regional, and national law challenges**. The Web site must operate in conformance to a wide variety of laws that cover a variety of issues, including the protection of trademarks and patents, the sale of copyrighted material, the collection and safeguarding of personal or financial data, the payment of sales taxes and fees, and much more.

ELECTRONIC AND MOBILE COMMERCE APPLICATIONS

E-commerce and m-commerce are being used in innovative and exciting ways. This section examines a few of the many B2B, B2C, C2C, and m-commerce applications in retail and wholesale, manufacturing, marketing, advertising, bartering, retargeting, price comparison, couponing, investment and finance, banking, and e-boutiques. As with any new technology, m-commerce will succeed only if it provides users with real benefits. Companies involved in e-commerce and m-commerce must think through their strategies carefully and ensure that they provide services that truly meet customers' needs.

Retail and Wholesale

electronic retailing (e-tailing): The direct sale of products or services by businesses to consumers through electronic storefronts, typically designed around the familiar electronic catalog and shopping cart model.

cybermall: A single Web site that offers many products and services at one Internet location.

E-commerce is being used extensively in retailing and wholesaling. **Electronic retailing**, sometimes called *e- tailing*, is the direct sale of products or services by businesses to consumers through electronic storefronts, which are typically designed around the familiar electronic catalog and shopping cart model. Companies such as Office Depot, Walmart, and many others have used the same model to sell wholesale goods to employees of corporations. Tens of thousands of electronic retail Web sites sell everything from soup to nuts.

Cybermalls are another means to support retail shopping. A **cybermall** is a single Web site that offers many products and services at one Internet location—similar to a regular shopping mall. An Internet cybermall pulls multiple buyers and sellers into one virtual place, easily reachable through a Web browser. For example, the Super Cyber Mall (*www.zutronic.com/Super_Cyber_Mall_EWZB.html*) provides direct links to Office Depot, Starbucks, 1800ProFlowers, Open Sky, Camping World, Foot Locker, and dozens of other online shopping sites.[30]

A key sector of wholesale e-commerce is spending on manufacturing, repair, and operations (MRO) goods and services—from simple office supplies to mission-critical equipment, such as the motors, pumps, compressors, and instruments that keep manufacturing facilities running smoothly. MRO purchases often approach 40 percent of a manufacturing company's total revenues, but the purchasing system can be haphazard, without automated controls. In addition to these external purchase costs, companies face significant internal costs resulting from outdated and cumbersome MRO management processes. For example, studies show that a high percentage of manufacturing downtime is often caused by not having the right part at the right time in the right place. The result is lost productivity and capacity. E-commerce software for plant operations provides powerful comparative searching capabilities to enable managers to identify functionally equivalent items, helping them spot opportunities to combine purchases for cost savings. Comparing various suppliers, coupled with consolidating more spending with fewer suppliers, leads to decreased costs. In addition, automated workflows are typically based on industry best practices, which can streamline processes.

The Maldron Holding Company's MRO Center (*www.mrocenter.com*) is a global supplier of over 18 million unique MRO items from some 18,000 manufacturers in 62 countries. Manufacturers can place their entire product catalog in MRO Center's database for access by potential purchasers.[31]

The MRO Manager eProcurement solution is an e-commerce supply chain system that replaces paper purchase orders, invoices, and checks with their electronic counterparts. Multifamily housing owners and operators use this solution to standardize on suppliers and products as well as to monitor their spending by region, community, or unit.[32]

Manufacturing

electronic exchange: An electronic forum where manufacturers, suppliers, and competitors buy and sell goods, trade market information, and run back-office operations.

One approach taken by many manufacturers to raise profitability and improve customer service is to move their supply chain operations onto the Internet. Here, they can form an **electronic exchange**, an electronic forum where manufacturers, suppliers, and competitors buy and sell goods, trade market information, and run back-office operations, such as inventory control, as shown in Figure 8.6. This approach has greatly speeded up the movement of raw materials and finished products among all members of the business community and has reduced the amount of inventory that must be maintained. It has also led to a much more competitive marketplace and lower prices.

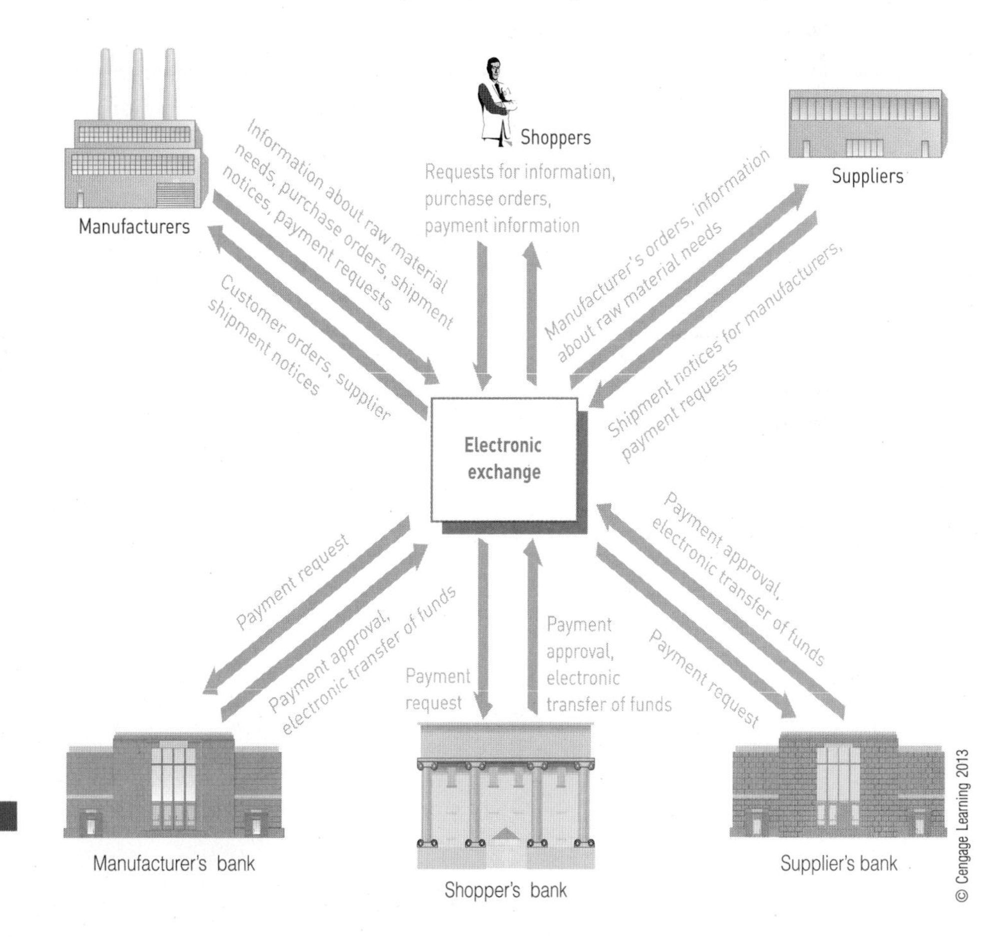

FIGURE 8.6
Model of an electronic exchange

Companies can join one of three types of exchanges based on who operates the exchange. Private exchanges are owned and operated by a single company. The owner uses the exchange to trade exclusively with established business partners. Walmart's Retail Link is such an exchange. Consortium-operated exchanges are run by a group of traditionally competing companies with common procurement needs. For example, Covisint was developed to serve the needs of the big three auto makers. Independent exchanges are open to any set of buyers and sellers within a given market. They provide services and a common technology platform to their members and are open, usually for a fee, to any company that wants to use them.

Several strategic and competitive issues are associated with the use of exchanges. Many companies distrust their corporate rivals and fear they might lose trade secrets through participation in such exchanges. Suppliers worry that online marketplaces will drive down the prices of goods and favor buyers. Suppliers also can spend a great deal of money setting up to participate in multiple exchanges. For example, more than a dozen new exchanges have appeared in the oil industry, and the printing industry is up to more than 20 online marketplaces. Until a clear winner emerges in particular industries, suppliers are more or less forced to sign on to several or all of them. Yet another issue is potential government scrutiny of exchange participants: When competitors get together to share information, it raises questions of collusion or antitrust behavior.

Many companies that already use the Internet for their private exchanges have no desire to share their expertise with competitors. At Walmart, the world's number-one retail chain, executives turned down several invitations to join exchanges in the retail and consumer goods industries. Walmart is pleased with its in-house exchange, Retail Link, which connects the company to 7,000 worldwide suppliers that sell everything from toothpaste to furniture.

Marketing

The nature of the Web enables firms to gather more information about customer behavior and preferences as customers and potential customers gather their own information and make their purchase decisions. Analysis of this data is complicated because of the Web's interactivity and because each visitor voluntarily provides or refuses to provide personal data such as name, address, e-mail address, telephone number, and demographic data. Internet advertisers use the data to identify specific markets and target them with tailored advertising messages. This practice, called **market segmentation**, divides the pool of potential customers into subgroups usually defined in terms of demographic characteristics, such as age, gender, marital status, income level, and geographic location.

market segmentation: The identification of specific markets to target them with tailored advertising messages.

In the past, market segmentation has been difficult for B2B marketers because firmographic data (addresses, financials, number of employees, and industry classification code) was difficult to obtain. Now, however, Nielsen, the marketing and media information company, has developed its Business-Facts database, which provides this information for more than 12 million businesses. Using this data, analysts can estimate potential sales for each business and rank the business against all other prospects and customers.[33]

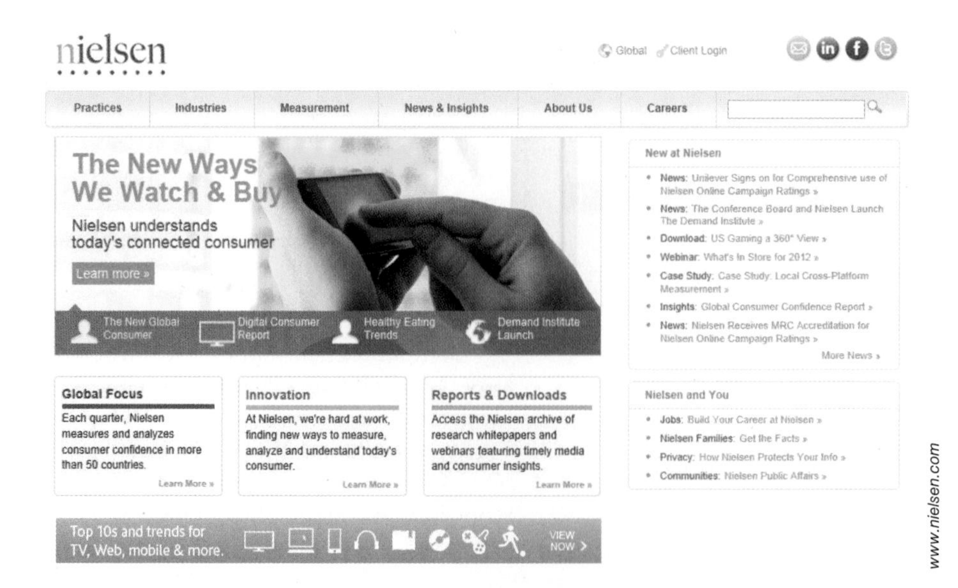

Nielsen is a major marketing company that measures and analyzes how consumers acquire information, consume media, and buy goods and services.

www.nielsen.com

Advertising

Mobile ad networks distribute mobile ads to publishers such as mobile Web sites, application developers, and mobile operators. Mobile ad impressions are generally bought at a cost per thousand (CPM), cost per click (CPC), or cost per action (CPA), in which the advertiser pays only if the customer clicks through and then buys the product or service. The main measures of success are the number of users reached, click through rate (CTR), and the number of actions users take, such as the number of downloads prompted by the ad. The advertiser is keenly interested in this data to measure the effectiveness of its advertising spending and may pay extra to purchase the data from the mobile ad network or a third party. Generally, there are three types of mobile ad networks—blind, premium blind, and premium networks—though no clear lines separate them. The characteristics of these mobile advertising networks are summarized in Table 8.4.

TABLE 8.4 Characteristics of three types of mobile advertising networks

Characteristic	Blind Networks	Premium Blind Networks	Premium Networks
Degree to which advertisers can specify where ads are run	An advertiser can specify country and content channel (e.g. news, sports, or entertainment) on which the add will run but not a specific Web site.	Most advertising is blind, but for an additional charge, the advertiser can buy a specific spot on a Web site of its choice.	Big brand advertisers can secure elite locations on top-tier destinations.
Predominant pricing model and typical rate	CPC ($.01 per click)	CPM ($20 per thousand impressions)	CPM ($40 per thousand impressions)
Examples	Admoda/Adultmoda	Jumptap	Advertising.com/AOL
	AdMob	Madhouse	Hands
	BuzzCity	Millennial Media	Microsoft Mobile Advertising
	InMobi	Quattro Wireless	Nokia Interactive Advertising
			Pudding Media
			YOC Group

AdMob is a mobile advertising provider that serves up ads for display on mobile devices and in applications like those that run on the Android and iPhone. With AdMob, smartphone application developers can distribute their apps for free and recover their costs over time through payments from advertisers. AdMob was acquired by Google and is now part of that firm, earning it close to $1 billion in advertising revenue.[34]

Because m-commerce devices usually have a single user, they are ideal for accessing personal information and receiving targeted messages for a particular consumer. Through m-commerce, companies can reach individual consumers to establish one-to-one marketing relationships and communicate whenever it is convenient—in short, anytime and anywhere.

Bartering

During the recent economic downturn, many people and businesses have turned to bartering as a means of gaining goods and services. A number of Web sites have been created to support this activity, as shown in Table 8.5. Businesses are willing to barter to reduce excess inventory, gain new customers, or avoid paying cash for necessary raw materials or services. Cash-strapped customers find bartering to be an attractive alternative to paying

© iStockphoto/Martin McCarthy

Consumers are increasingly using mobile phones to purchase goods and perform other transactions online.

scarce dollars. Generally, bartering transactions have tax-reporting, accounting, and other record-keeping responsibilities associated with them. Indeed, the IRS hosts a Bartering Tax Center Web site that provides details about the tax laws and responsibilities for bartering transactions.[35]

TABLE **8.5** Popular bartering Web sites

Web Site	Purpose
Craiglist.org	Includes a section where users can request an item in exchange for services or exchange services for services.
Goozez.com	Allows users to exchange games and movies.
SwapAGift.com	Enables users to sell unused gift cards for a percentage of their remaining balance.
SwapHog.com	Bartering site that offers a third-party service that first receives all items and inspects them before finalizing the transaction to eliminate fraud and ensure a successful transaction.
Swapstyle.com	Users can swap, sell, or buy direct women's accessories, clothes, cosmetics, and shoes.
Swaptree.com	Users trade books, DVDs, and video games on a one-for-one basis.
TradeAway.com	Enables users to exchange a wide variety of items including jewelry, landscape work, and vacation packages.

Source: Wagner, Vivian, "Trading Out: The Rise of Bartering," *E-Commerce Times*, *www.ecommercetimes.com/rsstory/74088.html*, January 4, 2012.

Retargeting

Over 95 percent of Web site visitors leave a shopping site without making a purchase.[36] Retargeting is a technique used by advertisers to recapture these shoppers by using targeted and personalized ads to direct shoppers back to a retailer's site. For example, a visitor who viewed the men's clothing portion of

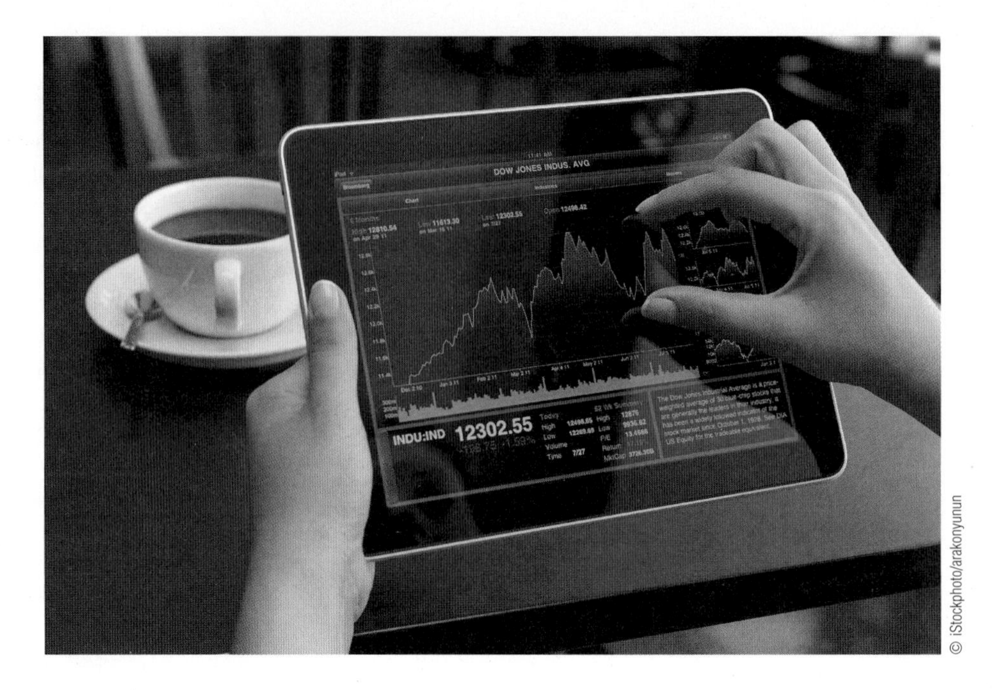

a retailer's Web site and then abandoned the Web site would be targeted with banner ads showing various men's clothing items from that retailer. The banner ads might even display the exact items the visitor viewed, such as men's casual slacks. The retargeting could be even further enhanced to include comments and recommendations from other consumers who purchased the same items. Thus, retargeting ensures that potential consumers see relevant, targeted ads for products they've already expressed interest in.

Price Comparison

An increasing number of companies provide mobile phone apps that enable shoppers to compare prices and products on the Web. Amazon's Price Check and Google's Shopper enable shoppers to do a quick price comparison by simply scanning the product's barcode or by taking a picture of a book, DVD, CD, or video game cover. Barcode Scanner allows shoppers to scan UPC or Quick Response codes to perform a price comparison.[37]

Couponing

Over 300 billion coupons are distributed each year in North America.[38] These coupons have an estimated value of $485 billion.[39] Surprisingly, only about 1.1 percent of those coupons were redeemed even during tough economic times for many people.[40]

Many manufacturers and retailers now send mobile coupons directly to consumers' smartphones. Unfortunately, the standard red laser scanners used at checkout stands have difficulty reading information displayed on a smartphone without special smartphone apps. Therefore, for many shoppers, current technology requires that the consumer print out the coupon, have it scanned, and present it to the clerk to enter the numbers from the coupon manually.

Procter and Gamble, the world's largest consumer products company, and mobeam, inc., will pilot a new approach to enable mobile coupons to be read directly by the standard bar code scanners at the checkout. This will enable consumers to receive electronic coupons, sort and organize them on their smartphone, and bring them to the store where they can be scanned directly. The mobeam approach converts the barcode into a beam of light that can be read by the typical barcode scanner. The mobeam application must be loaded onto the smartphone before scanning.[41]

Groupon is an innovative approach to couponing. Discount coupons for consumers are valid only if a predetermined minimum number of people sign up for them. Merchants do not pay any money up front to participate in Groupon but must pay Groupon half of whatever the customer pays for the coupon.

A recent study by Juniper Research estimates that regular mobile coupon users will grow to more than 600 million worldwide and the mobile coupon redemption rate will grow to about 8 percent worldwide by 2016. This growth means that mobile coupons are about eight times more likely to be redeemed than paper ones. Total redemption value is estimated to be over $43 billion worldwide.[42]

Investment and Finance

The Internet has revolutionized the world of investment and finance. Perhaps the changes have been so significant because this industry had so many built-in inefficiencies and so much opportunity for improvement.

The brokerage business adapted to the Internet faster than any other arm of finance. The allure of online trading that enables investors to do quick, thorough research and then buy shares in any company in a few seconds and at a fraction of the cost of a full-commission firm has brought many investors to the Web. Scottrade offers a mobile trading app for investors to monitor their investments, view streaming quotes, generate investment ideas from research and analysts' opinions, and execute trades from their Apple, Blackberry, or Android mobile devices.[43]

Investors can use mobile trading apps on tablets or smartphones.

© iStockphoto/arakonyunun

Banking

Online banking customers can check balances of their savings, checking, and loan accounts; transfer money among accounts; and pay their bills. These customers enjoy the convenience of not writing checks by hand, of tracking their current balances, and of reducing expenditures on envelopes and stamps. In addition, paying bills online is good for the environment because it reduces the amount of paper used, thus saving trees and reducing greenhouse gases.

All of the major banks and many of the smaller banks in the United States enable their customers to pay bills online; many support bill payment via cell phone or other wireless device. Banks are eager to gain more customers who pay bills online because such customers tend to stay with the bank longer, have higher cash balances, and use more of the bank's products and services. To encourage the use of this service, many banks have eliminated all fees associated with online bill payment.

Consumers who have enrolled in mobile banking and downloaded the mobile application to their cell phones can check their credit card balances before making major purchases and can avoid credit rejections. They can also transfer funds from savings to checking accounts to avoid an overdraft.

Visa-affiliated financial institutions offer their account holders an entire suite of applications that enable them to monitor account history and balances, transfer money between accounts, and receive transaction alerts on their mobile devices. These mobile services are managed by Visa and can be accessed with any mobile device such as a smartphone or tablet computer. Additional services planned for the future include mobile check deposit, mobile payments, and mobile offers.[44]

E-Boutiques

An increasing number of Web sites offer personalized shopping consultations for shoppers interested in upscale, contemporary clothing—dresses, sportswear, denim, handbags, jewelry, shoes, and gifts. Key to the success of Web sites such as Charm Boutique and ShopLaTiDa is a philosophy of high customer service and strong, personal client relationships. Online boutique shoppers complete a personal shopping profile by answering questions about body measurements, profession, interests, preferred designers, and areas of shopping where they would welcome assistance. Shoppers are then given

ETHICAL & SOCIETAL ISSUES

AbilityOne: E-Commerce Not Just for the Sighted

In the United States, 7 out of 10 working-age people who are blind are not employed. National Industries for the Blind aims to reduce that number. Their mission is "to enhance the opportunities for economic and personal independence of persons who are blind, primarily through creating, sustaining and improving employment."

In the United States, the Javits-Wagner-O'Day Act of 1971 has a similar goal. It requires the Federal government to purchase certain supplies and services from nonprofit organizations that employ people who are blind or have other significant disabilities. This law is administered by the U.S. AbilityOne Commission, a government agency that operates in conjunction with National Industries for the Blind (NIB) and National Industries for the Severely Handicapped (NISH). These two organizations, in turn, are authorized to provide supplies and services under this act. They coordinate offerings from many small nonprofits that produce items or provide services that the government needs and sell them through AbilityOne.

Today, people expect to be able to obtain goods and services via e-commerce. AbilityOne, therefore, needed an e-commerce Web site that would let government agencies order from them. The commission's challenges in developing its site were to make it accessible to the visually impaired while basing it on standard software to reduce development effort and keeping it easily usable by people with full vision. (Most external site users, those who purchase goods and services for government agencies, have full vision, but accessibility was still required for two reasons: Many internal users at NIB are vision-impaired, and accessibility is part of NIB's mission.)

To address these challenges, AbilityOne decided to work with NIB to develop its site. The agencies began with a commitment to meet all the requirements of Section 508c of the Rehabilitation Act of 1998. That act specifies accommodations to make information systems accessible to people with disabilities, such as ensuring that all graphics on a site have speakable text descriptions. Section 508c compliance was a requirement throughout the vendor selection process for the new site. Ultimately, NIB decided to base its site on Oracle's E-Business Suite and its iStore e-commerce application. NIB found that iStore offered sufficient user interface flexibility to meet the accessibility requirements. For example, iStore allows each user to customize the site's colors to black and white, high contrast, or a mix of settings that optimizes the site for his or her vision. NIB also chose BizTech, an IT services firm and Oracle developer, to lead the implementation project.

After confirming that the AbilityOne site met Section 508c requirements, NIB tested the site further against the Web Content Accessibility Guidelines of the World Wide Web Consortium (W3C) and against NIB's own internal usability guidelines. Only after passing all these tests was the site made available to AbilityOne's customers.

The result was a site that is fully usable by sighted and visually impaired people, that is visually attractive to sighted people, that allows government agencies to obtain any of the 8,000 products that are available through AbilityOne, and that facilitates the employment of visually impaired people as AbilityOne customer representatives. Commercial e-commerce sites can lose nothing, but can only gain, by making their sites equally accessible.

Discussion Questions

1. How can a company with a general e-commerce site benefit from conforming to accessibility guidelines such as those mentioned above?

2. As this case study states, NIB selected BizTech (*www.biztech.com*) to customize the software. BizTech had a great deal of experience with Oracle software but relatively little with accessibility. It would also have been possible for NIB to select a software partner that had more experience with accessibility but less with Oracle software. Given that the work includes both using Oracle software and ensuring accessibility, do you think NIB made the right choice?

Critical Thinking Questions

1. U.S. Federal government agencies are required by law to purchase certain supplies and services through AbilityOne, thus ensuring that visually impaired and other severely handicapped people have a major role in producing or providing them. Do you agree with this law? Why or why not?
2. With some exceptions, complying with accessibility standards is voluntary on the part of companies that sell software. However, it costs money in added development effort. This cost must be offset, or companies that comply with these standards will be at a competitive disadvantage. How can a software vendor use compliance with accessibility standards to increase its revenue?

SOURCES: AbilityOne Web site, *www.abilityone.com*, accessed February 29, 2012; National Industries for the Blind, "AbilityOne eCommerce Site," *Computerworld* case study, *www.eiseverywhere.com/file_uploads/13271db4a95e3264477277f0829035c6_National_Industries_for_the_Blind_-_Ability One_eCommerce_Site.pdf*, accessed February 29, 2012; National Industries for the Blind Web site, *www.nib.org*, accessed February 29, 2012; U.S. Government, "Resources for Understanding and Implementing Section 508," *www.section508.gov*, accessed February 29, 2012; "Javits–Wagner–O'Day Act," *Wikipedia*, *en.wikipedia.org/wiki/Javits–Wagner–O'Day_Act*, September 18, 2011; "Web Content Accessibility Guidelines (WCAG) 2.0," World Wide Web Consortium, *www.w3.org/TR/WCAG20*, December 11, 2008.

suggestions on what styles and designers might work best and where they can be found—online or in brick-and-mortar shops.

Quintessentially Gifts is a luxury gifts and shopping service whose researchers and editorial stylists can find the rarest and most exquisite gifts for the affluent shopper. From a McQueen Luxury Dive Toy's underwater scooter to a Hermes Birkin handbag sans the usual two-year wait, the gift team can get it for you.[45]

Quintessentially Gifts is an online shopping service that features unusual luxury gifts.

www.quintessentiallygifts.com

STRATEGIES FOR SUCCESSFUL E-COMMERCE AND M-COMMERCE

With all the constraints to e-commerce just covered, a company must develop an effective Web site, one that is easy to use and accomplishes the goals of the company yet is safe, secure, and affordable to set up and maintain. The next sections examine several issues for a successful e-commerce site.

Defining an Effective E-Commerce Model and Strategy

The first major challenge is for the company to define an effective e-commerce model and strategy. Although companies can select from a number of approaches, the most successful e-commerce models include three basic components: community, content, and commerce, as shown in Figure 8.7. Message boards and chat rooms can build a loyal *community* of people who are interested in and enthusiastic about the company and its products and services. Providing useful, accurate, and timely *content*, such as industry and economic news and stock quotes, is a sound approach to encourage people to return to your Web site time and again. *Commerce* involves consumers and businesses paying to purchase physical goods, information, or services that are posted or advertised online.

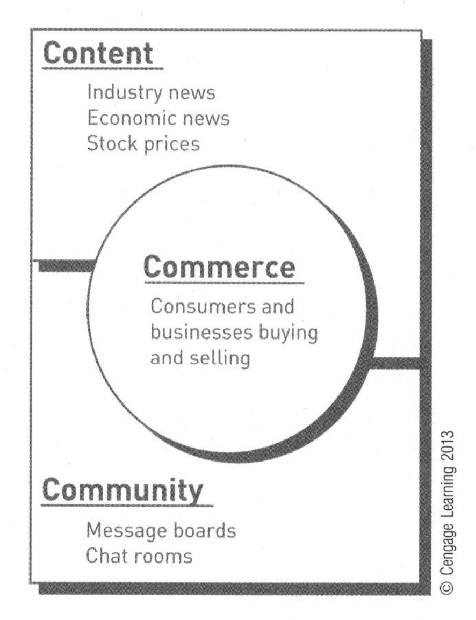

FIGURE 8.7

Three basic components of a successful e-commerce model

Defining the Web Site Functions

When building a Web site, you should first decide which tasks the site must accomplish. Most people agree that an effective Web site is one that creates an attractive presence and that meets the needs of its visitors, as with the following:

- Obtaining general information about the organization
- Obtaining financial information for making an investment decision in the organization
- Learning the organization's position on social issues
- Learning about the products or services that the organization sells
- Buying the products or services that the company offers
- Checking the status of an order
- Getting advice or help on effective use of the products
- Registering a complaint about the organization's products
- Registering a complaint concerning the organization's position on social issues
- Providing a product testimonial or an idea for product improvement or a new product

- Obtaining information about warranties or service and repair policies for products
- Obtaining contact information for a person or department in the organization

After a company determines which objectives its site should accomplish, it can proceed to the details of developing the site.

As the number of e-commerce shoppers increases and they become more comfortable—and more selective—making online purchases, a company might need to redefine the basic business model of its site to capture new business opportunities. For example, consider the major travel sites such as Expedia, Travelocity, CheapTickets, Orbitz, and Priceline. These sites used to specialize in one area of travel—inexpensive airline tickets. Now they offer a full range of travel products, including airline tickets, auto rentals, hotel rooms, tours, and last-minute trip packages. Expedia provides in-depth hotel details to help comparison shoppers and even offers 360-degree visual tours and expanded photo displays. It also entices flexible travelers to search for rates, compare airfares, and configure hotel and air prices at the same time. Expedia has developed numerous hotel partnerships to reduce costs and help secure great values for consumers. Meanwhile, Orbitz has launched a special full-service program for corporate business travelers.

Establishing a Web Site

Companies large and small can establish Web sites. Some companies elect to develop their sites in-house, but this decision requires learning HTML, Java, and Web design software. Many firms, especially those with few or no experienced Web developers, have decided that to outsource the building of their Web sites gets the Web sites up and running faster and cheaper than doing the job themselves. Web site hosting companies such as HostWay and BroadSpire make it possible to set up a Web page and conduct e-commerce within a matter of days and with little up-front cost.

These companies can also provide free hosting for your store, but to allow visitors to pay for merchandise with credit cards, you need a merchant account with a bank. If your company doesn't already have one, it must establish one.

Web development firms can provide organizations with prebuilt templates and Web site builder tools to enable customers to construct their own Web sites. Businesses can custom design a new Web site or redesign an existing Web site. Such firms have worked with thousands of customers to help them get their Web sites up and running.

storefront broker: A company that acts as an intermediary between your Web site and online merchants who have the products and retail expertise.

Another model for setting up a Web site is the use of a **storefront broker**, a business that serves as an intermediary between your Web site and online merchants who have the actual products and retail expertise. The storefront broker deals with the details of the transactions, including who gets paid for what, and is responsible for bringing together merchants and reseller sites. The storefront broker is similar to a distributor in standard retail operations, but in this case, no product moves—only electronic data flows back and forth. Products are ordered by a customer at your site, orders are processed through a user interface provided by the storefront broker, and the product is shipped by the merchant.

ProStores is a service firm that, among other things, helps clients to set up electronic storefronts. It helped BiggSports, a small retailer with three brick-and-mortar locations, to launch a Web site that enabled cost-effective expansion of sales nationally.[46]

Building Traffic to Your Web Site

The Internet includes hundreds of thousands of e-commerce Web sites. With all those potential competitors, a company must take strong measures to ensure that the customers it wants to attract can find its Web site. The first step is to obtain and register a domain name, which should say something about your business. For instance, stuff4u might seem to be a good catchall, but it doesn't describe the nature of the business—it could be anything. If you want to sell soccer uniforms and equipment, then you'd try to get a domain name such as *www.soccerstuff4u.com*, *www.soccerequipment.com*, or *www.stuff4soccercoaches.com*. The more specific the Web address, the better.

The next step to attracting customers is to make your site search-engine friendly by improving its rankings. Following are several ideas on how to accomplish this goal:

meta tag: An HTML code, not visible on the displayed Web page, that contains keywords representing your site's content, which search engines use to build indexes pointing to your Web site.

- Include a meta tag in your store's home page. A **meta tag** is an HTML code, not visible on the displayed Web page, that contains keywords representing your site's content. Search engines use these keywords to build indexes pointing to your Web site. Again, keywords are critical to attracting customers, so they should be chosen carefully. They should clearly define the scope of the products or services you offer.
- Use Web site traffic data analysis software to turn the data captured in the Web log file into useful information. This data can tell you the URLs from which your site is being accessed, the search engines and keywords that find your site, and other useful information. Using this data can help you identify search engines to which you need to market your Web site, allowing you to submit your Web pages to them for inclusion in the search engine's index.
- Provide quality, keyword-rich content. Be careful not to use too many keywords, as search engines often ban sites that do this. Judiciously place keywords throughout your site, ensuring that the Web content is sensible and easy to read by humans as well as search engines.
- Add new content to the Web site on a regular basis. Again, this makes the site attractive to humans as well as search engines.
- Acquire links to your site from other reputable Web sites that are popular and related to your Web site. Avoid the use of low-quality links, as they can actually hurt your Web site's rating.

The use of the Internet is growing rapidly in markets throughout Europe, Asia, and Latin America. Obviously, companies that want to succeed on the Web cannot ignore this global shift. A company must be aware that consumers outside the United States will access sites with a variety of devices, and the firm should modify its site design accordingly. In Europe, for example, closed-system iDTVs (integrated digital televisions) are becoming popular for accessing online content, with more than 50 percent of the population now using them.[47] Because such devices have better resolution and more screen space than the PC monitors that U.S. consumers use to access the Internet, iDTV users expect more ambitious graphics. Successful global firms operate with a portfolio of sites designed for each market, with shared sourcing and infrastructure to support the network of stores and with local marketing and business development teams to take advantage of local opportunities. Service providers continue to emerge to solve the cross-border logistics, payments, and customer service needs of these global retailers.

Maintaining and Improving Your Web Site

Web site operators must constantly monitor the traffic to their sites and the response times experienced by visitors. AMR Research, a Boston-based, independent research analysis firm, reports that Internet shoppers expect service to be better than or equal to their in-store experience.

Nothing will drive potential customers away faster than experiencing unreasonable delays while trying to view or order your products or services. To keep pace with technology and increasing traffic, it might be necessary to modify the software, databases, or hardware on which the Web site runs to ensure acceptable response times.

Staples, Inc., spent considerable time and effort to revamp its B2B Web site to include new search and navigation features so customers can find products more quickly. Customers can see alternative product recommendations and sort and compare products on cost and other features. Customers also have quicker and easier access to previously ordered items so they can streamline the reordering of frequently used items.[48]

Web site operators must also continually be alert to new trends and developments in the area of e-commerce and be prepared to take advantage of new opportunities. For example, recent studies show that customers more frequently visit Web sites they can customize. **Personalization** is the process of tailoring Web pages to specifically target individual consumers. The goal is to meet the customer's needs more effectively, make interactions faster and easier, and consequently, increase customer satisfaction and the likelihood of repeat visits. Building a better understanding of customer preferences also can aid in cross-selling related products and more expensive products. The most basic form of personalization involves using the consumer's name in an e-mail campaign or in a greeting on the Web page. Amazon uses a more advanced form of personalization in which the Web site greets each repeat customer by name and recommends a list of new products based on the customer's previous purchases.

personalization: The process of tailoring Web pages to specifically target individual consumers.

Businesses use two types of personalization techniques to capture data and build customer profiles. *Implicit personalization* techniques capture data from actual customer Web sessions—primarily based on which pages were viewed and which weren't. *Explicit personalization* techniques capture user-provided information, such as information from warranties, surveys, user registrations, and contest-entry forms completed online. Data can also be gathered through access to other data sources such as the Bureau of Motor Vehicles, Bureau of Vital Statistics, and marketing affiliates (firms that share marketing data). Marketing firms aggregate this information to build databases containing a huge amount of consumer behavioral data. During each customer interaction, powerful algorithms analyze both types of data in real time to predict the consumer's needs and interests. This analysis makes it possible to deliver new, targeted information before the customer leaves the site. Because personalization depends on gathering and using personal user information, privacy issues are a major concern.

iGoDigital is a sales data program used by Amazon, Best Buy, Dell, Sears, Walmart, and many other e-commerce retailers. The software recommends items to Web site visitors that might interest them based on their purchases and Web-browsing habits. iGoDigital collects information about individual shoppers such as what items they searched for, placed in their shopping basket, or purchased. This data is then stored in a central personalization engine and is used to make shopping recommendations to individual consumers on any Web site they visit that uses the software.[49]

These tips and suggestions are only a few ideas that can help a company set up and maintain an effective e-commerce site. With technology and competition changing constantly, managers should read articles in print and on the Web to keep up to date on ever-evolving issues.

TECHNOLOGY INFRASTRUCTURE REQUIRED TO SUPPORT E-COMMERCE AND M-COMMERCE

Now that we've examined how to establish e-commerce effectively, let's look at some of the technical issues related to e-commerce systems and the technology that makes it possible. Successful implementation of e-business requires significant changes to existing business processes and substantial investment in

IS technology. These technology components must be chosen carefully and be integrated to support a large volume of transactions with customers, suppliers, and other business partners worldwide. Online consumers complain that poor Web site performance (e.g., slow response time, inadequate customer support, and lost orders) drives them to abandon some e-commerce sites in favor of those with better, more reliable performance. This section provides a brief overview of the key technology infrastructure components (see Figure 8.8).

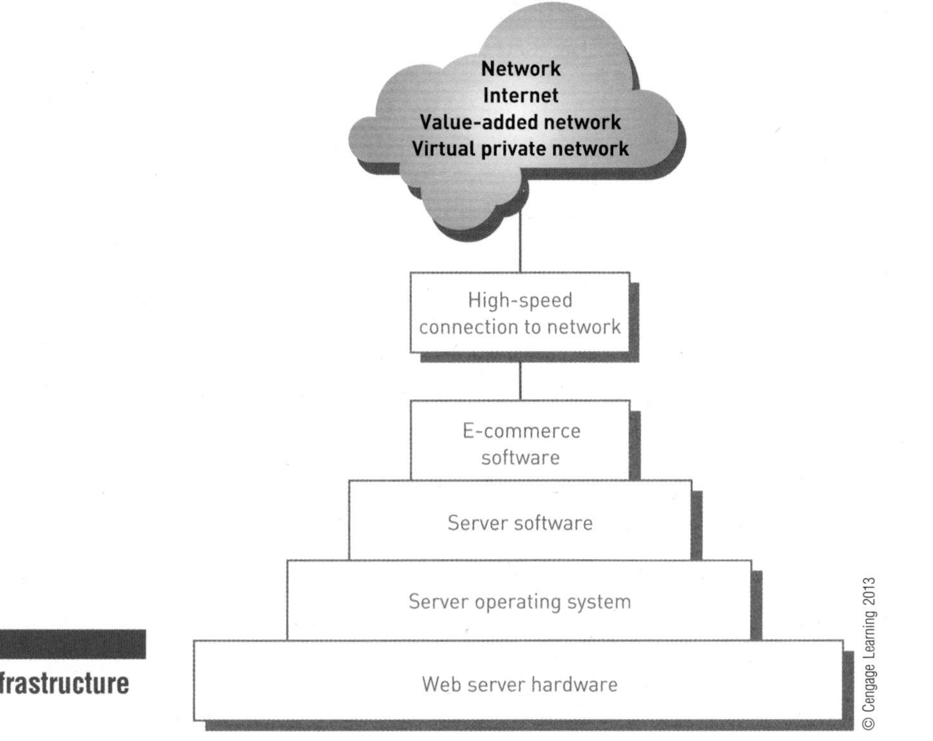

FIGURE 8.8
Key technology infrastructure components

Hardware

A Web server hardware platform complete with the appropriate software is a key ingredient to e-commerce infrastructure. The amount of storage capacity and computing power required of the Web server depends primarily on two things: the software that must run on the server and the volume of e-commerce transactions that must be processed. The most successful e-commerce solutions are designed to be highly scalable so that they can be upgraded to meet unexpected user traffic.

Key Web site performance measures include response time, transaction success rate, and system availability. Table 8.6 shows the values for the key measures for four popular online retailers for one week.

TABLE **8.6** Key performance measures for popular retail Web sites

Retail Apparel Firm	Response Time (seconds)	Success Rate	Outage Time During One Week
Abercrombie	4.64	98.6%	1 hour
Macy's	6.81	99.5%	0 hours
Sears	12.90	99.1%	1 hour
J Crew	7.89	97.7%	1 hour
Saks Fifth Avenue	10.59	95.7%	2 hours

Source: "Keynote Online Retail Transaction Indices," *e-Commerce Times*, *www.ecommercetimes.com/web-performance*, February 10, 2012.

A key decision facing a new e-commerce company is whether to host its own Web site or to let someone else do it. Many companies decide that using a third-party Web service provider is the best way to meet initial e-commerce needs. The third-party company rents space on its computer system and provides a high-speed connection to the Internet, thus minimizing the initial out-of-pocket costs for e-commerce start-up. The third party can also provide personnel trained to operate, troubleshoot, and manage the Web server.

Web Server Software

In addition to the Web server operating system, each e-commerce Web site must have Web server software to perform fundamental services, including security and identification, retrieval and sending of Web pages, Web site tracking, Web site development, and Web page development. The two most widely used Web server software packages are Apache HTTP Server and Microsoft Internet Information Services.

E-Commerce Software

After you have located or built a host server, including the hardware, operating system, and Web server software, you can begin to investigate and install e-commerce software to support five core tasks: catalog management to create and update the product catalog, product configuration to help customers select the necessary components and options, shopping cart facilities to track the items selected for purchase (see Figure 8.9), e-commerce transaction processing, and Web traffic data analysis to provide details to adjust the operations of the Web site.

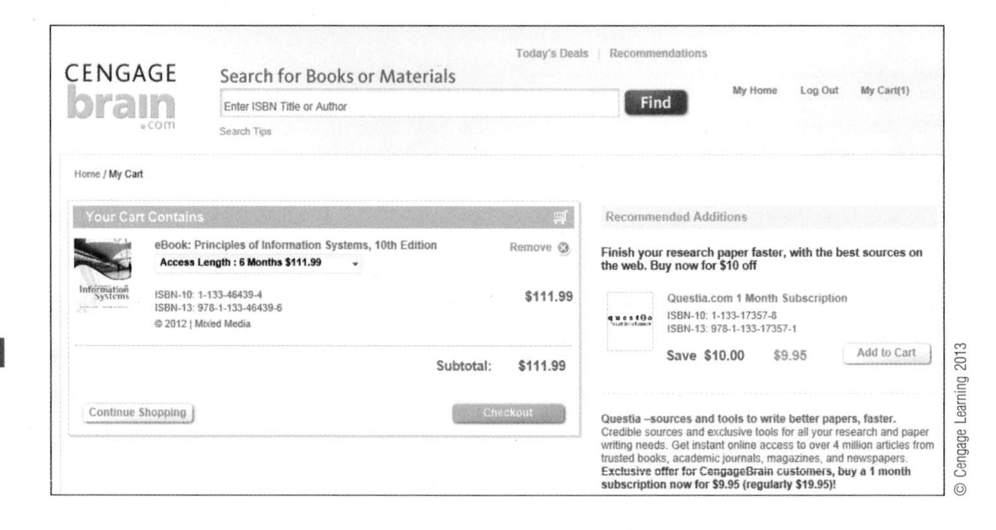

FIGURE **8.9**

Electronic shopping cart

An electronic shopping cart allows online shoppers to view their selections and add or remove items.

Mobile Commerce Hardware and Software

For m-commerce to work effectively, the interface between the wireless, handheld device and its user must improve to the point that it is nearly as easy to purchase an item on a wireless device as it is to purchase it on a PC. In addition, network speed must improve so that users do not become frustrated. Security is also a major concern, particularly in two areas: the security of the transmission itself and the trust that the transaction is being made with the intended party. Encryption can provide secure transmission. Digital certificates, discussed later in this chapter, can ensure that transactions are made between the intended parties.

The handheld devices used for m-commerce have several limitations that complicate their use. Their screens are small, perhaps no more than a few square inches, and might be able to display only a few lines of text. Their input capabilities are limited to a few buttons, so entering data can be tedious and error prone. They also have less processing power and less bandwidth than desktop computers, which are usually connected to a high-speed LAN. They also operate on limited-life batteries. For these reasons, it is currently impossible to directly access many Web sites with a handheld device. Web developers must rewrite Web applications so that users with handheld devices can access them.

To address the limitations of wireless devices, the industry has undertaken a standardization effort for their Internet communications. The Wireless Application Protocol (WAP) is a standard set of specifications for Internet applications that run on handheld, wireless devices. It effectively serves as a Web browser for such devices.

Electronic Payment Systems

Electronic payment systems are a key component of the e-commerce infrastructure. Current e-commerce technology relies on user identification and encryption to safeguard business transactions. Actual payments are made in a variety of ways, including electronic cash, electronic wallets, and smart, credit, charge, and debit cards. Web sites that accept multiple payment types convert more visitors to purchasing customers than merchants who offer only a single payment method.

digital certificate: An attachment to an e-mail message or data embedded in a Web site that verifies the identity of a sender or Web site.

certificate authority (CA): A trusted third-party organization or company that issues digital certificates.

Authentication technologies are used by many organizations to confirm the identity of a user requesting access to information or assets. A **digital certificate** is an attachment to an e-mail message or data embedded in a Web site that verifies the identity of a sender or Web site. A **certificate authority (CA)** is a trusted third-party organization or company that issues digital certificates. The CA is responsible for guaranteeing that the people or organizations granted these unique certificates are in fact who they claim to be. Digital certificates thus create a trust chain throughout the transaction, verifying both purchaser and supplier identities.

Many organizations that accept credit cards to pay for items purchased via e-commerce have adopted the Payment Card Industry (PCI) security standard. This standard spells out measures and security procedures to safeguard the card issuer, the cardholder, and the merchant. Some of the measures include installing and maintaining a firewall configuration to control access to computers and data, never using software or hardware vendor-supplier defaults for system passwords, and requiring merchants to protect stored data, encrypt transmission of cardholder information across public networks, use and regularly update antivirus software, and restrict access to sensitive data on a need-to-know basis.

Various measures are being implemented to increase the security associated with the use of credit cards at the time of purchase. The Address Verification System is a check built into the payment authorization request that compares the address on file with the card issuer to the billing address provided by the cardholder. The Card Verification Number technique is a check of the additional digits printed on the back of the card. Visa has Advanced Authorization, a Visa-patented process that provides an instantaneous rating of that transaction's potential for fraud to the financial institution that issued the card. The card issuer can then send an immediate response to the merchant regarding whether to accept or decline the transaction. The technology is now being applied to every Visa credit and check card purchase today. Visa estimates that this technique will reduce fraudulent credit card charges by 40 percent.

The Federal Financial Institutions Examination Council has developed a new set of guidelines called "Authentication in an Internet Banking Environment," which recommend two-factor authorization. This approach adds another identity check along with the password system. A number of multifactor authentication schemes can be used, such as biometrics, one-time passwords, or hardware tokens that plug into a USB port on the computer and generate a password that matches the ones used by a bank's security system. Currently, the use of biometric technology to secure online transactions is rare for both cost and privacy reasons. It can be expensive to outfit every merchant with a biometric scanner, and it is difficult to convince consumers to supply something as personal and distinguishing as a fingerprint. In spite of these problems, a growing number of financial service firms from large (e.g., Citibank) to small (e.g., Purdue Employees Federal Credit Union) are considering biometric systems. Bank Leumi uses a biometric speaker verification system as part of its multifactor authentication process to safeguard its customers from fraud and to provide a positive customer experience.[50]

Secure Sockets Layer

Secure Sockets Layer (SSL): A communications protocol used to secure sensitive data during e-commerce.

All online shoppers fear the theft of credit card numbers and banking information. To help prevent this type of identity theft, the **Secure Sockets Layer (SSL)** communications protocol is used to secure sensitive data. The SSL communications protocol includes a handshake stage, which authenticates the server (and the client, if needed), determines the encryption and hashing algorithms to be used, and exchanges encryption keys. Following the handshake stage, data might be transferred. The data is always encrypted, ensuring that your transactions are not subject to interception or "sniffing" by a third party. Although SSL handles the encryption part of a secure e-commerce transaction, a digital certificate is necessary to provide server identification.

Electronic Cash

electronic cash: An amount of money that is computerized, stored, and used as cash for e-commerce transactions.

Electronic cash is an amount of money that is computerized, stored, and used as cash for e-commerce transactions. Typically, consumers must open an account with an electronic cash service provider by providing identification information. When the consumers want to withdraw electronic cash to make a purchase, they access the service provider via the Internet and present proof of identity—a digital certificate issued by a certification authority or a username and password. After verifying a consumer's identity, the system debits the consumer's account and credits the seller's account with the amount of the purchase. PayPal, BillMeLater, MoneyZap, and TeleCheck are four popular forms of electronic cash.

PayPal enables any person or business with an e-mail address to securely, easily, and quickly send and receive payments online. To send money, you enter the recipient's e-mail address and the amount you want to send. You can pay with a credit card, debit card, or funds from a checking account. The recipient gets an e-mail that says, "You've Got Cash!" Recipients can then collect their money by clicking a link in the e-mail that takes them to *www.paypal.com*. To receive the money, the user also must have a credit card or checking account to accept fund transfers. To request money for an auction, invoice a customer, or send a personal bill, you enter the recipient's e-mail address and the amount you are requesting. The recipient gets an e-mail and instructions on how to pay you using PayPal. PayPal has 100 million active accounts in 190 markets and makes payments in 25 currencies around the world.[51]

Redcats USA is an online retailer of men's and women's plus-size apparel, home and lifestyle products, sporting goods, and outdoor gear. The firm recently added the Express Checkout service from PayPal to streamline its consumer checkout process. Consumers who have a PayPal account can specify that they want to use this payment service at the start of their checkout process. They are then directed to a PayPal-operated site where they sign in with their PayPal user name and password. PayPal then retrieves the consumers' shipping, billing, and payment details and passes this data to Redcats, thus saving consumers the effort of entering this information. Redcats accepts other methods of payment including all major credit cards, any Redcard's stored-branded credit card, and eBillme (an alternative to PayPal).[52]

Credit, Charge, Debit, and Smart Cards

Many online shoppers use credit and charge cards for most of their Internet purchases. A credit card, such as Visa or MasterCard, has a preset spending limit based on the user's credit history, and each month the user can pay all or part of the amount owed. Interest is charged on the unpaid amount. A charge card, such as American Express, carries no preset spending limit, and the entire amount charged to the card is due at the end of the billing period. Charge cards do not involve lines of credit and do not accumulate interest charges. American Express became the first company to offer disposable credit card numbers in 2000. Other banks, such as Citibank, protect the consumer by providing a unique number for each transaction. Debit cards look like credit cards, but they operate like cash or a personal check. Credit, charge, and debit cards currently store limited information about you on a magnetic strip. This information is read each time the card is swiped to make a purchase. All credit card customers are protected by law from paying more than $50 for fraudulent transactions.

smart card: A credit card–sized device with an embedded microchip to provide electronic memory and processing capability.

The **smart card** is a credit card–sized device with an embedded microchip to provide electronic memory and processing capability. Smart cards can be used for a variety of purposes, including storing a user's financial facts, health insurance data, credit card numbers, and network identification codes and passwords. They can also store monetary values for spending.

Smart cards are better protected from misuse than conventional credit, charge, and debit cards because the smart-card information is encrypted. Conventional credit, charge, and debit cards clearly show your account number on the face of the card. The card number, along with a forged signature, is all that a thief needs to purchase items and charge them against your card. A smart card makes credit theft practically impossible because a key to unlock the encrypted information is required, and there is no external number that a thief can identify and no physical signature a thief can forge. Table 8.7 compares various types of payment systems.

TABLE 8.7 Comparison of payment systems

Payment System	Description	Advantages	Disadvantages
Credit card	Carries preset spending limit based on the user's credit history	Each month the user can pay all or part of the amount owed	Unpaid balance accumulates interest charges—often at a high rate of interest
Charge card	Looks like a credit card but carries no preset spending limit	Does not involve lines of credit and does not accumulate interest charges	The entire amount charged to the card is due at the end of the billing period
Debit card	Looks like a credit card or automated teller machine (ATM) card	Operates like cash or a personal check	Money is immediately deducted from user's account balance
Smart card	Is a credit card device with embedded microchip capable of storing facts about card holder	Better protected from misuse than conventional credit, charge, and debit cards because the smart card information is encrypted	Not widely used in the United States

The Peninsula Taxi Association in Cape Town, South Africa, has a fleet of 250 cabs and introduced a smart card payment system that allows riders to use the card on local buses as well as to pay for cab fare.[53] Austin, Texas, and Salt Lake City, Utah, are participating in a pilot project to encourage consumers to use smart cards to pay for their purchases. The test includes loading data from consumers' loyalty cards onto their smart card to reduce the number of cards consumers must carry.[54]

P-Card

p-card (procurement card or purchasing card): A credit card used to streamline the traditional purchase order and invoice payment processes.

A **p-card (procurement card or purchasing card)** is a credit card used to streamline the traditional purchase order and invoice payment processes. The p-card is typically issued to selected employees who must follow company rules and guidelines that may include a single purchase limit, a monthly spending limit, or merchant category code restrictions. Due to an increased risk of unauthorized purchases, each p-card holder's spending activity is reviewed periodically by someone independent of the cardholder to ensure adherence to the guidelines.

The city government of Richmond Heights, Ohio, had such success using procurement cards that it passed an ordinance outlining the use of the cards for government officials.[55] Washington State University adopted procurement cards as the preferred means of paying for purchases under $3000.[56]

Payments Using Cell Phones

The use of cell phones has become commonplace to make purchases and transfer funds between consumers. Two options are available: payments linked to your bank account and payments added to your phone bill. The goals are to make the payment process as simple and secure as possible and for it to work on many different phones and through many different cell phone service providers—not simple tasks. Fortunately, the intelligence built into the iPhone, BlackBerry, and other smartphones can make this all possible.

You can use several services (e.g. Phone Transact iMerchant Pro, Square, RoamData RoamPay, and PayWare Mobile) to plug a credit card reader device into the headphone jack on a cell phone to accept credit card payments. Intuit's GoPayment service does not require a credit card reader but provides software that lets you enter the credit card number.

With Xipwire, consumers can text someone with a special code to place a purchase on their monthly phone bill and bypass any credit card system altogether. With Apple's Bump iPhone application, two iPhone users can link their devices together to transfer cash from one credit card funded PayPal account to another. A free Starbucks Card Mobile app that runs on iPhones, iPod touches, and some BlackBerry smartphones enables customers to pay for their java by holding their mobile device in front of a scanner that reads the app's on-screen barcode. Registered customers link their credit card information to their Starbucks.com account.[57]

SUMMARY

Principle:

Electronic and mobile commerce are evolving, providing new ways of conducting business that present both potential benefits and problems.

Electronic commerce is the conducting of business activities (e.g. distribution, buying, selling, marketing, and servicing of products or services) electronically over computer networks. Business-to-business (B2B) e-commerce allows manufacturers to buy at a low cost worldwide, and it offers enterprises the chance to sell to a global market. B2B e-commerce is currently the largest type of e-commerce. Business-to-consumer (B2C) e-commerce enables organizations to sell directly to consumers, eliminating intermediaries. In many cases, this practice squeezes costs and inefficiencies out of the supply chain and can lead to higher profits and lower prices for consumers. Consumer-to-consumer (C2C) e-commerce involves consumers selling directly to other consumers. Online auctions are the chief method by which C2C e-commerce is currently conducted. e-Government involves the use of information and communications technology to simplify the sharing of information, speed formerly paper-based processes, and improve the relationship between citizens and government.

A successful e-commerce system must address the many stages consumers experience in the sales life cycle. At the heart of any e-commerce system is the ability of the user to search for and identify items for sale; select those items; negotiate prices, terms of payment, and delivery date; send an order to the vendor to purchase the items; pay for the product or service; obtain product delivery; and receive after-sales support.

From the perspective of the provider of goods or services, an effective e-commerce system must be able to support the activities associated with supply chain management and customer relationship management.

A firm must overcome three key challenges to convert its business processes from the traditional form to e-commerce processes: (1) it must deal effectively with consumer privacy concerns, (2) it must successfully overcome consumers' lack of trust, and (3) overcome global issues.

Mobile commerce is the use of wireless devices such as cell phones and smartphones to facilitate the sale of goods or services—anytime and anywhere. The market for m-commerce in North America is expected to mature much later than in Western Europe and Japan. Numerous retailers have established special Web sites for users of mobile devices.

Principle:

E-commerce and m-commerce can be used in many innovative ways to improve the operations of an organization.

Electronic retailing (e-tailing) is the direct sale from a business to consumers through electronic storefronts designed around an electronic catalog and shopping cart model.

A cybermall is a single Web site that offers many products and services at one Internet location.

Manufacturers are joining electronic exchanges, where they can work with competitors and suppliers to use computers and Web sites to buy and sell goods, trade market information, and run back-office operations, such as inventory control. They are also using e-commerce to improve the efficiency of the selling process by moving customer queries about product availability and prices online.

The Web allows firms to gather much more information about customer behavior and preferences than they could using other marketing approaches. This new technology has greatly enhanced the practice of market segmentation and enabled companies to establish closer relationships with their customers.

The Internet has revolutionized the world of investment and finance, especially online stock trading and online banking. The Internet has also created many options for electronic auctions, where geographically dispersed buyers and sellers can come together.

The numerous m-commerce applications include advertising, bartering, retargeting, price comparison, couponing, investment and finance, banking, and e-boutiques.

Principle:

E-commerce and m-commerce offer many advantages yet raise many challenges.

Businesses and people use e-commerce and m-commerce to reduce transaction costs, speed the flow of goods and information, improve the level of customer service, and enable the close coordination of actions among manufacturers, suppliers, and customers.

E-commerce and m-commerce also enable consumers and companies to gain access to worldwide markets. They offer great promise for developing countries, enabling them to enter the prosperous global marketplace and hence helping to reduce the gap between rich and poor countries.

Because e-commerce and m-commerce are global systems, they face cultural, language, time and distance, infrastructure, currency, product and service, and state, regional, and national law challenges.

Principle:

Organizations must define and execute a strategy to be successful in e-commerce and m-commerce.

Most people agree that an effective Web site is one that creates an attractive presence and meets the needs of its visitors. E-commerce start-ups must decide whether they will build and operate the Web site themselves or outsource this function. Web site hosting services and storefront brokers provide alternatives to building your own Web site.

To build traffic to your Web site, you should register a domain name that is relevant to your business, make your site search-engine friendly by including a meta tag in your home page, use Web site traffic data analysis software to attract additional customers, and modify your Web site so that it supports global commerce. Web site operators must constantly monitor the traffic and response times associated with their sites and adjust software, databases, and hardware to ensure that visitors have a good experience when they visit.

Principle:

E-commerce and m-commerce require the careful planning and integration of a number of technology infrastructure components.

A number of infrastructure components must be chosen and integrated to support a large volume of transactions with customers, suppliers, and other business partners worldwide. These components include hardware, Web server software, and e-commerce software.

M-commerce presents additional infrastructure challenges, including improving the ease of use of wireless devices, addressing the security of wireless transactions, and improving network speed. The Wireless Application Protocol (WAP) is a standard set of specifications to enable development of m-commerce software for wireless devices. The development of WAP and its derivatives addresses many m-commerce issues.

Electronic payment systems are a key component of the e-commerce infrastructure. A digital certificate is an attachment to an e-mail message or data embedded in a Web page that verifies the identity of a sender or a Web site. To help prevent the theft of credit card numbers and banking information, the Secure Sockets Layer (SSL) communications protocol is used to secure all sensitive data. Several electronic cash alternatives require the purchaser to open an account with an electronic cash service provider and to present proof of identity whenever payments are to be made. Payments can also be made by credit, charge, debit, smart cards, and p-cards. Retail and banking industries are developing means to enable payments using the cell phone like a credit card.

CHAPTER 8: SELF-ASSESSMENT TEST

Electronic and mobile commerce are evolving, providing new ways of conducting business that present both potential benefits and problems.

1. ________________ e-commerce activities include identifying and comparing competitive suppliers and products, negotiating and establishing prices and terms, ordering and tracking shipments, and steering organizational buyers to preferred suppliers and products.
2. The revenue generated by U.S. B2C e-commerce is about ________________ as/than B2C revenue.
   a. five times larger
   b. three times smaller
   c. the same
   d. thirteen times larger
3. The region of the world with the least forecasted B2C e-commerce spending for 2014 is ________________.
4. The largest B2C retailer in the United States is ________________.
   a. Amazon
   b. Staples
   c. Apple
   d. Walmart
5. E-Government simplifies the sharing of information, speeds up formerly paper-based processes, and improves the relationship between citizens and government. True or False?

E-commerce and m-commerce can be used in many innovative ways to improve the operations of an organization.

6. B2C systems must support negotiation between a buyer and the selected seller over the final price, delivery date, delivery costs, and any extra charges. True or False?
7. About ________________ of all adult Internet users will not buy anything online primarily because they have privacy concerns or lack trust in online merchants.
   a. one-half
   b. one-third
   c. one-fourth
   d. fewer than one-fourth
8. The Internet Corporation for Assigned Names and Numbers (ICANN) created a domain called ________________ to attract mobile users to the Web.

E-commerce and m-commerce offer many advantages yet raise many challenges.

9. Conversion to an e-commerce or m-commerce system enables organizations to achieve many benefits including reducing the cost of business, speeding the flow of goods and information, increasing the accuracy of order processing and order fulfillment, and improving the level of customer service. True or False?
10. ________________ is the direct sale of products or services by businesses to consumers through electronic storefronts, which are typically designed around the familiar electronic catalog and shopping cart model.
11. There are ________________ kinds of electronic exchanges based on who operates the exchange.
12. Businesses are willing to ________________ as a means to reduce excess inventory, gain new customers, or avoid paying cash for necessary raw materials or services.
13. The ________________ security standard spells out measures and security procedures to safeguard the card issuer, the cardholder, and the merchant.

Organizations must define and execute a strategy to be successful in e-commerce and m-commerce.

14. When building a Web site, you should first decide ________________.
    a. what Web software you will use
    b. what Web hardware is necessary
    c. what tasks the site must accomplish
    d. what products and services you will sell
15. Web site hosting companies make it possible to set up a Web page and conduct e-commerce within a matter of days and with few up-front costs. True or False?

E-commerce and m-commerce require the careful planning and integration of a number of technology infrastructure components.

16. The amount of storage capacity and computing power required of a Web server depends primarily on ________________.
    a. the geographical location of the server and number of different products sold
    b. the software that must run on the server and the volume of e-commerce transactions
    c. the size of the business organization and the location of its customers
    d. the number of potential customers and average dollar value of each transaction
17. Key Web site performance measures include response time, transaction success rate, and system availability. True or False?

CHAPTER 8: SELF-ASSESSMENT TEST ANSWERS

1. Buy-side
2. d.
3. Africa and the Middle East
4. a.
5. true
6. false
7. b.
8. .mobi
9. true
10. Electronic retailing or e-tailing
11. three
12. barter
13. Payment Card Industry or PCI
14. c.
15. true
16. b.
17. true

REVIEW QUESTIONS

1. Briefly define the term electronic commerce, and identify five forms of electronic commerce based on the parties involved in the transactions.
2. Compare the U.S. dollar volume of B2B, B2C, and mobile e-commerce. How does the volume of U.S. B2C compare to the global volume?
3. What is disintermediation? What advantages does it present?
4. What tools are available to help shoppers compare prices, features, and values and check other shoppers' opinions?
5. What are Web-influenced sales?
6. Identify the six stages consumers experience in the sales life cycle that must be supported by a successful e-commerce system.
7. Identify three key challenges that an organization faces in creating a successful e-commerce operation.
8. Outline at least three specific trust-building strategies for an organization to gain the trust of consumers.
9. What is electronic couponing and how does it work? What are some of the issues with electronic couponing?
10. What is an electronic exchange? Identify and briefly describe three types of exchanges based on who operates the exchange.
11. What is the two-step authentication associated with items purchased via e-commerce?
12. Why is it necessary to continue to maintain and improve an existing Web site?
13. Identify the key elements of the technology infrastructure required to successfully implement e-commerce within an organization.

DISCUSSION QUESTIONS

1. Briefly discuss three models for selling mobile ad impressions. What are the primary measures for the success of mobile advertising?
2. Why are many manufacturers and retailers outsourcing the physical logistics of delivering merchandise to shoppers? What advantages does such a strategy offer? What are the potential issues or disadvantages?
3. What is retargeting? What are some strategies used to retarget Web site visitors?
4. What are some of the privacy concerns that shoppers have with e-commerce?
5. Identify three specific actions an organization can take to overcome consumers' lack of trust.
6. The volume of mobile commerce in the United States is expected to reach only about 7 percent of total B2C sales by 2016. Do you think that mobile commerce is a significant capability for shoppers? Why or why not?
7. Identify and briefly describe three m-commerce applications you have used.
8. Discuss the use of e-commerce to improve spending on manufacturing, repair, and operations (MRO) of goods and services.
9. Outline the key steps in developing a corporate global e-commerce strategy.
10. Identify three kinds of business organizations that would have difficulty in becoming a successful e-commerce organization.
11. Identify five major global issues that deter the spread of e-commerce. What steps can an organization take to overcome these barriers?

PROBLEM-SOLVING EXERCISES

1. Develop a set of criteria you would use to evaluate various business-to-consumer Web sites based on factors such as ease of use, response time, availability, protection of consumer data, and security of payment process. Develop a simple spreadsheet containing these criteria. Evaluate three popular Web sites using the criteria you developed. What changes would you recommend to the Web developer of the site that scored lowest?
2. Use the charting capability of your spreadsheet software to plot the growth of B2C e-commerce and retail sales for the period 2004 to the present. Using current growth rates, predict the year that B2C e-commerce will exceed 10 percent of retail sales. Document any assumptions you make.
3. Your large screen TV just gave out and must be replaced within the week! Use your Web-enabled smartphone (or borrow a friend's) to perform a price and product comparison to identify the manufacturer and model that best meets your needs and the retailer with the lowest delivered cost. Obtain peer input to validate your choice. Write a brief summary of your experience, and identify the Web sites you found most useful.

TEAM ACTIVITIES

1. Imagine that your team has been hired as consultants to a large organization that has just suffered a major public relations setback due to a large-scale data breach that it handled poorly. What actions would you recommend for the firm to regain consumer confidence?
2. As a team, develop a set of criteria that you would use to evaluate the effectiveness of a mobile advertising campaign to boost the popularity of a candidate for an elected state government position. Identify the measures you would use and the data that must be gathered.

WEB EXERCISES

1. Do research on the Web to find out more about consumer privacy data breaches associated with e-commerce Web sites. Document the actions taken by the operators of the breached Web sites. Which operators did the best job of informing consumers and taking follow-up action? Which Web sites did the worst job? Document your findings in a brief report.
2. Do research on the Web to find a dozen Web sites that offer mobile coupons. Separate the sites into two groups: those that provide coupons for a single retailer and those that aggregate coupons for multiple retailers. Produce a table that summarizes your results and shows the approximate number of coupons available at each site.

CAREER EXERCISES

1. Do research and write a brief report on the impact of consumer privacy concerns on e-commerce.
2. For your chosen career field, describe how you might use or be involved with e-commerce. If you have not chosen a career yet, answer this question for someone in marketing, finance, or human resources.

CASE STUDIES

Case One

MobiKash: Bringing Financial Services to Rural Africa

Full participation in the twenty-first-century economy requires access to financial services. However, this access is a luxury for many citizens of African nations. Due to the long distances between bank branches and the lack of rapid, cost-effective transportation to the urban areas in which banks are typically found, fewer than 10 percent of Africans participate in formal banking. Those who do often face time-consuming inefficiencies.

A new company, MobiKash Afrika, hopes to change this by empowering people in Africa with a secure and independent mobile commerce system that is easy to use. In planning its system, MobiKash established several standards:

- The service must be independent of specific mobile telephone operators.
- The service must be independent of specific banks or financial institutions.
- The service must work with all bill issuers.
- The service must not require smartphones or high-end featurephones.

MobiKash offers its members five services, all accessible from a mobile phone: loading money into their MobiKash account from any bank account, paying bills, sending money to any other mobile phone user or bank account, managing a bank or MobiKash account, and obtaining or depositing cash. Only the last pair of services requires members to visit a physical location where cash can be handled, but that site doesn't have to be a bank. MobiKash agents in market towns, convenient to rural areas, can handle transactions that require cash. (As of the end of 2011, 3,000 MobiKash agents were operating in Kenya. The firm expected to cover all 47 Kenyan counties by mid-2012.) Account holders don't even need to visit a bank to set up their MobiKash accounts: in fact, anyone with a mobile phone to whom a MobiKash user sends money becomes a MobiKash user automatically.

MobiKash charges for some services. Withdrawing cash costs 25 to 75 Kenya shillings (Kshs), about U.S. $0.30 to $0.90, for withdrawals up to Kshs 10,000 (about U.S. $20), with higher fees for larger withdrawals. Paying bills from a mobile phone incurs a fixed fee of Kshs 25, no matter how large the bill is. The largest fee that MobiKash charges is Kshs 350, about U.S. $4, for cash withdrawals in excess of Kshs 75,000, about U.S. $900. This fee schedule is consistent with the financial resources of MobiKash users and the value those users place on each financial service.

The MobiKash system is based on Sybase 365 mCommerce software. Several factors contributed to this choice, including the local presence of Sybase in Africa with experience in similar applications, its understanding of how to integrate with African financial institutions, and the system's ability to work with any mobile telephone. It operates from an existing Sybase data center in Frankfurt, Germany.

In early 2012, MobiKash services were available only in Kenya. However, MobiKash is expanding in east, west, and southern Africa, starting with Zimbabwe, and is planning to cover at least nine countries by the end of 2012. It is working with Masary, an Egyptian e-wallet firm, to cover northern Africa as well. Work is also under way to support intercontinental fund transfers to and from North America, Europe, and the Middle East. As for the future, at the end of 2011, CEO Duncan Otieno said, "We see MobiKash in the next five years playing with the international or global mobile commerce space in at least 40 countries. The plans for building this network are already in progress."

Discussion Questions

1. Firms can base m-commerce systems on commercially available software, as MobiKash did here. Alternatively, they can write their own software. List three pros and cons of each approach. Do you think MobiKash made the right choice?
2. What is the value of each of the four MobiKash standards listed near the beginning of this case study?

Critical Thinking Questions

1. MobiKash is not the only mobile money system in Kenya. Safaricom M-Pesa is the oldest and largest of the other mobile money systems. However, the others are tied to specific banks or network operators. Using the competitive concepts you studied in Chapter 2, how can MobiKash compete against established firms in this market?
2. Contrast your m-commerce needs with those of a typical rural African. Would you find the MobiKash offering attractive in full, in part (which parts?), or not at all?

SOURCES: Masary Web site, *www.e-masary.com*, accessed March 1, 2012; MobiKash Afrika, "The First Intra-region Mobile Network and Bank Agnostic Mobile Commerce Solution," *Computerworld* case study, *www.eiseverywhere.com/file_uploads/e1bfbec2f385506b3890cbd7e b7e9dd9_MobiKash_Afrika_-_The_First_Intra-region_Mobile_Network_and_Bank_Agnostic_Mobile_Commerce_Solution.pdf*, accessed March 1, 2012; MobiKash Africa Web site, *www.mobikash.com*, accessed March 1, 2012; "Reaching the Unbanked in a MobiKash World," interview with CEO Duncan Otieno, *MobileWorld*, *www.mobileworldmag.com/reaching-the-unbanked-in-a-mobikash-world.html*, December 28, 2011; Sybase, "MobiKash Africa: Customer Case Study," *www.sybase.com/files/Success_Stories/Mobikash-CS.pdf*, accessed March 1, 2012.

Case Two

Nambé Moves to the Cloud

If you want a fruit bowl shaped like flower petals, a swooping wood-and-silver candle holder for six tea candles, or a bronze-finished three-bottle wine sling, Nambé is the place to go. Jeff Creecy, vice president of IT at this high-end housewares vendor, describes the company's products as "functional art, luxury you can afford."

However, when Creecy assumed his current position, he found that Nambé's Web site didn't measure up to the standard of its products. He didn't think the site was building brand awareness beyond the existing client base. He also wasn't confident that the site had the ability to handle increased visitor traffic or that it could provide an easy navigation experience for visitors.

"Visually, the old site was okay," Creecy concedes. "It wasn't awful. But it wasn't great either. It was very difficult to search and navigate. It tended to crash when traffic was heavy. And the integration and maintenance required to support it was really difficult."

Another problem was that the original software engineers who built the site had left the company. Having been a software developer early in his career, Creecy knew how difficult it is to understand and maintain a design built by others. "When I joined we were a couple of generations away from those who developed it," Creecy says. Nor was Nambé in a position to invest the resources required to maintain a site properly.

Creecy's background is unusual for a corporate CIO. In addition to having been a programmer, he was educated in marketing. At Nambé, he oversees direct sales as well as heading the IT department. Thus, he feels strongly that "an e-commerce system should really be in the hands of marketers and merchants, not IT. I focus first on solving the business problem, then see where the technology takes me."

Following that approach, he and his team looked at several ways to build an e-commerce system that was on-premise, open-source, and software-as-a-service (SaaS). They selected an SaaS solution, CDC Software's eCommerce Platform. With SaaS, Nambé doesn't run the software itself. Instead, another company runs the software for Nambé. Nambé and its customers access the site over the Internet—"in the cloud," in today's terminology—using Web browsers. SaaS offers several advantages. One, according to Creecy, is that "with SaaS you get a product that is constantly being improved." Another is that the service provider can assign more or less hardware to Nambé's site as load demands; Nambé pays only for what it uses.

While the new site retains much of the old visual design, Creecy says that "The big changes we made were in search and navigation and the technology underneath that...[The new site] groups things into the attributes we think are relevant to the consumer. We worked very hard to enhance our customers' ability to find the products they want." For instance, the top navigation lets consumers see products made of crystal or those made of metal alloys. A later site improvement added zooming capabilities to product photos.

As a marketer, Creecy was also tuned in to the risks of channel conflict. The majority of Nambé's business comes through department and specialty stores that sell its products and through corporate gift programs. Recognizing that nambe.com couldn't compete with retail clients on price without jeopardizing its wholesale business, Nambé has refocused its site to operate primarily as a brand-building tool, although it continues to sell products. This involves changing the emphasis from "buy here now," with promotions to encourage that behavior, to "aren't these nice?" With his marketing background, Creecy understands how a subtle change in emphasis can affect subconscious customer perceptions.

Creecy estimates that only about 7 or 8 percent of total sales come from nambe.com. "One thing I'm committed to is to make sure we have a site that coexists properly with our selling channels," he says.

One of the benefits of outsourcing the operation of Nambé's site is peace of mind. It "enables me to sleep better," Creecy says. "I never worry about it crashing."

Discussion Questions

1. When you visit an e-commerce site, you usually have no way of knowing if it is operated by the firm whose name is on the top or by an SaaS supplier on behalf of that firm. Does this inability bother you? If you could find out who operates the site, would you? As a consumer considering buying something, would you care?
2. The case gives reasons why Nambé selected SaaS, a cloud-based approach, for its e-commerce site. What *disadvantages* does the cloud approach have?

Critical Thinking Questions

1. Nambé is careful not to promote its products too strongly on nambe.com, lest it damage its relationships with its primary sales channels. It could, instead, say "We'll sell as much as we can online, and if stores drop our products, so be it." What are the pros and cons of this alternate strategy?
2. While it is good for a CIO to have a background in marketing, as Nambé's Creecy does, do you think the benefits justify the compromise of having less education or experience in his or her own professional field? If you had to choose between two candidates who appear equally qualified in other respects, but one has a degree in IS while the other has a degree in marketing, which would you pick and why? If your answer depends on factors in that CIO's specific job, what are they, and how would they affect your preference?

SOURCES: CDC Software, "Nambé Case Study," 2011, *www.cdcsoftware.com/en/Resource-Library/Documents/Case-Studies/NA-ECM-GB-CS-NambeCaseStudy-PM-11252011* (free registration required), downloaded March 2, 2012; Enright, A., "Outsourcing Its Web Platform Lets Nambé Focus on Brand-Building," *Internet Retailer, www.internetretailer.com/2010/09/02/outsourcing-its-web-platform-lets-nambe-focus-brand-building,* September 2, 2010; Nambé Web site, *www.nambe.com,* accessed March 13, 2012; "Nambe Puts E-Commerce in the Cloud," *Consumer Goods Technology, consumergoods.edgl.com/case-studies/Nambe-Puts-E-Commerce-in-the-Cloud74115,* July 6, 2011.

Questions for Web Case

See the Web site for this book to read about the Altitude Online case for this chapter. The following are questions concerning this Web case.

Altitude Online: Electronic and Mobile Commerce

Discussion Questions

1. How does Altitude Online's Web site contribute to the company's commerce?
2. How will the new ERP system impact Altitude Online's Web presence?

Critical Thinking Questions

1. How can companies like Altitude Online, which sell services rather than physical products, use e-commerce to attract customers and streamline operations?
2. Consider a company like Fluid by reviewing its site, *www.fluid.com*. Fluid is similar to Altitude Online in the services it offers. What site features do you think are effective for e-commerce? How might you design the site differently?

NOTES

Sources for the opening vignette: Heiler Software AG, "Kramp Focuses on Long Tail and Efficient Customer Response in its E-Commerce Strategy", *www.heiler.com/international/pdf/Case_Study_Kramp_EN.pdf,* August 11, 2011; IBM, "Kramp Group Cuts Transaction Costs and Enhances Customer Service," *www14.software.ibm.com/webapp/iwm/web/signup.do?source=swg-smartercommerce&S_PKG=500007153&S_CMP=web_ibm_ws_comm_cntrmid_b2b* (free registration required), downloaded March 1, 2012; Kramp Group Web site (English), *www.kramp.com/shop/action/start_60_-1,* accessed March 1, 2012.

1. "U.S. Census Bureau E-Stats," *www.census.gov/estats,* May 21, 2011.
2. Brohan, Mark, "The Web Grows in Importance at W. W. Grainger," *Internet Retailer, www.internetretailer.com/2011/07/20/web-grows-importance-w-w-grainger,* July 20, 2011.
3. "Global B2C E-Commerce Players Report 2011," *Market Watch, www.marketwatch.com/story/global-b2c-e-commerce-players-report-2011-2011-12-27,* December 27, 2011.
4. "Quarterly Retail e-Commerce Sales 3rd Quarter 2011," *U.S. Census Bureau News,* November 17, 2011.
5. Ibid.
6. "Forecast for Global e-Commerce: Growth", *Source, www.capturecommerce.com/blog/general/forecast-for-global-ecommerce-growth,* December 2010/January 2011.
7. "Investor Relations," *www.eastonbellsports.com/investor-relations,* accessed February 5, 2012.
8. "Online Shopping at Easton Bell Sports," Sybit CRM Success Story, *www.sybit.de/en/crm/success-stories/success-online-shopping-at-easton-bell-sports.html,* accessed February 5, 2012.
9. Rueter, Thad, "Wrigley Gives Online Shoppers the Chance to Personalize Pack of Sugar-Free Gum," *Internet Retailer, http://ecommerce-news.internetretailer.com/search?w=Wrigley+Gives+Online+Shoppers+the+Chance+to+Personalize+Pack+of+Sugar-Free+Gum,* March 28, 2011.
10. Ribeiro, John, "Walmart's Shopycat Uses Facebook to Recommend Gifts," *PC World, www.pcworld.com/businesscenter/article/245258/walmarts_shopycat_uses_facebook_to_recommend_gifts.html,* December 1, 2011.
11. Abraham, Jack, "Commerce 3.0: Online Research, Offline Buying," *E-Commerce Times, www.ecommercetimes.com/story/73066.html,* August 13, 2011.
12. DePaul, Jennifer, "The $23 Billion 'Amazon Tax' Loophole," *The Fiscal Times, www.thefiscaltimes.com/Articles/2011/11/28/The-23-Billion-Amazon-Tax-Loophole.aspx#page1,* November 28, 2011.
13. Stone, Andrea, "Illegal Gun Sales Run Rampant on Internet," *The Huffington Post, www.huffingtonpost.com/2011/12/14/illegal-internet-gun-sales_n_1148497.html,* December 14, 2011.
14. Parker Ashley, "City Files Suit over Internet Cigarette Sales," *The New York Times,* February 20, 2011.
15. Higgins, John N., "New Challenges in Store for Customer Facing Gov Websites," *CRM Buyer, www.crmbuyer.com/story/74154.html,* January 11, 2012.
16. Hardawar, Devindra, "Forrester: U.S. Mobile Commerce Will Hit $31B, by 2016, Still a Tiny Sliver of eCommerce," Venture Beat, *http://venturebeat.com/2011/06/17/ecommerce-16b-2016,* June 17, 2011.
17. "Reducing Shipping Costs for Low-Volume Merchants," *Practical eCommerce, www.practicalecommerce.com/articles/3333-Reducing-Shipping-Costs-for-Low-Volume-Merchants,* February 1, 2012.
18. Frook, John Evan, "Investors Pump $1.75 Million Into Bypass Lane's Mobile 'Beer-Me Application,'" *Mobile Sports Report, www.mobilesportsreport.com/2011/12/investors-pump-1-75-million-into-bypass-lanes-mobile-beer-me-application,* December 6, 2011.
19. "Progressive Automotive Systems Expands e-Commerce Offerings by Integrating Its R.O. Writer Software with DST TurboParts," AutoChannel, *www.theautochannel.com/news/2012/01/18/021456-progressive-automotive-systems-inc-expands-e-commerce-offerings-by-integrating.html,* January 20, 2012.
20. "Ryder Introduces First Mobile Application to Help Customers Find Truck Rental, Maintenance, and Fuel Locations, Ryder Newsroom, *http://investors.ryder.com/newsroom/News-Release-Details/2012/Ryder-Introduces-First-Mobile-Application-to-Help-Customers-Find-Truck-Rental-Maintenance-and-Fueling-Locations1128179/default.aspx,* February 8, 2012.
21. Wahba, Phil, "Borders Files for Bankruptcy, to Close Stores," *The Huffington Post,* February 16, 2011.
22. "Chronology of Data Breaches," Privacy Rights Clearing House, *www.privacyrights.org/data-breach/new,* accessed January 23, 2011.
23. "Sony Hacked Yet Again," Fox News, *www.foxnews.com/scitech/2011/05/20/sony-hacked-playstation-so-net-isp,* May 21, 2011.
24. Sherman, Erik, "Zappos Hacked: 4 Million Accounts at Risk," CBS News, *www.cbsnews.com/8301-505124_162-57359700/zappos-hacked-24-million-accounts-at-risk,* January 12, 2012.
25. McGrane, Victoria and Smith, Randall, "Hacking at Citi is Latest Data Scare," *The Wall Street Journal,* June 9, 2011.
26. Kubota, Yoko, "Sega Says 1.3 Million Users Affected by Cyber Attack," Reuters, *www.reuters.com/article/2011/06/19/us-sega-hackers-idUSL3E7HJ01520110619,* June 19, 2011.
27. "About Identity Theft," Federal Trade Commission Web site, *www.ftc.gov/bcp/edu/microsites/idtheft//consumers/about-identity-theft.html,* accessed January 23, 2012.
28. Ritter, Ken, "Zappos, Amazon Sued Over Customer Data Breach," KomoNews.com, *www.komonews.com/news/business/Zappos-Amazon-sued-over-customer-data-breach–137620588.html,* January 18, 2012.
29. Demery, Paul, "Fly Fishing Retailer Hooks Higher Conversions," *Internet Retailer, www.internetretailer.com/2011/07/07/fly-fishing-retailer-hooks-higher-conversions,* July 7, 2011.
30. "Online Shopping at Open Sky, at the Zutronic.com Super Cybermall Online Shopping Mall," blog, *https://blog.zutronic.com/2012/02/online-shopping-at-open-sky-at-the-zutronic-com-super-cyber-mall-online-shopping-mall,* February 11, 2012.

31. "About Us," MRO Web site, *www.mrocenter.com/aboutus.aspx*, accessed February 13, 2012.
32. Staff, "NWP and eSupply Systems Announce Agreement to Introduce MRO Manager eProcurement Solution," PRLOG Press Release, *www.prlog.org/11554576-nwp-and-esupply-systems-announce-agreement-to-introduce-mro-manager-eprocurement-solution.html*, June 22, 2011.
33. "Nielsen Business Facts," *http://tetrad.com/demographics/usa/nielsen/businessfacts.html*, accessed February 17, 2011.
34. "About Us," *www.admob.com/home/about*, accessed February 17, 2012.
35. IRS Bartering Tax Center. *www.irs.gov/businesses/small/article/0,,id=187920,00.html*, accessed February 16, 2012.
36. Rudelle, J. B., "Old-School Retailing Just Doesn't Cut it," *CRM Buyer*, *www.crmbuyer.com/story/72105.html*, March 21, 2011.
37. Bergen, Jennifer, "The 10 Best Shopping Apps to Compare Prices," *PC Magazine*, November 25, 2011.
38. "Mobeam Partners with Procter & Gamble to Reinvent the Coupon," mobeam Web site, *www.mobeam.com/uncategorized/mobeam-partners-with-procter-gamble-to-reinvent-the-coupon*, December 19, 2011.
39. "Use Mobile Grocery Coupons for the Holidays," *Mobile Grocery Coupons* (blog), *http://mobilegrocerycoupons.wordpress.com*, November 15, 2011.
40. "Coupon Trend Reports," January 20, 2011 at *www.santella.com/Trends.htm*.
41. Clancy, Heather, "Proctor and Gamble Brings Coupons to the Mobile Phone," *Smart Planet* (blog), *www.smartplanet.com/blog/business-brains/proctor-gamble-brings-coupons-to-the-mobile-phone/20641?tag=search-river*, December 19, 2011.
42. Tarnowski, Joseph, "Global Mobile Coupon Redemption to Be 8 Times Paper by 2016," *Progressive Grocer*, *www.progressivegrocer.com/top-stories/headlines/technology/id34517/global-mobile-coupon-redemption-to-be-8-times-paper-by-2016*, January 5, 2012.
43. Barrett, Larry, "Scottrade Rolls Out Mobile Trading App for Investors," *Financial Planning*, *www.financial-planning.com/news/scottrade-mobile-trading-application-2674496-1.html*, August 3, 2011.
44. "Visa Introduces Suite of Mobile Services for U.S. Financial Institutions," *The Street*, *www.thestreet.com/story/11409239/1/visa-introduces-suite-of-mobile-services-for-us-financial-institutions.html*, February 8, 2012.
45. Hogg, Charlotte, "Lee Coleman, Founder of Quintessentially Gifts, Reveals Her Top Tip for Entrepreneurs," *The Next Women*, *http://thenextwomen.com/2011/10/10/lee-coleman-founder-quintessentially-gifts-reveals-her-top-tip-entrepreneurs*, October 10, 2011.
46. "BiggSports Scores Greater Profitability with ProStores," *www.prostores.com/online-store-Bigg-Sports.html*, accessed February 20, 2012.
47. "iDTV" *www.slideshare.net/Saky10/idtv*, accessed February 20, 2012.
48. Martinez, Juan, "Staples Revamps b-to-b Commerce Site, Reports Net Sales Increase," *Direct Marketing News*, *www.dmnews.com/staples-revamps-b-to-b-e-commerce-site-reports-net-sales-increase/article/216825*, November 16, 2011.
49. Deatsch, Katie, "Potpourri Group Senses Mores Sales with Personalization," *Internet Retailer*, *www.internetretailer.com/2011/03/31/potpourri-group-senses-more-sales-personalization*, March 31, 2011.
50. "Voice Biometrics," Nuance Web site, *www.nuance.com/for-business/by-solution/customer-service-solutions/solutions-services/inbound-solutions/voice-authentication/ voice-biometrics/index.htm*, accessed March 23, 2012.
51. "PayPal and x.Commerce Launch PayPal Access," PayPal News Room, *https://www.paypal-media.com/news#*, October 11, 2011.
52. Enright, Allison, "Redcats Adds PayPal Express Checkout to Speed Checkout," *Internet Retailer*, *www.internetretailer.com/2011/07/07/redcats-adds-paypal-express-checkout-speed-checkout*, July 7, 2011.
53. "Taxis Adopt Smart Card System," *The New Age*, *www.thenewage.co.za/42657-1011-53-Taxis_adopt_smart_card_system*, February 7, 2012.
54. Harvey, Tom, "Salt Lake a Testing Ground for New to Pay," *Salt Lake Tribune*, February 9, 2012.
55. "An Ordinance Enacting a Procurement Card Policy for City Purchases," *www.richmondheightsohio.org/legislation-2012/14-2012.pdf*, accessed February 22, 2012.
56. Washington State University Announcements for 1/13/2012, *http://announcements.wsu.edu/pages/newsletters.asp?Action=Detail&CampaignID=1970#08eb63c60632c540eb357e915aa52a1c*, accessed March 9, 2012.
57. Adhikari, Richard, "The Long Leap to Clear Mobile Payment Hurdles," *Tech News World*, *www.technewsworld.com/story/71694.html*, January 21, 2011.

9 Enterprise Systems

Principles	Learning Objectives
• An organization must have information systems that support routine, day-to-day activities and that help a company add value to its products and services.	• Identify the basic activities and business objectives common to all transaction processing systems. • Describe the transaction processing systems associated with the order processing, purchasing, and accounting business functions.
• An organization that implements an enterprise system is creating a highly integrated set of systems, which can lead to many business benefits.	• Discuss the advantages and disadvantages associated with the implementation of an enterprise resource planning system, customer resource management, and product lifecycle management system. • Identify the challenges that multinational corporations face in planning, building, and operating their enterprise systems.

Information Systems in the Global Economy

YIOULA GROUP, ATHENS, GREECE

From Stand-Alone to Integrated Applications

YIOULA Group is the largest glass manufacturer in the Balkans, producing over 625,000 glass containers annually as well as over 30,000 tons of tableware. Starting in the 1990s, the company expanded by acquiring other glassmaking firms in Romania, Bulgaria, and Ukraine. In 2012, it had seven factories in those four countries, about 2,100 employees, and net annual sales of about €180 million (about U.S. $240 million).

As a result of its growth through acquisition, by 2010 YIOULA Group found itself with a confusing variety of information systems. The group was unable to compare production costs for the same item across factories, could not improve efficiencies by coordinating purchasing and financial management across all its plants, and was not positioned for continued growth or expansion into new market areas. Clearly, its legacy stand-alone applications needed to be replaced.

YIOULA Group CIO Zacharias Maridakis had previous experience using integrated enterprise software when he worked at Mobil Oil's Greek subsidiary, Mobil Oil Hellas S.A., in the 1990s. Therefore, he was well acquainted with the advantages of such software. Under his direction, YIOULA Group investigated various software packages. They selected JD Edwards EnterpriseOne, named for a company that became part of Oracle Corporation in 2005. Part of the reason for this choice was that most other ERP packages, including the SAP software with which Maridakis had worked at Mobil, are designed primarily for much larger organizations. EnterpriseOne was always intended for medium-sized firms.

Since YIOULA Group had little experience with EnterpriseOne, it enlisted the help of Oracle partner Softecon to help configure the software to the company's needs, meet the legal requirements of each region in which it operates, and manage implementation in each area. Support for the Greek language (as well as English and eighteen others) is a standard JD Edwards EnterpriseOne capability available from Oracle; Softecon added the other languages that YIOULA Group needed to the user interface. YIOULA Group also added a specialized cost comparison module from Softecon to the basic EnterpriseOne package. This module helps the group choose the lowest-cost facility to manufacture a product.

The conversion to a single enterprise package gave YIOULA Group the expected benefits. Time from order to invoice, delivery time, and cash collection have all been accelerated. Financial data is now available two weeks after the end of a period versus one month previously. A consolidated view of inventory across all plants has enabled the group to manage inventory more efficiently and comprehensively and to use just-in-time purchasing methods.

Perhaps even more importantly, YIOULA Group is now positioned to grow. As Maridakis puts it, "Oracle's JD Edwards EnterpriseOne is a key enabler of our strategy to enhance market leadership in the Balkans, grow our business in the Ukraine, and continue to improve productivity, efficiency, and profitability as we expand into new markets."

As you read this chapter, consider the following:

- What advantages does using an integrated enterprise system offer an organization, compared to using separate systems for each application area?
- How does having multiple operations in different locations or different countries affect these advantages?

WHY LEARN ABOUT ENTERPRISE SYSTEMS?

© Andrey Burmakin/Shutterstock

Individuals and organizations today are moving from a collection of nonintegrated transaction processing systems to highly integrated enterprise systems to perform routine business processes and to maintain records about them. As mentioned in Chapter 1, these systems support a wide range of business activities associated with supply chain management and customer relationship management. Although they were initially thought to be cost effective only for very large companies, even small and midsized companies are now implementing these systems to reduce costs and improve service.

In our service-oriented economy, outstanding customer service has become a goal of virtually all companies. To provide good customer service, employees who work directly with customers—whether in sales, customer service, or marketing—require high-quality transaction processing systems and their associated information. Such workers might use an enterprise system to check the inventory status of ordered items, view the production planning schedule to tell the customer when the item will be in stock, or enter data to schedule a delivery to the customer.

No matter what your role, it is very likely that you will provide input to or use the output from your organization's enterprise systems. Your effective use of these systems will be essential to raise the productivity of your firm, improve customer service, and enable better decision making. Thus, it is important that you understand how these systems work and what their capabilities and limitations are.

enterprise system: A system central to the organization that ensures information can be shared across all business functions and all levels of management to support the running and managing of a business.

An **enterprise system** is central to individuals and organizations of all sizes and ensures that information can be shared across all business functions and all levels of management to support the running and managing of a business. Enterprise systems employ a database of key operational and planning data that can be shared by all. This eliminates the problems of lack of information and inconsistent information caused by multiple transaction processing systems that support only one business function or one department in an organization. Examples of enterprise systems include enterprise resource planning systems that support supply-chain processes, such as order processing, inventory management, and purchasing, and customer relationship management systems that support sales, marketing, and customer service-related processes.

FOCUS Bikes is an internationally known manufacturer of sport and racing bikes founded by cross-country bicycle racing champion and entrepreneur, Mike Kluge. While the manufacturing is done in Germany, the assembly and distribution to customers is done in Carlsbad, California. The operation is relatively small with about 8 to 10 bikes assembled per day. Yet, the firm effectively employs an enterprise system to handle inventory management, purchasing, warehousing, and sales orders to manage its business and to support sales and customer service.[1]

As demonstrated by YIOULA Group in the opening vignette, businesses rely on enterprise systems to perform many of their daily activities in areas such as product supply, distribution, sales, marketing, human resources, manufacturing, accounting, and taxation so that work is performed quickly without waste or mistakes. Without such systems, recording and processing business transactions would consume huge amounts of an organization's resources. This collection of processed transactions also forms a storehouse of data invaluable to decision making. The ultimate goal is to satisfy customers and provide a competitive advantage by reducing costs and improving service.

AN OVERVIEW OF TRANSACTION PROCESSING SYSTEMS

Every organization has many transaction processing systems (TPSs), which capture and process the detailed data necessary to update records about the fundamental business operations of the organization. These systems include order entry, inventory control, payroll, accounts payable, accounts receivable, and the general ledger, to name just a few. The input to these systems includes basic business transactions, such as customer orders, purchase orders, receipts, time cards, invoices, and customer payments. The processing activities include data collection, data editing, data correction, data manipulation, data storage, and document production. The result of processing business transactions is that the organization's records are updated to reflect the status of the operation at the time of the last processed transaction. For example, Najrain Dairy Company, Ltd., in Saudi Arabia produces dairy, water, and juice products. It uses software called InventoryPower to meet its needs for inventory control reporting and stock balance management. With this transaction processing system, it keeps track of the stock level of dozens of products so that customer orders can be fulfilled accurately and without delay.[2]

A TPS also provides valuable input to management information systems, decision support systems, and knowledge management systems, first discussed in Chapter 1. A transaction processing system serves as the foundation for these other systems (see Figure 9.1).

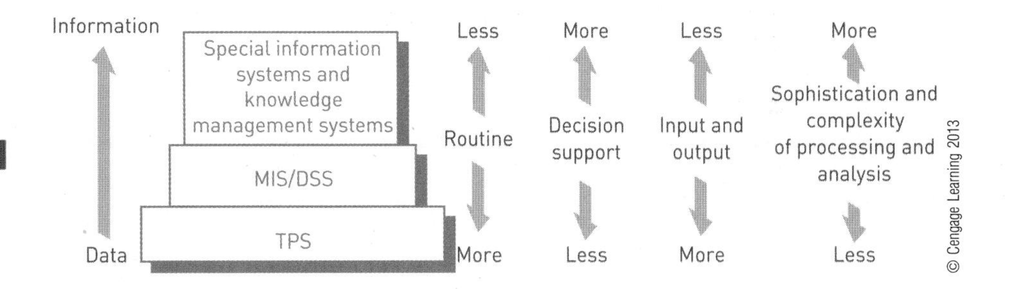

FIGURE 9.1
TPS, MIS/DSS, and special information systems in perspective

Transaction processing systems support routine operations associated with customer ordering and billing, employee payroll, purchasing, and accounting. TPSs, however, don't provide much support for decision making.

TPSs use a large amount of input and output data to update the official records of the company about orders, sales, customers, and so on. Heartland Payments Systems is a United States-based payments processor whose transaction processing systems deliver credit, debit, and prepaid card processing and payroll and related services to over 250,000 business locations nationwide. Heartland was the official concession payments processer for the XLVI Super Bowl in Indianapolis, Indiana. It processed 43,982 credit and debit card transactions for $924,500 during the event.[3]

Bianca USA manufactures home textiles for its trading partners, which include such firms as Bed, Bath, & Beyond; Saks, Inc., and Parisian. This medium-sized firm receives an average of 600 orders per day. The firm has created an integrated transaction processing system using QuickBooks for sales and accounting plus add-on software from eBridge to generate Electronic Data Interchange records such as invoices and advanced shipping notices.[4]

Because TPSs often perform activities related to customer contacts—such as order processing and invoicing—these information systems play a critical role in providing value to the customer. For example, by capturing and tracking the movement of each package, shippers such as Federal Express and DHL Express can provide timely and accurate data on the exact location of a package. Shippers and receivers can access an online database and, by providing the airbill number of a package, find the package's current location. If the package has been delivered, they can see who signed for it (a service that is especially useful in large companies where packages can become "lost" in internal distribution systems and mailrooms). Such a system provides the basis for added value through improved customer service.

Traditional Transaction Processing Methods and Objectives

batch processing system: A form of data processing whereby business transactions are accumulated over a period of time and prepared for processing as a single unit or batch.

With **batch processing systems**, business transactions are accumulated over a period of time and prepared for processing as a single unit or batch (see Figure 9.2a). Transactions are accumulated for the length of time needed to meet the needs of the users of that system. For example, it might be important to process invoices and customer payments for the accounts receivable system daily. On the other hand, the payroll system might receive time cards and process them biweekly to create checks, update employee earnings records, and distribute labor costs. The essential characteristic of a batch processing system is that there is some delay between an event and the eventual processing of the related transaction to update the organization's records.

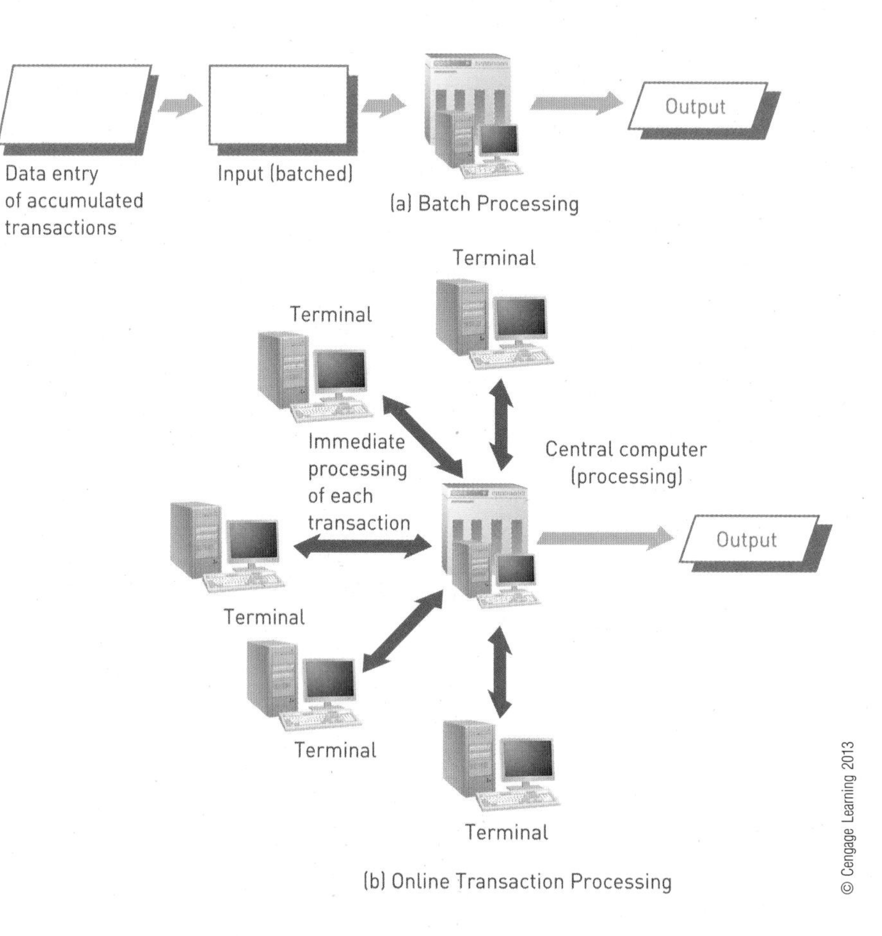

FIGURE 9.2

Batch versus online transaction processing

(a) Batch processing inputs and processes data in groups. (b) In online processing, transactions are completed as they occur.

Automatic Data Processing (ADP) is a major provider of business outsourcing solutions for payroll, human resources, tax, and employee benefits administration for over 570,000 clients worldwide. It uses a batch processing system to prepare the paychecks of one out of seven Americans.[5]

After years of working in the automotive repair parts business, entrepreneur Steve Kirby decided to start his own company. China Auto Group is a manufacturer of automotive repair parts in China. These parts are sold to customers in the United States, Europe, and South America including original equipment manufacturers such as General Motors and auto enthusiasts seeking restoration car parts. Initially, accounting and inventory control were handled using Excel spreadsheets. As the business grew and became more complicated, China Auto Group implemented QuickBooks to process accounting transactions and ACCTivate to perform inventory control processing. These two packages complement each other and work effectively for a company the size of China Auto Group.[6]

An electronic batch processing software module of the AdvancedMD software does a check on patients' medical insurance eligibility. This check, done within the week before a patient's appointment, avoids issues associated with treating patients who no longer have medical insurance coverage.[7]

online transaction processing (OLTP): A form of data processing where each transaction is processed immediately without the delay of accumulating transactions into a batch.

With **online transaction processing (OLTP)**, each transaction is processed immediately without the delay of accumulating transactions into a batch, as shown in Figure 9.2b. Consequently, at any time, the data in an online system reflects the current status. This type of processing is essential for businesses that require access to current data such as airlines, ticket agencies, and stock investment firms. Many companies find that OLTP helps them provide faster, more efficient service—one way to add value to their activities in the eyes of the customer. More and more companies are using the Internet to capture and process transaction data such as customer orders and shipping information from e-commerce applications. The following are a few examples of organizations using online transaction processing systems:

- Fit 4 Travel is a travel agency based in Australia that specializes in the planning of sports and adventure holidays. Typical customers are Australian athletes who plan to compete abroad in a marathon, swimming competition, or other sporting event. The firm implemented a Web-based online reservation transaction processing system capable of integrating all aspects of the trip, including flights, transfers, hotels, tours, car rental, services, and event logistics.[8]
- ResortCom International uses OLTP systems to manage a total value of over $500 million each year for timeshare payments, service loans, and credit card transactions for more than 1 million customers.[9]
- Westside Produce is an independent handler of cantaloupe and honeydew melons and contracts with melon growers to harvest, cool, market, and ship melons throughout North America. The firm uses an OLTP system to track products at the case level from a case's point of origin to its final point of sale. The goal is to protect consumers in the event of food-borne illness outbreaks.[10]
- Interactive entertainment software developer, LucasArts is building large online transaction processing systems that will support millions of gamers.[11]

Hospitality companies such as ResortCom International can use OLTP systems to manage timeshare payments and other financial transactions.

TPS applications do not always run using online processing. For many applications, batch processing is more appropriate and cost effective. Payroll transactions and billing are typically done via batch processing. Specific goals of the organization define the method of transaction processing best suited for the various applications of the company.

Figure 9.3 shows the traditional flow of key pieces of information from one TPS to another for a typical manufacturing organization. Transactions input to one system are processed and create new transactions that flow into another system.

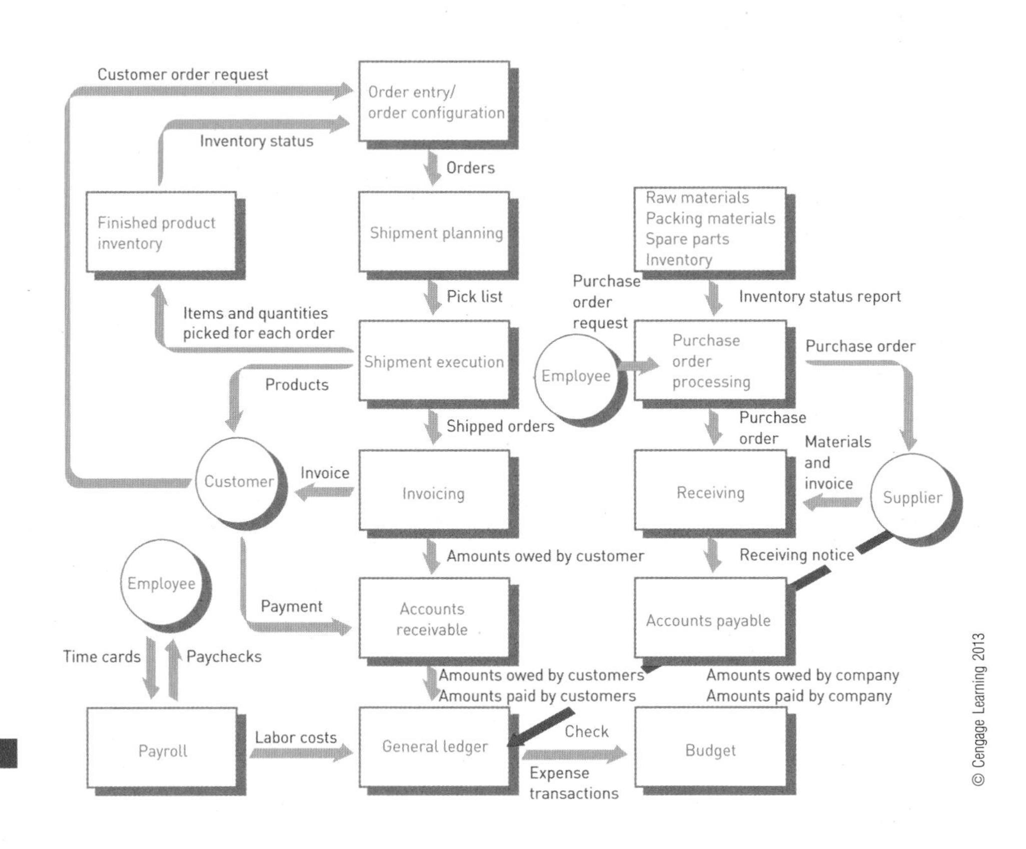

FIGURE 9.3
Integration of a firm's TPS

Because of the importance of transaction processing, organizations expect their TPSs to accomplish a number of specific objectives, including the following:

- Capture, process, and update databases of business data required to support routine business activities
- Ensure that the data is processed accurately and completely
- Avoid processing fraudulent transactions
- Produce timely user responses and reports
- Reduce clerical and other labor requirements
- Help improve customer service
- Achieve competitive advantage

A TPS typically includes the following types of systems:

- **Order processing systems.** Running these systems efficiently and reliably is so critical that the order processing systems are sometimes referred to as the lifeblood of the organization. The processing flow begins with the receipt of a customer order. The finished product inventory is checked to see if sufficient inventory is on hand to fill the order. If sufficient inventory is available, the customer shipment is planned to meet the customer's desired receipt date. A product pick list is printed at the warehouse from which the order is to be filled on the day the order is planned to be shipped. At the warehouse, workers gather the items needed to fill the order and enter the item identifier and quantity for each item to update the finished product inventory. When the order is complete and sent on its way, a customer invoice is created with a copy included in the customer shipment.

- **Accounting systems.** The accounting systems must track the flow of data related to all the cash flows that affect the organization. As mentioned earlier, the order processing system generates an invoice for customer orders to include with the shipment. This information is also sent to the accounts receivable system to update the customer's account. When the customer pays the invoice, the payment information is also used to update the customer's account. The necessary accounting transactions are sent to the general ledger system to keep track of amounts owed and amounts paid. Similarly, as the purchasing systems generate purchase orders and those items are received, information is sent to the accounts payable system to manage the amounts owed by the company. Data about amounts owed and paid by customers to the company and from the company to vendors and others are sent to the general ledger system that records and reports all financial transactions for the company.
- **Purchasing systems.** The traditional transaction processing systems that support the purchasing business function include inventory control, purchase order processing, receiving, and accounts payable. Employees place purchase order requests in response to shortages identified in inventory control reports. Purchase order information flows to the receiving system and accounts payable systems. A record of receipt is created upon receipt of the items ordered. When the invoice arrives from the supplier, it is matched to the original order and the receiving report, and a check is generated if all data is complete and consistent.

In the past, organizations knitted together a hodgepodge of systems to accomplish the transaction processing activities shown in Figure 9.3. Some of the systems might have been applications developed using in-house resources, some may have been developed by outside contractors, and others may have been off-the-shelf software packages. Much customization and modification of this diversity of software was necessary for all the applications to work together efficiently. In some cases, it was necessary to print data from one system and manually reenter it into other systems. Of course, this increased the amount of effort required and increased the likelihood of processing delays and errors.

The approach taken today by many organizations is to implement an integrated set of transaction processing systems from a single or limited number of software vendors that handle most or all of the transaction processing activities shown in Figure 9.3. The data flows automatically from one application to another with no delay or need to reenter data. For example, Denny's Automotive distributes replacement parts and accessories to auto dealers throughout South Africa. During its initial growth, the firm discovered that its information systems were not well integrated, creating problems in trying to order the correct parts and quantities. This problem slowed sales and reduced profits. Denny's implemented an integrated set of transaction processing systems to tie together its inventory control, general ledger, accounts payable, and accounts receivable processes. As a result, accurate data is now available for making sound purchasing decisions.[12]

Table 9.1 summarizes some of the ways that companies can use transaction processing systems to achieve competitive advantage.

TABLE 9.1 Examples of transaction processing systems for competitive advantage

Competitive Advantage	Example
Customer loyalty increased	Customer interaction system to monitor and track each customer interaction with the company
Superior service provided to customers	Tracking systems that customers can access to determine shipping status
Better relationship with suppliers	Internet marketplace to allow the company to purchase products from suppliers at discounted prices
Superior information gathering	Order configuration system to ensure that products ordered will meet customer's objectives
Costs dramatically reduced	Warehouse management system employing RFID technology to reduce labor hours and improve inventory accuracy
Inventory levels reduced	Collaborative planning, forecasting, and replenishing to ensure the right amount of inventory is in stores

Depending on the specific nature and goals of the organization, any of the objectives in Table 9.1 might be more important than others. By meeting these objectives, TPSs can support corporate goals such as reducing costs; increasing productivity, quality, and customer satisfaction; and running more efficient and effective operations. For example, overnight delivery companies such as FedEx expect their TPSs to increase customer service. These systems can locate a client's package at any time, from initial pickup to final delivery. This improved customer information allows companies to produce timely information and be more responsive to customer needs and queries.

Transaction Processing Systems for Entrepreneurs and Small and Medium-Sized Enterprises

Many software packages provide integrated transaction processing system solutions for small and medium-sized enterprises (SMEs), wherein small is an enterprise with 50 to 100 employees and medium-sized is one with 250 to 500 employees. These systems are typically easy to install and operate and usually have a low total cost of ownership, with an initial cost of a few hundred to a few thousand dollars. Such solutions are highly attractive to firms that have outgrown their current software but cannot afford a complex, high-end integrated system solution. Table 9.2 presents some of the dozens of such software solutions available.

TABLE 9.2 Sample of integrated TPS solutions for SMEs

Vendor	Software	Type of TPS Offered	Target Customers
AccuFund	AccuFund	Financial reporting and accounting	Nonprofit, municipal, and government organizations
OpenPro	OpenPro	Complete ERP solution, including financials, supply chain management, e-commerce, customer relationship management, and retail POS system	Manufacturers, distributors, and retailers
Intuit	QuickBooks	Financial reporting and accounting	Manufacturers, professional services, contractors, nonprofits, and retailers
Sage	Timberline	Financial reporting, accounting, and operations	Contractors, real estate developers, and residential builders
Redwing	TurningPoint	Financial reporting and accounting	Professional services, banks, and retailers

QuickBooks is accounting software from Intuit that small business owners use to easily maintain their accounting records. In India, hundreds of companies are subsidiaries of much larger foreign companies. Quite often, these subsidiaries adopt QuickBooks to maintain all their accounts in an accurate and consistent manner. Users can maintain a customer database and record customer payments as well as keep track of current balances. The software makes it easy to create a supplier database and write checks to pay for goods and services.[13]

Qvinci.web allows companies to collect QuickBooks data from many locations and format that data into the company's standardized chart of accounts. Financial managers at various SMEs, such as Aire Serv Heating and Air Conditioning; the Diocese of Lexington, Kentucky; Crunchy Logistics; Ignite Spot; and Hotze Health and Wellness Center, use QuickBooks to create predefined financial reports.[14]

TRANSACTION PROCESSING ACTIVITIES

Along with having common characteristics, all TPSs perform a common set of basic data-processing activities. TPSs capture and process data that describes fundamental business transactions. This data is used to update databases and to produce a variety of reports for people both within and outside the enterprise. The business data goes through a **transaction processing cycle** that includes data collection, data editing, data correction, data manipulation, data storage, and document production (see Figure 9.4).

transaction processing cycle: The process of data collection, data editing, data correction, data manipulation, data storage, and document production.

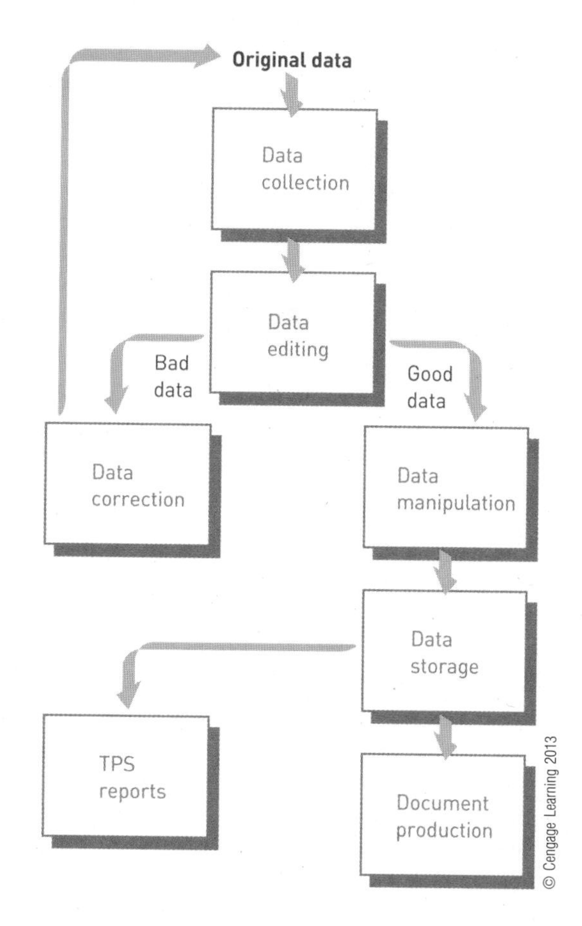

FIGURE **9.4**
Data processing activities common to transaction processing systems

Data Collection

data collection: Capturing and gathering all data necessary to complete the processing of transactions.

Capturing and gathering all data necessary to complete the processing of transactions is called **data collection**. In some cases, it can be done manually, such as by collecting handwritten sales orders or changes to inventory.

In other cases, data collection is automated via special input devices such as scanners, point-of-sale (POS) devices, and terminals.

Data collection begins with a transaction (e.g., taking a customer order) and results in data that serves as input to the TPS. Data should be captured at its source and recorded accurately in a timely fashion with minimal manual effort and in an electronic or digital form that can be directly entered into the computer. This approach is called source data automation. An example of source data automation is an automated device at a retail store that speeds the checkout process—either UPC codes read by a scanner or RFID signals picked up when the items approach the checkout stand. Using both UPC bar codes and RFID tags is quicker and more accurate than having a clerk enter codes manually at the cash register. The product ID for each item is determined automatically, and its price retrieved from the item database. The point-of-sale TPS uses the price data to determine the customer's bill. The store's inventory and purchase databases record the number of units of an item purchased, along with the price and the date and time of the purchase. The inventory database generates a management report notifying the store manager to reorder items that have fallen below the reorder quantity. The detailed purchases database can be used by the store or sold to marketing research firms or manufacturers for detailed sales analysis (see Figure 9.5).

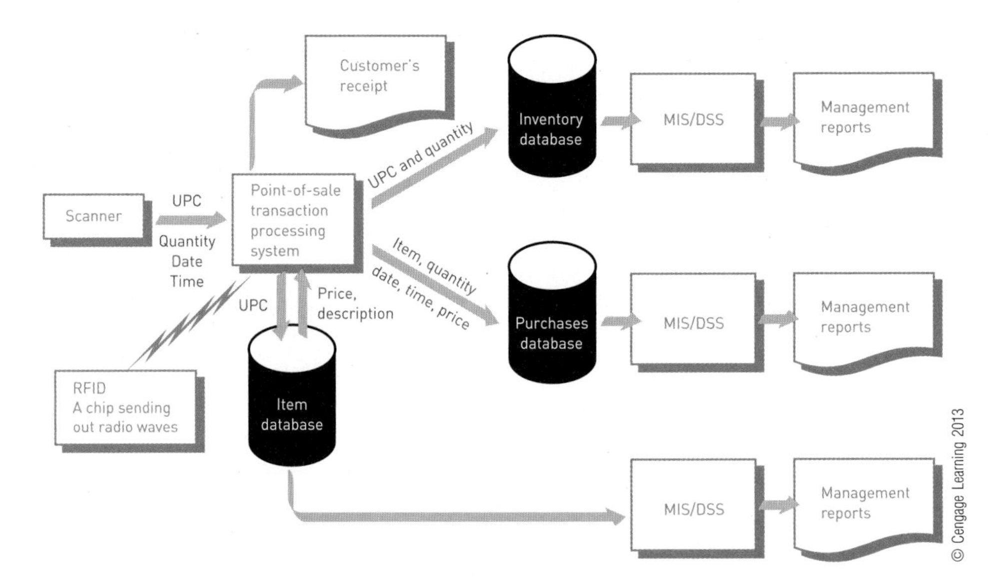

FIGURE 9.5

Point-of-sale transaction processing system

The purchase of items at the checkout stand updates a store's inventory database and its database of purchases.

Many grocery stores combine point-of-sale scanners and coupon printers. The systems are programmed so that each time a specific product—for example, a box of cereal—crosses a checkout scanner, an appropriate coupon, perhaps a milk coupon, is printed. Companies can pay to be promoted through the system, which is then reprogrammed to print those companies' coupons if the customer buys a competitive brand. These TPSs help grocery stores increase profits by improving their repeat sales and bringing in revenue from other businesses. For example, Teamwork Retail is a cloud-based retail application that supports POS systems operating in more than 100,000 stores. Sales clerks armed with iPads can access the application to check inventory in their store or anywhere in the supply chain. They can also process payments, including both cash and credit cards; accept credit cards online; and e-mail or wirelessly print customer receipts.[15]

Data Editing

data editing: The process of checking data for validity and completeness.

An important step in processing transaction data is to check data for validity and completeness to detect any problems, a task called **data editing**. For example, quantity and cost data must be numeric, and names must be alphabetic; otherwise, the data is not valid. Often, the codes associated with an individual transaction are edited against a database containing valid codes. If any code entered (or scanned) is not present in the database, the transaction is rejected.

Data Correction

data correction: The process of reentering data that was not typed or scanned properly.

It is not enough simply to reject invalid data. The system should also provide error messages that alert those responsible for editing the data. Error messages must specify the problem so proper corrections can be made. A **data correction** involves reentering data that was not typed or scanned properly. For example, a scanned UPC code must match a code in a master table of valid UPCs. If the code is misread or does not exist in the table, the checkout clerk is given an instruction to rescan the item or type the information manually.

Data Manipulation

data manipulation: The process of performing calculations and other data transformations related to business transactions.

Another major activity of a TPS is **data manipulation**, the process of performing calculations and other data transformations related to business transactions. Data manipulation can include classifying data, sorting data into categories, performing calculations, summarizing results, and storing data in the organization's database for further processing. In a payroll TPS, for example, data manipulation includes multiplying an employee's hours worked by the hourly pay rate. Overtime pay, federal and state tax withholdings, and deductions are also calculated.

Data Storage

data storage: The process of updating one or more databases with new transactions.

Data storage involves updating one or more databases with new transactions. After being updated, this data can be further processed and manipulated by other systems so that it is available for management reporting and decision making. Thus, although transaction databases can be considered a by-product of transaction processing, they have a pronounced effect on nearly all other information systems and decision-making processes in an organization.

Document Production

document production: The process of generating output records, documents, and reports.

Document production involves generating output records, documents, and reports. These can be hard-copy paper reports or displays on computer screens (sometimes referred to as soft copy). Printed paychecks, for example, are hard-copy documents produced by a payroll TPS, whereas an outstanding balance report for invoices might be a soft-copy report displayed by an accounts receivable TPS. Often, as shown earlier in Figure 9.5, results from one TPS flow downstream to become input to other systems, which might use the results of updating the inventory database to create the stock exception report (a type of management report) of items whose inventory level is below the reorder point.

In addition to major documents such as checks and invoices, most TPSs provide other useful management information, such as printed or on-screen reports that help managers and employees perform various activities. A report showing current inventory is one example; another might be a document listing items ordered from a supplier to help a receiving clerk check the order for completeness when it arrives. A TPS can also produce reports required by local, state, and federal agencies, such as statements of tax withholding and quarterly income statements.

ENTERPRISE RESOURCE PLANNING AND CUSTOMER RELATIONSHIP MANAGEMENT

As defined in Chapter 4, enterprise resource planning (ERP) is a set of integrated programs that manage a company's vital business operations for an entire organization, even a complex, multisite, global organization. Recall that a business process is a set of coordinated and related activities that takes one or more types of input and creates an output of value to the customer of that process. The customer might be a traditional external business customer who buys goods or services from the firm. An example of such a process is capturing a sales order, which takes customer input and generates an order. The customer of a business process might also be an internal customer, such as a worker in another department of the firm. For example, the shipment process generates the internal documents workers need in the warehouse and shipping departments to pick, pack, and ship orders. At the core of the ERP system is a database that is shared by all users so that all business functions have access to current and consistent data for operational decision making and planning, as shown in Figure 9.6.

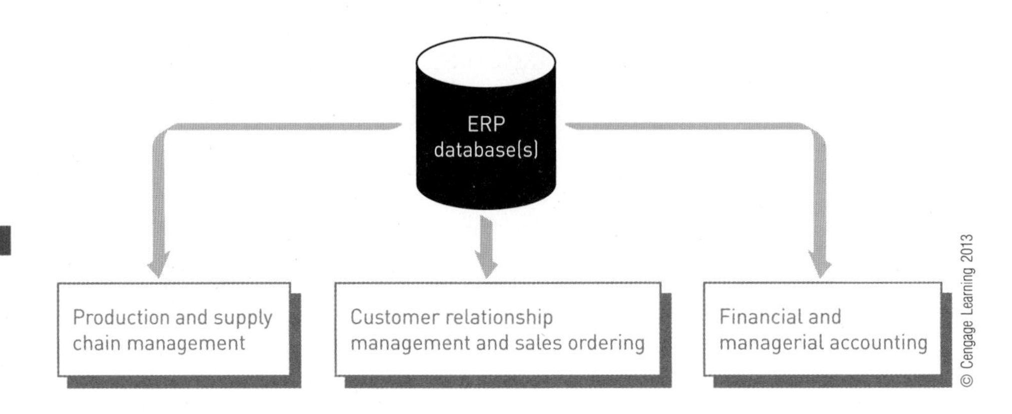

FIGURE **9.6**

Enterprise resource planning system

An ERP integrates business processes and the ERP database.

An Overview of Enterprise Resource Planning

ERP systems evolved from materials requirement planning systems (MRP) developed in the 1970s. These systems tied together the production planning, inventory control, and purchasing business functions for manufacturing organizations. During the late 1980s and early 1990s, many organizations recognized that their legacy TPSs lacked the integration needed to coordinate activities and share valuable information across all the business functions of the firm. As a result, costs were higher and customer service poorer than desired. Large organizations, specifically members of the *Fortune* 1000, were the first to take on the challenge of implementing ERP. As they did, they uncovered many advantages as well as some disadvantages summarized in the following sections.

Advantages of ERP

Increased global competition, new needs of executives for control over the total cost and product flow through their enterprises, and ever-more-numerous customer interactions drive the demand for enterprise-wide access to real-time information. ERP offers integrated software from a single vendor to help meet those needs. The primary benefits of implementing ERP include improved access to quality data for operational decision making, elimination of inefficient or outdated systems, improvement of work processes, and technology standardization. ERP vendors have also developed specialized systems that provide effective solutions for specific industries and market segments.

Improved Access to Quality Data for Operational Decision Making

ERP systems operate via an integrated database, using one set of data to support all business functions. The systems, for instance, can support decisions on optimal sourcing or cost accounting for the entire enterprise or business units from the start rather than gathering data from multiple business functions and then trying to coordinate that information manually or reconciling data with another application. The result is an organization that looks seamless, not only to the outside world but also to the decision makers who are deploying resources within the organization. The data is integrated to facilitate operational decision making and allows companies to provide greater customer service and support, strengthen customer and supplier relationships, and generate new business opportunities. It is essential that the data used for decision making is of high quality.

Mannington Mills is one of the largest and oldest flooring manufacturers in the United States. The firm recently implemented an ERP system and was able to improve data accuracy at the time of initial data entry. Increasing data accuracy improved operational decision making as well as eliminated incorrect tax filing and incorrect payment of employees. It also eliminated errors before transferring data to third parties such as suppliers and shippers.[16]

Elimination of Costly, Inflexible Legacy Systems

Adoption of an ERP system enables an organization to eliminate dozens or even hundreds of separate systems and replace them with a single integrated set of applications for the entire enterprise. In many cases, these systems are decades old, the original developers are long gone, and the systems are poorly documented. As a result, the systems are extremely difficult to fix when they break, and adapting them to meet new business needs takes too long. They become an anchor around the organization that keeps it from moving ahead and remaining competitive. An ERP system helps match the capabilities of an organization's information systems to its business needs—even as these needs evolve.

Syndiant is a small business of 50 employees that designs light-modulating panels for high-resolution displays used in ultraportable projectors small enough to be embedded in a cell phone. Initially, the firm used Intuit's QuickBooks to manage its finances. The firm recognized that it would need more sophisticated software once it brought its product to market, began buying from multiple suppliers, and selling the product through multiple distribution channels. The company also needed the capability to account for accumulated costs incurred throughout the manufacturing process to meet generally accepted accounting principles (GAAP). Syndiant decided to implement an ERP system to keep up with a rapid increase in sales while managing to maintain tight control over its costs. The availability of accurate and timely financial information on costs enables the firm to make accurate decisions on price quotes to customers.[17]

Improvement of Work Processes

best practices: The most efficient and effective ways to complete a business process.

Competition requires companies to structure their business processes to be as effective and customer oriented as possible. ERP vendors do considerable research to define the best business processes. They gather requirements of leading companies within the same industry and combine them with research findings from research institutions and consultants. The individual application modules included in the ERP system are then designed to support these **best practices**, the most efficient and effective ways to complete a business process. Thus, implementation of an ERP system ensures good work processes based on best practices. For example, for managing customer payments, the ERP system's finance module can be configured to reflect the most efficient practices of leading companies in an industry. This increased efficiency ensures that everyday business operations follow the optimal chain of activities, with all users supplied the information and tools they need to complete each step.

Eldorado Gold is an international gold producer with headquarters in Canada and mine operations at four locations in China and one in Turkey. The company implemented an ERP system to standardize on a set of best practices for financial management, control and materials management, maintenance, human resources, and payroll. Eldorado is now seeing benefits from integrating the system and is maintaining consistent business processes across the entire organization.[18]

Upgrade of Technology Infrastructure

When implementing an ERP system, an organization has an opportunity to upgrade the information technology (such as hardware, operating systems, and databases) that it uses. While centralizing and formalizing these decisions, the organization can eliminate the hodgepodge of multiple hardware platforms, operating systems, and databases it is currently using—most likely from a variety of vendors. Standardizing on fewer technologies and vendors reduces ongoing maintenance and support costs as well as the training load for those who must support the infrastructure.

One reason SAP purchased Sybase, a data management company, was to build a gateway to integrate smartphones with its enterprise resource planning application. The Sybase Unwired Platform-based suite enables users to access SAP ERP applications via their personal smartphone. As a result of developments such as this by SAP and other enterprise software developers, smartphones are being used in the supply chain to enable technical service personnel to enter information that is then updated on a back-end ERP application such as Plant Maintenance, Customer Service, or Quality Management. Users in the field can query their ERP system to check inventory, sales orders, and account balances, among others. Managers can even authorize changes to expedite specific work orders or customer deliveries. Many organizations scan barcodes with their smartphones.[19]

Mamut One Enterprise ERP software has been designed to meet the business needs of entrepreneurs and small and medium organizations. Over 400,000 European users of the software can perform accounting, invoicing, finance, and inventory functions. Many of these users access the software and data via their mobile phones. They can take their office with them wherever they are, thus increasing their productivity.[20]

Challenges in Implementing ERP Systems

Implementing an ERP system, particularly for a large organization, is extremely challenging and requires tremendous amounts of resources, the best IS and businesspeople, and plenty of management support. In spite of all this, many ERP implementations fail, and problems with an ERP implementation can require expensive solutions. The following is a sample of ERP projects that failed in 2011:

- A *Fortune* 100 company and one of the world's largest technology distributors participated in a global rollout of an ERP system that involved converting systems in a number of countries. The conversions went smoothly until the staff encountered problems with the conversion in Australia. Problems there led to business disruptions that affected sales and income, causing income to drop unexpectedly for two quarters in a row.[21]
- A state university sued an ERP system provider over problems associated with an ERP implementation at the university that eventually led to millions in costly overruns and months of project delays.[22]
- A small manufacturer of outdoor furniture with 20 employees tried to implement an ERP system. Unfortunately, the ERP system did not meet the manufacturer's expectations, and the firm blamed the poor implementation for causing a loss for the year. The manufacturer eventually sued the software provider.[23]
- A county government worked with a consulting firm and an ERP software provider to implement an ERP system. When problems surfaced during the implementation, government officials decided to terminate the

implementation, uninstall the software, and sue both the consulting firm and software provider for $35 million in damages.[24]

Half of nearly 200 ERP implementations worldwide evaluated by Panorama, an ERP consulting firm, were judged to be failures.[25] Table 9.3 lists and describes the most significant challenges to successful implementation of an ERP system.

TABLE **9.3** Challenges to successful ERP system implementation

Challenge	Description
Cost and disruption of upgrades	Most companies have other systems that must be integrated with the ERP system, such as financial analysis programs, e-commerce operations, and other applications that communicate with suppliers, customers, distributors, and other business partners. This integration takes even more effort and time.
Cost and long implementation lead time	The average ERP implementation cost is $5.5 million with an average project duration just over 14 months.
Difficulty in managing change	Companies often must radically change how they operate to conform to the ERP's work processes. These changes can be so drastic to long-time employees that they depart rather than adapt to the change, leaving the firm short of experienced workers.
Management of software customization	The base ERP system may need to be modified to meet mandatory business requirements. This modification can become extremely expensive and further delay implementation.
User frustration with the new system	Effective use of an ERP system requires changes in work processes and in the details of how work gets done. Many users initially balk at these changes and require much training and encouragement.

Source: Jutras, Cindy, "2011 ERP Solution Study Highlights," *Epicor*, September 2011.

The following list provides tips for avoiding many common causes for failed ERP implementations:

- Assign a full-time executive to manage the project.
- Appoint an experienced, independent resource to provide project oversight and to verify and validate system performance.
- Allow sufficient time for transition from the old way of doing things to the new system and new processes.
- Plan to spend considerable time and money training people; many project managers recommend that 30 to 60 days per employee be budgeted for training of personnel.
- Define metrics to assess project progress and to identify project-related risks.
- Keep the scope of the project well defined and contained to essential business processes.
- Be wary of modifying the ERP software to conform to your firm's business practices.

Leading ERP Systems

ERP systems are commonly used in manufacturing companies, colleges and universities, professional service organizations, retailers, and health care organizations. The business needs for each of these types of organizations varies greatly. Naturally, the needs of a large multinational organization are very different from the needs of a small, local organization. Thus, no one ERP software solution from a single vendor is "best" for all organizations.

SAP is the largest and most-recognized ERP solution provider among *Fortune* 1000 and Global 5000 organizations, with more than 35,000 customers in more than 120 countries. The scope of its ERP software encompasses accounting, distribution, financials, manufacturing, human resources, and payroll functions.

Oracle has accumulated an impressive customer list and set of ERP solutions through its acquisition of PeopleSoft (including JD Edwards) and Siebel systems. Its challenge is to integrate all those products through Project Fusion.

Infor is the third-largest ERP software manufacturer and has several different ERP systems that appeal to midsized organizations. The Microsoft Dynamics ERP solution is very popular among small businesses.

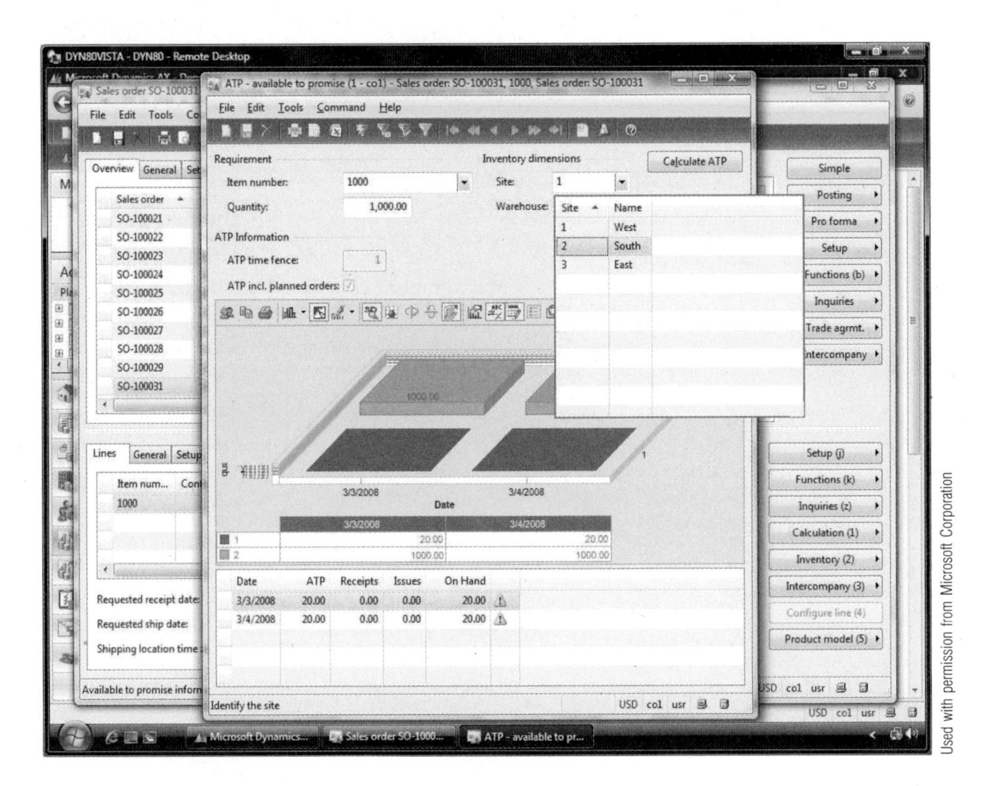

Microsoft Dynamics is an ERP solution that is very popular among small businesses.

Organizations that are successful in implementing ERP are not limited to large *Fortune* 1000 companies. Entrepreneurs and small and medium (both for-profit and not-for-profit) organizations can achieve real business benefits from their ERP efforts. Many organizations elect to implement open-source ERP systems from vendors such as Compiere. With open-source software, organizations can see and modify the source code to customize it to meet their needs. Such systems are much less costly to acquire and are relatively easy to modify to meet business needs. Frequent reasons for customization are to integrate other business systems to work with the ERP package, to add data fields or change field sizes from those in the standard system, and to meet regulatory requirements. A wide range of software service organizations can perform the system development and maintenance. Plex and NetSuite are cloud-computing-based ERP solutions that enable users to access the ERP application using a Web browser and avoid paying for and maintaining high-cost hardware.

Table 9.4 provides a list of highly rated ERP systems.

TABLE **9.4** Highly rated ERP systems, 2011 (alphabetical order)

ERP Systems for Large Organizations	ERP Systems for Mid-sized Organizations	ERP Systems for Small Organizations
Microsoft Dynamics	Epicor	ABAS
Oracle	Industrial and Financial Systems	Activant Solutions, Inc.
Oracle eBusiness Suite	Infor	Baan
Oracle JD Edwards	Lawson	Compiere
Oracle Peoplesoft	Plex	Netsuite
SAP	Sage	Syspro

Source: "2011 Guide to ERP Systems and Vendors," Panorama Consulting Group, 2011.

Organizations select ERP software solutions based on their business needs. For example, PEPCO Poland is a nationwide chain of discount clothing stores. As the firm gained in market share and sales increased, it decided to implement the Microsoft Dynamics ERP system to improve its sales, financial management, customer services, and product management in both its stores and warehouses.[26] Another example is Empired Ltd, an IT services provider, that implemented NetSuite ERP software to manage its core business operations including invoice generation, resource planning and scheduling, and full financial reporting. The new software replaced several legacy systems previously used for financial forecasting and project management.[27] Some organizations actually use software modules from more than one ERP vendor. For example, Hitachi uses Baan ERP to run its distribution and manufacturing operations and to support a seamless flow of data to the SAP financials software.[28]

Supply Chain Management (SCM)

supply chain management (SCM): A system that includes planning, executing, and controlling all activities involved in raw material sourcing and procurement, converting raw materials to finished products, and warehousing and delivering finished products to customers.

This section outlines the use of an ERP system within a manufacturing organization to support what is known as **supply chain management (SCM)**, which includes the planning, execution, and control of all activities involved in raw material sourcing and procurement, conversion of raw materials to finished products, and the warehousing and delivery of finished product to customers. The goal of SCM is to reduce costs and improve customer service, while at the same time reducing the overall investment in inventory in the supply chain.

Another way to think about SCM is that it is the management of materials, information, and finances as they move in a process from supplier to manufacturer to wholesaler to retailer to consumer. The material flow includes the inbound movement of raw materials from supplier to manufacturer as well as the outbound movement of finished product from manufacturer to wholesaler, retailer, and customer. The information flow involves the capture and transmission of orders and invoices among suppliers, manufacturer, wholesalers, retailers, and customers. The financial flow consists of payment transactions among suppliers, manufacturers, wholesalers, retailers, customers, and their financial institutions.

The ERP system for a manufacturing organization typically encompasses SCM activities and manages the flow of materials, information and finances. Manufacturing ERP systems follow a systematic process for developing a production plan that draws on the information available in the ERP system database.

The process starts with *sales forecasting* to develop an estimate of future customer demand. This initial forecast is at a fairly high level, with estimates made by product group rather than by each individual product item. The

INFORMATION SYSTEMS @ WORK

Project Management: A Vital Element

As you read this chapter, you surely realize that developing or implementing an enterprise information system is not a trivial activity. It takes lots of people with different skill sets to develop such a system. When a system is critical to the firm, as enterprise information systems often are, the firm can ill afford mistakes, schedule slips, or cost overruns. Someone has to coordinate the activities of all project team members to avoid such problems. This person, probably the most important person to the success of any information system development effort, is the *project manager*.

How necessary is good project management? The Standish Group surveys software development for a study it releases every two years. The survey released in 2012 collected data on 10,000 projects. It found that 37 percent of the software development projects in its study were successful, meaning that they came in on time and on budget and that the users accepted them. It categorized 42 percent of the projects as challenges. These projects had problems such as being late, being over budget, or not having all the features that users wanted. The remaining 21 percent were failures: they weren't completed or were rejected by the customer. What's more, the projects with the highest failure rates tended to be large ones, that is, enterprise systems. The failure can often be connected to poor project management.

Todd Little is a senior development manager for Landmark Graphics Corporation, which develops applications for the oil and gas industry. Regarding project problems, he says "very rarely are they technical challenges; almost always, the challenges are with people."

Because programmers are not trained to deal with human challenges, organizations recognize that the skills required to be a good project manager are not just an extension of the skills required to be a good systems analyst, systems designer, or programmer. The U.S. Social Security Administration (SSA), for example, "invests heavily in program management development." The SSA's training curriculum includes "over 50 courses available via classroom, internet, interactive video teletraining, video on demand, and university partnership program formats. Courses include all mandatory and basic certification requirements as well as tailored content specific to SSA process and tools, and a large number of additional courses that SSA deems important for full PM development."

Among other things, the SSA wants its program managers to be certified according to the requirements of the Federal Acquisition Certification (FAC) for Program and Project Managers. This certification was established by the Office of Management and Budget in 2007. Its head, Paul Denett, wrote at the time that "Well-trained and experienced program and project managers are critical to the acquisition process and the successful accomplishment of mission goals." The FAC process defines three levels of certification: Entry/Apprentice, Mid-Level/Journeyman, and Senior/Expert. For each level, it further establishes the essential competencies and training requirements for certification. The SSA reported in 2011 that it had "no program failures and few difficulties" since adopting this program. Given the difficulties that attend many enterprise information systems projects, as found in the Standish Group study, this outcome is impressive.

The Project Management Institute (*www.pmi.org*) was established to provide project management guidance to all organizations, not just those in the U.S. government. It offers six certification programs and a wide range of training programs to support them. Mark A. Langley, president and CEO of PMI, states that "more organizations are experiencing tangible long-term value from investing time, money and resources to build organizational project management expertise." Among other benefits, PMI's 2011 *Pulse of the Profession* report finds that organizations that standardize project management across the enterprise experience an average of 28 percent more projects meeting their intended goals; those with formal project management training programs increase the number of projects that meet goals, are on time, and are within budget; and organizations with a high percentage of Project Management Professional-credentialed project managers (35 percent or higher) have more successful project outcomes. To PMI, these findings more than justify these organizations' investment in professional project management.

Discussion Questions

1. Organizations can often choose either an experienced business manager with project management experience or an experienced information systems specialist to lead enterprise system development and implementation. Which of the two would you choose and why?
2. Suppose your school is hiring a project manager to supervise the development of a new schoolwide

(that is, enterprise-level) student information system. Examine the six certifications at *www.pmi.org/Certification/What-are-PMI-Certifications.aspx*. Which are most relevant to an enterprise IS development project at a school? Should your school require candidates to have PMI certification? If your school shouldn't require certification, should it consider certification to be a plus when comparing candidates for this job?

Critical Thinking Questions

1. Do you think it is a good idea to certify project managers rather than relying on each organization's upper management to verify their ability to do their job? Should laws require project managers in critical areas, such as nuclear power plant control, air traffic control, or medical system development, to be certified? Why or why not?
2. Consider the findings taken from PMI's *Pulse of the Profession* report. Can you consider PMI to be an unbiased source of information on this topic? If the organization is biased, do you think that could influence the findings of the group's 2011 report? If it could, can the findings still be of value? Explain your answers.

SOURCES: Claps, M., "Case Study: Montgomery County Deploys ERP and CRM Simultaneously," Gartner Group research note G00214740, *www.gartner.com/technology/streamReprints.do?id=1-16PRCGW&ct=110721,* August 19, 2011; Denett, P., "The Federal Acquisition Certification for Program and Project Managers," U.S. Office of Management and Budget, *www.whitehouse.gov/omb/procurement_index_workforce,* April 25, 2007; Thibodeau, P., "It's Not the Coding That's Hard, It's the People," *Computerworld, www.computerworld.com/s/article/9220591/It_s_not_the_coding_that_s_hard_it_s_the_people,* October 13, 2011; Staff, "PMI's Pulse of the Profession (2011)," Project Management Institute, *www.pmi.org/~/media/PDF/Home/Pulse of the Profession White Paper - FINAL.ashx,* accessed Oct. 23, 2011; Staff, "SSA Program Manager Development Practices," U.S. Social Security Administration, Office of the CIO, *www.cio.gov/modules/best_practices/bp_display.cfm/page/Best-Practices-SSA-Program-Manager-Development-Practices,* April 7, 2011.

sales forecast extends for months into the future; it might be developed using an ERP software module or produced by other means using specialized software and techniques. Many organizations are moving to a collaborative process with major customers to plan future inventory levels and production rather than relying on an internally generated sales forecast.

The *sales and operations plan* (*S&OP*) takes demand and current inventory levels into account and determines the specific product items that need to be produced as well as when to meet the forecast future demand. Production capacity and any seasonal variability in demand must also be considered.

Demand management refines the production plan by determining the amount of weekly or daily production needed to meet the demand for individual products. The output of the demand management process is the master production schedule, which is a production plan for all finished goods.

Detailed scheduling uses the production plan defined by the demand management process to develop a detailed production schedule specifying production scheduling details, such as which item to produce first and when production should be switched from one item to another. A key decision is how long to make the production runs for each product. Longer production runs reduce the number of machine setups required, thus reducing production costs. Shorter production runs generate less finished product inventory and reduce inventory holding costs.

Materials requirement planning (*MRP*) determines the amount and timing for placing raw material orders with suppliers. The types and amounts of raw materials required to support the planned production schedule are determined by the existing raw material inventory and the "bill of materials," or BOM—a sort of "recipe" of ingredients needed to make each item. The quantity of raw materials to order also depends on the lead time and lot sizing. Lead time is the amount of time it takes from the placement of a purchase order until the raw materials arrive at the production facility. Lot size has to do with discrete quantities that the supplier will ship and the amount that is economical for the producer to receive or store. For example, a supplier might ship a certain raw material in units of 80,000-pound rail cars. The producer might need 95,000 pounds of the raw material. A decision must be made to order one or two rail cars of the raw material.

The Newark Group is a producer of recycled paperboard and paperboard products with operations in North America and Europe. The firm implemented a materials requirement planning system that enables it to view, plan, execute, settle, and analyze all its inbound raw material and outbound customer shipments. The result is improved efficiency and lower transportation costs.[29]

Purchasing uses the information from MRP to place purchase orders for raw materials with qualified suppliers. Typically, purchase orders are released so that raw materials arrive just in time to be used in production and to minimize warehouse and storage costs. Often, producers will allow suppliers to tap into data via an extranet that enables them to determine what raw materials the supplier needs, minimizing the effort and lead time to place and fill purchase orders.

Production uses the high-level production schedule to plan the details of running and staffing the production operation. This more detailed schedule takes into account employee, equipment, and raw material availability along with detailed customer demand data. Bombardier Aerospace employs over 35,000 people as the manufacturer of a number of business and regional aircraft including the Learjet, Challenger, Globa, Flexjet, and Bombardier aircraft. The firm uses a production planning and detailed scheduling system to manage its aircraft assembly lines.[30]

Sales ordering is the set of activities that must be performed to capture a customer sales order. A few of the essential steps include recording the items to be purchased, setting the sales price, recording the order quantity, determining the total cost of the order including delivery costs, and confirming the customer's available credit. Should the item(s) the customer wants to order be out of stock, the sales order process should communicate this fact and suggest other items that could substitute for the customer's initial choice. Setting sales prices can be quite complicated and can include quantity discounts, promotions, and incentives. After the total cost of the order is determined, a company must check the customer's available credit to see if this order is within the credit limit. Figure 9.7 shows a sales order entry window in SAP business software.

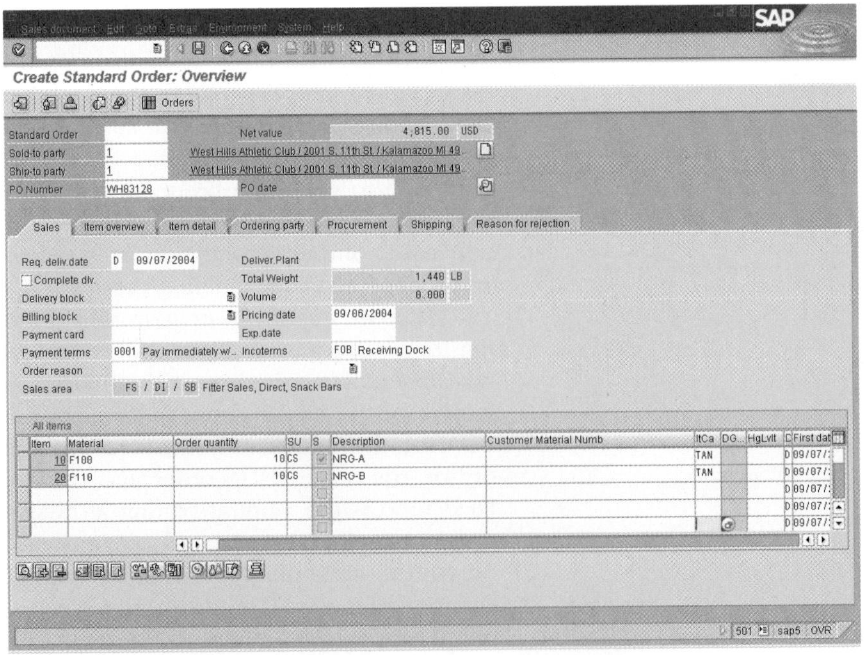

FIGURE **9.7**

Sales order entry window

ERP systems do not work directly with production machines, so they need a way to capture information about what was produced. This data must be passed to the ERP accounting modules to keep an accurate count of finished product inventory. Many companies have personal computers on the production floor that count the number of cases of each product item by scanning a UPC code on the packing material. Other approaches for capturing production quantities include the use of RFID chips and manually entering the data via a handheld computer.

Separately, production-quality data can be added based on the results of quality tests run on a sample of the product for each batch of product produced. Typically, this data includes the batch identification number, which identifies the production run and the results of various product quality tests.

Retailers as well as manufacturers use demand forecasting to match production to consumer demand and to allocate products to stores. The Body Shop sells a wide range of beauty care products in over 2,500 stores in 60 markets worldwide. The firm uses demand forecasting software to set and maintain profitable inventory levels across an ever-changing lineup of fragrances, lotions, and skin care products.[31]

Financial and Managerial Accounting and ERP

The general ledger is the main accounting record of a business. It is often divided into categories, including assets, liabilities, revenue, expenses, and equity. These categories, in turn, are subdivided into subledgers to capture details such as cash, accounts payable, accounts receivable, and so on. The business processes required to capture and report these accounting details are essential to the operation of any organization and are frequently included within the scope of an organization's ERP system. In the ERP system, input to the general ledger occurs simultaneously with the input of a business transaction to a specific module. The following are several examples of how this process occurs:

- An order clerk records a sale, and the ERP system automatically creates an accounts receivable entry indicating that a customer owes money for goods received.
- A buyer enters a purchase order, and the ERP system automatically creates an accounts payable entry in the general ledger registering that the company has an obligation to pay for goods that will be received at some time in the future.
- A dock worker enters a receipt of purchased materials from a supplier, and the ERP system automatically creates a general ledger entry to increase the value of inventory on hand.
- A production worker withdraws raw materials from inventory to support production, and the ERP system generates a record to reduce the value of inventory on hand.

Thus, the ERP system captures transactions entered by workers in all functional areas of the business. The ERP system then creates the associated general ledger record to track the financial impact of the transaction. This set of records is an extremely valuable resource that companies can use to support financial and managerial accounting.

Financial accounting consists of capturing and recording all the transactions that affect a company's financial state and then using these documented transactions to prepare financial statements to external decision makers, such as stockholders, suppliers, banks, and government agencies. These financial statements include the profit-and-loss statement, balance sheet, and cash-flow statement. They must be prepared in strict accordance to rules and guidelines of agencies such as the Securities and Exchange Commission, the Internal Revenue Service, and the Financial Accounting Standards Board. Data gathered for financial accounting can also form the basis for tax accounting because it involves external reporting of a firm's activities to the local, state, and federal tax agencies.

Managerial accounting involves using both historical and estimated data in providing information that management uses to conduct daily operations, plan future operations, and develop overall business strategies. Managerial accounting provides data to enable the firm's managers to assess the profitability of a given product line or specific product, identify underperforming sales regions, establish budgets, make profit forecasts, and measure the effectiveness of marketing campaigns.

All transactions that affect the financial state of the firm are captured and recorded in the database of the ERP system. This data is used in the financial accounting module of the ERP system to prepare the statements required by various constituencies. The data can also be used in the managerial accounting module of the ERP system along with various assumptions and forecasts to perform different analyses such as generating a forecasted profit-and-loss statement to assess the firm's future profitability. Using an ERP with financial and managerial accounting systems can contribute significantly to a company's success.

Customer Relationship Management

customer relationship management (CRM) system: A system that helps a company manage all aspects of customer encounters, including marketing and advertising, sales, customer service after the sale, and programs to retain loyal customers.

As discussed in Chapter 2, a **customer relationship management (CRM) system** helps a company manage all aspects of customer encounters, including marketing and advertising, sales, customer service after the sale, and programs to keep loyal customers (see Figure 9.8). The goal of CRM is to understand and anticipate the needs of current and potential customers to increase customer retention and loyalty while optimizing the way that products and

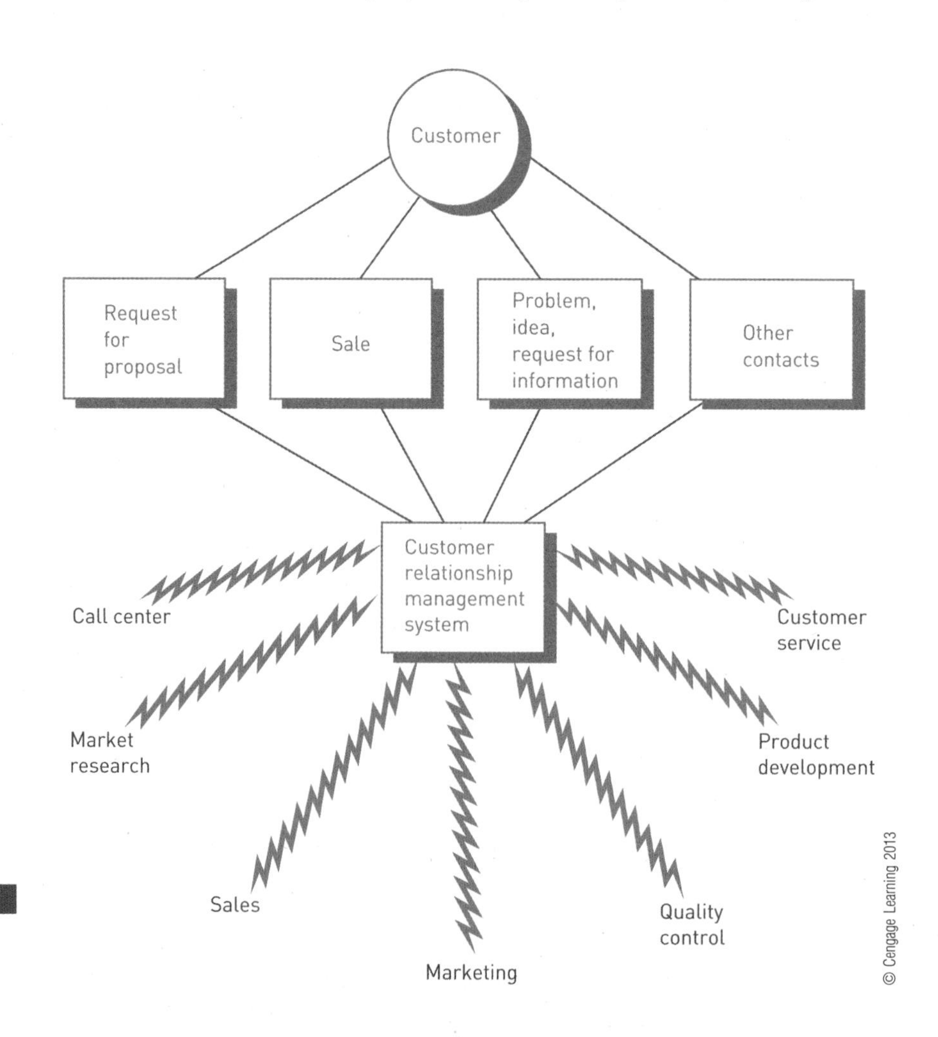

FIGURE 9.8
Customer relationship management system

services are sold. CRM is used primarily by people in the sales, marketing, and service organizations to capture and view data about customers and to improve communications. Businesses implementing CRM systems report benefits such as improved customer satisfaction, increased customer retention, reduced operating costs, and the ability to meet customer demand.

CRM software automates and integrates the functions of sales, marketing, and service in an organization. The objective is to capture data about every contact a company has with a customer through every channel and to store it in the CRM system so that the company can truly understand customer actions. CRM software helps an organization build a database about its customers that describes relationships in sufficient detail so that management, salespeople, customer service providers, and even customers can access information to match customer needs with product plans and offerings, remind them of service requirements, and know what other products they have purchased.

The Boston Red Sox baseball team uses a CRM system to capture data from various sources including its ticketing system and provide improved insight into the preferences of its fans. The organization analyzes this data to develop promotions and marketing programs that increase attendance and improve customer retention rates at Fenway Park.[32]

The key features of a CRM system include the following:

- **Contact management.** The ability to track data on individual customers and sales leads and access that data from any part of the organization (see Figure 9.9, which shows the SAP Contact Manager).
- **Sales management.** The ability to organize data about customers and sales leads and then to prioritize the potential sales opportunities and identify appropriate next steps.
- **Customer support.** The ability to support customer service representatives so that they can quickly, thoroughly, and appropriately address customer requests and resolve customers' issues while at the same time collecting and storing data about those interactions.
- **Marketing automation.** The ability to capture and analyze all customer interactions, generate appropriate responses, and gather data to create and build effective and efficient marketing campaigns.
- **Analysis.** The ability to analyze customer data to identify ways to increase revenue and decrease costs, identify the firm's "best customers," and determine how to retain them and find more of them.
- **Social networking.** The ability to create and join sites like Facebook, where salespeople can make contacts with potential customers.
- **Access by smartphones.** The ability to access Web-based customer relationship management software by devices such as the BlackBerry or Apple iPhone.
- **Import contact data.** The ability for users to import contact data from various data service providers such as Jigsaw, which offers company-level contact data that can be downloaded for free directly into the CRM application.

Figure 9.9 shows the SAP Contact Manager.

The focus of CRM involves much more than installing new software. Moving from a culture of simply selling products to placing the customer first is essential to a successful CRM deployment. Before any software is loaded onto a computer, a company must retrain employees. Who handles customer issues and when must be clearly defined, and computer systems need to be integrated so that all pertinent information is available immediately, whether a customer calls a sales representative or a customer service representative. Most CRM systems can now be accessed via wireless devices.

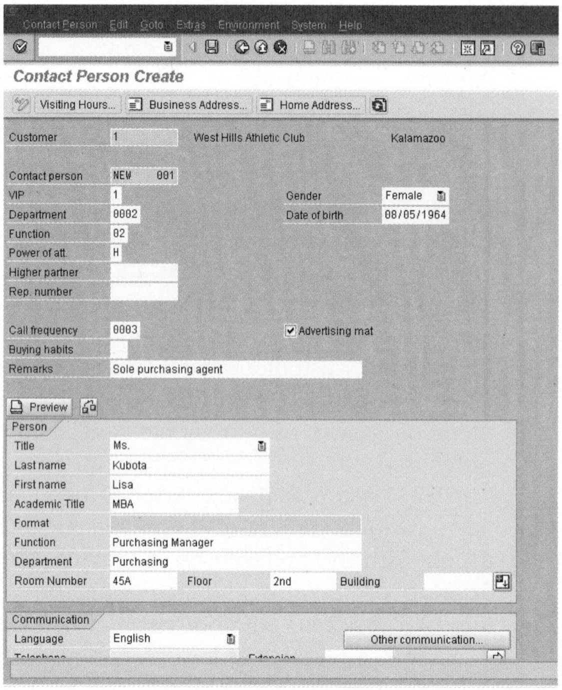

FIGURE **9.9**
SAP Contact Manager

© SAP AG

Organizations choose to implement CRM for a variety of reasons, depending on their needs. Consider the following examples:

- Stallergenes is a European biopharmaceutical company that specializes in developing products for the treatment of severe respiratory allergies. Each year nearly 500,000 patients benefit from this company's products.[33] Stallergenes chose Veeva CRM, a cloud-based CRM, to support its rapidly expanding international operations. This CRM software provides a high degree of flexibility and is easily expanded to support the firm's growing sales force.[34]
- Burns & McDonnell is an architecture, construction, engineering, and environmental firm with 3,000 employees and is headquartered in Kansas City, Missouri. The firm plans, designs, permits, builds, and manages facilities all over the world.[35] It recently chose Microsoft Dynamics CRM to improve employee productivity and increase client sales and service. The software will enable its employees to see an integrated view of each customer including key events, project efforts, and customer interactions across all of its 11 global practices, thus increasing sales potential.[36]
- The Department of Social Welfare and Development for the Philippines ensures that the poor and vulnerable are provided with quality social protection programs and services. The agency recently adopted Microsoft Dynamics CRM to ensure the accuracy and efficiency of identifying those who are eligible for services and providing them with seamless integration to all the many services and systems necessary to meet their needs.[37]

Table 9.5 lists the highest-rated CRM systems (in alphabetical order) for large, medium, and small organizations. NetSuite, Microsoft Dynamics DRM, and Salesforce.com are highly rated for all sizes of organizations.

Due to the popularity of mobile devices, shoppers can easily compare products and prices on their mobile phones and instantly tweet their experiences with a brand to dozens of friends. Savvy retailers today use

TABLE 9.5 Highest-rated CRM systems, 2011

CRM Systems for Large Organizations	CRM Systems for Mid-sized Organizations	CRM Systems for Small Organizations
Amdocs		
Microsoft Dynamics CRM	Microsoft Dynamics CRM	Microsoft Dynamics CRM
NetSuite	NetSuite	NetSuite
Oracle Siebel CRM	Oracle Siebel CRM	
	RightNow	
Salesforce.com	Salesforce.com	Salesforce.com
	Sage	
SAP	SugarCRM	SugarCRM
		Zoho

Source: Staff, "The 2011 CRM Market," *CRM Magazine, www.destinationcrm.com/Articles/Editorial/Magazine-Features/The-2011-CRM-Market-Leaders-76485.aspx*, August 9, 2011.

CRM systems to stay on top of what these customers are saying on social networks and to respond quickly to their comments.

Tasti D-Lite serves its creamy dairy dessert in more than 100 flavors through a network of 60 independently owned and operated centers, most in the greater New York City area. The firm has installed display screens so that as its customers check out, they can view social media comments about the Tasti D-Lite products and other customers' shopping experiences. They can also learn about the TastiRewards loyalty program through which they can earn points by connecting to the brand via Facebook, Foursquare, or Twitter.[38] Tasti D-Lite has licensed CRM software to support these and others of its social media initiatives.[39]

Tommy Bahama uses the data captured by its CRM system to dig in and learn more about its most passionate customers. As the company analyzed this data, management was surprised to learn that its core customers were much younger than they thought and that the company had many more customers in northern cities than assumed. It is now working to tie its CRM software into an improved customer loyalty program that distinguishes individuals and rewards them appropriately.[40]

Product Lifecycle Management (PLM)

Product lifecycle management (PLM) software provides a means for managing all the data associated with the product development, engineering design, production, support, and disposal of manufactured products. As products advance through these stages, product data is generated and distributed to various groups both within and outside the manufacturing firm. This data includes design and process documents, bill of material definitions, product attributes, and documents needed for FDA and environmental compliance.

PLM software and its data are used by both internal and external users. Internal users include engineering, operations and manufacturing, procurement and sourcing, manufacturing, marketing, quality, regulatory, and others. External users include the manufacturer's design partners, packaging suppliers, raw material suppliers, and contract manufacturers. These users must collaborate to define, maintain, update, and securely share product information throughout the life cycle of the product. Frequently, these external users are asked to sign nondisclosure agreements to reduce the risk of proprietary information being shared with competitors.

Diane von Furstenberg is a fashion designer for men's, women's, and children's clothing, footwear, bags, and sunglasses. The firm implemented a

Kick-Starting a Small Nonprofit with Online Fundraising

Prader-Willi syndrome (PWS) is a rare genetic disorder that occurs about once in 15,000 children, one percent of the frequency of autism spectrum disorders. It leads to life-threatening childhood obesity, in part because the brain is convinced that the body is in a perpetual state of starvation, as well as other conditions. PWS has historically received little attention or support from medical researchers. The Foundation for Prader-Willi Research (FPWR) sought to change that situation.

FPWR was established in 2003 by 40 families to fund research into PWS. In 2005, the organization's $100,000 income came from a few large donors. The founders realized that increasing donations substantially required changing the way FPWR operated. In particular, the foundation had to go online. However, the organization had few information technology resources, limited financial resources, and no experience with the Web.

Online giving is a chancy process at best. Research has found that nearly half of the people who visit a charity's Web site, intending to make a donation, don't follow through. What's more, the fraction of visitors who don't donate varies a great deal from site to site. Simon Norris, CEO of consulting firm Nomensa, suggests that "Nonprofits should take a lesson from successful e-commerce brands to understand and deliver an optimal donation experience."

The key to this experience is creating a relationship between the charity and the donor. The FPWR founders understood this principle. They knew, as you learned in this chapter, that customer relationship management (CRM) systems can "help a company manage all aspects of customer encounters." Replace "customer" by "donor," and that help was exactly what the foundation needed: software to manage its donor relationships in order to strengthen the donors' connection to FPWR.

Fortunately, FPWR could choose from many available CRM packages—even if the C is taken as Constituent rather than Customer, as is more appropriate when donors are part of the picture. Unfortunately, the very abundance of CRM packages makes it difficult to choose one. FPWR's limited budget was a critical factor in selecting a CRM package. When the foundation learned of CiviCRM, designed specifically for charitable organizations and available at no charge, that system became the obvious answer.

CiviCRM is designed specifically for donor tracking. It can record contributions of cash, items or services of value (in-kind), and volunteer time. It can handle one-time gifts, recurring gifts, pledges of future gifts, and more. It can track offline gifts to provide a complete picture of a donor's contributions through all channels. The system also differentiates grants (which obligate FPWR to do something in return) from contributions (which don't). It tracks household and workplace affiliations to indicate who is connected to whom. It also lets the organization manage volunteers by skills and availability and create membership levels with various criteria and benefits.

Three people from FPWR plus a hired developer set up CiviCRM, and FPWR now takes in over $700,000 annually through online donations.

In 2011, FPWR used CiviCRM to launch its OneSmallStep for Research initiative (OSS). This initiative brought together over 500 fundraisers in 53 cities around the world to raise money for PWS research. OSS organizers in each city used CiviCRM to establish their campaigns and recruit fundraisers who, in turn, solicit donations. CiviCRM handles multiple currencies, languages, and payment processors as well as manages the legal donation tracking requirements of different countries. Through the first four months of 2012, OSS raised $168,000 toward its goal of $1 million. Without CiviCRM, this achievement would not have been possible.

Discussion Questions

1. Other than a smaller budget, how are charities different from companies that sell products by e-commerce? How are they similar?
2. Your university almost certainly solicits donations from its graduates. How do its donor management requirements differ from those of FPWR? How are they similar?

Critical Thinking Questions

1. Anything involving children's health, such as PWS, has an emotional appeal to donors. Most products sold online have less emotional appeal but greater practical value. How does the emotional appeal of a product or service affect the information a CRM system should store in its database for each constituent (donor or customer)?
2. Using the relational database concepts you learned in Chapter 5, figure out what entities, attributes, and relationships the CiviCRM database design should include. Then, using the ERD concepts you learned there, draw a data model for it. Include all the entities and relationships that are needed for the CiviCRM features described in the case.

SOURCES: CiviCRM Web site, *civicrm.org*, accessed May 2, 2012; Foundation for Prader-Willi Research Web site, *www.fpwr.org*, accessed May 2, 2012; Nomensa, "47% of Donors Not Completing Their Journey to Give," *www.nomensa.com/about/news-items/47-donors-not-completing-their-journey-give*, November 1, 2011; Norris, S. and Potts, J., "Designing the Perfect Donation Experience," Nomensa Ltd., *www.nomensa.com/insights/designing-perfect-donation-process-part-1* (requires free registration), October 2011; Sheridan, A., "Getting to Know You: CRM for the Charity Sector," *Fundraising, www.civilsociety.co.uk/fundraising/opinion/content/8759/getting_to_know_you_crm_for_the_sector*, April 6, 2011.

PLM system to improve the speed to market for the company's latest designs and to improve worker productivity. The system enables workers to see key information much earlier in the design process, thus making it possible to implement changes earlier and at a lower cost.[41]

Brown-Foreman is a producer and distributor of fine spirits and wines, with over two dozen brands including Old Forrester, Early Times, Jack Daniels, and Southern Comfort. It distributes its products in over 135 countries worldwide. The firm implemented a PLM system to create, track, and analyze its existing product formulas. Using the system, product development people can build new formulas from verified information, thus preventing duplication of effort. The PLM system also helps ensure regulatory compliance with improved control over labeling for kosher, allergic, GMO, and vegan customers. Employees can also search the PLM database to determine how its products and their ingredients fare on topics such as allergens, and kosher, GMO, regulatory, and vegan standards, topics important to many of its customers.[42]

Hosted Software Model for Enterprise Software

Many business application software vendors are pushing the use of the hosted software model for SMEs. The goal is to help customers acquire, use, and benefit from the new technology while avoiding much of the associated complexity and high start-up costs. Applicor, Intacct, NetSuite, SAP, and Workday are among the software vendors who offer hosted versions of their ERP or CRM software at a cost of $50 to $200 per month per user.

This pay-as-you-go approach is appealing to SMEs because they can experiment with powerful software capabilities without making a major financial investment. Organizations can then dispose of the software without large investments if the software fails to provide value or otherwise misses expectations. Also, using the hosted software model means the small business firm does not need to employ a full-time IT person to maintain key business applications. The small business firm can expect additional savings from reduced hardware costs and costs associated with maintaining an appropriate computer environment (such as air conditioning, power, and an uninterruptible power supply).

Love Culture operates 60 stores in 26 states offering clothing, footwear, and accessories for savvy, fashion-forward young women. The firm implemented a hosted CRM solution because of its minimal initial cost and lack of dependency on site IS infrastructure and technical resources. The CRM solution helps decision makers to minimize inventory while simultaneously reducing stockouts and to execute the correct pricing strategy at each store location.[43]

Table 9.6 lists the advantages and disadvantages of hosted software.

TABLE 9.6 Advantages and disadvantages of hosted software model

Advantages	Disadvantages
Decreased total cost of ownership	Potential availability and reliability issues
Faster system start-up	Potential data security issues
Lower implementation risk	Potential problems integrating the hosted products of different vendors
Management of systems outsourced to experts	Savings anticipated from outsourcing may be offset by increased effort to manage vendor

SUMMARY

Principle:

An organization must have information systems that support routine, day-to-day activities and that help a company add value to its products and services.

Transaction processing systems (TPSs) are at the heart of most information systems in businesses today. A TPS is an organized collection of people, procedures, software, databases, and devices used to capture fundamental data about events that affect the organization (transactions) and that use that data to update the official records of the organization.

The methods of TPSs include batch and online processing. Batch processing involves the collection of transactions into batches, which are entered into the system at regular intervals as a group. Online transaction processing (OLTP) allows transactions to be entered as they occur.

Order processing systems capture and process customer order data from receipt of order through creation of a customer invoice.

Accounting systems track the flow of data related to all the cash flows that affect the organization.

Purchasing systems support the inventory control, purchase order processing, receiving, and accounts payable business functions.

Organizations today, including SMEs, typically implement an integrated set of TPSs from a single or limited number of software vendors to meet their transaction processing needs.

Organizations expect TPSs to accomplish a number of specific objectives, including processing data generated by and about transactions, maintaining a high degree of accuracy and information integrity, compiling accurate and timely reports and documents, increasing labor efficiency, helping provide increased and enhanced service, and building and maintaining customer loyalty. In some situations, an effective TPS can help an organization gain a competitive advantage.

All TPSs perform the following basic activities: data collection, which involves the capture of source data to complete a set of transactions; data editing, which checks for data validity and completeness; data correction, which involves providing feedback of a potential problem and enabling users to change the data; data manipulation, which is the performance of calculations, sorting, categorizing, summarizing, and storing data for further processing; data storage, which involves placing transaction data into one or more databases; and document production, which involves outputting records and reports.

Principle:

An organization that implements an enterprise system is creating a highly integrated set of systems, which can lead to many business benefits.

Enterprise resource planning (ERP) software supports the efficient operation of business processes by integrating activities throughout a business, including sales, marketing, manufacturing, logistics, accounting, and staffing.

Implementing an ERP system can provide many advantages, including providing access to data for operational decision making; elimination of costly, inflexible legacy systems; providing improved work processes; and creating the opportunity to upgrade technology infrastructure.

Some of the disadvantages associated with ERP systems are that they are time consuming, difficult, and expensive to implement; they can also be difficult to integrate with other systems.

The most significant challenges to successful implementation of an ERP system include the cost and disruption of upgrades, the cost and long implementation lead time, the difficulty in managing change, the management of software customization, and user frustration with the new system.

Many SMEs are implementing ERP systems to achieve organizational benefits. In many cases, they are choosing open-source systems because of the lower total cost of ownership and the ability of such systems to be easily modified.

No one ERP software solution is "best" for all organizations. SAP, Oracle, Infor, and Microsoft are among the leading ERP suppliers.

Although the scope of ERP implementation can vary, most firms use ERP systems to support financial and managerial accounting and business intelligence. Most manufacturing organizations use ERP to support the supply chain management activities of planning, executing, and controlling all tasks involved in raw material sourcing and procurement, conversion of raw materials to finished products, and the warehousing and delivery of finished product to customers.

The production and supply chain management process starts with sales forecasting to develop an estimate of future customer demand. This initial forecast is at a fairly high level, with estimates made by product group rather than by individual product item. The sales and operations plan takes demand and current inventory levels into account and determines the specific product items that need to be produced as well as when to meet the forecast future demand. Demand management refines the production plan by determining the amount of weekly or daily production needed to meet the demand for individual products. Detailed scheduling uses the production plan defined by the demand management process to develop a detailed production schedule specifying details, such as which item to produce first and when production should be switched from one item to another. Materials requirement planning determines the amount and timing for placing raw material orders with suppliers. Purchasing uses the information from materials requirement planning to place purchase orders for raw materials and transmit them to qualified suppliers. Production uses the detailed schedule to plan the logistics of running and staffing the production operation. The individual application modules included in the ERP system are designed to support best practices, the most efficient and effective ways to complete a business process.

Organizations are implementing CRM systems to manage all aspects of customer encounters, including marketing and advertising, sales, customer service after the sale, and programs to keep and retain loyal customers.

Manufacturing organizations are implementing product life cycle management (PLM) software to manage all the data associated with the product development, engineering design, production, support, and disposal of their products. These systems are used by both internal and external users to enable them to collaborate.

Business application software vendors are experimenting with the hosted software model to see if the approach meets customer needs and is likely to generate significant revenue. This approach is especially appealing to SMEs due to the low initial cost, which makes it possible to experiment with powerful software capabilities.

Numerous complications arise that multinational corporations must address in planning, building, and operating their enterprise systems. These challenges include dealing with different languages and cultures, disparities in IS infrastructure, varying laws and customs, and multiple currencies.

CHAPTER 9: SELF-ASSESSMENT TEST

An organization must have information systems that support routine, day-to-day activities and that help a company add value to its products and services.

1. A(n) ________________ is central to an organization and ensures that information can be shared across all business functions and all levels of management to support the running and managing of a business.
2. The result of processing business transactions is that the organization's records are updated to reflect the status of the operation at the time of the ________________.
3. Which of the following is *not* one of the basic components of a TPS?
   a. databases
   b. networks
   c. procedures
   d. analytical models
4. With ________________ processing systems, business transactions are accumulated over a period of time and prepared for processing as a single unit.
5. Online processing methods are most appropriate and cost effective for all forms of transaction processing systems. True or False?
6. Which of the following is not an objective of an organization's transaction processing system?
   a. Capture, process, and update databases of business data required to support routine business activities
   b. Ensure that data is processed immediately upon occurrence of a business transaction
   c. Avoid processing fraudulent transactions
   d. Produce timely user responses and reports

7. Business data goes through a(n) ________________ that includes data collection, data editing, data correction, data manipulation, data storage, and documentation production.
8. Unfortunately, there are few choices for software packages that provide integrated transaction processing system solutions for small and medium-sized enterprises. True or False?
9. Capturing and gathering all the data necessary to complete the processing of transactions is called ________________.

An organization that implements an enterprise system is creating a highly integrated set of systems, which can lead to many business benefits.

10. Which of the following is a primary benefit of implementing an ERP system?
    a. elimination of costly, inefficient legacy systems
    b. easing adoption of improved work processes
    c. improving access to quality data for operational decision making
    d. all of the above
11. The individual application modules included in an ERP system are designed to support ________________, the most efficient and effective ways to complete a business process.
12. ________________ software provides a means for managing all the data associated with the product development, engineering design, production, support, and disposal of manufactured products.
13. The hosted software model for enterprise software helps customers acquire, use, and benefit from new technology while avoiding much of the associated complexity and high start-up costs. True or False?
14. Many multinational companies roll out standard IS applications for all to use. However, standard applications often don't account for all the differences among business partners and employees operating in other parts of the world. Which of the following is a frequent modification that is needed for standard software?
    a. Software might need to be designed with local language interfaces to ensure the successful implementation of a new IS.
    b. Customization might be needed to handle date fields correctly.
    c. Users might also have to implement manual processes and overrides to enable systems to function correctly.
    d. all of the above

CHAPTER 9: SELF-ASSESSMENT TEST ANSWERS

1. enterprise system
2. last processed transaction
3. d
4. batch
5. False
6. b
7. transaction processing cycle
8. False
9. data collection
10. d
11. best practices
12. Product life cycle management
13. True
14. d

REVIEW QUESTIONS

1. Identify and discuss some of the issues that are common to the planning, building, and operation of an ERP, CRM, or PLM enterprise system whether for an SME or a large multinational organization.
2. What basic transaction processing activities are performed by all transaction processing systems?
3. Provide an example for which the use of a batch processing system to handle transactions is appropriate. Provide an example for which the use of online transaction processing is appropriate.
4. Discuss the benefits and potential risks of sharing product data with external users through use of a PLM system.
5. Identify and discuss some of the benefits that are common to the use of an ERP, CRM, or PLM enterprise system, whether it be for an SME or a large multinational organization.
6. How does materials requirement planning support the purchasing process? What are some of the issues and complications that arise in materials requirement planning?
7. What is the role of a CRM system? What sort of benefits can such a system produce for a business?
8. What are the business processes included with the scope of product life cycle management?

9. Identify and discuss at least three factors that are increasing the need for enterprise-wide access to real-time information.
10. What is the difference between managerial and financial accounting?
11. What is the role of the general ledger system in keeping track of the financial transactions of the organization? How is it used?
12. List and briefly describe the set of activities that must be performed by the sales-ordering module of an ERP system to capture a customer sales order.

DISCUSSION QUESTIONS

1. Assume that you are the owner of a small neighborhood bakery serving hundreds of customers in your area. Identify the kinds of customer information you would like your firm's CRM system to capture. How might this information be used to provide better service or increase revenue? Identify where or how you might capture this data.
2. Briefly describe the hosted software model for enterprise software and discuss its primary appeal for SMEs.
3. In what ways is the implementation of an ERP system simpler and less risky for an SME than for a large multinational corporation?
4. What are some of the challenges and potential problems of implementing a CRM system and CRM mindset in a firm's employees? How might you overcome these?
5. You are a member of the marketing organization for a consumer packaged goods company. The firm is considering the implementation of a CRM system. Make a convincing argument for the scope of the system to include gathering customer data from social networks.
6. Your friend has been appointed the project manager of your firm's ERP implementation system. What advice would you offer to help ensure the success of the project?
7. What benefits should the suppliers and customers of a firm that has successfully implemented an ERP system expect to see? What issues might arise for suppliers and customers during an ERP implementation?
8. Many organizations are moving to a collaborative process with their major suppliers to get their input on designing and planning future product modification or new products. Explain how a PLM system might enhance such a process. What issues and concerns might a manufacturer have in sharing product data?

PROBLEM-SOLVING EXERCISES

1. Imagine that you are a new employee in the marketing organization of a large national fast food restaurant chain. The company is considering implementing a CRM system to improve its understanding of its customers and their needs. You have been invited to a meeting to share your thoughts on how such a system might be used and what capabilities are most important. How would you prepare for this meeting? What points would you make?
2. Use a spreadsheet program to develop a sales forecasting system for a new car dealership that can estimate monthly sales for each make and model based on historical sales data and various parameters. Suggestion: Assume that this month's sales will be the same as the sales for this month last year except for adjustments due to the cost of gas and each make of car's miles per gallon. You can further refine the model to take into account change in interest rates for new cars or other parameters you wish to include. Document the assumptions you make in building your model.

TEAM ACTIVITIES

1. With your team members, interview several business managers at a firm that has implemented an enterprise system (ERP, CRM, or PLM system). Interview them to define the scope, cost, and schedule for the overall project. Make a list of what they see as the primary benefits of the

implementation. What were the biggest hurdles they had to overcome? With the benefit of 20-20 hindsight, is there anything they would have done differently?

2. As a team, develop a list of seven key criteria that a nonprofit charitable organization should consider in selecting a CRM system. Discuss each criterion and assign a weight representing the relative importance of that criterion. Develop a simple spreadsheet to use in scoring various CRM alternatives.

WEB EXERCISES

1. Do research on the Web and find a Web site that offers a demo of an enterprise system (ERP, CRM, or PLM system). View the demo, perhaps more than once. Write a review of the software based on the demo. What are its strengths and weaknesses? What additional questions about the software do you have? E-mail your questions to the vendor and document the response to your questions.
2. Using the Web, identify several software services firms that offer consulting services to help organizations to implement enterprise systems. Gain an understanding of what sort of services they offer and become familiar with several of their success stories. If you had to choose one of the software services firms to assist your SME organization, which one would you choose and why?

CAREER EXERCISES

1. Imagine that you are a construction equipment salesperson for a large equipment manufacturer. You make frequent sales calls on construction firms in a three-state area. The purpose of these sales calls is to acquaint the firms with your company's products and get them to consider purchase of your products. Describe the basic functionality you would want in your organization's CRM system for it to support you in preparing and making presentations to these people.
2. Enterprise system software vendors need business systems analysts that understand both information systems and business processes. Make a list of six or more specific qualifications needed to be a strong business systems analyst supporting the implementation and conversion to an enterprise system within an SME. Are there additional qualifications needed for someone who is doing similar work but for a large multinational organization?

CASE STUDIES

Case One

Kerry Group Is on Your Table

In business, *sourcing* is the set of activities involved in finding, evaluating, and then engaging suppliers of goods or services. Before a business can start to manage its supply chain, as described in this chapter, it must complete a sourcing process.

Ireland's Kerry Group, a supplier of food ingredients and flavors to the worldwide food industry and of consumer food products to the British Isles, requires a wide range of raw materials from many suppliers. With annual revenue of €5.3 billion (about U.S. $7 billion) in 2011, it needs a lot of those materials. With plants in 25 countries and 40 percent of revenue from outside Europe, it is impossible for the people in one plant to know about all possible suppliers worldwide, but making local sourcing decisions would reduce economies of scale. With the thin profit margins of the food industry, good sourcing decisions are vital to Kerry Group profitability. Software to manage the sourcing process is one way to help make those decisions.

Kerry Group was already a SAP customer when it chose SAP Sourcing OnDemand, having used SAP ERP systems since 2009. The advantage of obtaining a new system from its existing ERP supplier is assured compatibility with applications the company already uses. "What we needed was an intuitive sourcing system that would be completely integrated with our SAP back-office for an end-to-end procurement process," said Peter Fotios, Kerry Group's director of e-procurement services.

SAP Sourcing OnDemand uses the *cloud computing* concept. As its OnDemand name suggests, customers do not have to dedicate computing resources to the software. They use SAP resources *on demand* as their needs require, paying on a per-user, per-month subscription basis. Meanwhile, SAP is responsible for administrative tasks such as data backup and, if necessary, restoration.

Kerry Group implemented SAP Sourcing OnDemand by beginning with a pilot plant. "We rolled it out smoothly in Ireland first, then England and then throughout our global operations in 23 countries," explains Fotios. If any problems appeared in Ireland, the pilot site, Kerry Group could have focused all its problem-solving resources on that location. Fortunately, no major issues arose.

Another thing that Kerry Group did right at implementation time was training. Recognizing that it had competent in-house trainers and competent technical professionals, but few if any who were both, the firm engaged SAP's Irish training partner Olas to assist with that end of the project. Olas brought SAP expertise to the training team, completing the required set of capabilities.

Moving forward, Kerry Group has project plans extending into 2016 for the full roll-out of all its planned SAP ERP capabilities. The smoothness of its Sourcing OnDemand implementation, which took a total of four weeks elapsed time because the software was already running in the cloud when they began, is a good indication that the rest of the project (which is in many ways more complex) will probably go well. If Kerry Group is to carry out its mission statement, which includes being "the world leader in food ingredients and flavors serving the food and beverage industry," the roll-out will have to be smooth.

Discussion Questions

1. According to Peter Fotios, one of Kerry Group's top requirements was ease of integration with its existing SAP system. Why might a company select a sourcing system that was not as easy to integrate with its existing SCM software?
2. Kerry Group is taking a slow and methodical approach to implementing the parts of SAP ERP software they will use, extending that implementation over six years. What does the company gain and what does it lose by taking its time in this way?

Critical Thinking Questions

1. Why should Kerry Group standardize on one ERP package? Wouldn't it be simpler and less expensive to let each plant and sales operation choose its own software, as long as it can report its financial results to headquarters in a standard form?
2. As the largest ERP vendor, SAP can support Kerry Group in 45 countries where the latter has manufacturing facilities or sales operations, providing a local contact for training, problem-solving, and requesting new features. These factors are important. However, others are important also. Suppose another ERP supplier didn't have this depth of local support but offered software that met Kerry Group's needs better, cost less, or both. Rank those factors for importance and discuss how you would make this vendor selection decision.

SOURCES: Kerry Group Web site, *www.kerrygroup.com*, accessed May 2, 2012; Staff, "Kerry's SAP Transformation Measures Up to Their L&D Beliefs," Olas, *olas.ie/olas/Files/DRAFT%20KErry%20synopsis%20RB07.pdf*, May 18, 2011; Staff, "Kerry Group Transforms Its Global Procurement Group in Weeks With SAP Sourcing OnDemand Solution," SAP, *www.sap.com/news-reader/index.epx?pressid=18809*, May 1, 2012.

Case Two

GMC Puts ERP in the Cloud

GMC Global, part of the 4,000-employee, Australia-based management consultants SMEC Group, is a world leader in helping mining companies create efficient and effective operations. The company applies the same drive for efficiency and effectiveness to its own operations. "Our business depends on effective resource management and accurate tracking of time, so having access to timely data that we can easily convert into an invoice is critical for us," said Thomas Hynes, executive general manager of global operations.

Before 2012, GMC Global used a variety of functional applications. It had one system for inventory, invoicing, and financial management and one for time and expenses, as well as a home-grown spreadsheet-based systems in North and South America. None of these systems were integrated with each other. That led to inconsistent processes from location to location, which in turn, made it difficult to assign staff effectively or to ensure consistent client results. Forrester Research gives these additional dangers of poorly integrated applications in a project-oriented services business such as GMC Global's:

- With no clear picture of each project and the overall business, there is no solid information basis for decision making.
- Errors creep into project management data, making it more likely that the team will miss project milestones.
- Project managers who must work with multiple systems have less time to manage their projects actively and to engage with customers and prospects.
- Lack of consistent project metrics leads to lack of insight into staff and team performance, making it hard for a firm to optimize its resources.

Integrating applications requires those applications to share data. The best way to share data is usually by using a shared database. However, in early 2012, GMC Global had active projects in 14 countries, covering every continent except Europe and Antarctica. Over 80 percent of the firm's employees are consultants who spend the great majority of their time in remote locations. They could be in different places every year. Accessing a fixed database would be quick from wherever the database is located but could become a bottleneck elsewhere in the world because the consultants would have to access it remotely and share its physical connection to the Internet.

GMC Global's solution was *cloud computing*. It chose Professional Services Automation (PSA) integrated software from NetSuite, Inc.. After an eight-week period for installation and training, GMC Global went live in March 2012. Its 180 employees now use PSA for project tracking and resource management, time and expense accounting, billing, and reporting.

GMC Global isn't finished. The firm plans to upgrade to NetSuite's Services Resource Planning (SRP) software suite to provide even greater financial control of its global business. SRP is the service-oriented equivalent of ERP in production-based organizations. "Moving to NetSuite SRP will complete the picture for us and enable some of the things that are restricting us at the moment, like global accessibility and managing processes consistently across all regions," said Hynes. "Once we get there, we expect to have a far better view of our overall business."

Discussion Questions

1. You read about cloud computing in Chapter 3, which discussed it as it applies to all types of applications. What are the specific advantages and disadvantages of cloud computing for ERP?
2. How would moving to a single shared database, used by all applications, address the concerns that Forrester lists as dangers of poor integration?

Critical Thinking Questions

1. How can a university, which is also a service organization, benefit from integrated software? (Contrast integrated software with applications that are not integrated. List the specific applications you have in mind.)
2. GMC Global is a service organization. Do you think the applicability of cloud computing to its ERP requirements would be different for a manufacturing firm of comparable size? What about the applicability of cloud ERP to a chain of retail stores?

SOURCES: GMC Global Web site, *www.gmcglobal.com*, accessed April 30, 2012; Martens, C., "Market Overview: Project-Based ERP For Service Delivery Professionals, Forrester Research, January 13, 2012; NetSuite Web site, *www.netsuite.com*, accessed April 30, 2012; Savvas, A., "Mining Firm GMC Deploys Cloud-Based ERP," *Computerworld UK*, *www.computerworlduk.com/news/cloud-computing/3349590/mining-firm-gmc-deploys-cloud-based-erp*, April 10, 2012.

Questions for Web Case

See the Web site for this book to read about the Altitude Online case for this chapter. Following are questions concerning this Web case.

Altitude Online: Enterprise Systems Considerations

Discussion Questions

1. Judging from the ERP features, how important is an ERP to the functioning of a business? Explain.
2. What consideration do you think led Altitude Online to decide to host the ERP on its own servers rather than using SaaS? What are the benefits and drawbacks of both approaches?

Critical Thinking Questions

1. What challenges lay ahead for Altitude Online as it rolls out its new ERP system?
2. How might the ERP affect Altitude Online's future growth and success?

NOTES

Sources for the opening vignette: Oracle Corporation, JD Edwards EnterpriseOne Web site, *www.oracle.com/us/products/applications/jd-edwards-enterpriseone/index.html*, accessed April 30, 2012; "YIOULA Group Builds Scalable Platform for Sustained Growth, Profitability, and Market Leadership," Oracle Corporation, *www.oracle.com/us/corporate/customers/yioula-group-jde-snapshot-176618.pdf*, October 2010; Softecon Enterprise Web site (English), *www.softecon.com/site/en*, accessed April 29, 2012; YIOULA Group Web site, *www.yioula.com*, accessed April 27, 2012.

1. FOCUS Bikes Web site, *www.focus-bikes.com/int/en/home.html*, accessed May 10, 2012.
2. "InventoryPower 5 – Inventory Software for SME," InventorySoft Web site, *www.inventorysoft.com*, accessed March 10, 2012.
3. "Heartland Payment Systems Handles Nearly 44,000 Card Transactions As Official Concessions Processor of XLVI Super Bowl," Heartland Payment Systems Web site, *www.heartlandpaymentsystems.com/article/Heartland-Payment-Systems-Handles-Nearly-44-9244.aspx*, February 9, 2012.
4. "Customer Success Story, Bianca USA," EBridge Connections Web site, *www.ebridgesoft.com*, accessed May 2, 2012.
5. Staff, "ADP Sees Strong Demand from Mid-Sized Organizations Seeking a Rapidly Deployed Payroll Solution," press release, *www.adp.com/media/press-releases/2012-press-releases/adp-sees-strong-demand-from-mid-sized-organizations-seeking-a–rapidly-deployed-payroll-solution.aspx*, February 22, 2012.
6. "China Auto Group," Acctivate Web site,*www.acctivate.com/Customer/china-auto-group*, accessed May 3, 2012.
7. "Electronic Eligibility Software," ADP Web site, *www.advancedmd.com/products-solutions/practice-management/medical-billing-software/electronic-eligibility-medical-software*, accessed March 4, 2012.
8. "Case Study Fit 4 Travel," Rocket Software Web site, *www.rocketsoftware.com/u2*, accessed May 2, 2012.
9. "With RightNow, ResortCom International Supports Over 1 Million Customers and Processes Half a Billion Dollars in Transactions Annually," Oracle Customer Success Stories, Oracle Web site, *www.oracle.com/us/corporate/customers/customersearch/resortcom-international-rightnow-cs-1563704.html*, accessed March 25, 2012.
10. "Westside Produce Trusts IBM to Protect Its Produce Lifecycle," IBM Customer Success Stories, IBM Web site, *www-01.ibm.com/software/success/cssdb.nsf/CS/*

ARBN-8SBRLM?OpenDocument&Site=corp&cty=en_us, March 12, 2012.

11. "LucasArts Job Listings & Profile," Job Monkey Web site, *www.jobmonkeyjobs.com/employer/company/1213/Lucasarts*, March 25, 2012.
12. "Case Study: Denny's Automotive," Rocket Software Web site, *www.rocketsoftware.com/u2*, accessed May 2, 2012.
13. "QuickBooks–A Boon to SME's and Foreign Subsidiaries," Cogzidel Consultancy Services, *http://cogzidel.in/blog/2011/08/03/quickbooks-%E2%80%93-a-boon-to-smes-and-foreign-subsidiaries/*, August 3 2011.
14. "Qvinci Customer Testimonials," Qvinci Web site, *www.qvinci.com/Marketing/Testimonials*, accessed March 7, 2012.
15. Staff, "Paymentsite Drives Teamwork Retail's New iPad POS," press release, PRWeb, *www.prweb.com/releases/2011/12/prweb9069596.htm*, December 29, 2011.
16. Wisor, David C., "Case Study: Mannington Mills Achieves HR Master Data Accuracy," presented at SAP HR2012 Conference in Las Vegas, Nevada, March 2012.
17. "Syndiant Fabless Manufacturer Goes to Market with Support from a Cloud-Based ERP Solution," Microsoft Case Studies, Microsoft Web site, *www.microsoft.com/casestudies/Microsoft-Dynamics-GP/Syndiant/Fabless-Manufacturer-Goes-to-Market-with-Support-from-a-Cloud-Based-ERP-Solution/4000009416*, March 28, 2011.
18. Orbach, Dor, "Eldorado Gold Integrates Chinese Mines into Global SAP Business All-in-One Environment," *http://blog.illumiti.com/blog/bid/58062/Eldorado-Gold-integrates-Chinese-mines-into-global-SAP-Business-All-in-One-environment*, May 17, 2011.
19. Murray, Martin, "Smartphone and the Supply Chain," *http://logistics.about.com/od/supplychainsoftware/a/Smartphones-And-The-Supply-Chain.htm*, accessed May 6, 2012.
20. "Software for SME's," New Business Web site, *www.newbusiness.co.uk/articles/it-advice/software-smes*, August 25, 2011.
21. Kanaracus, Chris, "SAP Project Issues Hurt Ingram Micro Profits Again," *Tech World*, July 30, 2011.
22. Kanaracus, Chris, "University Accuses Oracle of Lies, Rigged Demo in Lawsuit," *InfoWorld*, December 6, 2011.
23. Kanaracus, Chris, "Customer Sues Epicor after ERP Software Project Attempt Ends in Big Mess," *PC World*, November 29, 2011.
24. Kanaracus, Chris, "County Alleges SAP, Deloitte, Engaged in Racketeering," *CIO*, February 3, 2011.
25. "Allen Bridges the Adoption Gap between ERP Technical Success and User Adoption Failure," Allen Communication Learning Services Web site, *www.allencomm.com/2012/01/allen-bridges-the-adoption-gap*, January 25, 2012.
26. "PEPCO Poland Sp. Z.o.o.," Microsoft Dynamics Customer Stories, Microsoft Web site, *www.microsoft.com/dynamics/en/gulf/products/ax-customer-stories.aspx*, accessed March 27, 2012.
27. "National IT Services Firm Moves from Six Disparate Software Systems to NetSuite OneWorld SRP," ERP-News, *www.erpnews.net/49/national-it-services-firm-moves-from-six-disparate-software-systems-to-netsuite-oneworld-srp*, November 24, 2011.
28. Vandana, Priti, "Merino Successfully Integrates Baan IV c4 with SAP for Hitachi India," *www.baanboard.com/baanboard/showthread.php?t=60452*, April 28, 2011.
29. "The Newark Group Improves the Bottom Line with a Completely Integrated, Comprehensive Supply Chain Solution," IBM Web site, *www-01.ibm.com/software/success/cssdb.nsf/CS/DJOY-8LW5AW?OpenDocument&Site=default&cty=en_us*, September 20, 2011.
30. Sandhu, Tandeep and Konuk, Emel, "Case Study: How Bombardier Leverages Production Planning and Detailed Scheduling to Perform Planning and Scheduling of its Aircraft Assembly Line," presented at Manufacturing 2012, Orlando, Florida, March 20, 2012.
31. "The Sweet Smell of Success," Demand Management Systems Web site, *www.demandmgmt.com*, accessed March 31, 2012.
32. Kolakowski, John, "Boston Red Sox Improves Business Functionality with Green Bacon Solutions," *CRM Software Blog, www.crmsoftwareblog.com/2012/01/boston-red-sox-improves-business-functionality-with-green-beacon-solutions/*, January 18, 2012.
33. "Stallergenes, Biopharmaceutical Company," Stallergenes Web site, *www.stallergenes.com/nc/en.html*, accessed March 11, 2012.
34. "Stallergenes Adopts Veeva CRM," *IT & Software, itsoftware.pharmaceutical-business-review.com/news/stallergenes-adopts-veeva-crm-121011*, October 12, 2011.
35. "About Us–Burns & McDonnell," Burns & McDonnell Web site, *www.burnsmcd.com/Company/About-Us*, accessed March 13, 2012.
36. Staff, "Burns & McDonnell Picks Microsoft Dynamics CRM," *Computer Business Review Online, www.cbronline.com/news/burns-mcdonnell-selects-microsoft-crm-software-210212*, February 12, 2012.
37. Staff, "Department of Social Welfare and Development (DSWD) Adopted Microsoft Dynamics CRM 2011," press release, Gurango Software Web site, *www.gurango.com/news/department-of-social-welfare-and-development-dswd-adopted-microsoft-dynamics-crm-2011.aspx*, February 28, 2012.
38. "Case Study–A Social Media Marketing at Point-of-Sale," Bematech Web site, *www.logiccontrols.com/web/news_041911.html*, April 19, 2011.
39. "Tasti D-Lite Adopts Engage 121 as Network-Wide Social CRM Solution," Engage 121 Web site, *www.engage121.com/node/73*, June 29, 2011.
40. Pillar, Matt, "CRM Sheds Light on Tommy Bahama Customers," *RetailSolutionsOnline.com*, October 25, 2011.
41. Staff, "Diane von Furstenberg Selects Computer Generated Solutions' Product Lifecycle Management System," Press release, PRWeb, *www.prweb.com/releases/2012/2/prweb9232535.htm*, February 28, 2012.

42. "A Case for PLM: Brown-Forman Reveals Innovation Transformation" (video), *https://consumergoods.webex.com/ec0605lc/eventcenter/recording/recordAction.do;jsessionid=5LgyP32J2vdNwV11qbFP9242PnKw7lfJLjQMp2nHMLrXtwwJJv4G!-1133797779?theAction=poprecord&actname=%2Feventcenter%2Fframe%2Fg.do&apiname=lsr.php&renewticket=0&renewticket=0&actappname=ec0605lc&entappname=url0107lc&needFilter=false&&isurlact=true&entactname=%2FnbrRecordingURL.do&rID=47386567&rKey=8eb861d17fd62a3e&recordID=47386567&rnd=5145083162&siteurl=consumergoods&SP=EC&AT=pb&format=short*, recorded July 20, 2011.
43. Staff, "Love Culture Adopts Epicor Retail SaaS to Support Growth," *Apparel, http://apparel.edgl.com/news/Love-Culture-Adopts-Epicor-Retail-SaaS-to-Support-Growth78098*, January 20, 2012.

10 Information and Decision Support Systems

Principles	Learning Objectives
• Good decision-making and problem-solving skills are the key to developing effective information and decision support systems.	• Define the stages of decision making. • Discuss the importance of implementation and monitoring in problem solving.
• The management information system (MIS) must provide the right information to the right person in the right format at the right time.	• Explain the uses of MISs and describe their inputs and outputs. • Discuss information systems in the functional areas of business organizations.
• Decision support systems (DSSs) are used when the problems are unstructured.	• List and discuss important characteristics of DSSs that give them the potential to be effective management support tools. • Identify and describe the basic components of a DSS.
• Specialized support systems, such as group support systems (GSSs) and executive support systems (ESSs), use the overall approach of a DSS in situations such as group and executive decision making.	• State the goals of a GSS and identify the characteristics that distinguish it from a DSS. • Identify the fundamental uses of an ESS and list the characteristics of such a system. • List and discuss other special-purpose systems.

Information Systems in the Global Economy

IRISH LIFE, IRELAND

Decision Makers Gain Insight from BI Dashboards

Irish Life, founded in 1939, is Ireland's largest life insurer. In addition, the company also handles pensions for 200,000 Irish workers and is Ireland's largest investment manager, with over €31 billion (about U.S. $41 billion) in assets.

However, Irish Life had a problem. It collected vast amounts of data through its three major lines of business. It had business intelligence (BI) software to help analyze all this data, but that software wasn't doing the job. Paul Egan, business intelligence IT manager at Irish Life, explains that "a lot of the tools were only IT tools and only IT people could use them, but [with those tools] we could never keep up with the appetite the business had for this." Irish Life needed software that its business managers could use in their decision making without having to become technical specialists.

After looking at the available BI packages from its incumbent supplier and other software vendors, Irish Life sought advice from consultants at the Gartner Group. The life insurance provider then chose software from Tableau Software of Seattle, Washington, and engaged Tableau partner MXI Computing to help implement that software.

Using the Tableau software, Irish Life can represent data graphically across the organization, mapping patterns and trends more clearly than it could before the company began to use it. It will make Tableau dashboards available to about 300 users. These users will be able to build their own dashboards, to publish on the Web, or distribute on mobile devices running Android or iOS software. The net result, Irish Life believes, will be improved decision making due to better availability of data and better insight into the data. The Intelligence and Design stages of decision making are especially well positioned to benefit from this insight.

For example, Irish Life recently announced the Personal Lifestyle Strategy program for customized retirement planning within the framework of a corporate pension plan. Making the decisions that were involved in developing this program required detailed analysis of workforce data—exactly what data visualization is suited for.

"Managers can come up with their own dashboards based on the numbers they know they need. There's less work for IT in the front end: IT now only have to worry about the data warehouse, which is where we can add value. We don't have to worry about the visuals as much," added Egan. Insights from the BI tool have already led to Irish Life moving its management team's focus in certain cases to product lines or customer accounts that needed closer attention. "It will make a difference in how effectively we manage the business," Egan said.

The results that Irish Life expected to achieve are being realized. Gerry Hassett, CEO at Irish Life Retail, said, "We can see that the return on investment is already being delivered. We can now see what is happening on a day-to-day and week-to-week basis, which means we can develop our business accordingly and further enhance our competitive edge." He added that his dashboards "took a few weeks to implement" and that it had been "a seamless operation."

As you read this chapter, consider the following:

- What sorts of decisions can BI systems and dashboards help companies make? Who makes those decisions?
- What data is needed to make each decision properly? How can that data be presented most effectively?

WHY LEARN ABOUT INFORMATION AND DECISION SUPPORT SYSTEMS?

© Andrey Burmakin/Shutterstock

You have seen throughout this book how information systems can make you more efficient and effective through the use of database systems, the Internet, e-commerce, enterprise system, and many other technologies. The true potential of information systems, however, is in helping you and your coworkers make more informed decisions. This chapter shows you how to slash costs, increase profits, and uncover new opportunities for your company using management information and decision support systems. A financial planner can use management information and decision support systems to find the best investments for clients. A loan committee at a bank or credit union can use a group support system to help determine who should receive loans. Store managers can use decision support systems to help them decide what and how much inventory to order to meet customer needs and increase profits. An entrepreneur who owns and operates a temporary storage company can use vacancy reports to help determine what price to charge for new storage units. Everyone can be a better problem solver and decision maker. This chapter shows you how information systems can help. It begins with an overview of decision making and problem solving.

As shown in the opening vignette, information and decision support are the lifeblood of today's organizations. Thanks to information and decision support systems, managers and employees can obtain useful information in real time. As discussed in Chapter 9, TPS and ERP systems capture a wealth of data. When this data is filtered and manipulated, it can provide powerful support for managers and employees. The ultimate goal of management information and decision support systems is to help managers and executives at all levels make better decisions and solve important problems. The result can be increased revenues, reduced costs, and the realization of corporate goals. No matter what type of information and decision support system you use, its primary goal should be to help you and others become better decision makers and problem solvers.

DECISION MAKING AND PROBLEM SOLVING

Every organization needs effective decision makers. In most cases, strategic planning and the overall goals of the organization set the course for decision making, helping employees and business units achieve their objectives and goals. Often, information systems also assist with problem solving, helping people make better decisions and save lives. Today, companies like Chevron are starting to use the latest technology, such as new tablet computers, to help managers make better decisions by allowing them to connect to corporate computer systems and applications.[1] According to the general manager of technology at Chevron, "New mobility strategies are going to change the average workflow in companies."

Decision-making approaches can differ dramatically from one person to the next. Nobel laureate Daniel Kahneman, for example, shows that some people have good decision-making approaches, while others do not.[2] Some people make quick decisions, while others take their time. In the opinion of the CIO of Shatlee, "A person won't make a decision until he identifies a familiar pattern in the facts, which makes him comfortable enough to act. This is often a slow, incremental process, which partly explains why many companies prod rather than sprint."[3]

Decision Making as a Component of Problem Solving

In business, one of the highest compliments you can receive is to be recognized by your colleagues and peers as a "real problem solver." According to Jimmy Wang, vice president and CIO of Teva Pharmaceuticals Americas,

"A career in IT can certainly be rewarding. You're viewed as a problem solver."[4] Problem solving is a critical activity for any business organization. After identifying a problem, the process of solving the problem begins with decision making. A well-known model developed by Herbert Simon divides the **decision-making phase** of the problem-solving process into three stages: intelligence, design, and choice. This model was later incorporated by George Huber into an expanded model of the entire problem-solving process (see Figure 10.1).

decision-making phase: The first part of problem solving, including three stages: intelligence, design, and choice.

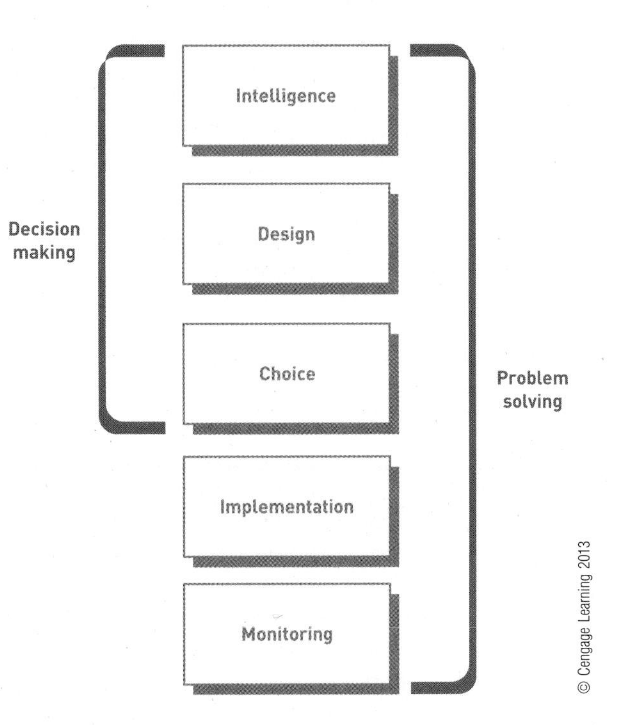

FIGURE **10.1**

How decision making relates to problem solving

The three stages of decision making—intelligence, design, and choice—are augmented by implementation and monitoring to result in problem solving.

The first stage in the problem-solving process is the **intelligence stage**. During this stage, you identify and define potential problems or opportunities. You also investigate resource and environmental constraints. For example, if you were a Hawaiian farmer, during the intelligence stage, you would explore the possibilities of shipping tropical fruit from your farm in Hawaii to stores in Michigan. The perishability of the fruit and the maximum price that consumers in Michigan are willing to pay for the fruit are problem constraints.

intelligence stage: The first stage of decision making in which you identify and define potential problems or opportunities.

In the **design stage**, you develop alternative solutions to the problem and evaluate their feasibility. In the tropical fruit example, you would consider the alternative methods of shipment, including the transportation times and costs associated with each. During this stage, you might determine that shipment by freighter to California and then by truck to Michigan is not feasible because the fruit would spoil. A priority of most farmers is to quickly deliver fresh food to the market. According to the CIO of Target, "From a decision-making standpoint, speed forces you to be crystal-clear on the priorities of the organization."[5]

design stage: The second stage of decision making in which you develop alternative solutions to the problem and evaluate their feasibility.

The last stage of the decision-making phase, the **choice stage**, requires selecting a course of action. In the tropical fruit example, you might select the method of shipping fruit by air from your Hawaiian farm to Michigan as the solution. The choice stage would then conclude with selection of an air carrier. As you will see later, various factors influence choice; the act of choosing is not as simple as it might first appear.

choice stage: The third stage of decision making, which requires selecting a course of action.

Problem solving includes and goes beyond decision making. It also includes the **implementation stage** when the solution is put into effect. For example, if your decision is to ship tropical fruit to Michigan as air freight

problem solving: A process that goes beyond decision making to include the implementation stage.

implementation stage: A stage of problem solving in which a solution is put into effect.

using a specific carrier, implementation involves informing your field staff of the new activity, getting the fruit to the airport, and actually shipping the product to Michigan.

monitoring stage: The final stage of the problem-solving process in which decision makers evaluate the implementation.

The final stage of the problem-solving process is the **monitoring stage**. In this stage, decision makers evaluate the implementation to determine whether the anticipated results were achieved and to modify the process in light of new information. Monitoring can involve feedback and adjustment. For example, after the first shipment of fruit from Hawaii to Michigan, you might learn that the flight of your chosen air freight firm routinely stops in Phoenix, Arizona, where the plane sits on the runway for a number of hours while loading additional cargo. If this unforeseen fluctuation in temperature and humidity adversely affects the fruit, you might have to readjust your solution to include a new carrier that does not make such a stop, or perhaps you would consider a change in fruit packaging. In a real-life example, American Airlines monitored its decision to use probability analysis to reduce inventory levels and shipping costs for airline maintenance equipment and in-flight service items.[6] The value of this inventory can be over $1 billion a year on average. The airline used a decision-making technique called decision tree analysis that diagrammed major decisions and possible outcomes from these decisions.

Programmed versus Nonprogrammed Decisions

programmed decision: A decision made using a rule, procedure, or quantitative method.

In the choice stage, various factors influence the decision maker's selection of a solution. One factor is whether the decision can be programmed. **Programmed decisions** are made using a rule, procedure, or quantitative method. For example, to say that inventory should be ordered when inventory levels drop to 100 units is a programmed decision because it adheres to a rule. Programmed decisions are easy to computerize using traditional information systems. For example, you can easily program a computer to order more inventory when levels for a certain item reach 100 units or less. Cisco Systems, a large computer-networking equipment and server-manufacturing company, controls its inventory and production levels using programmed decisions embedded into its computer systems.[7] The programmed decision-making process has improved forecasting accuracy and reduced the possibility of manufacturing the wrong types of inventory, which has saved money and preserved cash reserves. Management information systems can also reach programmed decisions by providing reports on problems that are routine and in which the relationships are well defined (in other words, they are structured problems).

Cisco Systems controls its inventory and production levels using programmed decisions embedded into its computer systems.

iStockphoto/rev606

nonprogrammed decision: A decision that deals with unusual or exceptional situations.

Nonprogrammed decisions deal with unusual or exceptional situations. In many cases, these decisions are difficult to quantify. Determining the appropriate training program for a new employee, deciding whether to develop a new type of product line, and weighing the benefits and drawbacks of installing an upgraded pollution control system are examples. Each of these decisions contains unique characteristics, and standard rules or procedures might not apply to them. Today, decision support systems help solve many nonprogrammed decisions in which the problem is not routine and rules and relationships are not well defined (unstructured or ill-structured problems). These problems can include deciding the best location for a manufacturing plant or whether to rebuild a hospital that was severely damaged from a hurricane or tornado.

Optimization, Satisficing, and Heuristic Approaches

optimization model: A process to find the best solution, usually the one that will best help the organization meet its goals.

In general, computerized decision support systems can either optimize or satisfice. An **optimization model** finds the best solution, usually the one that will best help the organization meet its goals.[8] For example, an optimization model can find the best route to ship products to markets, given certain conditions and assumptions. Laps Care from TietoEnatorAM is an information system that used optimization to assign medical personnel to home health care patients in Sweden while minimizing health care costs. The optimization system has improved the quality of medical care delivered to the elderly.[9] The system has also improved health care efficiency by 10 to 15 percent and lowered costs by more than 20 million Euros. In another case, Xerox developed an optimization system to increase printer productivity and reduce costs, which is called Lean Document Production (LDP) solutions.[10] The optimization routine reduced labor costs by 20 to 40 percent in some cases. Exxon Mobile used an optimization technique, called mixed-integer programming, to reduce costs in shipping vacuum gas oil (VGO) from its European operations to its refineries in the United States.[11] The optimization technique allowed Exxon Mobile to use existing shipping vessels more efficiently and reduce transportation costs. An automobile club used an optimization program to schedule personnel to service calls to minimize costs.[12] Kimberly-Clark used optimization to reduce inventory levels by more than 40 percent, drastically cutting costs.[13]

Optimization models use problem constraints. A limit on the number of available work hours in a manufacturing facility is an example of a problem constraint. Shermag, a Canadian furniture manufacturing company, used an optimization program to reduce raw materials costs, including wood, in its manufacturing operations.[14] The optimization program, which used the C++ programming language and CPLEX optimization software, helped the company reduce total costs by more than 20 percent. Some spreadsheet programs, such as Excel, have optimizing features, as shown in Figure 10.2. Optimization software also allows decision makers to explore various alternatives.[15]

satisficing model: A model that will find a good—but not necessarily the best—solution to a problem.

A **satisficing model** is one that finds a good—but not necessarily the best—solution to a problem.[16] Satisficing is used when modeling the problem properly to get an optimal decision would be too difficult, complex, or costly. Satisficing normally does not look at all possible solutions but only at those likely to give good results. Consider a decision to select a location for a new manufacturing plant. To find the optimal (best) location, you must consider all cities in the United States or the world. A satisficing approach is to consider only five or ten cities that might satisfy the company's requirements. Limiting the options might not result in the best decision, but it will likely result in a good decision without spending the time and effort to investigate all cities. Satisficing is a good alternative modeling method

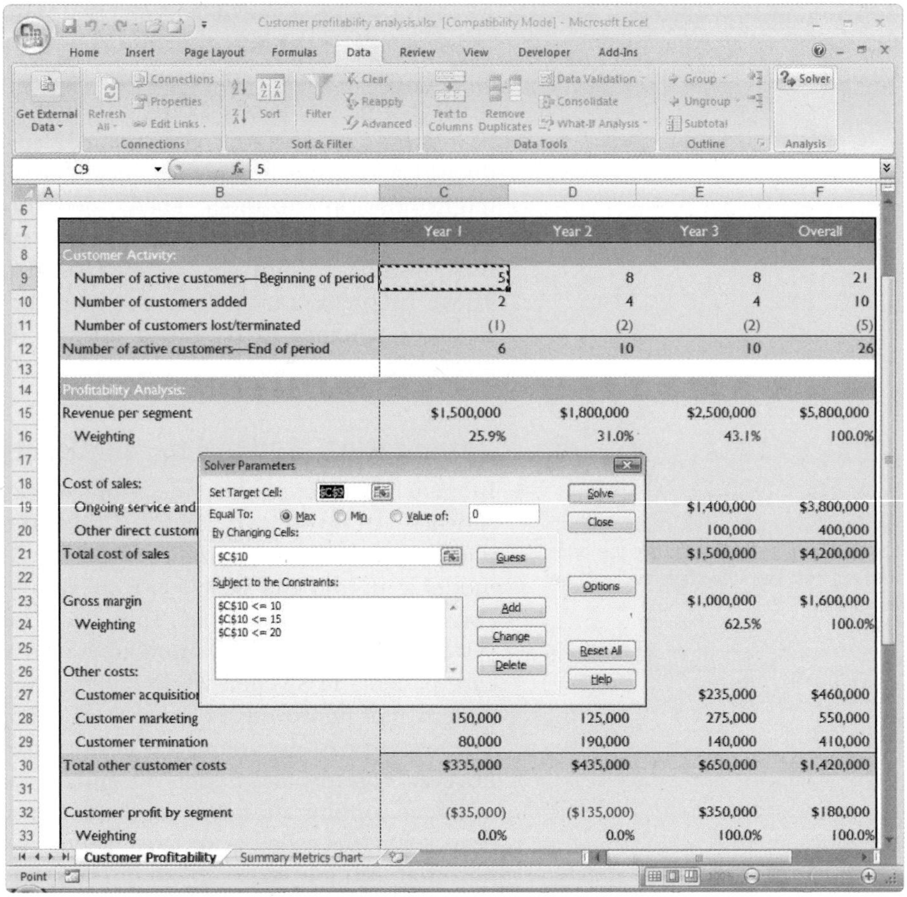

FIGURE 10.2
Optimization software
Some spreadsheet programs, such as Microsoft Excel, have optimizing routines. This figure shows Solver, which can find an optimal solution given certain constraints.

because it is sometimes too expensive to analyze every alternative to find the best solution.

heuristics: Commonly accepted guidelines or procedures that usually find a good solution.

Heuristics, also known as "rules of thumb," are commonly accepted guidelines or procedures that usually find a good solution.[17] A heuristic that baseball team managers use is to place batters most likely to get on base at the top of the lineup, followed by the power hitters who can drive them in to score. An example of a heuristic used in business is to order four months' supply of inventory for a particular item when the inventory level drops to 20 units or less; although this heuristic might not minimize total inventory costs, it can serve as a good rule of thumb to avoid stockouts without maintaining excess inventory. Symantec, a provider of antivirus software, has developed an antivirus product that is based on heuristics.[18] The software can detect viruses that are difficult to detect using traditional antivirus software techniques.

The Benefits of Information and Decision Support Systems

The information and decision support systems covered in this chapter and the next help individuals, groups, and organizations make better decisions, solve problems, and achieve their goals. These systems include management information systems, decision support systems, group support systems, executive support systems, knowledge management systems, and a variety of special-purpose systems. Information and decision support systems can help sell products and services. Coulomb Technologies, for example, has developed an application for smartphones that tells customers with electric cars where to find the closest charging stations.[19] The application is available on iPhones and BlackBerry phones. According to the chief technology officer, "Folks who buy an electric vehicle are probably the same folks who would buy an iPhone."

As shown in Figure 10.3, the benefits of these systems are a measure of increased performance of the systems versus the cost to deliver them. The plus sign (+) by the arrow from *performance* to *benefits* indicates that increased performance has a positive impact on benefits. The minus sign (–) from *cost* to *benefits* indicates that increased cost has a negative impact on benefits.

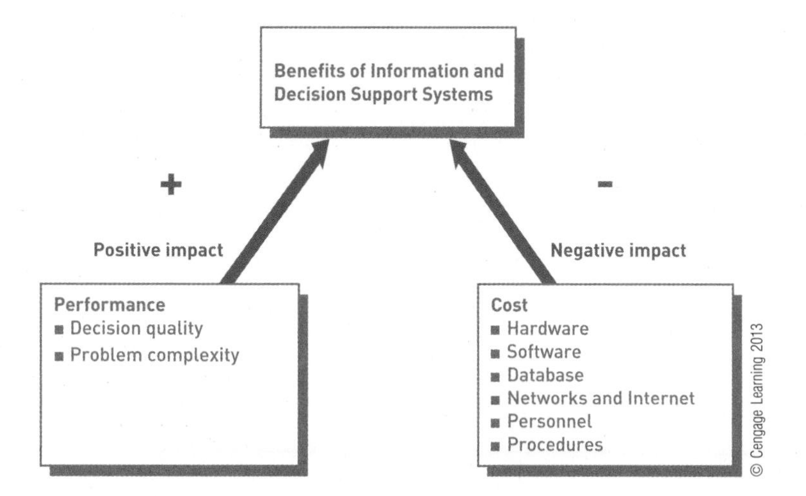

FIGURE **10.3**

Benefits of information and decision support systems

The performance of these systems is typically a function of decision quality and problem complexity. Decision quality can result in increased effectiveness, increased efficiency, higher productivity, and many other measures first introduced in Chapter 2. Problem complexity depends on how hard the problem is to solve and implement. The cost of delivering these systems are the expenditures of the information technology components covered in Part 2 of this book, including hardware, software, databases, networks and the Internet, people, and procedures. But how do these systems actually deliver benefits to the individuals, groups, and organizations that use them? It depends on the system. We begin our discussion with traditional management information systems.

AN OVERVIEW OF MANAGEMENT INFORMATION SYSTEMS

A management information system (MIS) is an integrated collection of people, procedures, databases, and devices that provides managers and decision makers with information to help achieve organizational goals. MISs can often give companies and other organizations a competitive advantage by providing the right information to the right people in the right format and at the right time.

Management Information Systems in Perspective

The primary purpose of an MIS is to help an organization achieve its goals by providing managers with insight into the regular operations of the organization so that they can control, organize, and plan more effectively. One important role of the MIS is to provide the right information to the right person in the right format at the right time. In short, an MIS provides managers with information, typically in reports, that supports effective decision making and provides feedback on daily operations. Figure 10.4 shows the role of MISs within the flow of an organization's information. Note that business transactions can enter the organization via traditional methods, via the Internet, or via an extranet connecting customers and suppliers to the firm's ERP or transaction processing systems. The use of MISs spans all levels of management, that is, they provide support to and are used by employees throughout the organization.

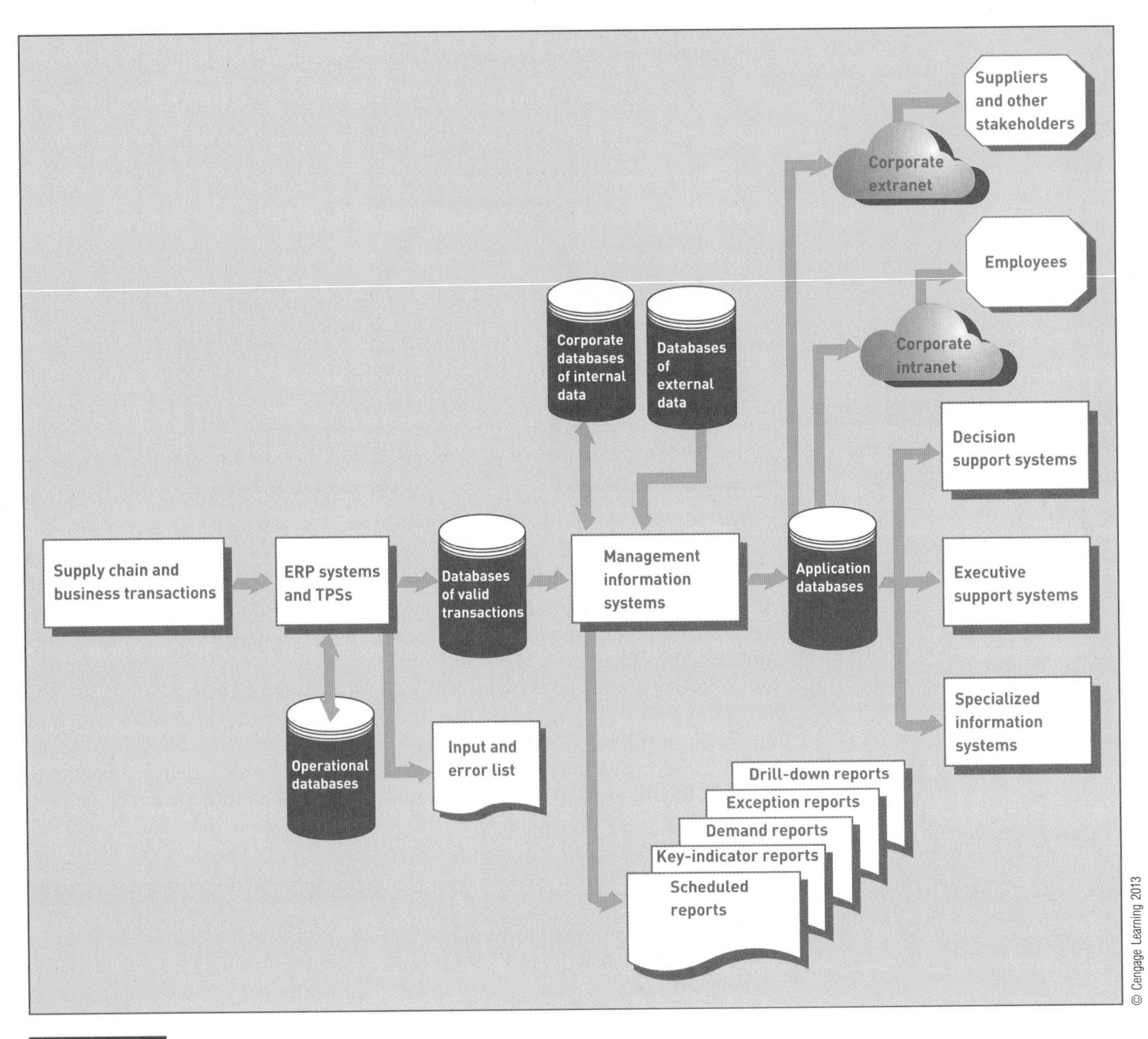

FIGURE 10.4

Sources of managerial information

The MIS is just one of many sources of managerial information. Decision support systems, executive support systems, and expert systems also assist in decision making.

Inputs to a Management Information System

As shown in Figure 10.4, data that enters an MIS originates from both internal and external sources, including a company's supply chain, first discussed in Chapter 2. The most significant internal data sources for an MIS are the organization's various TPS and ERP systems and related databases. External sources of data can include customers, suppliers, competitors, and stockholders, whose data is not already captured by the TPS and ERP systems, as well as other sources, such as the Internet. As discussed in Chapter 5, companies also use data warehouses and data marts to store valuable business information. Business intelligence, also discussed in Chapter 5, can be used to turn a database into useful information throughout the organization.

Outputs of a Management Information System

The output of most MISs is a collection of reports that are distributed to managers. Many MIS reports come from an organization's databases, first discussed in Chapter 5. These reports can be tailored for each user and can be delivered in a timely fashion. Providence Washington Insurance Company used the ReportNet module in business intelligence software from Cognos (*www.cognos.com*), an IBM company, to reduce the number of paper reports they produce and the associated costs.[20] The new reporting system creates an "executive dashboard" that shows current data, graphs, and tables to help managers make better real-time decisions (see Figure 10.5). Microsoft makes a reporting system called Business Scorecard Manager to give decision makers timely information about sales and customer information.[21] The software, which competes with Business Objects and IBM Cognos, can integrate with other Microsoft software products, including Microsoft Excel. Hewlett-Packard's OpenView Dashboard is another MIS package that can quickly and efficiently render pictures, graphs, and tables that show how a business is functioning.

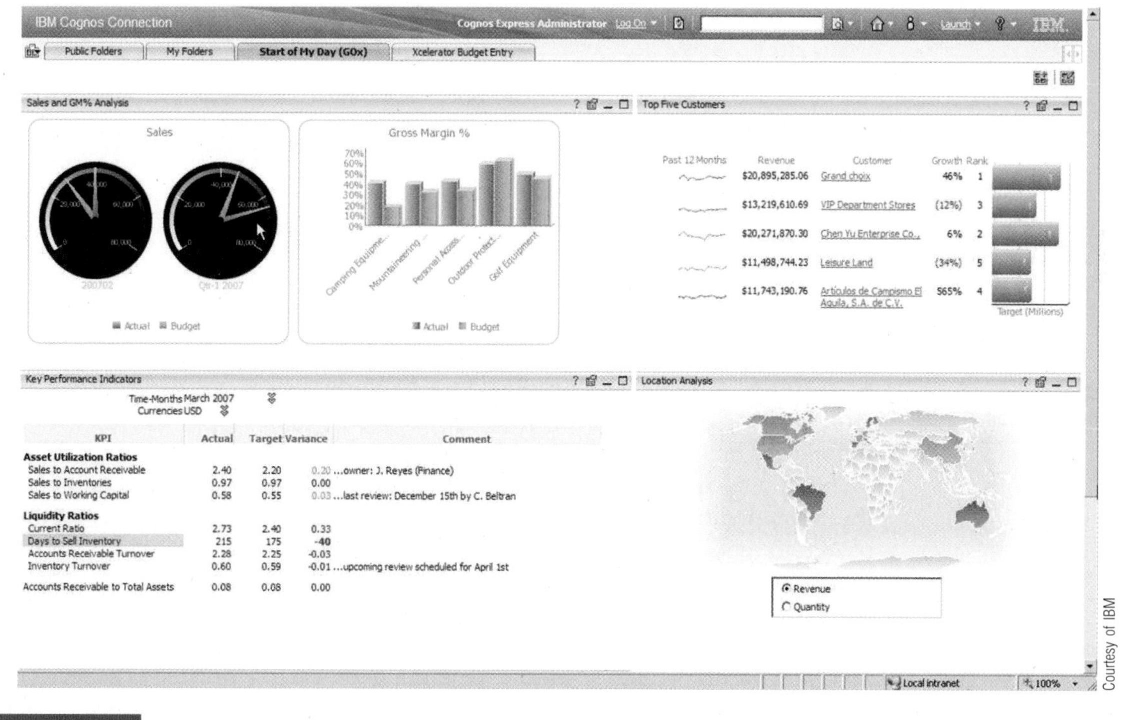

FIGURE **10.5**

Executive dashboard

IBM Cognos Connection software includes a dashboard with a graphic overview of how a business is functioning.

Management reports can come from various company databases, data warehouses, and other sources. These reports include scheduled reports, key-indicator reports, demand reports, exception reports, and drill-down reports (see Figure 10.6).

(a) Scheduled Report

Daily Sales Detail Report — Prepared: 08/10/08

Order #	Customer ID	Salesperson ID	Planned Ship Date	Quantity	Item #	Amount
P12453	C89321	CAR	08/12/08	144	P1234	$3,214
P12453	C89321	CAR	08/12/08	288	P3214	$5,660
P12454	C03214	GWA	08/13/08	12	P4902	$1,224
P12455	C52313	SAK	08/12/08	24	P4012	$2,448
P12456	C34123	JMW	08/13/08	144	P3214	$720
....	....	....	....	....	....	....

(b) Key-Indicator Report

Daily Sales Key-Indicator Report

	This Month	Last Month	Last Year
Total Orders Month to Date	$1,808	$1,694	$1,914
Forecasted Sales for the Month	$2,406	$2,224	$2,608

(c) Demand Report

Daily Sales by Salesperson Summary Report — Prepared: 08/10/08

Salesperson ID	Amount
CAR	$42,345
GWA	$38,950
SAK	$22,100
JWN	$12,350
....	....
....	....

(d) Exception Report

Daily Sales Exception Report—Orders Over $10,000 — Prepared: 08/10/08

Order #	Customer ID	Salesperson ID	Planned Ship Date	Quantity	Item #	Amount
P12345	C89321	GWA	08/12/08	576	P1234	$12,856
P22153	C00453	CAR	08/12/08	288	P2314	$28,800
P23023	C32832	JMN	08/11/08	144	P2323	$14,400
....	....	....	....	....	....	....
....	....	....	....	....	....	....

(e) First-Level Drill-Down Report

Earnings by Quarter (Millions)

		Actual	Forecast	Variance
2nd Qtr.	2008	$12.6	$11.8	6.8%
1st Qtr.	2008	$10.8	$10.7	0.9%
4th Qtr.	2008	$14.3	$14.5	-1.4%
3rd Qtr.	2008	$12.8	$13.3	-3.8%

(f) Second-Level Drill-Down Report

Sales and Expenses (Millions)

Qtr: 2nd Qtr. 2008	Actual	Forecast	Variance
Gross Sales	$110.9	$108.3	2.4%
Expenses	$ 98.3	$ 96.5	1.9%
Profit	$ 12.6	$ 11.8	6.8%

(g) Third-Level Drill-Down Report

Sales by Division (Millions)

Qtr: 2nd Qtr. 2008	Actual	Forecast	Variance
Beauty Care	$ 34.5	$ 33.9	1.8%
Health Care	$ 30.0	$ 28.0	7.1%
Soap	$ 22.8	$ 23.0	-0.9%
Snacks	$ 12.1	$ 12.5	-3.2%
Electronics	$ 11.5	$ 10.9	5.5%
Total	$110.9	$108.3	2.4%

(h) Fourth-Level Drill-Down Report

Sales by Product Category (Millions)

Qtr: 2nd Qtr. 2008 Division: Health Care	Actual	Forecast	Variance
Toothpaste	$12.4	$10.5	18.1%
Mouthwash	$ 8.6	$ 8.8	-2.3%
Over-the-Counter Drugs	$ 5.8	$ 5.3	9.4%
Skin Care Products	$ 3.2	$ 3.4	-5.9%
Total	$30.0	$28.0	7.1%

FIGURE 10.6

Reports generated by an MIS

The types of reports are (a) scheduled, (b) key indicator, (c) demand, (d) exception, and (e–h) drill down. (Source: George W. Reynolds, *Information Systems for Managers,* 3rd ed., St. Paul, MN: West Publishing, 1995.)

Scheduled Reports

scheduled report: A report produced periodically, such as daily, weekly, or monthly.

Scheduled reports are produced periodically, such as daily, weekly, or monthly. For example, a production manager might use a weekly summary report that lists total payroll costs to monitor and control labor and job costs. Monthly bills are also examples of scheduled reports. Using its monthly bills, the Sacramento Municipal Utility District compared how people use energy, trying to encourage better energy usage.[22] Other scheduled reports can help managers control customer credit, performance of sales representatives, inventory levels, and more.

key-indicator report: A summary of the previous day's critical activities, typically available at the beginning of each workday.

A **key-indicator report** summarizes the previous day's critical activities and is typically available at the beginning of each workday. These reports can summarize inventory levels, production activity, sales volume, and the like. Key-indicator reports are used by managers and executives to take quick, corrective action on significant aspects of the business.

Demand Reports

demand report: A report developed to give certain information at someone's request rather than on a schedule.

Demand reports are developed to provide certain information upon request. In other words, these reports are produced on demand rather than on a schedule. Like other reports discussed in this section, they often come from an organization's database system. For example, an executive might want to know the production status of a particular item—a demand report can be generated to provide the requested information by querying the company's database. Suppliers and customers can also use demand reports. FedEx, for example, provides demand reports on its Web site to allow customers to track packages from their source to their final destination.

Exception Reports

exception report: A report automatically produced when a situation is unusual or requires management action.

Exception reports are reports that are automatically produced when a situation is unusual or requires management action. For example, a manager might set a parameter that generates a report of all inventory items with fewer than the equivalent of five days of sales on hand. This unusual situation requires prompt action to avoid running out of stock on the item. The exception report generated by this parameter would contain only items with fewer than five days of sales in inventory.

As with key-indicator reports, exception reports are most often used to monitor aspects important to an organization's success. In general, when an exception report is produced, a manager or executive takes action. Parameters, or *trigger points*, should be set carefully for an exception report. Trigger points that are set too low might result in too many exception reports; trigger points that are too high could mean that problems requiring action are overlooked. For example, if a manager at a large company wants a report that contains all projects over budget by $100 or more, the system might retrieve almost every company project. The $100 trigger point is probably too low. A trigger point of $10,000 might be more appropriate.

Drill-Down Reports

drill-down report: A report providing increasingly detailed data about a situation.

Drill-down reports provide increasingly detailed data about a situation. Using these reports, analysts can see data at a high level first (such as sales for the entire company), then at a more detailed level (such as the sales for one department of the company), and then a very detailed level (such as sales for one sales representative). Companies and organizations of all sizes and types use drill-down reports.

Characteristics of a Management Information System

Scheduled, key-indicator, demand, exception, and drill-down reports have all helped managers and executives make better, more timely decisions. In general, MISs perform the following functions:

- Provide reports with fixed and standard formats.
- Produce hard-copy and soft-copy reports.
- Use internal data stored in the computer system.
- Allow users to develop custom reports.
- Require user requests for reports developed by systems personnel.

FUNCTIONAL ASPECTS OF THE MIS

Most organizations are structured along functional areas. This functional structure is usually apparent from an organization chart, which typically shows a hierarchy in roles or positions. Some traditional functional areas include finance, manufacturing, marketing, human resources, and other specialized information systems. The MIS can also be divided along those functional lines to produce reports tailored to individual functions (see Figure 10.7).

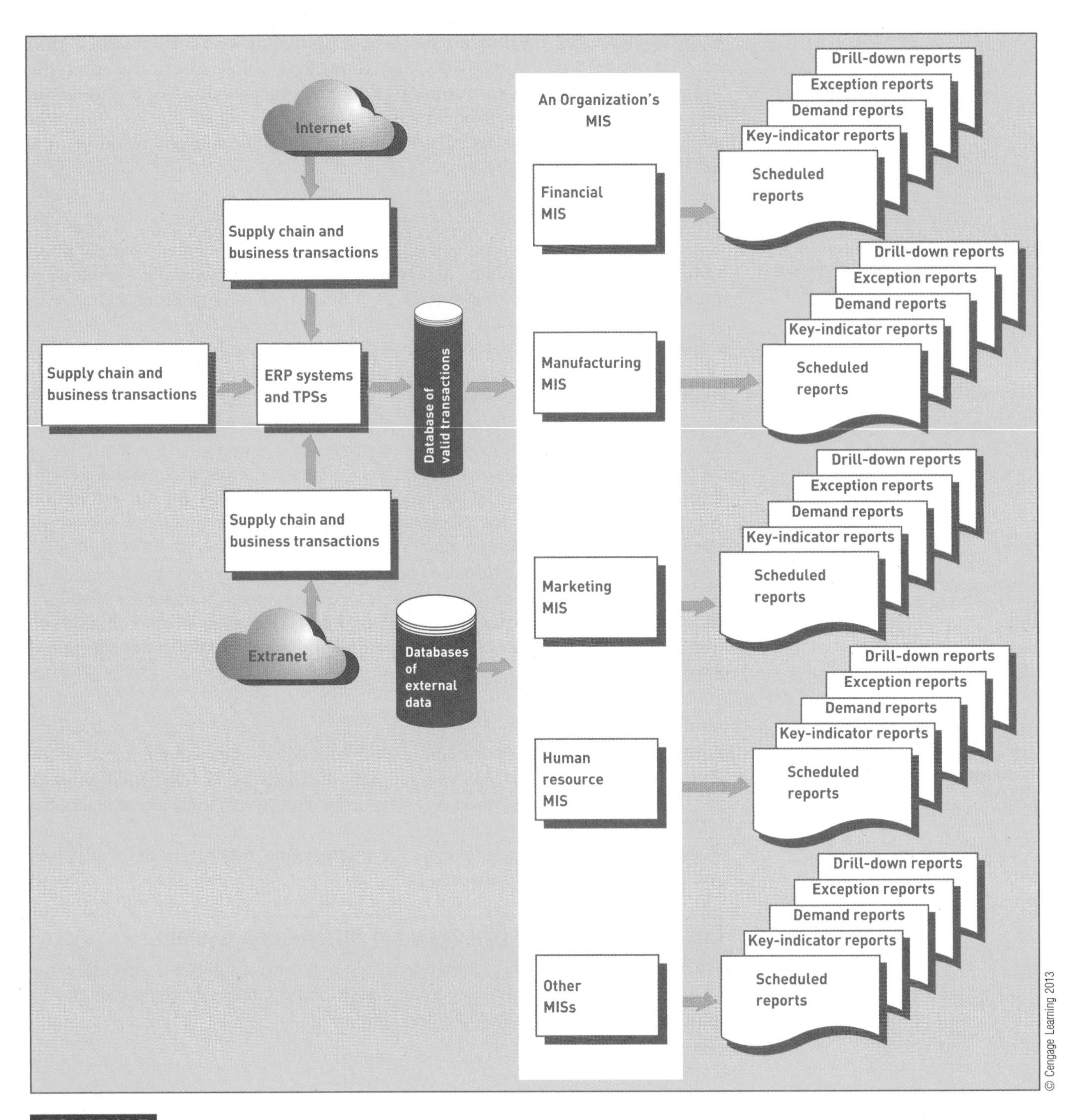

FIGURE 10.7

An organization's MIS

The MIS is an integrated collection of functional information systems, each supporting particular functional areas.

Financial Management Information Systems

financial MIS: An information system that provides financial information for executives and for a broader set of people who need to make better decisions on a daily basis.

A **financial MIS** provides financial information for executives and for a broader set of people who need to make better decisions on a daily basis. Thomson Reuters, for example, has developed an automated reporting system that scans articles about companies for its stock traders to determine if the news is favorable or unfavorable. The reports can result in buy orders if the news is positive or sell orders if the news is negative. Eventually, the system will be tied into machine trading that doesn't require trade orders generated by people.[23] Financial MISs can help companies raise funds through loans and stock offerings. In its initial public offering (IPO), LinkedIn, the professional Internet job and networking site, was valued at more than $3 billion.[24] Facebook was valued at $100 billion by some before its IPO.

Financial MISs are also important for countries. Using computer simulation and optimization, discussed later in this chapter, INDEVAL, Mexico's stock and securities institution, substantially improved its operations by developing new procedures for settling trades, which can total over $200 billion daily.[25] Most financial MISs perform the following functions:

- Integrate financial and operational information from multiple sources, including the Internet, into a single system.
- Provide easy access to data for both financial and nonfinancial users, often through the use of a corporate intranet to access corporate Web pages of financial data and information.
- Make financial data immediately available to shorten analysis turnaround time.
- Enable analysis of financial data along multiple dimensions—time, geography, product, plant, and customer.
- Analyze historical and current financial activity.
- Monitor and control the use of funds over time.

Figure 10.8 shows typical inputs, function-specific subsystems, and outputs of a financial MIS, including profit and loss, auditing, and uses and

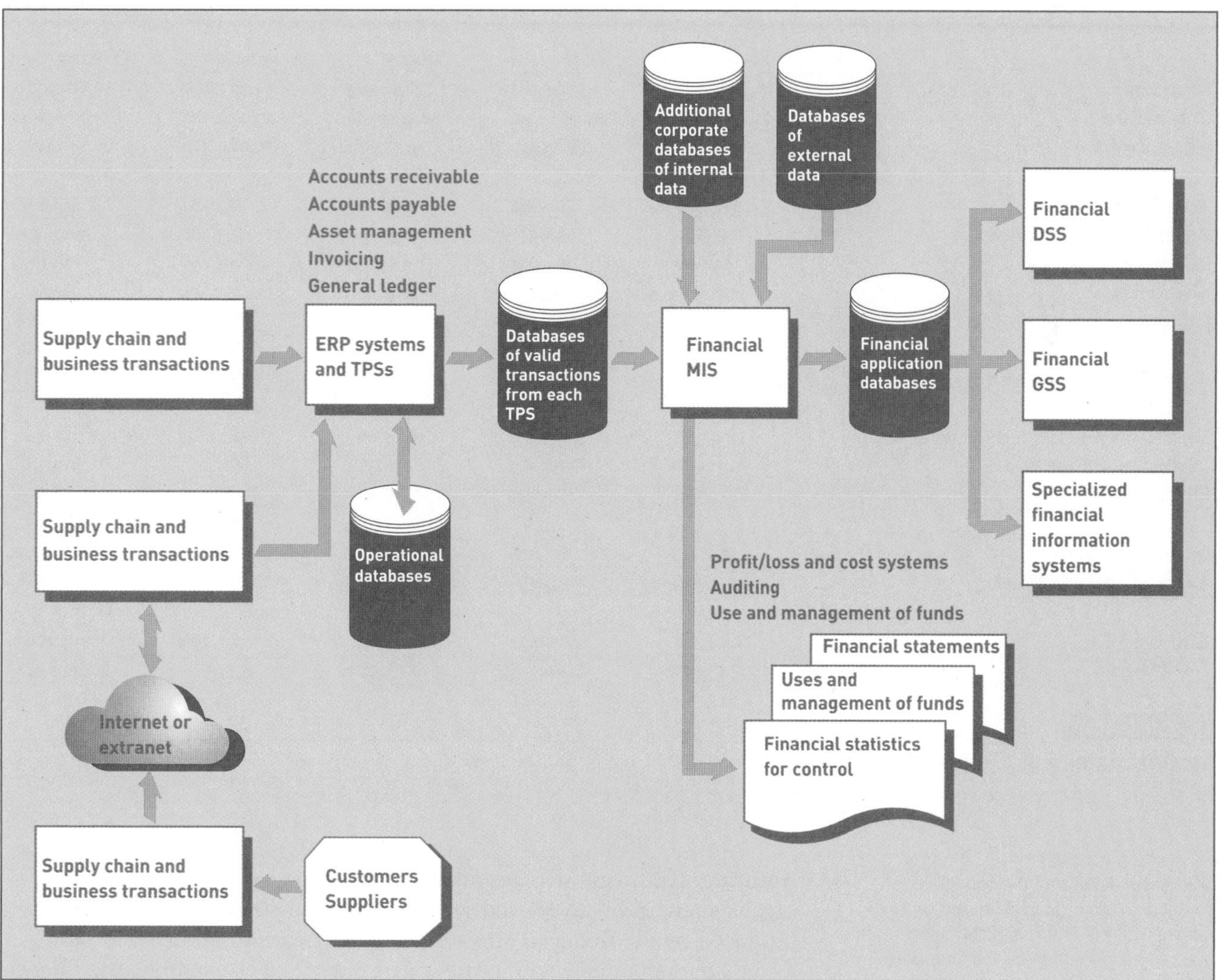

FIGURE 10.8
Overview of a financial MIS

management of funds. Some of the financial MIS subsystems and outputs follow:

profit center: A department within an organization that focuses on generating profits.

revenue center: A division within a company that generates sales or revenues.

cost center: A division within a company that does not directly generate revenue.

- **Profit/loss and cost systems.** Many departments within an organization are **profit centers**, which means that they focus on generating profits. An investment division of a large insurance or credit card company is an example of a profit center (see Figure 10.9). Other departments, such as a marketing or sales department, can be **revenue centers**, which are divisions within the company that focus primarily on generating sales or revenues. Still other departments, such as manufacturing or research and development, can be **cost centers**, which are divisions within a company that do not directly generate revenue. In most cases, information systems are used to compute revenues, costs, and profits.

Projected Five-Year Income Statement

Upland International
Recreational Products Division
Projected Five-Year Income Statement

Prepared: 4/29/2013

Tax Rate	33%
Cost of Goods	75%

	Year 1	Year 2	Year 3	Year 4	Year 5
Revenues	$3,200,000	$3,541,382	$3,919,184	$4,337,290	$4,800,000
Cost of Sales	$2,400,000	$2,656,037	$2,939,388	$3,252,967	$3,600,000
Gross Profit	$800,000	$885,346	$979,796	$1,084,322	$1,200,000
Accounting	$12,600	$14,112	$15,805	$17,702	$19,826
Advertising & Promotion	$37,800	$42,336	$47,416	$53,106	$59,479
Insurance	$3,600	$4,032	$4,516	$5,058	$5,665
Maintenance	$8,640	$9,677	$10,838	$12,139	$13,595
Utilities	$13,680	$15,322	$17,160	$19,219	$21,526
Miscellaneous	$4,300	$4,816	$5,394	$6,041	$6,766
Total General Expenses	$80,620	$90,294	$101,130	$113,265	$126,857
Earnings before Interest, Depr. & Tax	$719,380	$795,051	$878,666	$971,057	$1,073,143
Depreciation Expense	$141,550	$120,459	$102,511	$87,237	$74,238
Operating Profit	$577,830	$674,592	$776,156	$883,821	$998,905
Interest Expense	$65,733	$53,102	$39,490	$24,822	$9,016
Earnings Before Taxes	$512,097	$621,490	$736,665	$858,998	$989,889
Estimated Tax	$168,992	$205,092	$243,099	$283,469	$326,663
Net Income	$343,105	$416,398	$493,566	$575,529	$663,226

FIGURE 10.9
Income statement
An income statement shows a corporation's business results, including all revenues, earnings, expenses, costs, and taxes.

auditing: Analyzing the financial condition of an organization and determining whether financial statements and reports produced by the financial MIS are accurate.

- **Auditing.** **Auditing** involves analyzing the financial condition of an organization and determining whether financial statements and reports produced by the financial MIS are accurate. **Internal auditing** is performed by individuals within the organization.[26] For example, the finance department of a corporation might use a team of employees to perform an audit. **External auditing** is performed by an outside group, usually

internal auditing: Auditing performed by individuals within the organization.

external auditing: Auditing performed by an outside group.

an accounting or consulting firm such as PricewaterhouseCoopers, Deloitte & Touche, or one of the other major international accounting firms. Computer systems are used in all aspects of internal and external auditing. Even with internal and external audits, some companies can mislead investors and others with fraudulent accounting statements.

- **Uses and management of funds.** Internal uses of funds include purchasing additional inventory, updating plants and equipment, hiring new employees, acquiring other companies, buying new computer systems, increasing marketing and advertising, purchasing raw materials or land, investing in new products, and increasing research and development. External uses of funds are typically investment related. The recent economic downturn prompted some financial services companies and venture capitalists to use funds to finance acquisitions of other companies. According to one venture capitalist and cofounder of Netscape, an early Internet company, "We wanted to get these deals done because we had a strong feeling [the market] would heat up fast."[27] A number of powerful personal finance applications also help people manage and use their money. Mint.com, for example, can help people budget their expenditures.[28]

Mint.com provides personal finance software on its Web site.

Manufacturing Management Information Systems

Without question, advances in information systems have revolutionized manufacturing. As a result, many manufacturing operations have been dramatically improved over the last decade. The use of computerized systems is emphasized at all levels of manufacturing—from the shop floor to the executive suite. Dell Computer has used both optimization and heuristic software to help it manufacture a larger variety of products.[29] Dell was able to double its product variety, while saving about $1 million annually in manufacturing costs. Figure 10.10 gives an overview of some of the manufacturing MIS inputs, subsystems, and outputs.

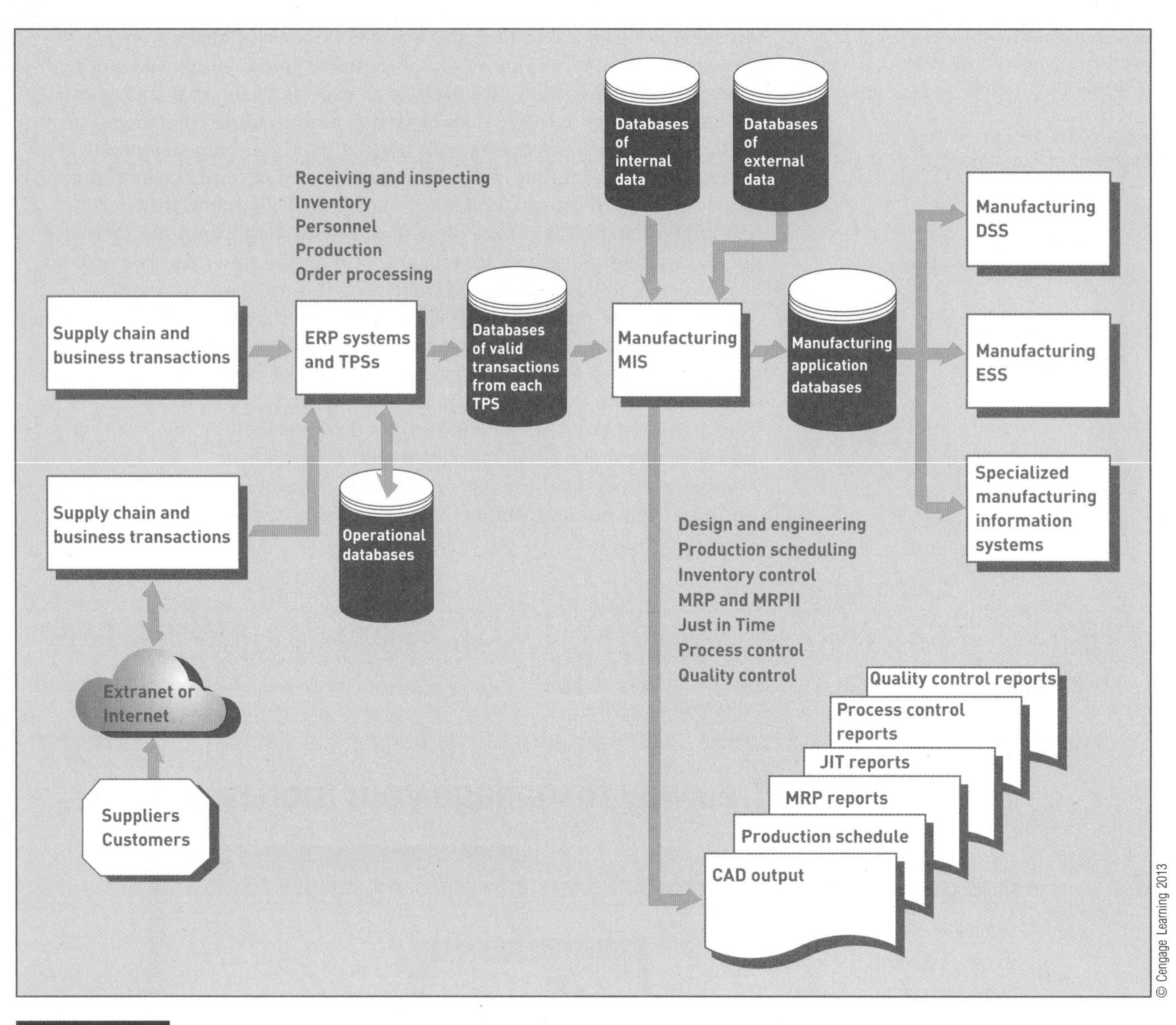

FIGURE **10.10**

Overview of a manufacturing MIS

The manufacturing MIS subsystems and outputs are used to monitor and control the flow of materials, products, and services through the organization. As raw materials are converted to finished goods, the manufacturing MIS monitors the process at almost every stage. The success of an organization can depend on the manufacturing function. Some common information subsystems and outputs used in manufacturing are provided in the following list:

- **Design and engineering.** Manufacturing companies often use computer-assisted design (CAD) with new or existing products. For example, Boeing uses a CAD system to develop a complete digital blueprint of an aircraft before it begins the manufacturing process.[30] As mock-ups are built and tested, the digital blueprint is constantly revised to reflect the most current design. Using such technology helps Boeing reduce manufacturing costs and the time to design a new aircraft.

ANTHONY BOLANTE/Reuters/Landov

Boeing uses computer-assisted design (CAD) in the development and design of its aircraft.

- **Master production scheduling.** Scheduling production and controlling inventory are critical for any manufacturing company. The overall objective of master production scheduling is to provide detailed plans for both short-term and long-range scheduling of manufacturing facilities. Production scheduling can become extremely difficult with natural disasters like the one that struck Japan in 2011.[31] The tsunami and resulting nuclear power plant problems in Japan have caused inventory shortages for Japanese companies and other manufacturing companies around the world that use raw materials and supplies from Japan, making production scheduling very difficult. In another example, Kimberly-Clark was able to slash inventory levels using scheduling software.[32]
- **Inventory control.** Most inventory control techniques are used to minimize inventory costs. Using spreadsheet models, Procter & Gamble reduced inventory levels and costs while streamlining its supply chain.[33] The large consumer products company was able to save over $1 billion using spreadsheet models and quantitative analysis techniques. Inventory control techniques determine when to restock and how much inventory to order. One method of determining the amount of inventory to order is called the **economic order quantity (EOQ)**. This quantity is calculated to minimize the total inventory costs. The "when to order" question is based on inventory usage over time. Typically, the question is answered in terms of a **reorder point (ROP)**, which is a critical inventory quantity level. When the inventory level for a particular item falls to the reorder point, or critical level, the system generates a report so that an order is immediately placed for the EOQ of the product. RODA, a Greek company that manufactures and distributes castors and wheels, had over 10,000 inventory items or stock keeping units (SKUs) to monitor and control.[34] The company used inventory reorder points and economic order quantities to manage its large number of inventory items and minimize inventory costs.
- Another inventory technique used when demand for one item depends on the demand for another is called **material requirements planning (MRP)**. The basic goals of MRP are to determine when finished products, such as automobiles or airplanes, are needed and then to work backward to determine deadlines and resources needed, such as engines and tires, to complete the final product on schedule. **Just-in-time (JIT) inventory** and manufacturing is an approach that maintains inventory at the lowest levels without sacrificing the availability of finished products. With this approach, inventory and materials are delivered just before they are used

economic order quantity (EOQ): The quantity that should be reordered to minimize total inventory costs.

reorder point (ROP): A critical inventory quantity level.

material requirements planning (MRP): A set of inventory-control techniques that help coordinate thousands of inventory items when the demand of one item is dependent on the demand for another.

just-in-time (JIT) inventory: An inventory management approach in which inventory and materials are delivered just before they are used in manufacturing a product.

in a product. JIT, however, can result in some organizations running out of inventory when demand exceeds expectations or there are problems with the manufacturing process.[35] The Japanese earthquake in 2011, for example, caused many manufacturing companies that used JIT in the auto and electronics industries to close.

- **Process control.** Managers can use a number of technologies to control and streamline the manufacturing process. Hawker Beechcraft, a maker of small airplanes and jets for individuals, corporations, and the military, used information systems to help make its manufacturing processes and assembly lines more efficient.[36] The aircraft company used optimization and project management programs to save about $30 million annually and streamline its assembly lines. In addition, computers can directly control manufacturing equipment, using systems called **computer-assisted manufacturing (CAM)**. CAM systems can control drilling machines, assembly lines, and more. **Computer-integrated manufacturing (CIM)** uses computers to link the components of the production process into an effective system. CIM's goal is to tie together all aspects of production, including order processing, product design, manufacturing, inspection and quality control, and shipping. A **flexible manufacturing system (FMS)** is an approach that allows manufacturing facilities to rapidly and efficiently change from making one product to another. In the middle of a production run, for example, the production process can be changed to make a different product or to switch manufacturing materials. By using an FMS, the time and cost to change manufacturing jobs can be substantially reduced, and companies can react quickly to market needs and competition. For example, Chrysler used a FMS to quickly change from manufacturing diesel minivans with right-hand drive to gasoline minivans with left-hand drive.[37]

computer-assisted manufacturing (CAM): A system that directly controls manufacturing equipment.

computer-integrated manufacturing (CIM): Using computers to link the components of the production process into an effective system.

flexible manufacturing system (FMS): An approach that allows manufacturing facilities to rapidly and efficiently change from making one product to making another.

Computer-assisted manufacturing systems control complex processes on the assembly line and provide users with instant access to information.

- **Quality control and testing.** With increased pressure from consumers and a general concern for productivity and high quality, today's manufacturing organizations are placing more emphasis on **quality control**, a process that ensures that the finished product meets the customers' needs. Information systems are used to monitor quality and take corrective steps to eliminate possible quality problems.

quality control: A process that ensures that the finished product meets the customers' needs.

Marketing Management Information Systems

A **marketing MIS** supports managerial activities in product development, distribution, pricing decisions, promotional effectiveness, and sales forecasting. Marketing functions are increasingly being performed on the Internet and

marketing MIS: An information system that supports managerial activities in product development, distribution, pricing decisions, promotional effectiveness, and sales forecasting.

mobile devices. Some large drug companies, for example, are investing in smartphones and tablet computers for their sales representatives to help market drugs to doctors and health care providers even before marketing applications have been fully developed for the devices.[38] According to one health analyst, "Several major pharmaceutical companies are putting the cart before the horse by purchasing iPads in large quantities prior to even owning a single application to run on the iPad." Others, however, believe that this type of investment in tablet computers and other mobile devices will make sales representatives more productive and increase revenues in the long run. Marketing MISs can also be used by attorneys who use Internet auction sites to market and advertise their services.[39]

Some grocery and retail stores are giving customers small mobile devices and then advertising product specials on the devices while customers are shopping.[40] The small devices can also determine a customer's total bill and speed checkout times. Because of the threat of online shopping, many traditional malls are using smartphone applications to help shoppers find deals and move through malls to increase mall traffic and revenues.[41] Figure 10.11 shows the inputs, subsystems, and outputs of a typical marketing MIS.

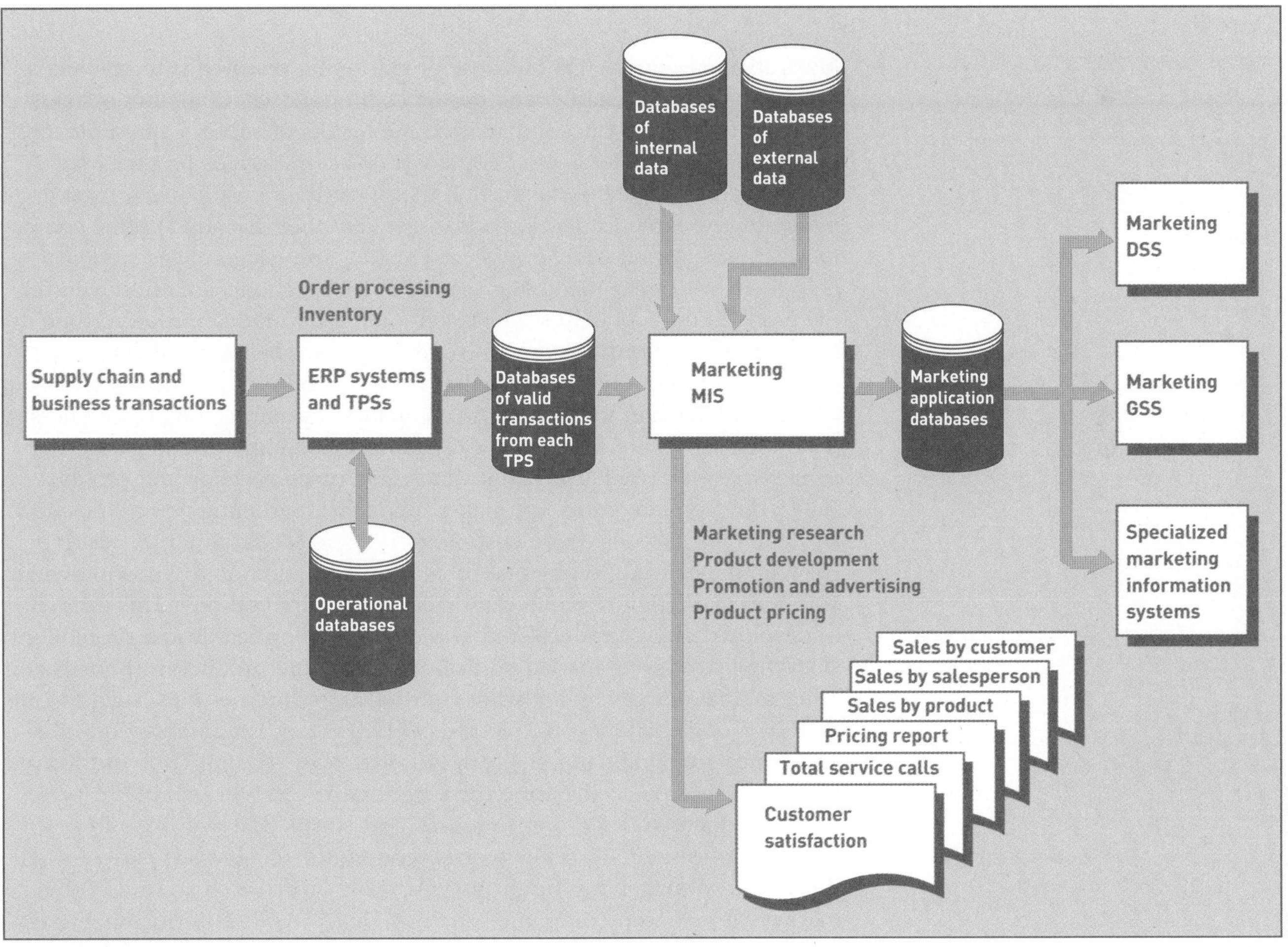

FIGURE 10.11
Overview of a marketing MIS

Subsystems for the marketing MIS and their outputs help marketing managers and executives increase sales, reduce marketing expenses, and develop plans for future products and services to meet the changing needs of customers. These subsystems include marketing research, product development, promotion and advertising, product pricing, and sales analysis:

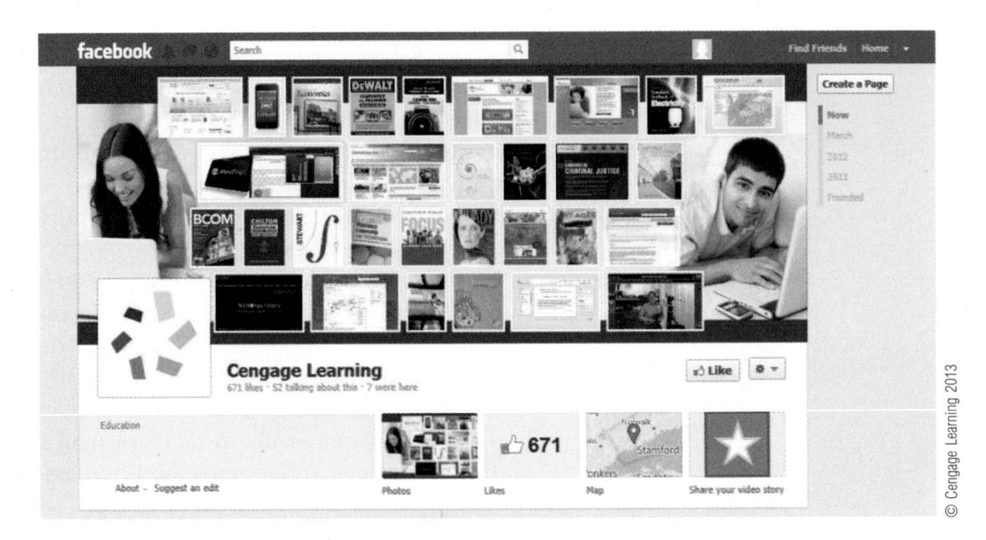

Corporate marketing departments use social networking sites, such as Facebook (*www.facebook.com*), to advertise their products and perform marketing research.

- **Marketing research.** The purpose of marketing research is to conduct a formal study of the market and customer preferences. Computer systems are used to help conduct and analyze the results of surveys, questionnaires, pilot studies, and interviews. People who place messages on social networks such as Facebook might find that what they post is being used in marketing research to develop ads targeted at their friends.[42] Some companies, such as Sprint, Levi Strauss, and Mattel, are using college students to help them with their marketing research.[43] The schools are often paid for the work, and it can help students get a job after graduation. According to an instructor at Northwestern University, "We are helping students to go out and get hired. They've done the work." To help create more excitement and interest in Internet sites, some companies hire marketing researchers to determine what Internet sites are the most appealing.[44] Exciting video games are often used. The result can be increased revenue and profits. BMW, the German luxury car maker, performs marketing research using search engines to determine customer preferences and to target ads to people who might want to buy one of its cars.[45] In addition to knowing what you buy, marketing research can determine where you buy. This can help in developing new products and services and tailoring ads and promotions. With the use of GPS, marketing firms can promote products to phones and other mobile devices by knowing your location. Internet sites, such as Latitude by Google, allow you to locate people, stores, restaurants, and other businesses and landmarks that are close to you.[46] Using GPS and location analysis from cell phone towers, advertisers will be able to promote products and services in stores and shops that are close to people with cell phones.[47] In other words, you could receive ads on your cell phone for a burger restaurant as you walk or drive close to it. Other marketing research companies such as Betawave are performing marketing research on customer engagement and attentiveness in responding to ads.[48] If successful, this marketing research might result in advertisers charging more for higher levels of customer engagement and attentiveness instead of charging for the number of viewers of a particular advertisement.
- **Product development and delivery.** Product development involves the conversion of raw materials into finished goods and services and focuses primarily on the various attributes of the product and its supply chain first

Google Latitude allows you to locate people, stores, restaurants, and other businesses and landmarks that are close to you.

Courtesy of Google

introduced in Chapter 2. Sysco, a Texas-based food distribution company, uses software and databases to prepare and ship over 20 million tons of meat, produce, and other food items to restaurants and other outlets every year.[49] The huge company supplies one out of three cafeterias, sports stadiums, restaurants, and other food outlets. Many factors, including plant capacity, labor skills, engineering issues, and materials, are important in product development decisions. In many cases, a computer program analyzes these various factors and selects the appropriate mix of labor, materials, plant and equipment, and engineering designs. Make-or-buy decisions can also be made with the assistance of computer programs. Product delivery involves determining the best way to get products to customers. With the Internet, customers can determine the best way to have products and services delivered to them.[50] They can specify the speed, cost, and convenience of the delivery system that best meets their needs.

- **Promotion and advertising.** One of the most important functions of any marketing effort is promotion and advertising. Product success is a direct function of the types of advertising and sales promotion done. Increasingly, organizations are using the Internet, smartphones, and other mobile devices to advertise and sell products and services. Many small businesses are effectively advertising their products and services using Internet sites such as Groupon.[51] With Groupon, users receive a daily advertisement for a deal from a local company, which can be a 50 percent discount over normal prices.[52] Some believe Groupon's annual revenues may exceed $700 million.[53] A BMW iPhone application called eVolve shows the advantages and fuel savings to potential owners of its electric cars, which are expected to hit the U.S. markets in a few years.[54] According to a BMW spokesperson, "We need to attempt to understand [driver] behavior and create a culture first, and we need to start early and broad." Companies are also trying to measure the effectiveness of different advertising approaches, such as TV and Internet advertising. Toyota, for example, used IAG Research to help it measure the effectiveness of TV and Internet advertising. Many companies are now promoting their products on games and other applications for Apple's iPhone and other devices. In some cases, the advertising is hidden within free gaming applications. When you download and start playing the game, the advertising pops up on the phone. Advertisers are increasingly using video and animation to create cartoons and videos to advertise products and services on the Internet and TV.[55] However, some people and organizations are not happy with Internet advertising. Some companies and individuals complain that popular Internet search programs often display ads for completely unrelated searches.[56] Some physicians and dentists, for example, claim that ads for their services are displayed to people searching for taxicab companies, barbers, and hair stylists, or other unrelated services.

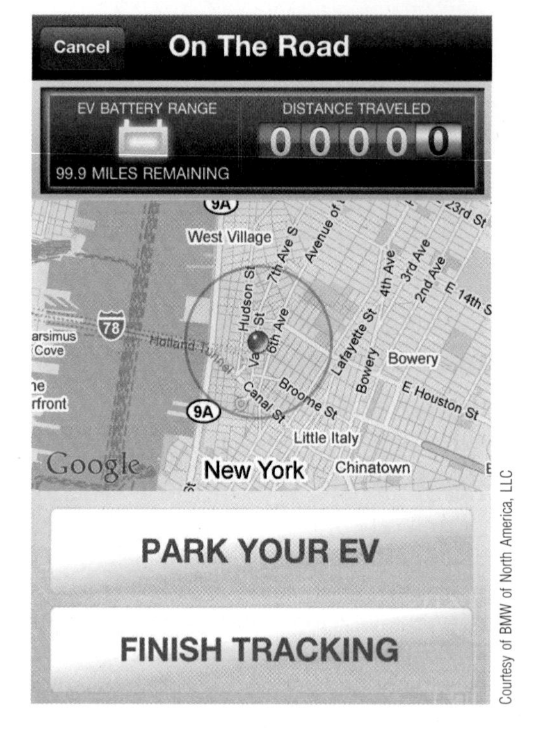

Courtesy of BMW of North America, LLC

The BMW Evolve app is designed for those without an electric vehicle (EV) to gauge their EV compatibility and savings in terms of fuel and CO^2 emissions.

- **Product pricing.** Product pricing is another important and complex marketing function. Retail price, wholesale price, and price discounts must be set. Shopkick, Inc., for example, makes smartphone applications that offer discounts to customers for entering a store.[57] According to the company, the smartphone application has drawn about 750,000 customers into stores. Target, Best Buy, and other stores have used this service to their advantage. Companies that have a unique brand, product, or service can often charge more. Not everyone, however, is happy with this pricing approach.[58] Companies often try to develop pricing policies that will maximize total sales revenues. Computers are often used to analyze the relationship between prices and total revenues. Pepsi, for example, used virtual grocery store software called SimuShop to test different pricing strategies.[59] The software helped Pepsi increase sales by more than 20 percent. According to one corporate vice president, "Pricing should be a CIO's issue by virtue of being an officer in the company, but also because the business may not know there are more advanced ways to look at it."[60] Some companies are using *Internet behavioral pricing,* where the price customers pay online depends on what they are willing to pay based on large databases of personal information that reveal individual shopping behaviors and practices.
- **Sales analysis.** Computerized sales analysis is important to identify products, sales personnel, and customers that contribute to profits and those

that do not. This analysis can be done for sales and ads that help generate sales. Engagement ratings, for example, show how ads convert to sales. IBM used the OnTarget sales analysis tool to identify new sales opportunities with existing customers.[61] The sales analysis tool helps IBM assign sales personnel to sales opportunities to improve both sales and profitability. Sales analysis can help companies determine the best approach for investing in Internet advertising.[62] Some sales analysis consultants specialize in Facebook and similar social network advertising. Today, Facebook represents about one-third of all Internet advertising. Yahoo! and other Internet companies make up the rest of Internet advertising. Several reports can be generated to help marketing managers make good sales decisions (see Figure 10.12). The sales-by-product report lists all major products and their sales for a specified period of time. This report shows which products are doing well and which need improvement or should be discarded altogether. The sales-by-salesperson report lists total sales for each salesperson for each week or month. This report can also be subdivided by product to show which products are being sold by each salesperson. The sales-by-customer report is a tool that can be used to identify high- and low-volume customers.

(a) Sales by Product

Product	August	September	October	November	December	Total
Product 1	34	32	32	21	33	152
Product 2	156	162	177	163	122	780
Product 3	202	145	122	98	66	633
Product 4	345	365	352	341	288	1,691

(b) Sales by Salesperson

Salesperson	August	September	October	November	December	Total
Jones	24	42	42	11	43	162
Kline	166	155	156	122	133	732
Lane	166	155	104	99	106	630
Miller	245	225	305	291	301	1,367

(c) Sales by Customer

Customer	August	September	October	November	December	Total
Ang	234	334	432	411	301	1,712
Braswell	56	62	77	61	21	277
Celec	1,202	1,445	1,322	998	667	5,634
Jung	45	65	55	34	88	287

FIGURE **10.12**

Reports generated to help marketing managers make good decisions

(a) This sales-by-product report lists all major products and their sales for the period from August to December. (b) This sales-by-salesperson report lists total sales for each salesperson for the same time period. (c) This sales-by-customer report lists sales for each customer for the period. Like all MIS reports, totals are provided automatically by the system to show managers at a glance the information they need to make good decisions.

Human Resource Management Information Systems

human resource MIS (HRMIS): An information system that is concerned with activities related to previous, current, and potential employees of an organization, also called a personnel MIS.

A **human resource MIS (HRMIS)**, also called the *personnel MIS*, is concerned with activities related to previous, current, and potential employees of the organization.[63] The HRMIS is being used more and more to oversee and manage part-time, virtual work teams, and job sharing in additional to traditional job titles and duties. Because the personnel function relates to all other functional areas in the business, the HRMIS plays a valuable role in ensuring organizational success. Some of the activities performed by this important MIS include workforce analysis and planning, hiring, training, job and task assignment, and many other personnel-related issues. An effective HRMIS allows a company to keep personnel costs at a minimum while serving the required business processes needed to achieve corporate goals. Although human resource information systems focus on cost reduction, many of today's HR systems concentrate on hiring and managing existing employees to get the total potential of the human talent in the organization. Figure 10.13 shows some of the inputs, subsystems, and outputs of the human resource MIS.

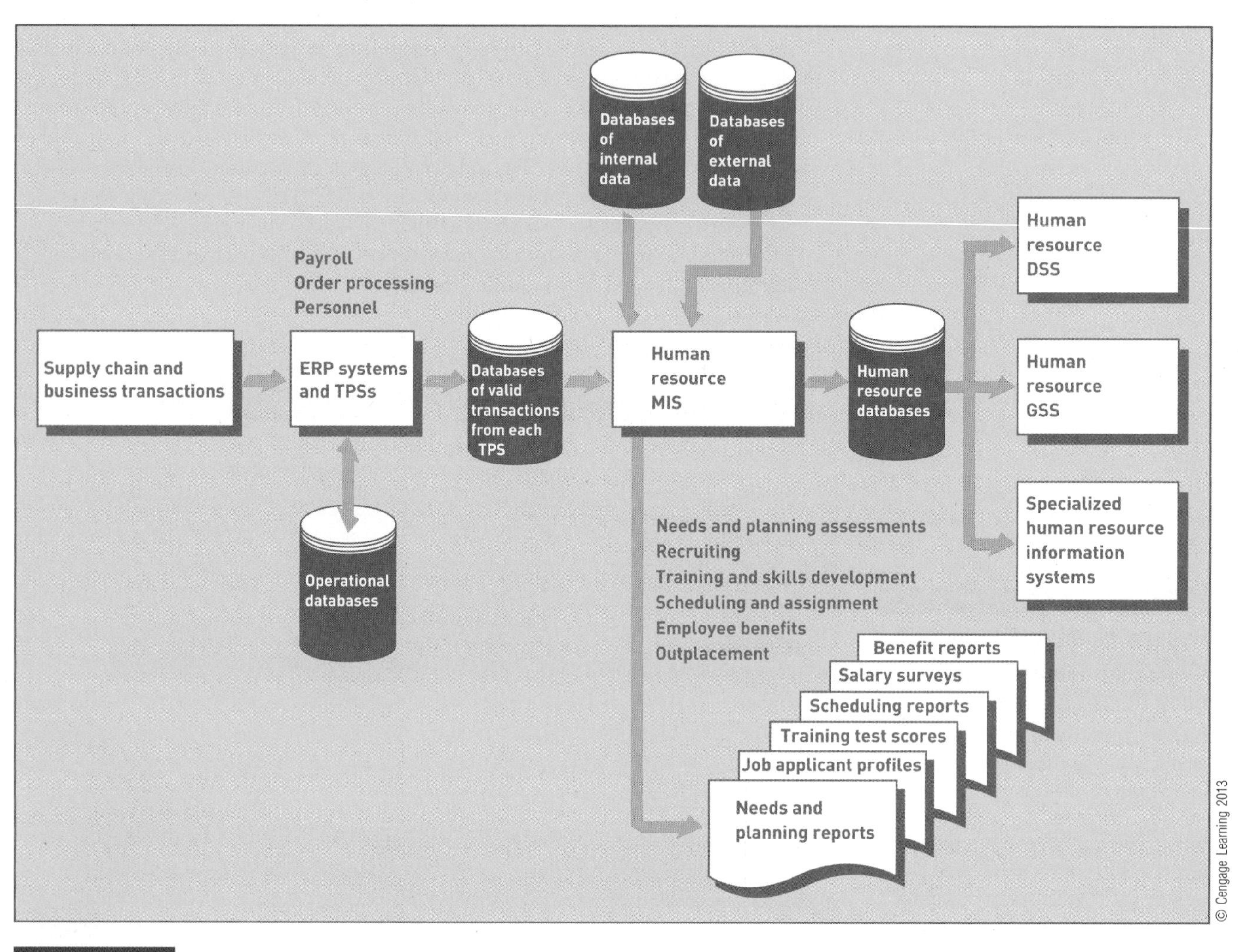

FIGURE **10.13**

Overview of a human resource MIS

Human resource subsystems and outputs range from the determination of human resource needs and hiring through retirement and outplacement. Most medium and large organizations have computer systems to assist with human resource planning, hiring, training and skills inventorying, and wage and salary administration. Outputs of the HRMIS include reports, such as human resource planning reports, job application review profiles, skills inventory reports, and salary surveys. Most human resource departments start with planning, discussed next.

- **Human resource planning.** One of the first aspects of any human resource MIS is determining personnel and human needs. The overall

Human resource MIS subsystems help to determine personnel needs and match employees to jobs.

purpose of this MIS subsystem is to put the right number and type of employees in the right jobs when needed, including internal employees who work exclusively for the organization and outside workers who are hired when they are needed. Determining the best use of existing employees is a key component of human resource planning. The Royal Dutch Touring Club, similar to AAA in the United States, was able to more efficiently use its personnel to service its 4 million automotive members.[64] The use of the Dutch club's information system resulted in outstanding customer satisfaction and retention. Its information system runs an optimization program about twice per minute to allocate personnel to over 1 million service calls from its club members every year. In addition to increasing customer satisfaction and retention, human resource management information systems can reduce personnel costs.

- **Personnel selection and recruiting.** If the human resource plan reveals that additional personnel are required, the next logical step is recruiting and selecting personnel. Companies seeking new employees often use computers to schedule recruiting efforts and trips and to test potential employees' skills. Many companies now use the Internet to screen for job applicants. Applicants use a template to load their résumés onto the Internet site. HR managers can then access these résumés and identify applicants they are interested in interviewing.
- **Training and skills inventory.** Some jobs, such as programming, equipment repair, and tax preparation, require very specific training for new employees. Other jobs may require general training about the organizational culture, orientation, dress standards, and expectations of the organization. When training is complete, employees often take computer-scored tests to evaluate their mastery of skills and new material.
- **Scheduling and job placement.** Employee schedules are developed for each employee, showing job assignments over the next week or month. Job placements are often determined based on skills inventory reports showing which employee might be best suited to a particular job. Sophisticated scheduling programs are often used in the airline industry, the military, and many other areas to get the right people

assigned to the right jobs at the right time. DB Schenker, a large European railway freight company, used its information system to minimize train crew costs while meeting all scheduling needs and work regulations.[65]

- **Wage and salary administration.** Another human resource MIS subsystem involves determining wages, salaries, and benefits, including medical payments, savings plans, and retirement accounts. Wage data, such as industry averages for positions, can be taken from the corporate database and manipulated by the HRMIS to provide wage information and reports to higher levels of management.
- **Outplacement.** Employees leave a company for a variety of reasons. Outplacement services are offered by many companies to help employees make the transition. *Outplacement* can include job counseling and training, job and executive search, retirement and financial planning, and a variety of severance packages and options. Many employees use the Internet to plan their future retirement or to find new jobs, using job sites such as *www.monster.com* and *www.linkedin.com*.

Other Management Information Systems

In addition to finance, manufacturing, marketing, and human resource MISs, some companies have other functional management information systems. For example, most successful companies have well-developed accounting functions and a supporting accounting MIS. Also, many companies use geographic information systems for presenting data in a useful form.

Accounting MISs

accounting MIS: An information system that provides aggregate information on accounts payable, accounts receivable, payroll, and many other applications.

In some cases, accounting works closely with financial management. An **accounting MIS** performs a number of important activities, providing aggregate information on accounts payable, accounts receivable, payroll, and many other applications. The organization's enterprise resource-planning and transaction-processing system captures accounting data, which is also used by most other functional information systems.

Some smaller companies hire outside accounting firms to assist them with their accounting functions. These outside companies produce reports for the firm using raw accounting data. In addition, many excellent integrated accounting programs are available for personal computers in small companies. Depending on the needs of the small organization and its staff's computer experience, using these computerized accounting systems can be a very cost-effective approach to managing information.

Geographic Information Systems

geographic information system (GIS): A computer system capable of assembling, storing, manipulating, and displaying geographic information, that is, data identified according to its location.

Increasingly, managers want to see data presented in graphical form. A **geographic information system (GIS)** is a computer system capable of assembling, storing, manipulating, and displaying geographically referenced information, that is, data identified according to its location. Google, for example, has developed a GIS that combines geothermal information with its mapping applications to help identify carbon emissions. Another Google GIS has been developed to map flu trends around the world by tracking flu inquiries on its Web site. Staples Inc., the large office supply store chain, used a GIS to select new store locations. Finding the best location is critical. It can cost up to $1 million for a failed store because of a poor location. Staples also used a GIS tool from Tactician Corporation along with software from SAS.[66] It is possible to use GIS to analyze customer preferences and shopping patterns in various locations using GPS and location analysis from cell phone towers.[67]

In addition, a number of applications follow the location of people and places using GPS, including Loopt (*www.loopt.com*), Where (*www.where.com*), and Google Latitude. Some people, however, are concerned about privacy concerns with these GPS applications.

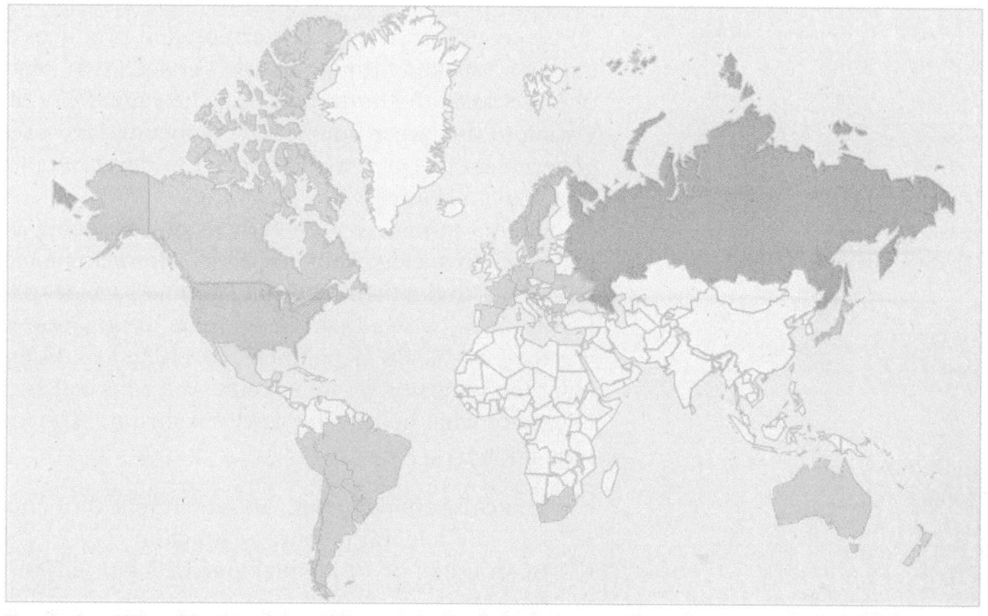

Google Flu Trends uses aggregated Google search data to estimate flu activity.

We saw earlier in this chapter that management information systems (MISs) provide useful summary reports to help solve structured and semistructured business problems. Decision support systems offer the potential to assist in solving both semistructured and unstructured problems. These systems are discussed next.

AN OVERVIEW OF DECISION SUPPORT SYSTEMS

A decision support system (DSS) is an organized collection of people, procedures, software, databases, and devices used to help make decisions that solve problems. The focus of a DSS is on decision-making effectiveness when faced with unstructured or semistructured business problems. Decision support systems offer the potential to generate higher profits, lower costs, and better products and services. TurboRouter, for example, is a decision support system developed in Norway to reduce shipping costs and cut emissions of merchant ships.[68] Operating a single ship can cost over $10,000 every day. TurboRouter efficiently schedules and manages the use of ships to transport oil and other products to locations around the world.

You Want to Put That *Where*?

Land use is often a contentious topic. Everyone wants the benefits that airports, electricity-generating plants, prisons, and all-night railroad freight car classification yards bring to society. Still, few people want to live next door to one of these operations.

Fortunately, geographic information systems (GISs) can help sort out the issues involved with siting these and other operations, including those that are not necessarily as objectionable. The government of Queensland, Australia, used a GIS to figure out the best locations for poultry farms in the southern part of that state.

As background, chicken passed beef and veal in 2007 to become the most popular form of meat for Australians. Queensland has about 20 percent of Australia's population and produces about 20 percent of its chicken. Chicken farming is split between two major centers: near the town of Mareeba in the north and near the capital city of Brisbane in the south. Owing to the larger number of competing land uses near Brisbane, the government of Queensland needed to develop objective ways to allocate this scarce resource.

Poultry farms are not nearly as objectionable as some other operations that modern society finds necessary, but many factors still determine the best places to put them. William Mortimer, senior spatial analyst at Queensland Government, writes that "Geographic information systems and spatial analysis tools enable the departmental decision makers to visualize and understand complex issues on a site specific and regional scale basis." Among these issues are what he calls *primary constraints*. The location of a poultry farm *may* not be:

- Too close (under 1 km, about 0.6 miles) to another poultry farm
- In a key mineral resource extraction area
- In an urban or residential area (a 2 km buffer, about 1.2 miles, is desirable)
- In an area of high ecological significance
- In a low-lying, flood-prone area
- In a koala conservation area
- In a designated water catchment area
- Within the Royal Australian Air Force base at Amberley

As secondary constraints, a poultry farm *should* not be:

- On land that is too steep (over 10 percent slope)
- Next to a watercourse
- On good quality agricultural land
- On land suitable for strategic crops
- In a national park or other protected area
- On an oil or gas pipeline
- On acid sulphate soil

Conversely, it is desirable for a poultry farm to be:

- Near poultry processing plants
- Near paved roads
- Near a reliable supply of clean water
- Near a supply of electricity
- Near poultry feed mills

Using these constraints and ESRI's ArcGIS software, the Queensland government was able to produce maps of the southern part of Queensland, showing areas that were suitable for new poultry farms, calculating automatically the amount of land available in each of them, and showing areas of different sizes in different colors. This mapping provides a good basis for future planning—both for the poultry industry and for those affected by it.

Discussion Questions

1. Assume the area under consideration for possible poultry farms is about 1,000 square miles, or 2,500 square km, and that the resulting maps have to be accurate to 100 feet, or 30 m. Estimate how long it would take to draw maps that reflect all 20 of the factors listed in this box by hand. You will have to make additional assumptions; be sure to state them.
2. How could this approach be used to help choose locations for new solar electricity-generating fields? What would have to be changed? What could stay essentially the same?

Critical Thinking Questions

1. Suppose your university outgrew its present facilities and needed to build a new campus. List primary constraints, secondary constraints, and desirable factors for its choice of location.
2. You have probably used a different type of geographic information system for navigation. Ignore its GPS aspect, which serves only to determine your location, and consider its maps. How do those maps differ from those that a land-use information system would have? How are they similar?

SOURCES: Australia Chicken Meat Federation Web site, *www.chicken.org.au*, accessed April 11, 2012; Department of Local Government and Planning, Queensland Government, "Rural Planning: The Identification and Constraint Mapping of Potential Poultry Farming Industry Locations within Southern Queensland," OZRI 2011 conference, *http://www10.giscafe.com/link/Esri-Australia-Rural-Planning-identification-constraint-mapping-potential-poultry-farming-industry-locations-within-Southern-Queensland./36838/view.html*, October 14, 2011; Queensland Government Web site, *www.qld.gov.au*, accessed April 11, 2012; ESRI ArcGIS software Web site, *www.esri.com/software/arcgis*, accessed April 11, 2012.

Decision support systems, although skewed somewhat toward the top levels of management, can be used at all levels. DSSs are also used in government, law enforcement, and nonprofit organizations (see Figure 10.14).

Characteristics of a Decision Support System

Decision support systems have many characteristics that allow them to be effective management support tools. Of course, not all DSSs work the same. The following list shows some important characteristics of a DSS:

- Provide rapid access to information. In Japan, for example, some farms are using DSSs to help determine when to plant crops and what crops to plant.[69] The Shinpuku Seika farm has placed sensors in the field that immediately transmit data on soil temperature and moisture over the Internet to a centralized site. The real-time data is then placed in an

FIGURE 10.14

Decision support systems are used by government and law enforcement professionals in many settings.

algorithm to determine the best time to plant and the best crop to plant to maximize profits.

- Handle large amounts of data from different sources. Private companies, for example, are collecting a vast amount of customer data. According to the vice president of a large e-book store, "Our Web logs on how customers are using e-readers and e-books have produced 35 TB (terabytes) of data and will load us up with another 20 TB this year."[70]
- Provide report and presentation flexibility.
- Offer both textual and graphical orientation.
- Support drill-down analysis. The FBI is developing a comprehensive DSS to allow its agents to drill down and analyze case histories, geospatial information, and other data to look for relationships to identify potential criminals and criminal acts.[71]
- Perform complex, sophisticated analyses and comparisons using advanced software packages. SAS Institute, for example, developed mathematical and statistical software to help businesses and nonprofit organizations make better decisions to achieve their goals.[72] It has provided decision support systems for casinos, communications companies, educational institutions, financial services companies, insurance companies, manufacturing companies, and many others. High Performance Analytics (HPA) is a mathematical modeling software product by SAS to provide even better decision support. According to the HPA product manager, "This gives statisticians an ad hoc model-building environment where they can use all their data with all their variables."
- Support optimization, satisficing, and heuristic approaches (see Figure 10.15).
- Perform simulation analysis. The DSS has the ability to duplicate the features of a real system, where probability or uncertainty is involved. Sasol, a South African energy and chemical company, used simulation to make its production facilities more efficient.[73] The innovative DSS helped Sasol increase its stock value by more than $200 million.
- Forecast a future opportunity or problem.[74]

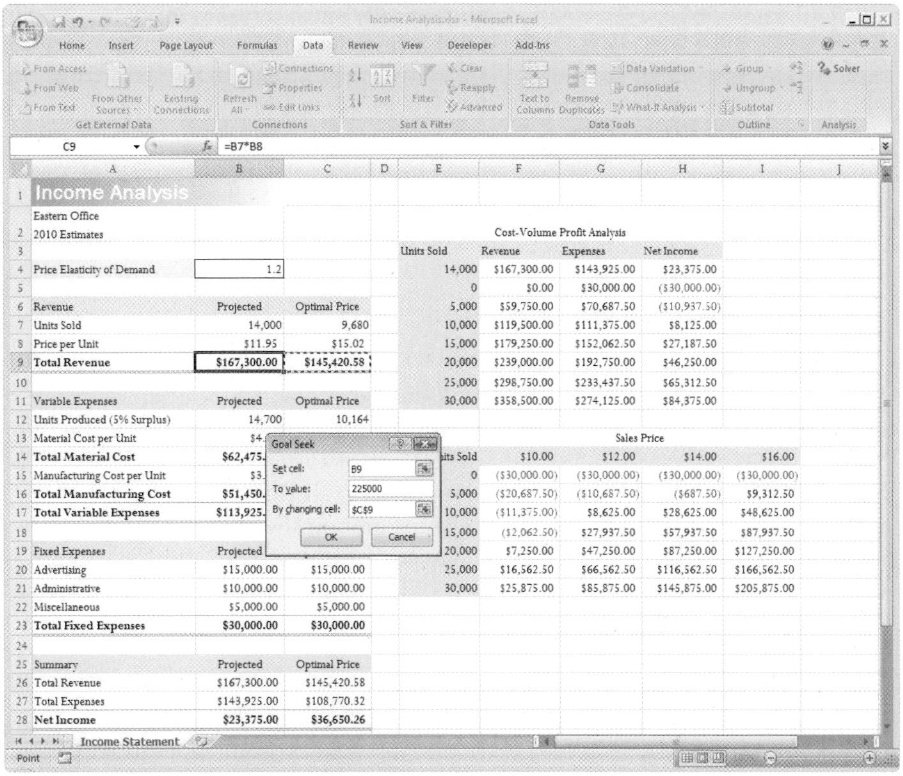

C9 =B7*B8

	A	B	C
1	Income Analysis		
2	Eastern Office 2010 Estimates		
3			
4	Price Elasticity of Demand	1.2	
5			
6	Revenue	Projected	Optimal Price
7	Units Sold	14,000	9,680
8	Price per Unit	$11.95	$15.02
9	**Total Revenue**	**$167,300.00**	**$145,420.58**
10			
11	Variable Expenses	Projected	Optimal Price
12	Units Produced (5% Surplus)	14,700	10,164
13	Material Cost per Unit	$4.	
14	**Total Material Cost**	**$62,475.**	
15	Manufacturing Cost per Unit	$3.	
16	**Total Manufacturing Cost**	**$51,450.**	
17	**Total Variable Expenses**	**$113,925.**	
18			
19	Fixed Expenses	Projected	
20	Advertising	$15,000.00	$15,000.00
21	Administrative	$10,000.00	$10,000.00
22	Miscellaneous	$5,000.00	$5,000.00
23	**Total Fixed Expenses**	**$30,000.00**	**$30,000.00**
24			
25	Summary	Projected	Optimal Price
26	Total Revenue	$167,300.00	$145,420.58
27	Total Expenses	$143,925.00	$108,770.32
28	**Net Income**	**$23,375.00**	**$36,650.26**

Goal Seek
Set cell: B9
To value: 225000
By changing cell: C9
OK Cancel

Cost-Volume Profit Analysis

Units Sold	Revenue	Expenses	Net Income
14,000	$167,300.00	$143,925.00	$23,375.00
0	$0.00	$30,000.00	($30,000.00)
5,000	$59,750.00	$70,687.50	($10,937.50)
10,000	$119,500.00	$111,375.00	$8,125.00
15,000	$179,250.00	$152,062.50	$27,187.50
20,000	$239,000.00	$192,750.00	$46,250.00
25,000	$298,750.00	$233,437.50	$65,312.50
30,000	$358,500.00	$274,125.00	$84,375.00

Sales Price

Units Sold	$10.00	$12.00	$14.00	$16.00
0	($30,000.00)	($30,000.00)	($30,000.00)	($30,000.00)
5,000	($20,687.50)	($10,687.50)	($687.50)	$9,312.50
10,000	($11,375.00)	$8,625.00	$28,625.00	$48,625.00
15,000	($2,062.50)	$27,937.50	$57,937.50	$87,937.50
20,000	$7,250.00	$47,250.00	$87,250.00	$127,250.00
25,000	$16,562.50	$66,562.50	$116,562.50	$166,562.50
30,000	$25,875.00	$85,875.00	$145,875.00	$205,875.00

Income Statement

FIGURE 10.15

With a spreadsheet program, a manager can enter a goal, and the spreadsheet will determine the input needed to achieve the goal.

Capabilities of a Decision Support System

Developers of decision support systems strive to make them more flexible than management information systems and to give them the potential to assist decision makers in a variety of situations. Jeppesen, a supplier of charts and navigational products to hundreds of airlines and thousands of pilots, needed a flexible decision support system to monitor and control its operations.[75] With constantly changing navigational charts and documents, Jeppesen found it difficult to ship accurate products in a timely fashion. As a result of its computerized decision support system, Jeppesen was able to reduce its late shipping percentage from 35 percent to almost 0 percent. In addition to being flexible, DSSs can assist with all or most problem-solving phases, decision frequencies, and varying degrees of problem structure. DSS approaches can also help at all levels of the decision-making process. A single DSS, however, might provide only a few of these capabilities, depending on its uses and scope.

Support for Problem-Solving Phases

The objective of most decision support systems is to assist decision makers with the phases of problem solving. As previously discussed, these phases include intelligence, design, choice, implementation, and monitoring. A specific DSS might support only one or a few phases. By supporting all types of decision-making approaches, a DSS gives the decision maker a great deal of flexibility in getting computer support for decision-making activities.

Support for Various Decision Frequencies

Decisions can range on a continuum from one-of-a-kind to repetitive decisions. One-of-a-kind decisions are typically handled by an ad hoc DSS. An **ad hoc DSS** is concerned with situations or decisions that come up only a few times during the life of the organization; in small businesses, they might

ad hoc DSS: A DSS concerned with situations or decisions that come up only a few times during the life of the organization.

happen only once. For example, a company might need to decide whether to build a new manufacturing facility in another area of the country. Repetitive decisions are addressed by an institutional DSS. An **institutional DSS** handles situations or decisions that occur more than once, usually several times per year or more. An institutional DSS is used repeatedly and refined over the years. Examples of institutional DSSs include systems that support portfolio and investment decisions and production scheduling. These decisions might require decision support numerous times during the year. Between these two extremes are decisions that managers make several times but not routinely.

institutional DSS: A DSS that handles situations or decisions that occur more than once, usually several times per year or more. An institutional DSS is used repeatedly and refined over the years.

Support for Various Problem Structures

As discussed previously, decisions can range from highly structured and programmed to unstructured and nonprogrammed. **Highly structured problems** are straightforward, requiring known facts and relationships. **Semistructured or unstructured problems**, on the other hand, are more complex. The relationships among the pieces of data are not always clear, the data might be in a variety of formats, and the data might be difficult to manipulate or obtain. In addition, the decision maker might not know the information requirements of the decision in advance.

highly structured problems: Problems that are straightforward and require known facts and relationships.

semistructured or unstructured problems: More complex problems in which the relationships among the pieces of data are not always clear, the data might be in a variety of formats, and the data is often difficult to manipulate or obtain.

Support for Various Decision-Making Levels

Decision support systems can provide help for managers at various levels within an organization. Operational managers can get assistance with daily and routine decision making. Tactical decision makers can use analysis tools to ensure proper planning and control. At the strategic level, DSSs can help managers by providing analysis for long-term decisions requiring both internal and external information (see Figure 10.16).

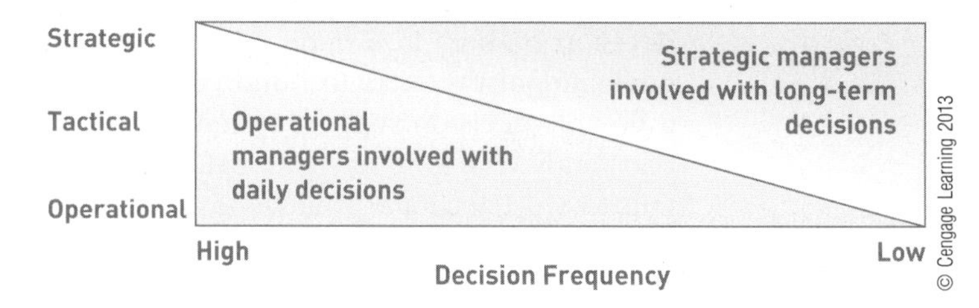

FIGURE **10.16**

Decision-making level

Strategic managers are involved with long-term decisions, which are often made infrequently. Operational managers are involved with decisions that are made more frequently.

A Comparison of DSS and MIS

A DSS differs from an MIS in numerous ways, including the type of problems solved, the support given to users, the decision emphasis and approach, and the type, speed, output, and development of the system used. Table 10.1 lists brief descriptions of these differences.

COMPONENTS OF A DECISION SUPPORT SYSTEM

At the core of a DSS are a database and a model base. In addition, a typical DSS contains a user interface, also called a **dialogue manager**, which allows decision makers to easily access and manipulate the DSS and to use common business terms and phrases. Finally, access to the Internet, networks, and other computer-based systems permits the DSS to tie into other powerful systems, including the TPS or function-specific subsystems. Figure 10.17 shows a conceptual model of a DSS, although specific DSSs might not have all the components shown in this figure.

dialogue manager: A user interface that allows decision makers to easily access and manipulate the DSS and to use common business terms and phrases.

TABLE 10.1 Comparison of DSSs and MISs

Factor	DSS	MIS
Problem type	Can handle unstructured problems that cannot be easily programmed.	Normally used only with structured problems.
Users	Supports individuals, small groups, and the entire organization. In the short run, users typically have more control over a DSS.	Supports primarily the organization. In the short run, users have less control over an MIS.
Support	Supports all aspects and phases of decision making; it does not replace the decision maker—people still make the decisions.	In some cases, makes automatic decisions and replaces the decision maker.
Emphasis	Emphasizes actual decisions and decision-making styles.	Usually emphasizes information only.
Approach	Serves as a direct support system that provides interactive reports on computer screens.	Typically serves as an indirect support system that uses regularly produced reports.
System	Uses computer equipment that is usually online (directly connected to the computer system) and related to real time (providing immediate results). Computer terminals and display screens are examples—these devices can provide immediate information and answers to questions.	Uses printed reports that might be delivered to managers once per week, so it cannot provide immediate results.
Speed	Is flexible and can be implemented by users, so it usually takes less time to develop and is better able to respond to user requests.	Provides response time usually longer than a DSS.
Output	Produces reports that are usually screen oriented, with the ability to generate reports on a printer.	Is oriented toward printed reports and documents.
Development	Has users who are usually more directly involved in its development. User involvement usually means better systems that provide superior support. For all systems, user involvement is the most important factor for the development of a successful system.	Is frequently several years old and often was developed for people who are no longer performing the work supported by the MIS.

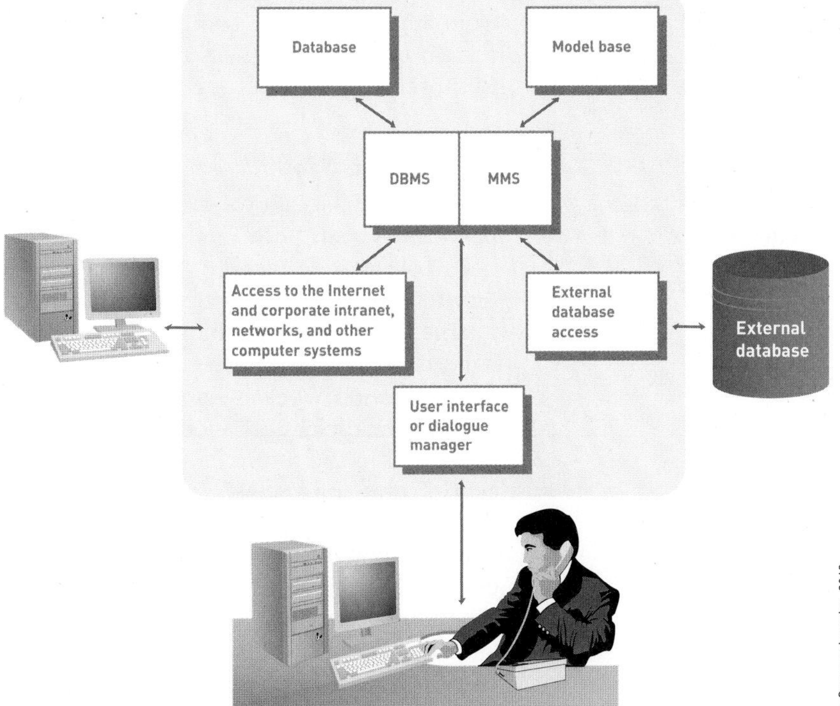

FIGURE **10.17**

Conceptual model of a DSS

DSS components include a model base; database; external database access; access to the Internet and corporate intranet, networks, and other computer systems; and a user interface or dialogue manager.

The Database

The database management system allows managers and decision makers to perform *qualitative analysis* on the company's vast stores of data in databases, data warehouses, and data marts, which are discussed in Chapter 5. A *data-driven DSS* primarily performs qualitative analysis based on the company's databases. One health care company, for example, offered a $3 million prize for the individual or group that could sift through a vast amount of data, analyze it, and most accurately predict when people go to hospitals for medical procedures and care.[76] Data-driven DSSs tap into vast stores of information contained in the corporate database, retrieving information on inventory, sales, personnel, production, finance, accounting, and other areas. Exeros, which was acquired by IBM, provides technology and software to help companies sift through a mountain of data and return valuable information and decision support. In another example, Tween Brands specialty retail store used the Oracle database to provide decision support to reduce inventory costs. Data mining and business intelligence, introduced in Chapter 5, are often used in a data-driven DSS.[77]

Data-driven DSSs can also be used by governmental agencies and nonprofit organizations. The Internal Revenue Service (IRS), for example, has asked some small businesses being audited to turn over their corporate data stored in accounting databases in hopes of generating more revenues from businesses that may not be paying what they owe.[78] Some researchers are searching Medicare's vast claims databases to uncover potential fraud.[79] In one case, the researchers uncovered doctors who performed six or more surgeries on the same patient, revealing potential fraud and Medicare abuses.

A database management system can also connect to external databases to give managers and decision makers even more information and decision support. External databases can include the Internet, libraries, and government databases, among others. Access to a combination of internal and external database can give key decision makers a better understanding of the company and its environment. Some drug companies, for example, search databases of individual TV-watching habits and databases containing drug purchases from pharmaceutical companies.[80] If they find that people who purchase certain types of drugs also watch certain TV programs, ads can be targeted to those programs.

The Model Base

model base: Part of a DSS that allows managers and decision makers to perform quantitative analysis on both internal and external data.

The **model base** allows managers and decision makers to perform *quantitative analysis* on both internal and external data. Once large databases have been collected and stored, companies use models (analytics) to turn the data into future products, services, and profits. According to the vice president of a large bookstore, "We have to decide how we capture the customer's imagination and how we move forward."[81] The senior vice president and director of Information Week's Global CIO unit believes seeing and shaping the future through mathematical modeling and analytics is the top priority for CIOs today and in the future.[82] A *model-driven DSS* primarily performs mathematical or quantitative analysis. The model base gives decision makers access to a variety of models so that they can explore different scenarios and see their effects. Ultimately, it assists them in the decision-making process. Organic, Chrysler's digital marketing agency, used a model-driven DSS to determine the best place to invest marketing funds.[83] Organic used a team of economists and statisticians to develop models that predicted the effectiveness of various advertising alternatives. According to Chrysler's director of media and events,

"As a marketer, it helps me be smarter about the dollars I need to reach the sales goals we are responsible for. It gives you some science." DHL, the largest express delivery business worldwide with over 500,000 employees in more than 200 countries, used mathematical models to improve and streamline its marketing and branding efforts. The improved advertising program helped the company increase its value by about $1.3 billion over a five-year period, representing an internal rate of return of more than 20 percent on investment.[84] To help stock traders effectively use model-driven DSSs, a number of programs and Internet sites let people try quantitative trading strategies before they make real investments.[85]

Not all model-based decision support systems, however, are accurate or provide advantages over more traditional decision-making approaches.[86] One investment firm was charged over $200 million by the Securities and Exchange Commission (SEC) for faulty and perhaps fraudulent quantitative investment models. According to one SEC official, "This is a wake-up call to quant managers who might otherwise rely on a secretive culture and complex computer models to keep material from investors."

model management software (MMS): Software that coordinates the use of models in a DSS, including financial, statistical analysis, graphical, and project management models.

Model management software (MMS) can coordinate the use of models in a DSS, including financial, statistical analysis, graphical, and project-management models. Depending on the needs of the decision maker, one or more of these models can be used (see Table 10.2). What is important is how the mathematical models are used, not the number of models that an organization has available. In fact, too many model-based tools can be a disadvantage. MMS can often help managers effectively use multiple models in a DSS.

TABLE 10.2 Model management software

DSSs often use financial, statistical, graphical, and project-management models.

Model Type	Description	Software
Financial	Provides cash flow, internal rate of return, and other investment analysis	Spreadsheet, such as Microsoft Excel
Statistical	Provides summary statistics, trend projections, hypothesis testing, and more	Statistical programs, such as SPSS or SAS
Graphical	Assists decision makers in designing, developing, and using graphic displays of data and information	Graphics programs, such as Microsoft PowerPoint
Project Management	Handles and coordinates large projects; also used to identify critical activities and tasks that could delay or jeopardize an entire project if they are not completed in a timely and cost-effective fashion	Project management software, such as Microsoft Project

The User Interface, or Dialogue Manager

The user interface, or dialogue manager, allows users to interact with the DSS to obtain information. It assists with all aspects of communications between the user and the hardware and software that constitute the DSS. In a practical sense, to most DSS users, the user interface is the DSS. Upper-level decision makers are often less interested in where the information came from or how it was gathered than that the information is both understandable and accessible.

INFORMATION SYSTEMS @ WORK

VÚB Decision Support Technology

VÚB (Všeobecná úverová banka) has operated as a retail and commercial bank in the Slovak Republic since 1990. A member of the international banking group Intesa Sanpaolo, VÚB has 207 branches and 11 mortgage centers. According to *Global Finance* magazine, it is "the safest bank in Central and Eastern Europe."

"Information and data are the treasure of our company," says Juraj Fehér, business intelligence architect at VÚB. "Almost every department in the bank has a system for data analysis and reporting." But Fehér found that the data underlying those systems was not always of high quality or structured to meet the bank's reporting needs. Therefore, VÚB began to look for a new database and reporting solution.

VÚB selected Microsoft SQL Server 2012 as the solution in early 2011. The company began implementing a prerelease version in June of that year. By February 2012, the system was supporting about 100 users. "By the end of the year we will have about 2,000 users," Fehér says.

A major strength of SQL Server 2012, in VÚB's evaluation, was its Reporting Services. These allow users to "create interactive, tabular, graphical, or free-form reports from relational, multidimensional, or XML-based data sources" with a minimum of training. The Report Builder can also render reports in the form of Excel files for further analysis using familiar tools or as Word files for inclusion in text-based memos and reports.

In addition to user-generated ad hoc reports, VÚB has developed 15 standard reports for corporate and retail branches. By the end of 2012, it hopes to expand that number to more than 100, replacing older reports. Some of these reports monitor risk management portfolios, track client behavior, and examine key performance indicators of branch efficiency and quality.

As input to its reports, VÚB uses SQL Server 2012 Analysis Services to create online analytical processing (OLAP) "data cubes." In early 2012, the bank had five cubes of about 18 gigabytes (GB) of data each. By December 2013, Fehér expects to have about 40 cubes, some of them with 60 GB of data.

With the insights it obtains from its database, VÚB can make better decisions. "We can better identify, extract, and analyze business data," Fehér says. "With faster and more effective information delivery … we can take better advantage of the treasure of our data, improving the quality of our business decisions."

Microsoft SQL Server is not VÚB's only decision support software. It also uses SAS for Enterprise Risk Management. For example, the bank uses it for risk managers to generate liquidity forecasts. "Thanks to our new look into the future, we can better decide whether to gather financial resources for five or 10 years," explains Andrej Hronec, VÚB head of Assets and Liabilities. In other words, the bank can make better decisions about what level of cash (liquidity) it should have.

The SAS software also allows VÚB to forecast net interest income. "Thanks to this [capability] we know our level of assets and liabilities, whether there will be a profit or a loss," says Hronec. He also notes that VÚB used this software to decide how to structure its bond portfolio. "If we did not have this view, we would have not bought them in such a structure as what we finally decided."

At the end of the day, information systems make VÚB both more profitable for its stockholders and safer for its depositors than it would be otherwise.

Discussion Questions

1. What are three data items that your school's top administration would like to forecast? Would a database such as VÚB's and similar report-building capabilities help it forecast them? Why or why not?
2. If any user can create the reports that he or she needs using SQL Server 2012 Reporting Services, why would VÚB bother creating standard reports for all its branches?

Critical Thinking Questions

1. Microsoft SQL Server 2012 meets VÚB's needs. Other software firms offer similar reporting products. Some might meet other users' needs better than SQL Server 2012. List at least five criteria you would use as a manager in a company making a software selection to choose one of them.
2. Suppose you were applying for a job at VÚB. How would you use the information in this case to improve your chances? What parts of your education would you point out in an interview as being especially helpful to them?

SOURCES: Global Finance, "World's Safest Banks in Central & Eastern Europe 2011," *www.gfmag.com/tools/best-banks/11337-worlds-safest-banks-in-central-a-eastern-europe-2011.html*, August 18, 2011; "Bank Helps Employees Generate Self-Service BI for Better Reporting and Decision Making," Microsoft Corporation, *www.microsoft.com/casestudies/Microsoft-SQL-Server-2012-Enterprise/VUB-Bank/Bank-Helps-Employees-Generate-Self-Service-BI-for-Better-Reporting-and-Decision-Making/690000000081*, March 6, 2012; "SQL Server 2012 Reporting Services (SSRS)," Microsoft Corporation, *msdn.microsoft.com/en-us/library/ms159106.aspx*, accessed April 18, 2012; "Risk Management: Driving Business Evolution with Enterprise Risk Management," SAS, *www.sas.com/solutions/riskmgmt*, accessed April 18, 2012; VÚB Web site (English), *www.vub.sk/en*, accessed April 18, 2012 (initially redirects to personal finance page).

GROUP SUPPORT SYSTEMS

group support system (GSS): Software application that consists of most of the elements in a DSS, plus software to provide effective support in group decision-making settings; also called *group support system* or *computerized collaborative work system.*

The DSS approach has resulted in better decision making for all levels of individual users. However, many DSS approaches and techniques are not suitable for a group decision-making environment. A **group support system (GSS)**, also called a *group decision support system* and a *computerized collaborative work system*, consists of most of the elements in a DSS, plus software to provide effective support in group decision-making settings (see Figure 10.18).

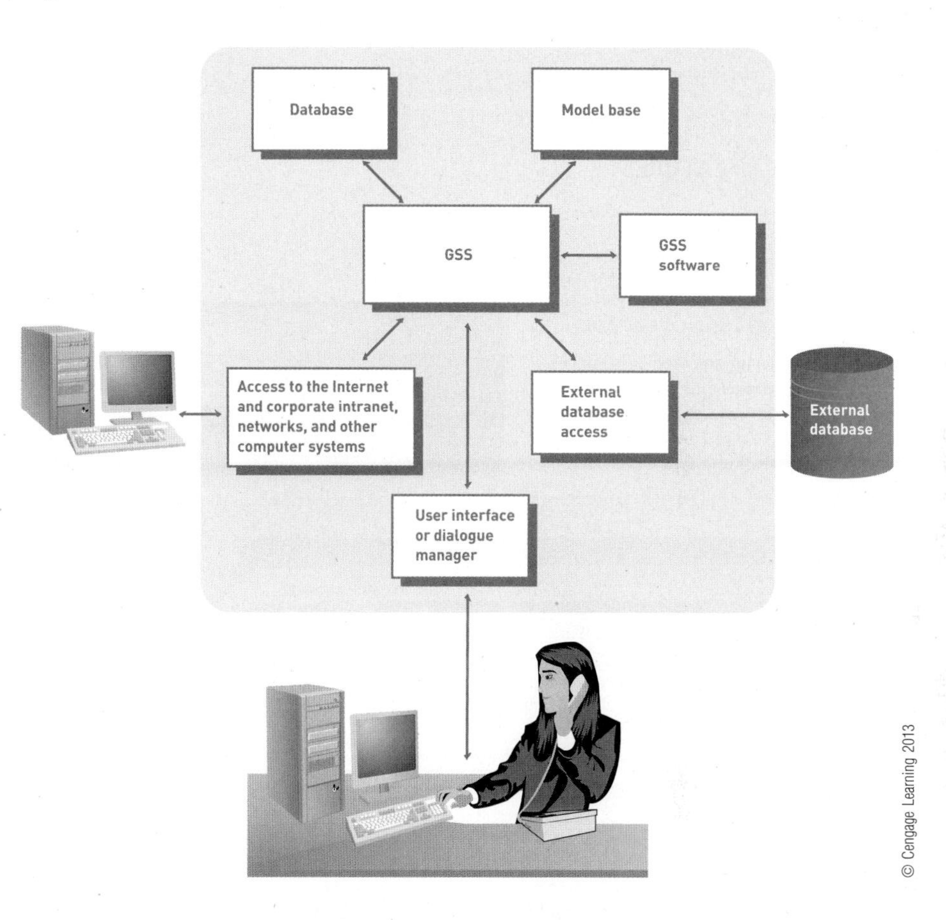

FIGURE **10.18**

Configuration of a GSS

A GSS contains most of the elements found in a DSS, plus software to facilitate group member communications.

Group support systems are used in most industries, governments, and the military.[87] They are also used between companies when the firms are involved in the same supply chain, as first discussed in Chapter 2.[88] For example, an organization might get raw materials and supplies from a separate company and use yet another company to distribute finished products to consumers. These separate companies involved in the same supply chain often use group support systems to coordinate joint forecasting, planning, and other activities critical to delivering finished products and services to customers. Engineers at MWH, a Colorado firm, used social-mapping software to show which employees effectively work together and collaborate on projects.[89] The software keeps track of employees who work together and then draws maps showing the relationships. The social-mapping software helped MWH identify key employees who collaborate on projects and foster innovation and creative thinking. Social networking Internet sites, such as Facebook, can be used to support group decision making.[90] Serena, a software company in California, used Facebook to collaborate on projects and exchange documents.[91] The company believes this type of collaboration is so important that it has instituted "Facebook Fridays" to encourage its employees to use the social networking site to collaborate and make group decisions. Facebook

Fridays also helped the company work with clients and recruit new employees. Many organizations have developed their own social networking sites to help their employees collaborate on important projects. With Ning, for example, companies and individuals can create their own social networking sites.[92] Some executives, however, believe that social networking Internet sites are a waste of time and corporate resources.[93]

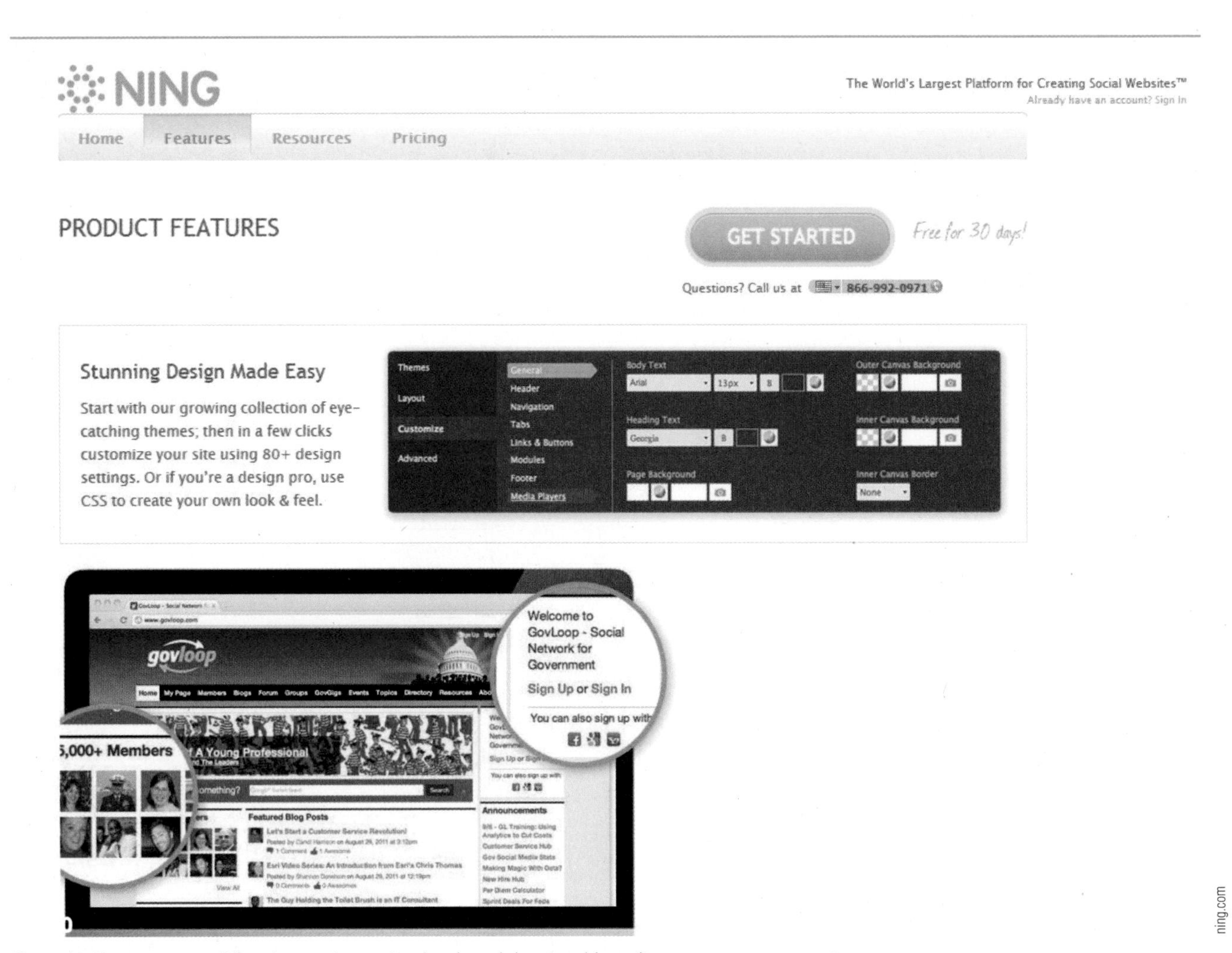

Organizations can use Ning to create customized social networking sites.

ning.com

Characteristics of a GSS That Enhance Decision Making

When it comes to decision making, a GSS's unique characteristics have the potential to result in better decisions. Developers of these systems try to build on the advantages of individual support systems while adding new approaches unique to group decision making. For example, some GSSs can allow the exchange of information and expertise among people without direct face-to-face interaction, although some face-to-face meeting time is usually beneficial. The following sections describe some characteristics that can improve and enhance decision making.

Special Design

The GSS approach acknowledges that special procedures, devices, and approaches are needed in group decision-making settings. These procedures must foster creative thinking, effective communications, and good group decision-making techniques.

Ease of Use

Like an individual DSS, a GSS must be easy to learn and use. Systems that are complex and hard to operate will seldom be used. Many groups have less tolerance than do individual decision makers for poorly developed systems.

Flexibility

Two or more decision makers working on the same problem might have different decision-making styles and preferences. Each manager makes decisions in a unique way, in part because of different experiences and cognitive styles. An effective GSS not only has to support the different approaches that managers use to make decisions but also must find a means to integrate their different perspectives into a common view of the task at hand. GSS flexibility is also important with customers and outside companies. Waste Management, for example, collaborates with customers to reduce waste and costs.[94] The company's collaborative efforts with customers can also help the environment. According to the CEO of Waste Management, "We need to be able to feed our waste materials collection and processing data into customers' systems so they can slice and dice it in order to understand how they're meeting their sustainability goals."

delphi approach: A decision-making approach in which group decision makers are geographically dispersed throughout the country or the world; this approach encourages diversity among group members and fosters creativity and original thinking in decision making.

brainstorming: A decision-making approach that consists of members offering ideas "off the top of their heads," fostering creativity and free thinking.

group consensus approach: A decision-making approach that forces members in the group to reach a unanimous decision.

Decision-Making Support

A GSS can support different decision-making approaches, including the **delphi approach** in which group decision makers are geographically dispersed throughout the country or the world. This approach encourages diversity among group members and fosters creativity and original thinking in decision making. In another approach, called **brainstorming**, members offer ideas "off the top of their heads," fostering creativity and free thinking. TD Bank, for example, used IBM's Connections software to allow over 85,000 workers to communicate and brainstorm.[95] According to the vice president and CIO of the company, "We needed a way for people to communicate and help each other." The **group consensus approach** forces members in the group to reach a unanimous decision. The Shuttle Project Engineering Office at the Kennedy Space Center has used the Consensus-Ranking Organizational-Support System

NASA engineers use the Consensus-Ranking Organizational-Support System (CROSS) to evaluate space projects in a group setting.

Courtesy of NASA

(CROSS) to evaluate space projects in a group setting. The group consensus approach analyzes the benefits of various projects and their probabilities of success. CROSS is used to evaluate and prioritize advanced space projects. With the **nominal group technique**, each decision maker can participate; this technique encourages feedback from individual group members, with the final decision being made by voting, similar to a system for electing public officials.

nominal group technique: A decision-making approach that encourages feedback from individual group members, with the final decision being made by voting, similar to a system for electing public officials.

Anonymous Input

Many GSSs allow anonymous input, where the person giving the input is not known to other group members. For example, some organizations use a GSS to help rank the performance of managers. Anonymous input allows the group decision makers to concentrate on the merits of the input without considering who gave it. In other words, input given by a top-level manager is given the same consideration as input from employees or other members of the group. Some studies have shown that groups using anonymous input can make better decisions and have superior results compared with groups that do not use anonymous input.

Reduction of Negative Group Behavior

One key characteristic of any GSS is the ability to suppress or eliminate group behavior that is counterproductive or harmful to effective decision making. In some group settings, dominant individuals can take over the discussion, thereby preventing other members of the group from participating. In other cases, one or two group members can sidetrack or subvert the group into areas that are nonproductive and do not help solve the problem at hand. At other times, members of a group might assume they have made the right decision without examining alternatives—a phenomenon called *groupthink*. If group sessions are poorly planned and executed, the result can be a tremendous waste of time. Today, many GSS designers are developing software and hardware systems to reduce these types of problems. Procedures for effectively planning and managing group meetings can be incorporated into the GSS approach. A trained meeting facilitator is often employed to help lead the group decision-making process and to avoid groupthink (see Figure 10.19).

FIGURE **10.19**

Using the GSS approach

A trained meeting facilitator can help lead the group decision-making process and avoid groupthink.

Parallel and Unified Communication

With traditional group meetings, people must take turns addressing various issues. One person normally talks at a time. With a GSS, every group member can address issues or make comments at the same time by entering them into a PC or workstation. These comments and issues are displayed on every group member's PC or workstation immediately. *Parallel communication* can speed meeting times and result in better decisions. Organizations are using unified communications to support group decision making. *Unified communications* ties together and integrates various communication systems, including traditional phones, cell phones, e-mail, text messages, the Internet, and more. With unified communications, members of a group decision-making team use a wide range of communications methods to help them collaborate and make better decisions.

Automated Record Keeping

Most GSSs can automatically keep detailed records of a meeting. Each comment that is entered into a group member's PC or workstation can be anonymously recorded. In some cases, literally hundreds of comments can be stored for future review and analysis. In addition, most GSS packages have automatic voting and ranking features. After group members vote, the GSS records each vote and makes the appropriate rankings.

GSS Hardware and Software Tools

GSS hardware includes computers, laptops, smartphones, and other devices to enhance collaboration and group decision making. Today, executives and corporate managers are collaborating with today's smartphones and tablet computers to a greater extent.[96] According to the CEO of Citrix, a software company that offers collaboration tools for organizations, "Our growth plan includes even richer collaboration experiences, partner app integrations, and expanded mobile device support, including GoToMyPC for the iPad." Advanced video devices are also being used to support group decision making. According to Diane Bryant, CIO of Intel, "New collaboration technologies such as video are almost a mandate, since Intel has employees in hundreds of factories and other operations around the world."[97]

GSS software, often called *groupware* or *workgroup software*, helps with joint work-group scheduling, communication, and management. One popular package, IBM's Lotus Notes, can capture, store, manipulate, and distribute memos and communications that are developed during group projects.[98] Some companies standardize on messaging and collaboration software, such as Lotus Notes. Lotus Connections is a feature of Lotus Notes that allows people to post documents and information on the Internet.[99] The feature is similar to popular social networking sites such as Facebook and LinkedIn, but it is designed for business use. Microsoft has invested billions of dollars in GSS software to incorporate collaborative features into its Office suite and related products. Office Communicator, for example, is a Microsoft product developed to allow better and faster collaboration.[100] Other GSS software packages include Collabnet, OpenMind, and TeamWare. All of these tools can aid in group decision making. *Shared electronic calendars* can be used to coordinate meetings and schedules for decision-making teams. Using electronic calendars, team leaders can block out time for all members of the decision-making team.

A number of additional collaborative tools are available on the Internet.[101] SharePoint (*www.microsoft.com*), WebOffice (*www.weboffice.com*), and BaseCamp (*www.basecamphq.com*) are just a few examples. Fuze Meeting (*www.fuzemeeting.com*) provides video collaboration tools on the Internet. The service can automatically bring participants into a live chat, allow workers to share information on their computer screens, and broadcast video content in high definition. Twitter (*www.twitter.com*) and Jaiku (*www.jaiku.com*) are Internet sites that some organizations use to help people and groups stay connected and coordinate work schedules. Yammer (*www.yammer.com*) is an Internet site that helps companies provide short answers to frequently asked questions. Managers and employees must first log into their private company network on Yammer to get their questions answered. Teamspace (*www.teamspace.com*) is yet another collaborative software package that assists teams to successfully complete projects. Many of these Internet packages embrace the use of Web 2.0 technologies. Some executives, however, worry about security and corporate compliance issues with any new technology.

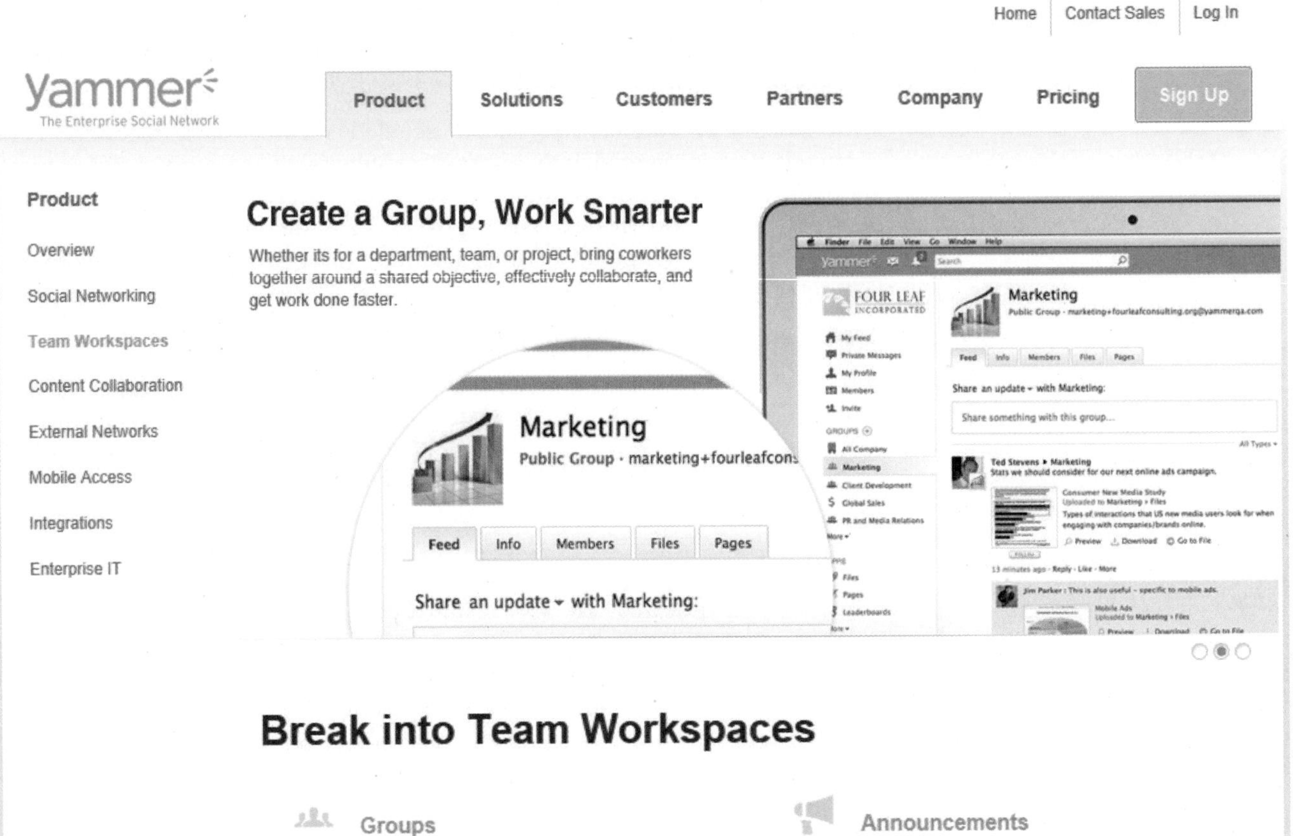

Yammer helps organizations provide short answers to frequently asked questions.

In addition to stand-alone products, GSS software is increasingly being incorporated into existing software packages. Microsoft is including a collaboration and communications tool called Lync into many of its applications.[102] Microsoft also developed the iLync application for Apple's iPhone. Today, some transaction-processing and enterprise resource-planning packages include collaboration software. In addition to groupware, GSSs use a number of tools discussed previously, including the following:

- E-mail, instant messaging (IM), and text messaging (TM)
- Video conferencing
- Group scheduling
- Project management
- Document sharing

GSS Alternatives

Group support systems can take on a number of network configurations, depending on the needs of the group, the decision to be supported, and the geographic location of group members. According to one study, GSS is about both the pace and quality of decision making.[103] GSS alternatives include a combination of decision rooms, local area networks, teleconferencing, and wide area networks:

decision room: A room that supports decision making, with the decision makers in the same building, and that combines face-to-face verbal interaction with technology to make the meeting more effective and efficient.

- The **decision room** is a room that supports decision making, with the decision makers in the same building, and that combines face-to-face verbal interaction with technology to make the meeting more effective and efficient. It is ideal for situations in which decision makers are located in the same building or geographic area and the decision makers are occasional users of the GSS approach. A typical decision room is shown in Figure 10.20.

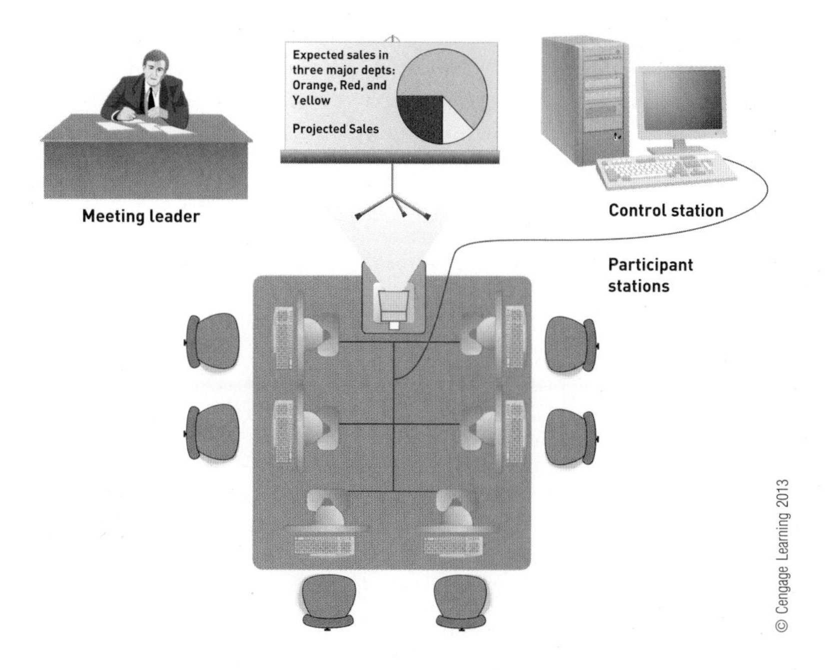

FIGURE **10.20**

GSS decision room

For group members who are in the same location, the decision room is an optimal GSS alternative. This approach can use both face-to-face and computer-mediated communication. By using networked computers and computer devices, such as project screens and printers, the meeting leader can pose questions to the group, instantly collect members' feedback, and with the help of the governing software loaded on the control station, process this feedback into meaningful information to aid in the decision-making process.

- The *local area decision network* can be used when group members are located in the same building or geographic area and under conditions in which group decision making is frequent. In these cases, the technology and equipment for the GSS approach is placed directly into the offices of the group members.
- *Teleconferencing* is used when the decision frequency is low and the location of group members is distant. These distant and occasional group meetings can tie together multiple GSS decision-making rooms across the country or around the world. Yum! Brands, owner of Kentucky Fried Chicken (KFC), Taco Bell, and Pizza Hut, uses a teleconferencing system by Tanberg to let employees have virtual meetings and make group decisions, reducing travel time and costs.[104] The Tandberg group support system uses high definition videos and group support software to help employees at distant locations collaborate and make group decisions. Teleconferencing, also called telepresence by some, can save companies thousands of dollars in travel costs. According to the Global Business Travel Association, businesses around the world take more than 400 million trips annually for a total cost of over $200 billion combined.[105] According to a partner in a venture capital firm, "A lot of what we do in venture is, we invest in people. You want to look somebody in the eye and say 'Do you trust this person enough to write them a multimillion dollar check?' We feel we can do that with telepresence."
- The *wide area decision network* is used when the decision frequency is high and the location of group members is distant. In this case, the decision makers require frequent or constant use of the GSS approach. This

virtual workgroups: Teams of people located around the world working on common problems.

GSS alternative allows people to work in **virtual workgroups**, where teams of people located around the world can work on common problems.

EXECUTIVE SUPPORT SYSTEMS

executive support system (ESS): Specialized DSS that includes all hardware, software, data, procedures, and people used to assist senior-level executives within the organization.

Because top-level executives often require specialized support when making strategic decisions, many companies have developed systems to assist executive decision making. This type of system, called an **executive support system (ESS)**, is a specialized DSS that includes all the hardware, software, data, procedures, and people used to assist senior-level executives within the organization. In some cases, an ESS, also called an *executive information system (EIS)*, supports decision making of members of the board of directors, who are responsible to stockholders. These top-level decision-making strata are shown in Figure 10.21.

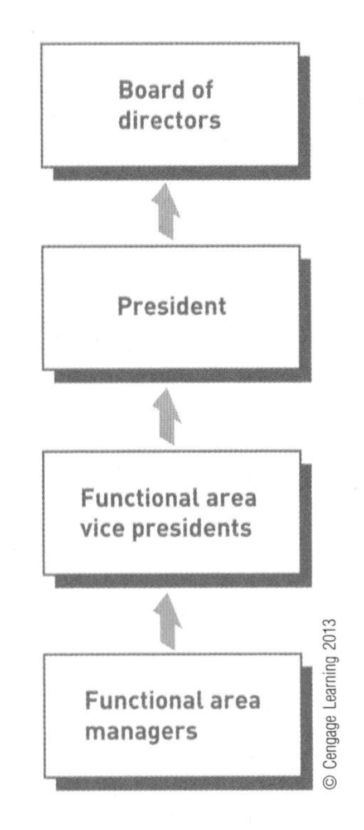

FIGURE **10.21**

Layers of executive decision making

An ESS can also be used by individuals at middle levels in the organizational structure. Once targeted at the top-level executive decision makers, ESSs are now marketed to and used by employees at other levels in the organization.

Executive Support Systems in Perspective

An ESS is a special type of DSS, and, like a DSS, is designed to support higher-level decision making in the organization. The two systems are, however, different in important ways. DSSs provide a variety of modeling and analysis tools to enable users to thoroughly analyze problems; that is, they allow users to *answer* questions. ESSs have the capability to present structured information about aspects of the organization that executives consider important.

Capabilities of Executive Support Systems

The responsibility given to top-level executives and decision makers brings unique problems and pressures to their jobs. This section discusses some of the characteristics of executive decision making that are supported through the ESS approach. ESSs take full advantage of data mining, the Internet, blogs, podcasts, executive dashboards, social networking sites, and many other technological innovations. As you will note, most of these decisions are related to an organization's overall profitability and direction. An effective ESS should have the capability to support executive decisions with components such as strategic planning and organizing and crisis management, among others.

Support for Defining an Overall Vision

One of the key roles of senior executives is to provide a broad vision for the entire organization. This vision includes the organization's major product lines and services, the types of businesses it supports today and in the future, and its overriding goals.

Support for Strategic Planning

strategic planning: Determining long-term objectives by analyzing the strengths and weaknesses of the organization, predicting future trends, and projecting the development of new product lines.

ESSs also support strategic planning. **Strategic planning** involves determining long-term objectives by analyzing the strengths and weaknesses of the organization, predicting future trends, and projecting the development of new product lines. It also involves planning the acquisition of new equipment, analyzing merger possibilities, and making difficult decisions concerning downsizing and the sale of assets if required by unfavorable economic conditions.

Support for Strategic Organizing and Staffing

Top-level executives are concerned with organizational structure. For example, decisions concerning the creation of new departments or downsizing the labor force are made by top-level managers. Overall direction for staffing decisions and effective communication with labor unions are also major decision areas for top-level executives. ESSs can help executives analyze the impact of staffing decisions, potential pay raises, changes in employee benefits, and new work rules.

Support for Strategic Control

Another type of executive decision relates to strategic control, which involves monitoring and managing the overall operation of the organization. Goal seeking can be done for each major area to determine what performance these areas need to achieve to reach corporate expectations. Effective ESS approaches can help top-level managers make the most of their existing resources and control all aspects of the organization.

Support for Crisis Management

Even with careful strategic planning, a crisis can occur. Major incidents, including natural disasters, fires, and terrorist activities, can totally shut down major parts of an organization. Handling these emergencies is another responsibility for top-level executives. In many cases, strategic emergency plans can be put into place with the help of an ESS. These contingency plans help organizations recover quickly if an emergency or crisis occurs.

Decision making is a vital part of managing businesses strategically. IS systems such as information and decision support, group support, and executive support systems help employees by tapping existing databases and providing them with current, accurate information. The increasing integration of all business information systems—from enterprise systems to MISs to DSSs—can help organizations monitor their competitive environment and make better-informed decisions. Organizations can also use specialized business information systems, discussed in the next chapter, to achieve their goals.

SUMMARY

Principle:

Good decision-making and problem-solving skills are the key to developing effective information and decision support systems.

Every organization needs effective decision making and problem solving to reach its objectives and goals. Problem solving begins with decision making. A well-known model developed by Herbert Simon divides the decision-making phase of the problem-solving process into three stages: intelligence, design, and choice. During the intelligence stage, potential problems or opportunities are identified and defined. Information is gathered that relates to the cause and scope of the problem. Constraints on the possible solution and the problem environment are investigated. In the design stage, alternative solutions to the problem are developed and explored. In addition, the feasibility and implications of these alternatives are evaluated. Finally, the choice stage involves selecting the best course of action. In this stage, the decision makers evaluate the implementation of the solution to determine whether the anticipated results were achieved and to modify the process in light of new information learned during the implementation stage.

Decision making is a component of problem solving. In addition to the intelligence, design, and choice steps of decision making, problem solving also includes implementation and monitoring. Implementation places the solution into effect. After a decision has been implemented, it is monitored and modified if necessary.

Decisions can be programmed or nonprogrammed. Programmed decisions are made using a rule, procedure, or quantitative method. Ordering more inventory when the level drops to 100 units or fewer is an example of a programmed decision. A nonprogrammed decision deals with unusual or exceptional situations. Determining the best training program for a new employee is an example of a nonprogrammed decision.

Decisions can use optimization, satisficing, or heuristic approaches. Optimization finds the best solution. Optimization problems often have an objective such as maximizing profits given production and material constraints. When a problem is too complex for optimization, satisficing is often used. Satisficing finds a good, although not necessarily the best, decision. Finally, a heuristic is a "rule of thumb" or common guideline or procedure used to find a good decision.

Principle:

The management information system (MIS) must provide the right information to the right person in the right format at the right time.

A management information system is an integrated collection of people, procedures, databases, and devices that provides managers and decision makers with information to help achieve organizational goals. An MIS can help an organization achieve its goals by providing managers with insight into the regular operations of the organization so that they can control, organize, and plan more effectively and efficiently. The primary difference between the reports generated by the TPS and ERP systems and those generated by the MIS is that MIS reports support managerial decision making at the higher levels of management.

Data that enters the MIS originates from both internal and external sources. The most significant internal sources of data for the MIS are an organization's various TPSs and ERP systems. Data warehouses and data marts also provide important input data for the MIS. External sources of data for the MIS include extranets, customers, suppliers, competitors, and stockholders.

The output of most MISs is a collection of reports that are distributed to managers. These reports include scheduled reports, key-indicator reports, demand reports, exception reports, and drill-down reports. Scheduled reports are produced periodically, such as daily, weekly, or monthly. A key-indicator report is a special type of scheduled report. Demand reports are developed to provide certain information at a manager's request. Exception reports are automatically produced when a situation is unusual or requires management action. Drill-down reports provide increasingly detailed data about situations.

Management information systems have a number of common characteristics, including producing scheduled, demand, exception, and drill-down reports; producing reports with fixed and standard formats; producing hard-copy and soft-copy reports; using internal data stored in organizational computerized databases; and having reports developed and implemented by IS personnel or end users. More and more MIS reports are being delivered over the Internet and through mobile devices, such as cell phones.

Most MISs are organized along the functional lines of an organization. Typical functional management information systems include financial, manufacturing, marketing, human resources, and other specialized systems. Each system is composed of inputs, processing subsystems, and outputs. The primary sources of input to functional MISs include the corporate strategic plan, data from the ERP system and TPS, information from supply chain and business transactions, and external sources including the Internet and extranets. The primary outputs of these functional MISs are summary reports that assist in managerial decision making.

Principle:

Decision support systems (DSSs) are used when the problems are unstructured.

A decision support system (DSS) is an organized collection of people, procedures, software, databases, and devices working to support managerial decision making. DSS characteristics include the ability to handle large amounts of data; obtain and process data from a variety of sources; provide report and presentation flexibility; support drill-down analysis; perform complex statistical analysis; offer textual and graphical orientations; support optimization, satisficing, and heuristic approaches; and perform what-if, simulation, and goal-seeking analysis.

DSSs provide support assistance through all phases of the problem-solving process. Different decision frequencies also require DSS support. An ad hoc DSS addresses unique, infrequent decision situations, and an institutional DSS handles routine decisions. Highly structured problems, semistructured problems, and unstructured problems can be supported by a DSS. A DSS can also support different managerial levels, including strategic, tactical, and operational managers. A common database is often the link that ties together a company's TPS, MIS, and DSS.

The components of a DSS are the database, model base, user interface or dialogue manager, and a link to external databases, the Internet, the corporate intranet, extranets, networks, and other systems. The database can use data warehouses and data marts. A data-driven DSS primarily performs qualitative analysis based on the company's databases. Data-driven DSSs tap into vast stores of information contained in the corporate database, retrieving information on inventory, sales, personnel, production, finance, accounting, and other areas. Data mining is often used in a data-driven DSS. The model base contains the models used by the decision maker, such as financial, statistical, graphical, and project-management models. A model-driven DSS primarily performs mathematical or quantitative analysis. Model management software (MMS) is often used to coordinate the use of models in a DSS. The user interface provides a dialogue management facility to assist in communications between the system and the user. Access to other computer-based systems permits the DSS to tie into other powerful systems, including the TPS or function-specific subsystems.

Principle:

Specialized support systems, such as group support systems (GSSs) and executive support systems (ESSs), use the overall approach of a DSS in situations such as group and executive decision making.

A group support system (GSS), also called a computerized collaborative work system, consists of most of the elements in a DSS, plus software to provide effective support in group decision-making settings. GSSs are typically easy to learn and use and can offer specific or general decision-making support. GSS software, also called groupware, is specially designed to help generate lists of decision alternatives and perform data analysis. These packages let people work on joint documents and files over a network. Newer Web 2.0 technologies are being used to a greater extent in delivering group decision-making support. Text messages and the Internet are also commonly used in a GSS.

The frequency of GSS use and the location of the decision makers will influence the GSS alternative chosen. The decision room alternative supports users in a single location who meet infrequently. Local area decision networks can be used when group members are located in the same geographic area and users meet regularly. Teleconferencing is used when decision frequency is low and the location of group members is distant. A wide area network is used when the decision frequency is high and the location of group members is distant.

Executive support systems (ESSs) are specialized decision support systems designed to meet the needs of senior management. They serve to indicate issues of importance to the organization, indicate new directions the company might take, and help executives monitor the company's progress. ESSs are typically easy to use, offer a wide range of computer resources, and handle a variety of internal and external data. In addition, the ESS performs sophisticated data analysis, offers a high degree of specialization, and provides flexibility and comprehensive communications capabilities. An ESS also supports individual decision-making styles. Some of the major decision-making areas that can be supported through an ESS are providing an overall vision, strategic planning and organizing, strategic control, and crisis management.

CHAPTER 10: SELF-ASSESSMENT TEST

Good decision-making and problem-solving skills are the key to developing effective information and decision support systems.

1. Developing decision alternatives is done during what decision-making stage?
   a. initiation stage
   b. intelligence stage
   c. design stage
   d. choice stage
2. Problem solving is one of the stages of decision making. True or False?
3. ______________ decisions deal with unusual or exceptional situations.
4. A decision that inventory should be ordered when inventory levels drop to 500 units is an example of a(n) ______________.
   a. synchronous decision
   b. asynchronous decision
   c. nonprogrammed decision
   d. programmed decision
5. A(n) ______________ model will find the best solution to help the organization meet its goals.
6. A satisficing model is one that will find a good problem solution, although not necessarily the best problem solution. True or False?

The management information system (MIS) must provide the right information to the right person in the right format at the right time.

7. What summarizes the previous day's critical activities and is typically available at the beginning of each workday?
   a. key-indicator report
   b. demand report
   c. exception report
   d. database report
8. MRP and JIT are subsystems of the ________________.
   a. marketing MIS
   b. financial MIS
   c. manufacturing MIS
   d. auditing MIS
9. ________________ involves analyzing the financial condition of an organization.

Decision support systems (DSSs) are used when the problems are unstructured.

10. The focus of a decision support system is on decision-making effectiveness when faced with unstructured or semistructured business problems. True or False?
11. The ________________ in a decision support system allows a decision maker to perform quantitative analysis.
12. What component of a decision support system allows decision makers to easily access and manipulate the DSS and to use common business terms and phrases?
    a. the knowledge base
    b. the model base
    c. the user interface or dialogue manager
    d. the expert system

Specialized support systems, such as group support systems (GSSs) and executive support systems (ESSs), use the overall approach of a DSS in situations such as group and executive decision making.

13. What decision-making technique allows voting group members to arrive at a final group decision?
    a. groupthink
    b. anonymous input
    c. nominal group technique
    d. delphi
14. A type of software that helps with joint work-group scheduling, communication, and management is called ________________.
15. The local area decision network is the ideal GSS alternative for situations in which decision makers are located in the same building or geographic area and the decision makers are occasional users of the GSS approach. True or False?
16. A(n) ________________ supports the actions of members of the board of directors, who are responsible to stockholders.

CHAPTER 10: SELF-ASSESSMENT TEST ANSWERS

1. c
2. False
3. Nonprogrammed
4. d
5. optimization
6. True
7. a
8. c
9. Auditing
10. True
11. model base
12. c
13. c
14. groupware or workgroup software
15. False
16. executive information system (EIS)

REVIEW QUESTIONS

1. What is optimization? Describe a situation when it should be used. What are the disadvantages of optimization?
2. What is the difference between intelligence and design in decision making?
3. What is the difference between a programmed decision and a nonprogrammed decision? Give several examples of each.
4. What are the basic kinds of reports produced by an MIS?
5. How can a social networking site be used in a DSS?
6. What are the functions performed by a financial MIS?
7. Describe the functions of a manufacturing MIS.
8. List and describe some other types of MISs.
9. What are the stages of problem solving?

10. What is the difference between decision making and problem solving?
11. How can location analysis be used in a marketing research MIS?
12. List some software tools used in group support systems.
13. Define *decision support system*. What are its characteristics?
14. Describe the difference between a data-driven and a model-driven DSS.
15. What is the difference between what-if analysis and goal-seeking analysis?
16. What are the components of a decision support system?
17. State the objective of a group support system (GSS) and identify three characteristics that distinguish it from a DSS.
18. How can social networking sites be used in a GSS?
19. How does an executive support system differ from a decision support system?
20. Identify three fundamental uses for an executive support system.

DISCUSSION QUESTIONS

1. Select an important problem you had to solve during the last two years. Describe how you used the decision-making and problem-solving steps discussed in this chapter to solve the problem.
2. Describe what types of applications for a tablet computer or smartphone could be used in a financial MIS.
3. Discuss how a social networking site can be used in an information and decision support system.
4. Describe the key features of a manufacturing MIS for an automotive company. What are the primary inputs and outputs? What are the subsystems?
5. How can a strong financial MIS provide strategic benefits to a firm?
6. Why is auditing so important in a financial MIS? Give an example of an audit that failed to disclose the true nature of the financial position of a firm. What was the result?
7. Describe two industries where a marketing MIS is critical to sales and success.
8. Describe how a social networking site, like Facebook, can be used in a group support system. What are the advantages and disadvantages of using this approach?
9. Pick a company and research its human resource management information system. Describe how the system works. What improvements could be made to the company's human resource MIS?
10. You have been hired to develop a DSS for a car company such as Ford or GM. Describe how you would use both data-driven and model-driven DSSs.
11. Describe how you would use Twitter in one of the MISs or DSSs discussed in this chapter.
12. What functions do DSSs support in business organizations? How does a DSS differ from a TPS and an MIS?
13. How is decision making in a group environment different from individual decision making, and why are information systems that assist in the group environment different? What are the advantages and disadvantages of making decisions as a group?
14. You have been hired to develop group support software for your university. Describe the features you would include in your new GSS software.
15. Imagine that you are the vice president of manufacturing for a Fortune 1000 manufacturing company. Describe the features and capabilities of your ideal ESS.

PROBLEM-SOLVING EXERCISES

1. Use the Internet to research the use of smartphones and tablet computers to facilitate group decision making. Use a word processor to describe what you discovered. Develop a set of slides using a graphics program to deliver a presentation on the use of smart phones and tablet computers in developing an information and decision support system.
2. Review the summarized consolidated statement of income for the manufacturing company whose data is shown here. Use graphics software to prepare a set of bar charts that shows the data for this year compared with the data for last year.
   a. This year, operating revenues increased by 3.5 percent, while operating expenses increased 2.5 percent.

b. Other income and expenses decreased to $13,000.
c. Interest and other charges increased to $265,000.

Operating results (in millions)	
Operating Revenues	$2,924,177
Operating Expenses (including taxes)	2,483,687
Operating Income	440,490
Other Income and Expenses	13,497
Income before Interest and Other Charges	453,987
Interest and Other Charges	262,845
Net Income	191,142
Average Common Shares Outstanding	147,426
Earnings per Share	1.30

If you were a financial analyst tracking this company, what detailed data might you need to perform a more complete analysis? Write a brief memo summarizing your data needs.

3. As the head buyer for a major supermarket chain, you are constantly being asked by manufacturers and distributors to stock their new products. Over 50 new items are introduced each week. Many times, these products are launched with national advertising campaigns and special promotional allowances to retailers. To add new products, the amount of shelf space allocated to existing products must be reduced or items must be eliminated altogether. Develop a marketing MIS that you can use to estimate the change in profits from adding or deleting an item from inventory. Your analysis should include input such as estimated weekly sales in units, shelf space allocated to stock an item (measured in units), total cost per unit, and sales price per unit. Your analysis should calculate total annual profit by item and then sort the rows in descending order based on total annual profit.

TEAM ACTIVITIES

1. Review the section on human resource MIS and have your team develop a job description for a new university president. Develop another job description for a new team member for your group.
2. Have your team make a group decision about how to solve the most frustrating aspect of college or university life. Appoint one or two members of the team to disrupt the meeting with negative group behavior. After the meeting, have your team describe how to prevent this negative group behavior. What GSS software features would you suggest to prevent the negative group behavior your team observed?
3. Have your team design a marketing MIS for a medium-sized retail store. Describe the features and characteristics of your marketing MIS. How could you achieve a competitive advantage over a similar retail store with your marketing MIS?

WEB EXERCISES

1. Use the Internet to explore applications for smartphones and tablet computers that can be used in decision making. You might be asked to develop a report or send an e-mail message to your instructor about what you found.
2. Use the Internet to explore the use of GPS and location analysis on cell phones and other mobile devices to market and sell products and services. Summarize your findings in a report.
3. Software, such as Microsoft Excel, is often used to find an optimal solution to maximize profits or minimize costs. Search the Internet using Yahoo!, Google, or another search engine to find other software packages that offer optimization features. Write a report describing one or two of the optimization software packages. What are some of the features of the packages?

CAREER EXERCISES

1. What decisions are critical for success in a career that interests you? What specific types of reports could help you make better decisions on the job? Give three specific examples.
2. Describe the features of an executive support system you would want, assuming you are the chief executive officer of a medium-sized company in an industry of your choice.

CASE STUDIES

Case One

Sports: Not Just Fun and Games

U.S. major league baseball and major league football each bring in annual revenue approaching $10 billion. Millions of dollars are paid for the services of people who can throw, hit, kick, or carry a ball better than most of us can. Today, decisions in sports must be made as carefully as decisions in any other business.

Sports professionals didn't always analyze their decisions. Traditionally, decisions were made on "gut feel." Experienced people knew what to do and trusted their hard-earned knowledge. They used data to justify their instinct, not to go against it. Anyone who disagreed with that approach didn't fit the culture and was ignored.

The 2011 movie *Moneyball* reflects a changing perception. Based closely on reality, it tells the story of how the 2002 Oakland Athletics baseball team used analytical methods to determine who the best players were, not by the traditional measures but instead by measures that related more closely to winning games. This approach enabled the organization to assemble a team that won its division in 2002 and 2003 (though they lost in the league playoffs both years).

One truism in sports is to pay top talent top dollar. However, shelling out large amounts of money for "superstars" may not be smart. Researchers Christopher Annala and Jason Winfree found that baseball team performance is negatively correlated with the level of salary inequality across its players. Teams in which all players were paid comparable amounts did better—all else being equal—than teams in which a few stars accounted for most of the payroll. Have baseball executives changed their behavior as a result? Not yet. Will they, if today's high-salary signings don't lead to championships? Almost certainly, yes. Behavior that leads to championships is ultimately rewarded for baseball executives as well as players.

In professional basketball, the Portland Trailblazers commissioned Protrade Sports (since renamed Citizen Sports and acquired by Yahoo!) in 2005 to create a model for the college player draft. The model used NCAA box score data, historical draft information, and performance of former college players in the NBA. The output of the model was an estimated probability that the college player would be a contributing pro in the NBA. Did that immediately turn the Trailblazers into champions? No, but their bottom-of-the league position in the 2005–2006 season improved during the next two years, and they were in the league playoffs in the three seasons after that.

Analytical methods are useful on the playing field as well. Since 2006, SportVision's PITCHf/x service has recorded the trajectory of every major league baseball pitch. Three cameras track each pitch, and a computer calculates the location, speed, and trajectory. The Tampa Bay Rays used this data to see how opposing pitchers release the ball at different points for different types of pitches, enabling the Rays's batters to anticipate pitches earlier than they could otherwise. This is considered to be one factor in the team's progress from a record of 66 wins, 96 losses in 2007 to the American League championship in 2008.

Ten years after the events recounted in *Moneyball*, and despite other successes, sports analytics are still not fully accepted. Some barriers are cultural. Some are due to a lack of trained analysts who understand sports and can communicate effectively with sports executives. These barriers will surely be overcome in time, with far-reaching implications for the world of professional sports.

Discussion Questions

1. How could you apply the concepts of predictive analytics to a sport of your choice, other than baseball, to improve a team's performance? What data would you use and what would you try to predict from that data?
2. The mathematical methods described in this case have been known for generations and used in business for decades. Why do you think it took so long for these methods to earn even a little acceptance in sports?

Critical Thinking Questions

1. Consider the factors you use to select a section of a required course that is offered at several times with different instructors. Could you use methods such as those found in this case study to improve your chances of making a decision you're eventually happy with?
2. Suppose you suggest using the methods in this study on the job (which is not in sports). Your manager replies, "We've always done it this way. It was good enough then, and it's good enough now." You are sure, however, that using the methods you learned in college will improve your department's performance. What does the experience of trying to introduce predictive analytics to sports suggest you should do?

SOURCES: Alamar, B. and Mehrotra, V., "Beyond 'Moneyball': The Rapidly Evolving World of Sports Analytics," *Analytics*, *www.analytics-magazine.org/special-articles/391-beyond-moneyball-the-rapidly-evolving-world-of-sports-analytics-part-i*, September/October 2011 (parts 2 and 3 of the article appear in the November-December 2011 and March-April 2012 issues of *Analytics*, respectively); Annala, C. and Winfree, J., "Salary Distribution and Team Performance in Major League Baseball," *Sport Management Review*, vol. 14, no. 2, May 2011; Keri, J., *The Extra 2%: How Wall Street Strategies Took a Major League Baseball Team from Worst to First*, New York: ESPN Books, 2011; *Moneyball* (movie) Web site, *www.moneyball-movie.com/site*, accessed April 20, 2012; Sportvision Web site, *www.sportvision.com*, accessed April 25, 2012.

Case Two

Mando: Inventory Management and More

Mando Corporation is South Korea's largest manufacturer of automobile steering, brake, and suspension components. Originally a division of automobile manufacturer Hyundai, it is now separate, though both are in the same *chaebol* (conglomerate). Mando now supplies many other automobile firms as well, including Chinese auto makers and GM. Its 2011 annual revenue was about U.S. $4 billion, with investment firm KDB Daewoo forecasting 14 percent growth for 2012. With plants in China, India, Malaysia, Turkey, Poland, and Brazil as well as South Korea and with a wide range of mechanical and electronic products, inventory management is critical to its success.

With inventory management (and more) in mind, Mando chose Oracle's E-Business Suite, version 12.1, as an integrated ERP system to connect all its divisions. Using a single enterprise-wide database reduced errors. For example, it enabled Mando to standardize on a common numbering system, eliminating inventory-tracking errors due to part number differences when applying design changes.

As you read in this chapter, inventory management decisions use a variety of reports. Therefore, if you saw "Mando Achieves 99.9% Accuracy in Inventory Tracking," you'd probably assume that management information systems and their reports were part of the reason. You'd be right. The E-Business Suite software can produce a wide variety of reports of all types.

Tracking inventory, knowing what you have, is only part of the answer. You have to have the *right* inventory. Inventory management decisions must be based on good inventory tracking data, but decisions are based on more than good inventory data. Inventory management decisions depend on reports as well.

Some inventory management decisions can be programmed. When stock drops to the reorder point, an order is placed for the reorder quantity. In this instance, management uses reports to make sure the programmed procedures are operating properly and meet the organization's needs.

Other inventory management decisions are less structured. New products have no usage history on which to base reorder points or quantities. Inventory of products being replaced must be managed to ensure proper phase-out. Management doesn't want to be left with a stock of parts that have no current use or run out of a key component before production ends. The transition from mechanical to electronic controls involves more than just replacing one part with a slightly different one. In making these inventory decisions, reports must be used along with sales forecasts and careful analysis to ensure that the right amounts of the right items are on hand.

In addition to reports, Mando used the capabilities of Oracle Business Intelligence software to create a real-time decision-making environment. Inventory information and other data are presented to senior managers through a dashboard on a daily and monthly basis. The dashboards deliver key information in an easy-to-view format and help managers determine business trends.

Park ByoungOk, Mando's CIO, is pleased with these software capabilities. He says, "The [ERP] system enabled us to standardize more than 200 processes globally, which gave senior managers an integrated, enterprise wide view of sales, financials, inventory, and quality management." Giving managers a good view of the company is, in the final analysis, the purpose of any management information system.

Discussion Questions

1. As a manager, you must choose between two inventory management software packages. One is a stand-alone package that only manages inventory. It allows users to define their own reports without much training. The other requires a professional programmer for new reports, but it is part of an ERP system that can handle much more than inventory management. Describe how you would choose between the two packages.
2. Mando's inventory at the end of 2011 was valued at approximately $300 million. Consider a small bicycle store whose inventory is valued at $30,000, about 1/10,000 of Mando's figure. In what ways are its inventory reporting needs similar? In what ways are they different?

Critical Thinking Questions

1. Sketch examples of the five major report types as they could apply to inventory management for an auto parts manufacturer. Invent data for your examples.
2. Explain why it would be more difficult for Mando to manage its inventory if the databases for all sixteen of its factories were different and used different numbers for the same part. Give a specific example of a problem that could arise.

SOURCES: "Mando: New Orders Are Winning Over Investors," KDB Daewoo Securities, downloaded from *www.kdbdw.com/bbs/download/82746.pdf?attachmentId=82746*, December 20, 2011; Mando Corporation English Web site, *www.mando.com/200909_mando/eng/main.asp*, accessed April 15, 2012; "Mando Corporation Achieves 99.9% Accuracy in Inventory Tracking," Oracle Corporation, *Information for Success*, p. 50, downloaded from *innovative.com.br/wp-content/uploads/2011/06/ebsr12referencebooklet-354227.pdf*, March 2011.

Questions for Web Case

See the Web site for this book to read about the Altitude Online case for this chapter. The following are questions concerning this Web case.

Altitude Online: Information and Decision Support Systems Considerations

Discussion Questions

1. What functional areas of Altitude Online are supported by MISs?
2. How do MISs and DSSs provide a value add to Altitude Online's products?

Critical Thinking Questions

1. How do you think MISs and DSSs assist Altitude Online's top executives in guiding the direction of the company?
2. How can the quality of information systems affect Altitude Online's ability to compete in the online marketing industry?

NOTES

Sources for the opening vignette: Irish Life Web site, *www.irishlife.ie*, accessed April 4, 2012; Savvas, A., "Irish Life Deploys New BI System," *Computerworld UK*, *www.computerworlduk.com/news/applications/3321944/irish-life-deploys-new-bi-system*, November 30, 2011; Smith, G., "Irish Life Chooses Tableau to Deliver Business Intelligence Dashboards," *Silicon Republic*, *www.siliconrepublic.com/strategy/item/25782-irish-life-chooses-tableau*, February 14, 2012; Tableau Software Web site, *www.tableausoftware.com*, accessed April 4, 2012.

1. Hamblen, Matt, "Chevron, TD Bank Hope to Tap Tablets' Potential," *Computerworld*, February 21, 2011, p. 6.
2. Lowenstein, Roger, "Better Think Twice," *Bloomberg Businessweek*, October 31, 2011, p. 97.
3. Nash, Kim, "The Thought That Counts," *CIO*, April 1, 2011, p. 28.
4. Wang, Jimmy, "CIO Profiles," *Information Week*, February 28, 2011, p. 8.
5. King, Julia, "The Rewards of Risk Taking," *Computerworld*, February 27, 2012, p. 22.
6. Bailey, M., et al, "American Airlines Uses Should-Cost Modeling," *Interfaces*, March-April 2011, p. 194.
7. Cisco Web site, *www.cisco.com*, accessed June 2, 2011.
8. Gnanlet, A., et al, "Sequential and Simultaneous Decision Making for Optimizing Health Care," *Decision Sciences*, May 2009, p. 295.
9. Tieto Web site, *www.tieto.com*, accessed June 5, 2011.
10. News and Features, Xerox Web site, *http://news.xerox.com/pr/xerox/xerox-lean-document-production-services-158574.aspx*, accessed June 1, 2011.
11. Furman, Kevin, et al, "Feedstock Routing in the ExxonMobile Downstream Sector," *Interfaces*, March-April 2011, p. 149.
12. Huigenbosch, Peter von, et al, "ANWB Automates and Improves Service Personnel Dispatching," *Interfaces*, March-April 2011, p. 123.
13. Shaikh, N., et al, "Kimberly-Clark Latin America Builds an Optimization-Based System for Machine Scheduling," *Interfaces*, September 2011, p. 455.
14. Shermag Web site, *www.shermag.com/eng/shermag.html*, accessed June 2, 2011.
15. CSIRO Web site, *www.cmis.csiro.au*, accessed June 3, 2011.
16. Brown, Reva, "Consideration of the Origin of Herbert Simon's Theory of Satisficing," *Management Decision*, vol. 42, no. 10, 2004, p. 1240.
17. Balakrishnan, R., et al, "Evaluating Heuristics Used When Designing Product Costing Systems," *Management Science*, March 2011, p. 520.
18. Symantec Web site, *www.symantec.com/norton/ps/3up_nl_nl_navnis360t1.html?om_sem_cid=hho_sem_sy:us:ggl:en:e|kw0000006084|7265273116*, accessed May 15, 2011.
19. Waxer, C., "Juicing the Market for Electric Cars," *CIO*, April 1, 2001, p. 20.
20. Cognos Web site, *www.cognos.com*, accessed June 4, 2011.
21. Microsoft Web site, *www.microsoft.com/dynamics/product/business_scorecard_manager.mspx*, accessed June 3, 2011.
22. Sacramento Municipal Utility District Web site, *https://www.smud.org/en/Pages/index.aspx*, accessed May 16, 2011.
23. Reuters Web site, *www.reuters.com*, accessed June 2, 2011.
24. Cowan, L., "LinkedIn Sees $3 Billion Value," *The Wall Street Journal*, May 10, 2011, p. B3.
25. Munoz, D., et al, "INDEVAL Develops a New Operating and Settlement System Using Operations Research," *Interfaces*, January-February, 2011, p. 8.
26. Ebaid, Ibrahim, "Internet Audit Function," *International Journal of Law and Management*, vol. 53, no. 2, 2011, p. 108.
27. Tam, Pui-Wing, et al, "A Venture-Capital Newbie Shakes up Silicon Valley," *The Wall Street Journal*, May 10, 2011, p. A1.
28. Mint Web site, *https://www.mint.com/what-is-mint*, accessed June 3, 2011.
29. Dell Web site, *www.dell.com*, accessed June 3, 2011.
30. Boeing Web site, *www.boeing.com*, accessed June 14, 2011.
31. Einhorn, B., et al, "Now a Weak Link in the Global Supply Chain," *Bloomberg Businessweek*, March 27, 2011, p. 18.
32. Shaikh, N., et al, "Kimberly-Clark Latin America Builds an Optimization-Based System for Machine Scheduling," *Interfaces*, September 2011, p. 455.
33. Farasyn, I., et al, "Inventory Optimization at Procter & Gamble," *Interfaces*, January-February, 2011, p. 66.
34. Nenes, G., et al, "Customized Inventory Management," *OR/MS Today*, April 2011, p. 22.
35. Black, T. and Ray, S., "The Downside of Just-In-Time Inventory," *BusinessWeek*, March 28, 2011, p. 1.
36. Abdinnour, Sue, "Hawker Beechcraft Uses a New Solution Approach to Balance Assembly Lines," *Interfaces*, March-April 2011, p. 164.
37. Chrysler Web site, *www.chrysler.com*, accessed June 2, 2011.
38. Betts, M., "Big Pharma Gobbles Up iPads for Sales," *Computerworld*, May 9, 2011, p. 4.
39. O'Connell, V., "Building eBay for Lawyers," *The Wall Street Journal*, March 23, 2011, p. B7.
40. Zimmerman, A., "Check Out the Future of Shopping," *The Wall Street Journal*, May 18, 2011, p. D1.
41. Hudson, Kris, "Malls Test Apps to Aid Shoppers," *The Wall Street Journal*, April 26, 2011, p. B5.
42. Fowler, G., "Facebook Friends Used in Ads," *The Wall Street Journal*, January 26, 2011, p. B9.
43. Rosman, Katherine, "Here, Tweeting Is a Class Requirement," *The Wall Street Journal*, March 9, 2011, p. D1.
44. MacMillian, D., "Creating Web Addicts for $10,000 a Month," *Bloomberg Businessweek*, January 24, 2011, p. 35.
45. BMW Web site, *www.bmwusa.com*, accessed June 2, 2011.
46. Google Web site, *www.google.com/latitude/intro.html*, accessed April 12, 2012.
47. Devries, J., "Latest Treasure Is Location Data," *The Wall Street Journal*, May 10, 2011, B1.
48. Betawave Web site, *www.betawave.com*, August 2, 2011.
49. Sysco Web site, *www.sysco.com*, accessed June 2, 2011.
50. Vascellaro, J., "Building Loyalty on the Web," *The Wall Street Journal*, March 28, 2011, p. B6.
51. Das, A. and Ghon, G., "Advertisers Vie to Take on Groupon IPO," *The Wall Street Journal*, January 15, 2001, p. B3.
52. Stone, B. and MacMillan, D., "Are Four Words Worth $25 Billion?" *Bloomberg Business Week*, March 21, 2011, p. 69.

53. Hickins, M., "Groupon Revenue Hit $760 Million, CEO Memo Shows," *The Wall Street Journal*, February 26, 2001, p. B3.
54. Fuhrmans, Vanessa, "BMW Wheels Out Apps to Drive Electric-Car Buzz," *The Wall Street Journal*, April 20, 2011, p. B1.
55. Gamerman, E., "Animation Nation," *The Wall Street Journal*, February 11, 2011, p. D1.
56. Efrati, A., "The Price of Unwanted Ad Clicks," *The Wall Street Journal*, January 12, 2011, p. B5.
57. Fowler, G., "Mobile Apps Drawing in Shoppers, Marketers," *The Wall Street Journal*, January 31, 2011, p. B4.
58. Peers, M., "Apple Risks App-lash on iPad," *The Wall Street Journal*, February 27, 2011, p. C12.
59. Waxer, Cindy, "One-Stop Virtual Shop," *CIO*, March 1, 2012, p. 13.
60. Nash, Kim, "The Art and Science of Pricing," *CIO*, February 1, 2012, p. 20.
61. IBM Web site, *www-304.ibm.com/partnerworld/gsd/solutiondetails.do?solution=14409&expand=true&lc=en*, accessed June 2, 2011.
62. Steel, E. and Fowler, S., "Facebook Gets New Friends," *The Wall Street Journal*, April 4, 2011, p. B4.
63. Wright, P. and McMahan, G., "Exploring Human Capital," *Human Resource Management Journal*, April 2011, p. 93.
64. Huigenbosch, Peter von, et al, "ANWB Automates and Improves Service Personnel Dispatching," *Interfaces*, March-April 2011, p. 123.
65. Jutte, Silke, et al, "Optimizing Railway Crew Scheduling at DB Schenker," *Interfaces*, March-April 2011, p. 109.
66. Tactician Web site, *www.tactician.com*, accessed August 2, 2011.
67. Ibid.
68. Turborouter Web site, *www.sintef.no/turborouter*, accessed June 3, 2011.
69. Wakabayashi, D., "Japanese Farms Look to the Cloud," *The Wall Street Journal*, January 18, 2011, p. B5.
70. Mearian, Lucas, "IT Looks for New Tools to Exploit Big Data," *Computerworld*, April 4, 2011, p. 6.
71. Fulgham, C., "It Is All About Helping Us Conduct Investigations Faster," *CIO*, February 1, 2011, p. 36.
72. Henschen, D., "SAS's New Big Data Approach," *Information Week*, April 25, 2011, p. 15.
73. Meer, M. et al, "Innovative Decision Support in a Petrochemical Production Environment," *Interfaces*, January-February, 2011, p. 79.
74. Sanna, N., "Getting Predictive About IT," *CIO*, February 1, 2011, p. 39.
75. Karisch, S., "Jessesen Soars with O.R.," *OR/MS Today*, April 2011, p. 16.
76. De-Vries, J., "May the Best Algorithm Win...," *The Wall Street Journal*, March 16, 2011, p. B4.
77. Howson, C., "Predictions of What's in Store for the BI Market," *Information Week*, January 31, 2011, p. 22.
78. Saunders, Laura, "Small Concerns Fight IRS Over Data," *The Wall Street Journal*, May 26, 2011, p. C1.
79. Carreyrou, J. and McGinty, T., "Medicare Records Reveal Troubling Trail of Surgeries," *The Wall Street Journal*, March 29, 2011, p. A1.
80. Vascellaro, J., "TVs Next Wave: Tuning in to You," *The Wall Street Journal*, March 7, 2011, p. A1.
81. Mearian, Lucas, "IT Looks for New Tools to Exploit Big Data," *Computerworld*, April 4, 2011, p. 6.
82. Evans, Bob, "The Top 10 CIO Priorities and Issues for 2011," *Information Week*, January 31, 2011, p. 12.
83. Chrysler Web site, *www.chrysler.com*, accessed June 5, 2011.
84. Fischer, M., et al, "Managing Global Brand Investments at DHL," *Interfaces*, January-February, 2011, p. 35.
85. Wealthfront Web site, *https://www.wealthfront.com*, accessed June 5, 2011.
86. Eaglesham, J. and Strasburg, J., "SEC Sues Over Bug in Quant Program," *The Wall Street Journal*, February 4, 2011, p. C1.
87. Ackerman, F. and Eden, C., "Negotiation in Strategy Making Teams: Group Support Systems and the Process of Cognitive Change," *Group Decision and Negotiation*, May 2011, p. 293.
88. Kim, D. and Lee, R., "Systems Collaboration and Strategic Collaboration: Their Impacts on Supply Chain Responsiveness and Market Performance," *Decision Sciences*, November 2010, p. 955.
89. MWH Web site, *www.mwhglobal.com*, accessed June 5, 2011.
90. Nash, K., "Collaborate with Ease," *CIO*, February 1, 2011, p. 13.
91. Serena Web site, *www.serena.com*, accessed June 5, 2011.
92. Ning Web site, *www.ning.com*, accessed June 14, 2011.
93. Vance, A., "Trouble at the Virtual Water Cooler," *Bloomberg Businessweek*, May 2, 2011, p. 31.
94. Bhasin, P., "Keep on Tracking," *CIO*, February 1, 2011, p. 43.
95. Gaudin, Sharon, "TD Bank Unites 85,000-plus Workers," *Computerworld*, February 13, 2012, p. 4.
96. Evans, Bob, "Why CIOs Must Have a Tablet Strategy," *Information Week*, March 14, 2001, p. 10.
97. Hamblen, Matt, "For a Sprawling IT Group, Collaboration Technologies are Imperative," *Computer World*, February 21, 2001, p. 28.
98. Carr, David, "IBM Lotus Notes Gets a Social Overhaul," *Information Week*, January 30, 2012, p. 14.
99. IBM Web site, *www-01.ibm.com/software/lotus/products/connections*, accessed June 14, 2011.
100. Microsoft Web site, *www.microsoft.com/unifiedcommunications*, accessed August 2, 2011.
101. Audy, J., et al, "Why Should We Work Together?" *OR/MS Today*, April 2011, p. 48.
102. Green, Tim, "Microsoft Preps iPhone, VoIP Conferencing Client," *Network World*, March 7, 2011, p. 19.
103. Nash, Kim, "The Thought That Counts," *CIO*, April 1, 2011, p. 28.
104. Betts, Mitch, "IT Leader Builds a Know-how Network," *Computerworld*, April 18, 2011, p. 4.
105. Bennett, D., "I'll Have My Robots Talk to Your Robots," *Bloomberg Businessweek*, February 21, 2011, p. 53.

11 Knowledge Management and Specialized Information Systems

Principles	Learning Objectives
• Knowledge management allows organizations to share knowledge and experience among managers and employees.	• Discuss the differences between data, information, and knowledge. • Describe the role of the chief knowledge officer (CKO). • List some of the tools and techniques used in knowledge management.
• Artificial intelligence systems form a broad and diverse set of systems that can replicate human decision making for certain types of well-defined problems.	• Define the term *artificial intelligence* and state the objective of developing artificial intelligence systems. • List the characteristics of intelligent behavior and compare the performance of natural and artificial intelligence systems for each of these characteristics. • Identify the major components of the artificial intelligence field and provide one example of each type of system.
• Expert systems can enable a novice to perform at the level of an expert but must be developed and maintained very carefully.	• List the characteristics and basic components of expert systems. • Outline and briefly explain the steps for developing an expert system. • Identify the benefits associated with the use of expert systems.
• Multimedia and virtual reality systems can reshape the interface between people and information technology by offering new ways to communicate information, visualize processes, and express ideas creatively.	• Discuss the use of multimedia in a business setting. • Define the term *virtual reality* and *augmented reality* and provide three examples of these applications.
• Specialized systems can help organizations and individuals achieve their goals.	• Discuss examples of specialized systems for organizational and individual use.

Information Systems in the Global Economy

REPSOL, SPAIN

Repsol Deploys Knowledge Management

Repsol is a Spanish oil and gas company with 2010 revenues of about 60 billion euros (approximately U.S. $80 billion in early 2012). Repsol's business ranges from exploration through retail marketing at over 7,000 service stations.

Repsol recognizes that "the key assets in an R&D organization like Repsol are the researchers and technologists, who have the training, experience, knowledge, creativity and motivation: the necessary ingredients for discovering, improving and assimilating new technologies." Without such *knowledge workers*, every stage in Repsol's value chain would dry up.

The value of such people is in what they know. China's Hilong Group, a supplier to the energy industry, lists many examples: geologists and geophysicists who use their knowledge to determine what rocks are beneath the earth's surface that can contain oil or gas, drilling engineers who plan well locations for efficient oil extraction, platform designers who must reduce costs to improve the efficiency of drilling platforms, and others. All of these people are in professions that call for specialized knowledge. Ideally, that knowledge should be available to anyone in the company who needs it.

Consider oil exploration, one of Repsol's core objectives. Geologists set off vibrations at one point and record them at another. They measure how strong the vibrations are at the receiving end, how long it takes the vibrations to get there, and how strength and delay vary with vibration frequency. From those measurements, the geologists infer how much oil is beneath the surface and how hard extracting it will be. Oil companies use their conclusions to decide how much to bid for the rights to drill in an area. Companies that understand these measurements well can make reasonable predictions of how much oil will come from an area, will create accurate bids, and will be profitable. Companies that do not understand the measurements will bid too much for a barren oil field and lose money, or they will not bid enough for a rich field and be outbid by others who figure out its true value.

Knowing how to interpret these measurements—or how to do specialized work in any other field—comes from years of study and experience. Even the best expert has knowledge gaps: the expert might not know a fact or process, for example, while someone else at the company probably does. That's where knowledge management (KM) comes in. With a KM system, experts throughout the company can tap into what their colleagues know, no matter where they or those colleagues happen to be.

Prior to 2011, knowledge management at Repsol was haphazard; it depended on each person's knowing who else might know something plus a few localized knowledge repositories. In 2011, Repsol chose IDOL and Virage software from Autonomy (Cambridge, U.K.) to underpin its corporate knowledge management efforts. The company explains that "Autonomy's enterprise search enable[s] Repsol employees to search across different departments and operating systems, geographic locations and languages, in order to find timely, relevant information, regardless of data type or format. [It can] deliver personalized information through Agents, which understand users' interests and monitor business-critical information to provide automated alerts [to help] Repsol work more efficiently and quickly identify market trends, risks or opportunities."

Domingo Valhondo, knowledge management chief at Repsol, said, "Autonomy allows us to harness all our electronic information on one platform—including our rich media data—and use it to its best advantage in order to improve the productivity of our employees across multiple business units." In the final analysis, employee productivity is important to any company. KM can help achieve it.

As you read this chapter, consider the following:

- How do knowledge management systems, such as the one Repsol uses, differ from other types of information systems you have read about in this book?
- What sort of knowledge do people use in your chosen career field? How can a knowledge management system help your employer make full use of everyone's knowledge?

WHY LEARN ABOUT KNOWLEDGE MANAGEMENT AND SPECIALIZED INFORMATION SYSTEMS?

Knowledge management and specialized information systems are used in almost every industry. As a manager, you might use a knowledge management system to support decisive action to help you correct a problem. As an executive at an automotive company, you might oversee robots that attach windshields to cars or paint body panels. As a young stock trader, you might use a special system called a *neural network* to uncover patterns and make millions of dollars trading stocks and stock options. As a marketing manager for a PC manufacturer, you might use virtual reality on a Web site to show customers your latest laptop and desktop computers. As a member of the military, you might use computer simulation as a training tool to prepare you for combat. As an employee of a petroleum company, you might use an expert system to determine where to drill for oil and gas. You will see many additional examples of using these specialized information systems throughout this chapter. Learning about these systems will help you discover new ways to use information systems in your day-to-day work.

Like other aspects of an information system, the overall goal of knowledge management and the specialized systems discussed in this chapter is to help people and organizations achieve their goals. In this chapter, we explore knowledge management, artificial intelligence, and many other specialized information systems, including expert systems, robotics, vision systems, natural language processing, learning systems, neural networks, genetic algorithms, intelligent agents, multimedia, virtual reality, and augmented reality.

KNOWLEDGE MANAGEMENT SYSTEMS

Chapter 1 defines and discusses data, information, and knowledge. Recall that *data* consists of raw facts, such as an employee number, number of hours worked in a week, inventory part numbers, or sales orders. A list of the quantity available for all items in inventory is an example of data. When these facts are organized or arranged in a meaningful manner, they become information. You may recall from Chapter 1 that *information* is a collection of facts organized so that they have additional value beyond the value of the facts themselves. An exception report of inventory items that might be out of stock in a week because of high demand is an example of information. *Knowledge* is the awareness and understanding of a set of information and the ways that information can be made useful to support a specific task or reach a decision. Knowing the procedures for ordering more inventory to avoid running out is an example of knowledge. In a sense, information tells you what has to be done (low inventory levels for some items), while knowledge tells you how to do it (make two important phone calls to the right people to get the needed inventory shipped overnight). See Figure 11.1.

FIGURE 11.1

Differences between data, information, and knowledge

Data	There are 20 PCs in stock at the retail store.
Information	The store will run out of inventory in a week unless more is ordered today.
Knowledge	Call 800-555-2222 to order more inventory.

A *knowledge management system (KMS)* is an organized collection of people, procedures, software, databases, and devices used to create, store, share, and use the organization's knowledge and experience. KMSs cover a wide range of systems, from software that contains some KMS components to dedicated systems designed specifically to capture, store, and use knowledge.

Overview of Knowledge Management Systems

Like the other systems discussed throughout the book, including information and decision support systems, knowledge management systems attempt to help organizations achieve their goals.[1] For businesses, these goals usually mean increasing profits or reducing costs. Advent, a San Francisco company that develops investment applications for hedge funds and financial services companies, used a KMS to help its employees locate and use critical information.[2] In one study, workers with more knowledge management experience were able to benefit from knowledge management faster and to a greater extent than workers with less experience.[3] For nonprofit organizations, KM can mean providing better customer service or providing special needs to people and groups.

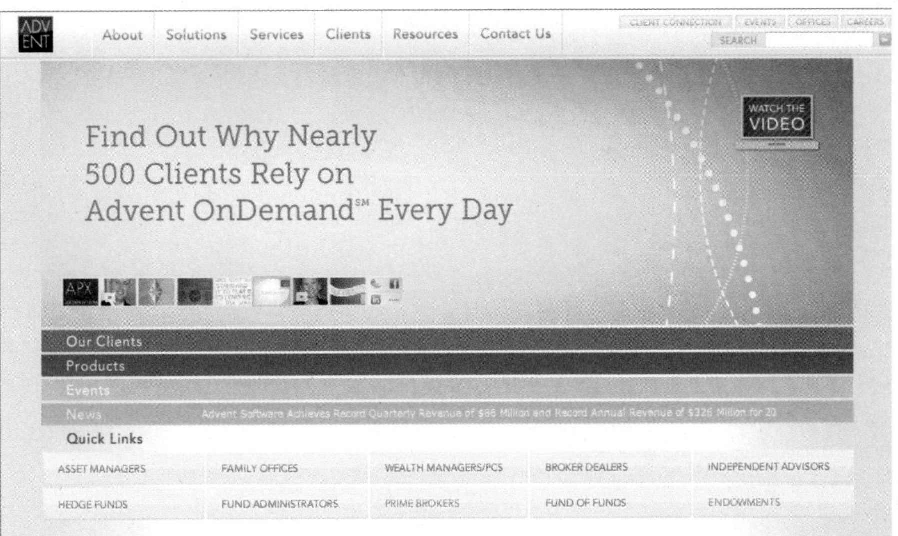

Advent Software uses a knowledge management system to help its employees find critical investment information.

A KMS can involve different types of knowledge. *Explicit knowledge* is objective and can be measured and documented in reports, papers, and rules. For example, knowing the best road to take to minimize drive time from home to the office when a major highway is closed is explicit knowledge. It can be documented in a report or a rule, as in "If I-70 is closed, take Highway 6 to the office." *Tacit knowledge*, on the other hand, is hard to measure and document and typically is not objective or formalized. Knowing the best way to negotiate with a foreign government about nuclear disarmament or a volatile hostage situation often requires a lifetime of experience and a high level of skill. These are examples of tacit knowledge. It is difficult to write a detailed report or a set of rules that would always work in every hostage situation. Many organizations actively attempt to convert tacit knowledge to explicit knowledge to make the knowledge easier to measure, document, and share with others.

Data and Knowledge Management Workers and Communities of Practice

The personnel involved in a KMS include data workers and knowledge workers. Secretaries, administrative assistants, bookkeepers, and similar data-entry personnel are often called *data workers*. As mentioned in Chapter 1, *knowledge workers* are people who create, use, and disseminate knowledge.[4] They are usually professionals in science, engineering, or business, and they usually work in offices and belong to professional organizations (see Figure 11.2). Other examples of knowledge workers include writers, researchers, educators, and corporate designers. Yum Brands, for example, connects its 1.6 million employees around the world to help them share and use knowledge.[5] The vice president of global IT for Yum Brands believes that this type of knowledge sharing can help employees "break out of silos and share know-how." This knowledge-sharing approach uses an internal social network called iChing, an enterprise search facility developed by Coveco, an online learning system by Saba, and a high-definition videoconferencing system from Tandberg.[6]

FIGURE 11.2
Knowledge worker
Knowledge workers are people who create, use, and disseminate knowledge and include professionals in science, engineering, business, and other areas.

chief knowledge officer (CKO): A top-level executive who helps the organization work with a KMS to create, store, and use knowledge to achieve organizational goals.

The **chief knowledge officer (CKO)** is a top-level executive who helps the organization work with a KMS to create, store, and use knowledge to achieve organizational goals. The CKO is responsible for the organization's KMS and typically works with other executives and vice presidents, including the chief executive officer (CEO) and chief financial officer (CFO), among others.

Some organizations and professions use *communities of practice* (*COP*) to create, store, and share knowledge. A COP is a group of people dedicated to a common discipline or practice, such as open-source software, auditing, medicine, or engineering. A group of oceanographers investigating climate change or a team of medical researchers looking for new ways to treat lung cancer are examples of COPs. COPs excel at obtaining, storing, sharing, and using knowledge.

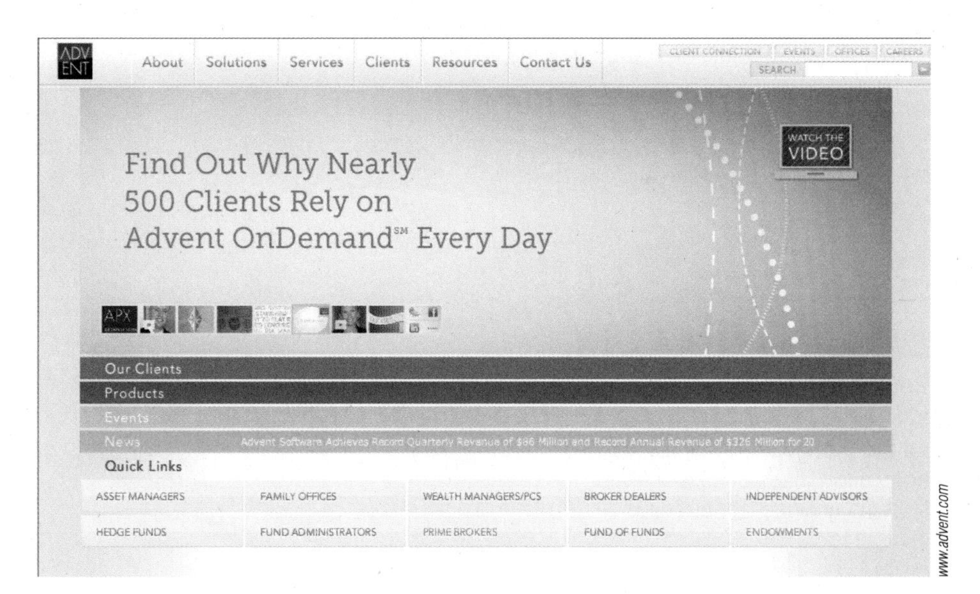

Obtaining, Storing, Sharing, and Using Knowledge

Obtaining, storing, sharing, and using knowledge is the key to any KMS. MWH Global, located in Colorado, uses a KMS to create, disseminate, and use knowledge specializing in environmental engineering, construction, and management activities worldwide.[7] The company has about 7,000 employees and 170 offices around the world. Using a KMS often leads to additional knowledge creation, storage, sharing, and usage. Drug companies and medical researchers invest billions of dollars in creating knowledge on cures for diseases. Knowledge management systems can also diminish the reliance on paper reports and thus reduce costs and help protect the environment. Although knowledge workers can act alone, they often work in teams to create or obtain knowledge. The Technology, Education, and Design (TED) conference is an annual event that brings together experts in a variety of areas to share ideas and approaches to solving problems.[8] Presenters give talks, which are limited to 18 minutes each. Some leaders in computer and technology companies often attend TED conferences to help them think about the future and collaborate with other industry leaders. When discussing the importance of collaboration, the technology officer of WPP, a marketing and communications firm says, "When it's part of the everyday process, you solve the problem of KM. You capture knowledge as it happens."[9] See Figure 11.3.

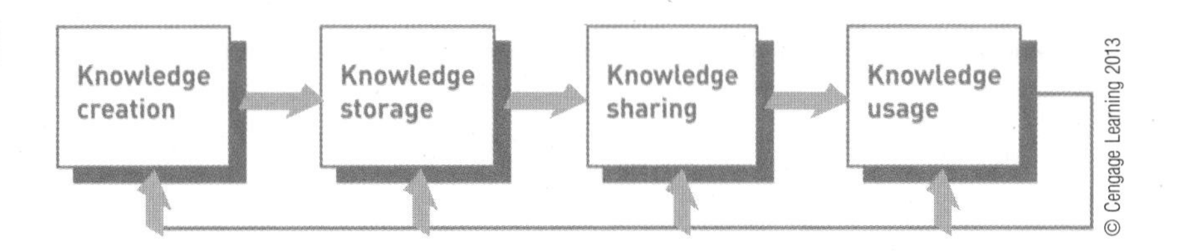

FIGURE 11.3
Knowledge management system
Obtaining, storing, sharing, and using knowledge is the key to any KMS.

After knowledge is created, it is often stored in a *knowledge repository* that includes documents, reports, files, and databases. The knowledge repository can be located both inside the organization and outside. Some types of software can store and share knowledge contained in documents and reports. Adobe Acrobat PDF files, for example, allow you to store corporate reports, tax returns, and other documents and send them to others over the Internet.[10] The publisher and the authors of this book used PDF files to store, share, and edit each chapter. Traditional databases, data warehouses, and data marts, discussed in Chapter 5, often store the organization's knowledge. Specialized knowledge bases in expert systems, discussed later in this chapter, can also be used.

Because knowledge workers often work in groups or teams, they can use collaborative work software and group support systems (discussed in Chapter 10) to share knowledge, such as groupware, meeting software, and collaboration tools. Intranets and password-protected Internet sites also provide ways to share knowledge. Many businesses, however, use patents, copyrights, trade secrets, Internet firewalls, and other measures to keep prying eyes from seeing important knowledge that is expensive and hard to create.

Using a knowledge management system begins with locating the organization's knowledge. This procedure is often done using a *knowledge map* or directory that points the knowledge worker to the needed knowledge. Medical researchers, university professors, and even textbook authors use Lexis-Nexis to locate important knowledge. Corporations often use the Internet or corporate Web portals to help their knowledge workers find knowledge stored in documents and reports.

Technology to Support Knowledge Management

KMSs use a number of tools discussed throughout the book. In Chapter 2, for example, we explored the importance of *organizational learning* and *organizational change*. An effective KMS is based on learning new knowledge and changing procedures and approaches as a result. A manufacturing company, for example, might learn new ways to program robots on the factory floor to improve accuracy and reduce defective parts. The new knowledge will likely cause the manufacturing company to change how it programs and uses its robots. In Chapter 5, we investigated the use of *data mining* and *business intelligence*. These powerful tools can be important in capturing and using knowledge. Enterprise resource planning tools, such as SAP, include knowledge management features. In Chapter 10, we showed how *groupware* can improve group decision-making and collaboration. Groupware can also be used to help capture, store, and use knowledge. Of course, hardware, software, databases, telecommunications, and the Internet, discussed in Part 2, are important technologies used to support most knowledge management systems.

Hundreds of organizations provide specific KM products and services (see Figure 11.4). In addition, researchers at colleges and universities have developed tools and technologies to support knowledge management. American companies spend billions of dollars on knowledge management technology every year. Companies such as IBM have many knowledge management tools in a variety of products, including Lotus Notes, discussed in Chapter 10. Microsoft offers a number of knowledge management tools, including Digital Dashboard, which is based on the Microsoft Office suite. Digital Dashboard integrates information from a variety of sources, including personal, group, enterprise, and external information and documents. Other tools from Microsoft include Web Store Technology, which uses wireless technology to deliver knowledge to any location at any time; Access Workflow Designer, which helps database developers create effective systems to process transactions and keep work flowing through the organization; and related products. Some additional knowledge management organizations and resources are summarized in Table 11.1. In addition to these tools, several artificial intelligence and special-purpose technologies and tools, discussed next, can be used in a KMS.

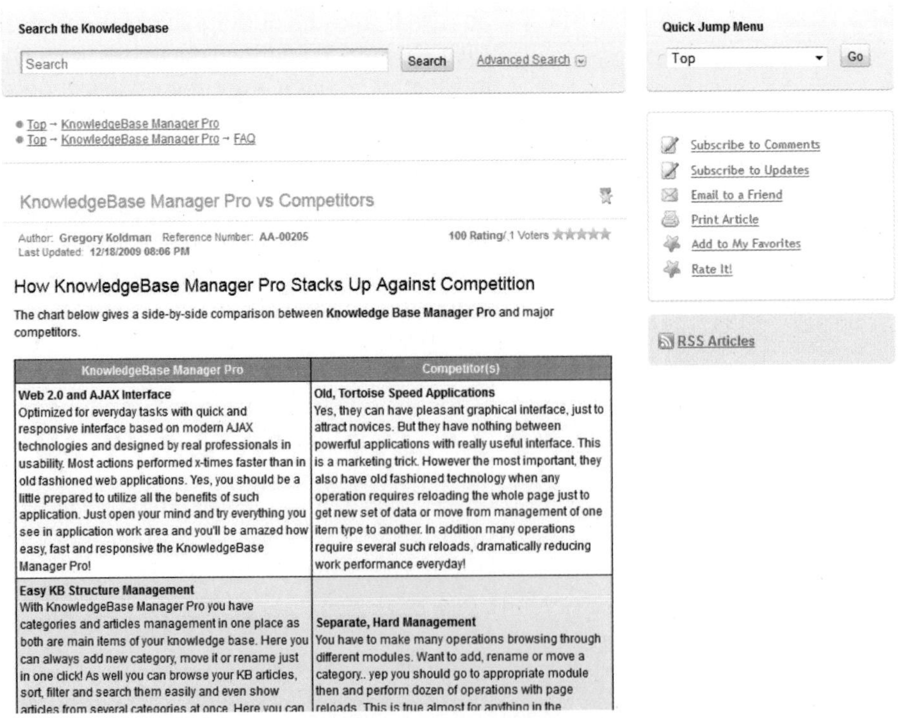

FIGURE 11.4

Knowledge management technology

Knowledgebase Manager Pro is designed for helping organizations create knowledgebases.

Courtesy of Web Site Scripts

TABLE 11.1 Additional knowledge management organizations and resources

Company	Description	Web Site
Knowledge Management World	Knowledge management publications, conferences, and information.	*http://www.kmworld.com/*[11]
Knowledge Management Online	Provides online information, articles, and blogs on knowledge management	*http://www.knowledge-management-online.com/*[12]
CortexPro	Knowledge management collaboration tools	*www.cortexpro.com*[13]
Delphi Group	A knowledge management consulting company	*www.delphigroup.com*[14]
KM Knowledge	Knowledge management sites, products and services, magazines, and case studies	*www.kmknowledge.com*[15]
Knowledge Management Solutions, Inc.	Tools to create, capture, classify, share, and manage knowledge	*www.kmsi.us*[16]
KnowledgeBase	Content creation and management	*www.knowledgebase.com*[17]

AN OVERVIEW OF ARTIFICIAL INTELLIGENCE

At a Dartmouth College conference in 1956, John McCarthy proposed the use of the term *artificial intelligence (AI)* to describe computers with the ability to mimic or duplicate the functions of the human brain. Many AI pioneers attended this first conference; a few predicted that computers would be as "smart" as people by the 1960s. The prediction has not yet been realized, but the application of artificial intelligence can be seen today, and research continues. Watson, a supercomputer developed by IBM with artificial intelligence capabilities, was able to soundly defeat two prior champions of the popular TV game show, *Jeopardy*.[18] The artificial intelligence computer could process human speech, search its vast databases for possible responses, and reply in a human voice. Watson took about four years to build and consists of 90 computer servers, 360 computer chips, and sophisticated software that

Watson is an AI system that can answer questions posed in natural language over a nearly unlimited range of knowledge.

fills a small room.[19] While Watson's game-playing prowess is impressive, the application of artificial intelligence and supercomputers such as Watson could improve our daily lives.[20] Doctors, for example, could use this type of artificial intelligence to make faster, more accurate diagnoses for patients. Medical researchers could use Watson-like artificial intelligence to make medical breakthroughs. Some have called Watson the computer advance of the century.[21]

Artificial Intelligence in Perspective

artificial intelligence systems: People, procedures, hardware, software, data, and knowledge needed to develop computer systems and machines that demonstrate the characteristics of intelligence.

Artificial intelligence systems include the people, procedures, hardware, software, data, and knowledge needed to develop computer systems and machines that demonstrate characteristics of intelligence.[22] Artificial intelligence can be used by most industries and applications. Researchers, scientists, and experts on how human beings think are often involved in developing these systems. Sonia Schulenburg, an ex-bodybuilder who also holds a doctorate in artificial intelligence, started a company called Level E Capital that uses artificial intelligence to pick and trade stocks.[23] Her trading system makes as many as 1,000 trades a day, and her company often outperforms popular stock indexes, such as the FTSE 100. According to Schulenburg, "My children say Mummy makes robots, the robots live in the computer, and the computer trades stocks."

The Nature of Intelligence

From the early AI pioneering stage, the research emphasis has been on developing machines with the ability to "learn" from experiences and apply knowledge acquired from those experiences; to handle complex situations; to solve problems when important information is missing; to determine what is important and to react quickly and correctly to a new situation; to understand visual images, process and manipulate symbols, be creative and imaginative; and to use heuristics, which together is considered **intelligent behavior**. In a book called *The Singularity Is Near* and in articles by and about him, Ray Kurzweil predicts computers will have humanlike intelligence in 20 years.[24] Kurzweil also foresees that by 2045 human and machine intelligence might merge. Machine intelligence, however, is hard to achieve.

intelligent behavior: The ability to learn from experiences and apply knowledge acquired from those experiences; to handle complex situations; to solve problems when important information is missing; to determine what is important and to react quickly and correctly to a new situation; to understand visual images, process and manipulate symbols, be creative and imaginative; and to use heuristics.

The *Turing Test* attempts to determine whether the responses from a computer with intelligent behavior are indistinguishable from responses from a human being. No computer has passed the Turing Test, developed by Alan Turing, a British mathematician.[25] The Loebner Prize offers money and a gold medal for anyone developing a computer that can pass the Turing Test.[26] Some of the specific characteristics of intelligent behavior include the ability to do the following:

- **Learn from experience and apply the knowledge acquired from experience**. Learning from past situations and events is a key component of intelligent behavior and is a natural ability of humans, who learn by trial and error. This ability, however, must be carefully programmed into a computer system. Today, researchers are developing systems that can "learn" from experience. The 20 questions (20Q) Web site, *www.20q.net*, is an example of a system that learns.[27] The Web site is an artificial intelligence game that learns as people play.
- **Handle complex situations**. In a business setting, top-level managers and executives must handle a complex market, challenging competitors, intricate government regulations, and a demanding workforce. Even human experts make mistakes in dealing with these matters. Very careful planning and elaborate computer programming are necessary to develop systems that can handle complex situations.

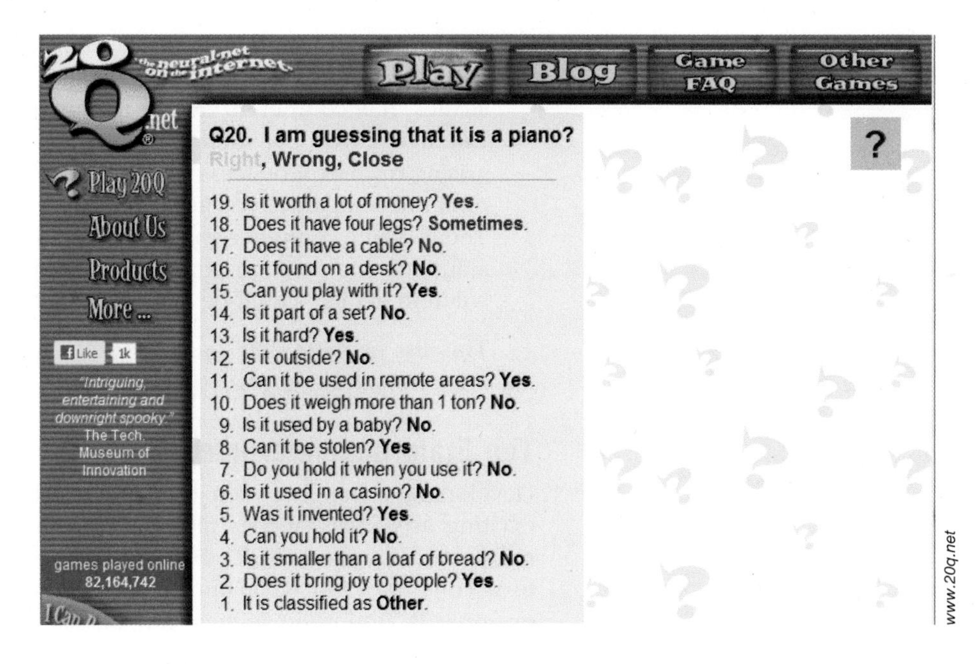

20Q is an online game where users play the popular game, Twenty Questions, against an artificial intelligence foe.

- **Solve problems when important information is missing**. An integral part of decision making is dealing with uncertainty. Often, decisions must be made with little or inaccurate information because obtaining complete information is too costly or impossible. Today, AI systems can make important calculations, comparisons, and decisions even when information is missing.
- **Determine what is important**. Knowing what is truly important is the mark of a good decision maker. Developing programs and approaches to allow computer systems and machines to identify important information is not a simple task.
- **React quickly and correctly to a new situation**. A small child, for example, can look over an edge and know not to venture too close. The child reacts quickly and correctly to a new situation. Computers, on the other hand, do not have this ability without complex programming.
- **Understand visual images**. Interpreting visual images can be extremely difficult, even for sophisticated computers. Moving through a room of chairs, tables, and other objects can be trivial for people but extremely complex for machines, robots, and computers. Such machines require an extension of understanding visual images, called a **perceptive system**. Having a perceptive system allows a machine to approximate the way a person sees, hears, and feels objects.

perceptive system: A system that approximates the way a person sees, hears, and feels objects.

- **Process and manipulate symbols**. People see, manipulate, and process symbols every day. Visual images provide a constant stream of information to our brains. By contrast, computers have difficulty handling symbolic processing and reasoning. Although computers excel at numerical calculations, they aren't as good at dealing with symbols and three-dimensional objects. Recent developments in machine-vision hardware and software, however, allow some computers to process and manipulate some symbols.
- **Be creative and imaginative**. Throughout history, some people have turned difficult situations into advantages by being creative and imaginative. For instance, when defective mints with holes in the middle arrived at a candy factory, an enterprising entrepreneur decided to market these new mints as LifeSavers instead of returning them to the manufacturer. Ice cream cones were invented at the St. Louis World's Fair when an imaginative store owner decided to wrap ice cream with a waffle from his grill for portability. Developing new products and services from an

existing (perhaps negative) situation is a human characteristic. While software has been developed to enable a computer to write short stories, few computers can be imaginative or creative in this way.

- **Use heuristics**. For some decisions, people use heuristics (rules of thumb arising from experience) or even guesses. In searching for a job, you might rank the companies you are considering according to profits per employee. Today, some computer systems, given the right programs, obtain good solutions that use approximations instead of trying to search for an optimal solution, which would be technically difficult or too time consuming.

This list of traits only partially defines intelligence. Another challenge is linking a human brain to a computer.

The Brain Computer Interface

Developing a link between the human brain and the computer is another exciting area that touches all aspects of artificial intelligence. Called *Brain Computer Interface* (*BCI*), the idea is to directly connect the human brain to a computer and have human thought control computer activities.[28] One example is BrainGate, which can be used to connect a human brain to a computer.[29] If successful, the BCI experiment could allow people to control computers and artificial arms and legs through thought alone. The objective is to give people without the ability to speak or move (called Locked-in Syndrome) the capability to communicate and move artificial limbs using advanced BCI technologies. Honda Motors has developed a BCI system that allows a person to complete certain operations, such as bending a leg, with 90 percent accuracy.[30] The new system uses a special helmet that can measure and transmit brain activity to a computer.

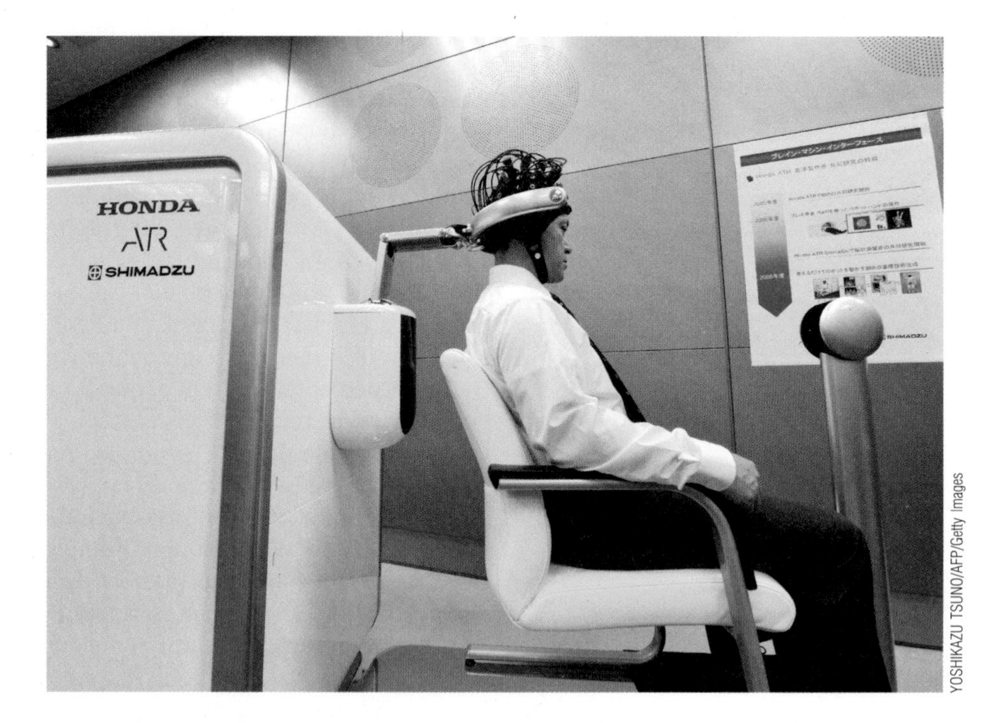

YOSHIKAZU TSUNO/AFP/Getty Images

Honda Motors has developed a brain-machine interface that measures electrical current and blood flow change in the brain and uses the data to control ASIMO, the Honda robot.

The Major Branches of Artificial Intelligence

AI is a broad field that includes several specialty areas, such as expert systems, robotics, vision systems, natural language processing, learning systems, and neural networks (see Figure 11.5). Many of these areas are related; advances in one can occur simultaneously with or result in advances in others.

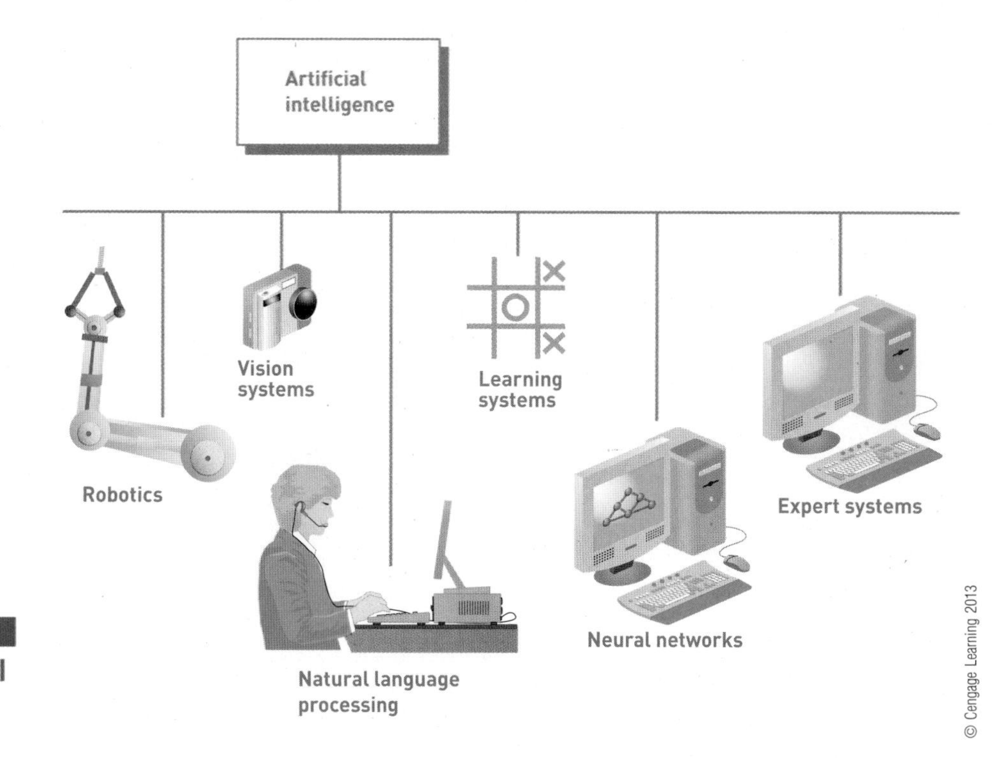

FIGURE 11.5
Conceptual model of artificial intelligence

Expert Systems

An expert system consists of hardware and software that stores knowledge and makes inferences, similar to those of a human expert. Because of their many business applications, expert systems are discussed in more detail in the next several sections of the chapter.

Robotics

robotics: Mechanical or computer devices that perform tasks requiring a high degree of precision or that are tedious or hazardous for humans.

Robotics involves developing mechanical or computer devices that can paint cars, make precision welds, and perform other tasks that require a high degree of precision or are tedious or hazardous for human beings. The word "robot" comes from a play by Karel Capek in the 1920s, when he used the word "robota" to describe factory machines that do drudgery work and revolt.[31] The use of robots has expanded and is likely to increase in the future. For many businesses, robots are used to do the three Ds—dull, dirty, and dangerous jobs, although some people fear that robots will increasingly take other jobs from employees.[32] For example, manufacturers use robots to locate, assemble, and paint products.[33] Some robots, such as the ER series by Intelitek, can be used for training or entertainment.[34] Contemporary robotics combines both high-precision machine capabilities and sophisticated controlling software. The controlling software in robots is what is most important in terms of AI.

The field of robotics has many applications, and research into these unique devices continues.[35] Even NASA uses robots. The Robonaut, also called R2, is a humanlike robot used on the International Space Station.[36] According to NASA's payload controller, "I'd like to introduce you to the newest member of our crew. We're going to see what Robonaut can do." Robonaut can work in front of a computer workstation like human astronauts on the International Space Station. The following are a few additional examples of robots in use:

- In 2011, a marine robot found the flight data recorder of an Air France jet that crashed in 2009 into the ocean on a trip from Rio de Janiero to Paris.[37] The flight data recorder was used to help determine the cause of the crash and avoid future ones.

To School by Remote Control

Public schools do as much as possible to allow students with handicaps to participate in educational activities. However, if a student can be harmed just by exposure to other students, what can a school do?

That was the problem facing the high school in Knox City, Texas. Knox City has a population of 1,130, with 70 students in its high school. With limited resources, it found itself needing to educate 15-year-old Lyndon Baty, who was born with polycystic kidney disease and had a transplant at age 7 but whose body began to reject the new kidney in the summer of 2010. Another transplant is in his future, but meanwhile his immune system cannot tolerate any risk of infection, which he could easily contract from teachers and other students. His parents tried home schooling, but they found that difficult. Lyndon says "I had no social interaction with anybody, no friends to talk to, nobody to be there for me."

Today, Lyndon attends school through a VGo robot. The robot, which he calls his "Batybot," resembles a skinny four-foot-tall penguin on wheels. A computer display where the penguin's head would be shows Lyndon's face. A camera enables Lyndon to see what he would see in the robot's place, a microphone lets him hear what he would hear, and a speaker lets him talk to those around him. Lights let him "raise his hand" in class. Lyndon controls the robot's motion through his computer. The only thing he can't do from home is open doors: someone must open them for the robot. Here's what some people say about the Batybot:

- **Sheri Baty, Lyndon's mother**. "From the first day that Lyndon got to use this technology, the next day there was just an immediate change in his life."
- **Math teacher Becky Jones**. "I can show him on the board, he can work things out, he can ask questions about what he doesn't understand."
- **Classmate Kelsey Vasquez**. "It's like him being there with us in class."
- **Lyndon**. "It's the most wonderful thing that's happened to me since my transplant."
- **Louis Baty, Lyndon's father and superintendent of the Knox City-O'Brien School District**. "The teachers have been very supportive of the equipment. To start with it was a little bit different, but now they're very supportive and consider it just another student in the class."

The VGo costs about $6,000 to purchase, plus about $1,200 per year for a service contract. The school must provide Wi-Fi coverage everywhere the robot has to go, but by 2012, most schools have a Wi-Fi network. In return for these investments, Lyndon can attend school with his classmates, with no additional expenses for tutors or other special equipment.

Lyndon Baty is not the only student to take advantage of the VGo. Cris Colaluca of Mohawk Junior High School in Lawrence County, Pennsylvania, is another student whose life was changed by a VGo robot. Cris, born with spina bifida, developed a rare seizure condition and requires 16 medications daily; however, his mind is intact. Since first grade, his body has not been able to tolerate school. His mother discusses the change when the robot entered Cris's life: "There was an old Cris, the boy that existed before the seizures hit. The seizures changed his health and his personality. He's still a happy-go-lucky kid, but because he had no peer interaction, he became subdued. When VGo came into his life, some of that spark came back. Some of his personality is back. It's an enthusiasm I haven't seen in a long time."

Discussion Questions

1. Would it be practical for a student to attend your college or university via VGo? What obstacles do you think the student and robot would encounter? How easy would it be to remove those obstacles or to develop ways to work around them?
2. Lyndon Baty is in a small school. He knew everyone at school before he could no longer attend in person. Cris Colaluca attends a larger school, where nobody can know all the other students, and never attended it in person. How do you think those differences affect their two VGo experiences?

Critical Thinking Questions

1. What business applications can you think of for VGo? In what industries would VGo be especially useful? If you were a manager in one of those industries (pick one), what would you urge your company to do regarding VGo robots?
2. You read about videoconferencing in Chapter 6. A conference in which some participants are present in person but others appear on computer screens is also sometimes called *telepresence*. Compare and contrast that type of telepresence with the type discussed in this case.

SOURCES: VGo Communications, "VGo Robotic Telepresence," *Computerworld* 2011 case study, *https://www.eiseverywhere.com/file_uploads/a578961052a0afaa64f4836662a98127_VGo_Robotic_Telepresence.pdf*, downloaded February 5, 2012; VGo Communications Web site, *www.vgocom.com*, accessed February 5, 2012; Shamlian, J., "Robot Helps 'Bubble Boy' Make the Grade" (video), *NBC Today*, *today.msnbc.msn.com/id/26184891/vp/41641714*, February 17, 2011; Weaver, R., "New Castle Teen Uses Robot to Continue Learning from Home," *Pittsburgh Tribune-Review*, *www.pittsburghlive.com/x/pittsburghtrib/news/regional/s_779061.html*, January 30, 2012.

- Robots are seeing increased use in corporate offices.[38] The HRP-4 robot from Japan, for example, costs about $350,000, has the general appearance of a human, and has excellent mobility and sensors that can recognize faces and other objects. The PR2, made in the United States at a cost of about $400,000, can travel around the office and do some routine tasks. The ANYBOTD QB also from the United States costs about $15,000, can travel around an office on two wheels, and has a video camera.
- The Envirobot can remove old coatings and clean surfaces on the decks of ships.[39]
- The Robot Learning Laboratory, part of the computer science department and the Robotics Institute at Carnegie Mellon University (*www.ri.cmu.edu*), conducts research into the development and use of robotics.[40]
- IRobot (*www.irobot.com*) is a company that builds a number of robots, including the Roomba Floorvac and Scooba for vacuuming and cleaning floors and the PackBot, an unmanned vehicle used to assist and protect soldiers.[41]
- Robots are used in a variety of ways in medicine. Service robots such as the TUG transport meals, medications, and medical equipment in hospitals.[42] The Porter Adventist Hospital (*www.porterhospital.org*) in Denver, Colorado, uses a $1.2-million Da Vinci Surgical System to perform surgery on prostate cancer patients.[43] The robot has multiple arms that hold surgical tools, and according to one doctor at Porter, "The biggest advantage is it improves recovery time. Instead of having an eight-inch incision, the patient has a 'band-aid' incision. It's much quicker." The Heart-Lander is a very small robot that is inserted below the rib cage and used to perform delicate heart surgery. Cameron Riviere at the Carnegie Mellon Robotics Institute developed the robot with help from Johns Hopkins University.[44]

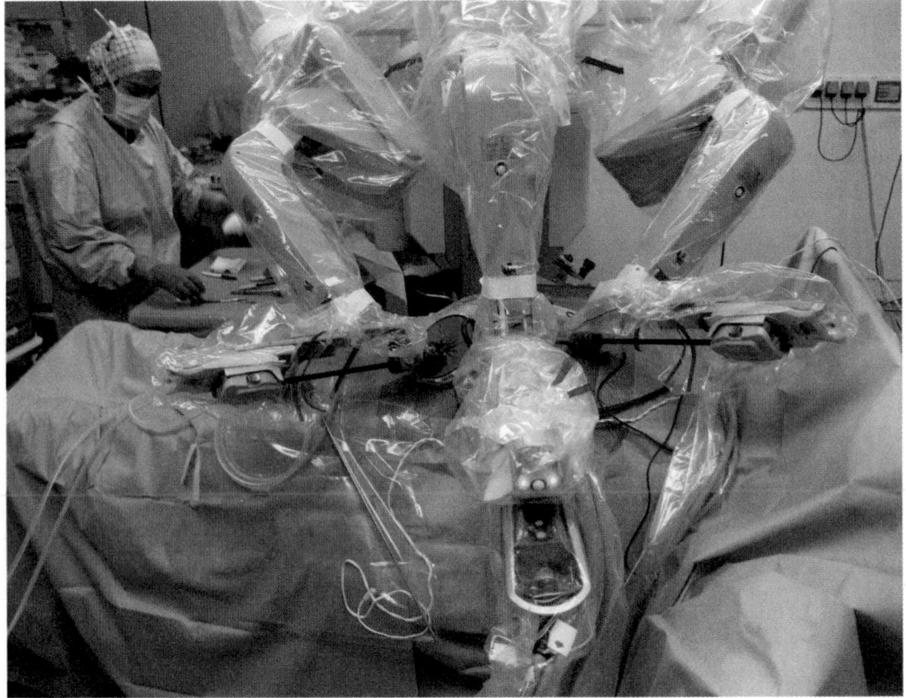

The arms of the Da Vinci robot assist in a kidney transplant. A surgeon controls the robot remotely from a corner of the operating room.

- DARPA (The Defense Advanced Research Project Agency) sponsors the DARPA Grand Challenge (*www.darpagrandchallenge.com*), a 132-mile race over rugged terrain for computer-controlled cars. The agency also sponsors other races and challenges.[45]
- Big Dog, made by Boston Dynamics, is a robot that can carry up to 200 pounds of military gear in field conditions.[46]

Although most of today's robots are limited in their capabilities, future robots will find wider applications in banks, restaurants, homes, doctors' offices, and hazardous working environments such as nuclear stations. The Pepliee and Simroid robots from Japan are ultra-humanlike robots or androids that can blink, gesture, speak, and even appear to breathe.[47] See Figure 11.6.

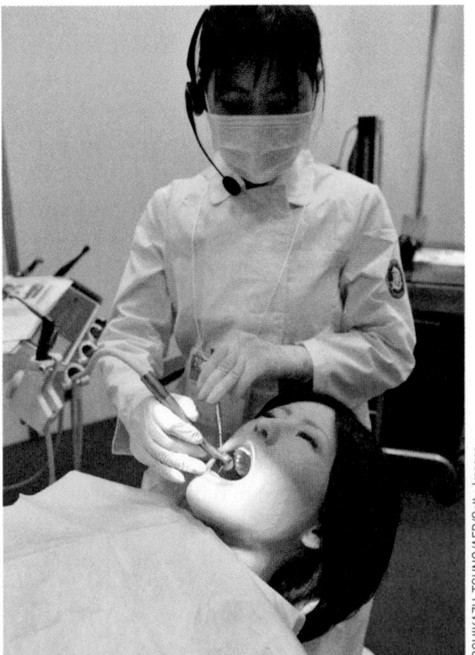

FIGURE **11.6**

Simroid, the humanoid robot

The Simroid robot was developed by Kokoro, a Japanese robot venture, for dentists and students to practice on.

INFORMATION SYSTEMS @ WORK

Allied Vision Technologies Is Watching the Pavement

More than 4 million miles (6.5 million km) of roads cross the United States, with millions more in other countries. Proper maintenance is essential to ensure the safety of the users of these roads. Frequent repaving to completely prevent problems is too expensive, but word of mouth, resident complaints, and spot checks are not a dependable basis for planning repair work. As a result, systematic inspection of all roads is necessary—but that's expensive, too. To make things worse, systematic inspection needs to be done on a regular basis with the changes tracked and analyzed over time to predict when road repair will be necessary and to schedule the work in advance. Cash-strapped city and regional governments, therefore, need an inexpensive way to manage ongoing road maintenance and repair.

Fortunately, vision systems are coming to the rescue. City, town, and regional governments use vision systems to help municipalities maintain their share of the world's roadways. Allied Vision Technologies (AVT) is a leading manufacturer of high-performance vision cameras. According to AVT's Web site, vision systems can "collect field data and assess the condition of all roadway and pavement features such as longitudinal cracks, transverse cracks, alligator cracks, edge cracks, potholes and rutting. Image-based systems offer a less labor-intensive and more reliable solution than traditional manual surveys, and allow the data to be stored for future referencing."

A system developed by AVT and used in the state of Florida incorporates two cameras from AVT and software from NorPix Inc. The cameras, mounted on the roof of a vehicle with one facing forward and one to the rear, capture images of the roadway every 5 to 10 feet (1.5 to 3 meters) as the vehicle drives over it, for a total of up to 180,000 images per day. GPS data is linked to the images to reference images to actual locations. Images are geometrically flattened to eliminate perspective distortion from the angle of the photograph, superimposed on a map for analysis, and compared with images of the same location from previous years to determine the rate of change. The AVT system still relies on people for the final data interpretation, though its image-processing capability simplifies their task.

Other vision systems can analyze the pavement as well. Pavemetrics Systems of Québec, Canada, offers systems that use lasers and high-speed cameras with custom optics to detect cracks, ruts, and surface deterioration at speeds of over 60 mph (100 km/hr), day or night, on all types of road surfaces. Pavemetrics systems classify cracks into three categories and evaluate their severity. The systems can also measure and report on the condition of lane markings.

The use of vision systems in pavement inspection is expected to become much more widespread in next few years. However, pavement inspection imposes unique challenges on vision systems. Inspection must take place quickly so that the inspecting vehicle won't obstruct traffic. The system must be able to tell the difference between a crack and other surface imperfections such as oil stains under conditions of low contrast between a crack and an undamaged surface. All in all, vision systems in pavement inspection have a tall order, but filling it saves government agencies a great deal of money by making road repairs at the right time—not too early, not too late—and on a planned rather than an emergency basis.

Vision systems are evolving rapidly. Improvements to the underlying hardware and software technologies, combined with research into the use of those components in a variety of vision-based applications, contribute to better acquisition, storage, and analysis of images.

Discussion Questions

1. You work for a city highway department. Your job is to drive over its roads, noting their condition to determine which must be repaired, which must be monitored, and which can be left alone for a while. Your mayor suggests that the city should buy a vision system and asks for your opinion. Options include (a) no new system, (b) a system such as the AVT system above, and (c) at higher cost, a system such as the Pavemetrics system. Write a memo to the mayor making and justifying your recommendation.
2. About 1.4 million of the 4 million miles of road in the United States are unpaved, mostly in rural areas, through forests, and in areas with very little traffic. Those roads must also be maintained. Would vision systems be of value in evaluating them?

Critical Thinking Questions

1. Vision systems use several underlying technologies. Some of them are mentioned in this sidebar. Which technologies do you think are most critical in terms of their rate of improvement to the future development of pavement monitoring systems?
2. A van equipped with a pavement inspection system costs about as much as a full-time highway

department employee for a year. The highway department employee can use a car that the department already has. In addition, there are ongoing costs for using and maintaining the inspection system. Compare, as best you can from the available information, the costs of a pavement inspection system with its possible benefits.

SOURCES: Allied Vision Technologies, "Mobile Machine Vision System Featuring AVT GigE Cameras Surveys Pavement Condition," *www.alliedvisiontec.com/emea/products/applications/application-case-study/article/mobile-machine-vision-system-featuring-avt-gige-cameras-surveys-pavement-condition.html*, July 7, 2011; Allied Vision Technologies, GC 1350 camera information, *www.alliedvisiontec.com/emea/products/cameras/gigabit-ethernet/prosilica-gc/gc1350.html*, accessed February 10, 2012; Chambon, S. and Moliard, J.-M., "Automatic Road Pavement Assessment with Image Processing: Review and Comparison," *International Journal of Geophysics*, *www.hindawi.com/journals/ijgp/2011/989354*, 2011; Norpix Web site, *www.norpix.com*, accessed February 6, 2012; Pavemetrics Systems, Inc., "LCMS—Laser Crack Measurement System," *www.pavemetrics.com/en/lcms.html*, accessed February 11, 2012; Salari, E. and Bao, G., "Automated Pavement Distress Inspection Based on 2D and 3D Information," 2011 IEEE International Conference on Electro/Information Technology, Mankato, MN, May 15–17, 2011; SSMC, "Pavement Mapping/Condition Assessment," *www.southeastern surveying.com/pavement_mapping.html*, February 11, 2011.

Vision Systems

vision systems: The hardware and software that permit computers to capture, store, and manipulate visual images.

Another area of AI involves vision systems. **Vision systems** include hardware and software that permit computers to capture, store, and manipulate visual images.[48] A video designer, for example, decided to develop a pocket translator application on a trip to Germany with his girlfriend.[49] The application, called Word Lens, uses optical character recognition to read text by holding a smartphone or other mobile device up to a restaurant menu, books, signs, and other text and taking a picture of the text. In a few seconds or less, the application can translate the camera image from one language to another, such as German to English. Vision systems are also effective at identifying people based on facial features. Another vision system is Nvidia's GeForce 3D, software that can display images on a computer screen that are three-dimensional when viewed with special glasses.[50]

Natural Language Processing and Voice Recognition

natural language processing: Processing that allows a computer to understand and react to statements and commands made in a "natural" language such as English.

As discussed in Chapter 4, **natural language processing** allows a computer to understand and react to statements and commands made in a "natural" language such as English. Google, for example, has a service called Google Voice Local Search that allows you to dial a toll-free number and search for local businesses using voice commands and statements.[51] Many companies provide natural language-processing help over the phone. When you call the help phone number, you are typically given a menu of options and asked to speak your responses. Many people, however, are frustrated talking to a machine instead of a human.

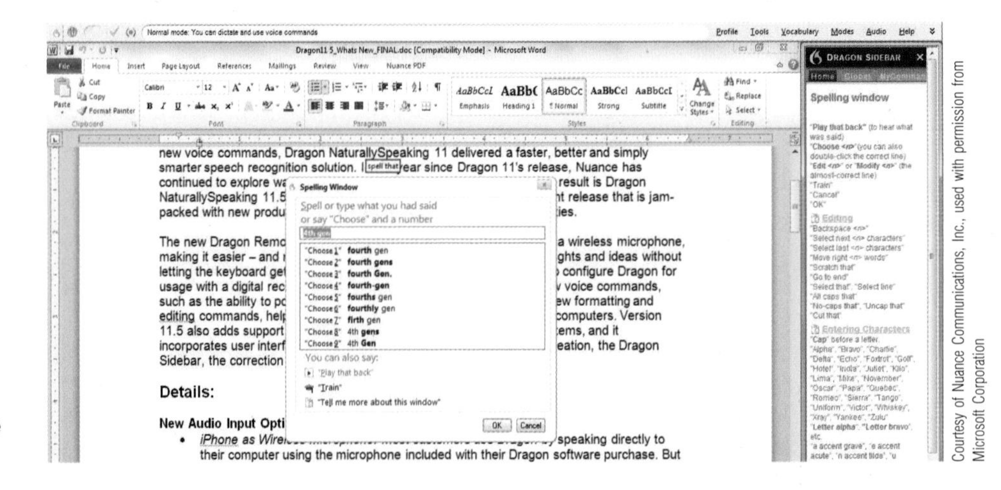

Courtesy of Nuance Communications, Inc., used with permission from Microsoft Corporation

Dragon Systems's Naturally Speaking uses continuous voice recognition, or natural speech, that allows the user to speak to the computer at a normal pace without pausing between words. The spoken words are transcribed immediately onto the computer screen.

In some cases, voice recognition is used with natural language processing. *Voice recognition* involves converting sound waves into words. After converting sounds into words, natural language-processing systems react to the words or commands by performing a variety of tasks. Brokerage services are a perfect fit for voice recognition and natural language-processing technology to replace the existing "press 1 to buy or sell a stock" touchpad telephone menu system. Using voice recognition to convert recordings into text is also possible. Some companies claim that voice recognition and natural language-processing software is so good that customers forget they are talking to a computer and start discussing the weather or sports scores.

Learning Systems

learning systems: A combination of software and hardware that allows a computer to change how it functions or how it reacts to situations based on feedback it receives.

Another part of AI deals with **learning systems**, a combination of software and hardware that allows a computer to change how it functions or how it reacts to situations based on feedback it receives. For example, some computerized games have learning abilities. If the computer does not win a game, it remembers not to make the same moves under the same conditions again. *Reinforcement Learning* is a learning system involving sequential decisions with learning taking place between each decision. Reinforcement learning often involves sophisticated computer programming and optimization techniques, first discussed in Chapter 10. The computer makes a decision, analyzes the results, and then makes a better decision based on the analysis. The process, often called *dynamic programming*, is repeated until it is impossible to make improvements in the decision.

Learning systems software requires feedback on the results of actions or decisions. At a minimum, the feedback needs to indicate whether the results are desirable (winning a game) or undesirable (losing a game). The feedback is then used to alter what the system will do in the future.

Neural Networks

neural network: A computer system that can act like or simulate the functioning of a human brain.

An increasingly important aspect of AI involves neural networks, also called neural nets. A **neural network** is a computer system that can act like or simulate the functioning of a human brain.[52] The systems can use massively parallel processors in an architecture that is based on the human brain's own meshlike structure. In addition, neural network software simulates a neural network using standard computers. Neural networks can process many pieces of data at the same time and learn to recognize patterns.

AI Trilogy, available from the Ward Systems Group (*www.wardsystems.com*), is a neural network software program that can run on a standard PC.[53] The software can make predictions with NeuroShell Predictor and classify information with NeuroShell Classifier (see Figure 11.7). The software package also contains GeneHunter, which uses a special type of algorithm called a genetic algorithm to get the best result from the neural network system. (Genetic algorithms are discussed next.) Some pattern recognition software uses neural networks to analyze hundreds of millions of bank, brokerage, and insurance accounts involving a trillion dollars to uncover money laundering and other suspicious money transfers.

Other Artificial Intelligence Applications

genetic algorithm: An approach to solving large, complex problems in which many related operations or models change and evolve until the best one emerges; also called a *genetic program*.

A few other artificial intelligence applications exist in addition to those just discussed. A **genetic algorithm**, also called a genetic program, is an approach to solving large, complex problems in which many repeated operations or models change and evolve until the best one emerges.[54] This approach is based on the theory of evolution that requires (1) variation and (2) natural selection. The first step is to change or vary competing solutions to the problem. This alteration can be done by changing the parts of a program or by combining different program segments into a new program, thus mimicking the evolution of species in which the genetic makeup of a plant or animal mutates or changes over time. The second step is to select only the best models or algorithms, which continue to evolve. Programs or program segments that are not as good as others are discarded—a process similar to what happens in natural selection in which only the best species survive and continue to evolve. This process of variation and selection continues until the genetic algorithm yields the best possible solution to the original problem. A genetic algorithm can be used to help schedule airline crews to meet flight requirements while minimizing total costs.

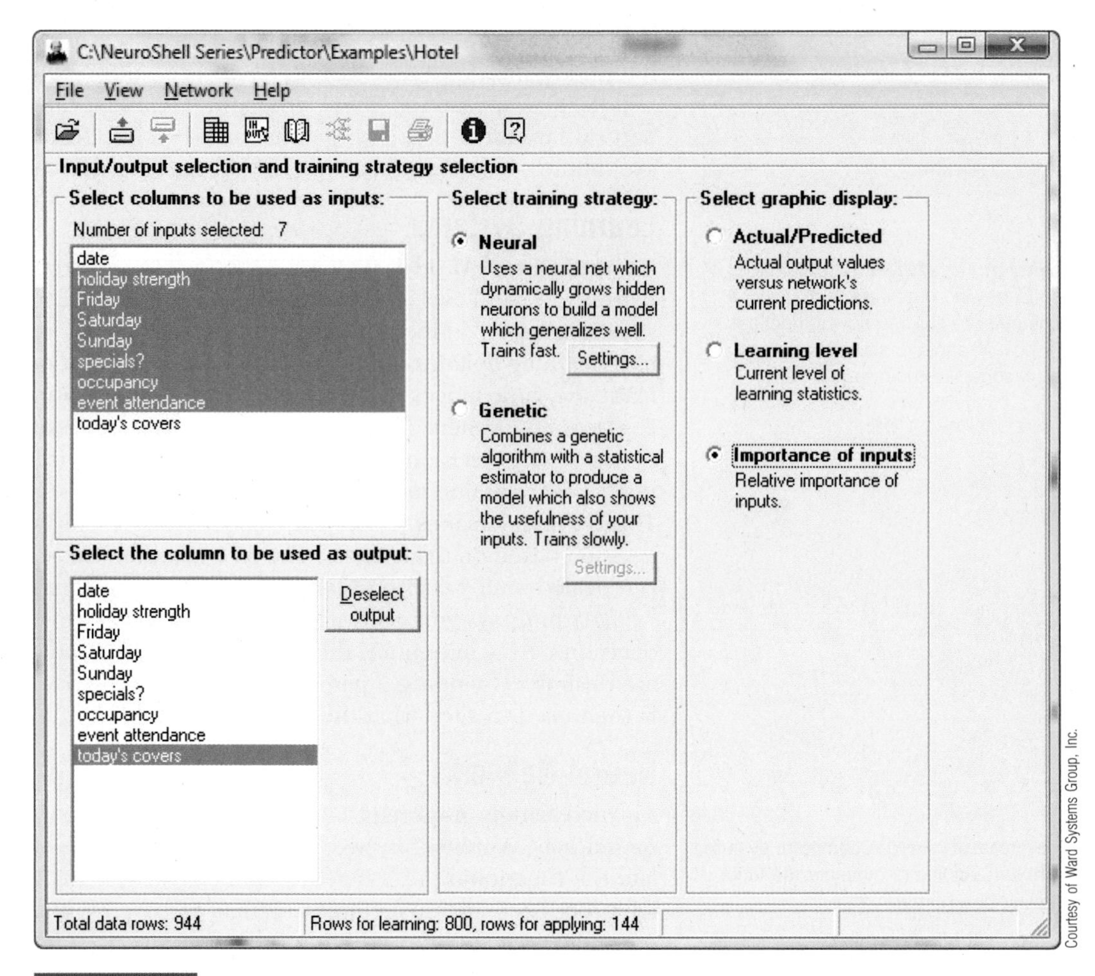

FIGURE 11.7

Neural network software

NeuroShell Predictor uses recognized forecasting methods to look for future trends in data.

intelligent agent: Programs and a knowledge base used to perform a specific task for a person, a process, or another program; also called an *intelligent robot* or *bot*.

An **intelligent agent** (also called an *intelligent robot* or *bot*) consists of programs and a knowledge base used to perform a specific task for a person, a process, or another program.[55] Like a sports agent who searches for the best endorsement deals for a top athlete, an intelligent agent often searches to find the best price, schedule, or solution to a problem. The programs used by an intelligent agent can search large amounts of data as the knowledge base refines the search or accommodates user preferences. Often used to search the vast resources of the Internet, intelligent agents can help people find information on any topic, such as the best price for a new digital camera.

AN OVERVIEW OF EXPERT SYSTEMS

As mentioned earlier, an expert system behaves similarly to a human expert in a particular field. Like human experts, computerized expert systems use heuristics, or rules of thumb, to arrive at conclusions or make suggestions. The Lantek expert system, for example, can be used to cut and fabricate metal into finished products for the automotive, construction, and mining industries.[56] The expert system can help reduce raw material waste and increase profits. Since expert systems can be difficult, expensive, and time consuming to develop, they should be developed when there is a high potential payoff or when they significantly reduce downside risk and the organization wants to capture and preserve irreplaceable human expertise.

Expert systems are used in metal fabrication plants to aid in decision making.

Components of Expert Systems

An expert system consists of a collection of integrated and related components, including a knowledge base, an inference engine, an explanation facility, a knowledge base acquisition facility, and a user interface. A diagram of a typical expert system is shown in Figure 11.8. In this figure, the user interacts with the interface, which interacts with the inference engine. The inference engine interacts with the other expert system components. These components must work together to provide expertise. This figure also shows the inference engine coordinating the flow of knowledge to other components of the expert system. Note that different knowledge flows can exist, depending on what the expert system is doing and on the specific expert system involved.

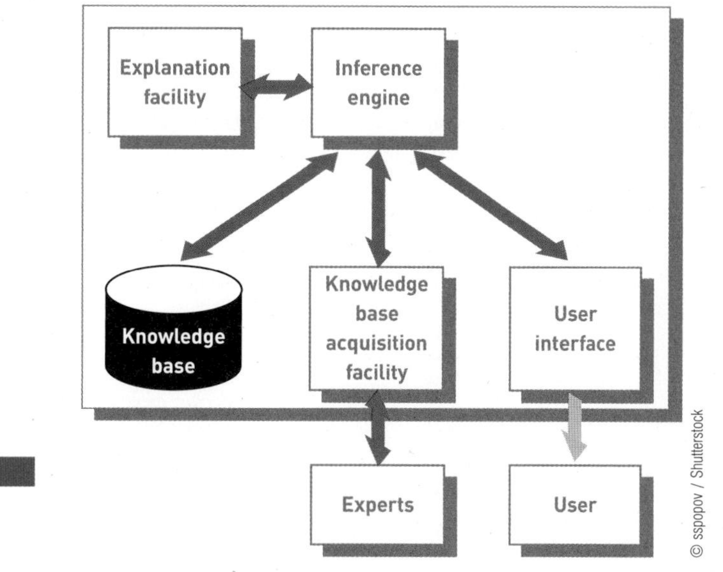

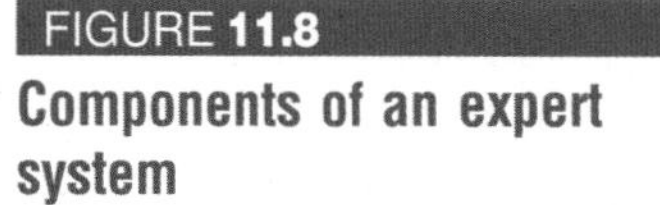
FIGURE 11.8

Components of an expert system

The Knowledge Base

The knowledge base stores all relevant information, data, rules, cases, and relationships that the expert system uses. As shown in Figure 11.9, a knowledge base is a natural extension of a database (presented in Chapter 5) and an information and decision support system (presented in Chapter 10). A knowledge base must be developed for each unique application. For example, a medical expert system contains facts about diseases and symptoms. The following are some tools and techniques that can be used to create a knowledge base:

Database
raw
facts
Information and decision support
information
Knowledge base
patterns and
relationships
Increasing understanding

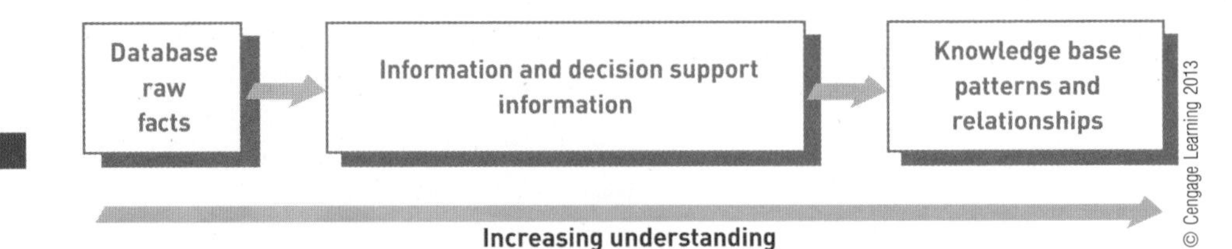

FIGURE **11.9**

Relationships between data, information, and knowledge

rule: A conditional statement that links conditions to actions or outcomes.

IF-THEN statements: Rules that suggest certain conclusions.

- **Using rules**. A **rule** is a conditional statement that links conditions to actions or outcomes. In many instances, these rules are stored as **IF-THEN statements**, which are rules that suggest certain conclusions. For example, "if a certain set of network conditions exists, then a certain network problem diagnosis is appropriate." In an expert system for a weather forecasting operation, for example, the rules could state that if certain temperature patterns exist with a given barometric pressure and certain previous weather patterns over the last 24 hours, then a specific forecast will be made, including temperatures, cloud coverage, and wind-chill factor. IBM has used a rule-based expert system to help detect execution errors in its large mainframe computers.[57] Figure 11.10 shows how to use expert system rules in determining whether a person should receive a mortgage loan from a bank. These rules can be placed in almost any standard programming language (discussed in Chapter 4) using IF-THEN statements or into special expert systems shells and products, discussed later in the chapter. In general, as the number of rules that an expert system knows increases, the precision of the expert system also increases.

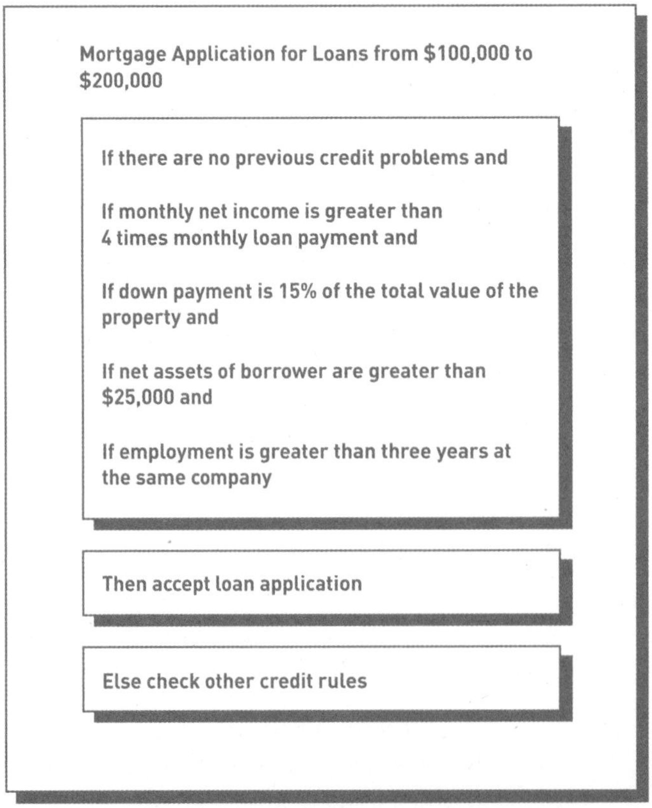

Mortgage Application for Loans from $100,000 to $200,000
If there are no previous credit problems and
If monthly net income is greater than 4 times monthly loan payment and
If down payment is 15% of the total value of the property and
If net assets of borrower are greater than $25,000 and
If employment is greater than three years at the same company
Then accept loan application
Else check other credit rules

FIGURE **11.10**

Rules for a credit application

- **Using cases.** An expert system can use cases in developing a solution to a current problem or situation. This process involves (1) finding cases stored in the knowledge base that are similar to the problem or situation at hand and (2) modifying the solutions to the cases to fit or accommodate the current problem or situation. For example, a company might use an expert system to determine the best location for a new service facility in the state of New Mexico. The expert system might identify two previous cases involving the location of a service facility where labor and transportation costs were also important—one in the state of Colorado and the other in the state of Nevada. The expert system can modify the solution to these two cases to determine the best location for a new facility in New Mexico.

The Inference Engine

inference engine: Part of the expert system that seeks information and relationships from the knowledge base and provides answers, predictions, and suggestions similar to the way a human expert would.

The overall purpose of an **inference engine** is to seek information and relationships from the knowledge base and to provide answers, predictions, and suggestions similar to the way a human expert would. In other words, the inference engine is the component that delivers the expert advice. Consider the expert system that forecasts future sales for a product. One approach is to start with a fact such as "The demand for the product last month was 20,000 units." The expert system searches for rules that contain a reference to product demand. For example, "IF product demand is over 15,000 units, THEN check the demand for competing products." As a result of this process, the expert system might use information on the demand for competitive products. Next, after searching additional rules, the expert system might use information on personal income or national inflation rates. This process continues until the expert system can reach a conclusion using the data supplied by the user and the rules that apply in the knowledge base.

The Explanation Facility

explanation facility: Component of an expert system that allows a user or decision maker to understand how the expert system arrived at certain conclusions or results.

An important part of an expert system is the **explanation facility**, which allows a user or decision maker to understand how the expert system arrived at certain conclusions or results. A medical expert system, for example, might reach the conclusion that a patient has a defective heart valve given certain symptoms and the results of tests on the patient. The explanation facility allows a doctor to find out the logic or rationale of the diagnosis made by the expert system. The expert system, using the explanation facility, can indicate all the facts and rules that were used in reaching the conclusion, which the doctors can look at to determine whether the expert system is processing the data and information correctly and logically.

The Knowledge Acquisition Facility

A difficult task in developing an expert system is the process of creating and updating the knowledge base.[58] In the past, when more traditional programming languages were used, developing a knowledge base was tedious and time consuming. Each fact, relationship, and rule had to be programmed into the knowledge base. In most cases, an experienced programmer had to create and update the knowledge base.

knowledge acquisition facility: Part of the expert system that provides a convenient and efficient means of capturing and storing all the components of the knowledge base.

Today, specialized software allows users and decision makers to create and modify their own knowledge bases through the knowledge acquisition facility, as shown in Figure 11.11. The purpose of the **knowledge acquisition facility** is to provide a convenient and efficient means of capturing and storing all components of the knowledge base. Knowledge acquisition software can present users and decision makers with easy-to-use menus. After entering the appropriate attributes, the knowledge acquisition facility correctly stores information and relationships in the knowledge base, making the knowledge base easier and less expensive to set up and maintain.

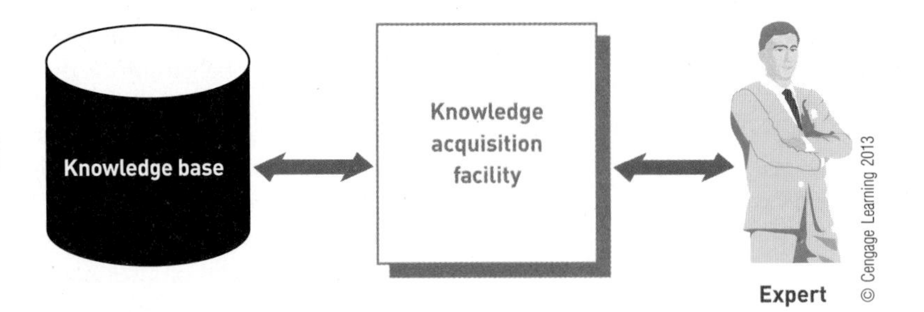

FIGURE 11.11
Knowledge acquisition facility
The knowledge acquisition facility acts as an interface between experts and the knowledge base.

The User Interface

The main purpose of the user interface is to make an expert system easier for users and decision makers to develop and use. At one time, skilled computer personnel created and operated most expert systems; today, the user interface permits decision makers to develop and use their own expert systems. A typical user interface between a user and the expert system for the credit application shown in Figure 11.10 follows:

EXPERT SYSTEM: Are there any previous credit problems?

USER: No.

EXPERT SYSTEM: Is the applicant's net income greater than four times the monthly loan payment?

USER: Yes.

EXPERT SYSTEM: Is the down payment greater than 15% of the property value?

USER: Yes.

EXPERT SYSTEM: Are the net assets of the borrower greater than $25,000?

USER: Yes.

EXPERT SYSTEM: Has the borrower worked more than three years at the same company?

USER: Yes.

EXPERT SYSTEM: **Accept the loan application.**

Participants in Developing and Using Expert Systems

Typically, several people are involved in developing and using an expert system (see Figure 11.12).

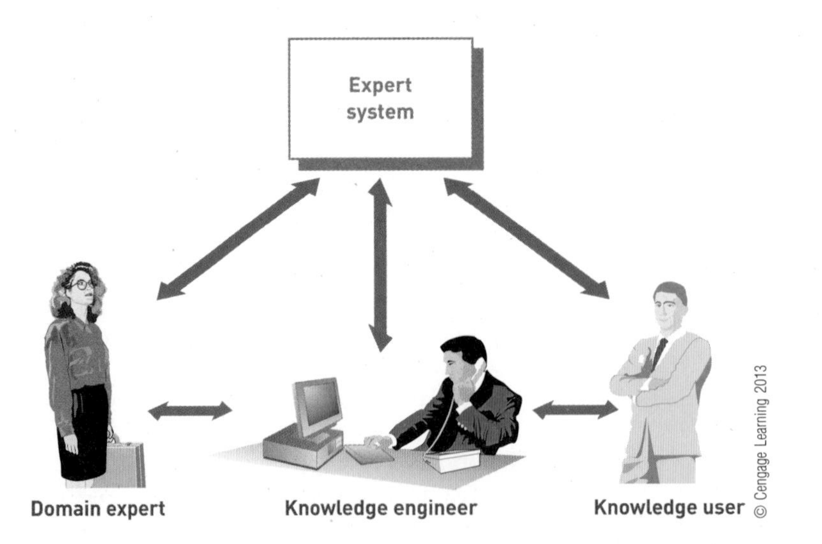

FIGURE 11.12
Participants in expert systems development and use

domain expert: The person or group with the expertise or knowledge the expert system is trying to capture (domain).

The **domain expert** is the person or group with the expertise or knowledge the expert system is trying to capture (domain). In most cases, the domain expert is a group of human experts. Research has shown that good domain

knowledge engineer: A person who has training or experience in the design, development, implementation, and maintenance of an expert system.

knowledge user: The person or group who uses and benefits from the expert system.

experts can increase the overall quality of an expert system. A **knowledge engineer** is a person who has training or experience in the design, development, implementation, and maintenance of an expert system, including training or experience with expert system shells. Knowledge engineers can help transfer the knowledge from the expert system to the knowledge user. The **knowledge user** is the person or group who uses and benefits from the expert system. Knowledge users do not need any previous training in computers or expert systems.

Expert Systems Development Tools and Techniques

Theoretically, expert systems can be developed from any programming language. Since the introduction of computer systems, programming languages have become easier to use, more powerful, and better able to handle specialized requirements. In the early days of expert systems development, traditional high-level languages, including Pascal, FORTRAN, and COBOL, were used, as shown in Figure 11.13. LISP was one of the first special languages developed and used for artificial intelligence applications, and PROLOG was also developed for AI applications. Since the 1990s, however, other expert system products (such as shells) have become available that remove the burden of programming, allowing nonprogrammers to develop and benefit from the use of expert systems.

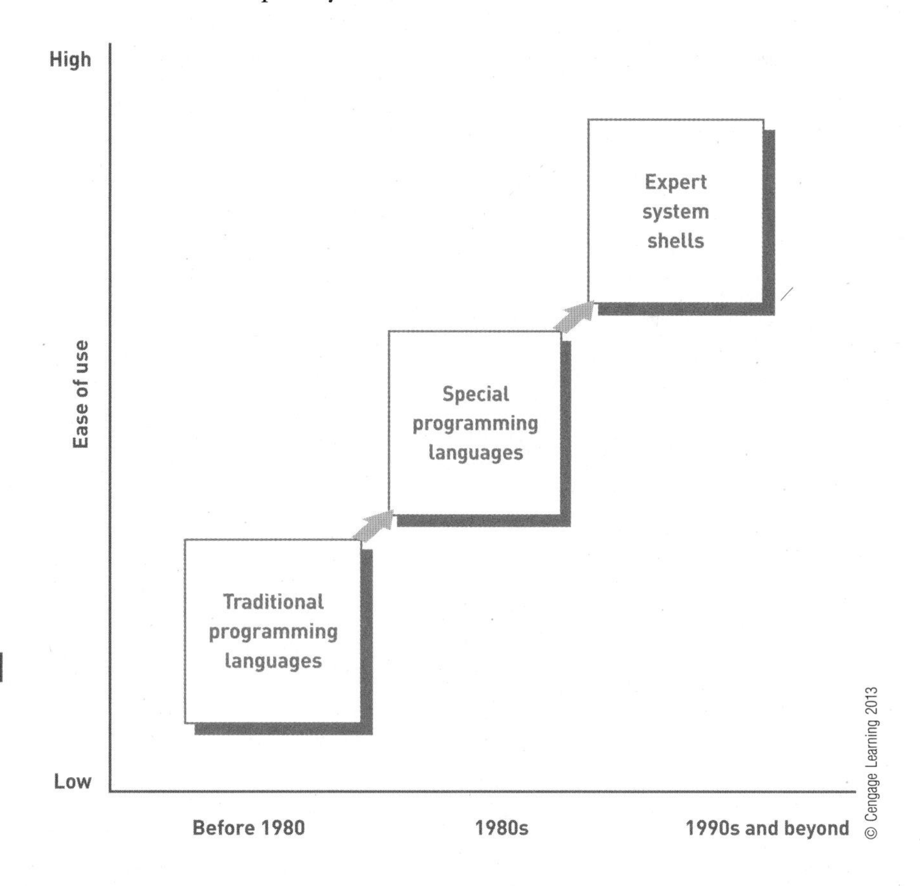

FIGURE **11.13**

Expert systems development

Software for expert systems development has evolved greatly since 1980, from traditional programming languages to expert system shells.

Expert System Shells and Products

An *expert system shell* is a collection of software packages and tools used to design, develop, implement, and maintain expert systems. Expert system shells are available for both personal computers and mainframe systems, with some shells being inexpensive, costing less than $500. In addition, off-the-shelf expert system shells are complete and ready to run. The user enters the appropriate data or parameters, and the expert system provides output to the problem or situation. Table 11.2 lists a few expert system products.

TABLE **11.2** Popular expert system products

Name of Product	Application and Capabilities
Clips	A tool for building expert systems on PCs.[59]
Cogito	Software by Expert System Semantic Intelligence that helps an organization extract knowledge from text in e-mails, articles, Web sites, documents, and other unstructured information.[60]
Exsys Corvid	An expert system tool that simulates a conversation with a human expert from Exsys (*www.exsys.com*).[61]
EZ-Xpert	A rule-based expert system that results in complete applications in the C++ or Visual Basic programming languages by EZ-Xpert (*www.ez-xpert.com*).[62]
Imprint Business Systems	An expert system that helps printing and packaging companies manage their businesses (*www.imprint-mis.co.uk*)[63]
Lantek Expert System	Software that helps metal fabricators reduce waste and increase profits (*www.lantek.es*).[64]

MULTIMEDIA AND VIRTUAL REALITY

The use of multimedia and virtual reality has helped many companies achieve a competitive advantage and increase profits. The approach and technology used in multimedia is often the foundation of virtual reality systems, discussed later in this section. While these specialized information systems are not used by all organizations, they can play a key role for many. We begin with a discussion of multimedia.

Overview of Multimedia

multimedia: Text, graphics, video, animation, audio, and other media that can be used to help an organization efficiently and effectively achieve its goals.

Multimedia is text, graphics, video, animation, audio, and other media that can be used to help an organization efficiently and effectively achieve its goals. Multimedia can be used to create stunning brochures, presentations, reports, and documents. Many companies are also starting to use multimedia approaches to develop exciting cartoons and video games to help advertise products and services.[65] For example, insurance company Geico uses animation in some of its TV ads. Animation Internet sites, such as Xtranormal and GoAnimate, can help individuals and corporations develop these types of animations. Although not all organizations use the full capabilities of multimedia, most use text and graphics capabilities.

Text and Graphics

All large organizations and most small and medium-sized ones use text and graphics to develop reports, financial statements, advertising pieces, and other documents used internally and externally. Internally, organizations use text and graphics to communicate policies, guidelines, and much more to managers and employees. Externally, they use text and graphics to communicate to suppliers, customers, federal and state groups , and a variety of other stakeholders. Text can have different sizes, fonts, and colors, and graphics can include photographs, illustrations, drawings, a variety of charts, and other still images. Graphic images can be stored in a variety of formats, including JPEG (Joint Photographic Experts Group format) and GIF (Graphics Interchange Format).

While standard word-processing programs are an inexpensive and simple way to develop documents and reports that require text and graphics, most organizations use specialized software (see Figure 11.14). Adobe Illustrator, for example, can be used to create attractive and informative charts, illustrations, and brochures. The software can also be used to develop digital art, reference manuals, profit and loss statements, and a variety of reports required by state and federal governments. Adobe Photoshop is a sophisticated and popular software package that can be used to edit photographs and other visual images. Once created, these documents and reports can be saved in an Adobe PDF (Portable Document Format) file and sent over the Internet or saved on a CD or similar storage device.

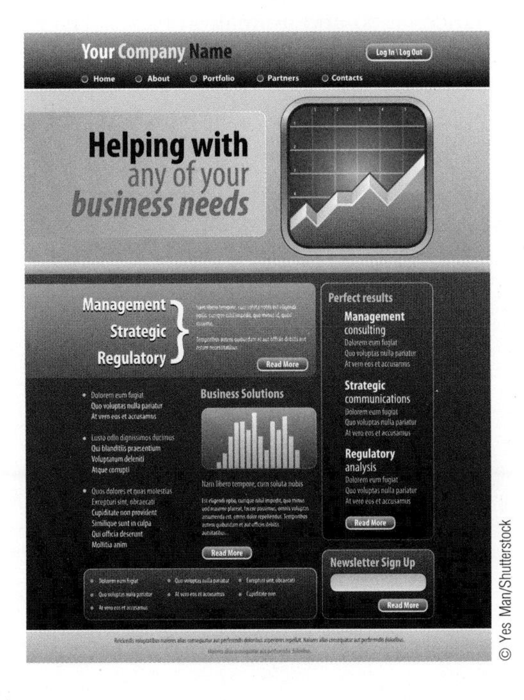

© Yes Man/Shutterstock

FIGURE **11.14**

Digital graphics

Businesses create graphics such as charts, illustrations, and brochures using software such as Adobe Photoshop or Adobe Illustrator.

Microsoft Silverlight can be used to add high-definition video and animation to Internet sites and other programming.[66] PowerPoint, also by Microsoft, can be used to develop a presentation that is displayed on a large viewing screen with sound and animation. There are many other graphics programs, including Paint and PhotoDraw by Microsoft, CorelDraw, and others.

Many graphics programs can also create 3D images.[67] James Cameron's movie *Avatar* used sophisticated computers and 3D imaging to create one of the most profitable movies in history. Once used primarily in movies, 3D technology can be employed by companies to design products, such as motorcycles, jet engines, bridges, and more. Autodesk, for example, makes exciting 3D software that companies can use to design large skyscrapers and other buildings.[68] The technology used to produce 3D movies will also be available with some TV programs. Nintendo has developed one of the first portable gaming devices that displays images in 3D.[69] The video game industry may generate as much as $50 billion annually.

Audio

Audio includes music, human voices, recorded sounds, and a variety of computer-generated sounds. It can be stored in a variety of file formats, including MP3 (Motion Picture Experts Group Audio Layer 3), WAV (wave format), MIDI (Musical Instrument Digital Interface), and other formats. When audio files are played while they are being downloaded from the Internet, it's called *streaming audio*.

Input to audio software includes audio recording devices: microphones, imported music or sound from CDs or audio files, MIDI instruments that can create music and sounds directly, and other audio sources. Once stored, audio files can be edited and augmented using audio software, including Apple's QuickTime, Microsoft's Sound Recorder, Adobe's Audition, and other software. Once edited, audio files can also be used to enhance presentations, create music, broadcast satellite radio signals, develop audio books, record podcasts for iPods and other audio players, provide realism to movies, and enhance video and animation (discussed next).

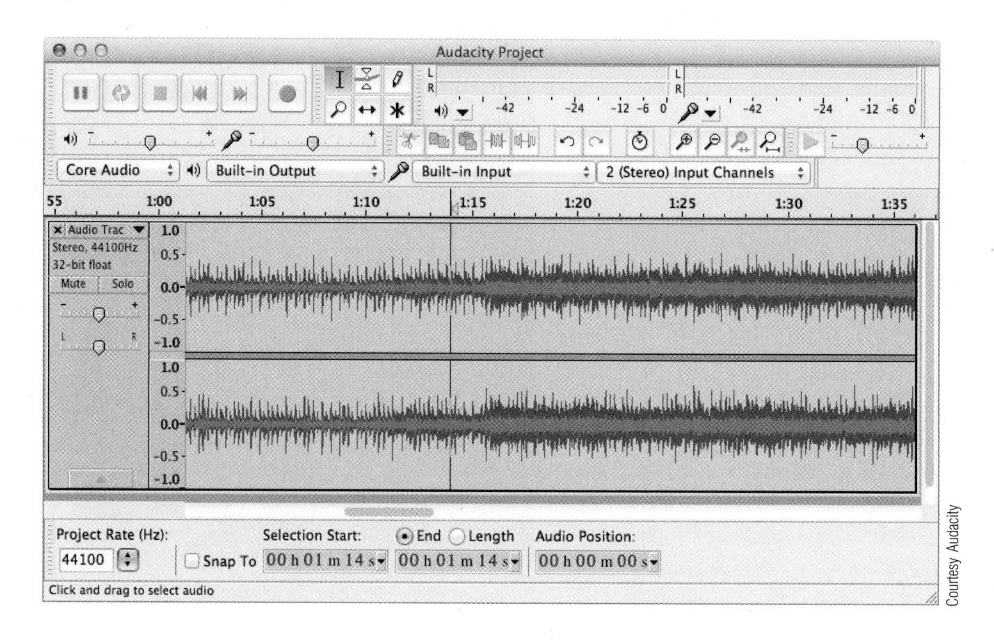

Courtesy Audacity

Audacity provides editing tools for editing and producing audio files in a variety of formats.

Video and Animation

The moving images of video and animation are typically created by rapidly displaying one still image after another. Video and animation can be stored in AVI (Audio Video Interleave) files used with many Microsoft applications, MPEG (Motion Picture Experts Group format) files, and MOV (QuickTime format) files used with many Apple applications. When video files are played while they are being downloaded from the Internet, it's called *streaming video*. For example, Netflix, which allows people to stream movies and TV programs to their TV, is becoming a popular alternative to renting DVDs at a video store.[70] On the Internet, Java applets (small downloadable programs) and animated GIF files can be used to animate or create "moving" images.

A number of video and animation software products can be used to create and edit video and animation files. Many video and animation programs can create realistic 3D moving images. Adobe's Premiere and After Effects and Apple's Final Cut Pro can be used to edit video images taken from cameras and other sources. Final Cut Pro, for example, has been used to edit and produce full-length motion pictures shown in movie theaters. Adobe Flash and LiveMotion can be used to add motion and animation to Web pages.

Video and animation have many business uses. Companies that develop computer-based or Internet training materials often use video and audio software. An information kiosk at an airport or shopping mall can use animation to help customers check-in for a flight or to get information. Further, Pixar uses sophisticated animation software to create dazzling 3D movies, with the exact process presented at Pixar's Web site.[71] See Figure 11.15.

FIGURE 11.15
Creating animation for Pixar
Pixar uses sophisticated proprietary animation software called Render-Man to create cutting-edge 3D movies such as those in the Harry Potter series.

File Conversion and Compression

Most multimedia applications are created, edited, and distributed in a digital file format, such as the ones discussed above. Older inputs to these applications, however, can be in an analog format from old home movies, magnetic tapes, vinyl records, or similar sources. In addition, there are older digital formats that are no longer popular or used. In these cases, the analog and older digital formats must be converted into a newer digital format before they can be edited and processed by today's multimedia software. This conversion can be done with a program or specialized hardware. Some of the multimedia software discussed above, such as Adobe Premium, Adobe Audition, and many others, have this analog-to-digital conversion capability. Standalone software and specialized hardware can also be used. Grass Valley, for example, is a hardware device that can be used to convert analog video to digital video or digital video to analog video. With this device, you can convert old VHS tapes to digital video files or digital video files to an analog format.

Because multimedia files can be large, it's sometimes necessary to compress files in order to make them easier to download from the Internet or send as e-mail attachments. Many of the multimedia software programs discussed above can be used to compress multimedia files. In addition, there are standalone file conversion programs, such as WinZip, that can be used to compress many file formats.

Designing a Multimedia Application

Designing multimedia applications requires careful thought and a systematic approach. The overall approach to modifying any existing application or developing a new one is discussed in the next chapters on systems development. There are, however, some additional considerations in developing a multimedia application. Multimedia applications can be printed on beautiful brochures, placed into attractive corporate reports, uploaded to the Internet, or displayed on large screens for viewing. Because these applications are typically more expensive than preparing documents and files in a word-processing program, it is important to spend time designing the best possible multimedia application. Designing a multimedia application requires that the end use of the document or file be carefully considered. For example, some text styles and fonts are designed for Internet display. Because different computers and Web browsers display information differently, it is a good idea to select styles, fonts, and presentations based on computers and browsers that are likely to display the multimedia application. Because large files can take much longer to load into a Web page, smaller files are usually preferred for Web-based multimedia applications.

Overview of Virtual Reality

The term *virtual reality* was initially coined in 1989 by Jaron Lanier, founder of VPL Research. Originally, the term referred to *immersive virtual reality* in which the user becomes fully immersed in an artificial, 3D world that is completely generated by a computer. Through immersion, the user can gain a deeper understanding of the virtual world's behavior and functionality. The Media Grid at Boston College, for example, has a number of initiatives in the use of immersive virtual reality in education.[72]

virtual reality system: A system that enables one or more users to move and react in a computer-simulated environment.

A **virtual reality system** enables one or more users to move and react in a computer-simulated environment. Virtual reality simulations require special interface devices that transmit the sights, sounds, and sensations of the simulated world to the user. These devices can also record and send the speech and movements of the participants to the simulation program, enabling users to sense and manipulate virtual objects much as they would real objects. This natural style of interaction gives the participants the feeling that they are immersed in the simulated world. For example, an auto manufacturer can use virtual reality to help it simulate and design factories.

Interface Devices

To see in a virtual world, the user often wears a head-mounted display (HMD) with screens directed at each eye. The HMD also contains a position tracker to monitor the location of the user's head and the direction in which the user is looking. Employing this information, a computer generates images of the virtual world—a slightly different view for each eye—to match the direction in which the user is looking and displays these images on the HMD. Many companies sell or rent virtual-reality interface devices, including Virtual Realities (*www.vrealities.com*), Amusitronix (*www.amusitronix.com*), and I-O Display Systems (*www.i-glassesstore.com*), among others.

The Electronic Visualization Laboratory at the University of Illinois at Chicago introduced a room constructed of large screens on three walls and a floor on which the graphics are projected. The CAVE, as this room is called, provides the illusion of immersion by projecting stereo images on the walls and floor of a room-sized cube (*www.evl.uic.edu*). Several persons wearing lightweight stereo glasses can enter and walk freely inside the CAVE. A head-tracking system continuously adjusts the stereo projection to the current position of the leading viewer.

Courtesy of Mechdyne Corporation

A CAVE provides the illusion of immersion in a virtual environment.

Users hear sounds in the virtual world through earphones, with information reported by the position tracker also being used to update audio signals. When a sound source in virtual space is not directly in front of or behind the user, the computer transmits sounds to arrive at one ear a little earlier or later than at the other and to be a little louder or softer and slightly different in pitch.

The *haptic* interface, which relays the sense of touch and other physical sensations in the virtual world, is the least developed and perhaps the most challenging to create. One virtual reality company has developed a haptic interface device that can be placed on a person's fingertips to give an accurate feel for game players, surgeons, and others. Currently, with the use of a glove and position tracker, the computer locates the user's hand and measures finger movements. The user can reach into the virtual world and handle objects; still, it is difficult to generate the sensations of a person tapping a hard surface, picking up an object, or running a finger across a textured surface. Touch sensations also have to be synchronized with the sights and sounds users experience. Today, some virtual reality developers are even trying to incorporate taste and smell into virtual reality applications.

Forms of Virtual Reality

Aside from immersive virtual reality, virtual reality can also refer to applications that are not fully immersive, such as mouse-controlled navigation through a 3D environment on a graphics monitor, stereo viewing from the monitor via stereo glasses, and stereo projection systems. *Augmented reality,* a newer form of virtual reality, has the potential to superimpose digital data over real photos or images. Augmented reality is being used in a variety of settings.[73] Some luxury car manufacturers, for example, display dashboard information, such as speed and remaining fuel, on windshields. The application is used in some military aircraft and is often called heads-up display. First down yellow lines displayed on TV screens during football games is another example of augmented reality, where computer-generated yellow lines are superimposed onto real images of a football field. GPS maps can be combined with real pictures of stores and streets to help you locate your position or find your way to a new destination. Using augmented reality, you can point a smartphone camera at a historic landmark, such as a castle, museum, or other building, and have information about the landmark appear on your screen, including a brief description of the landmark, admission price, and hours of operation. Although still in its early phases of implementation, augmented reality has the potential to become an important feature of tomorrow's smartphones and similar mobile devices.

Kyodo/Landov

Augmented reality technology shows additional information when it captures images through a camera on devices such as a smartphone.

Virtual Reality Applications

You can find thousands of applications of virtual reality, with more being developed as the cost of hardware and software declines and as people's imaginations are opened to the potential of virtual reality. Virtual reality applications are being used in medicine, education and training, business, and entertainment, among other fields.

Medicine

Barbara Rothbaum, the director of the Trauma and Recovery Program at Emory University School of Medicine and cofounder of Virtually Better, uses an immersive virtual reality system to help in the treatment of anxiety disorders.[74] Another VR program, called SnowWorld, is developed by the University of Washington to help treat burn patients.[75] Using VR, the patients can navigate through icy terrain and frigid waterfalls. VR helps because it takes a patient's mind off the pain.

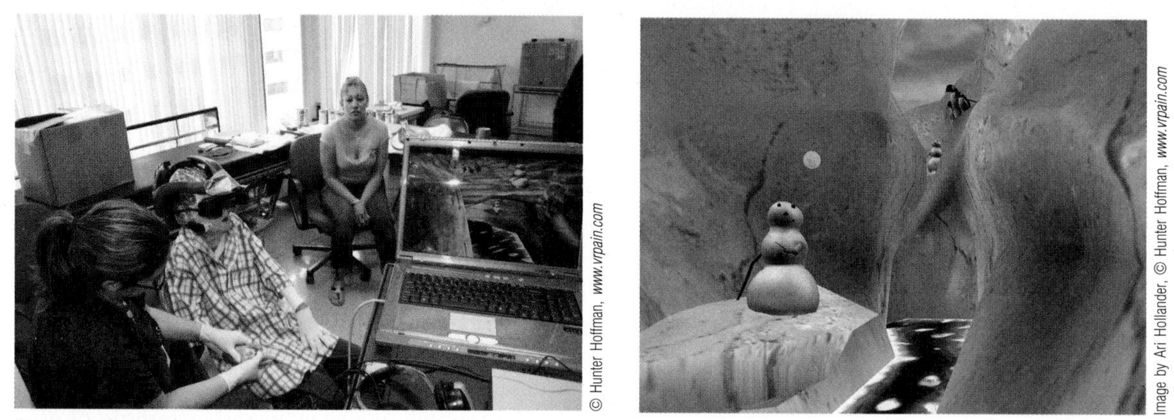

On the left, a patient is using SnowWorld VR pain distraction at a treatment center. On the right is an image of SnowWorld itself.

Education and Training

Virtual environments are used in education to bring exciting new resources into the classroom. In development for more than 10 years, *3D Rewind Rome* is a virtual reality show about ancient Rome developed at a virtual reality lab at University of California–Los Angeles (UCLA).[76] The show is historically accurate, with over 7,000 reconstructed buildings on a background of realistic landscape. At North Dakota State University, the Archaeology Technologies Laboratory has developed a 3D virtual reality system that displays an eighteenth-century American Indian village.

3D Rewind Rome is based on more than 10 years of research by archaeologists and historians coordinated by the University of California–Los Angeles.

Virtual technology has also been applied by the military. To help with aircraft maintenance, a virtual reality system has been developed to simulate an aircraft and give a user a sense of touch, while computer graphics provide a sense of sight and sound. The user sees, touches, and manipulates the various parts of the virtual aircraft during training. Also, the Pentagon is using a virtual reality training lab to prepare for a military crisis. The virtual reality system simulates various war scenarios.

Business and Commerce

Virtual reality has been used in all areas of business. Boeing used virtual reality to help it design and manufacture airplane parts and new planes, including the 787 Dreamliner. Boeing used 3D PLM from Dassault Systems.[77] One health care institution used Second Life to create a virtual hospital when it started construction of a real multimillion-dollar hospital. The purpose of the Second Life virtual hospital was to show clients and staff the layout and capabilities of the new hospital. Second Life has also been used in business and recruiting.[78] It also allows people to play games, interact with avatars, and build structures, such as homes.

A number of companies are using VR in advertising. Pizza chain Papa John's used VR as an advertising tool by placing a VR image on many of its pizza boxes. When the image is viewed by a Web camera on a computer, a standard keyboard can be used to manipulate images of a Chevrolet Camaro on the computer screen. It is a moving image of the Camaro that the founder of Papa John's sold to start his pizza company.

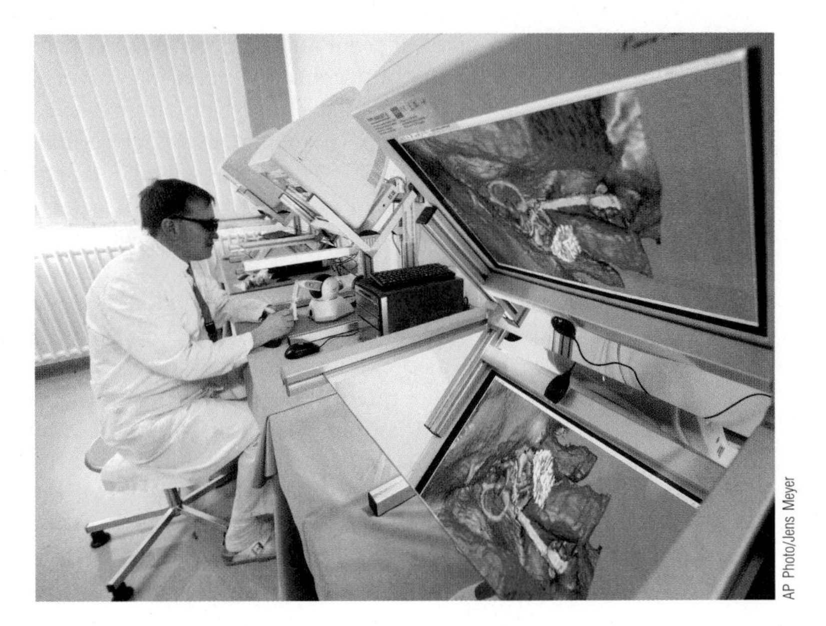

AP Photo/Jens Meyer

A surgeon uses Voxel-Man, a VR simulator to train surgical access to the middle ear.

Entertainment

Computer-generated imagery, or CGI, is a technology that has been around since the 1970s. Many movies use this technology to bring realism to the silver screen, including *Avatar, Finding Nemo*, *Spider-Man II*, and *Star Wars Episode II—Attack of the Clones*. A team of artists rendered the roiling seas and crashing waves of *Perfect Storm* almost entirely on computers using weather reports, scientific formulas, and their imagination. Other films that have used CGI technology include *Dinosaur* with its realistic talking reptiles, *Titan A.E.*'s beautiful 3D spacescapes, and the casts of computer-generated crowds and battles in *Gladiator* and *The Patriot*. CGI can also be used for sports simulation to enhance the viewers' knowledge and enjoyment of a game.

SPECIALIZED SYSTEMS

In addition to artificial intelligence, expert systems, and virtual reality, other interesting specialized systems have appeared. MIT's Fab Labs project, for example, has a goal to let anyone manufacture almost anything using specialized equipment and computers.[79] It is hoped that Fab Labs will be able to produce a wide range of products, from computers to roof panels.

Many special-purpose systems help overcome disabilities or improve health, often called *assistive technology*. IBM researchers, along with researches from the Institute of Bioengineering and Nanotechnology based in Singapore, have developed a nanoparticle 50,000 times smaller than the thickness of a human hair.[80] If successful, the nanoparticle could destroy bacteria, which threatens human health and life. According to one IBM researcher, "It's like the north pole and south pole. The particles disrupt the membrane, generate holes in it and empty out the bacteria." Eagle Eyes, a Boston College project, allows people with physical disabilities to control a computer by moving their head or eyes.[81] Using electrodes placed on a person's head, a camera connected to a computer detects head or eye movements and controls the operation of the computer. Today, more hospitals and health care facilities are using the Internet to connect doctors to patients in distant locations. In one case, a physician used Internet video to check on the treatment of a stroke patient located 15 miles away to make sure the drugs being used weren't increasing the chance of bleeding in the brain. After reviewing CT scans and the behavior of the patient, the doctor made specific drug recommendations.

Government agencies and the military use special-purpose systems to give them an edge. The U.S. Navy has tested an unmanned aircraft, the X-47B, that can take off from and land on an aircraft carrier.[82] Landing on an aircraft carrier is one of the most difficult maneuvers for skilled pilots, which makes this new experimental drone very impressive. Discussing this difficult task, the Navy's program manager says, "We're making that challenge with unmanned vehicles today." The U.S. military is also experimenting with very small drones that can move like a bird or insect.[83] One experimental drone weighs about the same as a AA battery with a 6.5-inch wing span mechanism that moves like the wings of a hummingbird. The small drone has video and audio devices used to send intelligence data back to a central command. The small drones have possible uses in police and fire departments. The U.S. military is also experimenting with remote-controlled helicopters such as the Fire Scout that can spy on enemy positions, locate drug smugglers, and monitor pirates attempting to hijack ships.[84] The robo-copters are about 30 feet long and 10 feet high. They can be launched from ships or land bases.

Research by universities and governmental agencies is also being conducted into using advanced radar, satellite imaging, and advanced software to help predict tornadoes and minimize their damage, like the ones that struck the United States in the spring and early summer of 2011.[85] According to the director of the Georgia Institute of Technology's Severe Storms Research Center, "The hope is research like this will help reduce the loss of life in such storms."

A number of special-purpose systems are now available in vehicles. Toyota uses hardware and software to help control its antilock brakes, keyless entry systems, transmission, engine, the accelerator, and other vehicle components.[86] Toyota and Microsoft are working on applications that can be downloaded from the Internet and used in cars and trucks.[87] Called Toyota Media Service, the Internet-based service will have the potential to perform a number of useful functions, including being able to heat or cool a house and turn lights on while driving home and helping drivers of electric vehicles find recharging stations. Research in Motion (RIM), maker of Blackberry phones and PlayBook tablet computers, is developing software to allow its devices to also be used in cars and trucks.[88] A recent acquisition by RIM should allow the mobile device maker to turn a vehicle's dashboard into an entertainment and control center that can read e-mails, play music stored on mobile devices, and control basic car and truck functions. In the future, automotive manufacturers and technology companies will explore ways to have highways and cars communicate with each other, avoiding accidents, finding the best route to avoid traffic and save time, and automatically pay highway tolls.[89] The Advanced Warning System by Mobileye warns drivers to keep a safe distance from other vehicles and drivers.[90]

Improvements are also being made in bridges and roadways. The St. Anthony Falls Bridge in Minneapolis, for example, has smart sensors in its concrete and steel that measure and report on any potential structural problems.[91] The sensors also measure temperatures on the road surface and help determine if anti-icing is needed during cold winter months.

Companies use special-purpose tracking devices, chips, and bar codes for a variety of purposes. As mentioned previously, *Radio Frequency Identification (RFID)* tags that contain small chips with information about products or packages can be quickly scanned to perform inventory control or trace a package as it moves from a supplier to a company to its customers. RFID systems can also be used to provide a better customer experience. According to the managing director of information systems at Aspen Skiing Company, "Guests simply walk through a turnstile to gain access while their RFID card remains in their pocket. They appreciate the reduced wait times on the slopes and the ease of renting skis or buying a burger."[92] Vail Resorts used RFID tags to reduce wait times at lift lines and prevent people from using chair lifts without paying for lift tickets.[93] RFID tags are starting to be used in footwear and apparel to a greater extent.[94] According to one industry analyst, "The growth in retail item-level tagging is huge, both in shipments and in total spending. The average growth rate is close to 60% for the next three years." Airline companies are now starting to use RFID tags and sophisticated software to improve baggage handling and processing. It is hoped that the new tags will reduce lost luggage, which is expensive for the airlines and very frustrating for passengers. Another technology is being used to create "smart containers" for ships, railroads, and trucks. NaviTag and other companies are developing communications systems that allow containers to broadcast the contents, location, and condition of shipments to shipping and cargo managers.[95] A railroad company can use standard radio messages to generate shipment and tracking data for customers and managers.

Toenisvorst, Germany

Innovative payment detectors scan the contents of a shopping cart by using RFID technology.

game theory: The use of information systems to develop competitive strategies for people, organizations, or even countries.

One special application of computer technology is derived from a branch of mathematics called **game theory**, which involves the use of information systems to develop competitive strategies for people, organizations, or even countries. Two competing businesses in the same market can use game theory to determine the best strategy to achieve their goals. In other applications, game theory was used to develop a better security system for the Los Angeles airport.[96] The game theory application used ARMOR (Assistant for Randomized Monitoring over Routes) software to help determine the best monitoring and patrolling strategies to enhance airport security. The military can also use game theory to determine the best military strategy to win a conflict against another country, and individual investors can use game theory to determine the best strategies when competing against other investors in a government auction of bonds. Groundbreaking work on game theory was pioneered by John Nash, the mathematician whose life was profiled in the book and film *A Beautiful Mind*.

informatics: A specialized system that combines traditional disciplines, such as science and medicine, with computer systems and technology.

Informatics, another specialized system, combines traditional disciplines, such as science and medicine, with computer systems and technology. *Bioinformatics*, for example, combines biology and computer science. Also called *computational biology*, bioinformatics has been used to help map the human genome and conduct research on biological organisms. Using sophisticated databases and artificial intelligence, bioinformatics helps unlock the secrets of the human genome, which could eventually prevent diseases and save lives. Some universities have courses on bioinformatics and offer bioinformatics certification. Medical informatics combines traditional medical research with computer science. Journals, such as *Healthcare Informatics*, report current research on applying computer systems and technology to reduce medical errors and improve health care.[97] Informatics can also be used in finance to develop sophisticated and profitable modeling programs that can analyze market risks and potential.

There are many other examples of specialized systems. A Carnegie Mellon professor is experimenting with electronic ink that conducts electricity.[98] The research could lead to display screens for computers, phones, and TVs almost as thin as a coat of paint. An MIT professor has designed a chip that uses an approach he calls sloppy arithmetic.[99] The chip, which is allowed to take short cuts and not be completely accurate, could be up to 100,000 faster than the typical chip as a result. The credit card company Visa is experimenting with an electronic wallet that contains credit card information that can be stored in a smartphone or other mobile device.[100] The electronic wallet can be used to pay for goods and services purchased online or from traditional stores. Banks such as Bancorp, TD Bank Group, and BB&T support this approach by Visa. Some grocery and retail stores are using devices like Scan It to allow customers to determine the total of all their purchases as they shop.[101] About the size of a smartphone, these scanning devices can alert shoppers to coupons and special offers, deduct these coupons and special offers from the total bill, and reduce checkout time at the cash register. Some believe that these scanning devices motivate shoppers to buy more. Tobii makes the *eye-tracker*, which can be used to locate a person's gaze on a computer screen.[102] The device works using two infrared cameras located on a display screen of a computer. After the eye-tracker is calibrated, it can determine where on the computer screen you are looking. The device has a number of applications. Game players can destroy an enemy in the game by just looking at it on the computer screen, and marketing companies can use it to determine what image on the computer screen attracts your attention. Some people, however, believe that this type of device could result in an invasion of privacy. Wi-Fi can also be used to help companies locate people and things. Royal Caribbean Cruises, for example, uses iPhones, iPads, and other mobile devices to determine the location of people and things onboard. If necessary, corrective action can be taken to make sure everything is in its proper location.[103] Segway is an electric scooter that uses sophisticated software, sensors, and gyro motors to transport people through warehouses, offices, downtown sidewalks, and other spaces.[104] Originally designed to transport people around a factory or around town, more recent versions are being tested by the military for gathering intelligence and transporting wounded soldiers to safety. Segway has also been used to play Polo instead of using horses.[105] Called Segway Polo, the sport has attracted many players, including one of the founders of Apple Computer. According to the head of Germany's Funky-Move Turtle's Polo team, "They couldn't drive very well, and they crashed into each other a lot, but their ball handling was amazing."

SUMMARY

Principle:

Knowledge management allows organizations to share knowledge and experience among managers and employees.

Knowledge is an awareness and understanding of a set of information and the ways that information can be made useful to support a specific task or reach a decision. A knowledge management system (KMS) is an organized collection of people, procedures, software, databases, and devices used to create, store, share, and use the organization's knowledge and experience. Explicit knowledge is objective and can be measured and documented in reports, papers, and rules. Tacit knowledge is hard to measure and document and is typically not objective or formalized.

Knowledge workers are people who create, use, and disseminate knowledge. They are usually professionals in science, engineering, business, and other areas. The chief knowledge officer (CKO) is a top-level executive who helps an organization use a KMS to create, store, and employ knowledge to achieve organizational goals. Some organizations and professions use communities of practice (COP) to create, store, and share knowledge. A COP is a group of people or a community dedicated to a common discipline or practice, such as open-source software, auditing, medicine, and engineering.

Obtaining, storing, sharing, and using knowledge is the key to any KMS, with the employment of a KMS often leading to additional knowledge creation, storage, sharing, and usage. Many tools and techniques can be used to create, store, and use knowledge. These tools and techniques are available from IBM, Microsoft, and other companies and organizations.

Principle:

Artificial intelligence systems form a broad and diverse set of systems that can replicate human decision making for certain types of well-defined problems.

The term artificial intelligence is used to describe computers with the ability to mimic or duplicate the functions of the human brain. The objective of building AI systems is not to replace human decision making but to replicate it for certain types of well-defined problems.

Intelligent behavior encompasses several characteristics, including the abilities to learn from experience and apply this knowledge to new experiences, handle complex situations and solve problems for which pieces of information might be missing, determine relevant information in a given situation, think in a logical and rational manner and give a quick and correct response, and understand visual images and process symbols. Computers are better than people at transferring information, making a series of calculations rapidly and accurately, and making complex calculations, but human beings are better than computers at all other attributes of intelligence.

Artificial intelligence is a broad field that includes several key components, such as expert systems, robotics, vision systems, natural language processing, learning systems, and neural networks. An expert system consists of the hardware and software used to produce systems that behave as a human expert would in a specialized field or area (e.g., credit analysis). Robotics uses mechanical or computer devices to perform tasks that require a high degree of precision or are tedious or hazardous for humans (e.g., stacking cartons on a pallet). Vision systems include hardware and software that permit computers to capture, store, and manipulate images and pictures (e.g., face-recognition software). Natural language processing allows the computer to understand and react to statements and commands made in a "natural" language, such as English. Learning systems use a combination of software and hardware to allow a computer to change how it functions or reacts to situations based on feedback it receives (e.g., a computerized chess game). A neural network is a computer system that can simulate the functioning of a human brain (e.g., a disease diagnostics system). A genetic algorithm is an approach to solving large, complex problems in which a number of related operations or models change and evolve until the best one emerges. The approach is based on the theory of evolution, which requires variation and natural selection. Intelligent agents consist of programs and a knowledge base used to perform a specific task for a person, a process, or another program.

Principle:

Expert systems can enable a novice to perform at the level of an expert but must be developed and maintained very carefully.

An expert system consists of a collection of integrated and related components, including a knowledge base, an inference engine, an explanation facility, a knowledge acquisition facility, and a user interface. The knowledge base is an extension of a database, discussed in Chapter 5, and an information and decision support system, discussed in Chapter 10. It contains all the relevant data, rules, and relationships used in the expert system. The rules are often composed of IF-THEN statements, which are used for drawing conclusions.

The inference engine processes the rules, data, and relationships stored in the knowledge base to provide answers, predictions, and suggestions similar to the way a human expert would. The explanation facility of an expert system allows the user to understand what rules were used in arriving at a decision. The knowledge acquisition facility helps the user add or update knowledge in the knowledge base. The user interface makes it easier to develop and use the expert system.

The people involved in the development of an expert system include the domain expert, the knowledge engineer, and the knowledge users. The domain expert is the person or group who has the expertise or knowledge being captured for the system. The knowledge engineer is the developer whose job is to extract the expertise from the domain expert. The knowledge user is the person who benefits from the use of the developed system.

Expert systems can be implemented in several ways. Previously, traditional high-level languages, including Pascal, FORTRAN, and COBOL, were used. LISP and PROLOG are two languages specifically developed for creating expert systems from scratch. A faster and less expensive way to acquire an expert system is to purchase an expert system shell or existing package. The shell program is a collection of software packages and tools used to design, develop, implement, and maintain expert systems.

Principle:

Multimedia and virtual reality systems can reshape the interface between people and information technology by offering new ways to communicate information, visualize processes, and express ideas creatively.

Multimedia is text, graphics, video, animation, audio, and other media that can be used to help an organization efficiently and effectively achieve its goals. Multimedia can be used to create stunning brochures, presentations, reports, and documents. Although not all organizations use the full capabilities of multimedia, most use text and graphics capabilities. Other applications of multimedia include audio, video, and animation. File compression and conversion are often needed in multimedia applications to import or export analog files and to reduce file size when storing multimedia files and sending them to others. Designing a multimedia application requires careful thought to get the best results and achieve corporate goals.

A virtual reality system enables one or more users to move and react in a computer-simulated environment. Virtual reality simulations require special interface devices that transmit the sights, sounds, and sensations of the simulated world to the user. These devices can also record and send the speech and movements of the participants to the simulation program. Thus, users can sense and manipulate virtual objects much as they would real objects. This natural style of interaction gives the participants the feeling that they are immersed in the simulated world.

Virtual reality can also refer to applications that are not fully immersive, such as mouse-controlled navigation through a three-dimensional environment on a graphics monitor, stereo viewing from the monitor via stereo glasses, and stereo projection systems. Some virtual reality applications allow views of real environments with superimposed virtual objects. Augmented reality, a newer form of virtual reality, has the potential to superimpose digital data over real photos or images. Virtual reality applications are found in medicine, education and training, real estate and tourism, and entertainment.

Principle:

Specialized systems can help organizations and individuals achieve their goals.

A number of specialized systems have recently appeared to assist organizations and individuals in new and exciting ways. Segway, for example, is an electric scooter that uses sophisticated software, sensors, and gyro motors to transport people through warehouses, offices, downtown sidewalks, and other spaces. Originally designed to transport people around a factory or around town, more recent versions are being tested by the military for gathering intelligence and transporting wounded soldiers to safety. Radio Frequency Identification (RFID) tags are used in a variety of settings. Game theory involves the use of information systems to develop competitive strategies for people, organizations, and even countries. Informatics combines traditional disciplines, such as science and medicine, with computer science. Bioinformatics and medical informatics are examples.

CHAPTER 11: SELF-ASSESSMENT TEST

Knowledge management allows organizations to share knowledge and experience among managers and employees.

1. ________________ knowledge is objective and can be measured and documented in reports, papers, and rules.
2. What type of person creates, uses, and disseminates knowledge?
   a. knowledge worker
   b. information worker
   c. domain expert
   d. knowledge engineer
3. The ________________ is a top-level executive who helps the organization work with a KMS to create, store, and use knowledge to achieve organizational goals.

Artificial intelligence systems form a broad and diverse set of systems that can replicate human decision making for certain types of well-defined problems.

4. The Brain Computer Interface (BCI) can directly connect the human brain to a computer and have human thought control computer activities. True or False?

5. ________________ are rules of thumb arising from experience or even guesses.
6. What is *not* an important attribute for artificial intelligence?
   a. the ability to use sensors
   b. the ability to learn from experience
   c. the ability to be creative
   d. the ability to make complex calculations
7. ________________ involves mechanical or computer devices that can paint cars, make precision welds, and perform other tasks that require a high degree of precision or that are tedious or hazardous for human beings.
8. What branch of artificial intelligence involves a computer understanding and reacting to statements in English or another language?
   a. expert systems
   b. neural networks
   c. natural language processing
   d. vision systems
9. A(n) ________________ is a combination of software and hardware that allows the computer to change how it functions or reacts to situations based on feedback it receives.

Expert systems can enable a novice to perform at the level of an expert but must be developed and maintained very carefully.

10. What is a disadvantage of an expert system?
    a. the inability to solve complex problems
    b. the inability to deal with uncertainty
    c. limitations to relatively narrow problems
    d. the inability to draw conclusions from complex relationships
11. A(n) ________________ is a collection of software packages and tools used to develop expert systems that can be implemented on most popular PC platforms to reduce development time and costs.
12. A heuristic consists of a collection of software and tools used to develop an expert system to reduce development time and costs. True or False?
13. What stores all relevant information, data, rules, cases, and relationships used by the expert system?
    a. the knowledge base
    b. the data interface
    c. the database
    d. the acquisition facility
14. A disadvantage of an expert system is the inability to provide expertise needed at a number of locations at the same time or in a hostile environment that is dangerous to human health. True or False?
15. What allows a user or decision maker to understand how the expert system arrived at a certain conclusion or result?
    a. the domain expert
    b. the inference engine
    c. the knowledge base
    d. the explanation facility
16. The purpose of the ________________ is to provide a convenient and efficient means of capturing and storing all components of the knowledge base.
17. In an expert system, the domain expert is the individual or group who has the expertise or knowledge one is trying to capture in the expert system. True or False?

Multimedia and virtual reality systems can reshape the interface between people and information technology by offering new ways to communicate information, visualize processes, and express ideas creatively.

18. ________________ has the potential to superimpose digital data over real photos or images.
19. What type of virtual reality is used to make human beings feel as though they are in a three-dimensional setting, such as a building, an archaeological excavation site, the human anatomy, a sculpture, or a crime scene reconstruction?
    a. cloud
    b. relative
    c. immersive
    d. visual

Specialized systems can help organizations and individuals achieve their goals.

20. ________________ involves the use of information systems to develop competitive strategies for people, organizations, or even countries.

CHAPTER 11: SELF-ASSESSMENT TEST ANSWERS

1. explicit
2. a
3. chief knowledge officer (CKO)
4. True
5. Heuristics
6. d
7. Robotics
8. c
9. learning system
10. c
11. expert system shell
12. False

13. a
14. False
15. d
16. knowledge acquisition facility
17. True
18. augmented reality
19. c
20. Game theory

REVIEW QUESTIONS

1. What is a *knowledge repository?*
2. What is a *community of practice?*
3. What is a *chief knowledge officer?* What are his or her duties?
4. What is a vision system? Discuss two applications of such a system.
5. What is natural language processing? What are the three levels of voice recognition?
6. Describe three examples of the use of robotics. How can a microrobot be used?
7. What is a learning system? Give a practical example of such a system.
8. Briefly describe a special-purpose system discussed at the end of this chapter. How could this system be used in a business setting?
9. Under what conditions is the development of an expert system likely to be worth the effort?
10. Identify the basic components of an expert system and describe the role of each.
11. Describe several business uses of multimedia.
12. What is augmented reality? How can it be used?
13. Expert systems can be built based on rules or cases. What is the difference between the two?
14. Describe the roles of the domain expert, the knowledge engineer, and the knowledge user in expert systems.
15. What is informatics? Give three examples.
16. Describe game theory and its use.
17. Identify three special interface devices developed for use with virtual reality systems.
18. Identify and briefly describe three specific virtual reality applications.
19. What is a knowledge base? How is it used?
20. Give three examples of other specialized systems.

DISCUSSION QUESTIONS

1. What are the requirements for a computer to exhibit human-level intelligence? How long will it be before we have the technology to design such computers? Do you think we should push to accelerate such a development? Why or why not?
2. You work for an insurance company as an entry-level manager. The company contains both explicit and tacit knowledge. Describe the types of explicit and tacit knowledge that might exist in your insurance company. How would you capture each type of knowledge?
3. How could you use a community of practice at a college or university?
4. What are some of the tasks at which robots excel? Which human tasks are difficult for robots to master? What fields of AI are required to develop a truly perceptive robot?
5. Describe how natural language processing could be used in a university setting.
6. Discuss how learning systems can be used in a military war simulation to train future officers and field commanders.
7. You have been hired to develop an expert system for a university career placement center. Develop five rules a student could use in selecting a career.
8. What is the relationship between a database and a knowledge base?
9. Imagine that you are developing the rules for an expert system to select the strongest candidates for a medical school. What rules or heuristics would you include?
10. Describe how game theory can be used in a business setting.
11. Describe how a university might use multimedia.
12. Describe how augmented reality can be used in a classroom. How could it be used in a work setting?
13. Describe a situation in which RFID could be used in a business setting.

PROBLEM-SOLVING EXERCISES

1. You are a senior vice president of a company that manufactures kitchen appliances. You are considering using robots to replace up to 10 of your skilled workers on the factory floor. Using a spreadsheet, analyze the costs of acquiring several robots to paint and assemble some of your products versus the cost savings in labor. How many years would

it take to pay for the robots from the savings in fewer employees? Assume that the skilled workers make $20 per hour, including benefits.

2. Develop an expert systems to predict the weather for the next few days. Use a word-processing program to list and describe the IF-THEN rules you would use.
3. Use a graphics program, such as PowerPoint, to develop a brochure for a small restaurant. Contrast your brochure to one that could have been developed using a specialized multimedia application used to develop brochures. Write a report using a word-processing application on the advantages of a multimedia application compared to a graphics program.

TEAM ACTIVITIES

1. Do research with your team to identify KMSs in three different businesses or nonprofit organizations. Describe the types of tacit and explicit knowledge that would be needed by each organization or business.
2. Have your team develop an expert system to predict how many years it will take a typical student to graduate from your college or university given the major the student selects, the number of courses taken each semester, the number of parties or social activities the student attends each month, and other factors. Each factor should be placed in one or more IF-THEN rules.
3. Have your team members explore the use of a special-purpose system in an industry of your choice. Describe the advantages and disadvantages of this special-purpose system.

WEB EXERCISES

1. Use the Internet to find information about the use of augmented reality. Describe what you found.
2. This chapter discussed several examples of expert systems. Search the Internet for two examples of the use of expert systems. Which one has the greatest potential to increase profits for a medium-sized firm? Explain your choice.
3. Use the Internet to find information about two of the special-purpose applications discussed in the chapter. Write a report about what you found.

CAREER EXERCISES

1. Describe how a COP can be used to help advance your career.
2. Describe how you could use multimedia in a career of your choice. Include applications in text and graphics, audio, video, and animation. Using a word processor, briefly describe how you would design each multimedia application to help your career.

CASE STUDIES

Case One

Google and BMW Train Cars to Drive Themselves

Artificial intelligence is defined in this chapter as computers having "the ability to mimic or duplicate the functions of the human brain." Driving a car in traffic is such an activity. By that definition vehicles that have begun to show the ability to drive in traffic have started to show intelligence.

Intelligent vehicle behavior, a step beyond the systems discussed in the Specialized Systems section of this chapter, is expected to become commonplace over the next few years. Alan Taub, General Motors vice president for research and development, predicts that self-driving will be a standard feature by 2020.

In October 2010, Google disclosed its Autonomous Car program. The firm's cars are easily spotted in and around San Francisco: the base Toyota Prius has a funnel-shaped device on the roof to hold a 64-beam laser rangefinder. Additional inputs to the control computer come from high-resolution maps of the area, two radar units on each bumper, a forward-facing camera to detect traffic lights, and GPS, inertial measurement unit, and wheel rotation counter to track vehicle motion.

Google's cars always travel with a licensed driver in the driver's seat. In the first 160,000 miles of operation, the driver had to take over twice: once when the car in front stopped and began to back into a parking space and once when a bicycle rider entered an intersection despite a red light. It would be fairly easy to program the cars' computers to check for cars starting to park but more difficult, though not impossible, to program them to handle cyclists running red lights. Situations in which a person has to take control should become rarer as time goes on.

The only accident involving a Google car in over 200,000 miles, by the way, came when one of the cars was rear ended while stopped at a traffic light. How many humans drive that far with only one minor fender-bender?

Google is not the only company studying autonomous driving technology. Vehicle manufacturers are investigating it, too. BMW is probably the furthest along: in January 2012, it demonstrated a self-driving car on Germany's no-speed-limits autobahns. The company's car looks like any other BMW since its radars, cameras, laser scanners, and distance sensors are all inside the body. Nico Kaempchen, project manager of Highly Automated Driving at BMW Group Research and Technology, says "the system works on all freeways that we have mapped out beforehand."

Programming autonomous cars means studying driver behavior to a detailed level. How do drivers alternate at an intersection with four stop signs? If other drivers don't yield to the driverless car, how should it inch into the intersection to show that it wants its turn? How large must an animal be before a car takes evasive action to avoid it?

Society must also resolve legal issues about this technology. How can police officers pull over driverless cars? Do they even have the right to? How can driverless cars recognize police officers directing traffic and ignore traffic lights over those officers' heads?

Despite such concerns, driverless vehicles were legalized in Nevada in 2011. Similar legislation is pending in Florida and Hawaii and may soon be introduced in California.

Like it or not, though, we will soon give our cars more control than our ancestors ever gave their most intelligent horses. "It won't truly be an autonomous vehicle," said Brad Templeton, a software designer and a consultant for the Google project, "until you instruct it to drive to work and it heads to the beach instead."

Discussion Questions

1. Suppose you saw a Google autonomous car driving along in front of you and going about 5 mph (or 10 km/hr) slower than you believe it is safe to go. There are two travel lanes in your direction. Would you pass the autonomous car? Do you think that is any riskier than passing a car controlled by a human driver?
2. As noted in the case, Google's autonomous cars are easily recognized but BMW's are not. Do you think this matters? If it does, which do you prefer and why?

Critical Thinking Questions

1. Would you buy an autonomous car in the first year such cars are on the market? (Assume you were going to buy a new car anyhow, that the car itself appeals to you, and that its price is reasonable.) If not, when do you think you would?
2. You are a lawyer. An autonomous car injures your client. What do you recommend your client do? If you recommend suing, who do you sue? What are some arguments you might use? What might the defendants argue?

SOURCES: Guizzo, E., "How Google's Self-Driving Car Works," *IEEE Spectrum, spectrum.ieee.org/automaton/robotics/artificial-intelligence/how-google-self-driving-car-works*, October 18, 2011; Hachman, M., "Google's Self-Driving Car Challenge: 1 Million Miles, by Itself," *PC Magazine, www.pcmag.com/article2/0,2817,2395049,00.asp*, October 20, 2011; Kelly, T., BMW Self Driving Car: Carmaker Shows off Hands-Free Car on Autobahn," *Huffington Post, www.huffingtonpost.com/2012/01/26/bmw-self-driving-car_n_1234362.html*, January 26, 2012; Trei, M., "BMW Challenges Google's Self-Driving Car," *NBC Bay Area News, www.nbcbayarea.com/blogs/press-here/BMW-Challenges-Googles-Self-Driving-Car-137892303.html*, January 23, 2012; Vanderbilt, T., "Five Reasons the Robo-Car Haters Are Wrong," *Wired, www.wired.com/autopia/2012/02/robo-car-haters-are-wrong*, February 9, 2012; Vanderbilt, T., "Let the Robot Drive: The Autonomous Car of the Future Is Here," *Wired, www.wired.com/magazine/2012/01/ff_autonomouscars*, January 20, 2012.

Case Two

Knowledge Management Improves Customer Support at Canon

Millions of U.S. consumers own Canon digital cameras, copiers, printers, binoculars, fax machines, camcorders, and calculators. Canon Information Technology Services (CITS) in Chesapeake, Virginia, fields support requests at a current rate of 200,000 calls, 50,000 e-mails, and 1,000 letters per month: a total of about three million contacts per year. CITS employs about 550 people to handle these contacts.

Canon's problem is that, until recently, they had no central knowledge repository. Product information was scattered over a CITS intranet, the Canon USA internet, hard-copy manuals, and an internally developed knowledge system. CITS could not ensure that all of the content was correct and did not conflict with the manuals or another system. Customers could use only the knowledge system, and it was not searchable. Support agents needed to check multiple sources of information on any product. This process was cumbersome, annoyed the agents, and wasted valuable time.

To address this problem, Canon installed Consona's Knowledge-Driven Support (KDS). KDS integrates knowledge management with case management software, a type of Customer Relationship Management software that you studied in Chapter 9.

As an example of integration, KDS supports *in-process authoring*. A representative who has just written a customer a long, complex explanation of how to solve a problem can enter that explanation directly into the knowledge base without having to recreate it or even copy and paste it. Agents don't have to take time after a call to create new knowledge when they could be improving their performance reports by taking another call. As Consona puts it, "knowledge isn't something that you do in addition to solving problems—it becomes the way you solve problems."

The results are that, during the first six months that the system was in full use, the fraction of customer questions resolved online without a phone call increased from 51 to 71 percent. This saved agent time while providing customers with better service. Another measure of the need for follow-up, e-mail escalation rate, dropped 47 percent from the same period of the previous year. Overall customer satisfaction scores were up from 6.5 to 7.1 on a scale of 1 to 10, and customer resolution rates rose from 50 to 60 percent.

"The Consona CRM knowledge base has been a great help to our service agents and to our customers. It lets the customers get the answers to the 'easy' questions themselves, while freeing up the agents to focus on the more difficult problems," says Jay Lucado, CITS assistant director of knowledge management and delivery.

CITS also leveraged the knowledge base to improve agent training. Its new training curriculum focuses on teaching agents to find the answers in the system rather than how to fix any problem a customer might have. In addition, system-based training is remotely available, which works well with CITS's work-at-home program. Agents works from their homes four days each week and are able to complete their training remotely as well.

Discussion Questions

1. Besides Canon, this knowledge management technology might also be useful to other companies. What are the characteristics of Canon that make it useful? In other words, what characteristics of a company suggest that it might find this technology useful for customer support? Conversely, what characteristics of a company might suggest that it would *not* find this technology useful?
2. In Canon's situation, a "case" is a single customer problem. Efforts to solve that problem may involve several contacts. The case management system tracks those contacts so that a representative who is new to the case can see its history at a glance. How could your college or university use a case management system that includes knowledge management?

Critical Thinking Questions

1. How would the need for a system such as KDS change if each of Canon's product lines (cameras, printers, etc.) was sold by a different company?
2. Discuss two reasons the cost-benefit ratio of a knowledge management system such as KDS goes up as the company using it gets larger. Which of these reasons apply to other applications besides knowledge management?

SOURCES: Briggs, M., "New Consona Report Uncovers Best Practices for Easier and More Effective Knowledge Management," Consona press release, *www.consona.com/news/km-report-best-practices.aspx*, August 4, 2011; Canon ITS Web site, *www.cits.canon.com*, accessed February 11, 2012; Canon USA Web site, *www.usa.canon.com/cusa/home*, accessed February 11, 2012; Consona, Inc., Knowledge-Driven Support Web site, *crm.consona.com/software/products/knowledge-driven-support.aspx*, accessed February 11, 2012; Johnson, S., "Canon Information Technology Services, Inc./Consona Knowledge Management," *Office Product News*, *www.officeproductnews.net/case_studies/canon_information_technology_services_inc_consona_knowledge_management*, October 24, 2011.

Questions for Web Case

See the Web site for this book to read about the Altitude Online case for this chapter. Following are questions concerning this Web case.

Altitude Online: Knowledge Management and Specialized Information Systems

Discussion Questions

1. Why do you think it is a good idea for Altitude Online to maintain records of all advertising projects?
2. How can social networks and blogs serve as knowledge management systems?

Critical Thinking Questions

1. What challenges lie in filling a wiki with information provided by employees?
2. What other tools could Altitude Online use to capture employee knowledge, build community, and reward productive employees?

NOTES

Sources for the opening vignette: Hilong Group, "Why Do So Many Oil Companies Walk in the Forefront of Knowledge Management?" *www.hilonggroup.net/en/news/showNews.aspx?classid=144959613005987840&id=68*, February 18, 2011; Orton, E., "Autonomy Selected by Repsol to Transform Knowledge Management System," *www.autonomy.com/content/News/Releases/2011/0719.en.html*, July 19, 2011; Savvas, A., "Repsol Deploys Knowledge Management," *Computerworld UK*; *www.computerworlduk.com/news/it-business/3292947/repsol-deploys-knowledge-management*, July 23, 2011; Repsol Web site, *www.repsol.com/es_en*, accessed February 3, 2012; Repsol Knowledge Management, *www.repsol.com/es_en/corporacion/conocer-repsol/canal-tecnologia/ctr_investigadores/gestion-conocimiento*, accessed February 3, 2012.

1. Ravishankar, M., et al, "Examining the Strategic Alignment and Implementation Success of a KMS," *Information Systems Research*, March 2011, p. 39.
2. Advent Web site, *www.advent.com*, accessed June 5, 2011.
3. Ko, D. and Dennis, A., "Profiting from Knowledge Management: The Impact of Time and Experience," *Information Systems Research*, March 2011, p. 134.
4. Staff, "Handwritten Notes Essential to Knowledge Workers," *Business Wire*, May 23, 2011.

5. Betts, Mitch, "IT Leader Builds a Know-how Network," *Computerworld*, April 18, 2011, p. 4.
6. VSGi Web site, *www.vsgi.com*, accessed June 15, 2011.
7. MWH Web site, *www.mwhglobal.com*, accessed June 15, 2011.
8. TED Web site, *www.ted.com*, accessed April 17, 2011.
9. WPP Web site, *www.wpp.com*, accessed on May 6, 2011.
10. "Adobe Creative Suite CS5.5," *www.adobe.com/products/creativesuite*, accessed June 5, 2011.
11. KM World Web site, *www.kmworld.com*, accessed June 15, 2011.
12. Knowledge Management Online Web site, *www.knowledge-management-online.com*, accessed June 15, 2011.
13. CortexPro Web site, *www.cortexpro.com*, accessed June 5, 2011.
14. Delphi Group Web site, *www.delphigroup.com*, accessed June 5, 2011.
15. KM Knowledge Web site, *www.kmknowledge*, accessed February 15, 2012.
16. KMSI Web site, *www.kmsi.us*, accessed June 5, 2011.
17. Knowledge Base Web site, *www.knowledgebase.com*, accessed June 5, 2011.
18. Baker, S., "Watson is Far from Elementary," *The Wall Street Journal*, March 14, 2001, p. A17.
19. Ante, S., "Computer Conquers Jeopardy," *The Wall Street Journal*, January 14, 2011, p. B5.
20. Henschen, D., "Jeopardy Challenge Is More Than a Game," *Information Week*, January 31, 2011, p. 26.
21. Staff, "Jeopardy Whiz Deemed Digital Milestone," *The Tampa Tribune*, February 17, 2001, p. 2.
22. Kurzweil, R., "When Computers Beat Humans on Jeopardy," *The Wall Street Journal*, February 17, 2011, p. A19.
23. Staff, "A Bodybuilder's Brainy Hedge Fund," *Bloomberg Businessweek*, May 9, 2011, p. 52.
24. O'Keefe, Brian, "The Smartest, the Nuttiest Futurist on Earth," *Fortune*, May 14, 2007, p. 60.
25. Turing Web site, *www.turing.org.uk/turing/scrapbook/test.html*, accessed June 15, 2011.
26. Christian, B., "More Than a Machine," *The Wall Street Journal*, March 8, 2011, p. A17.
27. 20 Q Web site, *www.20q.net*, accessed June 5, 2011.
28. Alzheimer Research Forum Web site, *www.alzforum.org/new/detail.asp?id=2173*, accessed June 5, 2011.
29. BrainGate Web site, *www.braingate2.org*, accessed August 4, 2011.
30. Honda Web site, *http://automobiles.honda.com*, accessed June 5, 2011.
31. Abate, Tom, "Future Moving from I, Robot to My Robot," *Rocky Mountain News*, February 26, 2007, p. 8.
32. Lynch, David, "Did That Robot Take My Job?" *Bloomberg Businessweek*, January 9, 2012, p. 15.
33. iRobot Web site, *www.irobot.com*, accessed June 5, 2011.
34. Intelitek Web site, *www.intelitek.com*, accessed June 16, 2011.
35. Bremmer, B., "Rise of the Machines (Again)," *Bloomberg Businessweek*, March 7, 2011, p. 32.
36. Staff, "Space Station Assistant Unpacked," *The Tampa Tribune*, March 17, 2011, p. 2.
37. Michaels, D. and Pasztor, A., "Robot Finds Air France Recorder," *The Wall Street Journal*, May 2, 2011, p. B1.
38. Spitznagel, E., "The Robot Revolution Is Coming," *Bloomberg Businessweek*, January 17, 2001, p. 69.
39. Bremmer, B., "Rise of the Machines (Again)," *Bloomberg Businessweek*, March 7, 2011, p. 32.
40. Robot Learning Laboratory Web site, *www.cs.cmu.edu/~rll*, accessed June 5, 2011.
41. Boehret, K., "The Little Robot That Could Clean the Icky Spots," *The Wall Street Journal*, April 6, 2011, p. D5.
42. Bremmer, B., "Rise of the Machines (Again)," *Bloomberg Businessweek*, March 7, 2011, p. 32.
43. Porter Adventist Hospital Web site, *www.porterhospital.org*, accessed June 5, 2011.
44. Carnegie Mellon Robotics Institute Web site, *www.ri.cmu.edu*, accessed June 16, 2011.
45. DARPA Grand Challenge Web site, *www.darpagrandchallenge.com*, accessed June 5, 2011.
46. Boston Dynamics Web site, *www.bostondynamics.com*, accessed June 16, 2011.
47. Intelligent Robotics Laboratory Web site, *www.is.sys.es.osaka-u.ac.jp/development/0006/index.en.html*, accessed August 4, 2011.
48. Staff, "SA Photonics Develops an Advanced Digital Night Vision System," *Business Wire*, April 19, 2011.
49. MacMillan, D., "Pocket Translator," *Bloomberg Businessweek*, May 9, 2011, p. 44.
50. Nvidia Web site, *www.nvidia.com/object/geforce_family.html*, accessed June 2, 2011.
51. Staff, "AlchemyAPI Announces Major Updates to Natural Language Processing Service," *PR Newswire*, January 11, 2011.
52. Staff, "Intelligence—The Future of Computing," *Business Wire*, January 21, 2011.
53. Ward Systems Group Web site, *www.wardsystems.com/learnmore.asp*, accessed June 16, 2011.
54. Barrios, A., et al, "A Double Genetic Algorthim for the MRCOSO/max," *Computers & Operations Research*, January 2011, p. 33.
55. Xu, Mark, et al, "Intelligent Agent System for Executive Information Scanning, Filtering, and Interpretation," *Information Processing & Management*, March 2011, p. 186.
56. Lantek Web site, *www.lanteksms.com/uk/lantek_expert3_fabricacion.asp*, accessed June 16, 2011.
57. IBM Web site, *www.ibm.com/us/en*, accessed June 5, 2011.
58. Chen, A., et al, "Knowledge Life Cycle, Knowledge Inventory, and Knowledge Acquisition," *Decision Sciences*, February 2010, p. 21.
59. CLIPS Web site, *http://clipsrules.sourceforge.net*, accessed June 16, 2011.
60. Expert System Web site, *www.expertsystem.net*, accessed June 16, 2011.
61. Exsys Web site, *www.exsys.com*, accessed June 6, 2011.
62. EZ-Xpert Expert System Web site, *www.ez-xpert.com*, accessed June 6, 2011.
63. Imprint Web site, *www.imprint-mis.co.uk*, accessed June 16, 2011.
64. Lantek Web site, *www.lantek.es*, accessed June 16, 2011.
65. Gamerman, E., "Animation Nation," *The Wall Street Journal*, February 11, 2011, p. D1.

66. Microsoft Silverlight Web site, *www.silverlight.net*, accessed June 6, 2011.
67. Wakabayashi, D. and Osawa, J., "Odd Couple behind Nintendo 3-D Push," *The Wall Street Journal*, March 2, 2011, p. A1.
68. Autodesk Web site, *http://usa.autodesk.com*, accessed June 6, 2011.
69. Edwards, C. and Alpeyev, P., "Nintendo Brings 3D to the Really Small Screen," *Bloomberg Businessweek*, January 24, 2011, p. 38.
70. Netflix Web site, *www.netflix.com*, accessed August 4, 2011.
71. Staff, "How We Do It," *www.pixar.com/howwedoit/index.html#*, accessed on June 6, 2011.
72. Media Grid Web site, *www.mediagrid.org*, accessed on June 6, 2011.
73. Boehret, K., "Why Smart Phones Can See More Than We Can," *The Wall Street Journal*, May 4, 2011, p. D3.
74. Emory University online archives, *www.emory.edu/EMORY_MAGAZINE/winter96/rothbaum.html*, accessed June 6, 2011.
75. HIT Lab Web site, *www.hitl.washington.edu/projects/vrpain/*, accessed June 6, 2011.
76. 3D Rewind Web site, *www.3drewind.com*, accessed June 6, 2011.
77. 3DS Web site, *www.3ds.com/home*, accessed June 6, 2011.
78. Second Life Web site, *www.secondlife.com*, accessed August 4, 2011.
79. Fab Central Web site, *http://fab.cba.mit.edu*, accessed August 4, 2011.
80. Winslow, R. and Tibken, S., "Big Blu's Tiny Bug Zapper," *The Wall Street Journal*, April 4, 2011, p. A3.
81. Staff, "Eagle Eyes Project," *www.bc.edu/schools/csom/eagleeyes*, accessed June 6, 2011.
82. Hodge, N., "Drone Will Call Aircraft Carriers Home," *The Wall Street Journal*, February 8, 2011, p. A7.
83. Staff, "That's No Bird," *The Tampa Tribune*, March 1, 2011, p. 1.
84. Hodge, N., "Robo-Copters Eye Enemies," *The Wall Street Journal*, May 17, 2011, p. A3.
85. McWhirter, C., et al, "Technology Offers Hope for Detecting Tornadoes Sooner," *The Wall Street Journal*, May 6, 2011, p. A5.
86. Searcey, D., "Toyota Maneuvers to Protect Crown Jewels," *The Wall Street Journal*, March 22, 2011, p. B1.
87. Murphy, C., "Why Toyota-Microsoft Pact Is a Very Big Deal," *Information Week*, April 25, 2011, p. 8.
88. Dvorak, P. and Weinberg, S., "RIM Hopes Cars Drive PlayBook Sales," *The Wall Street Journal*, May 28, 2011, p. B3.
89. White, J., "Car Talk and Talk and…," *The Wall Street Journal*, May 23, 2011, p. R8.
90. Mobileye Web site, *www.mobileye.com/consumer-products/product-line*, accessed June 6, 2011.
91. Minnesota Department of Transportation, *http://projects.dot.state.mn.us/35wbridge*, accessed June 6, 2011.
92. Fanning, Ellen, "Building Customer Loyalty," *Computerworld*, February 21, 2011, p. 42.
93. Worthen, B., "Getting a Lift," *The Wall Street Journal*, April 25, 2011, p. R7.
94. Betts, Mitch, "Footwear, Fashion Driving RFID Growth," *Computerworld*, February 21, 2011, p. 2.
95. NaviTag Web site, *http://navitag.com*, accessed August 4, 2011.
96. Teamcore Research Group Web site, *http://teamcore.usc.edu/projects/security*, accessed June 6, 2011.
97. Health Informatics Journal Web site, *http://jhi.sagepub.com*, accessed August 4, 2011.
98. Frenkel, K., "The Bendable Future," *Bloomberg Businessweek*, May 23, 2011, p. 42.
99. Bennett, D., "Innovator," *Bloomberg Businessweek*, January 31, 2011, p. 40.
100. Staff, "Visa Backs Virtual Wallet," *The Tampa Tribune*, May 16, 2011, p. 4.
101. Zimmerman, A., "Check Out the Future of Shopping," *The Wall Street Journal*, May 18, 2011, p. D1.
102. Staff, "Laptop Has Eye on You," *The Tampa Tribune*, March 7, 2011, p. 6.
103. Staff, "Cruise Control," *CIO*, April 1, 2011, p. 22.
104. Segway Web site, *www.segway.com*, accessed August 4, 2011.
105. Bay Area Segway Web site, *www.bayareaseg.com/Polo.htm*, accessed June 6, 2011.

PART 1 PART 2 PART 3 PART 4 PART 5

Systems Development

CHAPTERS

12 Systems Development: Investigation and Analysis

Principles	Learning Objectives
• Effective systems development requires a team effort from stakeholders, users, managers, systems development specialists, and various support personnel, and it starts with careful planning.	• Identify the key participants in the systems development process and discuss their roles. • Define the term *information systems planning* and list several reasons for initiating a systems project.
• Systems development often uses tools to select, implement, and monitor projects, including prototyping, rapid application development, CASE tools, and object-oriented development.	• Discuss the key features, advantages, and disadvantages of the traditional, prototyping, rapid application development, and end-user systems development life cycles. • Identify several factors that influence the success or failure of a systems development project. • Discuss the use of CASE tools and the object-oriented approach to systems development.
• Systems development starts with investigation and analysis of existing systems.	• State the purpose of systems investigation. • Discuss the importance of performance and cost objectives. • State the purpose of systems analysis and discuss some of the tools and techniques used in this phase of systems development.

Information Systems in the Global Economy

PRACTICE PLAN, UNITED KINGDOM

Requirements Analysis Leads to Effective System

Most citizens of the United Kingdom receive their dental care from the National Health Service (NHS). However, in contrast to medical care, NHS dental care is not free: a cleaning and examination costs £17, about US $26.50 in early 2012. Also, some dentists have opted for the higher earnings potential of private practice. As a result, many patients use private dental practices. To be competitive, these practices must structure their plans in such a way that patients can control their costs.

Practice Plan is the UK's leading provider of custom-branded dental plans. The company works with dental practices, helping them to develop and improve their own brands. In addition to providing collection and payment processing services, Practice Plan provides marketing, design, and business support functions. In mid- 2011, the company supported approximately 1,000 dental practices.

Practice Plan's growth wave "started in 2006," says IT Director John Cawrey. "A major change in the funding structure for dentistry under the National Health Service led to a wave of dentists taking the opportunity to move into private practice. This created an enormous demand for payment plan management services. As a result, our business went from handling around 250,000 patients' accounts to handling more than 450,000, almost overnight. Since then, we have continued to grow to around 600,000, and we're expecting further expansion."

That growth led to a business challenge: to develop a new application to combine Practice Plan's collections and payments into a single integrated solution that could cope with the growth and, the company hoped, more growth to come. Practice Plan had a few options for developing its collection and payment system. It could have written custom software or outsourced the project, a concept discussed in this chapter. It could have looked for an existing package and, if they found one, bought it. The firm could have combined parts of existing applications, either itself or with the help of an outside firm. Practice Plan needed to decide which path to take.

Instead of making that decision right away, Practice Plan began with requirements analysis. During this stage, the company realized that it could do more than just improve its collections. It could reduce duplication of effort in its systems, that resulted in multiple—and often inconsistent—copies of the data. As Cawrey says, "We realized that if we could consolidate all of these applications into a single system, we would have a much more efficient way of driving all our business processes."

Practice Plan then asked several companies, all of which had potentially useful packages or experience with high-volume payment projects, for suggestions. It accepted a proposal from Polymorph that would take advantage of programs they already knew: Notes and Domino, from IBM's Lotus division. The resulting system provided Practice Plan with the following benefits:

- The new system creates only one copy of the data, so that data inconsistency is a thing of the past.
- The new system can process collections for 600,000 patients in 70 minutes. Collections can, therefore, be processed closer to the end of the month, with this faster processing cycle reducing corrections that must be made due to events that took place after collections were processed.
- The system *scales linearly*: if collections for 600,000 patients can be processed in 70 minutes, those for 1,200,000 patients could be processed in 140 minutes.

"The solution is popular with our clients, popular with our own users, and has already made a dramatic difference to the efficiency of our business processes," concludes Cawrey. "As our business continues to grow, we are confident that it will grow with us, and adapt to the changing needs of the dental business in the UK."

As you read this chapter, consider the following:

- Suppose Practice Plan skipped the requirements analysis step of its development process and simply started to develop a collections application that could handle its growth. How would the results differ?
- Practice Plan's solution (a) used software that its staff already knew, (b) used the software integration firm of its choice, and (c) provided a fully integrated system for all its applications. It's rare that a company can achieve all three of these at the same time. If Practice Plan could meet only one or two of these objectives, how would you rank its priorities?

WHY LEARN ABOUT SYSTEMS INVESTIGATION AND ANALYSIS?

Throughout this book, you have seen many examples of the use of information systems in a variety of careers. But where do you start to acquire these systems or have them developed? How can you work with IS personnel, such as systems analysts and computer programmers, to get what you need to succeed on the job or in your own business? This chapter, the first of two chapters on systems development, provides the answers to these questions. You will see how you can initiate the systems development process and analyze your needs with the help of IS personnel. You will also see how you can use the systems development approach to start your own business. Systems investigation and systems analysis are the first two steps of the systems development process. This chapter provides specific examples of how new or modified systems are initiated and analyzed in a number of industries. In this chapter, you will learn how your project can be planned, aligned with corporate goals, rapidly developed, and much more. We start with an overview of the systems development process.

When an organization needs to accomplish a new task or change a work process, how does it do so? It develops a new system or modifies an existing one. Systems development is the activity of creating new systems or modifying existing systems. It refers to all aspects of the process—from identifying problems to solve or opportunities to exploit to implementing and refining the chosen solution. Systems development expenditures are expected to soar in the next few years, according to a *CIO* economic impact survey.[1] In addition, IS departments and systems developers will concentrate on creating more mobile applications for their businesses and organizations.

AN OVERVIEW OF SYSTEMS DEVELOPMENT

In today's businesses, managers and employees in all functional areas work together and use business information systems. As a result, they are helping with development and, in many cases, leading the way. Users might request that a systems development team determine whether they should purchase a few PCs or create an attractive Web site using the tools discussed in Chapter 7. In another case, an entrepreneur might use systems development to build an Internet site to compete with large corporations.

This chapter and the next provide you with a deeper appreciation of the systems development process. Systems development skills and techniques discussed in these two chapters can help people launch their own businesses. Corporations and nonprofit organizations also use systems development to achieve their goals. Hallmark, for example, successfully used systems development to create a new Web site to advertise its greeting cards and related products.[2] The company's new Web site was 300 percent faster than its old one and resulted in increased sales. According to the vice president of Hallmark Digital, "We were all standing around waiting for it to break, waiting for it to crash, and it didn't—it was a real confidence booster for the entire organization." In another example, Shopkick, Inc., created an application that offers discounts to customers for entering a store, including Target, Best Buy, and others.[3]

This chapter will also help you avoid systems development failures or projects that go over budget. A billion-dollar program called SBInet to build a high-tech fence on the border of Arizona and Mexico didn't work as expected.[4] As a result, the Department of Homeland Security halted the project. According to the Secretary of Homeland Security, "SBInet cannot meet its original objective of providing a single, integrated border-security technology solution." As another example, it took years for LCD displays to improve enough to allow the systems developers at Nintendo to create a mobile 3D gaming device.[5] According to one of the developers, "Our bar was set extra high, because we had tried 3D so many times in the past, and it wasn't successful." Further, some health care professionals believe that badly designed health care systems can kill more patients every year than using incorrect medications and inadequate medical devices.

Participants in Systems Development

Effective systems development requires a team effort. The team usually consists of stakeholders, users, managers, systems development specialists, and various support personnel. This team, called the *development team*, is responsible for determining the objectives of the information system and delivering a system that meets these objectives. Increasingly, companies seek members of the development team with training in mobile devices, Internet applications, and social networks.[6] Today, companies are setting up their own internal app stores.[7] According to the managing director of Enterprise Mobility Foundation, "Enterprise applications [on tablets] are an important and growing phenomenon. Organizations are realizing that a lot of applications that the company uses can be relevant on mobile devices." Selecting the best IS team for a systems development project is critical to project success. A *project* is a planned collection of activities that achieves a goal, such as constructing a new manufacturing plant or developing a new decision support system.[8]

All projects have a defined starting point and ending point, normally expressed as dates such as August 4 and December 11. Most IS budgets have a significant amount of funds allocated for new systems development efforts. Indeed, many CIOs, for example, believe an economic slump is an opportunity to invest for the future, not cut back on IS projects.[9]

A *project manager* is responsible for coordinating all people and resources needed to complete a project on time. According to the senior vice president of Sprint, "We want to dramatically improve project delivery cycle time to help our business partners deliver offerings to the market faster."[10] The project manager can make the difference between project success and failure. In systems development, the project manager can be an IS person inside the organization or an external consultant. Project managers need technical, business, and people skills. In addition to completing the project on time and within the specified budget, the project manager is usually responsible for controlling project quality, training personnel, facilitating communications, managing risks, and acquiring any necessary equipment, including office supplies and sophisticated computer systems.

stakeholders: People who, either themselves or through the organization they represent, ultimately benefit from the systems development project.

users: People who will regularly interact with the system.

In the context of systems development, **stakeholders** are people who, either themselves or through the organization they represent, ultimately benefit from the systems development project. **Users** are people who will regularly

© Yuri Arcurs/Shutterstock

Because stakeholders ultimately benefit from the systems development project, they often work with others in developing a computer application.

interact with the system. They can be employees, managers, or suppliers. For large-scale systems development projects in which the investment in and value of a system can be high, it is common for senior-level managers, including the functional vice presidents (of finance, marketing, and so on), to be part of the development team.

systems analyst: A professional who specializes in analyzing and designing business systems.

programmer: A specialist responsible for modifying or developing programs to satisfy user requirements.

Depending on the nature of the systems project, the development team might include systems analysts and programmers, among others. A **systems analyst** is a professional who specializes in analyzing and designing business systems. Systems analysts play various roles while interacting with the stakeholders and users, management, vendors and suppliers, external companies, programmers, and other IS support personnel (see Figure 12.1). Like an architect developing blueprints for a new building, a systems analyst develops detailed plans for the new or modified system. The **programmer** is responsible for modifying or developing programs to satisfy user requirements. Like a contractor constructing a new building or renovating an existing one based on an architect's drawings, the programmer takes the plans from the systems analyst and builds or modifies the necessary software.

The other support personnel on the development team are mostly technical specialists, including database and telecommunications experts, hardware engineers, and supplier representatives. It is becoming more common for companies to open IS departments with a team of systems analysts and other IS personnel in foreign countries. For small businesses, the development team might consist only of a systems analyst and the business owner as the primary stakeholder. For larger organizations, formal IS staff can include hundreds of people involved in a variety of activities, including systems development.

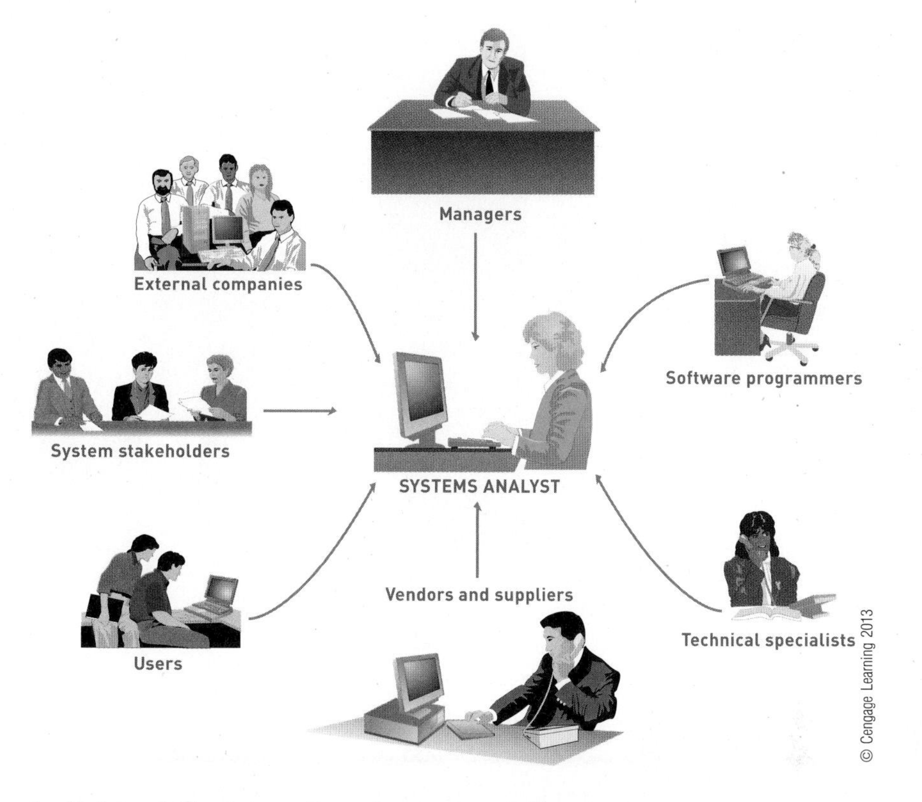

FIGURE 12.1

Role of the systems analyst

The systems analyst plays an important role in the development team and is often the only person who sees the system in its totality. The one-way arrows in this figure do not mean that there is no direct communication between other team members. These arrows just indicate the pivotal role of the systems analyst—a person who is often called on to be a facilitator, moderator, negotiator, and interpreter for development activities.

Individual Systems Developers and Users

For decades, systems development was oriented toward corporations and corporate teams or groups, whose major participants were discussed earlier in the chapter. While such teams continue to be an important part of systems development, we are seeing individual systems developers and users to a greater extent.

An *individual systems developer* is a person who performs all of the systems development roles, including systems analyst, programmer, technical specialist, and the other roles described earlier. While individual systems developers can create applications for a group or an entire organization, many specialize in developing applications for individuals. A large number of these applications are available for smartphones and other handheld computing devices. Revenues from mobile applications from all sources are expected to be over $15 billion annually according to Gartner, Inc.[11] An individual systems designer, for example, created an application called Word Lens that uses optical character recognition to read text by holding a smartphone or other mobile device up to restaurant menus, books, signs, and other text and taking a picture of the text. In a few seconds, the application can translate the camera image from one language to another, such as from German to English.[12]

Individual developers from around the world are using the steps of systems development to create unique applications for the iPhone and other mobile devices.[13] Before an individual developer can have his or her application placed or sold on Apple's application store, however, Apple must approve or select the application. Apple computer has sold more than $4 billion worth of applications on its App Store.[14] This vast array of applications is one reason for the popularity of Apple's smartphones, tablet computers, and other mobile devices. Individual systems developers that create applications for Apple's iTunes store may not be completely happy with the fees they pay to Apple, which can be as high as 30 percent of sales in some cases.[15] With the popularity of Apple's tablet computer, however, many individual systems developers feel they can't ignore this important market even with its high fees.

Other companies also have application stores. BlackBerry has an application store called App World, and Google has the Android Market store. Some applications, such as Google's Secure Data Connector, allow data to be

© iStockphoto/Sjoerd van der Wal

An iPad application developed to provide information about cars is featured at a motor show in Amsterdam, The Netherlands.

downloaded from secure corporate databases, including customer and supplier information. Mobile devices, however, can pose serious security risks to businesses and nonprofit organizations that allow their workers and managers to use these devices at work.[16] According to a report from a large computer consulting vendor, "Malicious software on the devices can be used to spy on users, access sensitive information on the phones, and reach back into corporate networks." Securing and managing smartphones and other mobile devices in a corporate setting is often called *mobile device management (MDM)*.[17] A number of companies are making MDM products, including Enterprise Mobility Management by McAfee, Afaria by Sybase, MDM by Tangoe, and others.

Individual users acquire applications for both personal and professional use. Cisco, the large networking company, has developed an iPhone application to help individual security personnel respond to IS- and computer-related threats.[18] The application, called Security Intelligence Operations To Go, can instantly notify security professionals of security attacks as they occur and can help them recover if they do occur. Individual applications can be used to compare prices of products, analyze loans, locate organic food, find reliable repair services, and search for an apartment. You can also turn a smartphone into a powerful scientific or financial calculator. Although most people purchase individual applications from authorized Web sites, unauthorized application stores that are not supported by the smartphone or cellular company can also be used to purchase or acquire useful applications.

end-user systems development: Any systems development project in which the primary effort is undertaken by a combination of business managers and users.

It is also possible for one person to be both an individual developer and user. The term **end-user systems development** describes any systems development project in which business managers and users assume the primary effort. User-developed systems range from the very small (such as a software routine to merge form letters) to those of significant organizational value (such as customer contact databases for the Web). Even if you develop your own applications, you will likely want to have an IS department develop applications for you that are too complex or time consuming to develop on your own. In this case, you will be involved in initiating systems development, which is discussed next.

Initiating Systems Development

Systems development initiatives arise from all levels of an organization and are both planned and unplanned. The Code for America (CFA) organization, for example, has initiated systems development efforts in Boston and other American cities to help cities and municipalities solve their problems.[19] As a result of a meeting with Boston firefighters, CFA launched a new systems

Many end users today are demonstrating their systems development capability by designing and implementing their own PC-based systems.

development effort to help firefighters locate fire hydrants that might be completely covered with snow in the winter. The programmers for CFA developed a Web site to pinpoint the location of every fire hydrant. CFA used open-source software and made its efforts free to other cities and municipalities.

Other organizations are also initiating systems development to take advantage of new technologies, such as smartphones and tablet computers, by developing unique and powerful applications for these newer devices.[20] Systems development projects are initiated for many reasons, as shown in Figure 12.2.

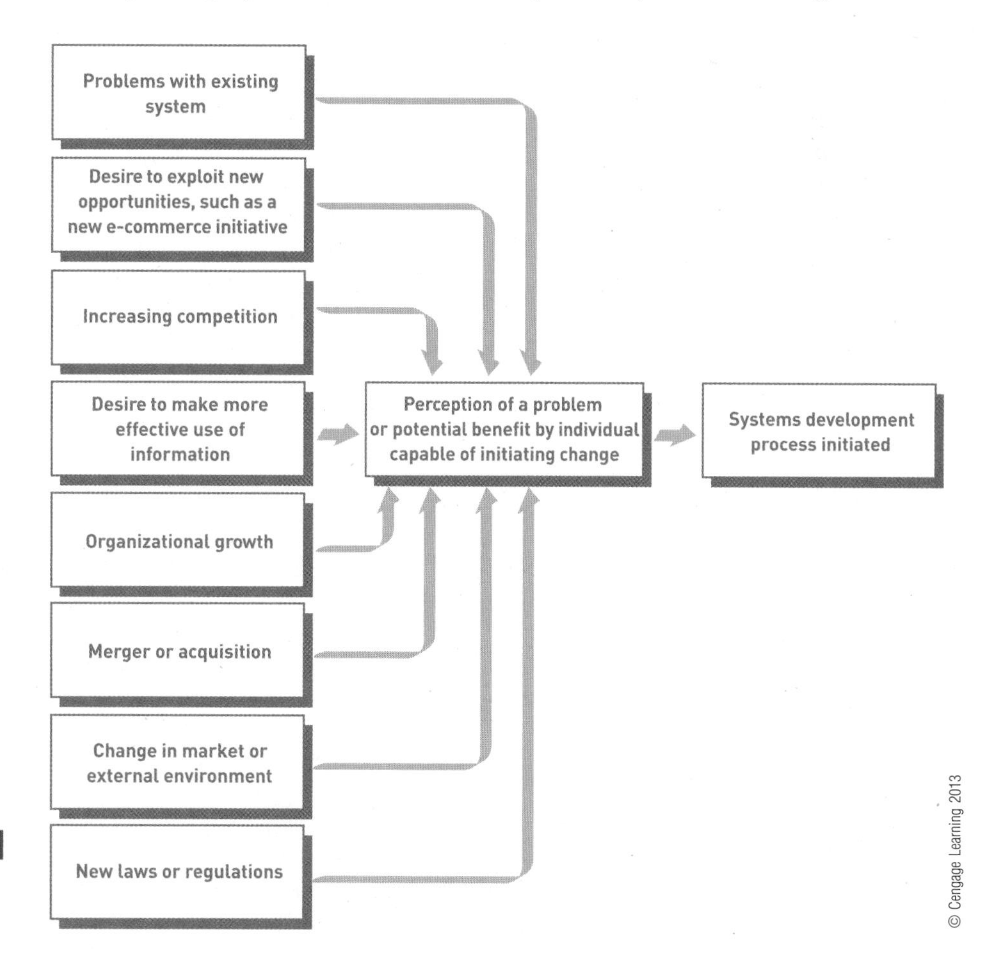

FIGURE **12.2**

Typical reasons to initiate a systems development project

Mergers and acquisitions, for example, can trigger many systems development projects. Companies that merge with or acquire other companies can be a challenge for the IS departments that are often integrated into a single IS function.[21] Consequently, getting IS departments and CIOs involved early in the merger or acquisition process can result in better deals and smoother transitions. According to one IS manager, "I've gone from not being included to having [a role] fairly early in the deal." Even with similar information systems, the procedures, culture, training, and management of the information systems are often different, requiring a realignment of the IS departments. Of course, systems development can be initiated because existing systems and procedures failed or caused problems.

Systems development can also be initiated when a vendor no longer supports an older system or older software. When this support is no longer available, companies are often forced to upgrade to new software and systems, which can be expensive and require additional training. This lack of support is a dilemma for many companies trying to keep older systems operational.

Likewise, governments can foster new systems development projects in the public and private sectors. Federal, state, and local tax breaks have resulted in new systems development efforts.[22] A depreciation tax benefit enacted by the U.S. Congress in 2010, for example, has caused some companies to purchase hardware and related computer equipment in 2011. According to the president of a computer research firm, "Buyers who are looking out 18 months now may move acquisitions into 2011 to take advantage of the accelerated depreciation." The Federal Trade Commission has called for Internet companies to add a "do-not-call" feature to their Web browser to protect customer privacy, and some Internet companies have complied. Firefox, for example, has a do-not-track feature as part of its Internet browser.[23]

Information Systems Planning and Aligning Corporate and IS Goals

Information systems planning and aligning corporate and IS goals are important aspects of any systems development project. Achieving a competitive advantage is often the overall objective of systems development.

Information Systems Planning

information systems planning: Translating strategic and organizational goals into systems development initiatives.

The term **information systems planning** refers to translating strategic and organizational goals into systems development initiatives (see Figure 12.3). Proper IS planning ensures that specific systems development objectives support organizational goals. Long-range planning can also be important and can result in getting the most from a systems development effort. It can also align IS goals with corporate goals and culture, which is discussed next.

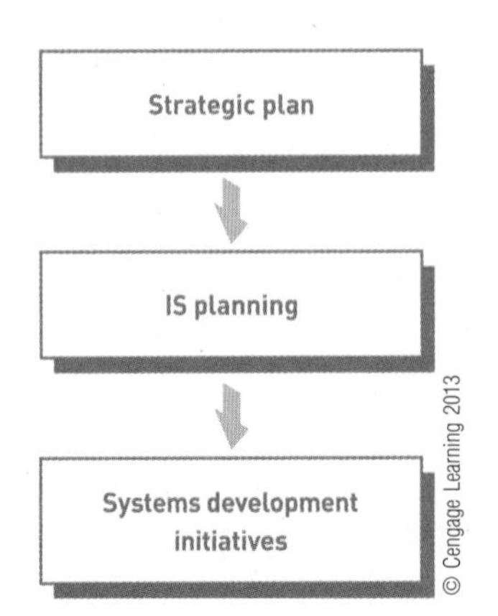

FIGURE **12.3**

Information systems planning

Information systems planning transforms organizational goals outlined in the strategic plan into specific systems development activities.

Privacy by Design at Hydro One

It's not hard to get managers to agree that an information system should protect privacy. It's harder to make privacy a priority so that the system meets all its privacy requirements.

Hydro One, a Toronto-based supplier of electricity to the Canadian province of Ontario, faced this problem in developing its Advanced Distribution System (ADS) pilot project. They decided to adopt the principles of Privacy by Design (PbD).

The PbD approach is based on seven principles covering information systems, business practices, and infrastructure. A development team begins to apply these principles in the first stages of system development, where respect for privacy is a core foundational requirement, and continues through the stages of system development covered in this chapter and the next. The PbD idea is that it is not necessary to sacrifice privacy to design an effective information system; rather, organizations can and should design a system that provides both privacy and effectiveness. In 2010, the 32nd International Conference of Data Protection and Privacy Commissioners approved resolutions that recommended organizations adopt PbD principles as a fundamental concept and encouraged governments to incorporate those principles into future privacy policies and legislation in their respective countries.

Hydro One's ADS had four business objectives when it was conceived: to optimize power distribution, optimize network planning, improve distribution reliability, and optimize outage restoration. Early in the systems analysis stage, the analysts broke those four objectives into 30 specific capabilities that ADS must have to support 30 business processes, such as maintenance reporting and outage-related customer communications. The processes communicated, to varying degrees, with shareholders, customers, suppliers, and regulatory authorities. In these interfaces, privacy requirements were most at risk. For example, the Grid domain will not retain any data about an individual customer's identity.

To make sure privacy requirements weren't violated, Hydro One adopted 12 security design principles such as "Compartmentalize elements with common security and privacy requirements." The company also looked at 60 threat scenarios, ranking each for seriousness (insignificant to catastrophic) and likelihood (rare to almost certain), to see if the design principles would protect against them. The likelihood of a threat was based on three factors: motive, means, and opportunity. If all three are low, the likelihood is *rare*. If all are high, it is *almost certain*. Other combinations lead to *unlikely*, *possible*, and *likely*.

The team then came up with 28 privacy requirements such as "The ADS information system produces audit records for each event that contains the following information: Date and time of the event, the component of the ADS information system where the event occurred, type of event, user/subject identity, and the outcome of the event." By making sure that system design conformed to these privacy requirements, Hydro One would be at the forefront of privacy practices.

The result of this design process was a clear statement of design requirements for the system's three business domains: the Grid, services, and customers. As a result of using PbD, Hydro One managers are confident that ADS will protect private information while achieving its business objectives.

When management consulting firm Ernst & Young reported the top 11 privacy trends for 2011, Privacy by Design was on that list. In their report, Ernst & Young concluded: "Protecting personal information can no longer be an afterthought that is bolted onto an existing privacy or security program. As Privacy by Design suggests, it needs to be a series of much-needed policies that embed privacy protection into new technologies and business practices at the outset. The focus on privacy will enhance the business performance of leading organizations."

Discussion Questions

1. Instead of developing 28 specific requirements that designers could follow, suppose the system planners simply said, "Be sure private information is protected." System designers might prefer this single statement because it gives them more freedom in their work. Making this statement also means less work for the planners. Since Hydro One didn't take this approach, making a blanket statement must have some drawbacks. What are they?
2. The term "Privacy by Design" might seem inaccurate, since this work is done during the system analysis phase rather than the later system design phase. Do you think it makes sense anyhow? If so, why? Can you suggest a better term?

Critical Thinking Questions

1. Consider the four ADS business objectives as given in the case. The first two have to do with planning company's network. The last two have to do with operating it. Do you think the privacy issues related to one category are different from those that arise with the other? Why or why not?
2. Think of four threats to the information system of a hospital emergency room. Evaluate their likelihood for the three factors in the case as low, medium, or high. For each threat, unless it scored low on all three, suggest a way to reduce at least one factor that you rated as medium or high.

SOURCES: Dougherty, S., "Privacy by Design: From Resolution to Reality," IBM case study, *www.cio.gov.bc.ca/local/cio/informationsecurity/documents/PS_2011_PDFs/Dougherty_Steven-WorkshopF.pdf*, downloaded January 23, 2012; Hydro One Web site, *www.hydroone.com*, accessed January 23, 2012; Hill, K., "Why Privacy by Design Is the New Corporate Hotness," *Forbes*, *www.forbes.com/sites/kashmirhill/2011/07/28/why-privacy-by-design-is-the-new-corporate-hotness*, July 28, 2011; Cavoukian, A., Privacy by Design Web site, *www.privacybydesign.ca*, Office of the Information and Privacy Commissioner, Ontario, Canada, accessed January 23, 2012; Ernst & Young, "Top 11 Privacy Trends for 2011," *www.ey.com/GL/en/Services/Advisory/IT-Risk-and-Assurance/Top-11-privacy-trends-for-2011*, January 2011.

Aligning Corporate and IS Goals

Aligning organizational goals and IS goals is critical for any successful systems development effort.[24] As mentioned in Chapter 2, firms that achieve a competitive advantage often emphasize the alignment of organizational and IS goals and strategies.[25] In other words, these organizations make sure their IS departments are totally supportive of the broader goals and strategies of the organization.[26] According to the senior vice president and CIO of the Associated Press, "When developing your priority list, give No. 1 ranking to understanding the business and business expectations."[27]

Specific systems development initiatives can spring from the IS plan, but the IS plan must also provide a broad framework for future success. The IS plan should guide development of the IS infrastructure over time. Another benefit of IS planning is that it ensures better use of IS resources, including funds, personnel, and time for scheduling specific projects. The steps of IS planning are shown in Figure 12.4.

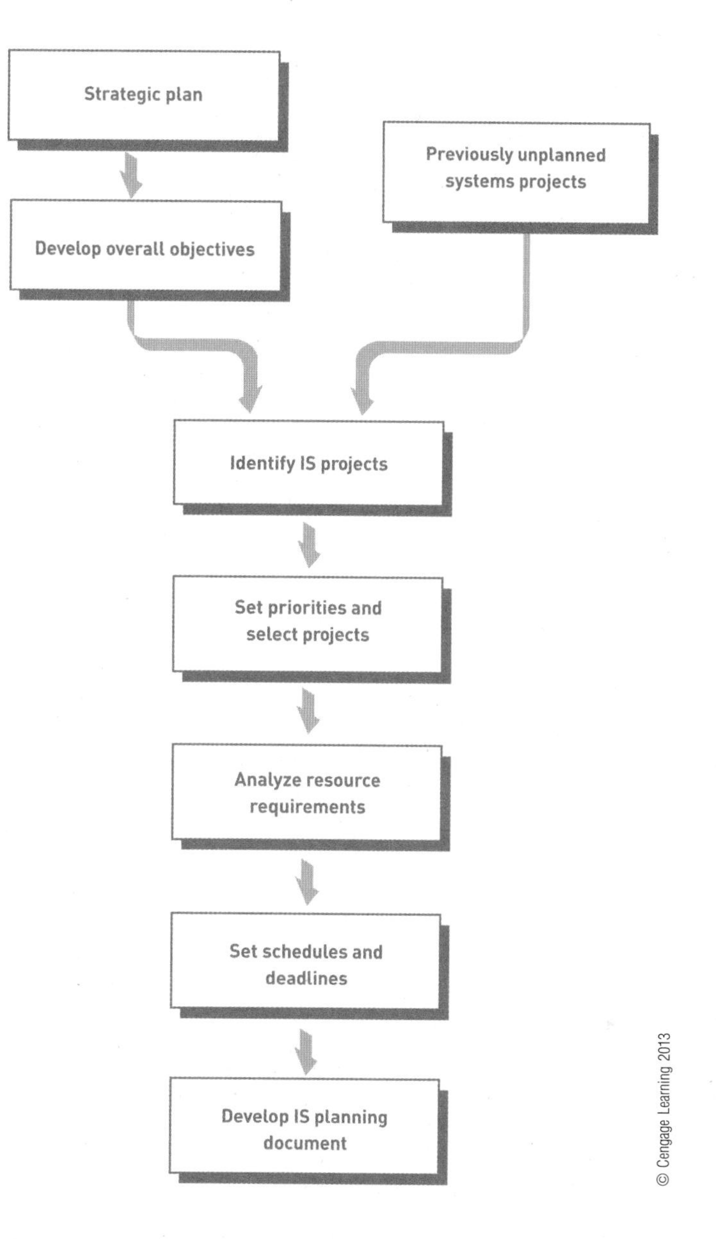

FIGURE 12.4

The steps of IS planning

Some projects are identified through overall IS objectives, whereas additional projects, called *unplanned projects*, are identified from other sources. All identified projects are then evaluated in terms of their organizational priority.

Developing a Competitive Advantage

In today's business environment, many companies seek systems development projects that will provide them with a competitive advantage. Thinking competitively usually requires creative and critical analysis. By looking at problems in new or different ways and by introducing innovative methods to solve them, many organizations have gained significant competitive advantage.

creative analysis: The investigation of new approaches to existing problems.

Creative analysis involves investigating new approaches to existing problems. By looking at problems in new or different ways and by introducing innovative methods to solve them, many firms have gained a competitive advantage. Typically, these new solutions are inspired by people and events not directly related to the problem. Sir Isaac Newton, for example, watched something as simple as an apple fall from a tree as he developed the laws of gravity. Albert Einstein modified Newton's laws of gravity to create special and general relativity watching trains and thinking about the constant speed of light from different vantage points. Today, physicists are once again using creative analysis to dream about nine or more dimensions that could modify Einstein's theories with an approach called "string theory." It is hoped that string theory will explain everything from the gravity of large planets and stars to the movement of the smallest particles known to today's scientists.

critical analysis: The unbiased and careful questioning of whether system elements are related in the most effective ways.

Critical analysis requires unbiased and careful questioning of whether system elements are related in the most effective ways. It involves considering the establishment of new or different relationships among system elements and perhaps introducing new elements into the system. Critical analysis in systems development involves the following actions:

- **Questioning statements and assumptions**. Questioning users about their needs and clarifying their initial responses can result in better systems and more accurate predictions.
- **Identifying and resolving objectives and orientations that conflict**. Each department in an organization can have different objectives and orientations. The buying department might want to minimize the cost of spare parts by always buying from the lowest-cost supplier, but engineering might want to buy more expensive, higher-quality spare parts to reduce the frequency of replacement. These differences must be identified and resolved before a new purchasing system is developed or an existing one modified.

Establishing Objectives for Systems Development

The overall objective of systems development is to achieve business goals, not technical goals, by delivering the right information to the right person at the right time. The impact a particular system has on an organization's ability to meet its goals determines the true value of that system to the organization. Although all systems should support business goals, some systems are more pivotal in continued operations and goal attainment than others. These systems are called **mission-critical systems**. An order-processing system, for example, is usually considered mission-critical. Without it, few organizations could continue daily activities, and they clearly would not meet set goals.

mission-critical systems: Systems that play a pivotal role in an organization's continued operations and goal attainment.

The goals defined for an organization also define the objectives that are set for a system. A manufacturing plant, for example, might determine that minimizing the total cost of owning and operating its equipment is critical to meeting production and profit goals. **Critical success factors (CSFs)** are factors that are essential to the success of certain functional areas of an organization. The CSF for manufacturing—minimizing equipment maintenance and operating costs—would be converted into specific objectives for a proposed system. One specific objective might be to alert maintenance planners when a piece of equipment is due for routine preventative maintenance (e.g., cleaning and lubrication). Another objective might be to alert the maintenance planners when the necessary cleaning materials, lubrication oils, or spare parts inventory levels are below specified limits. These objectives could be accomplished either through automatic stock replenishment via electronic data interchange or through the use of exception reports.

critical success factors (CSFs): Factors that are essential to the success of a functional area of an organization.

Regardless of the particular systems development effort, the development process should define a system with specific performance and cost objectives. The success or failure of the systems development effort will be measured against these objectives. New York, for example, awarded a $63 million systems development project to an outside firm.[28] When the project ran into large cost overruns, the city sued the outside firm for $600 million.

Performance Objectives

The extent to which a system performs as desired can be measured through its performance objectives. System performance is usually determined by factors such as the following:

- **The quality or usefulness of the output.** Is the system generating the right information to the right people in a timely fashion?

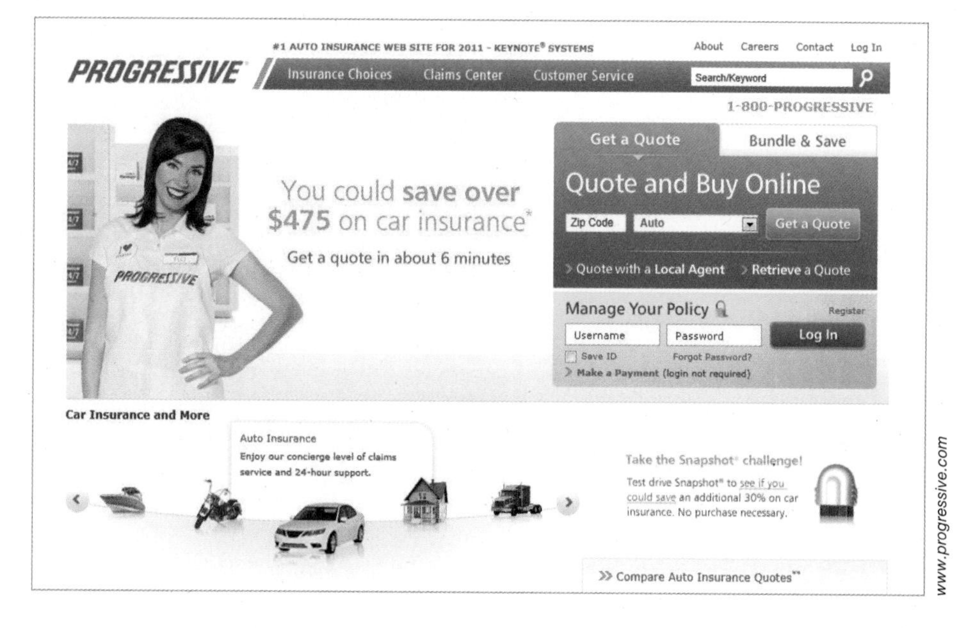

The Progressive Web site invites customers to get fast quotes on auto insurance and related insurance products.

www.progressive.com

- **The accuracy of the output.** Is the output accurate, and does it reflect the true situation?
- **The speed at which output is generated.** Is the system generating output in time to meet organizational goals and operational objectives? Progressive, a large insurance company, has developed a Web site that allows customers to get fast quotes on auto insurance and related insurance products with few inputs.[29] They hope the increased speed of outputting quotes will translate into more customers. According to the CIO of the FBI, concerning one of its IS projects, "It is all about helping us conduct investigations faster."[30]
- **The flexibility of the system.** Is the information system flexible and adaptable enough to produce a variety of reports and documents, depending on current conditions and the needs of the organization?
- **The ease of use of the application.** Developing applications that can be easily used by managers and employees is an important goal for any systems development process. Genetech, for example, developed about 15 applications for its employees that can be downloaded into company-owned iPhones.[31] Downloading the corporate applications from its Web site is as easy as downloading music and other content from Apple's App Store. According to Genetech's vice president of technology, "We're cutting down on the drag of technology so people can focus more of their intellectual energy and creativity on the important stuff."
- **The scalability of the resulting system.** *Scalability* allows an information system to handle business growth and increased business volume. The number of trades processed at the Chicago Mercantile Exchange, for example, has grown from 30 million a year in 2004 to more than 6 billion today.[32] This incredible growth requires a highly scalable information system.
- **The risk of the system.** One important objective of many systems development projects is to reduce risk, such as increased cost or delays in the project.[33] For one company implementing a new customer relationship management system, the cost of the systems development effort was twice what was projected.[34] According to one outside observer, "It's been a significant drag on that company's performance."

The Chicago Mercantile Exchange needs a highly scalable information system to process a dramatic increase in trades.

Cost Objectives

Organizations can spend more than is necessary during a systems development project. The benefits of achieving performance goals should be balanced with all costs associated with the system, including the following:

- **Development costs.** All costs required to get the system up and running should be included. Some computer vendors give cash rewards to companies using their systems to reduce costs and act as an incentive.
- **Costs related to the uniqueness of the system application.** A system's uniqueness has a profound effect on its cost. An expensive but reusable system might be preferable to a less costly system with limited use.
- **Fixed investments in hardware and related equipment.** Developers should consider the costs of such items as computers, network-related equipment, and environmentally controlled data centers in which to operate the equipment.
- **Ongoing operating costs of the system.** Operating costs include costs for personnel, software, supplies, and resources such as the electricity required to operate the system.

SYSTEMS DEVELOPMENT LIFE CYCLES AND APPROACHES

The systems development process is also called a *systems development life cycle (SDLC)* because the activities associated with it are ongoing. As each system is built, the project has timelines and deadlines until at last the system is installed and accepted. Many hospitals, for example, are facing tight U.S. government deadlines to complete systems development projects to better track medical diagnoses and treatments.[35] A key fact of systems development is that the later in the SDLC an error is detected, the more expensive it is to correct (see Figure 12.5). One reason for the mounting costs is that if an error is found in a later phase of the SDLC, the previous phases must be reworked to some extent. Thus, experienced systems developers prefer an approach that will catch errors early in the project life cycle.

Common systems development life cycles include traditional, prototyping, rapid application development (RAD), and individual development. In addition, companies can outsource the systems development process. With some companies, these approaches are formalized and documented so that systems developers have a well-defined process to follow; other companies use less formalized approaches. The systems development life cycle and approach that organizations use depends on the culture and approach of the IS department

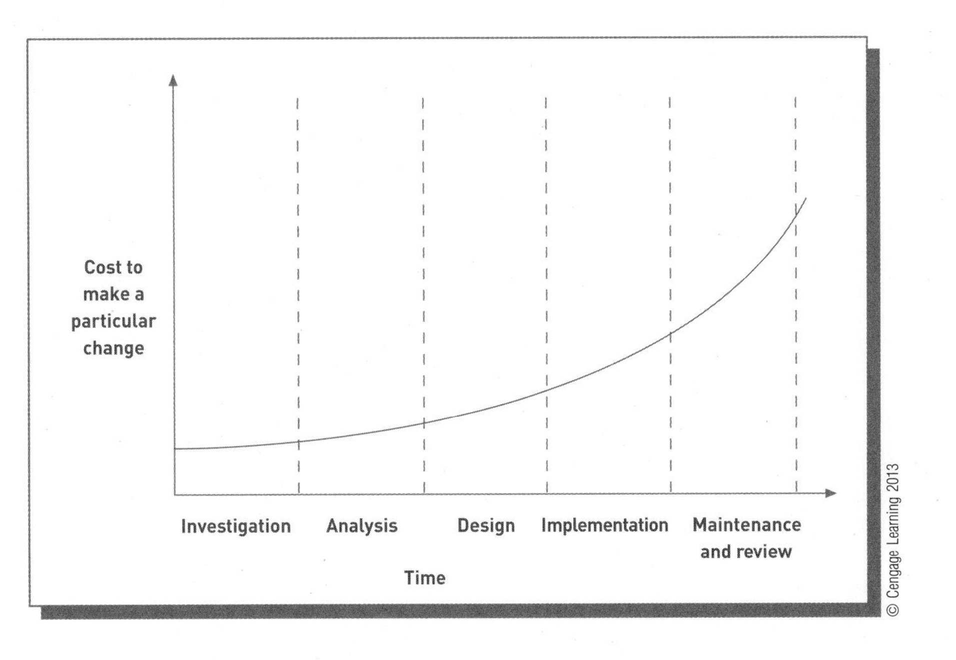

FIGURE 12.5
Relationship between timing of errors and costs
The later that system changes are made in the SDLC, the more expensive these changes become.

and the characteristics of the specific systems development project.[36] Keep Figure 12.5 in mind as you are introduced to alternative SDLCs in the next section.

The Traditional Systems Development Life Cycle

Traditional systems development efforts can range from a small project, such as purchasing an inexpensive computer program, to a major undertaking, such as installing a large computer system at a corporation or university. The steps of traditional systems development might vary from one company to the next, but most approaches have five common phases: investigation, analysis, design, implementation, and maintenance and review (see Figure 12.6).

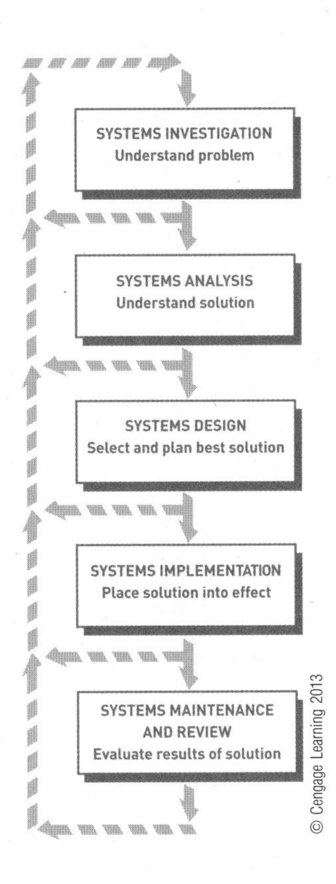

FIGURE 12.6
Traditional systems development life cycle
Sometimes, information learned in a particular phase requires cycling back to a previous phase.

systems investigation: The systems development phase during which problems and opportunities are identified and considered in light of the goals of the business.

systems analysis: The systems development phase involving the study of existing systems and work processes to identify strengths, weaknesses, and opportunities for improvement.

systems design: The systems development phase that defines how the information system will do what it must do to obtain the solution.

systems implementation: The systems development phase involving the creation or acquisition of various system components detailed in the systems design, assembling them, and placing the new or modified system into operation.

systems maintenance and review: The systems development phase that ensures the system operates and modifies the system so that it continues to meet changing business needs.

prototyping: An iterative approach to the systems development process in which at each iteration requirements and alternative solutions to a problem are identified and analyzed, new solutions are designed, and a portion of the system is implemented.

In the **systems investigation** phase, potential problems and opportunities are identified and considered in light of the goals of the business. Systems investigation attempts to answer the questions "What is the problem, and is it worth solving?" The primary result of this phase is a defined development project for which business problems or opportunity statements have been created, to which some organizational resources have been committed, and for which systems analysis is recommended. **Systems analysis** attempts to answer the question "What must the information system do to solve the problem?" This phase involves studying existing systems and work processes to identify strengths, weaknesses, and opportunities for improvement. The major outcome of systems analysis is a list of requirements and priorities. **Systems design** seeks to answer the question "How will the information system do what it must do to obtain the solution?" The primary result of this phase is a technical design that either describes the new system or describes how existing systems will be modified. The system design details system outputs, inputs, and user interfaces; specifies hardware, software, database, telecommunications, personnel, and procedure components; and shows how these components are related. **Systems implementation** involves creating or acquiring the various system components detailed in the systems design, assembling them, and placing the new or modified system into operation. An important task during this phase is to train the users. Systems implementation results in an installed, operational information system that meets the business needs for which it is developed. It can also involve phasing out or removing old systems, a process that can be difficult for existing users, especially when the systems are free.

The purpose of **systems maintenance and review** is to ensure that the system operates as intended and to modify the system so that it continues to meet changing business needs. As shown in Figure 12.6, a system under development moves from one phase of the traditional SDLC to the next.

The traditional SDLC allows for a large degree of management control. However, a major problem is that the user does not use the solution until the system is nearly complete. Table 12.1 lists advantages and disadvantages of the traditional SDLC.

TABLE **12.1** Advantages and disadvantages of traditional SDLC

Advantages	Disadvantages
Formal review at the end of each phase allows maximum management control.	Users get a system that meets the needs as understood by the developers; this might not be what the users really needed.
This approach creates considerable system documentation.	Documentation is expensive and time consuming to create. It is also difficult to keep current.
Formal documentation ensures that system requirements can be traced back to stated business needs.	Often, user needs go unstated or are misunderstood.
It produces many intermediate products that can be reviewed to see whether they meet the users' needs and conform to standards.	Users cannot easily review intermediate products and evaluate whether a particular product (e.g., a data flow diagram) meets their business requirements.

Prototyping

Prototyping takes an iterative approach to the systems development process. During each iteration, requirements and alternative solutions to the problem

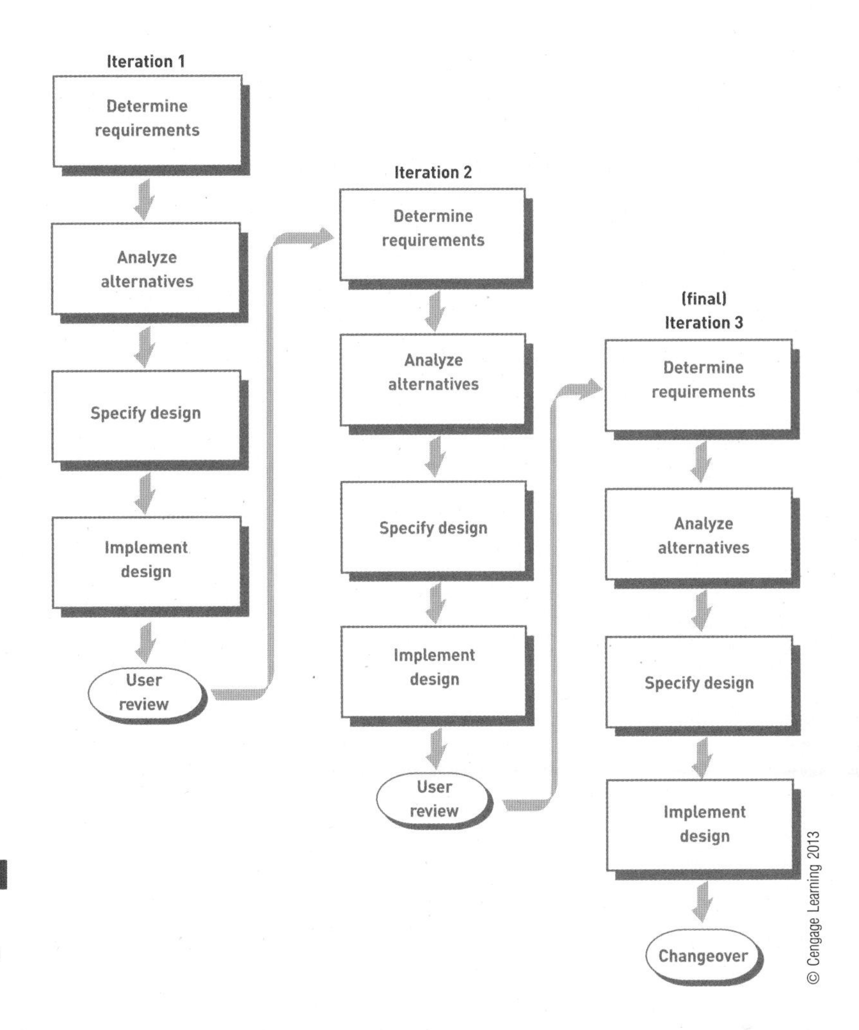

FIGURE 12.7
Prototyping
Prototyping is an iterative approach to systems development.

are identified and analyzed, new solutions are designed, and a portion of the system is implemented.[37] Users are then encouraged to try the prototype and provide feedback (see Figure 12.7). Prototyping begins by creating a preliminary model of a major subsystem or a scaled-down version of the entire system. For example, a prototype might show sample report formats and input screens. After they are developed and refined, the prototypical reports and input screens are used as models for the actual system. The first preliminary model is refined to form the second- and third-generation models and so on until the complete system is developed (see Figure 12.8).

Prototypes can be classified as operational or nonoperational. An *operational prototype* is a prototype that works, that is, accesses real data files, edits input data, makes necessary computations and comparisons, and produces real output. A *nonoperational prototype* is a mock-up or model that includes output and input specifications and formats. The advantages and disadvantages of prototyping are summarized in Table 12.2.

Rapid Application Development, Agile Development, and Other Systems Development Approaches

rapid application development (RAD): A systems development approach that employs tools, techniques, and methodologies designed to speed application development.

Rapid application development (RAD) employs tools, techniques, and methodologies designed to speed application development. The CIO of Hewlett-Packard has slashed project completion time to just six months using RAD and says, "We ought to be changing the productivity of everybody in our organization every 30 days. If we're going to keep up with the growth in front of us, we better figure out how to do that."[38] According to the vice

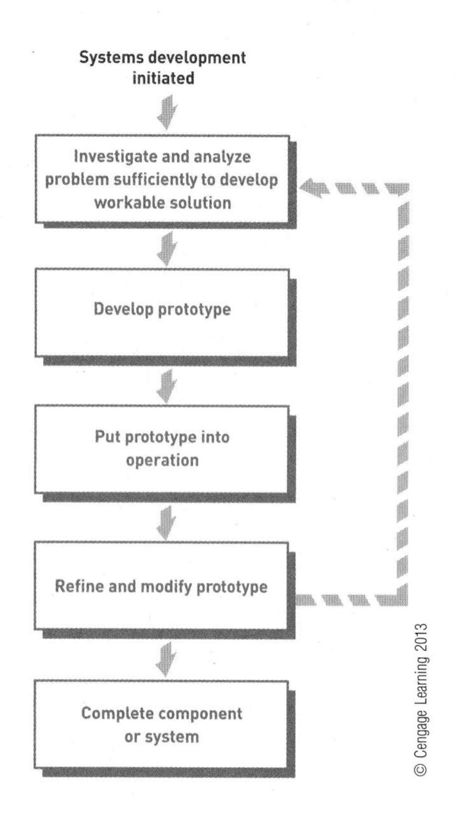

FIGURE 12.8
Refining during prototyping
Each generation of prototype is a refinement of the previous generation based on user feedback.

TABLE 12.2 Advantages and disadvantages of prototyping

Advantages	Disadvantages
Users can try the system and provide constructive feedback during development.	Each iteration builds on the previous one. The final solution might be only incrementally better than the initial solution.
An operational prototype can be produced in weeks.	Formal end-of-phase reviews might not occur. Thus, it is very difficult to contain the scope of the prototype, and the project never seems to end.
As solutions emerge, users become more positive about the process and the results.	System documentation is often absent or incomplete because the primary focus is on development of the prototype.
Prototyping enables early detection of errors and omissions.	System backup and recovery, performance, and security issues can be overlooked in the haste to develop a prototype.

president and CIO of State Street, a financial services company, "Our goal is to optimize the time from idea to solution. The idea is when the stopwatch starts. Really, it's about the customer—when does their stopwatch start."

RAD tools can also be used to make systems development projects more flexible and agile so that they are able to rapidly change with changing conditions and environments. Vendors such as Computer Associates International, IBM, and Oracle market products targeting the RAD market. Rational Software, a division of IBM, has a RAD tool called Rational Rapid Developer to make developing large Java programs and applications easier and faster.[39] Rational allows both systems developers and users to collaborate on systems development projects using Team Concert, which is like a social networking site for IBM developers and users.[40] Locus Systems, a program developer, used a RAD tool called OptimalJ from Compuware to generate more than 60 percent of the computer code for three applications it developed.[41] Advantage Gen, formerly known as COOL:Gen, is a RAD tool from Computer Associates International.[42] It can be used to rapidly generate computer code from business models and specifications.

RAD should not be used on every software development project. In general, it is best suited for DSSs and MISs and less suited for TPSs. During a RAD project, the level of participation of stakeholders and users is much higher than in other approaches. Table 12.3 lists advantages and disadvantages of RAD.

TABLE **12.3** Advantages and disadvantages of RAD

Advantages	Disadvantages
For appropriate projects, this approach puts an application into production sooner than any other approach.	This intense SDLC can burn out systems developers and other project participants.
Documentation is produced as a byproduct of completing project tasks.	This approach requires systems analysts and users to be skilled in RAD systems development tools and RAD techniques.
RAD forces teamwork and lots of interaction between users and stakeholders.	RAD requires a larger percentage of stakeholders' and users' time than other approaches.

Other approaches to rapid development, such as *agile development* or *extreme programming* (*XP*), allow the systems to change as they are being developed. Agile development requires cooperation and frequent face-to-face meetings with all participants, including systems developers and users, as they modify, refine, and test how the system meets users' needs and what its capabilities are. Organizations are using agile development to a greater extent today to improve the results of systems development, including global systems development projects requiring IS resources distributed in different locations. The FBI used agile development to save time and reduce the number of people needed for a project called Sentinel.[43] Agile development was able to reduce the number of people involved from about 200 to 50.[44]

Extreme programming uses pairs of programmers who work together to design, test, and code parts of the systems they develop. Research has shown that a pair of programmers usually outperforms the average individual programmer and on a level with the organization's best programmers. The iterative nature of XP helps companies develop robust systems with fewer errors. Sabre Airline Solutions, a $2-billion computer company serving the airline travel industry, used XP to eliminate programming errors and shorten program development times.[45]

In addition to the systems development approaches discussed previously, a number of other agile and innovative systems development approaches have been created by computer vendors and authors of systems development books. These approaches all attempt to deliver better systems in a shorter amount of time. A few agile development tools and approaches are the following:

- **Adaptive Software Development.** Adaptive Software Development (ASD) grew out of rapid application development techniques and stresses an iterative process that involves analysis, design, and implementation at each cycle or iteration. The approach was primarily developed by James Highsmith.[46]

- **Lean Software Development and Lean User Experience.** Lean Software Development came from a book with the same title by Mary and Tom Poppendieck.[47] The approach comes from lean manufacturing practices used by Toyota and stresses continuous learning, just-in-time decision making, empowering systems development teams, and the elimination of waste.[48] Lean User Experience (Lean UX) is also based on Toyota's lean manufacturing.[49] The approach attempts to rapidly convert user requirements into information systems using daily meetings and delivering results in a short time frame.
- **Rational Unified Process.** Rational Unified Process (RUP) is an iterative systems development approach developed by IBM and includes a number of tools and techniques that are typically tailored to fit the needs of a specific company or organization. RUP uses an iterative approach to software development that stresses quality as the software is changed and updated over time.[50] Many companies have used RUP to their advantage.
- **Feature-Driven Development.** Originally used to complete a systems development project at a large bank, Feature-Driven Development (FDD) is an iterative systems development approach that stresses the features of the new or modified system and involves developing an overall model, creating a list of features, planning by features, designing by features, and building by features.[51]
- **Crystal Methodologies.** Crystal Methodologies is a family of systems development approaches developed by Alistair Cockburn that concentrates on effective team work and the reduction of paperwork and bureaucracy to make development projects faster and more efficient.[52]
- **Scrum.** Scrum is a systems development approach that stresses agile, incremental development. A ScrumMaster is the person who coordinates all Scrum activities, and a Scrum team consists of a dozen or fewer people who perform all systems development activities from investigation to testing. The Product Owner consists of people or stakeholders who benefit from the Scrum systems development effort. The Scrum Alliance offers training for its Certified ScrumMaster (CSM) designation.[53]

Outsourcing, On-Demand Computing, and Cloud Computing

Many companies hire an outside consulting firm or computer company that specializes in systems development to take over some or all of its development and operations activities. The drug company Pfizer, for example, used outsourcing to allow about 4,000 of its employees to outsource some of their jobs to other individuals or companies around the globe.[54] Steel Technologies outsourced many of its computer operations and infrastructure to ERP Suites instead of spending almost $1 million on computers, storage devices, and power supplies.[55] This approach is often called *Infrastructure as a Service (IaaS.)* According to the CIO of Steel Technologies, "They own the data center. They own the hardware. They own the applications and operating system. And they provide the managed service that takes care of it all." As discussed in Chapter 7, cloud computing allows companies to run their applications on the Internet. Cloud computing, however, has its disadvantages, including security and service outages. After a service outage at a large cloud computing provider, the chief technology officer for the city of Lynwood, Washington, said, "The recent outage confirmed, for us, that cloud services are not yet ready for prime time."[56] Some CIOs are waiting to fully implement cloud computing until this approach has matured and some of its potential risks have been resolved. Other organizations are implementing less critical cloud computing applications to test the approach before it is used to a greater extent.[57]

Many companies are looking to outsource the acquisition of mobile applications or apps.[58] As with acquiring other information systems components, organizations have to decide if they want to build their own mobile applications or purchase mobile applications from other companies.[59] Table 12.4 describes the circumstances in which outsourcing is a good idea.

TABLE **12.4** When to use outsourcing for systems development

Reason	Example
When a company believes it can cut costs	PacifiCare outsourced its IS operations to IBM and Keane, Inc., in the hope that the outsourcing will save it about $400 million over ten years.
When a firm has limited opportunity to distinguish itself competitively through a particular IS operation or application	Kodak outsourced its IS operations, including mainframe processing, telecommunications, and personal computer support, because it had limited opportunity to distinguish itself through these IS operations. Kodak kept application development and support in-house because it thought that these activities had competitive value.
When outsourcing does not strip the company of technical know-how required for future IS innovation	Firms must ensure that their IS staffs remain technically up-to-date and have the expertise to develop future applications.
When the firm's existing IS capabilities are limited, ineffective, or technically inferior	A company might use outsourcing to help it make the transition from a centralized mainframe environment to a distributed client/server environment.
When a firm is downsizing	First Fidelity, a major bank, used outsourcing as part of a program to reduce the number of employees by 1,600 and slash expenses by $85 million.

A number of companies and nonprofit organizations offer outsourcing and on-demand computing services—from general systems development to specialized services. IBM's Global Services, for example, is one of the largest full-service outsourcing and consulting services.[60] IBM has consultants located in offices around the world and generates billions of dollars in revenues each year. Electronic Data Systems (EDS) is another large company that specializes in consulting and outsourcing.[61] EDS, which was acquired by Hewlett-Packard, has approximately 100,000 employees around the world. Accenture is another company that specializes in consulting and outsourcing.[62] Wipro Technologies, headquartered in India, is another worldwide outsourcing company with more than $4 billion in annual revenues.[63] Tata, a large outsourcing firm based in India, is now targeting smaller businesses by using cloud computing and a new service is called iON.[64] With cloud computing, smaller businesses can download software that allows Tata to manage its client's programs and processes from a remote location. Infosys, a consulting and outsourcing company, helped Hallmark Cards generate new revenues from digital greeting cards and other digital products.[65] Amazon offers Elastic Compute Cloud for organizations and individuals that pay only for the computing resources they use.[66] Some companies hire nontraditional outsourcing companies to develop software and other systems for them. Norfolk Southern, for example, hired General Electric to develop software to manage its train traffic.[67] See Figure 12.9.

Arindam Mukherjee/Landov

FIGURE 12.9

Outsourcing

Wipro Technologies, headquartered in India, is a worldwide outsourcing company.

Outsourcing has some disadvantages, however. Internal expertise can be lost, and loyalty can suffer under an outsourcing arrangement. For some companies, it can be difficult to achieve a competitive advantage when competitors are using the same computer or consulting company. When the outsourcing or on-demand computing is done offshore or in a foreign country, some people raise security concerns. U.S. federal authorities often investigate defense contractors for improper outsourcing.

Mobile Application Development

Today, more organizations are developing or buying mobile applications for their managers and workers.[68] Indeed, one company that makes charging stations for electric vehicles has developed an application for iPhones and BlackBerry phones that alerts customers how to find the closest charging station.[69] Corporate applications should be able to work on a wide variety of devices, including iPhones, BlackBerry phones, tablet computers, and other mobile devices.[70] According to an analyst with Nucleus Research, Inc., "Enterprise vendors are recognizing that devices don't matter when you're accessing an application. If I'm out in the field, I should be able to access information to do my job." Making corporate information available on mobile devices is becoming increasingly important as more workers and managers use these devices.[71] According to a vice president for Dice.com, an Internet site for IS jobs, "Virtually all companies are figuring out how to make use of mobile apps and don't know how to do it as well as they need to." In one survey, 14 percent of people purchasing tablet computers planned to use them for work-related activities. According to the CIO of Pizza Hut, "If your mobile strategy is stalled because you're looking for the business case and the ROI, put your big-picture hat on and realize that mobile devices are on people all day—they are on, they are connected, and when you press a button they are ready to go."[72]

While the overall approach of systems development is the same for mobile devices compared to traditional systems development projects, there are some important differences.[73] The user interface is not the typical graphical user interface discussed in Chapter 4. Instead, most mobile devices use a touch user interface, called a *natural user interface (NUI)*, or multitouch interface by some. The systems development teams for mobile devices are typically smaller, allowing them to be more flexible and agile.[74] It can also be difficult to find IS personnel with the skills and experience to develop good mobile applications.[75] Having the application communicate with the Internet or corporate computers is another issue for mobile devices that must be resolved. How to handle phone calls in the middle of running an application also needs to be considered.

A number of systems development tools are also available for mobile applications. RehabCare, a large health care company with close to 19,000 workers, for example, developed a mobile application in four days using Force.com development tools.[76] The application helps about 8,000 therapists prescreen patients using the Apple iPad. The process used to take several people and up to five hours to process all the paperwork for a single patient. Now, everything is stored electronically. According to the CIO of the company, "It has really paid off beautifully for us in fewer lost opportunities."

In addition to developing innovative applications, organizations are using innovative approaches to deliver these applications to workers and managers. Some CIOs, for example, are investigating the use of application stores to deliver corporate information and applications to workers and executives.[77] As with users of smartphones, tablet computers, and other mobile devices, workers and executives could go to a corporate applications store and download the latest programs, decision support systems, or other work-related applications. According to a consultant at the research firm Gartner, "The idea has legs." Even the U.S. Army has an application store, called Army Marketplace, for its soldiers.[78]

Allowing workers and executives to use their own mobile devices at work, however, is a difficult issue.[79] Most workers and executives want to use their own mobile devices at work, but it can be very difficult for systems developers to create and maintain applications for a wide range of mobile devices with different operating systems.

FACTORS AFFECTING SYSTEMS DEVELOPMENT SUCCESS

Successful systems development means delivering a system that meets user and organizational needs—on time and within budget. Some systems development efforts have been so successful that companies have sold their systems to other companies. Union Pacific, a large railroad company, developed workforce-management software for itself and then decided to sell it to other companies, thus possibly generating an additional $60 million for the company.[80] According to the CIO of Union Pacific, "We looked around at things and thought, 'Hey, we can make some money on that.'"

Getting users and stakeholders involved in systems development is critical for most systems development projects. Having the support of top-level managers is also important. According to the CIO of Northrop Grumman Corporation, "Once I understood what needed to be done, I solicited the support of the top echelons of the company. Once I had the top covered, then I began to engage my own staff."[81] In addition to user involvement and top management support, other factors can contribute to successful systems development efforts at a reasonable cost. These factors are discussed next.

Degree of Change

A major factor that affects the quality of systems development is the degree of change associated with the project. The scope can vary from enhancing an existing system to major reengineering. The project team needs to recognize where it is on this spectrum of change. The ability to manage change is also critical to the success of systems development because new systems inevitably involve change. Unfortunately, not everyone adapts easily, and the increasing complexity of systems can multiply the problems. It is essential to recognize existing or potential problems (particularly the concerns of users) and to deal with them before they become a serious threat to the success of the new or modified system.

The Importance of Planning

The bigger the project, the more likely that poor planning will lead to significant problems. Many companies find that large systems projects fall behind schedule, go over budget, and do not meet expectations. Although proper planning cannot guarantee that these types of problems will be avoided, it can minimize the likelihood of their occurrence. Good systems development is not automatic, and certain factors contribute to the failure of systems development projects. These factors and the countermeasures to eliminate or alleviate the problems are summarized in Table 12.5.

TABLE **12.5** Project planning issues that frequently contribute to project failure

Factor	Countermeasure
Solving the wrong problem	Establish a clear connection between the project and organizational goals.
Poor problem definition and analysis	Follow a standard systems development approach.
Poor communication	Set up communications procedures and protocols.
Project is too ambitious	Narrow the project focus to address only the most important business opportunities.
Lack of top management support	Identify the senior manager who has the most to gain from the success of the project and recruit this person to champion the project.
Lack of management and user involvement	Identify and recruit key stakeholders to be active participants in the project.

Organizational experience with the systems development process is also an important factor for systems development success.[82] The *Capability Maturity Model (CMM)* is one way to measure this experience. It is based on research done at Carnegie Mellon University and work done by the Software Engineering Institute (SEI). CMM grades an organization's systems development maturity using five levels: initial, repeatable, defined, managed, and optimized.

Use of Project Management Tools

Project management involves planning, scheduling, directing, and controlling human, financial, and technological resources for a defined task whose result is achievement of specific goals and objectives. Corporations and nonprofit organizations use the following important tools and techniques.

project schedule: A detailed description of what is to be done.

project milestone: A critical date for the completion of a major part of the project.

project deadline: The date the entire project is to be completed and operational.

A **project schedule** is a detailed description of what is to be done. In the schedule, each project activity, the use of personnel and other resources, and expected completion dates are described. A **project milestone** is a critical date for the completion of a major part of the project, such as program design, coding, testing, and release (for a programming project). The **project deadline** is the date the entire project is to be completed and operational—when the organization can expect to begin to reap the benefits of the project.

critical path: Activities that, if delayed, would delay the entire project.

In systems development, each activity has an earliest start time, earliest finish time, and slack time, which is the amount of time an activity can be delayed without delaying the entire project. The **critical path** consists of all activities that, if delayed, would delay the entire project. These activities have zero slack time. Any problems with critical path activities will cause problems for the entire project. To ensure that critical path activities are completed in a timely fashion, formalized project management approaches have been developed. Tools such as Microsoft Project are available to help compute these critical project attributes.

Program Evaluation and Review Technique (PERT): A formalized approach for developing a project schedule that creates three time estimates for an activity.

Gantt chart: A graphical tool used for planning, monitoring, and coordinating projects.

Although the steps of systems development seem straightforward, larger projects can become complex, requiring hundreds or thousands of separate activities. For these systems development efforts, formal project management methods and tools are essential. A formalized approach called **Program Evaluation and Review Technique (PERT)** creates three time estimates for an activity: shortest possible time, most likely time, and longest possible time. A formula is then applied to determine a single PERT time estimate. A **Gantt chart** is a graphical tool used for planning, monitoring, and coordinating projects; it is essentially a grid that lists activities and deadlines. Each time a task is completed, a marker such as a darkened line is placed in the proper grid cell to indicate the completion of a task (see Figure 12.10).

PROJECT PLANNING DOCUMENTATION — Page 1 of 1

System: Warehouse Inventory System (Modification) — Date 12/10

System: — Scheduled activity; ▬ Completed activity — Analyst: Cecil Truman — Signature

Activity*	Individual assigned	Week 1	2	3	4	5	6	7	8	9	10	11	12	13	14
R — Requirements definition															
R.1 Form project team	VP, Cecil, Bev														
R.2 Define obj. and constraints	Cecil														
R.3 Interview warehouse staff for requirements report	Bev														
R.4 Organize requirements	Team														
R.5 VP review	VP, Team														
D — Design															
D.1 Revise program specs.	Bev														
D. 2. 1 Specify screens	Bev														
D. 2. 2 Specify reports	Bev														
D. 2. 3 Specify doc. changes	Cecil														
D. 4 Management review	Team														
I — Implementation															
I. 1 Code program changes	Bev														
I. 2. 1 Build test file	Team														
I. 2. 2 Build production file	Bev														
I. 3 Revise production file	Cecil														
I. 4. 1 Test short file	Bev														
I. 4. 2 Test production file	Cecil														
I. 5 Management review	Team														
I. 6 Install warehouse**															
I. 6. 1 Train new procedures	Bev														
I. 6. 2 Install	Bev														
I. 6. 3 Management review	Team														

*Weekly team reviews not shown here
**Report for warehouses 2 through 5

FIGURE **12.10**

Sample Gantt chart

A Gantt chart shows progress through systems development activities by putting a bar through appropriate cells.

Both PERT and Gantt techniques can be automated using project management software. Project management software helps managers determine the best way to reduce project completion time at the least cost. The software packages include OpenPlan by Deltek, Microsoft Project, and Unifier by Skire.

computer-aided software engineering (CASE): Tools that automate many of the tasks required in a systems development effort and encourage adherence to the SDLC.

Computer-aided software engineering (CASE) tools automate many of the tasks required in a systems development effort and encourage adherence to the SDLC, thus instilling a high degree of rigor and standardization to the entire systems development process. Oracle Designer by Oracle (*www.oracle.com*) and Visible Analyst by Visible Systems Corporation (*www.visible.com*) are examples of CASE tools. Oracle Designer is a CASE tool that can help systems analysts automate and simplify the development process for database systems. Other CASE tools include Embarcadero Describe (*www.embarcadero.com*), Popkin Software (*www.popkin.com*), Rational Software (part of IBM), and Visio (a charting and graphics program) from Microsoft.

Object-Oriented Systems Development

The success of a systems development effort can depend on the specific programming tools and approaches used. As mentioned in Chapter 4, object-oriented (OO) programming languages allow the interaction of programming objects (recall that an object consists of both data and the actions that can be performed on the data). So, an object could be data about an employee and all the operations (such as payroll, benefits, and tax calculations) that might be performed for that employee.

Chapter 4 discusses a number of programming languages that use the object-oriented approach, including Visual Basic, C++, and Java. These languages allow systems developers to take the OO approach, making program development faster and more efficient and resulting in lower costs. Modules can be developed internally or obtained from an external source. After a company has the programming modules, programmers and systems analysts can modify them and integrate them with other modules to form new programs.

object-oriented systems development (OOSD): An approach to systems development that combines the logic of the systems development life cycle with the power of object-oriented modeling and programming.

Object-oriented systems development (OOSD) combines the logic of the systems development life cycle with the power of object-oriented modeling and programming. OOSD follows a defined systems development life cycle, much like the SDLC. The life cycle phases are usually completed with many iterations. Object-oriented systems development typically involves the following tasks:

- **Identifying potential problems and opportunities within the organization that would be appropriate for the OO approach.** This process is similar to traditional systems investigation. Ideally, these problems or opportunities should lend themselves to the development of programs that can be built by modifying existing programming modules.
- **Defining what kind of system users require.** This analysis means defining all the objects that are part of the user's work environment (object-oriented analysis). The OO team must study the business and build a model of the objects that are part of the business (such as a customer, an order, or a payment). Many of the CASE tools discussed in the previous section can be used, starting with this step of OOSD.
- **Designing the system.** This process defines all the objects in the system and the ways they interact (object-oriented design). Design involves developing logical and physical models of the new system by adding details to the object model started in analysis.
- **Programming or modifying modules.** This implementation step takes the object model begun during analysis and completed during design and turns it into a set of interacting objects in a system. Object-oriented programming languages are designed to allow the programmer to create

classes of objects in the computer system that correspond to the objects in the actual business process. Objects such as customer, order, and payment are redefined as computer system objects—a customer screen, an order entry menu, or a dollar sign icon. Programmers then write new modules or modify existing ones to produce the desired programs.

- **Evaluation by users.** The initial implementation is evaluated by users and improved. Additional scenarios and objects are added, and the cycle repeats. Finally, a complete, tested, and approved system is available for use.
- **Periodic review and modification.** The completed and operational system is reviewed at regular intervals and modified as necessary.

SYSTEMS INVESTIGATION

As discussed earlier in the chapter, systems investigation is the first phase in the traditional SDLC of a new or modified business information system. The purpose is to identify potential problems and opportunities and consider them in light of the goals of the company. For example, cost savings was the primary reason for a library system in the state of Washington with over 20 branches to initiate systems investigation. The primary objective of its investigation was to determine if the use of DSL lines for communications and Internet connection could save money.[83] As a result of systems investigation, the library system projected cost savings to be approximately $400,000 in the first year. In general, systems investigation attempts to uncover answers to the following questions:

- What primary problems might a new or enhanced system solve?
- What opportunities might a new or enhanced system provide?
- What new hardware, software, databases, telecommunications, personnel, or procedures will improve an existing system or are required in a new system?
- What are the potential costs, both variable and fixed?
- What are the associated risks?

Initiating Systems Investigation

Because systems development requests can require considerable time and effort to implement, many organizations have adopted a formal procedure for initiating systems development, beginning with systems investigation. The **systems request form** is a document filled out by someone who wants the IS department to initiate systems investigation. This form typically includes the following information:

systems request form: A document filled out by someone who wants the IS department to initiate systems investigation.

- Problems in or opportunities for the system
- Objectives of systems investigation
- Overview of the proposed system
- Expected costs and benefits of the proposed system

The information in the systems request form helps to rationalize and prioritize the activities of the IS department. Based on the overall IS plan, the organization's needs and goals, and the estimated value and priority of the proposed projects, managers make decisions regarding the initiation of each systems investigation for such projects.

Participants in Systems Investigation

After a decision has been made to initiate systems investigation, the first step is to determine what members of the development team should participate in the investigation phase of the project. Members of the development

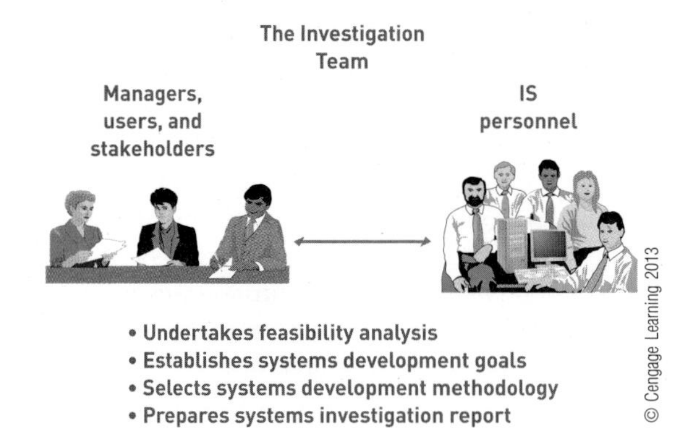

FIGURE 12.11
Systems investigation team
The team consists of upper- and middle-level managers, a project manager, IS personnel, users, and stakeholders.

team change from phase to phase (see Figure 12.11). The systems investigation team can be diverse, with members located around the world. Cooperation and collaboration are keys to successful investigation teams in these cases. The members of the development team who participate in investigation are then responsible for gathering and analyzing data, preparing a report justifying systems development, and presenting the results to top-level managers.

feasibility analysis: Assessment of the technical, economic, legal, operational, and schedule feasibility of a project.

technical feasibility: Assessment of whether the hardware, software, and other system components can be acquired or developed to solve the problem.

Feasibility Analysis

A key step of the systems investigation phase is **feasibility analysis**, which assesses technical, economic, legal, operational, and schedule feasibility (see Figure 12.12). **Technical feasibility** is concerned with whether the hardware, software, and other system components can be acquired or developed to solve the problem.

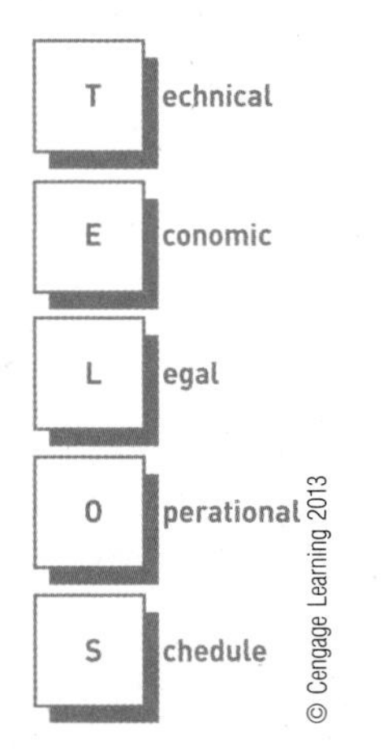

FIGURE 12.12
Technical, economic, legal, operational, and schedule feasibility

economic feasibility: The determination of whether the project makes financial sense and whether predicted benefits offset the cost and time needed to obtain them.

legal feasibility: The determination of whether laws or regulations may prevent or limit a systems development project.

Economic feasibility determines whether the project makes financial sense and whether predicted benefits offset the cost and time needed to obtain them. Economic feasibility can involve cash flow analysis such as that done in internal rate of return (IRR) or total cost of ownership (TCO) calculations, first discussed in Chapter 2. Spreadsheet programs, such as Microsoft Excel, have built-in functions to compute internal rate of return and other cash flow measures.

Legal feasibility determines whether laws or regulations may prevent or limit a systems development project. Legal feasibility involves an analysis of existing and future laws to determine the likelihood of legal action against the systems development project and the possible consequences of such action.

operational feasibility: The measure of whether the project can be put into action or operation.

Operational feasibility is a measure of whether the project can be put into action or operation. It can include logistical and motivational (acceptance of change) considerations. Motivational considerations are important because new systems affect people, and data flows can have unintended consequences. As a result, power and politics might come into play, and some people might resist the new system.

schedule feasibility: The determination of whether the project can be completed in a reasonable amount of time.

Schedule feasibility determines whether the project can be completed in a reasonable amount of time. This process involves balancing the time and resource requirements of the project with other projects.

Object-Oriented Systems Investigation

The object-oriented approach can be used during all phases of systems development, from investigation to maintenance and review. Consider a kayak rental business in Maui, Hawaii, where the owner wants to computerize its operations, including renting kayaks to customers and adding new kayaks into the rental program (see Figure 12.13). As you can see, the kayak rental clerk rents kayaks to customers and adds new kayaks to the current inventory available for rent. The stick figure is an example of an *actor*, and the ovals each represent an event, called a *use case*. In our example, the actor (the kayak rental clerk) interacts with two use cases (rent kayaks to customers and add new kayaks to inventory). The use case diagram is part of the Unified Modeling Language (UML) that is used in object-oriented systems development.

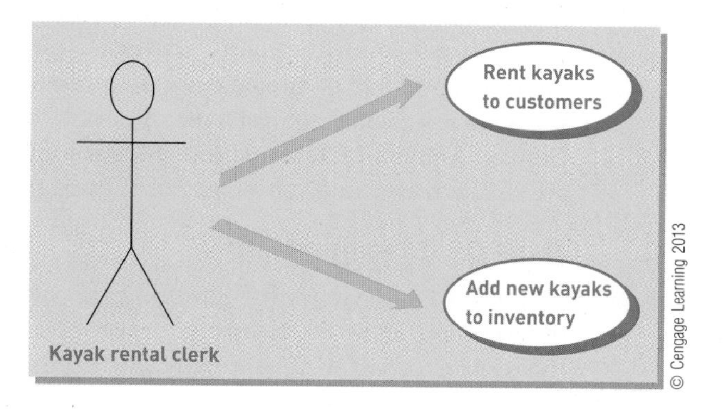

FIGURE **12.13**
Use case diagram for a kayak rental application

The Systems Investigation Report

systems investigation report: A summary of the results of the systems investigation and the process of feasibility analysis and recommendation of a course of action.

The primary outcome of systems investigation is a **systems investigation report**, also called a *feasibility study*. This report summarizes the results of systems investigation and the process of feasibility analysis and recommends a course of action: continue on into systems analysis, modify the project in some manner, or drop it. A typical table of contents for the systems investigation report is shown in Figure 12.14.

steering committee: An advisory group consisting of senior management and users from the IS department and other functional areas.

The systems investigation report is reviewed by senior management, often organized as an advisory committee, or **steering committee**, consisting of senior management and users from the IS department and other functional areas. These people help IS personnel with their decisions about the use of information systems in the business and give authorization to pursue further systems development activities. After review, the steering committee might agree with the recommendation of the systems development team, or it might suggest a change in project focus to concentrate more directly on meeting a specific company objective. Another alternative is that everyone might decide that the project is not feasible and thus cancel the project.

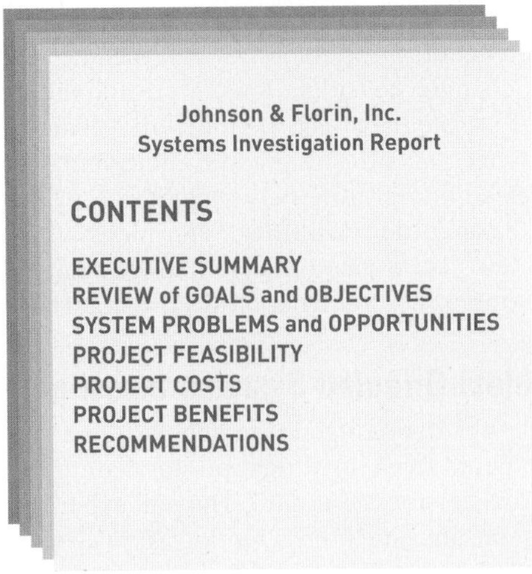

Johnson & Florin, Inc.
Systems Investigation Report

CONTENTS

EXECUTIVE SUMMARY
REVIEW of GOALS and OBJECTIVES
SYSTEM PROBLEMS and OPPORTUNITIES
PROJECT FEASIBILITY
PROJECT COSTS
PROJECT BENEFITS
RECOMMENDATIONS

FIGURE 12.14
Typical table of contents for a systems investigation report

SYSTEMS ANALYSIS

After a project has been approved for further study, the next step is to answer the question "What must the information system do to solve the problem?" The overall emphasis of analysis is gathering data on the existing system, determining the requirements for the new system, considering alternatives within these constraints, and investigating the feasibility of the solutions. The primary outcome of systems analysis is a prioritized list of systems requirements. A number of hospitals, for example, are scrambling to meet federally mandated systems analysis deadlines.[84] According to the vice president of General Dynamics Information Technology, "A large percentage of hospitals are in the heavy analysis stage, or they're just starting."

General Considerations

Systems analysis starts by clarifying the overall goals of the organization and determining how the existing or proposed information system helps meet them. A manufacturing company, for example, might want to reduce the number of equipment breakdowns. This goal can be translated into one or more informational needs. One need might be to create and maintain an accurate list of each piece of equipment and a schedule for preventative maintenance. Another need might be a list of equipment failures and their causes.

Analysis of a small company's information system is usually straightforward. On the other hand, evaluating an existing information system for a large company can be a long, tedious process. As a result, large organizations evaluating a major information system normally follow a formalized analysis procedure that involves these steps:

1. Assembling the participants for systems analysis
2. Collecting appropriate data and requirements
3. Analyzing the data and requirements
4. Preparing a report on the existing system, new system requirements, and project priorities

Participants in Systems Analysis

The first step in formal analysis is to assemble a team to study the existing system. This group includes members of the original investigation team—from users and stakeholders to IS personnel and management. Most organizations usually allow key members of the development team to not only analyze

INFORMATION SYSTEMS @ WORK

Nottingham Trent University Oversees Its Information Systems

You read in this chapter that information systems steering committees "help IS personnel with their decisions about the use of information systems in the business and give authorization to pursue further systems development activities." This is an important aspect of *information systems governance*: oversight of how the organization uses information systems to further its business objectives.

Nottingham Trent University (NTU) in Nottingham, England, was formed in the 1990s by a merger of several other institutions, some dating to the mid- 19th century. As a result, it had to deal with several different information systems and technology platforms. In 2004, NTU brought its resources together into a central IS department. When this centralization was followed by the departure of their IS director in 2006, the stage was set to rethink NTU's approach to managing information resources.

Richard Eade, information technology software manager at NTU, writes that "Balancing risks and opportunities [was] the major driver for the introduction of IT governance at NTU. One significant risk identified is associated with the university culture, ... a perception that IT staff can 'do what they like.' This has arisen from poor management. Consequently, IT governance has been introduced to bring in systems of control, without repressing initiative and enthusiasm."

Anarchy, while it sounds attractive to those who can do as they please, does not yield the best information systems or the best use of resources. NTU's challenge was to get control of its IS efforts in ways that faculty and staff would accept. In addition, the university wanted to move from spending 80 percent of its time operating existing systems, with only the remaining 20 percent of its efforts going to developing new ones, to a 50:50 ratio.

Achieving these goals was a big task. Eade says NTU approached it the same way one approaches eating an elephant: "One bite at a time."

Managing new system development was not the first area NTU addressed. Some top-priority areas had to be tackled first, including risk analysis, financial audits, security, and legal issues such as data protection. Once those urgent matters were taken care of, though, NTU could address project priorities and approval. In the long run, getting control of development priorities would have a greater payoff.

To improve the IS department's focus on meaningful projects, NTU created the position of *business relationship manager*. The person in this position was the IS department's "eyes and ears" to the rest of the university. This manager's role was to provide communication links in both directions.

Next, the department strengthened its steering group. Previously, it had focused on the allocation of capital funding but had no controls to ensure the money was spent as the steering group specified. Now, both a sponsor and a statement of business benefits were required for all new projects.

Finally, NTU recognized that project management was vital to project success. The goals of the IS department were to deliver projects on time, on budget, and at an agreed-upon quality level. In the past, all three goals were missed consistently. NTU now trained project managers in the PRINCE2 project management methodology. This methodology provides proven processes for every project, from start to finish. It defines the required competencies, duties, and behaviors of eight types of people involved in a project. In sum, it changes project management from a "seat of the pants" activity to a systematic one. It helps projects stay on track and ensures that scope changes were well documented and properly agreed upon, and it provides evidence that risks and issues are well managed. NTU uses project boards (committees) for high-level control, with weekly or monthly highlight reports circulated to the IS management team.

Eade summarizes NTU's new governance with "Ensuring that IT systems are fit for the purpose, are well managed, and can be relied upon, means that we must undertake more effective measures to identify that appropriate strategies, policies, procedures, and controls are in place to bring risks and problems to our attention. NTU senior managers have identified the key role the IT systems play—and will continue to play—in taking the university forward. They now expect the principles of IT governance to be in place to ensure that IT remains strong."

Discussion Questions

1. NTU's governance group first addressed urgent (but perhaps less important) issues and then later confronted more important (but less urgent) issues. What is the danger in taking this approach? Why do you think it worked for NTU?
2. How can NTU's use of formal IS governance help them achieve its goal of redirecting about 30 percent of its resources from operations to development?

Critical Thinking Questions

1. Consider Richard Eade's definition of *IS governance* in the last paragraph. Suppose you have just started a new job as head of IS governance for a university and are sending an e-mail to one of your grandparents explaining what you do. Rewrite what he said in your own words, for this purpose.
2. Suppose NTU, instead of arising from a recent merger, had been one institution for over a century but had the same problems. How might that difference affect its approach to information systems governance?

SOURCES: Eade, R., "IT Governance: How We Are Making It Work," UCISA (Universities and Colleges Information Systems Association) 2011 conference, March 23, 2011, *www.ucisa.ac.uk/events/2011/conference2011/~/media/Files/events/ucisa2011/presentations/richard_eade.ashx*, downloaded January 23, 2012; Eade, R., "Making IT Governance Work," ECAR Research Bulletin, no. 20, October 5, 2010; *www.educause.edu/Resources/MakingITGovernanceWork/214687*, downloaded January 23, 2012; Nottingham Trent University Web site, *www.ntu.ac.uk*, accessed January 23, 2012; Prince2 Methodology Web site, *www.prince-officialsite.com*, accessed January 23, 2012.

the condition of the existing system but also to perform other aspects of systems development, such as design and implementation.

After the participants in systems analysis are assembled, this group develops a list of specific objectives and activities. A schedule for meeting the objectives and completing the specific activities is also devised, along with deadlines for each stage. The group also draws up a statement of the resources required at each stage, such as clerical personnel, supplies, and so forth. Major milestones are normally established to help the team monitor progress and determine whether problems or delays occur in performing systems analysis.

Data Collection

The purpose of data collection is to seek additional information about the problems or needs identified in the systems investigation report. During this process, the strengths and weaknesses of the existing system are emphasized.

Identifying Sources of Data

Data collection begins by identifying and locating the various sources of data, including both internal and external sources (see Figure 12.15).

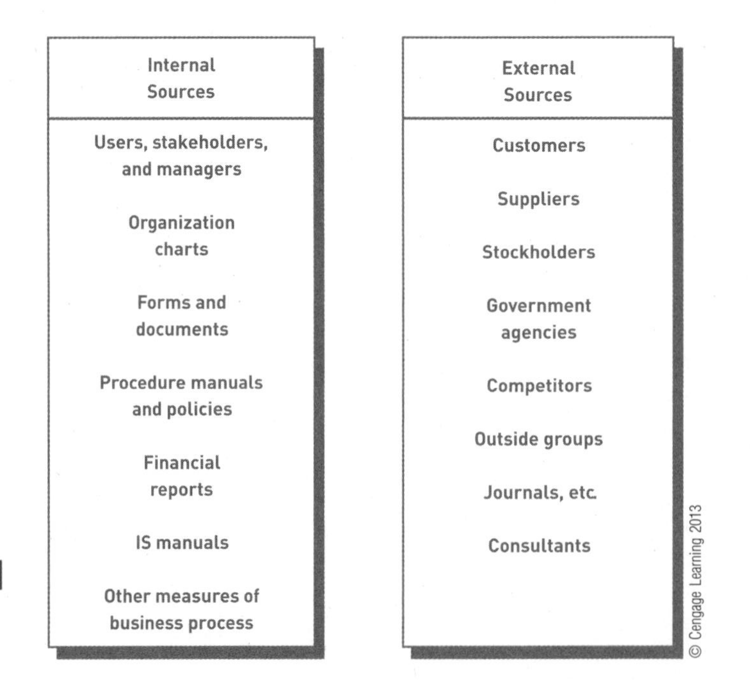

Internal Sources
Users, stakeholders, and managers
Organization charts
Forms and documents
Procedure manuals and policies
Financial reports
IS manuals
Other measures of business process
External Sources
Customers
Suppliers
Stockholders
Government agencies
Competitors
Outside groups
Journals, etc
Consultants
© Cengage Learning 2013

FIGURE **12.15**

Internal and external sources of data for systems analysis

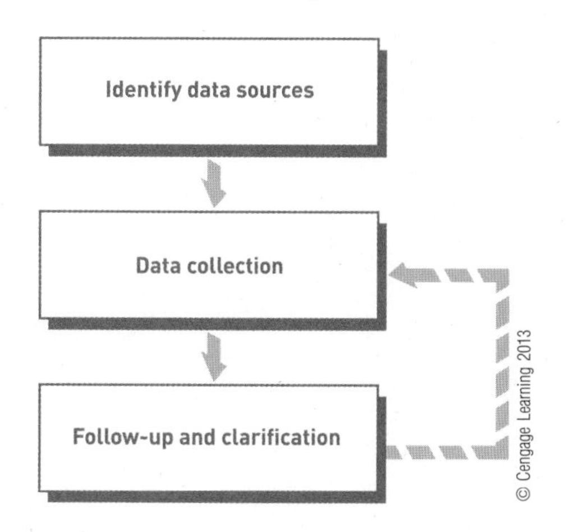

FIGURE 12.16
The steps in data collection

Collecting Data

After data sources have been identified, data collection begins. Figure 12.16 shows the steps involved. Data collection might require a number of tools and techniques, such as interviews, direct observation, and questionnaires.

structured interview: An interview in which the questions are written in advance.

unstructured interview: An interview in which the questions are not written in advance.

Interviews can either be structured or unstructured. In a **structured interview**, the questions are written in advance. In an **unstructured interview**, the questions are not written in advance; instead, the interviewer relies on experience in asking the best questions to uncover the inherent problems of the existing system and the needs of the employee or department. An advantage of the unstructured interview is that it allows the interviewer to ask follow-up or clarifying questions immediately.

direct observation: Directly observing the existing system in action by one or more members of the analysis team.

With **direct observation**, one or more members of the analysis team directly observe the existing system in action. One of the best ways to understand how the existing system functions is to work with the users to discover how data flows in certain business tasks. Determining the data flow entails direct observation of users' work procedures, their reports, current screens (if automated already), and so on. From this observation, members of the analysis team determine which forms and procedures are adequate and which need improvement. Direct observation requires a certain amount of skill. The observer must be able to see what is really happening and not be influenced by attitudes or feelings. This approach can reveal important problems and opportunities that would be difficult to obtain using other data collection methods. An example would be observing the work procedures, reports, and computer screens associated with an accounts payable system being considered for replacement.

questionnaires: A method of gathering data when the data sources are spread over a wide geographic area.

When many data sources are spread over a wide geographic area, **questionnaires** might be the best method. Like interviews, questionnaires can be either structured or unstructured. In most cases, a pilot study is conducted to fine tune the questionnaire. A follow-up questionnaire can also capture the opinions of those who do not respond to the original questionnaire.

statistical sampling: Selecting a random sample of data and applying the characteristics of the sample to the whole group.

Other data collection techniques can also be employed. In some cases, telephone calls work well. Activities can also be simulated to see how the existing system reacts. Thus, fake sales orders, stockouts, customer complaints, and data-flow bottlenecks can be created to see how the existing system responds to these situations. **Statistical sampling**, which involves selecting a random sample of data, is another technique. For example, suppose that you want to collect data that describes 10,000 sales orders received over the last few years. Because it is too time consuming to analyze each of the sales orders, you can collect a random sample of 100 to 200 sales orders from the entire batch. You can assume that the characteristics of this sample apply to all 10,000 orders.

© iStockphoto/Steve Cole

Direct observation is a method of data collection. One or more members of the analysis team directly observes the existing system inaction.

Data Analysis

The data collected in its raw form is usually not adequate to determine the effectiveness of the existing system or the requirements for the new system. The next step is to manipulate the collected data so that the development team members who are participating in systems analysis can use the data. This manipulation is called **data analysis**. Data and activity modeling and using data-flow diagrams and entity-relationship diagrams show the relationships among various objects, associations, and activities. Other common tools and techniques for data analysis include application flowcharts, grid charts, CASE tools, and the object-oriented approach.

data analysis: The manipulation of collected data so that the development team members who are participating in systems analysis can use the data.

Data Modeling

Data modeling, first introduced in Chapter 5, is a commonly accepted approach to modeling organizational objects and associations that employ both text and graphics. How data modeling is employed, however, is governed by the specific systems development methodology.

Data modeling is most often accomplished through the use of entity-relationship (ER) diagrams. Recall from Chapter 5 that an entity is a generalized representation of an object type—such as a class of people (employee), types of events (sales), categories of things (desks), or different locations (city). Recall also that entities possess certain attributes. Objects can be related to other objects in many ways. An ER diagram, such as the one shown in Figure 12.17a, describes a number of objects and the ways they are associated. An ER diagram (or any other modeling tool) cannot by itself fully describe a business problem or solution because it lacks descriptions of the related activities. It is, however, a good place to start because it describes object types and attributes about which data might need to be collected for processing.

Activity Modeling

To fully describe a business problem or solution, the related objects, associations, and activities must be described. Activities in this sense are events or items that are necessary to fulfill the business relationship or that can be associated with the business relationship in a meaningful way.

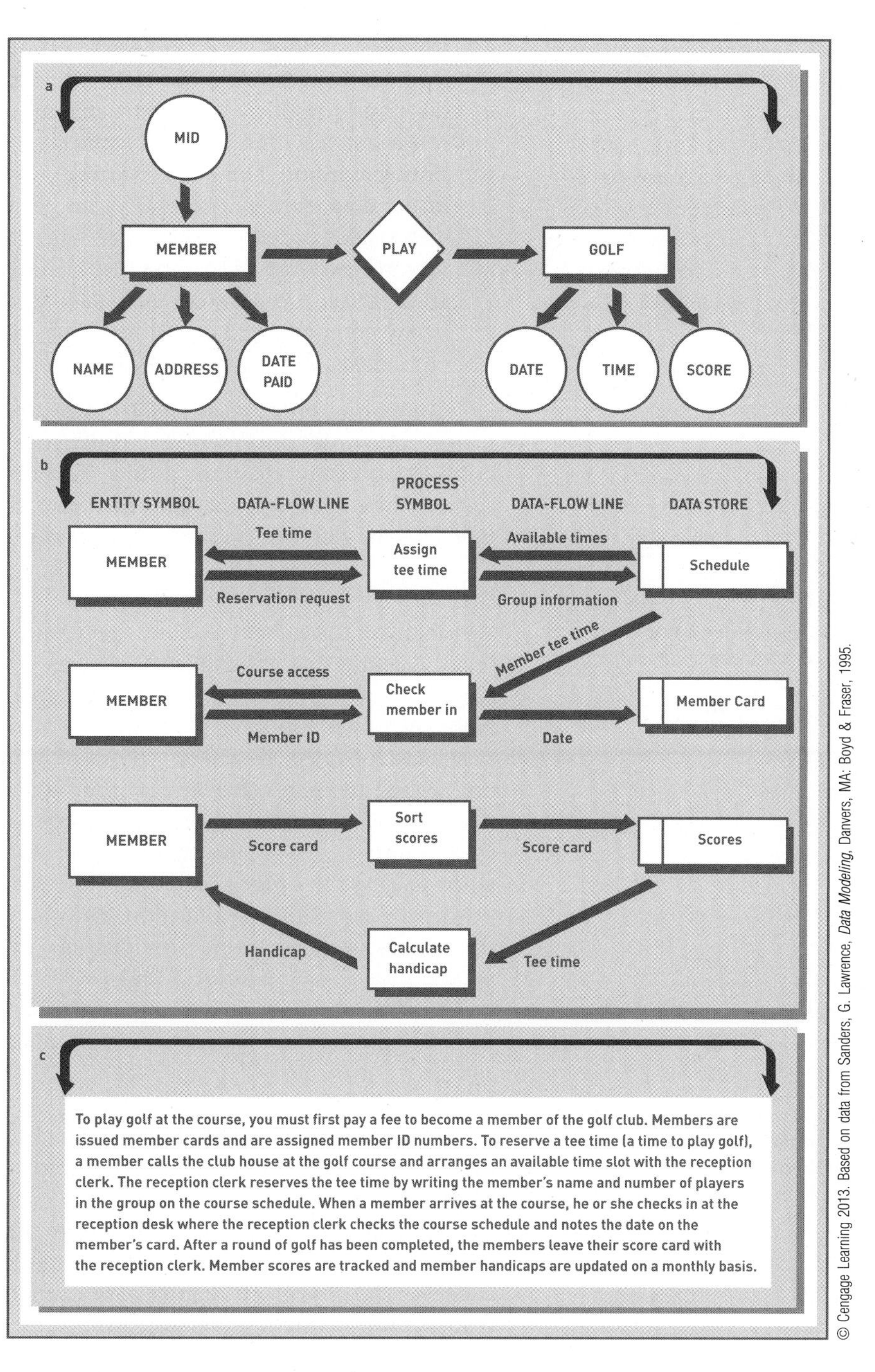

FIGURE **12.17**

Data and activity modeling

(a) An entity-relationship diagram. (b) A data-flow diagram. (c) A semantic description of the business process.

data-flow diagram (DFD): A model of objects, associations, and activities that describes how data can flow between and around various objects.

Activity modeling is often accomplished through the use of data-flow diagrams. A **data-flow diagram (DFD)** models objects, associations, and activities by describing how data can flow between and around various objects. DFDs work on the premise that every activity involves some communication, transference, or flow that can be described as a data element. DFDs describe the activities that fulfill a business relationship or accomplish a business task, not how these activities are to be performed. That is, DFDs show the logical sequence of associations and activities, not the physical processes. A system modeled with a DFD could operate manually or could be computer based; if computer based, the system could operate with a variety of technologies.

DFDs are easy to develop and easily understood by nontechnical people. Data-flow diagrams use four primary symbols, as illustrated in Figure 12.17b:

data-flow line: A line with arrows that show the direction of data element movement.

- **Data flow.** The **data-flow line** includes arrows that show the direction of data element movement.

process symbol: Representation of a function that is performed.

- **Process symbol.** The **process symbol** reveals a function that is performed. Computing gross pay, entering a sales order, delivering merchandise, and printing a report are examples of functions that can be represented with a process symbol.

entity symbol: Representation of either a source or destination of a data element.

- **Entity symbol.** The **entity symbol** shows either the source or destination of the data element. An entity can be, for example, a customer who initiates a sales order, an employee who receives a paycheck, or a manager who receives a financial report.

data store: Representation of a storage location for data.

- **Data store.** A **data store** reveals a storage location for data. A data store is any computerized or manual data storage location, including magnetic tape, disks, a filing cabinet, or a desk.

Comparing entity-relationship diagrams with data-flow diagrams provides insight into the concept of top-down design. Figure 12.17a and b show an entity-relationship diagram and a data-flow diagram for the same business relationship, namely a member of a golf club playing golf. Figure 12.17c provides a brief description of the business relationship for clarification.

Application Flowcharts

application flowcharts: Diagrams that show relationships among applications or systems.

Application flowcharts show the relationships among applications or systems. Assume that a small business has collected data about its order processing, inventory control, invoicing, and marketing analysis applications. Management is thinking of modifying the inventory control application. The raw facts collected, however, do not help in determining how the applications are related to each other and to the databases required for each. These relationships are established through data analysis with an application flowchart (see Figure 12.18). Using this tool for data analysis makes clear the relationships among the order processing functions.

In the simplified application flowchart in Figure 12.18, you can see that the telephone order clerk provides important data to the system about items such as versions, quantities, and prices. The system calculates sales tax and order totals. Any changes made to this order processing system could affect the company's other systems, such as inventory control and marketing.

Grid Charts

grid chart: A table that shows relationships among the various aspects of a systems development effort.

A **grid chart** is a table that shows relationships among various aspects of a systems development effort. For example, a grid chart can reveal the databases used by the various applications (see Figure 12.19).

The simplified grid chart in Figure 12.19 shows that the customer database is used by the order processing, marketing analysis, and invoicing applications. The inventory database is used by the order processing, inventory control, and marketing analysis applications. The supplier database is used by the inventory control application, and the accounts receivable database is used by the invoicing application. This grid chart shows which applications use common databases and reveals that, for example, any changes to the inventory control application must investigate the inventory and supplier databases.

CASE Tools

CASE repository: A database of system descriptions, parameters, and objectives.

As discussed earlier, many systems development projects use CASE tools to complete analysis tasks. Most computer-aided software engineering tools have generalized graphics programs that can generate a variety of diagrams and figures. Entity-relationship diagrams, data-flow diagrams, application flowcharts, and other diagrams can be developed using CASE graphics programs to help describe the existing system. During the analysis phase, a **CASE repository**—a database of system descriptions, parameters, and objectives—will be developed.

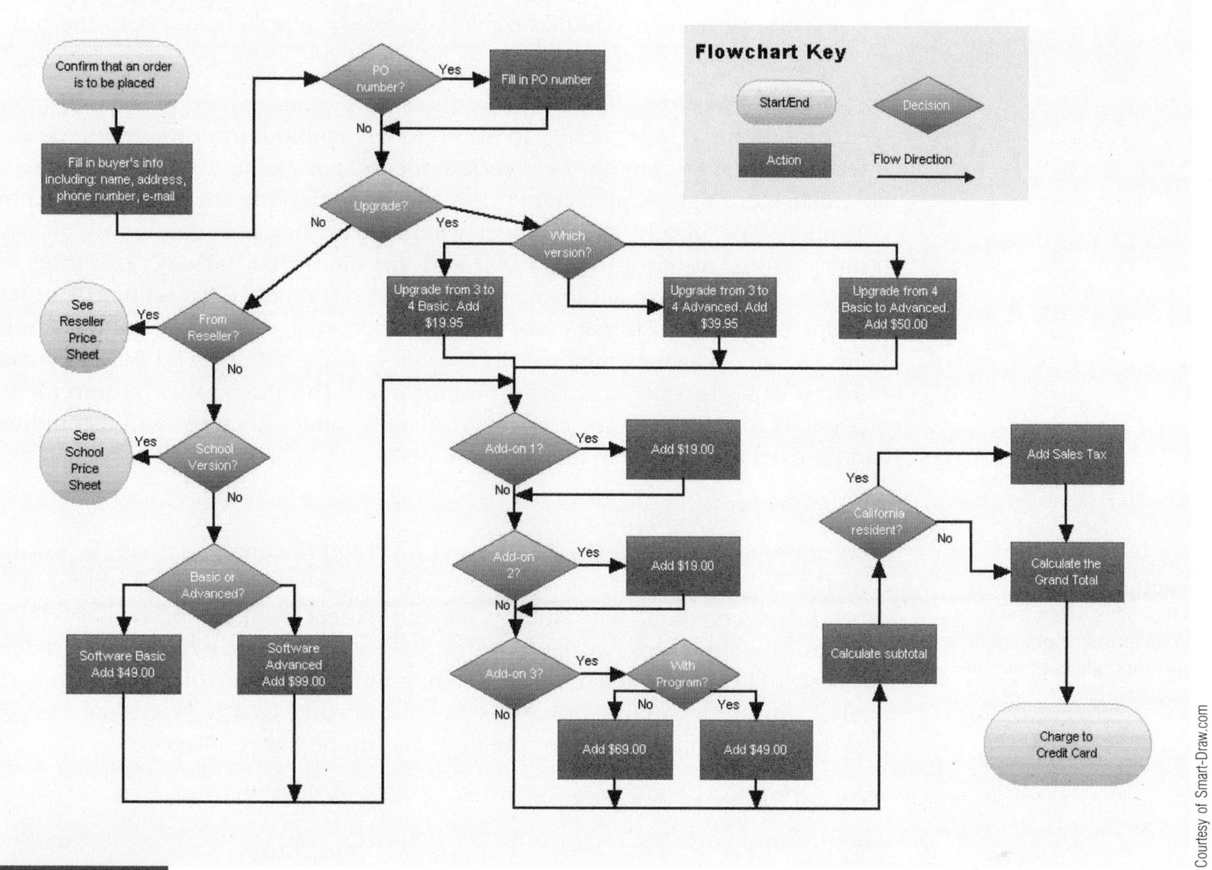

FIGURE **12.18**

Telephone order process application flowchart

The flowchart shows the relationships among various processes.

Databases → / Applications ↓	Customer database	Inventory database	Supplier database	Accounts receivable database
Order processing application	X	X		
Inventory control application		X	X	
Marketing analysis application	X	X		
Invoicing application	X			X

FIGURE **12.19**

Grid chart

The chart shows the relationships among applications and databases.

Requirements Analysis

requirements analysis: The determination of user, stakeholder, and organizational needs.

The overall purpose of **requirements analysis** is to determine user, stakeholder, and organizational needs.[85] For an accounts payable application, the stakeholders could include suppliers and members of the purchasing department. Questions that should be asked during requirements analysis include the following:

- Are these stakeholders satisfied with the current accounts payable application?
- What improvements could be made to satisfy suppliers and help the purchasing department?

One of the most difficult procedures in systems analysis is confirming user or systems requirements. In some cases, communications problems can interfere with determining these requirements. For example, an accounts payable manager might want a better procedure for tracking the amount owed by customers. Specifically, the manager wants a weekly report that shows all customers who owe more than $1,000 and are more than 90 days past due on their accounts. A financial manager might need a report that summarizes total amount owed by customers to consider whether to loosen or tighten credit limits. A sales manager might want to review the amount owed by a key customer relative to sales to that same customer. The purpose of requirements analysis is to capture these requests in detail. Numerous tools and techniques can be used to capture systems requirements.

Asking Directly

asking directly: An approach to gather data that asks users, stakeholders, and other managers about what they want and expect from the new or modified system.

One the most basic techniques used in requirements analysis is asking directly. **Asking directly** is an approach that asks users, stakeholders, and other managers about what they want and expect from the new or modified system. This approach works best for stable systems in which stakeholders and users clearly understand the system's functions. The role of the systems analyst during the analysis phase is to critically and creatively evaluate needs and define them clearly so that the systems can best meet them.

Critical Success Factors

Another approach uses critical success factors (CSFs). As discussed earlier, managers and decision makers are asked to list only the factors that are critical to the success of their areas of the organization. A CSF for a production manager might be adequate raw materials from suppliers, while a CSF for a sales representative could be a list of customers currently buying a certain type of product. Starting from these CSFs, the system inputs, outputs, performance, and other specific requirements can be determined.

The IS Plan

As we have seen, the IS plan translates strategic and organizational goals into systems development initiatives. The IS planning process often generates strategic planning documents that can be used to define system requirements. Working from these documents ensures that requirements analysis will address the goals set by top-level managers and decision makers (see Figure 12.20). There are unique benefits to applying the IS plan to define systems requirements. Because the IS plan takes a long-range approach to using information systems within the organization, the requirements for a system analyzed in terms of the IS plan are more likely to be compatible with future systems development initiatives.

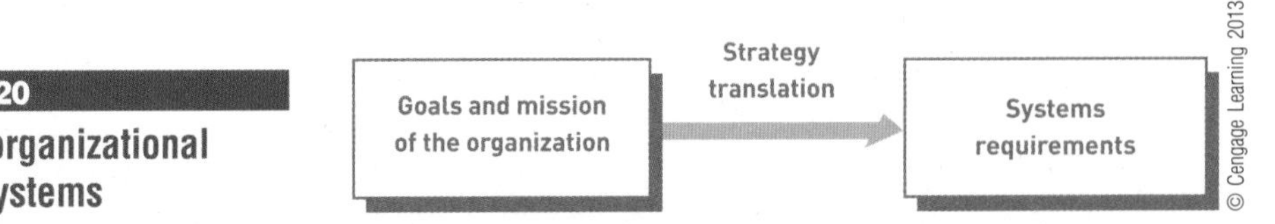

FIGURE 12.20

Converting organizational goals into systems requirements

Screen and Report Layout

Developing formats for printed reports and screens to capture data and display information are some of the common tasks associated with developing systems. Screens and reports relating to systems output are specified first to verify that the desired solution is being delivered. Manual or computerized screen and report layout facilities are used to capture both output and input requirements.

screen layout: A technique that allows a designer to quickly and efficiently design the features, layout, and format of a display screen.

Using a **screen layout**, a designer can quickly and efficiently design the features, layout, and format of a display screen. In general, users who interact with the screen frequently can be presented with more data and less descriptive information; infrequent users should have more descriptive information presented to explain the data that they are viewing (see Figure 12.21).

report layout: A technique that allows designers to diagram and format printed reports.

Report layout allows designers to diagram and format printed reports. Reports can contain data, graphs, or both. Graphic presentations allow managers and executives to quickly view trends and take appropriate action, if necessary.

a

ORDER ENTRY

ORDER NO.	CUSTOMER NO.	SALES PERSON	REGION	COMMISSION	NET DOLLARS
XXXXX	XXXXX	XXXXX	XXX	XXX	XXXXX

ITEM NO	QTY	UNIT	PRICE	DOLLARS	DISCOUNTS
XXXXXXXXX	XXXX	XX	XXXXX	XXXXXXX	XX XX XX
XXXXXXXXX	XXXX	XX	XXXXX	XXXXXXX	XX XX XX
XXXXXXXXX	XXXX	XX	XXXXX	XXXXXXX	XX XX XX
XXXXXXXXX	XXXX	XX	XXXXX	XXXXXXX	XX XX XX
XXXXXXXXX	XXXX	XX	XXXXX	XXXXXXX	XX XX XX
XXXXXXXXX	XXXX	XX	XXXXX	XXXXXXX	XX XX XX
XXXXXXXXX	XXXX	XX	XXXXX	XXXXXXX	XX XX XX
XXXXXXXXX	XXXX	XX	XXXXX	XXXXXXX	XX XX XX
XXXXXXXXX	XXXX	XX	XXXXX	XXXXXXX	XX XX XX
XXXXXXXXX	XXXX	XX	XXXXX	XXXXXXX	XX XX XX
XXXXXXXXX	XXXX	XX	XXXXX	XXXXXXX	XX XX XX
XXXXXXXXX	XXXX	XX	XXXXX	XXXXXXX	XX XX XX

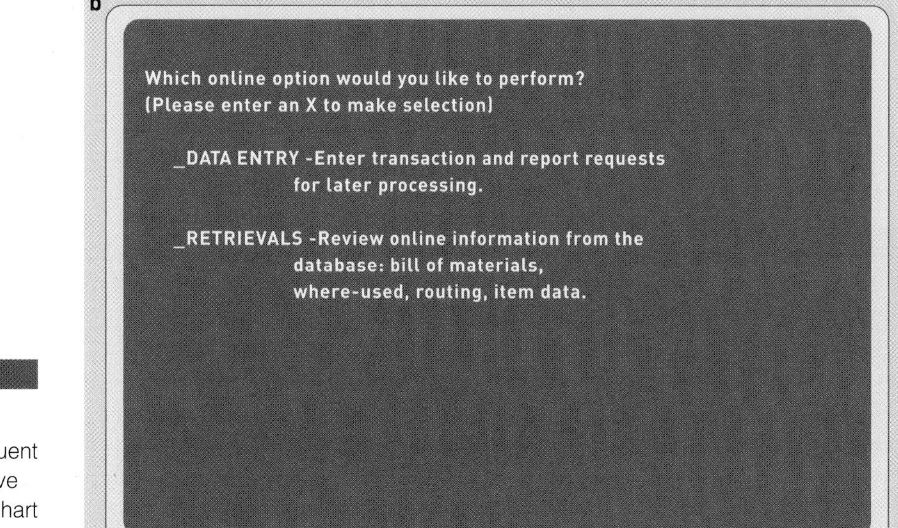

FIGURE **12.21**

Screen layouts

(a) A screen layout chart for frequent users who require little descriptive information. (b) A screen layout chart for infrequent users who require more descriptive information.

Screen layout diagrams can document the screens users' desire for the new or modified application. Report layout charts reveal the format and content of various reports that the application will prepare. Other diagrams and charts can be developed to reveal the relationship between the application and outputs from the application.

Requirements Analysis Tools

A number of tools can be used to document requirements analysis, including CASE tools. As requirements are developed and agreed on, entity-relationship diagrams, data-flow diagrams, screen and report layout forms, and other types of documentation are stored in the CASE repository. These requirements might also be used later as a reference during the rest of systems development or for a different systems development project.

Object-Oriented Systems Analysis

The object-oriented approach can also be used during systems analysis. Like traditional analysis, problems or potential opportunities are examined and key participants and collecting data are identified during object-oriented analysis. But instead of analyzing the existing system using data-flow diagrams and flowcharts, the team uses an object-oriented approach.

The section "Object-Oriented Systems Investigation" introduced a kayak rental example. A more detailed analysis of that business reveals that there are two classes of kayaks: single kayaks for one person and tandem kayaks that can accommodate two people. With the OO approach, a class is used to describe different types of objects, such as single and tandem kayaks. The classes of kayaks can be shown in a generalization/specialization hierarchy diagram as in Figure 12.22. KayakItem is an object that will store the kayak identification number (ID) and the date the kayak was purchased (datePurchased).

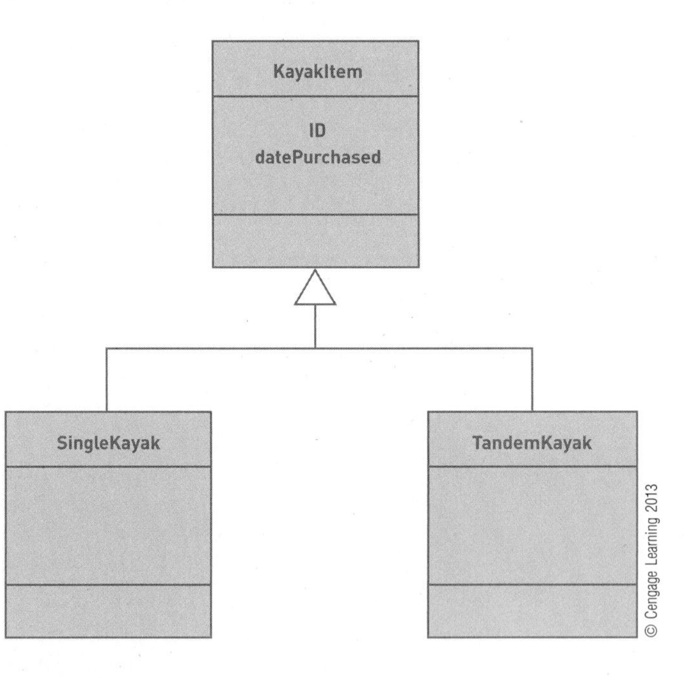

FIGURE **12.22**

Generalization/specialization hierarchy diagram for single and tandem kayak classes

Of course, there could be subclasses of customers, life vests, paddles, and other items in the system. For example, price discounts for kayak rentals could be given to seniors (people over 65 years) and students. Thus, the Customer class could be divided into regular, senior, and student customer subclasses.

The Systems Analysis Report

Systems analysis concludes with a formal systems analysis report. It should cover the following elements:

- The strengths and weaknesses of the existing system from a stakeholder's perspective
- The user/stakeholder requirements for the new system (also called the *functional requirements*)
- The organizational requirements for the new system
- A description of what the new information system should do to solve the problem

Suppose analysis reveals that a marketing manager thinks a weakness of the existing system is its inability to provide accurate reports on product availability. These requirements and a preliminary list of the corporate objectives for the new system will be in the systems analysis report. Particular attention is placed on areas of the existing system that could be improved to meet user requirements. The table of contents for a typical report is shown in Figure 12.23.

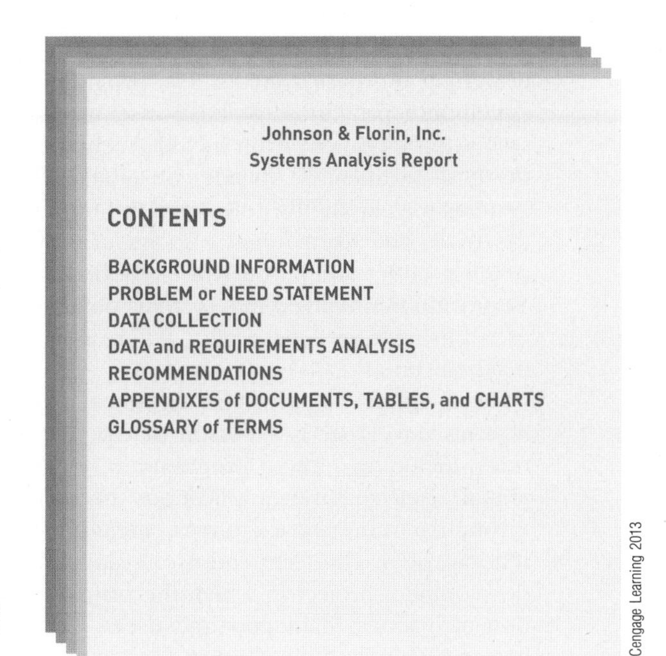
Johnson & Florin, Inc.
Systems Analysis Report

CONTENTS

BACKGROUND INFORMATION
PROBLEM or NEED STATEMENT
DATA COLLECTION
DATA and REQUIREMENTS ANALYSIS
RECOMMENDATIONS
APPENDIXES of DOCUMENTS, TABLES, and CHARTS
GLOSSARY of TERMS

FIGURE 12.23
Typical table of contents for a report on an existing system

The systems analysis report gives managers a good understanding of the problems and strengths of the existing system. If the existing system is operating better than expected or the necessary changes are too expensive relative to the benefits of a new or modified system, the systems development process can be stopped at this stage. If the report shows that changes to another part of the system might be the best solution, the development process might start over, beginning again with systems investigation. Or, if the systems analysis report shows that it will be beneficial to develop one or more new systems or to make changes to existing ones, then systems design, which is discussed in Chapter 13, begins.

SUMMARY

Principle:

Effective systems development requires a team effort from stakeholders, users, managers, systems development specialists, and various support personnel, and it starts with careful planning.

The systems development team consists of stakeholders, users, managers, systems development specialists, and various support personnel. The development team determines the objectives of the information system and delivers to the organization a system that meets its objectives.

Stakeholders are people who, either themselves or through the area of the organization they represent, ultimately benefit from the systems development project. Users are people who will interact with the system regularly. They can be employees, managers, customers, or suppliers. Managers on development teams are typically representative of stakeholders or can be stakeholders themselves. In addition, managers are most capable of initiating and maintaining change. For large-scale systems development projects for which the investment in and value of a system can be quite high, it is common to have senior-level managers be part of the development team.

A systems analyst is a professional who specializes in analyzing and designing business systems. The programmer is responsible for modifying or developing programs to satisfy user requirements. Other support personnel on the development team include technical specialists, either employees from the IS department or outside consultants. Depending on the magnitude of the systems development project and the number of IS development specialists on the team, the team might also include one or more IS managers. At some point in your career, you will likely be a participant in systems development. You could be involved in a systems development team as a user, as a manager of a business area or project team, as a member of the IS department, or maybe even as a CIO.

Individuals are involved as systems developers and users. An individual systems developer is a person that performs all of the systems development roles, including analyst, programmer, technical specialist, and so on. An individual user acquires applications for both personal and professional use through a number of sources, including Apple's App Store, popular with iPhone users. The term end-user systems development describes any systems development project in which the primary effort is undertaken by a combination of business managers and users. Today, more individuals are becoming entrepreneurs and developing applications for cell phones and other devices.

Systems development projects are initiated for many reasons, including the need to solve problems with an existing system, to exploit opportunities to gain competitive advantage, to increase competition, to make use of effective information, to spur organizational growth, to settle a merger or corporate acquisition, and to address a change in the market or external environment. External pressures, such as potential lawsuits or terrorist attacks, can also prompt an organization to initiate systems development.

Information systems planning refers to the translation of strategic and organizational goals into systems development initiatives. Benefits of IS planning include a long-range view of information technology use and better use of IS resources. Planning requires developing overall IS objectives; identifying IS projects; setting priorities and selecting projects; analyzing resource requirements; setting schedules, milestones, and deadlines; and developing the IS planning document. IS planning can result in a competitive advantage through creative and critical analysis.

Establishing objectives for systems development is a key aspect of any successful development project. Critical success factors (CSFs) can identify important objectives. Systems development objectives can include performance goals (quality and usefulness of the output and the speed at which output is generated) and cost objectives (development costs, fixed costs, and ongoing investment costs).

Principle:

Systems development often uses tools to select, implement, and monitor projects, including prototyping, rapid application development, CASE tools, and object-oriented development.

The five phases of the traditional SDLC are investigation, analysis, design, implementation, and maintenance and review. Systems investigation identifies potential problems and opportunities and considers them in light of organizational goals. Systems analysis seeks a general understanding of the solution required to solve the problem; the team studies the existing system in detail and identifies weaknesses. Systems design creates new or modifies existing system requirements. Systems implementation encompasses programming, testing, training, conversion, and operation of the system. Systems maintenance and review entails monitoring the system and performing enhancements or repairs.

Advantages of the traditional SDLC include the following: It provides for maximum management control, creates considerable system documentation, ensures that system requirements can be traced back to stated business needs, and produces many intermediate products for review. Its disadvantages include the following: Users may get a system that meets the needs as understood by the developers, the documentation is expensive and difficult to maintain, users' needs go unstated or might not be met, and users cannot easily review the many intermediate products produced.

Prototyping is an iterative approach that involves defining the problem, building the initial version, having users work with and evaluate the initial version, providing feedback, and incorporating suggestions into the second version. Prototypes can be fully operational or nonoperational, depending on how critical the system under development is and how much time and money the organization has to spend on prototyping.

Rapid application development (RAD) uses tools and techniques designed to speed application development. Its use reduces paper-based documentation, automates program source code generation, and facilitates user participation in development activities. RAD can use newer programming techniques, such as agile development or extreme programming.

Many companies hire an outside consulting firm that specializes in systems development to take over some or all of its systems development activities. This approach is called *outsourcing*. Reasons for outsourcing include companies' belief that they can cut costs, achieve a competitive advantage without having the necessary IS personnel in-house, obtain state-of-the-art technology, increase their technological flexibility, and proceed with development despite downsizing. Many companies offer outsourcing services, including computer vendors and specialized consulting companies.

A number of factors affect systems development success. The degree of change introduced by the project, continuous improvement and reengineering, organizational experience with systems development, the use of project management tools, and the use of CASE tools and the object-oriented approach are all factors that affect the success of a project. The greater the amount of change a system will endure, the greater the degree of risk and often the greater the amount of reward. Continuous improvement projects do not require significant business process or IS changes, while reengineering involves fundamental changes in how the organization conducts business and completes tasks. Successful systems development projects often involve such factors as support from top management, strong user involvement, use of a proven methodology, clear project goals and objectives, concentration on key problems and straightforward designs, staying on schedule and within budget, good user training, and solid review and maintenance programs. Today, more organizations are developing or buying mobile applications for their managers and workers.

The use of automated project management tools enables detailed development, tracking, and control of the project schedule. Effective use of a quality assurance process enables the project manager to deliver a high-quality system and to make intelligent trade-offs among cost, schedule, and quality. CASE tools automate many of the systems development tasks, thus reducing an analyst's time and effort while ensuring good documentation. Object-oriented systems development (OOSD) can also be an important success factor. With the OOSD approach, a project can be broken down into a group of objects that interact. Instead of requiring thousands or millions of lines of detailed computer instructions or code, the systems development project might require a few dozen or maybe a hundred objects.

Principle:

Systems development starts with investigation and analysis of existing systems.

In most organizations, a systems request form initiates the investigation process. Participants in systems investigation can include stakeholders, users, managers, employees, analysts, and programmers. The systems investigation is designed to assess the feasibility of implementing solutions for business problems, including technical, economic, legal, operational, and schedule feasibility. Internal rate of return (IRR) analysis is often used to help determine a project's economic feasibility. An investigation team follows up on the request and performs a feasibility analysis that addresses technical, economic, legal, operational, and schedule feasibility.

If the project under investigation is feasible, major goals are set for the system's development, including performance, cost, managerial goals, and procedural goals. Many companies choose a popular methodology so that new IS employees, outside specialists, and vendors will be familiar with the systems development tasks set forth in the approach. A systems development methodology must be selected. Object-oriented systems investigation is being used to a greater extent today. The use case diagram is part of the Unified Modeling Language that is used to document object-oriented systems development. As a final step in the investigation process, a systems investigation report should be prepared to document relevant findings.

Systems analysis is the examination of existing systems, a process that begins after a team receives approval for further study from management. Additional study of a selected system allows those involved to further understand the system's weaknesses and potential areas for improvement. An analysis team assembles to collect and analyze data on the existing system.

Data collection methods include observation, interviews, questionnaires, and statistical sampling. Data analysis manipulates the collected data to provide information. The analysis includes grid charts, application flowcharts, and CASE tools. The overall purpose of requirements analysis is to determine user and organizational needs.

Data analysis and modeling is used to model organizational objects and associations using text and graphical diagrams. Analysts do this through the use of entity-relationship (ER) diagrams. Activity modeling often uses data-flow diagrams (DFDs), which model objects, associations, and activities by describing how data can flow between and around various objects. DFDs use

symbols for data flows, processing, entities, and data stores. Application flowcharts, grid charts, and CASE tools are also used during systems analysis.

Requirements analysis determines the needs of users, stakeholders, and the organization in general. Asking directly, using critical success factors, and determining requirements from the IS plan can be employed. Often, screen and report layout charts are used to document requirements during systems analysis.

Like traditional analysis, teams often identify problems or potential opportunities during object-oriented analysis. Object-oriented systems analysis can involve using diagramming techniques, such as a generalization/specialization hierarchy diagram.

CHAPTER 12: SELF-ASSESSMENT TEST

Effective systems development requires a team effort from stakeholders, users, managers, systems development specialists, and various support personnel, and it starts with careful planning.

1. ________________ is the activity of creating or modifying existing business systems. It refers to all aspects of the process—from identifying problems to be solved or opportunities to be exploited to the implementation and refinement of the chosen solution.
2. Which of the following people ultimately benefit from a systems development project?
   a. computer programmers
   b. systems analysts
   c. stakeholders
   d. senior-level manager
3. ________________ is a person who performs all the systems development roles.
4. Like a contractor constructing a new building or renovating an existing one, the chief information officer (CIO) takes the plans from the systems analyst and builds or modifies the necessary software. True or False?
5. The term ________________ refers to the translation of strategic and organizational goals into systems development initiatives.
6. What involves investigating new approaches to existing problems?
   a. critical success factors
   b. systems analysis factors
   c. creative analysis
   d. critical analysis

Systems development often uses tools to select, implement, and monitor projects, including, prototyping, rapid application development, CASE tools, and object-oriented development.

7. What kind of development uses tools, techniques, and methodologies designed to speed application development?
   a. rapid application development
   b. joint optimization
   c. prototyping
   d. extended application development
8. Traditional systems development encourages systems to constantly change as they are being developed. True or False?
9. ________________ takes an iterative approach to the systems development process. During each iteration, the team identifies and analyzes requirements and alternative solutions to the problem, designs new solutions, and implements a portion of the system.
10. Rapid application development (RAD) employs tools, techniques, and methodologies designed to speed application development. True or False?
11. What consists of all activities that, if delayed, would delay the entire project?
    a. deadline activities
    b. slack activities
    c. RAD tasks
    d. the critical path

Systems development starts with investigation and analysis of existing systems.

12. The systems request form is a document that is filled out during systems analysis. True or False?
13. Feasibility analysis is typically done during which systems development stage?
    a. investigation
    b. analysis
    c. design
    d. implementation

14. Data modeling is most often accomplished through the use of ________________, whereas activity modeling is often accomplished through the use of ________________.
15. The overall purpose of requirements analysis is to determine user, stakeholder, and organizational needs. True or False?

CHAPTER 12: SELF-ASSESSMENT TEST ANSWERS

1. Systems development
2. c
3. Individual systems developer
4. False
5. information systems planning
6. c
7. a
8. False
9. Prototyping
10. True
11. d
12. False
13. a
14. entity-relationship (ER) diagrams, data-flow diagrams
15. True

REVIEW QUESTIONS

1. What is an individual systems developer?
2. What is the goal of IS planning? What steps are involved in IS planning?
3. What are the typical reasons to initiate systems development?
4. What is the difference between creative analysis and critical analysis?
5. What is the difference between a programmer and a systems analyst?
6. What is the difference between a Gantt chart and PERT?
7. Describe what is involved in feasibility analysis.
8. What is end-user systems development? What are the advantages and disadvantages of end-user systems development?
9. List factors that have a strong influence on project success.
10. What is the purpose of systems analysis?
11. What are the steps of object-oriented systems development?
12. Define the different types of feasibility that systems development must consider.
13. What are the objectives of agile development?
14. What is the result or outcome of systems analysis? What happens next?

DISCUSSION QUESTIONS

1. Why is it important for business managers to have a basic understanding of the systems development process?
2. Briefly describe the role of a system user in the systems investigation and systems analysis stages of a project.
3. Briefly describe what is involved in developing applications for mobile devices.
4. You have decided to become an IS entrepreneur and develop applications for the iPhone and other mobile devices. Describe what applications you would develop and how you would do it.
5. Your company wants to develop or acquire a new sales program to help sales representatives identify new customers. Describe what factors you would consider in deciding whether to develop the application in-house or outsource the application to an outside company.
6. You have been hired by your university to find an outsourcing company to perform the payroll function. What are your recommendations? Describe the advantages and disadvantages of the outsourcing approach for this application.
7. You have been hired as a project manager to develop a new Web site in the next six months for a store that sells music and books online. Describe how project management tools, such as a Gantt chart and PERT, might be used.
8. For what types of systems development projects might prototyping be especially useful? What are the characteristics of a system developed with a prototyping technique?

9. Assume that you work for a financial services company. Describe three applications that would be critical for your clients. What systems development tools would you use to develop these applications?
10. How important are communications skills to IS personnel? Consider this statement: "IS personnel need a combination of skills—one-third technical skills, one-third business skills, and one-third communications skills." Do you think this is true? How would this affect the training of IS personnel?
11. You have been hired to perform systems investigation for a French restaurant owner in a large metropolitan area. She is thinking of opening a new restaurant with a state-of-the-art computer system that would allow customers to place orders on the Internet or at kiosks at restaurant tables. Describe how you would determine the technical, economic, legal, operational, and schedule feasibility for the restaurant and its proposed computer system.
12. Discuss three reasons why aligning overall business goals with IS goals is important.
13. You are the chief information officer for a medium-sized retail store and would like to develop an extranet to allow your loyal customers to see and buy your products on the Internet. Describe how you would determine the requirements for the new system.

PROBLEM-SOLVING EXERCISES

1. For a business of your choice, use a graphics program to develop one or more Use Case diagrams and one or more Generalized/Specialized Hierarchy diagrams for your new business, using the object-oriented approach.
2. You have been hired to develop new mobile applications for an insurance company. Using a word processing program, describe the applications you would develop. Briefly describe how you would develop these applications.

TEAM ACTIVITIES

1. Your team should interview people involved in systems development in a local business or at your college or university. Describe the process used. Identify the users, analysts, and stakeholders for a systems development project that has been completed or is currently under development.
2. Your team has been hired to determine the feasibility of a new Internet site that contains information about upcoming entertainment, sports, and other events for students at your college or university. Use a word processing program to describe your conclusions for the different types of feasibility and your final recommendations for the new Internet site.
3. Your team has been hired to use systems development to produce a billing program for a small heating and air conditioning company. Develop a flowchart that shows the major features and components of your billing program.

WEB EXERCISES

1. Cloud computing in which applications such as word processing and spreadsheet analysis are delivered over the Internet is becoming more popular. You have been hired to analyze the potential of a cloud computing application that performs payroll and invoicing over the Internet from a large Internet company. Describe the systems development steps and procedures you would use to analyze the feasibility of this approach.
2. Using the Internet, explore the most useful mobile applications for a business or industry of your choice. Also explore mobile applications for this business or industry that are not currently available. Write a report describing what you found.

CAREER EXERCISES

1. Pick a career that you are considering. What type of information system would help you on the job? Perform technical, economic, legal, operational, and schedule feasibility for an information system you would like developed.
2. Using the Internet, research career opportunities in which you would develop applications for new tablet computers. Describe two individuals or companies that have successfully developed applications for tablet computers.

CASE STUDIES

Case One

Grain Hill Selects New ERP System

In this chapter, you saw that systems investigation and analysis lead to a detailed description of how a new system will solve a business problem. The next step can be designing and developing that system. Alternatively, the company that wants a new system can try to identify commercial products that will meet its needs.

Such was the situation for Grain Hill Corporation. Based in Panama City, Panama, Grain Hill is the leading rice producer in Central America with major national brands such as Arrocera San Francisco in El Salvador, Agricorp in Nicaragua, and Corporación Arrocera de Costa Rica. Grain Hill's products are distributed internationally under brand names such as Doña Lisa and are imported to the United States through All Foods Inc. Grain Hill also acts as a food distributor for other products within Central America.

As a result of its history of mergers and acquisitions, Grain Hill found itself with five computing platforms in three countries. It decided to replace them with a new enterprise resource planning (ERP) system for production and distribution so that it could better manage its companies and optimize its processes.

Grain Hill began by assuming that it would purchase a software package. This assumption is common when a company's needs are not unusual and the company knows, through experience and reading, that some packages could meet its need. However, Grain Hill had trouble building consensus and creating an evaluation method.

As Grain Hill CIO Benjamin Altamirano puts it, "We were deciding whether or not to build upon the systems we already had or pursue discussions with SAP and Oracle. It then became apparent to us that we had to find the right evaluation method before we found the right software." That job was given to information systems manager Diego Muñoz.

Muñoz did what many people in his situation do: he called an advisor who was familiar with the technical area in which he needed help. He chose Technology Evaluation Centers (TEC) of Montreal, Canada.

TEC project manager Denis Rousseau traveled to Nicaragua to assess Grain Hill's business firsthand. In previous telephone conversations, Rousseau determined that an ERP system for distribution was best suited for Grain Hill's current and future needs. During his visit, he helped Grain Hill stakeholders identify specific capabilities that this system should have to support the direction the company wanted to take.

"Prior to TEC, we had difficulties defining future requirements," says Muñoz. "We had an idea of where we wanted our company to go, but we didn't have the expertise to identify the [software] functionality needed to get there."

Like other vendor selection advisory companies, TEC uses a standard methodology in its work. After Grain Hill and TEC identified Grain Hill's requirements and ranked them in order of importance, TEC loaded them into its decision support system to generate a preliminary list of software packages that matched Grain Hill's needs. Grain Hill was able to narrow the list to three vendors and study their capabilities closely. TEC and Grain Hill then developed a request for proposal (RFP) for those three vendors, detailing the terms of the final selection process, which included vendor demonstration guidelines as well as the scoring method.

At the end of this exercise, Grain Hill was able to make an impartial selection of the best software package for its ERP needs. The company saved 48 percent on the total cost of ownership it had initially estimated, while also expanding the capabilities of its system beyond what the firm had initially envisaged.

Discussion Questions

1. Suppose Grain Hill wanted to evaluate ERP offerings from SAP, Oracle, and others on its own, without the help of TEC or another advisor. List the steps it should follow.
2. An in-depth evaluation such as TEC performed for Grain Hill can easily cost $25,000 to $50,000. How can this expense be justified?

Critical Thinking Questions

1. As this case suggests, TEC is not the only company that provides vendor selection advice. Using an advisor, therefore, doesn't eliminate the problem; it changes the problem from selecting a software vendor to selecting an advisor. Is this a better problem to have? An easier problem to solve? Why or why not?
2. Suppose you are a sales representative for a software vendor. You learn that one of your sales prospects, such as Grain Hill, is working with a vendor selection advisor instead of making the software selection decision itself. How does that affect your approach to selling your software to Grain Hill? If Grain Hill decides not to buy your software, what would you change? If

Grain Hill had made its decision without the help of this advisor, is your approach different?

SOURCES: Agricorp, Nicaragua Web site, *www.agricorp.com.ni* (in Spanish), accessed January 26, 2012; All Foods Inc. Web site, *www.allfoodsusa.com*, accessed January 26, 2012; Arrocera San Francisco, El Salvador Web site, *www.arrocerasanfrancisco.com* (in Spanish), accessed January 26, 2012; Technology Evaluation Centers, "Grain Hill Corporation Selects a New ERP System," *www.technologyevaluation.com/pdf/case-study/26168/grain-hill-corporation-selects-a-new-erp-system.pdf* (free registration required), September 7, 2011, Technology Evaluation Centers Web site, *www.technologyevaluation.com*, accessed January 26, 2012.

Case Two

Requirements Tracking at Honeywell Technology Solutions Lab

As you read in this chapter, determining system requirements is a vital part of the development of any information system. Complex information systems have many sets of requirements. It is, therefore, essential to have a systematic way to determine them.

Honeywell Technology Solutions Lab (HTSL), through its IT Services and Solutions business unit, develops software solutions for other parts of Honeywell Inc. HTSL is based in Bengaluru (Bangalore), India, with centers in Beijing, Brno, Hyderabad, Madurai, and Shanghai.

In 2010, the company identified a problem: At HTSL, various groups such as requirement writers and development, quality assurance (QA), and project management teams worked independently in separate "silos." It was difficult to track project requirements and the status of their implementation. HTSL needed a system to manage the requirements and its relationships to each other.

Beyond managing the requirements, HTSL needed an application that could coordinate test cases, design elements, and defects. Requirement writers would create the requirements for software, and HTSL customers (other Honeywell divisions) would review and approve these requirements. Once approved, the development team would implement them, and the QA team would generate test cases based on them. Any defects found in executing the test cases would also be tracked.

HTSL had a great deal of experience in developing software for aerospace, automation and control, specialty materials, and transportation systems. However, they had no experience in developing software to manage the development process itself. The company recognized this deficit and turned to specialists.

Kovair, of Santa Clara, California, is such a specialist. Its Application Lifecycle Management (ALM) package is for "implementing a software development life cycle (SDLC) process, collaborating on the entire development cycle and tracing implementations back to original specs. [It] ensures that all developers are working from the same playbook … and that there are no costly last minute surprises."

One ALM module is Requirements Management. Using it, HTSL can gather requirements, rank them, manage their changes, and coordinate them with system test cases. The Requirements Management module can also produce a variety of reports, including formatted requirements specifications and reports showing the distribution of requirements by type, criticality, source, or any other descriptor.

Honeywell already had a formal development process called "Review, Approval, Baseline, Technical Design, Test Design, Implementation and Testing." Kovair's ALM solution was customized to fit into this process. When a requirement is entered into ALM, it is marked "Submitted," and the review process begins. ALM generates Review tasks for stakeholders, ensuring that they will give their views on the new requirement. When they approve it, perhaps after changes, its status is changed to "Approved," and a task is entered for its owner to add it to the baseline system design. When this step is completed, two new tasks are created: one for the development team to develop technical specifications and then the software and one for the quality assurance team to develop test cases. Development can then continue.

What were the results? HTSL has reduced rework due to incorrect requirements and sped up development. Development team productivity was improved by about 20 percent, and requirements-related defects were reduced by at least 1 percent.

Discussion Questions

1. What would have happened if HTSL tried to develop a system like ALM on its own instead of turning to Kovair? Discuss both pros and cons of the likely outcome.
2. The ALM software is intended to help companies manage the steps of software development. Software development is only one of the processes that businesses use every day. What are the characteristics of a process that make a software package like ALM useful? Identify criteria that can be used to decide whether a company should look for an ALM-like package for that process.

Critical Thinking Questions

1. ALM is designed to manage the entire SDLC, not just requirements. Suppose you had to choose between (a) a system that could manage the entire SDLC and (b) a system that only managed requirements but did that better. How would you choose between them?
2. This case is based in part on information from Kovair. Many organizations need to track software development projects, so other companies besides Kovair offer packages to do that. Suppose you were given the job of choosing such a package. List at least four criteria you would use in comparing different packages. Rank the items on your list from most to least important.

SOURCES: Kovair, Inc., "Requirements Management Case Study for Honeywell," *www.kovair.com/whitepapers/Requirements-Management-Case-Study-for-Honeywell.pdf*, August 2011; HTSL Web site, *www.honeywell.com/sites/htsl*, accessed January 26, 2012; Kovair ALM Web site, *www.kovair.com/alm/application-lifecycle-management-description.aspx*, accessed January 27, 2012.

Questions for Web Case

See the Web site for this book to read about the Altitude Online case for this chapter. Following are questions concerning this Web case.

Altitude Online: Systems Investigation and Analysis Considerations

Discussion Questions

1. What important activities did Jon's team engage in during the systems investigation stage of the systems development life cycle?
2. Why are all forms of feasibility considerations especially important for an ERP development project?

Critical Thinking Questions

1. Why is the quality of the systems analysis report crucial to the successful continuation of the project?
2. Why do you think Jon felt the need to travel to communicate with Altitude Online colleagues rather than using e-mail or phone conferencing? What benefit does face-to-face communication provide in this scenario?

NOTES

Sources for the opening vignette: IBM, "Practice Plan Keeps Dentists and Patients Smiling," *www-01.ibm.com/software/success/cssdb.nsf/CS/STRD-8JAKFM?OpenDocument*, July 1, 2011; Practice Plan Web site, *www.practiceplan.co.uk*, accessed January 22, 2012; Polymorph Web site, *www.polymorph.co.uk*, accessed January 22, 2012; United Kingdom National Health Service, "How Much Will I Pay for NHS Dental Treatment?," *www.nhs.uk/chq/Pages/1781.aspx*; March 31, 2011.

1. Brousell, Lauren, "IT Budgets Bulge in 2011," *CIO*, February 1, 2011, p. 8.
2. Thibodeau, P., "Hallmark's Fresh Start," *Computerworld*, May 9, 2011, p. 42.
3. Fowler, G., "Mobile Apps Drawing in Shoppers, Marketers," *The Wall Street Journal*, January 31, 2011, p. B4.
4. Johnson, K., "Homeland Security Scraps Border Fence," *The Wall Street Journal*, January 15, 2011, p. A3.
5. Wakabayashi, D. and Osawa, J., "Odd Couple behind Nintendo 3-D Push," *The Wall Street Journal*, March 2, 2011, p. A1.
6. Nash, Kim, "The Talent Advantage," *CIO*, March 1, 2011, p. 32.
7. Violino, B., "Grand Opening for App Stores," *Computerworld*, May 23, 2011, p. 24.
8. Swanborg, Rick, "Winning Ways to Project Success," *CIO*, August 1, 2011, p. 26.
9. Waghray, A., "CIO Profiles," *Information Week*, February 14, 2011, p. 10.
10. Campbell, Peter, "CIO Profiles," *Information Week*, August 15, 2011, p. 8.
11. Nash, Kim, "The Talent Advantage," *CIO*, March 1, 2011, p. 32.
12. MacMillan, D., "Pocket Translator," *Bloomberg Businessweek*, May 9, 2011, p. 44.
13. Tozzi, J., "A 911 System for the Mobile Era," *Bloomberg Businessweek*, May 16, 2011, p. 40.
14. Burrows, P., "How Apple Feeds Its Army of App Makers," *Bloomberg Businessweek*, June 13, 2011, p. 39.
15. Peers, M., "Apple Risks App-lash on iPad," *The Wall Street Journal*, February 27, 2011, p. C12.
16. Greene, T., "IBM Cautions of Mobile, Cloud Security Issues," *Network World*, April 4, 2011, p. 16.
17. Henderson, T. and Allen, B., "New Tools to Protect Mobile Devices," *Network World*, May 23, 2001, p. 27.
18. Cisco Web site, *www.cisco.com*, accessed July 17, 2011.
19. Matlin, C., "Innovator," *Bloomberg Businessweek*, April 11, 2011, p. 34.
20. Evans, Bob, "Why CIOs Must Have a Tablet Strategy," *Information Week*, March 14, 2001, p. 10.
21. Pratt, M., "Get into the M&A Game," *Computerworld*, May 23, 2011, p. 17.
22. Thibodeau, P., "Tax Law May Accelerate IT Purchases," *Computerworld*, February 7, 2011, p. 4.
23. Angwin, Julia, "Web Tool on Firefox to Deter Tracking," *The Wall Street Journal*, January 24, 2011, p. B1.
24. Leaver, S., "Beyond IT-Business Alignment," *CIO*, May 1, 2011, p. 28.
25. Tanriverdi, H., et al, "Reframing the Dominant Quests of Information Systems Strategy Research for Complex Adaptive Business Systems," *Information Systems Research*, December 2010, p. 822.
26. Ravishankar, M., et al, "Examining the Strategic Alignment and Implementation Success of a KMS," *Information Systems Research*, March 2011, p. 39.
27. Cichowski, E., "CIO Profiles," *Information Week*, January 31, 2011, p. 20.
28. Kanaracus, Chris, "NYC Seeks $600m Refund for IT Work," *www.computerworld.com*, accessed August 24, 2011.
29. Progressive Web site, *www.progressive.com*, accessed July 17, 2011.
30. Fulgham, C., "It Is All about Helping Us Conduct Investigations Faster," *CIO*, February 1, 2011, p. 36.
31. Genetech Web site, *www.gene.com*, accessed July 17, 2011.
32. CME Group Web site, *www.cmegroup.com*, accessed July 17, 2011.
33. Bataller, E., "Risk Avengers," *Information Week*, January 31, 2001, p. 32.
34. Pratt, Mary, "What CFOs Want from IT," *Computerworld*, January 24, 2011, p. 19.
35. Mearian, Lucas, "IT Faces Deadline on New Medical Codes," *Computerworld*, July 18, 2011, p. 4.
36. Kennaley, M., "Pragmatic Development," *Information Week*, April 25, 2011, p. 35.
37. Staff, "Development of a New Rapid Prototyping Process," *Rapid Prototyping Journal*, vol. 17, no. 2, 2011, p. 138.
38. Murphy, C., "IT Is Too Darn Slow," *Information Week*, February 28, 2011, p. 19.
39. IBM Support home page, *www-947.ibm.com/support/entry/portal/Overview/Software/Rational/Rational_Rapid_Developer*, accessed August 24, 2011.

40. Jazz projects Web page, *https://jazz.net/projects/rational-team-concert/*, accessed July 17, 2011.
41. Compuware Web site, *www.compuware.com*, accessed August 24, 2011.
42. CA Technologies Web site, *www.ca.com/us/products/detail/CA-Gen.aspx*, accessed August 24, 2011.
43. Foley, John, "FBI Recasts Sentinel as a Model of Agility," *InformationWeek*, May 30, 2011, p. 12.
44. Fulgham, C., "It Is All about Helping Us Conduct Investigations Faster," *CIO*, February 1, 2011, p. 36.
45. Sabre Airline Solutions Web site, *www.sabreairlinesolutions.com/home/*, accessed August 24, 2011.
46. Jim Highsmith Web site, *http://adaptivesd.com/*, accessed August 26, 2011.
47. Lean Software Development Web site, *www.poppendieck.com/*, accessed August 26, 2011.
48. West, Dave, "Agile Processes Go Lean," *Information Week*, April 27, 2009, p. 32.
49. Cyrillo, Marcio, "Lean UX: Rethink Development," *InformationWeek*, November 14, 2011, p. 40.
50. Staff, "The Rational Unified Process," *www.ibm.com/rational*, accessed July 17, 2011.
51. Feature Driven Development Web site, *www.featuredrivendevelopment.com/*, accessed August 26, 2011.
52. Alistair Web site, *http://alistair.cockburn.us/Crystal+methodologies*, accessed August 26, 2011.
53. ScrumAlliance Web site, *www.scrumalliance.org/pages/certified_scrummaster_csm*, accessed May 28, 2011.
54. Pfizer Web site, *www.pfizer.com*, accessed July 17, 2011.
55. Overby, S., "Restaffing for the Cloud," *CIO*, June 1, 2011, p. 24.
56. Thibodeau, P. and Vijayan, J., "Amazon Service Outage Reinforces Cloud Doubts," *Computerworld*, May 9, 2011, p. 8.
57. Budge, J. and Wang, R., "Which Applications Belong on the Cloud," *CIO*, May 1, 2011, p. 49.
58. Efrati, A., "Google Searches for Mobile Apps Experts," *The Wall Street Journal*, January 31, 2001, p. B1.
59. McMahan, T., "The Ins and Outs of Mobile Apps," *The Wall Street Journal*, June 13, 2011, p. R8.
60. IBM Web site, *www.ibm.com*, accessed on July 17, 2011.
61. EDS Web site, *www.eds.com*, accessed July 17, 2011.
62. Accenture Web site, *www.accenture.com*, accessed July 17, 2011.
63. WiPro Web site, *www.wipro.com*, accessed July 17, 2011.
64. Bahree, M., "Tata Targets Smaller Clients," *The Wall Street Journal*, February 22, 2011, p. B7.
65. Nash, Kim, "Rise to the Occasion," *CIO*, July 1, 2011, p. 9.
66. Amazon Web Services Web site, *http://aws.amazon.com/ec2/*, accessed July 17, 2011.
67. Linebaugh, Kate, "GE Makes Big Bet on Software Development," *The Wall Street Journal*, November 17, 2011, p. B8.
68. Sheridan, B., "The Apps Class of 2010," *Bloomberg Businessweek*, January 3, 2011, p. 80.
69. Waxer, C., "Juicing the Market for Electric Cars," *CIO*, April 1, 2001, p. 20.
70. Jackson, J., "Business Apps Target Consumer Devices," *Computerworld*, January 10, 2011, p. 6.
71. Light, J., "How's Your HTML5? App Skills in Demand," *The Wall Street Journal*, January 31, 2001, p. B7.
72. Henschen, D., "They're Ready," *Information Week*, February 14, 2001, p. 21.
73. Esposito, D., "A Whole New Ball Game," *Information Week*, March 14, 2011, p. 42.
74. Murphy, Chris, "App Dev Must Get Agile Enough for Mobile," *InformationWeek*, October 17, 2011, p. 10.
75. Stackpole, Beth, "The Mobile App Gold Rush," *Comuterworld*, August 22, 2011, p. 18.
76. Kaneshige, T., "RehabCare's Two-Year Apple, Mobile Makeover," *CIO*, February 1, 2011, p. 24.
77. Betts, M., "Enterprise Apps Stores: A Good Idea?" *Computerworld*, January 24, 2011, p. 4.
78. Hoover, Nicholas, "Army Readies Mobile App Store," *InformationWeek*, October 31, 2011, p. 18.
79. Miller, Lloyd, "Should Employees Be Allowed to Use Their Own Devices for Work?" *The Wall Street Journal*, November 15, 2011, p. B8.
80. Nash, K., "Market Makers," *CIO*, June 1, 2011, p. 37.
81. Pratt, Mary, "An Enterprise Perspective Yields IT Innovations," *Computerworld*, February 21, 2011, p. 44.
82. Capability Maturity Model for Software home page, *www.sei.cmu.edu*, accessed July 17, 2011.
83. Greene, Tim, "Library System Shushes MPLS for Cheaper DSL," *Network World*, January 24, 2011, p. 14.
84. Mearian, Lucas, "IT Faces Deadline on New Medical Codes," *Computerworld*, July 18, 2011, p. 4.
85. Jayanth, R., et al, "Vendor and Client Interaction for Requirements Assessment in Software Development," *Information Systems Research*, June 2011, p. 289.

13 Systems Development: Design, Implementation, Maintenance, and Review

Principles	Learning Objectives
• Designing new systems or modifying existing ones should always help an organization achieve its goals.	• State the purpose of systems design and discuss the differences between logical and physical systems design. • Describe the process of design modeling and the diagrams used during object-oriented design. • Discuss the issues involved in environmental design. • Define the term *RFP* and discuss how this document is used to drive the acquisition of hardware and software. • Describe the techniques used to make systems selection evaluations.
• The primary emphasis of systems implementation is to make sure that the right information is delivered to the right person in the right format at the right time.	• State the purpose of systems implementation and discuss the activities associated with this phase of systems development. • List the advantages and disadvantages of purchasing versus developing software. • Discuss the software development process and list some of the tools used in this process, including object-oriented program development tools.
• Maintenance and review add to the useful life of a system but can consume large amounts of resources. These activities can benefit from the same rigorous methods and project management techniques applied to systems development.	• State the importance of systems and software maintenance and discuss the activities involved. • Describe the systems review process.

Information Systems in the Global Economy

IMMOBILIENSCOUT24, GERMANY

ImmobilienScout24 Uses Agile System Development Approach

After defining the functions of a new application—the problem it will solve and the approach it will use to solve it—it's time to work on its insides—its technical design and the actual code. The difference between one system development approach and another can make an application failure-prone, slow when it doesn't fail, and difficult to use. Another application, developed to the same specifications, can be reliable, responsive, and user-friendly. Companies that develop applications to sell or use would obviously prefer the second outcome.

This difference is crucial to ImmobilienScout24, which has the leading real estate Web site in Germany. It prides itself as an innovator that responds quickly to market demands. That response means frequent, rapid modifications to the company's Web site. "We have a very creative product management approach which continuously adapts our Internet platform to the market requirements in order to expand our market leadership in a competitive environment," says Katrin Jähn, head of their test automation team.

Besides its Web site, ImmobilienScout24 also develops programs for computers of all sizes. For example, it offers a free iPhone app for property hunting. With this app, "we meet the needs of our users by offering more flexibility and mobility. For example, if you are out on a walk and find a particularly nice area, you can use the ImmoScout application to display all of the available apartments and houses in that area," says ImmobilienScout24 CEO Marc Stilke.

To develop applications faster and improve their reliability, ImmobilienScout24 made two choices. First, the firm adopted the Scrum agile methodology for software development. This approach, which you read about in Chapter 12, is named for the high-powered scrum formation used by rugby teams. It is based on frequent team meetings in which a project team selects small tasks from its to-do list to work on next. Daily reviews and a clear focus, enforced by a so-called ScrumMaster, lead to rapid progress. The selection of small but tangible tasks leads to visible results in short periods of time. The daily reviews by the entire team lead to high-quality programs. If user requirements change during development, that simply means choosing different tasks to work on next; it doesn't force everyone to return to the specification stage. The business benefit of the Scrum system development approach is that it substantially reduces the time to make new software capabilities available to customers.

The next choice ImmobilienScout24 made was to select Hewlett-Packard's QuickTest Professional, now known as HP Functional Testing, for testing. ImmobilienScout24 uses this software to manage a total of 5,100 test cases, organized into 1,300 business process chains of activity. By automating the testing process, ImmobilienScout24 cut its release cycle from four weeks to three and hopes to reach two weeks in the future.

New approaches to software development, such as those discussed in Chapter 12 and this chapter, are being developed regularly. They promise improvements in both productivity and quality. By keeping informed of new approaches, all organizations can reap the same benefits that ImmobilienScout24 did.

As you read this chapter, consider the following:

- How do the development environment and the development tools affect the speed and effectiveness of software development?
- How important is testing to the software development process? How much emphasis should management place on programming, how much on testing what is programmed?

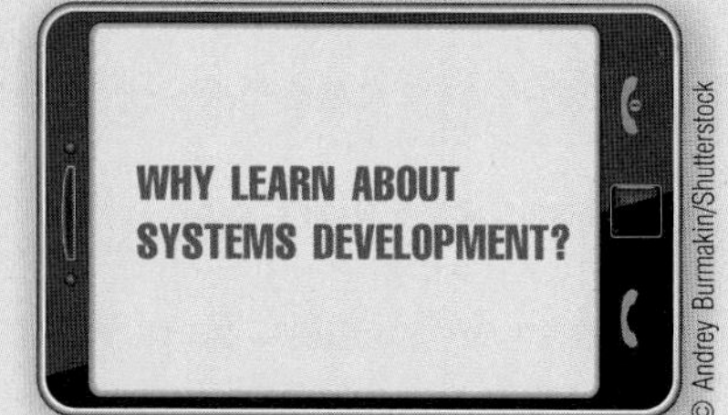

© Andrey Burmakin/Shutterstock

Information systems are designed and implemented for employees and managers every day. A manager at a hotel chain can use an information system to look up client preferences. An accountant at a manufacturing company can use an information system to analyze the costs of a new plant. A sales representative for a music store can use an information system to determine which CDs to order and which to discount because they are not selling. A computer engineer can use an information system to help determine why a computer system is running slowly. Information systems have been designed and implemented for almost every career and industry. An individual can use systems design and implementation to create applications for smartphones for profit or enjoyment. This chapter shows how you can be involved in designing and implementing an information system that will directly benefit you on the job. It also shows how to avoid errors and how to recover from disasters. This chapter starts with describing how systems are designed.

The way an information system is designed, implemented, and maintained profoundly affects the daily functioning of an organization. Like investigation and analysis covered in Chapter 12, the design, implementation, maintenance, and review covered in this chapter strive to achieve organizational goals, such as reducing costs, increasing profits, or improving customer service. We begin this chapter with a discussion of systems design.

SYSTEMS DESIGN

systems design: The stage of systems development that answers the question "How will the information system solve a problem?"

The purpose of **systems design** is to answer the question "How will the information system solve a problem?" The primary result of the systems design phase is a technical design that details system outputs, inputs, and user interfaces; specifies hardware, software, databases, telecommunications, personnel, and procedures; and shows how these components are related.

Systems design is typically accomplished using the tools and techniques discussed in Chapter 12. Depending on the specific application, these methods can be used to support and document all aspects of systems design. Two key aspects of systems design are logical and physical design.

Logical and Physical Design

logical design: A description of the functional requirements of a system.

As discussed in Chapter 5, design has two dimensions: logical and physical. The **logical design** refers to what the system will do; it describes the functional requirements of a system. Today, for example, many stock exchanges, large hedge funds, and institutional stock investors include speed as a critical logical design element for new computer trading systems. The objective is to increase profits by being faster in placing electronic trades than traditional computerized trading systems. Without logical design, the technical details of the system (such as which hardware devices should be acquired) often obscure the best solution. Logical design involves planning the purpose of each system element, independent of hardware and software considerations. The logical design specifications that are determined and documented include output, input, process, file and database, telecommunications, procedures, controls and security, and personnel and job requirements.

Security is always an important logical design issue for corporations and governments. To provide better security and privacy for its customers, Firefox has designed its new Web browser to include a "do-not-call" feature.[1] Rules published in September 2005 require that federal agencies incorporate security procedures in the design of new or modified systems. In addition, the Federal Information Security Management Act, enacted in 2002, requires federal agencies to make sure that security protection measures are incorporated into systems provided by outside vendors and contractors.

physical design: The specification of the characteristics of the system components necessary to put the logical design into action.

The physical design refers to how a computer system accomplishes tasks, including what each component does and how the components work together. **Physical design** specifies the characteristics of the system components necessary to put the logical design into action. In this phase, the characteristics of the hardware, software, database, telecommunications, personnel, and procedure and control specifications must be detailed. These physical design components were discussed in Part 2 on technology.

Object-Oriented Design

Logical and physical design can be accomplished using either the traditional approach or the object-oriented (OO) approach to systems development. Both approaches use a variety of design models to document the new system's features and the development team's understandings and agreements. Many organizations today are turning to OO development because of its increased flexibility. This section outlines a few OO design considerations and diagrams.

Using the OO approach, you can design key objects and classes of objects in the new or updated system. This process includes considering the problem domain, the operating environment, and the user interface. The problem domain involves the classes of objects related to solving a problem or realizing an opportunity. In our example of the Maui, Hawaii, kayak rental shop, first introduced in Chapter 12, KayakItem in Figure 12.23 is an example of a problem domain object that will store information on kayaks in the rental program. The operating environment for the rental shop's system includes objects that interact with printers, system software, and other software and hardware devices. The user interface for the system includes objects that users interact with, such as buttons and scroll bars in a Windows program.

During the design phase, you also need to consider the sequence of events that must happen for the system to function correctly. For example, you might want to design the sequence of events for adding a new kayak to the rental program. The event sequence is often called a *scenario*, and it can be diagrammed in a sequence diagram. See Figure 13.1.

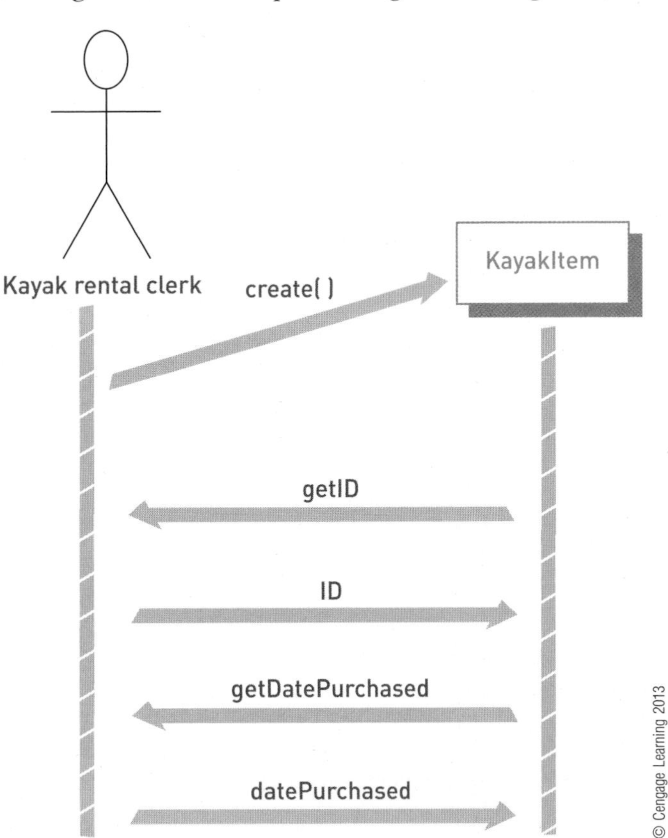

FIGURE **13.1**

Sequence diagram to add a new KayakItem scenario

You read a sequence diagram starting at the top and moving down:

1. The Create arrow at the top is a message from the kayak rental clerk to the KayakItem object to create information on a new kayak to be placed into the rental program.
2. The KayakItem object knows that it needs the ID for the kayak and sends a message to the clerk requesting the information (see the getID arrow).
3. The clerk then types the ID into the computer. This action is shown with the ID arrow. The data is stored in the KayakItem object.
4. Next, KayakItem requests the purchase date. This is shown in the get-DatePurchased arrow.
5. Finally, the clerk types the purchase date into the computer. The data is also transferred to KayakItem object, as shown in the datePurchased arrow at the bottom of Figure 13.1.

This scenario is only one example of a sequence of events. Other scenarios might include entering information about life jackets, paddles, suntan lotion, and other accessories. The same types of use case and generalization/specialization hierarchy diagrams discussed in Chapter 12 can be created for each event, and additional sequence diagrams will also be needed.

Interface Design and Controls

Designing a good interface for users leads to greater satisfaction with the system and better security. Programming languages and software development platforms often include user interface design and controls tools.[2] Microsoft, for example, has a number of interface designs and controls for Windows and its other software products available to systems developers. Other software companies also provide interface design and control tools. A *sign-on procedure* requiring identification numbers, passwords, and other safeguards is available with most systems to improve security and prevent unauthorized use. With a *menu-driven system* (see Figure 13.2), users simply pick what they want to do from a list of alternatives. Most people can easily operate these types of systems. In addition, many designers incorporate a *help facility* into the system or applications program. When users want to know more about a program or feature or what type of response is expected, they can activate the help facility. Computer programmers can develop *lookup tables* to simplify and shorten data entry. For example, if you are entering a sales order for a company, you can type its abbreviation, such as ABCO. The program searches the customer table, normally stored on a disk, the Internet, or other storage device, and looks up all the information pertaining to the company abbreviated ABCO that you need to complete the sales order.

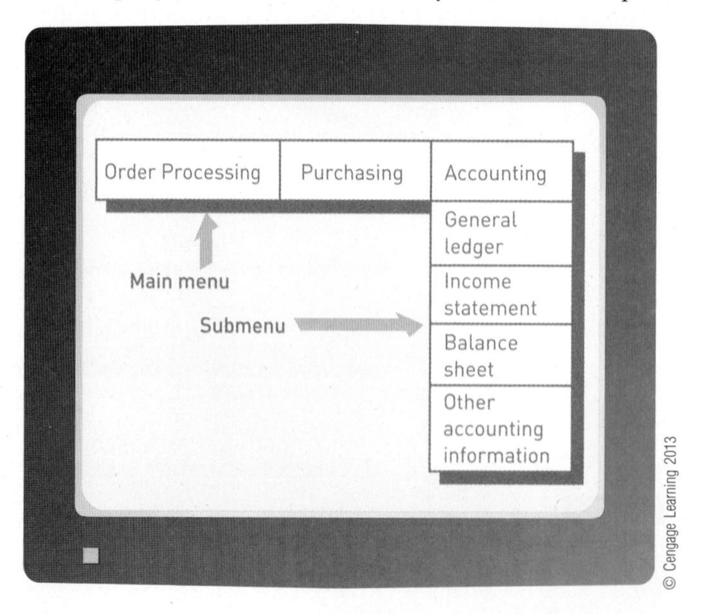

FIGURE **13.2**

Menu-driven system

A menu-driven system allows you to choose what you want from a list of alternatives.

Design of System Security and Controls

After specifying security features during the logical design phase discussed above, designers must develop specific system security and controls for all aspects of the IS, including hardware, software, database systems, telecommunications, and Internet operations. These key considerations involve error prevention, detection, and correction; disaster planning and recovery; and systems controls. Designing security systems is always a top concern for any systems development effort. You want secure systems without the burden of a large number of identification numbers and passwords for different applications. The U.S. Commerce Department is trying to avoid having its users remember many identification numbers and passwords. Instead, the organization hopes to design a security system that uses smart cards, fingerprint readers, or digital tokens to gain access to computers, phones, and other devices.[3] Once you are signed in, you will have access to all applications available on that device.

Designing security controls and procedures into the use of smartphones and other mobile devices can be a challenge for many organizations.[4] Employees want to be able to do their work using their smartphones, tablet computers, and other mobile devices at work and while traveling. Corporations want to make sure that the usage of these devices is secure. Stolen laptop computers and other mobile devices have been a major cause of identity theft for individuals. It has also resulted in the loss of corporate secrets and procedures. According to one survey, over 50 percent of respondents indicated that potential security problems with mobile devices have prevented them from using mobile devices to a greater extent to perform corporate work. To combat these problems, systems developers are installing software to encrypt the data on mobile devices and require IDs and passwords to gain access to them. Companies are also protecting their computers with software and firewalls to block unauthorized people from gaining access to corporate data and programs using corporate-issued mobile devices.

Preventing, Detecting, and Correcting Errors

The most cost-effective time to deal with potential errors is early in the design phase. Every possibility should be considered, even minor problems. One company, for example, had installed backup electrical generators in case of a power failure. When a fuel truck crashed near its facility and spilled its flammable cargo, the city shut down all power to the area and wouldn't let the company use its electrical generators, fearing they could cause an explosion or severe fire. This relatively minor incident of a truck crash completely shut down the company's IS center until the spill could be cleaned up. In addition to minor problems, other important security and control measures, including disaster planning and recovery and adequate backup procedures, must be considered.

Disaster Planning and Recovery

Disaster planning is the process of anticipating and providing for disasters.[5] The purpose of disaster planning is to ensure *business continuity*, where the organization's critical applications, data, and information systems are continuously available. The earthquake, tsunami, and resulting problems with several of Japan's nuclear plants were devastating, but technology helped people stay in touch and deal with the crisis.[6] Keeping its Ginza store open, Apple Computer was able to provide critical emergency communications to employees, customers, and others that were able to get to the computer store through email, Facebook, Twitter, and other Internet sites. Some people actually used the store as an emergency shelter. Disasters can range from a minor problem to a major catastrophe. A disaster can be an act of nature (a flood, fire, or earthquake) or a human act (terrorism, human error, labor unrest, or erasure of an important file). Disaster planning often focuses primarily on two issues: maintaining the integrity of corporate information and keeping the information system running until normal operations can be resumed.

disaster recovery: The implementation of the disaster plan.

Disaster recovery is the implementation of the disaster plan. Alabama's Troy University, for example, upgraded its disaster recovery system to help prevent and recover from potential disasters, including hurricanes.[7] According to Troy's chief security and technology officer, "We don't want our services to go down for a second." Organizations can spend from 5 to 10 percent of their total annual budget on disaster prevention and recovery.[8] According to a Forrester report, "It's much more likely that a CIO or other executive will approve a budget for an upgrade if you can explain that in the next five years there is a 20% probability that a severe winter storm will knock out power to the data center and cost $500,000 in lost revenue and employee productivity." According to a research analyst at Gartner, "Increasingly, there's an understanding and an appreciation that we have to protect more of our assets—such as branch offices, remotes, desktops, laptops, and even test and development data."[9] See Figure 13.3.

DENNIS M. SABANGAN/EPA/Landov

FIGURE 13.3
Disaster recovery efforts

The primary tools used in disaster planning and recovery are hardware, software, and database, telecommunications, and personnel backups.[10] Most of these systems were discussed in Part 2 on information technology concepts. According to a Gartner analyst, "Backup is really the insurance policy that you hope you never have to cash."[11] For some companies, personnel backup can be critical. Without IS employees, the IS department can't function. For hardware, hot and cold sites can be used as backup. A duplicate, operational hardware system that is ready for use (or immediate access to one through a specialized vendor) is an example of a **hot site**. If the primary computer has problems, the hot site can be used immediately as a backup. However, the hot site cannot be affected by the same disaster. Another approach is to use a **cold site**, also called a *shell*, which is a computer environment that includes rooms, electrical service, telecommunications links, data storage devices, and similar equipment. If a primary computer has

hot site: A duplicate, operational hardware system or immediate access to one through a specialized vendor.

cold site: A computer environment that includes rooms, electrical service, telecommunications links, data storage devices, and the like; it is also called a *shell*.

a problem, backup computer hardware is brought into the cold site, and the complete system is made operational. Files and databases can be protected by making a copy of all files and databases changed during the last few days or the last week, a technique called **incremental backup**. This approach to backup uses an **image log**, which is a separate file that contains only changes to applications. Whenever an application is run, an image log is created that contains all changes made to all files. If a problem occurs with a database, an old database with the last full backup of the data, along with the image log, can be used to recreate the current database. Organizations can also hire outside companies to help them perform disaster planning and recovery. EMC, for example, offers data backup in its RecoverPoint product.[12] For individuals and for some applications, backup copies of important files can be placed on the Internet. *Failover is* another approach to backup.[13] When a server, network, or database fails or is no longer functioning, failover automatically switches applications and other programs to a redundant or replicated server, network, or database so there is no interruption of service. SteelEye's LifeKeeper (*www.steeleye.com*) and Continuous Protection by NeverFail (*www.neverfailgroup.com*) are examples of failover software. Failover is especially important for applications that must be operational at all times.[14]

incremental backup: A backup copy of all files changed during the last few days or the last week.

image log: A separate file that contains only changes to applications.

Companies that suffer a disaster can employ a disaster recovery service, which can secure critical data backup information. These service companies can also provide a facility from which to operate and communications equipment to stay in touch with customers.

Systems Controls

Most IS departments establish tight **systems controls**, which are rules and procedures to maintain data security. Many companies are also developing systems controls to preserve important information in case a company faces legal action and is required by law to produce corporate e-mails, documents, and other important data, often called *electronic discovery* or *e-discovery*.[15] Many types of systems controls can be developed, documented, implemented, and reviewed. These controls touch all aspects of the organization (see Table 13.1).

systems controls: Rules and procedures to maintain data security.

After controls are developed, they should be documented in standards manuals that indicate how the controls are to be implemented. They should then be implemented and frequently reviewed. It is common practice to measure the extent to which control techniques are used and to take action if the controls have not been implemented. Organizations often have *compliance departments* to make sure the IS department is adhering to its systems controls along with all local, state, and federal laws and regulations.[16]

TABLE 13.1 Using systems controls to enhance security

Controls	Description
Input controls	Maintain input integrity and security. Their purpose is to reduce errors while protecting the computer system against improper or fraudulent input. Input controls range from using standardized input forms to eliminating data-entry errors and using tight password and identification controls.
Processing controls	Deal with all aspects of processing and storage. The use of passwords and identification numbers, backup copies of data, and storage rooms that have tight security systems are examples of processing and storage controls.
Output controls	Ensure that output is handled correctly. In many cases, output generated from the computer system is recorded in a file that indicates the reports and documents that were generated, the time they were generated, and their final destinations.
Database controls	Deal with ensuring an efficient and effective database system. These controls include the use of identification numbers and passwords, without which a user is denied access to certain data and information. Many of these controls are provided by database management systems.
Telecommunications controls	Provide accurate and reliable data and information transfer among systems. Telecommunications controls include firewalls and encryption to ensure correct communication while eliminating the potential for fraud and crime.
Personnel controls	Make sure that only authorized personnel have access to certain systems to help prevent computer-related mistakes and crime. Personnel controls can involve the use of identification numbers and passwords that allow only certain people access to particular data and information. ID badges and other security devices (such as smart cards) can prevent unauthorized people from entering strategic areas in the information systems facility.

A compliance department makes sure the IS department adheres to its systems controls and to all local, state, and federal laws and regulations.

ENVIRONMENTAL DESIGN CONSIDERATIONS

environmental design: Also called *green design*, it involves systems development efforts that slash power consumption, require less physical space, and result in systems that can be disposed of in a way that doesn't negatively affect the environment.

Developing new systems and modifying existing ones in an environmentally sensitive way are becoming important for many IS departments.[17] **Environmental design,** also called *green design,* involves systems development efforts that slash power consumption, require less physical space, and result in systems that can be disposed of in a way that doesn't negatively affect the environment.[18]

Today, companies are using innovative ways to design efficient systems and operations, including using virtual servers to save energy and space, pushing cold air under data centers to cool equipment, using software to efficiently control cooling fans, building facilities with more insulation, and even collecting rain water from roofs to cool equipment.[19] The large accounting firm KPMG was able to slash its energy usage for its servers by about 50 percent using some of these techniques.[20] Facebook, with the help of its business partners, developed a data center in rural Prineville, Oregon, to be more energy efficient by using the latest computer chips and servers that are lighter and easier to maintain.[21] According to the vice president of technical operations at Facebook, "These servers are 38% more efficient than the servers we were buying previously." The servers were also less expensive to purchase than the previous ones. Nissan used virtual server technology to reduce its number of servers from about 160 to about 30 in its Tennessee plants, saving energy and money.[22] Reducing the number of servers has slashed electricity costs by more than 30 percent for the automotive company. VistaPrint a graphics design and printing company, switched from traditional servers to virtual servers and saved about $500,000 in electricity costs over a three-year period, representing a 75 percent reduction in energy usage.[23]

The Leadership in Energy and Environmental Design (LEED) rating system was developed by the U.S. Green Building Council and includes a number of standards for the construction and operation of buildings.[24] Today, solar panels, gas turbines, and fuel cells are making IS departments and data centers more energy efficient. These energy-saving approaches can also give IS departments a significant return on investment. According to the CIO of the North County Transit District in San Diego, "Solar is what has made the [data center] project even have an ROI."[25]

Solar panels are helping to make IS departments and data centers more energy efficient and to provide IS departments a significant return on investment.

Many companies are developing products and services to help save energy. PC companies, such as Hewlett-Packard and others, are designing computers that use less power and are made from recycled materials.[26] One Japanese firm is designing a new computer chip that could potentially reduce power consumption by 50 percent and almost double battery life for mobile devices, including laptop computers, tablet computers, and smartphones.[27] According to the president of International Business Strategies, "The technology could have a very, very significant position." Voltaic Generator has developed a solar PC case that charges batteries from sunlight and other light sources.[28] The solar-powered bag can power computers, cell phones, and other electronic devices. Environmental design also involves developing software and systems that help organizations reduce power consumption for other aspects of their operations. Carbonetworks and Optimum Energy, for example, have developed software products that reduce energy costs by helping companies determine when and how to use electricity.[29]

Hewlett-Packard, Dell Computer, and others have developed procedures and machines to dispose of old computers and computer equipment in environmentally friendly ways. The city of Minneapolis recycled and refurbished more than 1,000 PCs, donating them to low-income families.[30] Venjuvo and other companies recycle old electronics equipment, offering cash in some cases, depending on the age and type of equipment.[31] Old computers and computer equipment are fed into machines that shred them into small pieces and sort them into materials that can be reused. The process is often called *green death*. The U.S. government has a tool called the *Electronic Product Environmental Assessment Tool (EPEAT)* to analyze the energy usage of new systems.[32] The U.S. Department of Energy rates products with the *Energy Star* designation to help people select products that save energy.[33] Today, utility companies are providing their corporate and individual customers with "smart meters" and specialized software that can help them reduce their power consumption and electric bills.

Utility companies provide customers with smart meters that can help reduce power consumption.

Generating Systems Design Alternatives

Generating systems design alternatives often involves getting the involvement of a single vendor or multiple vendors. If the new system is complex, the original development team might want to involve other personnel in generating alternative designs. In addition, if new hardware and software are to be acquired from an outside vendor, a formal request for proposal (RFP) can be made.

Request for Proposals

request for proposal (RFP): A document that specifies in detail required resources such as hardware and software.

The **request for proposal (RFP)** is a document that specifies in detail required resources such as hardware and software.[34] The RFP is an important document for many organizations involved with large, complex systems development efforts. Smaller, less-complex systems often do not require an RFP. A company that is purchasing an inexpensive piece of software that will run on existing hardware, for example, might not need to go through a formal RFP process.

In some cases, the RFP is part of the vendor contract. The Table of Contents for a typical RFP is shown in Figure 13.4.

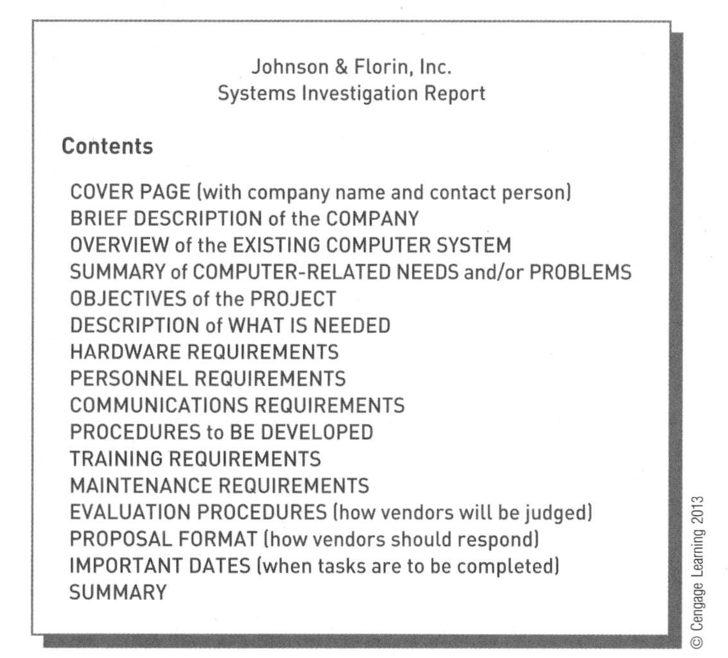

Johnson & Florin, Inc.
Systems Investigation Report

Contents

COVER PAGE (with company name and contact person)
BRIEF DESCRIPTION of the COMPANY
OVERVIEW of the EXISTING COMPUTER SYSTEM
SUMMARY of COMPUTER-RELATED NEEDS and/or PROBLEMS
OBJECTIVES of the PROJECT
DESCRIPTION of WHAT IS NEEDED
HARDWARE REQUIREMENTS
PERSONNEL REQUIREMENTS
COMMUNICATIONS REQUIREMENTS
PROCEDURES to BE DEVELOPED
TRAINING REQUIREMENTS
MAINTENANCE REQUIREMENTS
EVALUATION PROCEDURES (how vendors will be judged)
PROPOSAL FORMAT (how vendors should respond)
IMPORTANT DATES (when tasks are to be completed)
SUMMARY

FIGURE **13.4**

Typical table of contents for a request for proposal

Financial Options

When acquiring computer systems, several choices are available, including purchase, lease, or rent. Cost objectives and constraints set for the system play a significant role in the choice, as do the advantages and disadvantages of each. In addition, traditional financial tools, including net present value and internal rate of return, can be used. Table 13.2 summarizes the advantages and disadvantages of these financial options.

Determining which option is best for a particular company in a given situation can be difficult. Financial considerations, tax laws, the organization's policies, its sales and transaction growth, marketplace dynamics, and the organization's financial resources are all important factors. In some cases, lease or rental fees can amount to more than the original purchase price after a few years. Financial options should also include the use of virtualization, cloud computing, SaaS, DaaS, and similar approaches.

TABLE **13.2** Advantages and disadvantages of acquisition options

Renting (Short-Term Option)	
Advantages	**Disadvantages**
No risk of obsolescence	No ownership of equipment
No long-term financial investment	High monthly costs
No initial investment of funds	Restrictive rental agreements
Maintenance usually included	
Leasing (Intermediate to Long-Term Option)	
Advantages	**Disadvantages**
No risk of obsolescence	High cost of canceling lease
No long-term financial investment	Longer time commitment than renting
No initial investment of funds	No ownership of equipment
Less expensive than renting	
Purchasing (Long-Term Option)	
Advantages	**Disadvantages**
Total control over equipment	High initial investment
Can sell equipment at any time	Additional cost of maintenance
Can depreciate equipment	Possibility of obsolescence
Low cost if owned for a number of years	Other expenses, including taxes and insurance

Evaluating and Selecting a Systems Design

Evaluating and selecting the best design involves achieving a balance of system objectives that will best support organizational goals. Normally, evaluation and selection involve both a preliminary and a final evaluation before a design is selected.

preliminary evaluation: An initial assessment whose purpose is to dismiss the unwanted proposals; it begins after all proposals have been submitted.

final evaluation: A detailed investigation of the proposals offered by the vendors remaining after the preliminary evaluation.

A **preliminary evaluation** begins after all proposals have been submitted. The purpose of this evaluation is to dismiss unwanted proposals. Several vendors can usually be eliminated by investigating their proposals and comparing them with the original criteria. The **final evaluation** begins with a detailed investigation of the proposals offered by the remaining vendors. The vendors should be asked to make a final presentation and to fully demonstrate the system. The demonstration should be as close to actual operating conditions as possible. Figure 13.5 illustrates the evaluation process.

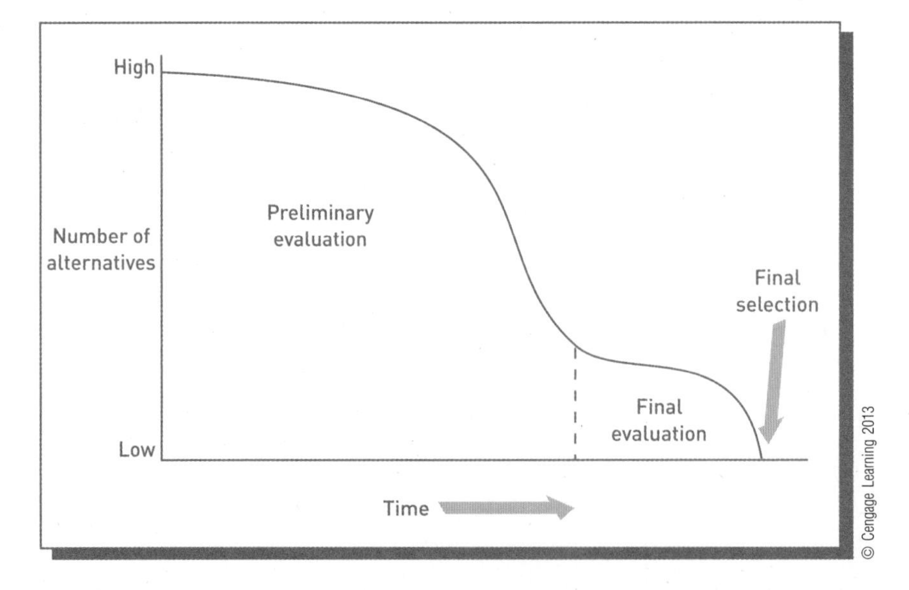

FIGURE **13.5**

Stages in preliminary and final evaluations

The number of possible alternatives decreases as the firm gets closer to making a final decision.

Evaluation Techniques

The exact procedure used to make the final evaluation and selection varies from one organization to the next. Some were first introduced in Chapter 2, including return on investment (ROI), earnings growth, market share, customer satisfaction, and total cost of ownership (TCO). In addition, four other approaches are commonly used: group consensus, cost/benefit analysis, benchmark tests, and point evaluation.

Group Consensus

group consensus: Decision making by a group that is appointed and given the responsibility of making the final evaluation and selection.

In **group consensus**, a decision-making group is appointed and given the responsibility of making the final evaluation and selection.[35] Usually, this group includes the members of the development team who participated in either systems analysis or systems design.

Cost/Benefit Analysis

cost/benefit analysis: An approach that lists the costs and benefits of each proposed system. After they are expressed in monetary terms, all the costs are compared with all the benefits.

Cost/benefit analysis is an approach that lists the costs and benefits of each proposed system.[36] After they are expressed in monetary terms, all the costs are compared with all the benefits. Table 13.3 lists some of the typical costs and benefits associated with the evaluation and selection procedure. This approach is used to evaluate options whose costs can be quantified, such as which hardware or software vendor to select.

TABLE **13.3** Cost/benefit analysis table

Costs	Benefits
Development costs	Reduced costs
Personnel	Fewer personnel
Computer resources	Reduced manufacturing costs
	Reduced inventory costs
	More efficient use of equipment
	Faster response time
	Reduced downtime or crash time
	Less spoilage
Fixed Costs	**Increased Revenues**
Computer equipment	New products and services
Software	New customers
One-time license fees for software and maintenance	More business from existing customers
	Higher price as a result of better products and services
Operating Costs	**Intangible Benefits**
Equipment lease and/or rental fees	Better public image for the organization
Computer personnel (including salaries, benefits, etc.)	Higher employee morale
	Better service for new and existing customers
Electric and other utilities	The ability to recruit better employees
Computer paper, tape, and disks	Position as a leader in the industry
Other computer supplies	System easier for programmers and users
Maintenance costs	
Insurance	

Benchmark Tests

benchmark test: An examination that compares computer systems operating under the same conditions.

A **benchmark test** is an examination that compares computer systems operating under the same conditions. Most computer companies publish their own benchmark tests, but some forbid disclosure of benchmark tests without prior written approval. Thus, one of the best approaches is for an organization to develop its own tests and then use them to compare the equipment it is considering. This approach might be used to compare the end-user system response time on two similar systems. Several independent companies and journals also rate computer systems.

Point Evaluation

point evaluation system: An evaluation process in which each evaluation factor is assigned a weight in percentage points based on importance. Then each proposed information system is evaluated in terms of this factor and given a score, such as one ranging from 0 to 100. The scores are totaled, and the system with the greatest total score is selected.

One of the disadvantages of cost/benefit analysis is the difficulty of determining the monetary values for all the benefits. An approach that does not use monetary values is a **point evaluation system**. Each evaluation factor is assigned a weight in percentage points based on importance. Then each proposed information system is evaluated in terms of this factor and given a score, such as one ranging from 0 to 100, where 0 means that the alternative does not address the feature at all and 100 means that the alternative addresses the feature perfectly. The scores are totaled, and the system with the greatest total score is selected. When using point evaluation, an organization can list and evaluate hundreds of factors. Figure 13.6 shows a simplified version of this process. This approach is used when there are many options to be evaluated, such as which software best matches the needs of a particular business.

FIGURE **13.6**

Illustration of the point evaluation system

In this example, software has been given the most weight (40 percent), compared with hardware (35 percent) and vendor support (25 percent). When system A is evaluated, the total of the three factors amounts to 82.5 percent. System B's rating, on the other hand, totals 86.75 percent, which is closer to 100 percent. Therefore, the firm chooses system B.

		System A			System B		
Factor's importance		*Evaluation*		*Weighted evaluation*	*Evaluation*		*Weighted evaluation*
Hardware	35%	95	35%	33.25	75	35%	26.25
Software	40%	70	40%	28.00	95	40%	38.00
Vendor support	25%	85	25%	21.25	90	25%	22.50
Totals	100%			82.5			86.75

Freezing Design Specifications

Near the end of the design stage, some organizations prohibit further changes in the design of the system. Freezing systems design specifications means that the user agrees in writing that the design is acceptable (see Figure 13.7). Other

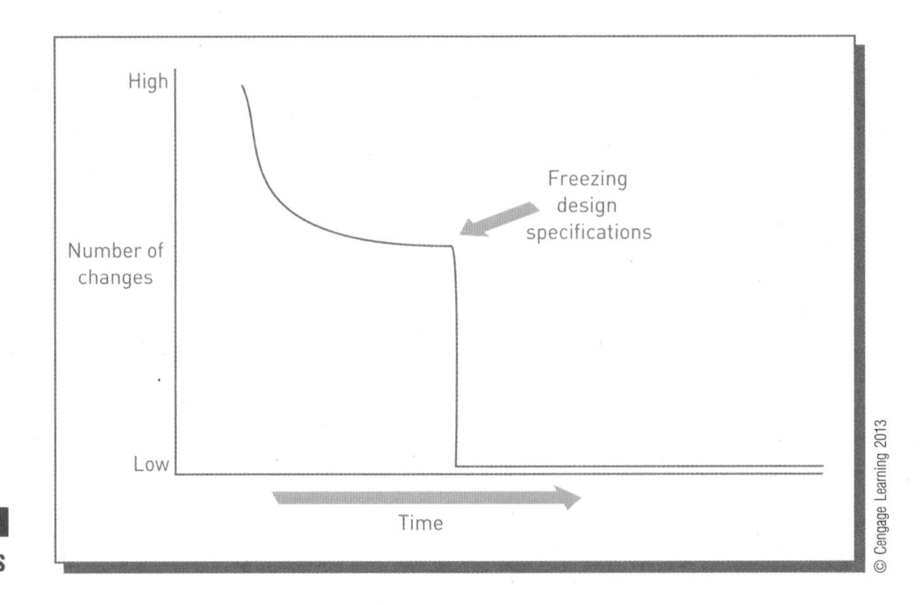

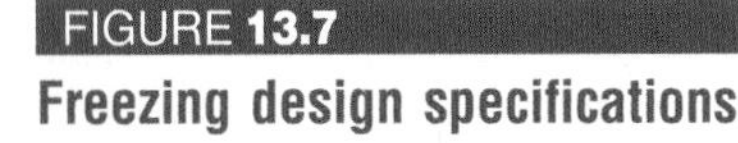
FIGURE **13.7**

Freezing design specifications

organizations, however, allow or even encourage design changes. These organizations often use agile or rapid systems development approaches, introduced in Chapter 12.

The Contract

One of the most important steps in systems design is to develop a good contract if new computer facilities are being acquired.[37] According to a technology lawyer and consultant, "Customers don't pay enough attention to the contract they're signing. In many cases, they ignore the very language that defines the scope of the software offering or implementation."[38] Organizations that use the cloud-computing approach need to take special precautions in signing contracts with cloud-computing providers, including how privacy is protected, how the organization can comply with various laws and regulations when using cloud computing, where the cloud-computing servers and computers are located in the world, how discovery is handled if there is a lawsuit, and the security of the data stored on cloud computers.[39] A good contract should have provisions for monitoring systems development progress, ownership and property rights of the new or modified system, contingency provisions in case something doesn't work as expected, and dispute resolution if something goes wrong. Typically, the request for proposal becomes part of the contract. This saves a considerable amount of time in developing the contract because the RFP specifies in detail what is expected from the vendors.

The Design Report

System specifications are the final results of systems design. They include a technical description that details system outputs, inputs, and user interfaces as well as all hardware, software, databases, telecommunications, personnel, and procedure components and the way these components are related. The specifications are contained in a **design report**, which is the primary result of systems design. The design report reflects the decisions made for systems design and prepares the way for systems implementation. The contents of the design report are summarized in Figure 13.8.

design report: The primary result of systems design, reflecting the decisions made for systems design and preparing the way for systems implementation.

Johnson & Florin, Inc.
Systems Design Report

Contents

PREFACE
EXECUTIVE SUMMARY of SYSTEMS DESIGN
REVIEW of SYSTEMS ANALYSIS
MAJOR DESIGN RECOMMENDATIONS
- Hardware design
- Software design
- Personnel design
- Communications design
- Database design
- Procedures design
- Training design
- Maintenance design

SUMMARY of DESIGN DECISIONS
APPENDICES
GLOSSARY of TERMS
INDEX

FIGURE 13.8
Typical table of contents for a systems design report

SYSTEMS IMPLEMENTATION

systems implementation: A stage of systems development that includes hardware acquisition, programming and software acquisition or development, user preparation, hiring and training of personnel, site and data preparation, installation, testing, start-up, and user acceptance.

After the information system has been designed, a number of tasks must be completed before the system is installed and ready to operate. This process, called **systems implementation**, includes hardware acquisition, programming and software acquisition or development, user preparation, hiring and training of personnel, site and data preparation, installation, testing, start-up, and user acceptance. According to a survey of CIOs, not being able to implement new or modified systems was the most important concern for today's IS departments.[40] The typical sequence of systems implementation activities is shown in Figure 13.9.

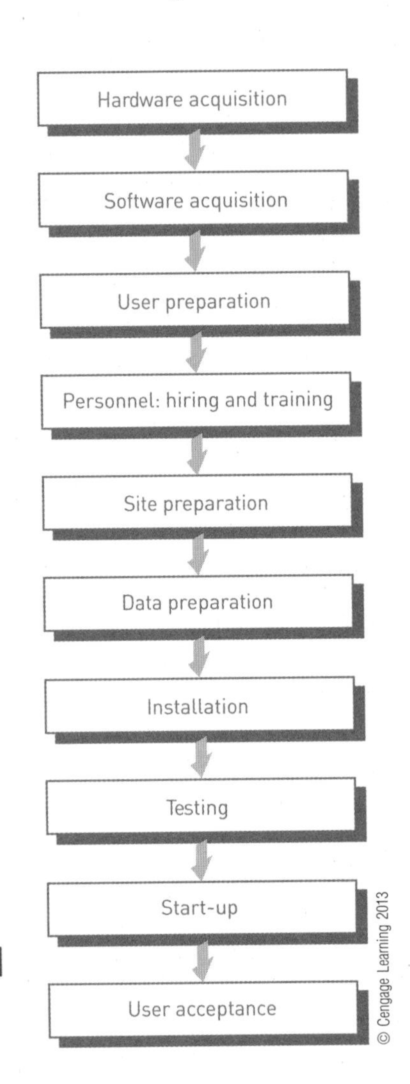

FIGURE **13.9**

Typical steps in systems implementation

Virtualization, first introduced in Chapter 3, has had a profound impact on many aspects of systems implementation. Virtualization is being used to acquire hardware, software, databases, and other capabilities. IBM has developed the VMControl tool to help companies implement virtual servers from various vendors.[41] Southern Company, a large electricity producer serving Alabama, Georgia, Mississippi, and Florida, ran a pilot project to acquire desktop computers that used virtualization to reduce costs and simplify the management and servicing of its desktop computers.[42] According to the director of client services for Southern Company, "I think it's a much better architecture. It's more stable. It's more secure. It's easier to manage." We start our discussion of systems implementation with hardware acquisition.

Acquiring Hardware from an IS Vendor

To obtain the components for an information system, organizations can purchase, lease, or rent computer hardware and other resources from an IS vendor. An *IS vendor* is a company that offers hardware, software, telecommunications systems, databases, IS personnel, or other computer-related resources. Types of IS vendors include general computer manufacturers (such as IBM and Hewlett-Packard), small computer manufacturers (such as Dell and Sony), peripheral equipment manufacturers (such as Epson and SanDisk), computer dealers and distributors (such as The Shack and Best Buy), and chip makers (such as Intel and AMD). Some of the most successful vendors include IBM (hardware and other services), Oracle (databases), Apple (personal and tablet computers), Microsoft (software), Accenture (IS consulting), and many others. In one survey, it was predicted that most companies will acquire and support tablet computers and applications by 2013.[43] Many companies have multiple hardware vendors, but managing them can be difficult. Different vendors must compete against each other to get a contract with an organization. Then, the selected vendors must work together to develop an effective information system at a good price. Indeed, open communications among the vendors is critical. Some CIOs put together vendor teams that work together to solve current and future problems for the organization.[44]

Spencer Platt/Getty Images

IBM is an IS vendor that offers hardware, software, and IS personnel.

In addition to buying, leasing, or renting computer hardware, companies can pay only for the computing services that it uses. Called "pay-as-you-go," "on-demand," or "utility" computing, this approach requires an organization to pay only for the computer power it uses, just as it would pay for a utility such as electricity.[45] Companies like IBM and Hewlett-Packard offer their clients a "capacity-on-demand" approach in which organizations pay according to the computer resources actually used, including processors, storage devices, and network facilities.[46]

Companies can also purchase used computer equipment. Popular Internet auction sites sometimes sell more than $1 billion of computer-related equipment annually, and companies can purchase equipment for about 20 or 30 cents on the dollar. However, buyers need to beware: prices are not always low, and equipment selection can be limited on Internet auction sites. Also, used equipment rarely comes with warranties or support. Organizations that use the cloud-computing approach for some applications don't have to worry about acquiring hardware for these applications.[47] They also don't have to worry about acquiring software for these applications, discussed next.

Acquiring Software: Make or Buy?

make-or-buy decision: The decision regarding whether to obtain the necessary software from internal or external sources.

As with hardware, application software can be acquired in several ways. As previously mentioned, it can be purchased from external developers or developed in-house.[48] This decision is often called the **make-or-buy decision**. A comparison of the two approaches is shown in Table 13.4. Today, most software is purchased.[49] Individuals and organizations can purchase software from a number of online application stores (app stores) or retail stores located around the country.[50] Market.Android.com, for example, is Google's application store. As mentioned in Chapter 4, companies can also purchase open-source software from Red Hat and many other open-source software companies, including programs that use the Internet and cloud-computing approaches.[51] Implementation is not complete, however, when the software code is finished. According to Sharon Donovan-Hart, CIO of State Street, "A lot of people think their job is complete once their code is in production. I see it as when users are getting benefit of a system."[52] Some companies, such as Secure by Design, offer automated software installation and updating services.[53] Using these tools, companies can easily install and update software designed to run on Windows, Linux, and other operating systems.

TABLE 13.4 Comparison of off-the-shelf and developed software

Factor	Developed (Make)	Off-the-Shelf (Buy)
Cost	High cost	Low cost
Needs	Custom software more likely to satisfy your needs	Might not get what you need
Quality	Quality can vary depending on the programming team	Can assess the quality before buying
Speed	Can take years to develop	Can acquire it now
Competitive advantage	Can develop a competitive advantage with good software	Other organizations can have the same software and same advantage

Externally Acquired Software and Software as a Service (SaaS)

A company planning to purchase or lease software from an outside company has many options. Commercial off-the-shelf development is often used. The *commercial off-the-shelf (COTS)* development process involves the use of commonly available products from software vendors. It combines software from various vendors into a finished system.[54] In many cases, it is necessary to write some original software from scratch and combine it with purchased or leased software. For example, a company can purchase or lease software from several software vendors and combine it into a single software program.

Organizations are also acquiring more virtualization software from software vendors, including operating systems and application software. Windows Server, for example, provides virtualization tools that allow multiple operating systems to run on a single server. Virtualization software such as VMware is being used by businesses to safeguard private data. Kindred Healthcare used VMware on its server to run hundreds of virtual Windows PC desktops that are accessed by mobile computers throughout the organization.[55]

Businesses are using virtualization software such as VMware to safeguard private data.

As mentioned in Chapter 4, *Software as a Service (SaaS)* allows businesses to subscribe to Web-delivered application software by paying a monthly service charge or a per-use fee.[56] Instead of acquiring software externally from a traditional software vendor, SaaS allows individuals and organizations to access needed software applications over the Internet. Instead of making its own software, Tidewell, which is a hospice that serves about 8,000 Florida families, used the SaaS approach to provide valuable information to its health care providers and clients.[57] This approach has been substantially less expensive than developing its own software in-house. The Humane Society of the United States used a SaaS product called QualysGuard by Qualys to obtain and process credit-card contributions from donors.[58] Companies such as Google are using the cloud-computing approach to deliver word processing, spreadsheet programs, and other software over the Internet.

In-House Developed Software

Another option is to make or develop software internally. Some advantages inherent with in-house developed software include meeting user and organizational requirements and having more features and increased flexibility in terms of customization and changes. Software programs developed within a company also have greater potential for providing a competitive advantage because competitors cannot easily duplicate them in the short term. In-house developed software, however, is not always successful.[59] One study estimated that about 20 percent of software development projects failed to be delivered or accepted by users. If software is to be developed internally, a number of tools and techniques can be used. A few of the tools and techniques used to develop in-house software are briefly discussed below:

cross-platform development: A software development technique that allows programmers to develop programs that can run on computer systems having different hardware and operating systems, or platforms.

- **CASE and object-oriented approaches.** As mentioned in Chapter 12, CASE tools and the object-oriented approach are often used during software development.
- **Cross-platform development.** One software development technique, called **cross-platform development**, allows programmers to develop programs that can run on computer systems having different hardware and operating systems, or platforms. Web service tools, such as .NET by

Microsoft, introduced in Chapter 7, are examples. With cross-platform development, the same program can run on both a personal computer and a mainframe or on two different types of PCs.

- **Integrated development environment.** The combination of the tools needed for programming with a programming language in one integrated package is called an **integrated development environment (IDE)**. An IDE allows programmers to use simple screens, customized pull-down menus, and graphical user interfaces. Visual Studio from Microsoft is an example of an IDE. Oracle Designer, which is used with Oracle's database system, is another example of an IDE. Eclipse Workbench (*www.eclipse-workbench.com*) supports IDEs that can be used with the Java, C, and C++ programming languages. Eclipse Workbench includes a debugger and a compiler, along with other tools. Increasingly, IDEs are being developed over the Internet, using cloud computing.[60]

integrated development environments (IDEs): A development approach that combines the tools needed for programming with a programming language in one integrated package.

- **Documentation.** With internally developed software, documentation is always important. **Technical documentation** is written details used by computer operators to execute the program and by analysts and programmers to solve problems or modify the program. In technical documentation, the purpose of every major piece of computer code is written out and explained. Key variables are also described. **User documentation** is developed for the people who use the program. In easy-to-understand language, this type of documentation shows how the program can and should be used. Incorporating a description of the benefits of the new application into user documentation can help stakeholders understand the reasons for the program and can speed user acceptance.

technical documentation: Written details used by computer operators to execute the program and by analysts and programmers to solve problems or modify the program.

user documentation: Written descriptions developed for people who use a program; in easy-to-understand language, it shows how the program can and should be used.

Acquiring Database and Telecommunications Systems

Because databases are a blend of hardware and software, many of the approaches discussed earlier for acquiring hardware and software also apply to database systems, including open-source databases. Thus, with its cool temperatures and low utility rates, Wyoming is hoping to lure database centers to the state.[61] *Virtual databases* and *Database as a Service* (*DaaS*) are popular ways to acquire database capabilities.[62] Sirius XM Radio, Bank of America, and Southwest Airlines, for example, use the DaaS approach to manage many of their database operations from the Internet. In another case, a brokerage company was able to reduce storage capacity by 50 percent using database virtualization.

With the increased use of e-commerce, the Internet, intranets, and extranets, telecommunications is one of the fastest-growing applications for today's organizations. Like database systems, telecommunications systems require a blend of hardware and software. For personal computer systems, the primary pieces of hardware are modems and routers. For client/server and mainframe systems, the hardware can include multiplexers, concentrators, communications processors, and a variety of network equipment. Communications software will also have to be acquired. Again, the earlier discussion on acquiring hardware and software also applies to the acquisition of telecommunications hardware and software. As discussed in Chapter 12 and previous chapters, individuals and organizations are using the Internet and cloud computing more than ever to implement many new systems development efforts. Systems analysts and programmers are also starting to use the Internet to develop applications.

User Preparation

User preparation is the process of readying managers, decision makers, employees, other users, and stakeholders for the new systems. This activity is an important but often ignored area of systems implementation. When a new operating

user preparation: The process of readying managers, decision makers, employees, other users, and stakeholders for the new system.

Kids and Programming

In this chapter, you read about computer programming as a professional activity. The great majority of the programs you use or interact with on a daily basis are produced by professional programmers. Your computer's operating system, the word processor you use to write term papers, the spreadsheet program you use to track your expenses, and the browser you use to surf the Web—all of these were written by pros.

What, however, would be the impact on society if everyone could develop his or her own apps? What if children were taught to program? Would it help their creativity, their career prospects, or anything else? Is Professor John Naughton, who teaches public understanding of technology at the U.K.'s Open University, right when he writes that, "Starting in primary school, children from all backgrounds and every part of the UK should have the opportunity to: learn some of the key ideas of computer science; understand computational thinking; learn to program; and have the opportunity to progress to the next level of excellence in these activities"? He concludes, "If we don't act now we will be short-changing our children.... They will grow up as passive consumers of closed devices and services, leading lives that are increasingly circumscribed by technologies created by elites working for huge corporations such as Google, Facebook and the like. We will, in effect, be breeding generations of hamsters for the glittering wheels of cages built by Mark Zuckerberg and his kind." Or, would computing become just another topic to stuff into school curricula, already overloaded with too much content and suffering from insufficient budgets?

While there may not yet be definitive answers to these questions, efforts to find those answers are under way. Many people and organizations are devoted to the cause of teaching children to program in the belief that the latter as well as society overall will benefit.

At the lowest level, being able to program requires a computer. Not all schools have enough computers for students in a classroom, let alone to ensure that children have them for homework. The Raspberry Pi project will bring a $35 computer to anyone who wants one. The project's computer runs a Linux operating system and connects to a TV set as a display and to any keyboard one happens to have. Storage comes in the form of inexpensive SD cards as used in most digital cameras. The SD cards that serious photographers discard as too small and outdated are more than sufficient.

The next requirement is for software. Programming needs a programming language. While children could in principle be taught a professional language such as C++, the amount of learning required until the pupil can do anything interesting makes such learning unsuited to all but the most highly motivated students. Instead, educators have designed languages such as Logo (the first such language), Simple, Kodu, and Scratch for this purpose. A pro wouldn't use these languages to code a word processor or a CRM application, but they let children develop simple games and animations soon after the children start using them. They teach, without the pupils realizing what's happening, the mental discipline of breaking a process down into logical components, planning the sequence of operations, and figuring out what data the program needs to accomplish its purpose.

Discussion Questions

1. This sidebar mentions two differences in teaching programming to adults and to children: the languages that are taught and the assignments that are given in teaching them. What other differences can you think of?

2. Do you think that knowing how to program helps you (if you know how) or would help you (if you don't) in your studies now? Why or why not?

Critical Thinking Questions

1. Do you agree with John Naughton that tomorrow's educated person will know how to program a computer?
2. Suppose a large school district decides to add programming to its elementary school curriculum. Few teachers can program. Few programmers can teach young children, and most of them now earn far more than schools can pay teachers. How could a school district deal with that problem?

SOURCES: Turtle Logo Web site, Codeplex, *logo.codeplex.com*, accessed April 1, 2012; Scratch Web site, Massachusetts Institute of Technology, *scratch.mit.edu*, accessed April 1, 2012; Kodu, Microsoft Research, *research.microsoft.com/en-us/projects/kodu*, accessed April 1, 2012; Naughton, J., "Why All Our Kids Should Be Taught How to Code," *The Guardian*, *www.guardian.co.uk/education/2012/mar/31/why-kids-should-be-taught-code*, March 31, 2012; Raspberry Pi Web site, *www.raspberrypi.org*, accessed March 31, 2012; Simple Web site, *www.simplecodeworks.com/website.html*, accessed April 1, 2012; Watters, A., "5 Tools to Introduce Programming to Kids," *Mind/Shift*, KQED, *blogs.kqed.org/mindshift/2011/05/5-tools-to-introduce-programming-to-kids*, May 18, 2011; Wayner, P., "Programming for Children, Minus Cryptic Syntax," *The New York Times*, *www.nytimes.com/2011/11/10/technology/personaltech/computer-programming-for-children-minus-cryptic-syntax.html*, November 9, 2011.

system or application software package is implemented, user training is essential. In some cases, companies decide not to install the latest software because the amount of time and money needed to train employees is too much. Because user training is so important, some companies provide training for their clients, including in-house, software, video, Internet, and other training approaches.

Providing users with proper training can help ensure that the information system is used correctly, efficiently, and effectively.

IS Personnel: Hiring and Training

Depending on the size of the new system, an organization might have to hire and, in some cases, train new IS personnel. An IS manager, systems analysts, computer programmers, data-entry operators, and similar personnel might be needed for the new or modified system.

The eventual success of any system depends on how it is used not only by the end users but also by the IS personnel within the organization. Training programs should be conducted for the IS personnel who will be using the computer system. These programs are similar to those for the users, although they can be more detailed in the technical aspects of the systems. Effective training will help IS personnel use the new system to perform their jobs and support other users in the organization. IBM and many other companies are using online and simulated training programs to cut training costs and improve effectiveness.

Site Preparation

site preparation: Preparation of the location of a new system.

The location of the new system needs to be prepared, a process called **site preparation**. For a small system, site preparation can be as simple as rearranging the furniture in an office to make room for a computer. With a larger system, this process is not so easy because it can require special wiring and air conditioning. A special floor, for example, might have to be built under which the cables connecting the various computer components are placed, and a new security system might be needed to protect the equipment. Today, developing IS sites that are energy efficient is important for most systems development implementations. Security is also important for site preparation. One company, for example, installed special security kiosks that lets visitors log on and request a meeting with a company employee. The employee can see the visitor on his or her computer screen and accept or reject the visitor. If the visitor is accepted, the kiosk prints a visitor pass.

Data Preparation

data preparation, or data conversion: Making sure that all files and databases are ready to be used with new computer software and systems.

Data preparation, or **data conversion**, involves making sure that all files and databases are ready to be used with new computer software and systems. If an organization is installing a new payroll program, the old employee-payroll data might have to be converted into a format that can be used by the new computer software or system. After the data has been prepared or converted, the computerized database system or other software will then be used to maintain and update the computer files.

Installation

Installation: The process of physically placing the computer equipment on the site and making it operational.

Installation is the process of physically placing the computer equipment on the site and making it operational. Although normally the manufacturer is responsible for installing computer equipment, someone from the organization (usually the IS manager) should oversee the process, making sure that all equipment specified in the contract is installed at the proper location. After the system is installed, the manufacturer performs several tests to ensure that the equipment is operating as it should.

Testing

unit testing: Testing of individual programs.

system testing: Testing the entire system of programs.

volume testing: Testing the application with a large amount of data.

integration testing: Testing all related systems together.

acceptance testing: Conducting any tests required by the user.

Good testing procedures are essential to make sure that the new or modified information system operates as intended. Inadequate testing can result in mistakes and problems.

Several forms of testing should be used, including testing each program (**unit testing**), testing the entire system of programs (**system testing**), testing the application with a large amount of data (**volume testing**), and testing all related systems together (**integration testing**), as well as conducting any tests required by the user (**acceptance testing**). Figure 13.10 lists the types of testing.

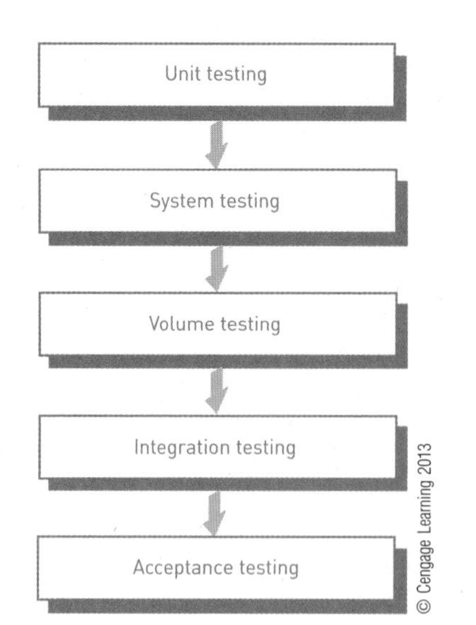

FIGURE 13.10
Types of testing

alpha testing: Testing an incomplete or early version of the system.

beta testing: Testing a complete and stable system by end users.

In addition to the previous forms of testing, there are different types of testing, among them alpha and beta testing. **Alpha testing** involves testing an incomplete or early version of the system; **beta testing** involves testing a complete and stable system by end users.[63] Alpha-unit testing, for example, is testing an individual program before it is completely finished.[64] Beta-unit testing, on the other hand, is performed after alpha testing when the individual program is complete and ready for use by end users. Vodafone used an innovative way to beta test its systems development efforts to deliver mobile applications.[65] It created a Web site called *www.betavine.com* that anyone, from customers to software professionals, can use to test mobile applications for Vodafone's networks and other wireless networks.

Unit testing is accomplished by developing test data that will force the computer to execute every statement in the program.[66] In addition, each program is tested with abnormal data to determine how it will handle problems.

System testing requires the testing of all the programs together. It is not uncommon for the output from one program to become the input for another. So, system testing ensures that program output can be used as input for another program within the system. Volume testing ensures that the entire system can handle a large amount of data under normal operating conditions. Integration testing ensures that the new programs can interact with other major applications. It also ensures that data flows efficiently and without error to other applications. For example, a new inventory control application might require data input from an older order-processing application. Integration testing would be done to ensure smooth data flow between the new and existing applications.

Finally, acceptance testing makes sure that the new or modified system is operating as intended. Run times, the amount of memory required, disk access methods, and more can be tested during this phase. Acceptance testing ensures that all performance objectives defined for the system are satisfied. Involving users in acceptance testing can help them understand and effectively interact with the new system. Acceptance testing is the final check of the system before start-up. In addition to the forms of testing described above, some companies are using *security testing* for critical software. Security testing makes sure that sensitive data remains protected from hackers and corporate spies.

Start-Up

start-up: The process of making the final tested information system fully operational.

Start-up, also called *cutover*, begins with the final tested information system. When start-up is finished, the system is fully operational. Start-up can be critical to the success of the organization because, if not done properly, the results can be disastrous. In one case, a small manufacturing company decided to stop an

INFORMATION SYSTEMS @ WORK

Implementation: It's about People, Too

ING "is a global financial institution of Dutch origin, offering banking, investments, life insurance and retirement services to ... a broad customer base." The company operates in 40 countries and has 85 million customers, over 100,000 employees, and €1,279 billion (about U.S. $1.7 trillion) in assets at the end of 2011. In short, it is one of the world's largest financial firms.

When ING decided to move to cloud computing, it realized that implementing such a major change would be a challenge. The company avoided a common mistake: too narrow a focus on the technology side of implementation. Instead, it recognized that, in addition to technology, people and business processes had to support the change to cloud computing. As Tony Kerison, ING's chief technical officer, put it, "Establishment of internal cloud competences is critical to be able to successfully leverage cloud services. Without these, and having the workforce understand them, we'll never be able to exploit the marketplace in the right way. Rolling out a comprehensive cloud and virtualization training program, tailored to ING's needs, is an integral part of our IT strategy."

Training users in cloud computing required ING's IT staff to work outside their comfort zone and deal with human issues such as the need to change motivations. Like most IT professionals, ING's staff didn't choose that profession because they wanted to deal with human issues or were competent in doing so. Unlike many organizations, though, ING was smart enough to recognize its shortcomings and to outsource its user training to an organization that specializes in training. The firm went to ITpreneurs of Rotterdam, The Netherlands, to help it through the transition to the cloud.

ITpreneurs and ING put together a multistep program to prepare ING for cloud computing. The program began with a video message from ING's chief technical officer. Senior management, including the head of ING Domestic Banking, also participated. Their participations showed that the move to the cloud was supported by the business side of the organization as well as by the company's technology experts.

ING then initiated internal marketing activities to announce the availability of the cloud training program. While the internal marketing campaign was under way, the workforce started taking the appropriate training. The training courses were integrated within the online ING learning platform, allowing learners from various locations to take the online courses at times most convenient to them.

ING professionals whose roles were affected by the cloud participated in the Cloud Competence Development Program. Those involved in the technology side of the cloud participated in further specialized courses. By September 2011, ING had trained close to 1,500 people to ensure that they could support ING's cloud strategy.

The result was that "ING's success in preparing the organization for the cloud lies in the fact that ING did not treat the journey towards the cloud as an IT-only issue, but rather involved the entire organization."

Sukhbir Jasuja, CEO of ITpreneurs, sums up the relationship: "Of course, having the opportunity to support an organization such as ING is great. But what is most inspiring is to see such a large organization being very focused and dedicated in making sure that before embarking on the cloud, the organization is ready for the cloud and that its workforce is able to support it."

Discussion Questions

1. Why was it important for ING's CIO to introduce the program to ING staff members? What did this step accomplish that ITpreneur trainers could not?
2. In this chapter, you learned about methods, such as pilot start-up and the phase-in approach, that decrease the technical risk of direct conversion. However, by lengthening the conversion process and/or requiring users to deal with parts of the old and new systems concurrently, these methods also increase the human risk of implementation: that users will not support the new system fully so the company will not reap its full benefits. How would you balance these two opposing considerations?

Critical Thinking Questions

1. Instead of outsourcing user training, ING could have trained some of its systems analysts in human issues and made them responsible for the training. Discuss the pros and cons of this approach compared with what the firm did.
2. Suppose you work for a company in ING's position and want to engage an outside firm to train your users in cloud computing. Many firms claim to be able to carry out this training. List several criteria you would use to choose among them.

SOURCES: ING Web site, *www.ing.com*, accessed March 31, 2012; "ING Building Cloud Competences," Cloud Credential Council, *www.cloudcredential.org/documents/Case-study.pdf*, September 2, 2011; ITpreneurs Web site, *www.itpreneurs.com*, accessed March 31, 2012.

accounting service it used to send out bills on the same day the firm was going to start its own program to send out bills to customers. The manufacturing company wanted to save money by using its own billing program developed by an employee. The new program didn't work, the accounting service wouldn't help because it was upset about being terminated, and the manufacturing company wasn't able to send out any bills to customers for more than three months. The company almost went bankrupt.

direct conversion: Stopping the old system and starting the new system on a given date; also called *plunge* or *direct cutover*.

Various start-up approaches are available (see Figure 13.11). **Direct conversion** (also called *plunge* or *direct cutover*) involves stopping the old system and starting the new system on a given date. Direct conversion is usually the least desirable approach because of the potential for problems and errors when the old system is shut off and the new system is turned on at the same instant.

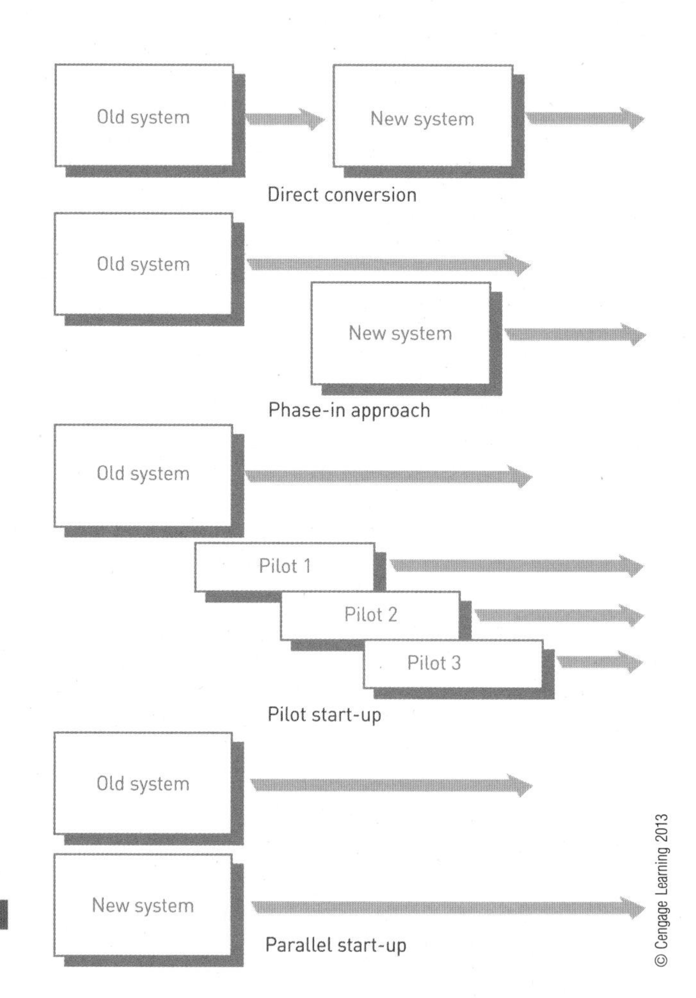

FIGURE **13.11**
Start-up approaches

phase-in approach: Slowly replacing components of the old system with those of the new one; this process is repeated for each application until the new system is running every application and performing as expected; it is also called a *piecemeal approach*.

The **phase-in approach** is a popular technique preferred by many organizations. In this approach, sometimes called a *piecemeal approach*, components of the new system are slowly phased in while components of the old one are slowly phased out. When everyone is confident that the new system is performing as expected, the old system is completely phased out. This gradual replacement is repeated for each application until the new system is running every application. In some cases, the phase-in approach can take months or years.

pilot start-up: Running the new system for one group of users rather than all users.

Pilot start-up involves running the new system for one group of users rather than all users. For example, a manufacturing company with many retail outlets throughout the country could use the pilot start-up approach and install a new inventory control system at one of the retail outlets. When this pilot retail outlet runs without problems, the new inventory control system can then be implemented at other retail outlets. Amazon, for example, used pilot testing for a new Web design that makes it easier for people using newer smartphones and tablet computers to access the company's Web site.[67] According to an Amazon representative, "We are continuing to roll out the new design to additional customers, but I can't speculate on when the new design will be live for everyone."

parallel start-up: Running both the old and new systems for a period of time and comparing the output of the new system closely with the output of the old system; any differences are reconciled. When users are comfortable that the new system is working correctly, the old system is eliminated.

Parallel start-up involves running both the old and new systems for a period of time. The output of the new system is compared closely with the output of the old system, and any differences are reconciled. When users are comfortable that the new system is working correctly, the old system is eliminated.

User Acceptance

user acceptance document: A formal agreement that the user signs stating that a phase of the installation or the complete system is approved.

Most mainframe computer manufacturers use a formal **user acceptance document**—a formal agreement the user signs stating that a phase of the installation or the complete system is approved. This is a legal document that usually removes or reduces the IS vendor's liability for problems that occur after the user acceptance document has been signed. Because this document is so important, many companies get legal assistance before they sign the acceptance document. Stakeholders can also be involved in acceptance testing to make sure that the benefits to them are indeed realized.

SYSTEMS OPERATION AND MAINTENANCE

systems operation: Use of a new or modified system in all kinds of operating conditions.

Systems operation involves all aspects of using the new or modified system in all kinds of operating conditions. Getting the most out of a new or modified system during its operation is the most important aspect of systems operations for many organizations. Throughout this book, we have seen many examples of information systems operating in a variety of settings and industries. Thus, we will not cover the operation of an information system in detail in this section. To provide adequate support, many companies use a formal help desk. A *help desk* consists of computer systems, manuals, people with technical expertise, and other resources needed to solve problems and give accurate answers to questions. If you are having trouble with your PC and call a toll-free number for assistance, you might reach a help desk in India, China, or another country.

systems maintenance: A stage of systems development that involves checking, changing, and enhancing the system to make it more useful in achieving user and organizational goals.

Systems maintenance involves checking, changing, and enhancing the system to make it more useful in achieving user and organizational goals. Maintenance, review, and support for mobile applications will become more important as more companies adopt the use of smartphones, tablet computers, and other mobile devices for their workers and managers.

Organizations can perform systems maintenance in-house, or they can hire outside companies to perform maintenance for them. Many companies that use database systems from Oracle or SAP, for example, often hire these companies to maintain their database systems.[68] Systems maintenance is important for individuals, groups, and organizations. Individuals, for example, can use the Internet, computer vendors, and independent maintenance companies, including YourTechOnline.com (*www.yourtechonline.com*), Geek Squad (*www.geeksquad.com*), PC Pinpoint (*www.pcpinpoint.com*), and others. Organizations often have personnel dedicated to maintenance.

Software maintenance for purchased software can be 20 percent or more of the purchase price of the software annually. The maintenance process can be especially difficult for older software. A *legacy system* is an old system that might have been patched or modified repeatedly over time. An old payroll program in COBOL developed decades ago and frequently changed is an example of a legacy system. Legacy systems can be very expensive to maintain, and it can be difficult to add new features to some legacy systems. Like many organizations, Crescent Healthcare, which provides drug treatments for cancer and other life threatening diseases, has a large investment in older legacy applications.[69] It is a challenge knowing which legacy systems to keep and which ones should be discarded for newer Internet or cloud applications. According to a company executive, "We went through the legacy systems to evaluate what true business processes they represented and how well we could replicate them with Force.com (cloud) services." British Airways had about 60 legacy systems that were becoming increasingly difficult to update and maintain.[70] According to the CIO of British Airways, "These 60-plus systems had been built up over many years—since IT was first applied to airlines. They were fit for purpose, but they were linked by legacy connections that were not easy to change and restrict[ed] our ability to improve business processes." British Airways hopes replacing these older legacy systems will make its applications easier to update and maintain.

At some point, it becomes less expensive to switch to new programs and applications than to repair and maintain the legacy system. Maintenance costs for older legacy systems can be 50 percent of total operating costs in some cases. According to the CIO of a large computer company, "CIOs around the world have told me that although the business is pressuring them to focus on innovation to increase flexibility, they're spending 70 percent to 80 percent of their budget just to keep legacy applications running."[71]

Software maintenance is a major concern for most organizations. In some cases, organizations encounter major problems that require recycling the entire systems development process. In other situations, minor modifications are sufficient to remedy problems. Hardware maintenance is also important. Companies such as IBM have developed *autonomic computing* in which computers are programmed to manage and maintain themselves.[72] The goal is for computers to be self-configuring, self-protecting, self-healing, and self-optimizing. Being self-configuring allows a computer to handle new hardware, software, or other changes to its operating environment. Being self-protecting means a computer can identify potential attacks, prevent them when possible, and recover from attacks that occur. Being self-healing means a computer can fix problems when they occur, and being self-optimizing allows a computer to run faster and get more done in less time.

Getting rid of old equipment is an important part of maintenance. The options include selling it on Web auction sites such as eBay, recycling the equipment at a computer-recycling center, and donating it to a charitable organization, such as a school, library, or religious organization. When discarding old computer systems, it is always a good idea to permanently remove sensitive files and programs. Companies such as McAfee and Blancco have software to help people remove data and programs from old computers and transfer them to new ones.[73] As mentioned in the section on environmental design, companies are disposing of old equipment in ways that minimize environmental damage.

Reasons for Maintenance

After a program is written, it will need ongoing maintenance. A Texas restaurant, for example, decided to make maintenance changes to its security system after its customer's credit card numbers were stolen. Experience shows that frequent, minor maintenance to a program, if properly done, can prevent major system failures later. Some of the reasons for program maintenance are the following:

- Changes in business processes
- New requests from stakeholders, users, and managers
- Bugs or errors in the program
- Technical and hardware problems

- Corporate mergers and acquisitions
- Government regulations
- Changes in the operating system or hardware on which the application runs
- Unexpected events, such as severe weather or terrorist attacks

Most companies modify their existing programs instead of developing new ones because existing software performs many important functions, and companies can have millions of dollars invested in their old legacy systems. So, as new systems needs are identified, the burden of fulfilling the needs most often falls on the existing system. Old programs are repeatedly modified to meet ever-changing needs. Yet, over time, repeated modifications tend to interfere with the system's overall structure, reducing its efficiency and making further modifications more burdensome.

Types of Maintenance

slipstream upgrade: A minor upgrade—typically a code adjustment or minor bug fix—not worth announcing. It usually requires recompiling all the code, and in so doing, it can create entirely new bugs.

patch: A minor change to correct a problem or make a small enhancement. It is usually an addition to an existing program.

release: A significant program change that often requires changes in the documentation of the software.

version: A major program change, typically encompassing many new features.

Software companies and many other organizations use four generally accepted categories to signify the amount of change involved in maintenance. A **slipstream upgrade** is a minor upgrade—typically a code adjustment or minor bug fix. Many companies don't announce to users that a slipstream upgrade has been made. A slipstream upgrade usually requires recompiling all the code, and so, it can create entirely new bugs. This maintenance practice can explain why the same computers sometimes work differently with what is supposedly the same software. A **patch** is a minor change to correct a problem or make a small enhancement. It is usually an addition to an existing program, that is, the programming code representing the system enhancement is usually "patched into," or added to, the existing code. Although slipstream upgrades and patches are minor changes, they can cause users and support personnel big problems if the programs do not run as before. Many patches come from off-the-shelf software vendors. A new **release** is a significant program change that often requires changes in the documentation of the software. Finally, a new **version** is a major program change, typically encompassing many new features.

The Request for Maintenance Form

request for maintenance form: A form authorizing modification of programs.

Because of the amount of effort that can be spent on maintenance, many organizations require a **request for maintenance form** to authorize modification of programs. This form is usually signed by a business manager, who documents the need for the change and identifies the priority of the change relative to other work that has been requested. The IS group reviews the form and identifies the programs to be changed, determines the programmer who will be assigned to the project, estimates the expected completion date, and develops a technical description of the change. A cost/benefit analysis might be required if the change requires substantial resources.

Performing Maintenance

maintenance team: A special IS team responsible for modifying, fixing, and updating existing software.

Depending on organizational policies, the people who perform systems maintenance vary. In some cases, the team that designs and builds the system also performs maintenance. This ongoing responsibility gives the designers and programmers an incentive to build systems well from the outset: if problems occur, they will have to fix them. In other cases, organizations have a separate **maintenance team**. This team is responsible for modifying, fixing, and updating existing software.

A number of vendors have developed tools to ease the software maintenance burden. Modernization Workbench from Micro Focus is a collection of tools that help organizations analyze the inner workings of legacy applications that are written in older programming languages such as COBOL. After analyzing the programming code, companies can update, or modernize, the application so it is easier to maintain.[74]

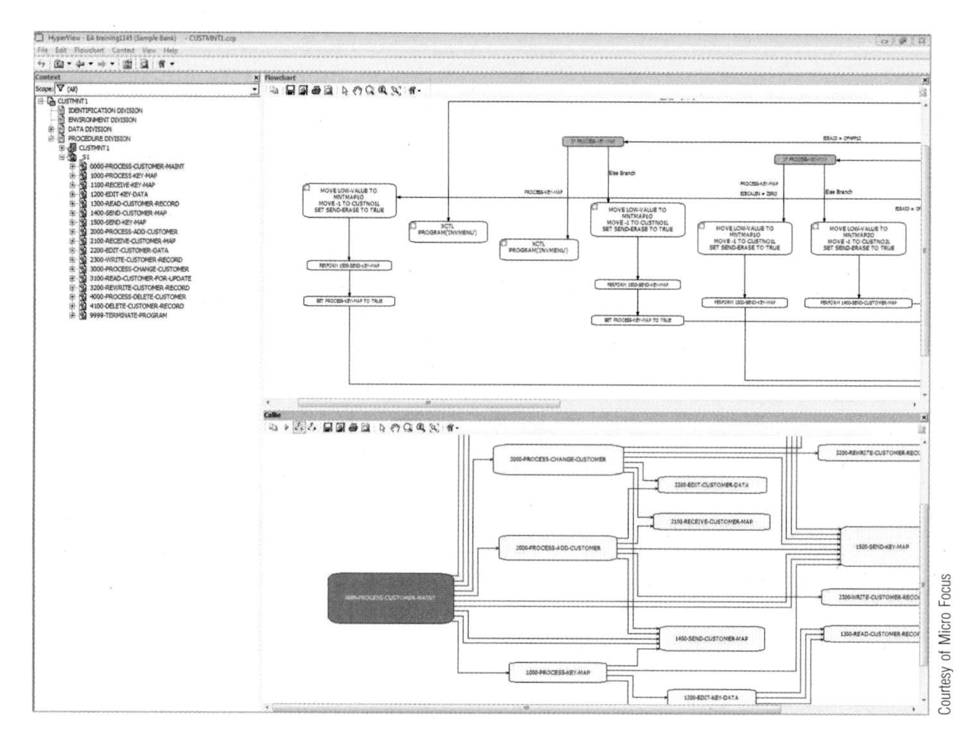

Modernization Workbench from Micro Focus provides business and technical insight into complex applications. Equipped with this insight, global development teams are more productive, modernization projects are accelerated, and managers can better govern the applications that run their business.

The Relationship between Maintenance and Design

Programs are expensive to develop, but they are even more expensive to maintain. For older programs, the total cost of maintenance can be up to five times greater than the total cost of development. A determining factor in the decision to replace a system is the point at which it is costing more to fix than to replace. Programs that are well designed and documented to be efficient, structured, and flexible are less expensive to maintain in later years. Thus, there is a direct relationship between design and maintenance. More time spent on design up front can mean less time spent on maintenance later.

In most cases, it is worth the extra time and expense to design a good system. Consider a system that costs $250,000 to develop. Spending 10 percent more on design would cost an additional $25,000, bringing the total design cost to $275,000. Maintenance costs over the life of the program could be $1,000,000. If this additional design expense can reduce maintenance costs by 10 percent, the savings in maintenance costs would be $100,000. Over the life of the program, the net savings would be $75,000 ($100,000–$25,000). This relationship between investment in design and long-term maintenance savings is shown in Figure 13.12.

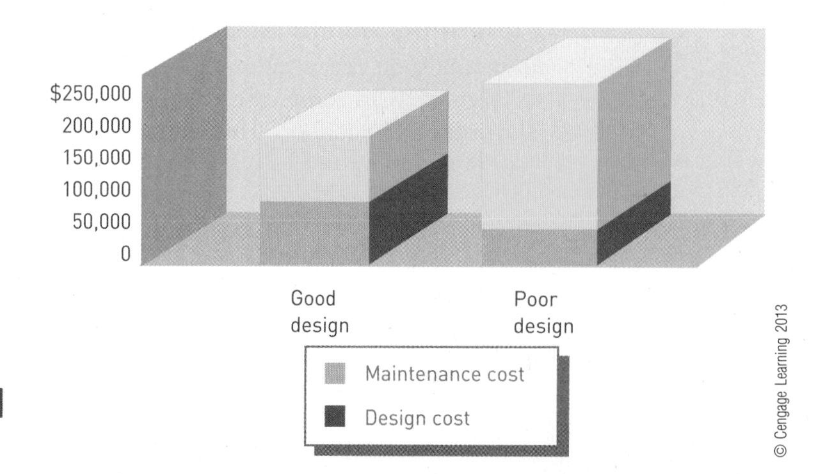

FIGURE 13.12
Value of investment in design

The need for good design goes beyond mere costs. Companies run a risk by ignoring small system problems when they arise because these small problems can become large in the future. As mentioned earlier, because maintenance programmers spend an estimated 50 percent or more of their time deciphering poorly written, undocumented program code, they have little time to spend on developing new, more effective systems. If put to good use, the tools and techniques discussed in this chapter will allow organizations to build longer-lasting, more reliable systems.

SYSTEMS REVIEW

systems review: The final step of systems development, involving analyzing systems to make sure that they are operating as intended.

Systems review, the final step of systems development, is the process of analyzing systems to make sure that they are operating as intended. The systems review process often compares the performance and benefits of the system as it was designed with the actual performance and benefits of the system in operation. After reviewing its Virtual Case File System, which some believe was over budget and didn't perform as expected, the FBI initiated a new systems development effort to create Sentinel, hardware and software used to store and analyze important information on its many cases.[75] The Transportation Security Agency (TSA) used an approach called the *Idea Factory* to review current information systems and recommend new ones or changes to existing systems.[76] The Idea Factory was featured on the White House Web site.[77] In some cases, a formal audit of the application can be performed, using internal and external auditors. Systems review can be performed during systems development, resulting in halting the new systems while they are being built because of problems.

Internal employees, external consultants, or both can perform systems review. When the problems or opportunities are industry-wide, people from several firms can get together. In some cases, they collaborate at an IS conference or in a private meeting involving several firms.

Types of Review Procedures

event-driven review: A review triggered by a problem or opportunity such as an error, a corporate merger, a new market for products, or other causes.

The two types of review procedures are event-driven and time-driven. See Table 13.5. An **event-driven review** is triggered by a problem or opportunity such as an error, a corporate merger, a new market for products, or other causes. The events that can trigger systems review can be highly complex or as simple as a broken cable, as was the case when a 75-year-old woman broke a fiber-optic cable with her shovel when digging for scrap metal in Armenia.[78] The shovel event disconnected most of the people in the Republic of Armenia from the Internet for about 12 hours. The U.S. State Department discovered its computer system used to run its annual lottery for visas had an internal programming error.[79] The lottery was supposed to pick 15,000 people for visas at random from the list of 15 million people who applied. As a result, the State Department was forced to fix the programming error and rerun the lottery for visas. The names of people that received a visa under the flawed lottery were put back into the lottery for a second chance.

TABLE **13.5** Examples of review types

Event-Driven	Time-Driven
Problem with an existing system	Monthly review
Merger	Yearly review
New accounting system	Review every few years
Executive decision that an upgraded Internet site is needed to stay competitive	Five-year review

According to one German man who had received a visa from the original, flawed program, "It's like you won $100,000 and then they just take it away from you and it's gone." In another case, a new cloud-computing company suddenly told its clients that it was closing all operations. This event-driven situation caused the companies using the cloud-computing service to make other arrangements in a short amount of time.

time-driven review: Review performed after a specified amount of time.

A **time-driven review** is performed after a specified amount of time. Many application programs are reviewed every six to twelve months. With this approach, an existing system is monitored on a schedule. If problems or opportunities are uncovered, a new systems development cycle can be initiated. A payroll application, for example, can be reviewed once a year to make sure that it is still operating as expected. If it is not, changes are made.

Many companies use both approaches. A billing application, for example, might be reviewed once a year for errors, inefficiencies, and opportunities to reduce operating costs. This is a time-driven approach. In addition, the billing application might be redone if corporations merge, if one or more new managers requires different information or reports, or if federal laws on bill collecting and privacy change. This is an event-driven approach.

System Performance Measurement

system performance measurement: Monitoring the system—the number of errors encountered, the amount of memory required, the amount of processing or CPU time needed, and other problems.

Systems review often involves monitoring the system, called **system performance measurement**. The number of errors encountered, the amount of memory required, the amount of processing or CPU time needed, and other problems should be closely observed. If a particular system is not performing as expected, it should be modified, or a new system should be developed or acquired. Comcast, the large cable provider, used Twitter to get user feedback on the performance of its information systems and all of its operations.[80] Some Comcast executives believe that using Twitter is like an "early-warning-system" that alerts Comcast to potential problems before they become serious and hurt system performance.

system performance products: Software that measures all components of the information system, including hardware, software, database, telecommunications, and network systems.

System performance products have been developed to measure all components of the information system, including hardware, software, database, telecommunications, and network systems. Microsoft Visual Studio, for example, has features that allow systems developers to monitor and review how applications are running and performing, thus permitting developers to make changes if needed.[81] IBM Tivoli OMEGAMON can monitor system performance in real time.[82] Precise Software Solutions has system performance products that provide around-the-clock performance monitoring for ERP systems, Oracle database applications, and other programs.[83] HP also offers a software tool called Business Technology Optimization (BTO) to help companies analyze the performance of their computer systems, diagnose potential problems, and take corrective action if needed.[84] When properly used, system performance products can quickly and efficiently locate actual or potential problems.

Measuring a system is, in effect, the final task of systems development. The results of this process can bring the development team back to the beginning of the development life cycle, where the process begins again.

SUMMARY

Principle:

Designing new systems or modifying existing ones should always help an organization achieve its goals.

The purpose of systems design is to prepare the detailed design needs for a new system or modifications to the existing system. Logical systems design refers to the way that the various components of an information system will work together. The logical design includes data requirements for output and input, processing, files and databases, telecommunications, procedures, personnel and job design, and controls and security design. Physical systems design refers to the specifications of the physical components. The physical design must specify characteristics for hardware and software design, database and telecommunications, and personnel and procedures design.

Logical and physical design can be accomplished using the traditional systems development life cycle or the object-oriented approach. Using the OO approach, analysts design key objects and classes of objects in the new or updated system. The sequence of events that a new or modified system requires is often called a scenario, which can be diagrammed in a sequence diagram.

A number of special design considerations should be taken into account during both logical and physical system design. Interface design and control relates to how users access and interact with the system. System security and control involves many aspects. Error prevention, detection, and correction should be part of the system design process. Causes of errors include human activities, natural phenomena, and technical problems. Designers should be alert to prevention of fraud and invasion of privacy.

Disaster recovery is an important aspect of systems design. Disaster planning is the process of anticipating and providing for disasters. A disaster can be an act of nature (a flood, fire, or earthquake) or a human act (terrorism, error, labor unrest, or erasure of an important file). The primary tools used in disaster planning and recovery are hardware, software, database, telecommunications, and personnel backup.

Security, fraud, and the invasion of privacy are also important design considerations. Most IS departments establish tight systems controls to maintain data security. Systems controls can help prevent computer misuse, crime, and fraud by employees and others. Systems controls include input, output, processing, database, telecommunications, and personnel controls.

Environmental design, also called green design, involves systems development efforts that slash power consumption, require less physical space, and result in systems that can be disposed of in a way that doesn't negatively affect the environment. A number of companies are developing products and services to help save energy. Environmental design also deals with how companies are developing systems to dispose of old equipment. The U.S. government is involved in environmental design. It has a plan to require federal agencies to purchase energy-efficient computer systems and equipment. The plan would require federal agencies to use the Electronic Product Environmental Assessment Tool (EPEAT) to analyze the energy usage of new systems. The U.S. Department of Energy rates products with the Energy Star designation to help people select products that save energy and are friendly to the environment.

Whether an individual is purchasing a personal computer or a large company is acquiring an expensive mainframe computer, the system could be obtained from one or more vendors. If new hardware or software will be purchased from a vendor, a formal request for proposal (RFP) is sometimes needed. The RFP outlines the company's needs; in response, the vendor provides a written reply. Financial options to consider include purchase, lease, and rent.

RFPs from various vendors are reviewed and narrowed down to the few most likely candidates. In the final evaluation, a variety of techniques—including group consensus, cost/benefit analysis, point evaluation, and benchmark tests—can be used. In group consensus, a decision-making group is appointed and given responsibility for making the final evaluation and selection. With cost/benefit analysis, all costs and benefits of the alternatives are expressed in monetary terms. Benchmarking involves comparing computer systems operating under the same conditions. Point evaluation assigns weights to evaluation factors, and each alternative is evaluated in terms of each factor and given a score from 0 to 100. After the vendor is chosen, contract negotiations can begin.

At the end of the systems design step, the final specifications are frozen, and no changes are allowed so that implementation can proceed. One of the most important steps in systems design is to develop a good contract if new computer facilities are being acquired. A final design report is developed at the end of the systems design phase.

Principle:

The primary emphasis of systems implementation is to make sure that the right information is delivered to the right person in the right format at the right time.

The purpose of systems implementation is to install the system and make everything, including users, ready for its operation. Systems implementation includes hardware acquisition, software acquisition or development, user preparation, hiring and training of personnel, site and data preparation, installation, testing, start-up, and user acceptance. Hardware acquisition requires purchasing, leasing, or renting computer resources from an IS vendor. Hardware is typically obtained from a computer hardware vendor.

Software can be purchased from vendors or developed in-house—a decision termed the *make-or-buy decision*. Virtualization, first introduced in Chapter 3, has had a profound impact on many aspects of systems implementation. A purchased software package usually has a lower cost, less risk regarding the features and performance, and easy installation. The amount of development effort is also less when software is purchased. Software as a Service (SaaS) is becoming a popular way to purchase software capabilities. Developing software can result in a system that more closely meets the business needs and has increased flexibility in terms of customization and changes. Developing software also has greater potential for providing a competitive advantage. However, such software is usually more expensive than purchased software. More companies are using service providers to acquire software, Internet access, and other IS resources.

Cross-platform development and integrated development environments (IDEs) make software development easier and more thorough. CASE tools are often used to automate some of these techniques. Technical and user documentation is always important in developing in-house software.

Database and telecommunications software development involves acquiring the necessary databases, networks, telecommunications, and Internet facilities. Companies have a wide array of choices, including newer object-oriented database systems. Virtual databases and Database as a Service (DaaS) are popular ways to acquire database capabilities.

Implementation must address personnel requirements. User preparation involves readying managers, employees, and other users for the new system. New IS personnel might need to be hired, and users must be well trained in the system's functions. Preparation of the physical site of the system must be done, and any existing data to be used in the new system will require conversion to the new format. Hardware installation is done during the implementation step, as is testing. Testing includes program (unit) testing, systems testing, volume testing, integration testing, and acceptance testing.

Start-up begins with the final tested information system. When start-up is finished, the system is fully operational. There are a number of different start-up approaches. Direct conversion (also called plunge or direct cutover) involves stopping the old system and starting the new system on a given date. With the phase-in approach, sometimes called a piecemeal approach, components of the new system are slowly phased in while components of the old one are slowly phased out. When everyone is confident that the new system is performing as expected, the old system is completely phased out. Pilot start-up involves running the new system for one group of users rather than all users. Parallel start-up involves running both the old and new systems for a period of time. The output of the new system is compared closely with the output of the old system, and any differences are reconciled. When users are comfortable that the new system is working correctly, the old system is eliminated. Many IS vendors ask the user to sign a formal user acceptance document that releases the IS vendor from liability for problems that occur after the document is signed.

Principle:

Maintenance and review add to the useful life of a system but can consume large amounts of resources. These activities can benefit from the same rigorous methods and project management techniques applied to systems development.

Systems operation is the use of a new or modified system. Systems maintenance involves checking, changing, and enhancing the system to make it more useful in obtaining user and organizational goals. Maintenance is critical for the continued smooth operation of the system. The costs of performing maintenance can well exceed the original cost of acquiring the system. Some major causes of maintenance are new requests from stakeholders and managers, enhancement requests from users, bugs or errors, technical or hardware problems, newly added equipment, changes in organizational structure, and government regulations.

Maintenance can be as simple as a program patch to correct a small problem to the more complex upgrading of software with a new release from a vendor. For older programs, the total cost of maintenance can be greater than the total cost of development. Increased emphasis on design can reduce maintenance costs. Requests for maintenance should be documented with a request for maintenance form, a document that formally authorizes modification of programs. The development team or a specialized maintenance team can then make approved changes. Maintenance can be greatly simplified with the object-oriented approach.

Systems review is the process of analyzing and monitoring systems to make sure that they are operating as intended. The two types of review procedures are the event-driven review and the time-driven review. An event-driven review is triggered by a problem or opportunity. A time-driven review is started after a specified amount of time.

Systems review involves measuring how well the system is supporting the mission and goals of the organization. System performance measurement monitors the system for number of errors, amount of memory and processing time required, and so on.

CHAPTER 13: SELF-ASSESSMENT TEST

Designing new systems or modifying existing ones should always help an organization achieve its goals.

1. ________________ details system outputs, inputs, and user interfaces; specifies hardware, software, databases, telecommunications, personnel, and procedures; and shows how these components are related.
2. Determining the hardware and software required for a new system is an example of ________________.
   a. logical design
   b. physical design
   c. interactive design
   d. object-oriented design
3. In cost/benefit analysis, a decision-making group is appointed and given the responsibility of making the final evaluation and selection. True or False?
4. ________________ involves systems development efforts that slash power consumption and require less physical space.

5. Scenarios and sequence diagrams are used with ________________.
   a. object-oriented design
   b. point evaluation
   c. incremental design
   d. nominal evaluation
6. A test that examines or compares computer systems operating under the same conditions is called ________________ testing.
7. The design report is the final result of systems design that provides technical and detailed descriptions of the new system. True or False?

The primary emphasis of systems implementation is to make sure that the right information is delivered to the right person in the right format at the right time.

8. Software as a Service (SaaS) allows an organization to subscribe to Web-based applications and pay for the software and services actually used. True or False?
9. ________________ is written details used by computer operators to execute a program and analysts and programmers to solve problems or modify the program.
10. What type of documentation is used by computer operators and by analysts and programmers?
    a. unit documentation
    b. integrated documentation
    c. technical documentation
    d. user documentation
11. ________________ testing involves testing the entire system of programs.
12. The phase-in approach to conversion involves running both the old system and the new system for three months or longer. True or False?

Maintenance and review add to the useful life of a system but can consume large amounts of resources. These activities can benefit from the same rigorous methods and project management techniques applied to systems development.

13. A(n) ________________ is a minor change to correct a problem or make a small enhancement to a program or system.
14. Many organizations require a request for maintenance form to authorize modification of programs. True or False?
15. A systems review that is caused by a problem with an existing system is called ________________.
    a. object review
    b. structured review
    c. event-driven review
    d. critical factors review
16. Corporate mergers and acquisitions can be a reason for systems maintenance. True or False?
17. Monitoring a system after it has been implemented is called ________________.

CHAPTER 13: SELF-ASSESSMENT TEST ANSWERS

1. Systems design
2. b
3. False
4. Environmental design
5. a
6. benchmark
7. True
8. True
9. Technical documentation
10. c
11. System
12. False
13. patch
14. True
15. c
16. True
17. system performance measurement

REVIEW QUESTIONS

1. What is the purpose of systems design?
2. Describe the design of system security and controls.
3. How can the object-oriented approach be used during systems design?
4. What is the difference between logical and physical design?
5. What is environmental design?
6. What are the advantages and disadvantages of in-house developed software?
7. What is disaster planning and recovery? What is the difference between a hot and cold site?
8. What is an RFP? What is typically included in one? How is it used?
9. What activities go on during the user preparation phase of systems implementation?
10. What is group consensus? How can it be used?
11. What are the major steps of systems implementation?
12. What are some tools and techniques for software development?
13. Give three examples of an IS vendor.
14. How can SaaS be used in software acquisition?

15. What are the steps involved in testing the information system?
16. What is the difference between an event-driven review and a time-driven review?
17. How is systems performance measurement related to the systems review?

DISCUSSION QUESTIONS

1. Describe the participants in the systems design stage. How do these participants compare with the participants of systems investigation?
2. Assume that you are the owner of a company that is about to start marketing and selling bicycles over the Internet. Describe what environmental design steps you could use to reduce power consumption with your information system.
3. Describe how you would create systems and security controls for smartphones and tablet computers for a medium-sized business.
4. You have been hired to design a computer system for a small business. Describe how you could use environmental design to reduce energy usage and the system's impact on the environment.
5. Identify some of the advantages and disadvantages of purchasing a database package instead of taking the DaaS approach.
6. Discuss the relationship between maintenance and systems design.
7. Is it equally important for all systems to have a disaster recovery plan? Why or why not?
8. Assume that you are the CIO of a medium-sized music company. The company president wants you to develop a Web site to advertise and sell the music the company produces. Describe the procedures you would use to hire several IS personnel to help you develop the needed Web site. What types of training would you make available to the new and existing IS personnel to help you in creating the needed Web site?
9. What are the advantages and disadvantages of the object-oriented approach to systems implementation?
10. You have been hired to oversee a major systems development effort to purchase a new accounting software package. Describe what is important to include in the contract with the software vendor.
11. You have been hired to purchase a new billing and accounting system for a medium-sized business. Describe how you would start up the new system and place it into operation.
12. Identify the various forms of testing. Why are there so many different types of tests?
13. What is the goal of conducting a systems review? What factors need to be considered during systems review?
14. Describe how you would select the best admissions software for your college or university. What features would be most important for school administrators? What features would be most important for students?
15. Assume that you have a personal computer that is several years old. Describe the steps you would use to perform systems review to determine whether you should acquire a new PC.

PROBLEM-SOLVING EXERCISES

1. You have been hired to develop a new student records and grade reporting system for your college or university. Describe how you would incorporate privacy and security measures into the design of the new system. Use a graphics program, such as PowerPoint, to develop a set of slides that shows how the different security and privacy measures will be included in the design. Write a brief report on the importance of including security and privacy concerns in your design.
2. A project team has estimated the costs associated with the development and maintenance of a new system. One approach requires a more complete design and will result in a slightly higher design and implementation cost but a lower maintenance cost over the life of the system. The second approach cuts the design effort, saving some dollars but with a likely increase in maintenance cost.
   a. Enter the following data in the spreadsheet. Print the result.

Benefits of good design

	Good Design	**Poor Design**
Design Costs	$14,000	$10,000
Implementation Cost	$42,000	$35,000
Annual Maintenance Cost	$32,000	$40,000

   b. Create a stacked bar graph that shows the total cost, including the design, implementation, and maintenance costs. Be sure that the chart has a title and that the costs are labeled on the chart.

c. Use your word-processing software to write a paragraph that recommends an approach to take and why.

3. Assume you have just started a campus bicycle rental business. Use a word-processing program to describe the logical and physical design of a computer application to purchase new bicycles for the rental program and another application to rent bicycles to students. Use a graphics program to develop one or more sequence diagrams to buy new bicycles and to rent them to students.

TEAM ACTIVITIES

1. Assume that your project team has been working for three months to complete the systems design of a new Web-based customer ordering system. Two possible options seem to meet all users' needs. The project team must make a final decision on which option to implement. The following table summarizes some of the key facts about each option.

Factor	**Option 1**	**Option 2**
Annual gross savings	$1.5 million	$3.0 million
Total development cost	$1.5 million	$2.2 million
Annual operating cost	$0.5 million	$1.0 million
Time required to implement	9 months	15 months
Risk associated with project (expressed in probabilities)		
Benefits will be 50% less than expected	20%	35%
Cost will be 50% greater than expected	25%	30%
Organization will not/cannot make changes necessary for system to operate as expected	20%	25%
Does system meet all mandatory requirements?	Yes	Yes

a. What process would you follow to make this important decision?
b. Who needs to be involved?
c. What additional questions need to be answered to make a good decision?
d. Based on the data, which option would you recommend and why?
e. How would you account for project risk in your decision making?

2. Your team has been hired by the owner of a new restaurant to explore word processing, graphics, database, and spreadsheet capabilities. The new owner has heard about cloud computing, SaaS, and DaaS. Your team should prepare a report on the advantages and disadvantages of using a traditional office suite from a company such as Microsoft as compared to other approaches.
3. Your team has been hired by your college or university to purchase desktop computers for a small computer lab consisting of 10 computers. Using the group consensus approach, have your team describe the hardware and software you would recommend. Using cost/benefit analysis, have your team make the same decision. Compare the advantages and disadvantages of group consensus versus cost/benefit analysis.

WEB EXERCISES

1. Use the Internet to find two different systems development projects that failed to meet cost or performance objectives. Summarize the problems and what should have been done. You might be asked to develop a report or send an e-mail message to your instructor about what you found.
2. Using the Web, search for information about disaster planning and recovery. Summarize what two or more organizations have done to prepare for or recover from a disaster.

CAREER EXERCISES

1. For a business or industry of your choice, describe how you would select a new CIO. What training opportunities would you offer a new CIO?
2. Research possible careers in developing applications for iPhones, other smartphones, and PDAs. Write a report that describes these opportunities. Include in your report applications that aren't currently available that you would find useful.

CASE STUDIES

Case One

IBM Grades Programmer Productivity

As you read in this chapter, computer programs are written by people. As with any type of work, some people are better at programming than others. Employers need a fair, objective way to find out who the excellent programmers are so as to recognize their superior work, to figure out what will help the others improve, and to learn what makes the excellent programmers better so the employers can try to duplicate this "secret sauce" throughout their workforce.

Superior programming is a composite of several measures. One is productivity: how much code a programmer produces. Another is quality: how error-free that code is. Other measures include the performance and security of the resulting program and its clarity for future modifications by people who were not involved in writing it.

Because it is difficult to measure these factors, most managers end up measuring the process by which software is built and the effort put into that process rather than its outcomes. As Jitendra Subramanyam, director of research at CAST, Inc., writes, "It's as if Michael Phelps tracks his time in the gym, the time it takes him to eat his meals, the time he spends on his Xbox, time walking his dog ... but bizarrely, not the time it takes him to swim the 100 meter butterfly!"

IBM recognizes the need to focus on people and their output. "At the end of the day, people are in the middle" of application development, says Pat Howard, vice president and cloud leader in IBM's global business services division. "It's really important to have great investments, great energy focused around the talent."

Howard's department uses the Applications Intelligence Platform from French software firm CAST to quantify performance. "Essentially it permitted our people to walk around with a scorecard. They could begin to earn points, based on the results or the value they were driving for the business," Howard says.

The program also helps to identify performance shortfalls and skill deficiencies. "We use it to identify where more training is needed," Howard says. Training budgets are tight, so "when you spend it, you've got to spend it really smartly, aim it at the right place."

Bank of New York Mellon is another CAST user. The bank uses CAST to control the quality of the software produced by offshore contract software developers. Vice president for systems and technology Robert-Michel Lejeune says, "You provide specifications, the offshorer has a process in place, but when they deliver, you don't know the level of quality. Using an automated tool provides you with facts and figures on the go."

A system such as this can never be the entire answer to employee or contractor performance evaluation. Systems cannot measure important employee efforts such as contributing ideas in team meetings, mentoring junior employees, and willingly taking on jobs that nobody else wants. However, CAST or something like it will be part of the answer at more and more companies in the future.

Discussion Questions

1. How would you feel as a programmer if your company announced that it was going to start using CAST Applications Intelligence Platform to measure your productivity and that of your colleagues? Would it matter to you if the firm said that 20 percent of your performance evaluation, which determines your salary increases, would be based on CAST's reports? What about 50 percent? 80 percent?
2. Consider other intellectual activities with a defined end product, such as writing a movie script based on a book or designing a university dormitory. Would such tools be useful for measuring the productivity of people who do those things and the quality of their output? If you don't think it would, how is programming different?

Critical Thinking Questions

1. Are there any drawbacks to using a programmer productivity measurement tool such as CAST's Applications Intelligence Platform? If there are, what are they?
2. Should software development managers be required to use a program such as CAST to measure the productivity of their teams?

SOURCES: Bednarz, A., "How IBM Started Grading Its Developers' Productivity," *Computerworld*, *www.computerworld.com/s/article/9221566/How_IBM_started_grading_its_developers_productivity*, November 7, 2011; CAST Web site, *www.castsoftware.com*, accessed March 4, 2012; Lejeune, R.-M., "Bank of New York Mellon Interview" (video), *www.youtube.com/watch?v=zLb7pCwA4rE*, February 7, 2012; Subramanyam, J., "5 Requirements for Measuring Application Quality," *Network World*, *www.networkworld.com/news/tech/2011/061611-application-quality.html*, June 7, 2011.

Case Two

COBOL: Not Going Away Anytime Soon

The COBOL programming language, which you read about in this chapter as an example of an older language found in legacy systems, was first developed in the 1960s using concepts that were even older than that. If any of your grandparents were computer programmers in the 1970s or 1980s, they might have used COBOL.

The problem businesses face today is that your grandparents want to retire. David Brown, managing director of the IT Transformation group at Bank of NY Mellon, is worried. "We have people we will be losing who have a lot of business knowledge. That scares me." But what really scares him is that nobody will understand how those programs work.

Why isn't COBOL used? It's perceived as outdated, not suited for modern needs such as mobile applications or the Web. However, IBM fellow Kevin Stoodley feels that perception is wrong. "COBOL has had lasting value, and it's not broken," he says.

A 2012 *Computerworld* survey of over 200 computer professionals confirms what Stoodley says. Forty-eight percent of the respondents said their organization uses

COBOL "a lot," more than any other language. The 64 percent whose organizations use it at least to some extent is higher than any language except JavaScript, which is nearly universal inside Web pages.

Adam Burden, global application modernization leader at IT consulting firm Accenture, says, "There's not a whole lot of new development going on. But our clients are enhancing their core applications and continue to maintain them." Some new development is going on, though: 37 percent of *Computerworld*'s survey respondents plan to use COBOL for at least some new applications in the future. That may not be a majority, but 37 percent of all organizations is a lot of organizations. Indeed, 69 percent of the respondents reported that at least some of their new software is written in COBOL today, and a whopping 95 percent use that language to maintain programs originally written in it.

Since the demand for COBOL programmers isn't going away and doesn't seem likely to decrease, new training programs in COBOL are still being developed. Manta Technologies, for example, will add COBOL to its set of training programs by releasing one module per month through 2012. "When we asked people for feedback, one of the things that customers have said consistently over the years is: 'You're perfect except … I use COBOL,'" said Bill Hansen, Manta president. "So the COBOL training has been on our list for development for some time." Hansen grows nostalgic: a training program for IBM's then-new COBOL compiler "was my first really big contract in 1982, and it helped get our training business going." Thirty years later, he's doing it again.

The long-term trend is still moving away from COBOL. Over time, applications will be rewritten in other languages as they move to new hardware environments. Automated software conversion programs such as OpenCOBOL, while not perfect, can help in this endeavor. Other COBOL applications will be replaced by packaged software. Yet it will take a very long time before all of this software makes much of a dent in today's inventory of COBOL software. Your grandchildren, when they use the 27th edition of this book several decades from now, may well learn the same thing.

Discussion Questions

1. Should today's programming students focus on developing expertise in COBOL, rather than trendier languages such as Java or Ruby? What are the pros and cons of learning COBOL versus learning newer languages?
2. Suppose you are considering purchasing an application that will manage an important part of your company's manufacturing process. One of the candidates on your short list of finalists is written in COBOL. The other two are not. Does this difference affect your selection?

Critical Thinking Questions

1. Many databases in the 1960s and 1970s stored dates with two digits for the year to save space. Their developers reasoned that those databases would be replaced long before they had to store years beginning with "20." That didn't happen. A major worldwide effort was needed in the late 1990s to deal with the so-called Y2K problem. Some databases were modified, others were replaced, and others received temporary fixes to keep working until 2020 or 2030. Does this experience suggest anything about how long COBOL programs will be around?
2. It has been suggested that COBOL programmers will soon belong to one of two groups: those who are about to retire and those who saw the opportunities created by those retirements—in other words, the very old (for active members of the workforce) and the very young. The large group in the middle considered COBOL "old hat" and didn't see the same opportunity, so they learned other languages. Should the middle group try to learn COBOL now? Should their employers try to train them to program in COBOL?

SOURCES: Staff, "COBOL Brain Drain: Survey Results," *Computerworld, www.computerworld.com/s/article/9225099/Cobol_brain_drain_Survey_results*, March 14, 2012; Manta Technologies, "Developing a COBOL Program," *www.mantatech.com/manta/PCOB01.htm*, accessed March 15, 2012; Mitchell, R., "Brain Drain: Where COBOL Systems Go from Here," *Computerworld, www.computerworld.com/s/article/9225079/Brain_drain_Where_Cobol_systems_go_from_here_*, March 14, 2012; OpenCOBOL Web site, *www.opencobol.org*, accessed March 15, 2012; Thomas, J., "Old School COBOL Gets New School Twist From Manta," The Four Hundred: iSeries and AS/400 Insight, *www.itjungle.com/tfh/tfh022012-story07.html*, February 20, 2012.

Questions for Web Case

See the Web site for this book to read about the Whitmann Price Consulting case for this chapter. Following are questions concerning this Web case.

Altitude Online: Systems Design: Design, Implementation, Maintenance, and Review

Discussion Questions

1. How did Jon's team coordinate with the vendor in the implementation stage of the systems development project?
2. What did Jon's team do in advance of contacting SAP that made the design and implementation systems proceed as smoothly as possible?

Critical Thinking Questions

1. What risks were involved in the systems development project?
2. What benefits were gained from this systems development project? Was it worth the risks?

NOTES

Sources for the opening vignette: Hewlett-Packard, "ImmobilienScout24 Increases Speed of Software Development," *h20195.www2.hp.com/V2/GetDocument.aspx?docname=4AA1-3475ENW*, October 2011;

Hewlett-Packard, "HP Unified Functional Testing Software," *www8.hp.com/us/en/software/software-product.html?compURI=tcm:245-936981,* 2012; ImmobilienScout24, "ImmobilienScout24: Germany's Largest Real Estate Market," *www.immobilienscout24.de/de/ueberuns/presseservice/press_releases/unternehmenstext_eng.jsp*, February 2012; Scrum Alliance Web site, *www.scrumalliance.org*, accessed March 5, 2012.

1. Angwin, Julia, "Web Tool on Firefox to Deter Tracking," *The Wall Street Journal*, January 24, 2011, p. B1.
2. Bruno, E., "Windows Ecosystem 2.0," *Information Week*, January 31, 2001, p. 51.
3. Staff, "Say Goodbye to All Those Passwords," *Bloomberg BusinessWeek*, *www.businessweek.com/magazine/content/11_06*, January 27, 2011.
4. Moerschel, G., "4 Strategies to Lower Mobility Risk," *Information Week*, January 31, 2011, p. 44.
5. Essex, Allen, "Valley Plans for Future Natural Disasters," *Tribune Business News*, January 8, 2011.
6. Tirone, Jonathan, "Crash iPads," *Bloomberg BusinessWeek*, March 21, 2011, p. 22.
7. Collett, Stacy, "Calculated Risk," *Computerworld*, February 7, 2011, p. 24.
8. Ibid.
9. Bednarz, A., "Will Cloud Backup Services Be the End for Tape?" *Network World*, February 21, 2011, p. 12.
10. Barbee, T., "Disaster Recovery on a Budget," *CIO*, May 1, 2011, p. 48.
11. Bednarz, A., "Will Cloud Backup Services Be the End for Tape?" *Network World*, February 21, 2011, p. 12.
12. EMC^2 Web site, *www.emc.com/products/detail/software/recoverpoint.htm*, accessed July 17, 2011.
13. Staff, "Symitar Announces Hosted Failover Service," *PR Newswire*, July 26, 2011.
14. Tirone, Jonathan, "Crash iPads," *Bloomberg BusinessWeek*, March 21, 2011, p. 22.
15. Ward, Burke, et al, "A United States Perspective on Electronic Discovery," *Emerald Group Publishing*, 2011, p. 268.
16. Rosso, Anne, "In Compliance," *Collector*, August, 2011, p. 60.
17. Varon, E., "Leaders vs. Laggards in Green IT," *CIO*, March 1, 2011, p. 24.
18. Staff, "BlueCross Headquarters Awarded Gold in Environmental Design," *Business Wire*, March 4, 2011.
19. Vinodh, S., "Environmental Conscious Product Design Using CAD and CAE," *Clean Technologies and Environmental Policy*, April 2011, p. 359.
20. Violino, Bob, "Top Green IT Organizations: KPMG," *Computerworld*, October 4, 2011, p. 20.
21. McMillan, R. and Gaudin, S., "Facebook Reveals Its Data Center Secrets," *Computerworld*, April 18, 2011, p. 8.
22. Nissan USA Web site, *www.nissanusa.com*, accessed July 17, 2011.
23. Vistaprint Web site, *www.vistaprint.com*, accessed August 8, 2011.
24. U.S. Green Building Council Web site, *www.usgbc.org*, accessed August 8, 2011.
25. Brandon, J., "Powerful Experiments," *Computerworld*, May 9, 2011, p. 38.
26. HP Web site, *www.hp.com*, accessed July 17, 2011.
27. Clark, Don, "Helping Chips to Sip Power," *The Wall Street Journal*, June 6, 2011, p. B5.
28. Voltaic Generator Web site, *www.voltaicsystems.com/bag_generator.shtml*, accessed July 17, 2011.
29. Optimum Energy Web site, *www.optimumenergyco.com*, accessed August 28, 2011.
30. Varon, Elana, "Recycling PCs," *CIO*, September 1, 2011, p. 16.
31. Venjuvo Web site, *www.venjuvo.com*, accessed July 17, 2011.
32. EPA Web site, *www.epa.gov/epp/pubs/products/epeat.htm*, accessed August 28, 2011.
33. Energy Star Web site, *www.energystar.gov*, accessed August 28, 2011.
34. Block, Janice, "Getting the Most Out of Your E-RFP Process," *Inside Council*, April 2011.
35. Alonso, S., et al, "A Web Based Consensus Support System for Group Decision Making Problems," *Information Sciences*, January, 2011, p. 4477.
36. Staff, "Improving the Practice of Cost Benefit Analysis in Transport," *Joint Transport Research Centre*, January, 2011.
37. Nash, K., "Highlight the Fine Print," *CIO*, March 1, 2011, p. 15.
38. Richard, K., "Enterprise Software: It's the Contract, Stupid," *Information Week*, March 28, 2011, p. 13.
39. Eisner, R., "Legal Risks in the Cloud," *Computerworld*, April 18, 2011, p. 24.
40. Murphy, Chris, "Create," *Information Week*, March 14, 2011, p. 23.
41. IBM Web site, *www-03.ibm.com/systems/management/director/about/announcement/20091020.html*, accessed July 17, 2011.
42. Pratt, M., "Desktop Dreams," *Computerworld*, June 6, 2011, p. 28.
43. Gartner Web site, *www.gartner.com/it/content/1462300/1462334/december_15_top_predictions_for_2011_dplummer.pdf*, accessed June 7, 2011.
44. Nash, Kim, "Forging the Vendor Collective," *CIO*, December 1, 2011, p. 20.
45. Staff, "Industry-Shaping Event Takes Utility Computing from Hype to Fact," *PR Newswire*, February 2, 2011.
46. IBM Web site, *www-03.ibm.com/systems/power/hardware/cod*, accessed August 29, 2011.
47. Henderson, T., "Will Cloud Computing Leave Traditional Vendor Relationships in the Dust?" *Network World*, May 23, 2011, p. 12.
48. Binstock, A., "In Praise of Small Code," *Information Week*, June 27, 2011, p. 41.
49. Dogan, K., et al, "Managing Versions of a Software Product," *Information Systems Research*, March 2011, p. 5.
50. Efrati, A., "Google Opens Online App Store," *The Wall Street Journal*, February 3, 2011, p. B2.
51. Binstock, A., ".NET Alternative in Transition," *Information Week*, June 13, 2011, p. 42.
52. King, Julia, "Pinning IT Projects More Firmly to Business Objectives," *Computerworld*, February 21, 2001, p. 32.
53. Gibbs, M., "Ninite Automates Installation of Windows, Linux Applications," *Network World*, May 23, 2011, p. 21.
54. Staff, "New General Dynamics Fortress Technologies' Commercial Off-the-Shelf Suite B," *PR Newswire*, August 23, 2011.

55. VMware Web site, *www.vmware.com*, accessed July 17, 2011.
56. Staff, "Omniture Co-founder Unveils New Software-as-a-Service Venture," *The Enterprise*, July 18, 2011, p. 3.
57. Schultz, B., "Florida Hospice Saves with SaaS," *Network World*, June 6, 2011, p. 24.
58. Qualys Web site, *www.qualys.com*, accessed August 29, 2011.
59. Thibodeau, P., "Software Development Still a Risky Business," *Computerworld*, October 24, 2012, p. 8.
60. Binstock, Andrew, "Developers Love IDEs," *InformationWeek*, January 30, 2012, p. 36.
61. Simon, S., "Wyoming Plays It Cool," *The Wall Street Journal*, March 8, 2011, p. A3.
62. IT Redux Web site, *http://itredux.com/office-20/database/?family=Database*, accessed July 17, 2011.
63. Bennett, Stephen, "Real-World Beta Testing," *Transport Topics,"* February 14, 2011, p. 12.
64. Staff, "Baidu Begins Alpha Testing on its Licensed Music Website," *TMT China Weekly*, May 6, 2011.
65. Vodafone Web site, *www.vodafone.com*, accessed July 17, 2011.
66. Esposito, Dino, ".NET Apps Put to the Test," *InformationWeek*, July 11, 2011, p. 40.
67. Woo, Stu, "Amazon.com Tests Redesign," *The Wall Street Journal*, September 6, 2011, p. B3.
68. Henschen, D., "SAP's ByDesign Moves from App to Platform," *Information Week*, February 28, 2011, p. 10.
69. Fogarty, K., "Modern Medicine," *CIO*, March 1, 2001, p. 30.
70. Overby, S., "Ground Control to Major Sales," *CIO*, February 1, 2011, p. 20.
71. Garcia, K., "Executive ViewPoint," *CIO*, December 1, 2011, p. 15.
72. IBM Web site, *www.research.ibm.com/autonomic*, accessed July 17, 2011.
73. Blancco Web site, *www.blancco.com*, accessed August 29, 2011.
74. Micro Focus Web site, *www.microfocus.com*, accessed July 17, 2011.
75. Foley, John, "FBI Recasts Sentinel As a Model of Agility," *InformationWeek*, May 30, 2011, p. 12.
76. TSA Web site, *www.tsa.gov/press/happenings/030209_weekly_story12.shtm*, accessed July 17, 2011.
77. The Idea Factory Web site, *www.whitehouse.gov/open/innovations/IdeaFactory*, accessed July 17, 2011.
78. Lomsadze, G., "A Shovel Cuts off Armenia's Internet," *The Wall Street Journal*, April 8, 2011, p. A10.
79. Staff, "You Get a U.S. Visa! Oops, No?" *The Tampa Tribune*, May 14, 2011, p. 1.
80. Comcast Web site, *www.comcast.com*, accessed August 29, 2011.
81. Microsoft Web site, *www.microsoft.com/visualstudio/en-us*, accessed August 29, 2011.
82. IBM Web site, *www.ibm.com/software/tivoli/products*, accessed July 17, 2011.
83. Precise Web site, *www.precise.com*, accessed July 17, 2011.
84. HP Web site, *https://h10078.www1.hp.com/cda/hpms/display/main/hpms_home.jsp?zn=bto&cp=1_4011_100__*, accessed July 17, 2011.

PART 1 | PART 2 | PART 3 | PART 4 | PART 5

Information Systems in Business and Society

CHAPTER

14 The Personal and Social Impact of Computers

Principles	Learning Objectives
• Policies and procedures must be established to avoid waste and mistakes associated with computer usage.	• Describe some examples of waste and mistakes in an IS environment, their causes, and possible solutions. • Identify policies and procedures useful in eliminating waste and mistakes. • Discuss the principles and limits of an individual's right to privacy.
• Computer crime is a serious and rapidly growing area of concern requiring management attention.	• Explain the types of computer crime and their effects. • Identify specific measures to prevent computer crime.
• Jobs, equipment, and working conditions must be designed to avoid negative health effects from computers.	• List the important negative effects of computers on the work environment. • Identify specific actions that must be taken to ensure the health and safety of employees.
• Practitioners in many professions subscribe to a code of ethics that states the principles and core values that are essential to their work.	• Outline criteria for the ethical use of information systems.

Information Systems in the Global Economy
BREDOLAB BOTNET, RUSSIA

One Master, 30 Million Slaves

On May 21, 2012, Gyorgi Avanesov was sentenced to four years in jail by the Court of First Instance of Armenia's Arabkir and Kanaker-Zeytun administrative districts. Avanesov was the first person to be jailed in Armenia for offenses related to cybercrime.

Thus ended, for the time being, the story of the Bredolab botnet. At its height, this network of slave computers—*bots*, in Internet parlance—consisted of 30 million computers infected with malware, controlled by 143 servers in the Netherlands and other countries. The botnet was used primarily to send spam e-mails and launch Distributed Denial of Service (DDoS) attacks.

In a regular Denial of Service (DoS) attack, a server is bombarded with so many requests for Web pages that it buckles under the strain. It can no longer handle valid requests and may crash, requiring restart. However, firewalls can protect against DoS attacks: as soon as they detect many requests from the same source, they block anything from that source. In a DDoS attack, the requests come from many sources, so blocking one or a few of them won't help. With 30 million computers under its control, most of them unknowingly, Bredolab DDoS attacks could not be blocked by this method.

For example, Avanesov was found guilty of attacking a Russian telecommunication company called Macomnet on October 1, 2010. Avanesov instructed 25 percent of his botnet to hit a Macomnet IP address. That brought down the company's Web site and deprived nearly 200 of its customers of telecommunications service.

The Bredolab virus infected a computer when its user visited an infected Web site or opened an infected e-mail attachment. E-mail attachments had file names such as "DHL invoice" that many users accepted as valid. Once installed, the virus gave Bredolab the ability to change or delete files on that computer, steal passwords, and monitor user activity. Bredolab could also send e-mails or Web page requests without the user's knowledge. At one point, Dutch investigators found that the virus was infecting 3 million computers per month, and that infected computers were sending out 3.6 million infected e-mail messages per day.

Avanesov earned an estimated €100,000 Euros (about U.S. $125,000) a month from his Bredolab botnet business by renting out access to the compromised computers to criminals who wanted to send out spam, spread malware, and fake antivirus attacks. This income allowed him to live lavishly, jetting off to the Seychelles with his girlfriend and fancying himself as a DJ. That life came to an end on October 25, 2010, when Dutch authorities disabled the servers that controlled the botnet. The next day, Armenian authorities arrested Avanesov at the Yerevan airport.

Despite the disabling of 143 servers, the publicity that accompanied these events, and the wide availability of instructions on cleansing infected computers, a few Bredolab servers remain active in Russia and Kazakhstan. Those servers can control the remaining infected computers. So, while the botnet is weaker than it was at its 2010 peak, it has not been eliminated. It is still capable of doing significant damage.

As you read this chapter, consider the following:

- Computers have brought great benefits to society. At the same time, they have created risks. How can you minimize the risks without reducing the benefits?
- For each computer-related risk you read about, what is its root cause: human error, human malice, or act of nature?

WHY LEARN ABOUT THE PERSONAL AND SOCIAL IMPACT OF COMPUTERS?

© Andrey Burmakin/Shutterstock

Both opportunities and threats surround a wide range of nontechnical issues associated with the use of information systems and the Internet. The issues span the full spectrum—from preventing computer waste and mistakes, to avoiding violations of privacy, to complying with laws on collecting data about customers, and to monitoring employees. If you become a member of a human resources, an information systems, or a legal department within an organization, you will likely be charged with leading the organization in dealing with these and other issues covered in this chapter. Also, as a user of information systems and the Internet, it is in your own self-interest to become well versed on these issues. You need to know about the topics in this chapter to help avoid becoming a victim of crime, fraud, privacy invasion, and other potential problems. This chapter begins with a discussion of preventing computer waste and mistakes.

Earlier chapters detailed the significant benefits of computer-based information systems in business, including increased profits, superior goods and services, and higher quality of work life. Computers have become such valuable tools that today's businesspeople have difficulty imagining work without them. Yet, the information age has also brought the following potential problems for workers, companies, and society in general:

- Computer waste and mistakes
- Computer crime
- Privacy issues
- Work environment problems
- Ethical issues

This chapter discusses some of the social and ethical issues as a reminder of these important considerations underlying the design, building, and use of computer-based information systems. No business organization, and hence no information system, operates in a vacuum. All IS professionals, business managers, and users have a responsibility to see that the potential consequences of IS use are fully considered. Even entrepreneurs, especially those who use computers and the Internet, must be aware of the potential personal and social impact of computers.

COMPUTER WASTE AND MISTAKES

Computer-related waste and mistakes are major causes of computer problems, contributing to unnecessarily high costs and lost profits. Examples of computer-related waste include organizations operating unintegrated information systems, acquiring redundant systems, and wasting information system resources. Computer-related mistakes refer to errors, failures, and other computer problems that make computer output incorrect or not useful; most of these are caused by human error. This section explores the damage that can be done as a result of computer waste and mistakes.

Computer Waste

Some organizations continue to operate their business using unintegrated information systems, which makes it difficult for decisions makers to collaborate and share information. This practice leads to missed opportunities, increased costs and lost sales. For example, until Casa Oliveira, a Venezuelan importer of liquor, wine, champagne, and other beverages, implemented an integrated ERP system, it was unable to share data from its disparate financial, sales, payroll, and purchasing systems. As a result, both costs and sales for the company were adversely affected due to missed opportunities to adjust stock levels to meet demand and target sales promotions more effectively.[1]

In some cases, organizations often unknowingly waste money to acquire systems in different organizational units that perform nearly the same functions. Implementation of such duplicate systems unnecessarily increases hardware and software costs. The U.S. government spends billions of dollars on information systems each year, with $79 billion spent in fiscal year 2011 alone. Some of this spending goes toward providing information systems that provide similar functions across the various branches and agencies of the government. Indeed, when the General Accounting Office (GAO) conducted an investigation of spending on information systems in the Department of Defense and the Department of Energy, it found 37 of its sample of 810 investments in information systems (amounting to $1.2 billion) were potential duplicates.[2]

A less dramatic, yet still relevant, example of waste is the amount of company time and money employees can spend playing computer games, sending personal e-mail, or browsing the Internet. When waste is identified, it typically points to one common cause: the improper use of information systems and resources. For example, Procter & Gamble employees complained for months that their computer Internet searches and e-mail were running slow. Following an investigation, it was discovered that over 50,000 YouTube videos were being played on the company's personal computers every day. In addition, computer users were listening to some 4,000 hours of music a day through Pandora and downloading films from Netflix. The demand for videos and music delivered to company personal computers exceeded the firm's Internet capacity and created a major bottleneck. The firm was forced to block employees' access to nonbusiness sites and remind them that the computers were to be used for work-related purposes.[3] Procter & Gamble is by no means unique in blocking access to nonwork related Web sites. Many other companies including Cintas, General Electric Aviation, Kroger, and TriHealth have all found it necessary to limit employee access to nonwork related Web sites.[4]

Computer-Related Mistakes

Despite many people's distrust of them, computers rarely make mistakes. If users do not follow proper procedures, however, even the most sophisticated hardware cannot produce meaningful output. Mistakes can be caused by unclear expectations and a lack of feedback. A programmer might also develop a program that contains errors, or a data-entry clerk might enter the wrong data. Unless errors are caught early and corrected, the speed of computers can intensify mistakes. As information technology becomes faster, more complex, and more powerful, organizations and computer users face increased risks of experiencing the results of computer-related mistakes. Consider these recent examples of computer-related mistakes:

- A programming error led to local Indiana governments being underpaid $206 million in income taxes collected by the state.[5]
- In a case that eventually was argued before the United States Supreme Court, a New Jersey man was arrested, held in jail for seven days, and strip searched twice based on an erroneous seven-year old arrest warrant that should have been purged.[6]
- Computer errors enabled some 19,000 ineligible Maine residents to receive state health care benefits.[7]
- Computer errors in the United Airlines reservation system led to delays in crediting mileage, mistakes in boarding passes, errors in frequent-flier status level, and problems with upgrades and redeeming flight credit certificates.[8]
- An astronaut on the International Space Station was forced to extend his spacewalk an additional half hour due to a computer glitch in one of the station's robotic arms used to carry astronauts where they need to go.[9]

PREVENTING COMPUTER-RELATED WASTE AND MISTAKES

To remain profitable in a competitive environment, organizations must use all resources wisely. To employ IS resources efficiently and effectively, employees and managers alike should strive to minimize waste and mistakes. This effort involves (1) establishing, (2) implementing, (3) monitoring, and (4) reviewing effective policies and procedures.

Establishing Policies and Procedures

The first step to prevent computer-related waste is to establish policies and procedures regarding efficient acquisition, use, and disposal of systems and devices. Computers permeate organizations today, and it is critical for organizations to ensure that systems are used to their full potential. As a result, most companies have implemented stringent policies on the acquisition of computer systems and equipment, including requiring a formal justification statement before computer equipment is purchased, definition of standard computing platforms (operating system, type of computer chip, minimum amount of RAM, and so on), and the use of preferred vendors for all acquisitions.

Prevention of computer-related mistakes begins by identifying the most common types of errors, of which there are surprisingly few. Types of computer-related mistakes include the following:

- Data-entry or data-capture errors
- Errors in computer programs
- Errors in handling files, including formatting a disk by mistake, copying an old file over a newer one, and deleting a file by mistake
- Mishandling of computer output
- Inadequate planning for and control of equipment malfunctions
- Inadequate planning for and control of environmental difficulties (such as electrical and humidity problems)
- Installing computing capacity inadequate for the level of activity
- Failure to provide access to the most current information by not adding new Web links and not deleting old links

To control and prevent potential problems caused by computer-related mistakes, companies have developed policies and procedures that cover the acquisition and use of computers. Training programs for individuals and work groups as well as manuals and documents covering the use and maintenance of computer systems also help prevent problems. The Process Improvement Institute offers a two-day course on preventing human errors that explains the underlying reasons that humans make mistakes and how these mistakes can be prevented.[10] Other preventive measures include approval of certain systems and applications before they are implemented and used to ensure compatibility and cost effectiveness and a requirement that documentation and descriptions of certain applications be filed or submitted to a central office, including all cell formulas for spreadsheets and a description of all data elements and relationships in a database system. Such standardization can ease access and use for all personnel.

Many organizations have established strong policies to prevent employees from wasting time using computers inappropriately at work. After companies have planned and developed policies and procedures, they must consider how best to implement them. In some cases, violating these policies can lead to termination. For example, an employee of the Emerson College Police Department was fired for allegedly visiting Netflix and an online dating Web site while on duty. Going one step further, the college administrators also fired the Chief of Police for allegedly delaying the reporting of the computer incident to the Human Resources Department.[11]

Implementing Policies and Procedures

Implementing policies and procedures to minimize waste and mistakes varies according to the business conducted. Most companies develop such policies and procedures with advice from the firm's internal auditing group or its external auditing firm. The policies often focus on the implementation of source data automation, the use of data editing to ensure data accuracy and completeness, and the assignment of clear responsibility for data accuracy within each information system. Some useful policies to minimize waste and mistakes include the following:

- Changes to critical tables, HTML, and URLs should be tightly controlled, with all changes documented and authorized by responsible owners.
- A user manual should be available covering operating procedures and documenting the management and control of the application.
- Each system report should indicate its general content in its title and specify the time period covered.
- The system should have controls to prevent invalid and unreasonable data entry.
- Controls should exist to ensure that data input, HTML, and URLs are valid, applicable, and posted in the right time frame.
- Users should implement proper procedures to ensure correct input data.

Training is another key aspect of implementation. Many users are not properly trained in using applications, and their mistakes can be very costly. Because more and more people use computers in their daily work, they should understand how to use them. Training is often the key to acceptance and implementation of policies and procedures. Because of the importance of maintaining accurate data and of people understanding their responsibilities, companies converting to ERP and e-commerce systems invest weeks of training for key users of the system's various modules.

It is critical that systems have correct input data so that they operate properly. When a controlled burn designed to reduce fire risk in Colorado got out of control, residents near the 4,500-acre fire were to receive reverse 911 calls warning them to evacuate. However, due to inaccurate data, it is estimated that roughly 12 percent of the homeowners in the fire zone never received a call, while some people living well outside the fire zone received calls in error. Unfortunately, one couple living inside the fire zone failed to evacuate and perished in the fire. Whether they received a call is still in dispute.[12]

Monitoring Policies and Procedures

To ensure that users throughout an organization are following established procedures, the next step is to monitor routine practices and take corrective action if necessary. By understanding what is happening in day-to-day activities, organizations can make adjustments or develop new procedures. Many organizations implement internal audits to measure actual results against established goals, such as percentage of end-user reports produced on time, percentage of data-input errors detected, number of input transactions entered per eight-hour shift, and so on.

Her Majesty's Revenue and Customs organization conducted an annual audit to ensure that the amount of tax deducted by employers matched its records. It discovered that 6 million people should receive tax rebates of averaging £300 ($475 USD) while 1 million others must pay an additional amount averaging £600 ($950 USD). Affected taxpayers were informed of the error detected by this monitoring.[13]

Reviewing Policies and Procedures

The final step is to review existing policies and procedures and determine whether they are adequate. During review, people should ask the following questions:

- Do current policies cover existing practices adequately? Were any problems or opportunities uncovered during monitoring?
- Does the organization plan any new activities in the future? If so, does it need new policies or procedures addressing who will handle them and what must be done?
- Are contingencies and disasters covered?

This review and planning allows companies to take a proactive approach to problem solving, which can enhance a company's performance, such as increasing productivity and improving customer service. During such a review, companies are alerted to upcoming changes in information systems that could have a profound effect on many business activities.

The results of failing to review and plan changes in policies and procedures can lead to disastrous consequences. For example, gaps in computer systems and poor procedures led to the erroneous release of an estimated 1,450 California prison inmates with either a high risk of violence or a high risk of committing crimes. These actions were taken under a program designed to reduce prison overcrowding by transferring thousands of low-level offenders from prison to county custody. While prisoners who pose a risk of reoffending were to be excluded from the program, prison officials lacked the necessary information to make this determination. They did not have access to the inmates' disciplinary history while they were incarcerated. In addition, prison officials employed a state Department of Justice system that records arrests but is missing nearly half of the state's arrests.[14]

Information systems professionals and users still need to be aware of the misuse of resources throughout an organization. Preventing errors and mistakes is one way to do so. Another is implementing in-house security measures and legal protections to detect and prevent a dangerous type of misuse: computer crime.

COMPUTER CRIME

Even good IS policies might not be able to predict or prevent computer crime. A computer's ability to process millions of pieces of data in less than one second makes it possible for a thief to steal data worth millions of dollars. Compared with the physical dangers of robbing a bank or retail store with a gun, computer crime is less dangerous as a computer criminal with the right equipment and know-how can steal large amounts of money without leaving his or her home.

The Internet Crime Computer Center is an alliance between the White Collar Crime Center and the Federal Bureau of Investigation that was formed in 2000. It provides a central site for Internet crime victims to report and to alert appropriate agencies of crimes committed. Just over 300,000 crimes were reported to this source in 2010 (the most recent year for which data is available).[15] Unfortunately, this number represents only a small fraction of total computer-related crimes, as many crimes go unreported because companies don't want the bad publicity or don't think that law enforcement can help. Such lack of publicity makes the job even tougher for law enforcement. Additionally, most companies that have been electronically attacked won't talk to the press. A big concern is loss of public trust and image—not to mention the fear of encouraging copycat hackers.

According to the Internet Crime Computer Center, the two most common online crimes reported included the nondelivery of payments or merchandise amounting to hundreds of millions of dollars and scams involving people posing as FBI agents. Victims of these crimes reported losing hundreds of millions of dollars.[16]

Many cases of fraud involve nondelivery of automobiles advertised for sale on the Internet. The hoaxer posts photos and a description of a vehicle offered for sale. When an interested buyer responds, the victim is told that the vehicle is located overseas. The fraudster then tells the victim to send a deposit via a wire transfer to initiate the shipping process. Once that transfer is done, the buyer hears nothing further from the fraudster. In a devious variation on this scam, the hoaxer advises the victim about a problem with the initial wire transfer. To correct the problem, the hoaxer sends the victim a cashier's check (counterfeit) and tells the victim to cash the check and resend a second wire to a different account. The victim is unaware that the cashier's check is counterfeit and follows the directions, getting stung a second time.[17]

Fraudsters frequently commit identity theft using the names and photos of U.S. government officials (including FBI agents) to set up sham social networking accounts and profiles. The fraudsters use this stolen identity to appear reputable and begin to "friend" potential victims trying to develop an online relationship with their victims. As the relationship grows, the fraudsters reveal that they are in a predicament and need cash. Perhaps they are overseas on urgent business and a family member has fallen ill. They need money to get home. Victims end up wiring funds (sometimes repeatedly) to the fraudster, believing they are involved in a genuine relationship.[18]

Today, computer criminals are a new breed—bolder and more creative than ever. With the increased use of the Internet, computer crime is now global. It's not just on U.S. shores that law enforcement has to battle cybercriminals. Regardless of its nonviolent image, computer crime is different only because a computer is used. It is still a crime. Part of what makes computer crime unique and difficult to combat is its dual nature—the computer can be both the tool used to commit a crime and the object of that crime.

THE COMPUTER AS A TOOL TO COMMIT CRIME

A computer can be used as a tool to gain access to valuable information and as the means to steal thousands or millions of dollars. It is, perhaps, a question of motivation—many people who commit computer-related crime claim they do it for the challenge, not for the money. Credit card fraud—whereby a criminal illegally gains access to another's line of credit with stolen credit card numbers—is a major concern for today's banks and financial institutions. In general, criminals need two capabilities to commit most computer crimes. First, the criminal needs to know how to gain access to the computer system. Sometimes, obtaining access requires knowledge of an identification number and a password. Second, the criminal must know how to manipulate the system to produce the desired result. Frequently, a critical computer password has been talked out of a person, a practice called **social engineering**. Or, the attackers simply go through the trash—**dumpster diving**—for important pieces of information that can help crack the computers or convince someone at the company to give them more access. In addition, over 2,000 Web sites offer the digital tools—often without charge—that will let people snoop, crash computers, hijack control of a machine, or retrieve a copy of every keystroke. While some of the tools were intended for legitimate use to provide remote technical support or monitor computer usage, hackers take advantage of them to gain unauthorized access to computers or data.

social engineering: Using social skills to get computer users to provide information that allows a hacker to access an information system or its data.

dumpster diving: Going through the trash of an organization to find secret or confidential information, including information needed to access an information system or its data.

A dumpster can be an excellent source of information for an identity thief. Data from discarded bills, credit card approval letters, or financial statements can provide all the information needed to rob you of your identity. Interestingly, the FBI's Domestic Investigations and Operations Guide encourages agents to consider dumpster diving as a technique to get information.[19]

Cyberterrorism

cyberterrorism: Any premeditated, politically motivated attack against information, computer systems, computer programs, and data which results in violence against non-combatant targets by sub-national groups or clandestine agents.[22]

In 2002, the FBI defined **cyberterrorism** as "any premeditated, politically motivated attack against information, computer systems, computer programs, and data which results in violence against non-combatant targets by sub-national groups or clandestine agents."[20] Cyberterrorism has been a concern for countries and companies around the globe.

The U.S. government considered the potential threat of cyberterrorism serious enough that in February 1998 it established the National Infrastructure Protection Center. This function was later transferred to the Homeland Security Department's Information Analysis and Infrastructure Protection Directorate to serve as a focal point for threat assessment of, warning of , investigation of, and response to threats or attacks against the country's critical infrastructure, which provides telecommunications, energy, banking and finance, water systems, government operations, and emergency services. Successful cyberattacks against the facilities that provide these services could cause widespread and massive disruptions to the normal function of American society. Janet Napolitano, secretary of the Department of Homeland Security, recently stated that "The U.S. has become 'categorically safer' since 9/11, but cyber-terrorism now tops the list of security concerns."[21]

cyberterrorist: Someone who intimidates or coerces a government or organization to advance his or her political or social objectives by launching computer-based attacks against computers, networks, and the information stored on them.

A **cyberterrorist** is someone who intimidates or coerces a government or organization to advance his or her political or social objectives by launching computer-based attacks against computers, networks, and the information stored on them. Following are some of the many recent examples of attacks by cyberterrorists from around the world:

- The Web sites for both the Tel Aviv Stock Exchange and El Al, the Israeli national airline, were brought down by cyberterrorists. In retaliation, Israeli hackers brought down the Saudi Arabian stock exchange.[23]
- Cyberterrorists attacked and defaced Croatia's NATO Web site to protest western intervention in Libya and other countries.[24]
- A large southern California water treatment facility hired a hacker to test the vulnerability of its computer systems. In just one day, the hacker and his team were able to gain control of the equipment that added chemicals to purify the drinking water. Had these been real cyberterrorists, they could have easily rendered the water undrinkable for millions of residents.[25]
- Six months following the California incident, a water pump failure at an Illinois water treatment facility was damaged in what may have been the result of a cyberattack by Russian-based cyberterrorists.[26]

Identity Theft

identify theft: A crime in which an imposter obtains key pieces of personal identification information, such as Social Security or driver's license numbers, to impersonate someone else.

Identity theft is a crime in which an imposter obtains key pieces of personal identification information, such as Social Security or driver's license numbers, to impersonate someone else. The information is then used to obtain credit, merchandise, and/or services in the name of the victim or to provide the thief with false credentials. The perpetrators of these crimes employ such an extensive range of methods that investigating them is difficult. A rapidly growing area of identity theft involves child identity theft, where the number of complaints reported to the Federal Trade Commission (FTC) has grown to over 19,000 in 2011 from just 6,000 in 2003.[27] Children's Social Security numbers are considered "clean," and their theft may not be detected for years. A North Carolina father could not list his son as a dependent because that identity had already been claimed on another person's tax return. As a result, he paid an extra $4,000 in federal taxes.[28]

In some cases, the identity thief uses personal information to open new credit accounts, establish cellular phone service, or open a new checking account to obtain blank checks. In other cases, the identity thief uses personal

Identity Theft and Gaming Consoles

As you've read here, ***identity theft*** is a crime in which one person obtains personal information about another in order to obtain a benefit, such as credit, by pretending to be that person.

Most of us if asked wouldn't think of gaming consoles as a source of identity information that could put us at risk. We would be wrong. A Drexel University research group, led by Assistant Professor Ashley Podhradsky, found that a used Xbox 360 console contains credit card and other information about its owner. (These consoles use credit cards to purchase games and accessories on the Xbox Store.) This information remains on the disk even if the user has followed the manufacturer's instructions to restore the console to "factory settings." Restoration does not erase all the data on the disk.

The file format used within the Xbox 360 is a streamlined version of the Windows FAT32 file format. Widely available disk analysis tools can read the data it contains. Speaking to a representative of gaming Web site Kotaku in a phone interview, Dr. Podhradsky said "A lot of [criminals] already know how to do all this. Anyone can freely download a lot of this software, pick up a discarded game console, and have someone's identity."

Dr. Podhradsky said Xbox publisher Microsoft does a "disservice" to its customers by not doing a better job of keeping personal data protected. "Microsoft does a great job of protecting their proprietary information," she said. "But they don't do a great job of protecting the user's data."

Most experienced gamers know that, when they erase data, nothing is really erased. A file is flagged to indicate that its space is available for use, but the data remains until the space is used for new data. Therefore, experienced gamers reformat their drives before sale. However, even reformatting doesn't destroy all the personal data on an Xbox 360 drive. Using disk inspection tools, the Drexel research team found user names and credit card numbers on drives that had supposedly been reformatted. Since the purchaser of a used system normally also has the seller's real name and address, the combination is sufficient to steal the seller's identity. The researchers were also able to find the previous owner's player list: the *gamer tags* of others with whom that user had played. Gamer tags can be used to search social networking and gamer sites to find personal information about that gamer.

Alex Garden, Microsoft's general manager of Xbox Live, wrote in response to these findings that "Security is an ongoing battle. No matter how well we work to improve security ... our work will never end. With every measure we put in place, ill-intentioned people will create new ways to attack." Microsoft is understandably reluctant to disclose specifics, lest those "ill-intentioned people" learn what will no longer work and, by implication, what might.

Meanwhile, what should people do before selling an Xbox 360? Drexel researchers urge them to remove its disk drive, attach it to a computer, and use a sanitation program such as Darik's Boot and Nuke (free for all major personal computing platforms) to erase it. Then users can reinstall the drive and sell the console.

Microsoft may have fixed this specific vulnerability by the time you read this. Other vulnerabilities, however, probably remain to be discovered and exploited. This type of vulnerability is not unique to Xbox among gaming consoles, nor is it unique to gaming consoles among smart devices. As the Drexel study says, "Five years from now, identities will be stolen on devices and technology that do not yet exist.... Consumers need to be diligent about protecting their own data, and not assume their technology is going to do it for them." Consumers need to be aware of all the places in which their personal data is stored and watch those places.

Discussion Questions

1. Have you ever sold a used gaming console, smartphone, tablet, or computer? How carefully did you erase all the data stored in it? Would you do the same if you were to sell a similar device now, having read this case? What would you do differently?
2. Have you ever bought a used gaming console, smartphone, tablet, or computer? Did you find any information about the previous owner on it? How hard did you have to look for it? What did you do with it? If you didn't find any, how hard did you look?

Critical Thinking Questions

1. Imagine that the additional cost to add personal information protection to a new game console will increase its cost by $25 per unit and delay its introduction into the market by six months. As the product manager for this new console, prepare an argument for the company's board of directors stating your position on adding this new protection. Be sure to consider the potential ethical and legal issues versus the impact on the company's bottom line and reputation with its customers.
2. It is the year 2023. Your freezer reads the RFID tag of the frozen peas you take out, updates its inventory database, tells your microwave how to prepare them, and places an order when the quantity of frozen peas falls below its reorder point. How does that situation resemble the one in this case? Should we start to worry about identity theft from kitchen appliances? What should appliance manufacturers, food shopping sites, consumers, and regulatory agencies start doing today to avoid this type of identity theft in the future?

SOURCES: Cunningham, A. (posting as humboldt111502), "Identity Theft and Used Gaming Consoles," Xbox Experts forum, *forums.xbox-experts.com/viewtopic.php?f=4&t=5640*, April 5, 2012; Darik's Boot and Nuke Web site, *www.dban.com*, accessed May 29, 2012; Plunkett, L., "Microsoft 'Cares Deeply' about Hijacked Accounts, Asks for 'Trust in Us,'" Kotaku, *kotaku.com/5883231/microsoft-cares-deeply-about-hijacked-accounts-asks-for-trust-in-us*, February 7, 2012; Podhradsky, A., et al., "Identity Theft and Used Gaming Consoles: Recovering Personal Information from Xbox 360 Hard Drives," AMCIS 2011 Proceedings—All Submissions, Paper 54, *aisel.aisnet.org/amcis2011_submissions/54*, August 5, 2011; Schreier, J., "Hackers Can Steal Credit Card Information from Your Old Xbox, Experts Tell Us [Update]," *kotaku.com/5897461/hackers-can-steal-credit-card-information-from-your-old-xbox-experts-tell-us*, March 29, 2012; Xbox 360 Web site, *www.xbox.com*, accessed May 29, 2012.

information to gain access to the person's existing accounts. Typically, the thief changes the mailing address on an account and runs up a huge bill before the person whose identity has been stolen realizes there is a problem. The Internet has made it easier for an identity thief to use the stolen information because transactions can be made without any personal interaction.

The IRS Web site provides the details of case after case of individuals who committed identify theft and used that information to falsely claim federal tax refunds. For example, a gang of four people obtained stolen identity information, including names and Social Security numbers, which they used to prepare several false federal tax returns. They claimed fraudulent refunds for more than $3 million during 2010 and more than $2 million for 2011. They were arrested, tried, found guilty, and are now serving time for their crime.[29]

Three neighbors on the same street in a suburb of Chicago were victims of identity theft and had several thousand dollars charged to their credit cards. Police think that the thief may have resorted to taking mail out of the mailboxes or to dumpster diving to steal the identity data.[30]

Internet Gambling

Many people enjoy Internet gambling as a recreational and leisure activity. Baccarat, bingo, blackjack, pachinko, poker, roulette, and sports betting are all readily available online. The size of the online gambling market is not known, but it is estimated that the global online gambling market is the neighborhood of $30 billion.[31]

The laws regarding the legality of online gambling are quite confusing. No federal law prohibits individuals engaging in online gambling. However, operating a casino, racebook, or sportsbook on a Web site located in the United States has, in some cases, been ruled illegal, which is why such Web sites run on servers located in other countries. It is against federal law for Web sites anywhere to take sports bets online. Although taking nonsports bets from U.S. residents by foreign Web sites is not against the law, it is against federal law for banks or other financial institutions to facilitate the transfer of funds to online casinos. So placing bets and using your credit card to process your wins and losses can get your bank in trouble. Naturally, all this is subject to changes in federal law and the interpretation of prosecutors and judges in ruling on specific cases. A handful of states also have their own laws concerning online gambling.

An example of online gambling Web sites running afoul of U.S. federal law are three popular online poker Web sites against which the U.S. Department of Justice filed bank fraud, money laundering, and illegal gambling charges. Prosecutors charged that the site operators arranged for billions of dollars in payments received from U.S. players to be disguised as payments to fictitious online merchants and to be processed through U.S. banks.[32]

Some members of Congress would like to legalize gambling and levy federal and state taxes on the proceeds as a means for increasing tax revenues. The increased revenue could easily exceed $10 billion over a 10-year period.

THE COMPUTER AS A TOOL TO FIGHT CRIME

The computer is also used as a tool to fight computer crime. Information systems are used to fight crime in many ways, including helping recover stolen property, monitoring sex offenders, and helping to better understand and diminish crime risks.

Recovery of Stolen Property

The LeadsOnline Web-based service system is one of several information systems used by law enforcement to recover stolen property. The system contains hundreds of millions of records in its database. Data is entered into the system from pawn brokers, secondhand dealers, and salvage yards. In some areas, state or local laws require that all such businesses register (with no charge to business owners) with LeadsOnline. The system allows law enforcement officers to search the database by item serial number or by individual. It even has a partnership with eBay that makes it possible to locate possible stolen merchandise that has been listed for sale or sold online.[33]

The LeadsOnline system has frequently helped to catch criminals and return stolen property to its rightful owners. Over $37,000 in stolen jewelry was recovered when a deputy in the sheriff's office of Hale County, Alabama, ran a suspect's name through LeadsOnline. Not only was the deputy able to recover the goods, but the pawn shop had a photo of the thief holding the jewelry in her hand.[34]

Monitoring Criminals

JusticeXchange is a Web-based data sharing system that provides law enforcement officials with fast, easy access to information about current and former offenders held in participating jails across the United States. The system receives data from agencies that book and house offenders through interfaces to their existing jail management systems. Users can search for historical and current information about prisoners, create a "watch" so that they are notified by e-mail of a specific offender's booking or release, and add behavioral information about currently incarcerated offenders to the database.[35]

Watch Systems is a technology partner and consultant to law enforcement organizations nationwide. Its Offender Watch product is a Web-based system used to track registered sex offenders. It stores the registered offender's address, physical description, and vehicle information. The public can access the information at *www.communitynotification.com*. The information available varies depending on the county and state. For example, in Hamilton County, Ohio, the data is provided by the sheriff's department and allows the user to search for registered sex offenders by township, school district, zip code, or within one mile of an entered address. The information displayed includes a photo of all registered sex offenders, their description, and current addresses. Law enforcement agencies can search the database based on full or partial license plate number or vehicle description.[36]

Assessing Crime Risk for a Given Area

The ready availability of personal computers, coupled with the development of mapping and analysis software, has led law enforcement agencies to use crime-related data, powerful analysis techniques, and geographic information systems (GIS) to better understand and even diminish crime risks. The use of such software enables law enforcement agencies, members of an organization's security department, and individuals to gain a quick overview of crime risk at a given address or in a given locale, as shown in Figure 14.1.

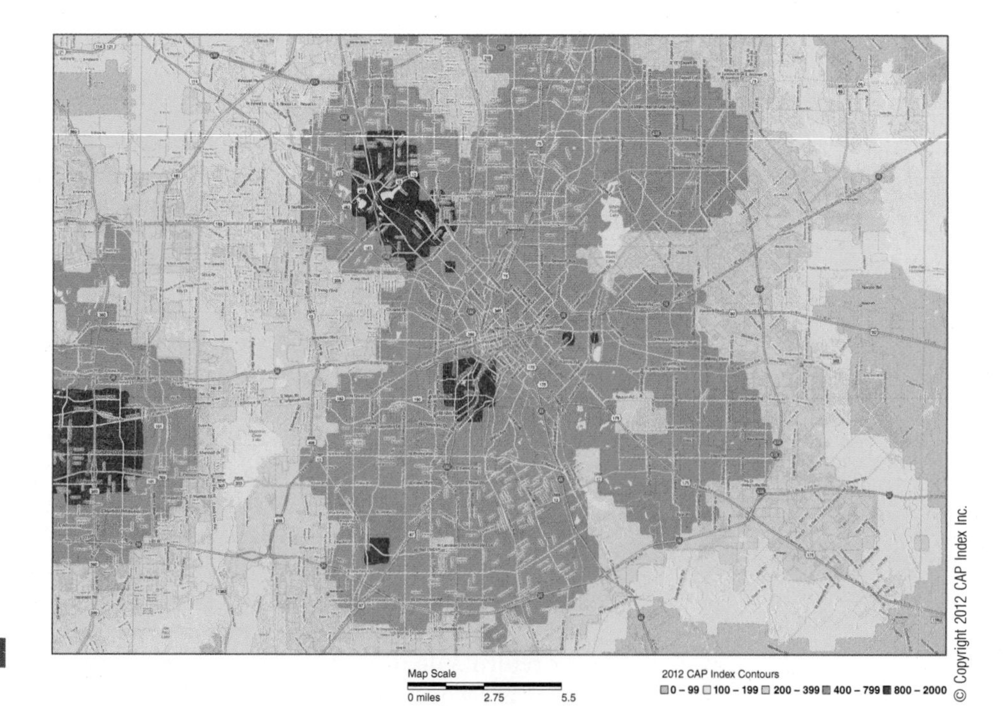

FIGURE **14.1**
Mapping crime risk

CAP Index Inc.'s CRIMECAST Reports provide a quick and thorough overview of the crime risk at any given location in the U.S., Canada and the U.K. A detailed map and spreadsheet of risk scores isolate and identify crime-related issues in the vicinity of a specific site. Figure 14.1 shows Dallas, Texas. CRIMECAST clients include more than 80 percent of FORTUNE 100 companies, including Bank of America, Kraft, Lowe's, and Marriott. Companies and government organizations use CRIMECAST data to assess crime risk levels at their facilities, for selection of new sites, for allocation of security resources, and to defend against litigation related to premises security.[37]

With GIS tools, law enforcement agencies can analyze crime data relative to other factors, including the locations of common crime scenes (such as convenience stores and gas stations) and certain demographic data (such as age and income distribution). Common GIS systems include the following:

- The National Equipment Registry maps mobile equipment thefts in areas where peak equipment thefts have occurred so that police and equipment owners can take appropriate action. It includes more than 15 million ownership records for construction and farm equipment.[38]
- The CompStat (short for computer statistics) program uses GIS software to map crime and identify problem precincts. The program has a proven track record of reducing crime in Boston, Philadelphia, Los Angeles, Miami, Newark, New Orleans, and New York. Other cities are exploring its use, including San Francisco.[39]
- CargoNet is a national database that helps law enforcement and the transportation industry track cargo crimes, identify cargo theft patterns, and improve stolen property recovery rates. The database can be accessed by traditional desktop and laptop computers, tablet computers, and even smartphones. The Tennessee Highway Patrol (THT) uses this system to fight cargo theft in transit in the state and surrounding states. Colonel Tracy Trout of the THT states that "with the information-sharing capabilities of cargo net, state troopers can conduct timely investigations, link recovered goods to owners, and remove criminals from interstate roads."[40]

THE COMPUTER AS THE OBJECT OF CRIME

A computer can also be the object of a crime rather than the tool for committing it. Tens of millions of dollars worth of computer time and resources are stolen every year. Each time system access is illegally obtained, data or computer equipment is stolen or destroyed, or software is illegally copied, the computer becomes the object of crime. These crimes fall into several categories: illegal access and use, data alteration and destruction, information and equipment theft, software and Internet piracy, computer-related scams, and international computer crime. See Table 14.1.

Illegal Access and Use

Crimes involving illegal system access and use of computer services are a concern to both government and business. Since the outset of information technology, computers have been plagued by criminal hackers. Originally, a **hacker** was a person who enjoyed computer technology and spent time learning and using computer systems. A **criminal hacker**, also called a **cracker**, is a computer-savvy person who attempts to gain unauthorized or illegal access to computer systems to steal passwords, corrupt files and programs, or even transfer money. In many cases, criminal hackers are people who are looking for excitement—the challenge of beating the system. Today, many people use the term hacker and cracker interchangeably. **Script bunnies** is a derogatory

hacker: A person who enjoys computer technology and spends time learning and using computer systems.

criminal hacker (cracker): A computer-savvy person who attempts to gain unauthorized or illegal access to computer systems to steal passwords, corrupt files and programs, or even transfer money.

script bunny: A derogatory term for inexperienced hackers who download programs called scripts that automate the job of breaking into computers.

TABLE **14.1** Common methods used to commit computer crimes

Methods	Examples
Add, delete, or change inputs to the computer system.	Delete records of absences from class in a student's school records.
Modify or develop computer programs that commit the crime.	Change a bank's program for calculating interest so it deposits rounded amounts in the criminal's account.
Alter or modify the data files used by the computer system.	Change a student's grade from C to A.
Operate the computer system in such a way as to commit computer crime.	Access a restricted government computer system.
Divert or misuse valid output from the computer system.	Steal discarded printouts of customer records from a company trash bin.
Steal computer resources, including hardware, software, and time on computer equipment.	Make illegal copies of a software program without paying for its use.
Offer worthless products for sale over the Internet.	Send e-mails requesting money for worthless hair growth product.
Blackmail executives to prevent release of harmful information.	Eavesdrop on organization's wireless network to capture competitive data or scandalous information.
Blackmail company to prevent loss of computer-based information.	Plant a logic bomb and send a letter threatening to set it off unless paid a considerable sum.

insider: An employee, disgruntled or otherwise, working solo or in concert with outsiders to compromise corporate systems.

term for inexperienced hackers who download programs called *scripts* that automate the job of breaking into computers. **Insiders** are employees, disgruntled or otherwise, working solo or in concert with outsiders to compromise corporate systems. The biggest threat for many companies is not external hackers but their own employees. Insiders have extra knowledge that makes them especially dangerous—they know logon IDs, passwords, and company procedures that help them evade detection.

Catching and convicting criminal hackers remains a difficult task. The method behind these crimes is often hard to determine, even if the method is known, and tracking down the criminals can take a lot of time.

Contractors often must be trusted with logon names and passwords to complete their job assignment and so can also be considered an insider threat. A contract programmer working for the Federal Reserve was accused of stealing software that tracks collections and payments made by the federal government. The software was used as a training tool in a side business the contractor was running.[41]

Data and information are valuable corporate assets. The intentional use of illegal and destructive programs to alter or destroy data is as much a crime as destroying tangible goods. The most common of these programs are viruses and worms, which are software programs that, when loaded into a computer system, will destroy, interrupt, or cause errors in processing. Such programs are also called *malware*, and the growth rate for such programs is epidemic. It is estimated that hundreds of previously unknown viruses and worms emerge each day. Table 14.2 describes the most common types of malware.

In some cases, a virus or a worm can completely halt the operation of a computer system or network for days until the problem is found and repaired. In other cases, a virus or a worm can destroy important data and programs. If backups are inadequate, the data and programs might never be fully functional again. The costs include the effort required to identify and neutralize the virus or worm and to restore computer files and data as well as the value of business lost because of unscheduled computer downtime.

TABLE **14.2** Common types of computer malware

Type of Malware	Description
Logic bomb	A type of Trojan horse that executes when specific conditions occur. Triggers for logic bombs can include a change in a file by a particular series of keystrokes or at a specific time or date.
Rootkit	A set of programs that enables its user to gain administrator level access to a computer or network. Once installed, the attacker can gain full control of the system and even obscure the presence of the rootkit from legitimate system administrators.
Trojan horse	A malicious program that disguises itself as a useful application or game and purposefully does something the user does not expect.
Variant	A modified version of a virus that is produced by the virus's author or another person by amending the original virus code.
Virus	Computer program file capable of attaching to disks or other files and replicating itself repeatedly, typically without the user's knowledge or permission.
Worm	Parasitic computer program that replicates, but unlike viruses, does not infect other computer program files. A worm can send the copies to other computers via a network.

Some criminal hackers use a new version of the SpyEye Trojan horse program that, when installed on a victim's computer, alters the text displayed by HTML code to either erase evidence of a money transfer transaction entirely or alter the amount of the money transfer and bank balances. When the victims access their account online, they find no trace of the transactions the hackers are using to drain a bank account.[42]

Nick Bradley, the senior manager for IBM global security operations, warns that organizations need to guard against new security threats that specifically target smartphones. Employees who use smartphones, such as the iPhone or Android phone, or tablet computers, such the Apple iPad or Samsung Electronics Galaxy Tab, to download and interact with corporate data must guard against malware just as computer users must.[43]

Spyware

spyware: Software that is installed on a personal computer to intercept or take partial control over the user's interaction with the computer without knowledge or permission of the user.

Spyware is software installed on a personal computer to intercept or take partial control over the user's interaction with the computer without the knowledge or permission of the user. Some forms of spyware secretly log keystrokes so that user names and passwords may be captured. Other forms of spyware record information about the user's Internet surfing habits and sites that have been visited. Still other forms of spyware change personal computer settings so that the user experiences slow connection speeds or is redirected to Web pages other than those expected. Spyware is similar to a Trojan horse in that users unknowingly install it when they download freeware or shareware from the Internet.

Win 7 Antispyware is malware that pretends to be antivirus software. Once on your computer, it runs automatically when your computer boots up. It runs a fake scan of your computer and pretends to check your computer for virus infections. The results of the scan are faked to show an alarming number of security risks. Win 7 Antispyware also disrupts the operation of your computer by generating various pop-up ads and bogus security notifications saying various programs cannot be started. The user is directed to a Web site that explains how to order the bogus antivirus software. Ordering the software will do nothing to improve the situation and your computer will continue to misbehave until other measures are taken to remove the Win 7 Antispyware.[44]

Information and Equipment Theft

Data and information are assets or goods that can also be stolen. People who illegally access systems often do so to steal data and information. To obtain illegal access, criminal hackers require identification numbers and passwords. Some criminals try various identification numbers and passwords until they find ones that work. Using password sniffers is another approach. A **password sniffer** is a small program hidden in a network or a computer system that records identification numbers and passwords. In a few days, a password sniffer can record hundreds or thousands of identification numbers and passwords. Using a password sniffer, a criminal hacker can gain access to computers and networks to steal data and information, invade privacy, plant viruses, and disrupt computer operations.

password sniffer: A small program hidden in a network or a computer system that records identification numbers and passwords.

In addition to theft of data and software, all types of computer systems and equipment have been stolen from homes, offices, schools, and vehicles. Two aides of a presidential candidate had two iPads, two handheld radios, and two laptops stolen from their rented vehicle.[45] Portable computers such as laptops and portable storage devices (and the data and information stored in them) are especially easy for thieves to take. This theft then raises the potential for the thieves to use this data to commit identity theft. For example, a file server holding patient information including name, address, telephone number, Social Security number, and diagnosis was stolen from a Baltimore-area rehabilitation center. In addition to notifying patients of the theft, the center recommended that patients place a fraud alert on their consumer credit files so that they would be alerted in the event anyone applies for credit using their name and personal information.[46]

Courtesy of Kensington Computer Products Group

To fight computer crime, many companies use devices that disable the disk drive or lock the computer to the desk.

In many cases, the data and information stored in these systems are more valuable than the equipment. Vulnerable data can be used in identity theft. In addition, the organization responsible receives a tremendous amount of negative publicity that can cause it to lose existing and potential future customers. Often, the responsible organization offers to pay for credit-monitoring services for those people affected in an attempt to restore customer goodwill and avoid law suits.

Patent and Copyright Violations

Works of the mind, such as art, books, films, formulas, inventions, music, and processes that are distinct and "owned" or created by a single person or group, are called intellectual property. Copyright law protects authored works such as art, books, film, and music. Patent laws protect processes, machines, objects made by humans or machines, compositions of matter, and new uses of these items. Software is considered intellectual property and may be protected by copyright or patent law.

Software piracy is the act of unauthorized copying, downloading, sharing, selling, or installing of software. When you purchase software, you are purchasing a license to use it; you do not own the actual software. The license states how many times you can install the software. If you make more copies of the software than the license permits, you are pirating.

software piracy: The act of unauthorized copying, downloading, sharing, selling, or installing of copyrighted software.

Businesses and consumers around the world purchased about $95 billion worth of legal software; however, another $59 billion was pirated.[47] The Business Software Alliance (BSA) has become a prominent software antipiracy organization. Software companies, including Adobe, Apple, Hewlett-Packard, IBM, Intel, and Microsoft, contribute funds to the operation of BSA. The BSA estimates that the 2010 global software piracy rate was 42 percent with Central and Eastern Europe having the highest piracy rate at 64 percent and North America having the lowest rate at 21 percent.[48]

digital rights management (DRM): Refers to the use of any of several technologies to enforce policies for controlling access to digital media, such as movies, music, and software.

Digital rights management (DRM) refers to the use of any of several technologies to enforce policies for controlling access to digital media, such as movies, music, and software. Many digital content publishers state that DRM technologies are needed to prevent revenue loss due to illegal duplication of their copyrighted works. While the costs of movie piracy can only be estimated imprecisely, the Motion Picture Association of America (MPAA) estimates that 29 million U.S. adults have watched illegal copies of movies or TV shows. It is estimated that 99 percent of the files available on bittorrent, a commonly used peer-to-peer file sharing service, were found to be infringing on copyright protection.[49] On the other hand, many digital content users argue that DRM and associated technologies lead to a loss of user rights. For example, users can purchase a music track online for less than a dollar through Apple's iTunes music store. They can then burn that song to a CD and transfer it to an iPod. However, the purchased music files are encoded in the AAC format supported by iPods and protected by FairPlay, a DRM technology developed by Apple. To the consternation of music lovers, many music devices are not compatible with the AAC format and cannot play iTunes' protected files.

Due to digital rights management (DRM) technology, music files that iTunes members purchase and download play only on iPods and other AAC-compatible devices.

Penalties for software piracy can be severe. If the copyright owner brings a civil action against someone, the owner can seek to stop the person from using its software immediately and can also request monetary damages. The copyright owner can then choose between compensation for actual damages—which includes the amount lost because of the person's infringement as well as any profits attributable to the infringement—and statutory damages, which can be as much as $150,000 for each program copied. In addition, the government can prosecute software pirates in criminal court for copyright infringement.

If convicted, they could be fined up to $250,000 or sentenced to jail for up to five years or both.

The owner and operator of several rogue Web sites that allowed users to download stolen software from Adobe, Autodesk, and Microsoft was sentenced to almost six years in prison and ordered to pay $400,000 in restitution. The person admitted to making more than half a million in illegal profits through software piracy.[50]

Another major issue in regards to copyright infringement is the downloading of copyright-protected music. Estimates vary widely as to how much music piracy is costing the recording industry. The Institute for Policy Innovation estimates that the U.S. global recording industry loses about $12.5 billion in revenue from music piracy every year. It is projected that this results in 70,000 lost jobs and $2 billion in lost wages for U.S. workers.[51]

The Recording Industry Association of America (RIAA) is a trade organization that works to protect the intellectual property and First Amendment rights of artists and music labels. RIAA members create, manufacture, or distribute approximately 85 percent of all legitimate recorded music produced and sold in the United States. The RIAA, acting on behalf of its member organizations, filed copyright infringements lawsuits against 750 individuals in April 2012.[52] The standard out-of-court settlement in such cases is a payment of $750 per work to the RIAA and an agreement not to engage in file sharing of music. If the case goes to court and the individual is found guilty of infringing copyright agreements, the damages can range from $750 and $150,000 per work if the infringement is deemed "willful."

The Stop Online Piracy Act (SOPA) and Preventing Real Online Threats to Economic Creativity and Theft of Intellectual Property Act (Protect Intellectual Property Act, or PIPA) are two bills proposed by Congress in 2011. Both SOPA (a House-sponsored bill) and PIPA (a Senate-sponsored bill) aim to stop online piracy and protect copyright holders.[53] Both bills have three main components: (1) they would allow the Justice Department and copyright holders to seek court orders against foreign Web sites accused of copyright infringement, (2) online advertising networks and credit card companies would be barred from doing business with foreign Web sites accused of copyright infringement, and (3) operators of search engines would be barred from linking to such sites.[54] Information systems industry companies including Internet service providers, Facebook, Google, Wikipedia, and the Business Software Alliance did not support the bills. While acknowledging that digital piracy is a real problem, they feel the bills would limit American innovation and detract from cybersecurity. Movie studios, music producers, and their lobbyist organizations (MPAA and RIAA) support the bills as means to increase their revenue and reduce piracy. Google sponsored a petition drive that attracted millions of participants, and Wikipedia, the online encyclopedia, held a one-day blackout in protest against the bills. Both the House and Senate postponed any vote on the bills in response to the strong opposition.[55] However, work continues to modify the bills, and they may resurface at some time in the near future.

Patent infringement is also a major problem for computer software and hardware manufacturers. It occurs when someone makes unauthorized use of another's patent. If a court determines that a patent infringement is intentional, it can award up to three times the amount of damages claimed by the patent holder. It is not unusual to see patent infringement awards in excess of $10 million. In a case that could be worth tens of millions of dollars, an inventor has sued Apple and Sony for patent infringement. At issue is the technology used in data-vending systems that enable users to store and manage digital content such as music, video, and software.[56]

In another major patent infringement lawsuit, Oracle filed a suit against Google alleging that Google's Android smartphone software infringes on patents and copyrights related to the Java programming language. Oracle acquired the Java technology when it bought out Sun Microsystems in early 2012. It seeks to block the alleged infringement and to be awarded damages. The suit could retard the sales of the highly popular Android and result in tens of millions in damages being awarded to Oracle.[57]

To obtain a patent or to determine if a patent exists in an area a company seeks to exploit requires a search by the U.S. Patent Office; these can last longer than 25 months. Indeed, the patent process is so controversial that manufacturing firms, the financial community, consumer and public interest groups, and government leaders are demanding patent reform.

Computer-Related Scams

People have lost hundreds of thousands of dollars on real estate, travel, stock, and other business scams. Today, many of these scams are being perpetrated with computers. Using the Internet, scam artists offer get-rich-quick schemes involving bogus real estate deals, tout "free" vacations with huge hidden costs, commit bank fraud, offer fake telephone lotteries, sell worthless penny stocks, and promote illegal tax-avoidance schemes.

Over the past few years, credit card customers of various banks have been targeted by scam artists trying to get personal information needed to use their credit cards. The scam works by sending customers an e-mail including a link that seems to direct users to their bank's Web site. At the site, they are greeted with a pop-up box asking them for their full debit card numbers, their personal identification numbers, and their credit card expiration dates. The problem is that the Web site is fake, operated by someone trying to gain access to customers' private information, a form of scam called *phishing*.

The current round of phishing scams are so sophisticated that they look like e-mails you would expect to receive from a major bank. Messages display a familiar bank logo and clicking it takes you to the real bank's Web site. The e-mail uses a real no-reply e-mail address from the bank itself rather than a clearly fake Yahoo or Hotmail address. However, the messages inform recipients that because the bank found a serious problem with their account, they must complete a form and provide account numbers, personal identification numbers, and other key information scammers need to impersonate you.[58] Phishing has become such a serious problem that the Bank of America, Facebook, Google, LinkedIn, Microsoft, PayPal, and Yahoo have formed the Domain-based Message Authentication, Reporting, and Conformance group to provide improved e-mail security and protection from phishing.[59]

Vishing is similar to phishing. However, instead of using the victim's computer, it uses the victim's phone. The victim is typically sent a notice or message to call to verify account information. If the victim returns the message, the caller asks for personal information, such as a credit card account number or name and address. The information gained can be used in identity theft to acquire and use credit cards in the victim's name. Vishing criminals can even use the Spoof Card, sold online for less than $5 dollars for 25 calls. It causes phones to display a caller ID number specified by the caller rather than the actual number of the caller.

One vishing scam is directed at diabetes patients. Callers claim to be from Medicare or a diabetes association and offer free testing or medical supplies. However, once they acquire personal information including Medicare data, they bill Medicare, often multiple times.[60]

International Computer Crime

Computer crime becomes more complex when it crosses borders. Money laundering is the practice of disguising illegally gained funds so that they seem legal. With the increase in electronic cash and funds transfer, some are concerned that terrorists, international drug dealers, and other criminals are using information systems to launder illegally obtained funds. For example, a pharmacist was sentenced to five years in prison for distribution of schedule III- and IV-controlled substances via the Internet without valid prescriptions and for money laundering. Customers of the Web site were not asked to provide medical records and answered only a few questions on a brief medical form. The Web site operator then "approved" these orders and asked the customer for a credit card for payment. The operator sent the order to a doctor or a lay person to "authorize" the order. The pharmacist and his workers then would receive the orders, fill them, and ship them to the customer. The doctors and lay people were paid for authorizing the orders. Over $3 million in payments were transferred from bank accounts in the Dominican Republic to accounts in the United States.[61]

PREVENTING COMPUTER-RELATED CRIME

Because of increased computer use today, greater emphasis is placed on the prevention and detection of computer crime. Although all states have passed computer crime legislation, some believe that these laws are not effective because companies do not always actively detect and pursue computer crime, security is inadequate, and convicted criminals are not severely punished. However, all over the United States, private users, companies, employees, and public officials are making individual and group efforts to curb computer crime, and recent efforts have met with some success.

Crime Prevention by State and Federal Agencies

State and federal agencies have begun aggressive attacks on computer criminals, including criminal hackers of all ages. In 1986, Congress enacted the Computer Fraud and Abuse Act, which mandates punishment based on the victim's dollar loss.

The Department of Defense also supports the Computer Emergency Response Team (CERT), which responds to network security breaches and monitors systems for emerging threats. Law enforcement agencies are also increasing their efforts to stop criminal hackers, and many states are now passing new, comprehensive bills to help eliminate computer crimes. Advice for providing good computer and network security as well as a complete listing of computer-related legislation by state can be found at the Online Security Web site at *www.onlinesecurity.com/forum/article46.php*. Recent court cases and police reports involving computer crime show that lawmakers are ready to introduce newer and tougher computer crime legislation.

Crime Prevention by Corporations

Companies are also taking crime-fighting efforts seriously. Many businesses have designed procedures and specialized hardware and software to protect their corporate data and systems. Specialized hardware and software, such as encryption devices, can be used to encode data and information to help prevent unauthorized use. Encryption is the process of converting an original electronic message into a form that can be understood only by the intended recipients. A key is a variable value that is applied using an algorithm to a string or block of unencrypted text to produce encrypted text or to decrypt encrypted text. Encryption methods rely on the limitations of computing power for their effectiveness: If breaking a code requires too much computing power, even the most determined code crackers will not be successful. The length of the key used to encode and decode messages determines the strength of the encryption algorithm.

Over 75 percent of U.S. organizations view data protection activities as a key component of enterprise risk management. Thus, the use of encryption to protect data stored on backup files and laptops and data transmitted externally is increasing. The primary justification for implementing encryption is to protect an organization's brand or reputation from the damage that would result from a serious data breach.[62]

Arch Chemicals is a United States-based company that provides solutions to control the growth of harmful microbes. The company employs 3,000 people and had recent annual sales in excess of $1.5 billion. Some 800 employees have laptop computers, and every year a few of those computers go missing. According to Damon Allen, manager, Distributed Systems Operations for North America, "The price of the laptop itself is insignificant when compared to the value of the data, so we wanted to be absolutely certain that unauthorized users could not gain access to valuable information via a lost or stolen device. A data breach can be catastrophic, and so you need to be 100 per cent certain that all data on all devices is protected at all times." As a result, the company implemented an encryption software solution to scramble the data on the hard drive of its laptops so that the data is unintelligible to any user without the proper authorization and the correct encryption key.[63]

As employees move from one position to another in a company, they can build up access to multiple systems if inadequate security procedures fail to revoke access privileges. It is clearly not appropriate for people who have changed positions and responsibilities to still have access to systems that they no longer use. To avoid this problem, many organizations create role-based system access lists so that only people filling a particular role (e.g., invoice approver) can access a specific system.

Hackers sometimes gain access to systems by exploiting inactive user accounts. It is critical that accounts of employees or contractors who have left the company or who are on extended disability be terminated. When an employee of Gucci was fired, the network administrator immediately disabled the former employee's accounts. However, unbeknownst to the administrator, the employee had created a fake account before he was fired. He used this account to delete files and e-mails, causing an estimated $20,000 in damages.[64]

separation of duties: The careful division of the tasks and responsibilities associated with a key process so that must be performed by more than one person.

In addition, a fundamental concept of good internal controls is the careful **separation of duties** associated with a key process so that they must be performed by more than one person. Separation of duties is essential for any process that involves the handling of financial transactions so that fraud requires the collusion of two or more parties. When designing an accounts receivable information system, for instance, separation of duties dictates that you separate responsibility for the receipt of customer payments, approving write-offs, depositing cash, and reconciling bank statements.

Proper separation of duties is frequently reviewed during any audit. The Utah State Auditor performed a review of the internal control and cash receipt process for the Medicaid Operations Division of the Utah Department of Health. The auditor found that the same individual had access to the cash and checks received in the mail and also had the ability to record the amounts received. This access would enable that individual to commit fraud by recording an amount less than the amount received and pocketing the difference.[65]

Fingerprint authentication devices provide security in the PC environment by using fingerprint recognition instead of passwords. Laptop computers from Lenovo, Toshiba, and others have built-in fingerprint readers used to log on and gain access to the computer system and its data. Some 2,000 mobile fingerprint scanner devices have been deployed to various law enforcement agencies throughout Los Angeles. The devices are used to capture a digital fingerprint and transmit it to an automated fingerprint identification system (AFIS). If a match is found, the individual's name, photograph, and most current arrest record are returned.[66]

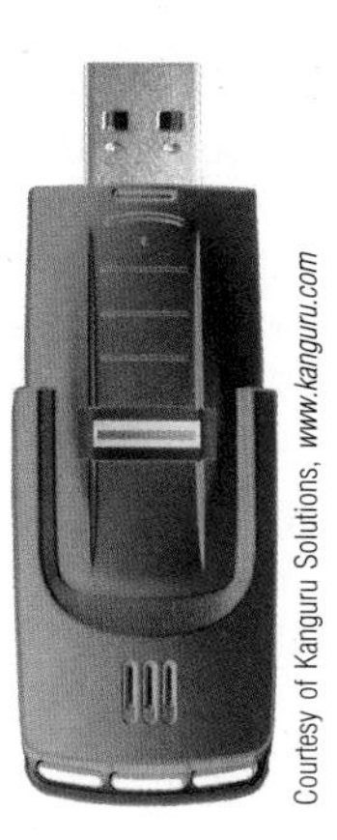
Courtesy of Kanguru Solutions, www.kanguru.com

Some USB flash drives have built-in fingerprint readers to protect the data on the device.

Crime-fighting procedures usually require additional controls on the information system. Before designing and implementing controls, organizations must consider the types of computer-related crime that might occur, the consequences of these crimes, and the cost and complexity of needed controls. In most cases, organizations conclude that the trade-off between crime and the additional cost and complexity weighs in favor of better system controls. Having knowledge of some of the methods used to commit crime is also helpful in preventing, detecting, and developing systems resistant to computer crime. Some companies actually hire former criminals to thwart other criminals.

The following list provides a set of useful guidelines to protect corporate computers from criminal hackers:

- Install strong user authentication and encryption capabilities on the corporate firewall.
- Install the latest security patches, which are often available at the vendor's Internet site.
- Disable guest accounts and null user accounts that let intruders access the network without a password.
- Do not provide overfriendly logon procedures for remote users (e.g., an organization that used the word "welcome" on its initial logon screen found it had difficulty prosecuting a criminal hacker).
- Restrict physical access to the server and configure it so that breaking into one server won't compromise the whole network.
- Dedicate one server to each application (e-mail, File Transfer Protocol, and domain name server). Turn audit trails on.
- Install a corporate firewall between your corporate network and the Internet.
- Install antivirus software on all computers and regularly download vendor updates.
- Conduct regular IS security audits.
- Verify and exercise frequent data backups for critical data.

Using Intrusion Detection System

intrusion detection system (IDS): Monitors system and network resources and traffic and notifies network security personnel when it senses a possible intrusion.

An **intrusion detection system (IDS)** monitors system and network resources and traffic and notifies network security personnel when it senses a possible intrusion. Examples of suspicious activities include repeated failed logon attempts, attempts to download a program to a server, and access to a system at unusual hours. Such activities generate alarms that are captured on log files. When they detect an apparent attack, intrusion detection systems send an alarm, often by e-mail or pager, to network security personnel. Unfortunately, many IDSs frequently provide false alarms that result in wasted effort. If the attack is real, network security personnel must make a decision about what to do to resist the attack. Any delay in response increases the probability of damage. Use of an IDS provides another layer of protection in case an intruder gets past the outer security layers—passwords, security procedures, and corporate firewall.

Rackspace provides cloud-computing hosting services for over 172,000 customers worldwide. It is critical that the firm maintain a highly reliable computing environment free from malicious attacks. The firm employs its Alert Logic Threat Manager intrusion detection system to guard its customers against threats such as worms, Trojans, and other malware.[67]

Security Dashboard

security dashboard: Software that provides a comprehensive display on a single computer screen of all the vital data related to an organization's security defenses, including threats, exposures, policy compliance and incident alerts.

Many organizations use **security dashboard** software to provide a comprehensive display on a single computer screen of all the vital data related to an organization's security defenses, including threats, exposures, policy compliance, and incident alerts. The goal is to reduce the effort required for monitoring and to identify threats earlier. Data comes from a variety of sources, including firewalls, applications, servers, and other software and hardware devices (see Figure 14.2).

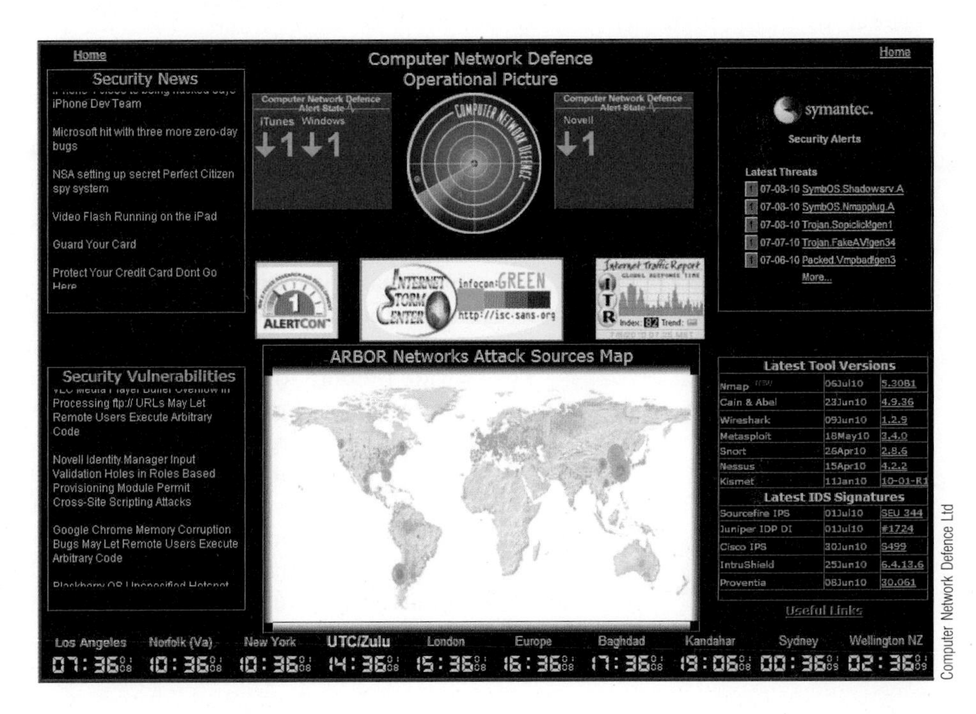

Computer Network Defence Ltd

FIGURE **14.2**

Computer Network Defence Internet Operational Picture

The Computer Network Defence Internet Operational Picture, a security dashboard designed for the United Kingdom government and military networks, displays near real-time information on new and emerging cyber threats.

Texas Children's Hospital is based in Houston and employs over 8,000 workers who need to access the Internet, electronic medical records, billing systems, diagnostic systems, and many other medical applications. The hospital needed to provide an information infrastructure that would enable its employees to collaborate without the risk of compromising patient data or exposing the system to malicious viruses and other malware. To that end, the hospital upgraded to software that provides protection against all forms of malware and can detect potential intrusions by unauthorized users. The hospital may also implement a security dashboard that provides a snapshot of the data needed to identify and respond to security incidents immediately.[68]

Using Managed Security Service Providers

Keeping up with computer criminals—and with new regulations—can be daunting for organizations. Criminal hackers are constantly poking and prodding, trying to breach the security defenses of companies. Also, such recent legislation as HIPAA, Sarbanes-Oxley, and the USA Patriot Act requires businesses to prove that they are securing their data. For most small and mid-sized organizations, the level of in-house network security expertise needed to protect their business operations can be quite costly to acquire and maintain. As a result, many are outsourcing their network security operations to managed security service providers (MSSPs), such as AT&T, CSC, Dell SecureWorks, IBM, Symantec, and Verizon. MSSPs monitor, manage, and maintain network security for both hardware and software. These companies provide a valuable service for IS departments drowning in reams of alerts and false alarms coming from virtual private networks (VPNs); antivirus, firewall, and intrusion detection systems; and other security-monitoring systems. In addition, some provide vulnerability scanning and Web blocking/filtering capabilities.

Coca-Cola, with over 90,000 employees and operations in more than 200 countries, outsourced its security monitoring and management to Symantec Managed Security Services. Coca-Cola gained top-notch human resources, hardware, and software to meet its information security objectives in a cost-effective manner.[69]

Guarding against Theft of Equipment and Data

Organizations need to take strong measures to guard against the theft of computer hardware and the data stored on it. Here are a few measures to be considered:

- Set clear guidelines on what kind of data (and how much of it) can be stored on vulnerable laptops. In many cases, private data or company confidential data may not be downloaded to laptops that leave the office.
- Require that data stored on laptops be encrypted and do spot checks to ensure that this policy is followed.
- Require that all laptops be secured using a lock and chain device so that they cannot be easily removed from an office area.
- Provide training to employees and contractors on the need for safe handling of laptops and their data. For example, laptops should never be left in a position where they can be viewed by the public, such as on the front seat of an automobile.
- Consider installing tracking software on laptops. The software sends messages via a wireless network to the specified e-mail address, pinpointing its location and including a picture of the thief (for those computers with an integrated Web cam).

Crime Prevention for Individuals and Employees

This section outlines actions that individuals can take to prevent becoming a victim of computer crime, including identity theft, malware attacks, theft of equipment and data, and computer scams.

Identity Theft

Consumers can protect themselves from identity theft by regularly checking their credit reports with major credit bureaus, following up with creditors if their bills do not arrive on time, not revealing any personal information in response to unsolicited e-mail or phone calls (especially Social Security numbers and credit card account numbers), and shredding bills and other documents that contain sensitive information.

Some consumers contract with a service company that provides fraud-monitoring services, helps file required reports, and disputes unauthorized transactions in accounts. Some services even offer identity theft guarantees of up to $1 million. Some of the more popular services include Trusted ID, Life Lock, Protect My ID, ID Watchdog, and Identity Guard. These services cost between $6 and $20 per month.

The U.S. Congress passed the Identity Theft and Assumption Deterrence Act of 1998 to fight identity theft. Under this act, the Federal Trade Commission (FTC) is assigned responsibility to help victims restore their credit and erase the impact of the imposter. It also makes identity theft a federal felony punishable by a prison term ranging from 3 to 25 years.

Malware Attacks

The number of personal computers infected with malware (viruses, worms, spyware, and so on) has reached epidemic proportions. As a result of the increasing threat of malware, most computer users and organizations have installed **antivirus programs** on their computers. Such software runs in the background to protect your computer from dangers lurking on the Internet and other possible sources of infected files. The latest virus definitions are downloaded automatically when you connect to the Internet, ensuring that your PC's protection is current. To safeguard your PC and prevent it from spreading malware to your friends and coworkers, some antivirus software scans and cleans both incoming and outgoing e-mail messages. Avast! Free Antivirus 7, AVG Anti-Virus Free 2012, Bitdefender Antivirus Plus 2012, Norton AntiVirus 2012, and Webroot Secure Anywhere Antivirus were among the top-rated antivirus programs for 2012 according to PC Magazine. They range in cost from nothing to $39.95 (see Figure 14.3).

antivirus program: Software that runs in the background to protect your computer from dangers lurking on the Internet and other possible sources of infected files.

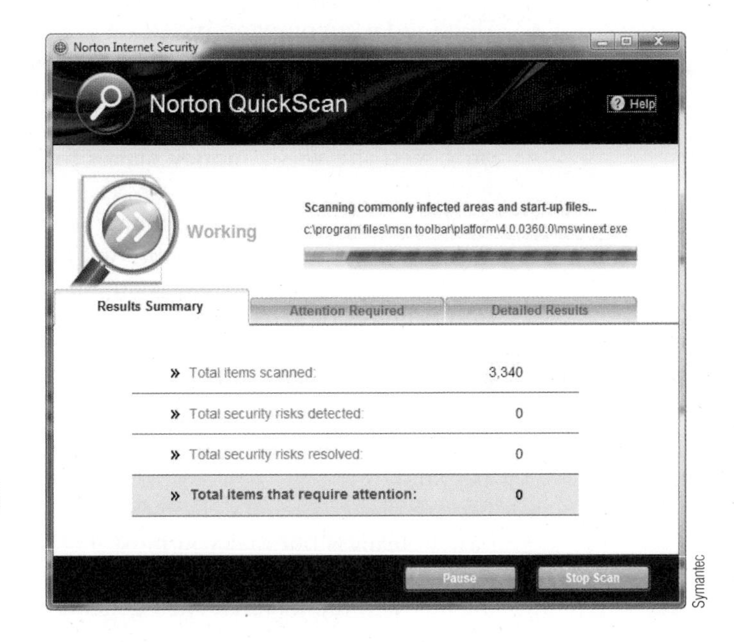

FIGURE **14.3**

Antivirus software

Antivirus software should be used and updated often.

Proper use of antivirus software requires the following steps:

1. **Install antivirus software and run it often**. Many of these programs automatically check for viruses each time you boot up your computer or insert a disk or CD, and some even monitor all e-mail, file transmissions, and copying operations.
2. **Update antivirus software often**. New viruses are created all the time, and antivirus software suppliers are constantly updating their software to detect and take action against these new viruses.
3. **Scan all removable media, including CDs, before copying or running programs from them**. Hiding on disks or CDs, viruses often move between systems. If you carry document or program files on removable media between computers at school or work and your home system, always scan them.
4. **Install software only from a sealed package or secure Web site of a known software company**. Even software publishers can unknowingly

distribute viruses on their program disks or software downloads. Most scan their own systems, but viruses might still remain.

5. **Follow careful downloading practices**. If you download software from the Internet or a bulletin board, check your computer for viruses immediately after completing the transmission.
6. **If you detect a virus, take immediate action**. Early detection often allows you to remove a virus before it does any serious damage.

Many e-mail services and ISP providers offer free antivirus protection. For example, AOL and MWEB (one of South Africa's leading ISPs) offer free antivirus software from McAfee.

Computer Scams

The following is a list of tips to help you avoid becoming a victim of a computer scam:

- Don't agree to anything in a high-pressure meeting or seminar. Insist on having time to think it over and to discuss your decision with someone you trust. If a company won't give you the time you need to check out an offer and think things over, you don't want to do business with it. A good deal now will be a good deal tomorrow; the only reason for rushing you is if the company has something to hide.
- Don't judge a company based on appearances. Flashy Web sites can be created and published in a matter of days. After a few weeks of taking money, a site can vanish without a trace in just a few minutes. You might find that the perfect money-making opportunity offered on a Web site was a money maker for the crook and a money loser for you.
- Avoid any plan that pays commissions simply for recruiting additional distributors. Your primary source of income should be your own product sales. If the earnings are not made primarily by sales of goods or services to consumers or sales by distributors under you, you might be dealing with an illegal pyramid scheme.
- Beware of shills—people paid by a company to lie about how much they've earned and how easy the plan was to operate. Check with an independent source to make sure that the company and its offers are valid.
- Beware of a company's claim that it can set you up in a profitable home-based business but that you must first pay up front to attend a seminar and buy expensive materials. Frequently, seminars are high-pressure sales pitches, and the material is so general that it is worthless.
- If you are interested in starting a home-based business, get a complete description of the work involved before you send any money. You might find that what you are asked to do after you pay is far different from what was stated in the ad. You should never have to pay for a job description or for needed materials.
- Get in writing the refund, buy-back, and cancellation policies of any company you deal with. Do not depend on oral promises.
- Do your homework. Check with your state attorney general and the National Fraud Information Center before getting involved, especially when the claims about a product or potential earnings seem too good to be true.

If you need advice about an Internet or online solicitation, or if you want to report a possible scam, use the Online Reporting Form or Online Question & Suggestion Form features on the Web site for the National Fraud Information Center at *http://fraud.org* or call the NFIC hotline at 1-800-876-7060.

PRIVACY ISSUES

Another important social issue in information systems involves privacy. In 1890, U.S. Supreme Court Justice Louis Brandeis stated that the "right to be left alone" is one of the most "comprehensive of rights and the most valued by civilized man." Basically, the issue of privacy deals with this right to be left alone or to be withdrawn from public view. With information systems, privacy deals with the collection and use or misuse of data. Data is constantly being collected and stored on each of us. This data is often distributed over easily accessed networks and without our knowledge or consent. Concerns of privacy regarding this data must be addressed. A difficult question to answer is "Who owns this information and knowledge?" If a public or private organization spends time and resources to obtain data on you, does the organization own the data, and can it use the data in any way it desires? Government legislation answers these questions to some extent for federal agencies, but the questions remain unanswered for private organizations. Today, many businesses have to handle many requests from law enforcement agencies for information about its employees, customers, and suppliers. Indeed, some phone and Internet companies have employees whose full-time role it is to deal with information requests from local, state, and federal law enforcement agencies.

Privacy and the Federal Government

The federal government has implemented a number of laws addressing personal privacy that are discussed in this section. The European Union has a data-protection directive that requires firms transporting data across national boundaries to have certain privacy procedures in place. This directive affects virtually any company doing business in Europe, and it is driving much of the attention being given to privacy in the United States.

Privacy at Work

The right to privacy at work is also an important issue. Employers are using technology and corporate policies to manage worker productivity and protect the use of IS resources. Employers are mostly concerned about inappropriate Web surfing, with over half of employers monitoring the Web activity of their employees. Organizations also monitor employees' e-mail, with more than half retaining and reviewing messages. Statistics such as these have raised employee privacy concerns. In many cases, workers claim their right to privacy trumps their companies' rights to monitor employee use of IS resources. However, most employers today have a policy that explicitly eliminates any expectation of privacy when an employee uses any company-owned computer, server, or e-mail system. The courts have ruled that, without a reasonable expectation of privacy, there is no Fourth Amendment protection for the employee. A California appeals court ruled in *Holmes v Petrovich Development Company* that e-mails sent by an employee to her attorney on the employer's computer were not "confidential communications between a client and lawyer." An Ohio federal district court in *Moore v University Hospital Cleveland Medical Center* ruled that an employee could be terminated for showing coworkers sexually explicit photos on his employer's computer. The court stated that the employee could have no expectation of privacy when accessing a hospital computer situated in the middle of a hospital floor within easy view of both patients and staff.[70]

Privacy and E-Mail

E-mail also raises some interesting issues about work privacy. Federal law permits employers to monitor e-mail sent and received by employees. Furthermore, e-mail messages that have been erased from hard disks can be retrieved and used in lawsuits because the laws of discovery demand that companies produce all relevant business documents. On the other hand, the use of e-mail among public officials might violate "open meeting" laws. These laws, which apply to many local, state, and federal agencies, prevent public officials from meeting in private about matters that affect the state or local area.

© iStockphoto/Daniel Laflor

E-mail has changed how workers and managers communicate in the same building or around the world. E-mail, however, can be monitored and intercepted. As with other services such as cell phones, the convenience of e-mail must be balanced with the potential of privacy invasion.

Privacy and Instant Messaging

Using instant messaging (IM) to send and receive messages, files, and images introduces the same privacy issues associated with e-mail. As with e-mail, federal law permits employers to monitor instant messages sent and received by employees. Employers' major concern involves IMs sent by employees over their employer's IM network or using employer-provided phones. To protect your privacy and your employer's property, do not send personal or private IMs at work. The following are a few other tips:

- Choose a nonrevealing, nongender-specific, unprovocative IM screen name (Sweet Sixteen, 2hot4u, UCLAMBA, all fail this test).
- Don't send messages you would be embarrassed to have your family members, colleagues, or friends read.
- Do not open files or click links in messages from people you do not know.
- Never send sensitive personal data such as credit card numbers, bank account numbers, or passwords via IM.

Privacy and Personal Sensing Devices

RFID tags, essentially microchips with antenna, are embedded in many of the products we buy, from medicine containers, clothing, and library books to computer printers, car keys, and tires. RFID tags generate radio transmissions that, if appropriate measures are not taken, can lead to potential privacy concerns. Once these tags are associated with the individual who purchased the item, someone can potentially track individuals by the unique identifier associated with the RFID chip.

A handful of states have reacted to the potential for abuse of RFID tags by passing legislation prohibiting the implantation of RFID chips under people's skin without their approval. Still, advocates for RFID chip implantation argue their potential value in tracking children or criminals and their value in carrying an individual's medical records.

Privacy and the Internet

Some people assume that there is no privacy on the Internet and that you use it at your own risk. Others believe that companies with Web sites should have strict privacy procedures and be held accountable for privacy invasion. Regardless of your view, the potential for privacy invasion on the Internet is

INFORMATION SYSTEMS @ WORK

Keeping E-Mail on Target

The U.S. Health Insurance Portability and Accountability Act (HIPAA) addresses (among other things) the privacy of health information. Its Title 2 regulates the use and disclosure of protected health information (PHI) by health care providers, insurance carriers, employers, and business associates such as billing services.

E-mail is often the best way for a hospital to communicate with off-site specialists and insurance carriers about a patient. Unfortunately, standard e-mail is insecure. It allows eavesdropping, later retrieval of messages from unprotected backups, message modification before it is received, invasion of the sender's privacy by providing access to information about the identity and location of the sending computer, and more. Since health care provider e-mail often carries PHI, health care facilities must be sure their e-mail systems meet HIPAA privacy and security requirements.

Children's National Medical Center (CNMC) of Washington, D.C., "The Nation's Children's Hospital," is especially aware of privacy concerns because all such concerns are heightened with children. CNMC did what many organizations do when faced with a specialized problem: rather than try to become specialists or hire specialists for whom the hospital has no long-term full-time need, it turned to a specialist firm.

CNMC chose Proofpoint of Sunnyvale, California, for its Security as a Service (SaaS) e-mail privacy protection service. Matt Johnston, senior security analyst at CNMC, says that children are "the highest target for identity theft. A small kid's record is worth its weight in gold on the black market. It's not the doctor's job to protect that information. It's *my* job."

Johnston likes several things about the Proofpoint service:

- "I don't have to worry about backups." Proofpoint handles those.
- "I don't have to worry about if a server goes down. [If it was a CNMC server, I would have to] get my staff ramped up and bring up another server. Proofpoint does that for us. It's one less headache."
- "We had a product in-house before. It required several servers which took a full FTE [full-time employee] just to manage this product. It took out too much time."
- "Spam has been on the rise. Since Proofpoint came in, we've seen a dramatic decrease in spam. It takes care of itself. The end user is given a digest daily."
- E-mail can be encrypted or not, according to rules that the end user need not be personally concerned with.
- "Their tech support has been great."

Proofpoint is not the only company that provides health care providers with e-mail security services. LuxSci of Cambridge, Massachusetts, also offers HIPAA-compliant e-mail hosting services, as do several other firms. They all provide the same basic features: user authentication, transmission security (encryption), logging, and audit. Software that runs on the provider's computers can also deliver media control and backup. Software that runs on a user organization's server necessarily relies on that organization to manage storage; for example, deleting messages from the server after four weeks as HIPAA requires.

As people become more aware of the privacy risks associated with standard e-mail, the use of secure solutions such as these will undoubtedly become more common in the future.

Discussion Questions

1. Universities use e-mail to communicate private information. For example, an instructor might send you an e-mail explaining what you must do in order to raise your grade. The regulations about protecting that information under the Family Educational Rights and Privacy Act (FERPA) are not as strict as those under HIPAA. Do you think they should be as strict as HIPAA's requirements? Why or why not?
2. You are a member of a student council considering the issue of student e-mail privacy. Prepare a statement of your position for or against increased e-mail security that includes the reasons behind your position.

Critical Thinking Questions

1. Consider the positive aspects of Proofpoint's service as Johnston listed them. Which could Proofpoint *not* offer without the SaaS approach, that is, if it sold software that ran on its customers' computers? Which benefit from the SaaS approach but do not absolutely require it? Which are unrelated to it?
2. Matt Johnston has a great deal of responsibility at CNMC. His job is technically challenging, has a lot of variety, and probably pays well. If he slips up, CNMC can be liable for heavy penalties under HIPAA, so he is also under a lot of pressure. Do you think you would enjoy being a security analyst at a hospital? Do you think you'd be good at it? For both answers, explain why.

SOURCES: Children's National Medical Center Web site, *www.childrensnational.org*, accessed May 21, 2012; LuxSci Web site, *www.luxsci.com*, accessed May 24, 2012; Proofpoint Web site, *www.proofpoint.com*, accessed May 21, 2012; Staff, "HIPAA Email Security Case Study: Children's National Medical Center," Proofpoint, *www.youtube.com/watch?v=RVaBaNvwkQE*, February 22.

huge. People wanting to invade your privacy could be anyone from criminal hackers to marketing companies to corporate bosses. Your personal and professional information can be seized on the Internet without your knowledge or consent. E-mail is a prime target, as discussed previously. Sending an e-mail message is like having an open conversation in a large room—people can listen to your messages. When you visit a Web site on the Internet, information about you and your computer can be captured. When this information is combined with other information, companies can find out what you read, what products you buy, and what your interests are.

Most people who buy products on the Web say it's very important for a site to have a policy explaining how personal information is used, and the policy statement must make people feel comfortable and be extremely clear about what information is collected and what will and will not be done with it. However, many Web sites still do not prominently display their privacy policy or implement practices completely consistent with that policy. The real issue that Internet users need to be concerned with is what do content providers want to do with that personal information? If a site requests that you provide your name and address, you have every right to know why and what will be done with it. If you buy something and provide a shipping address, will it be sold to other retailers? Will your e-mail address be sold on a list of active Internet shoppers? And if so, you should realize that this e-mail list is no different from the lists compiled from the orders you place with catalog retailers. You have the right to be taken off any mailing list.

The Children's Online Privacy Protection Act (COPPA) was passed by Congress in October 1998. This act was directed at Web sites catering to children, requiring site owners to post comprehensive privacy policies and to obtain parental consent before they collect any personal information from children under 13 years of age. Web site operators who violate the rule could be liable for civil penalties of up to $11,000 per violation. COPPA has made an impact in the design and operations of Web sites that cater to children. For example, the Web site Skid-e-kids violated the act by collecting personal information from some 5,600 children without obtaining prior parental consent. The Federal Trade Commission, responsible for enforcing the act, required that the information be deleted.[71]

A social network service employs the Web and software to connect people for whatever purpose. There are thousands of such networks, which have become popular among teenagers. Some of the more popular social networking Web sites include Bebo, Classmates, Facebook, Foursquare, LinkedIn, and MySpace. Most of these sites allow you to easily create a user profile that provides personal details, photos, and even videos that can be viewed by other visitors to the site. Some of the sites have age restrictions or require that a parent register his or her preteen by providing a credit card to validate the parent's identity. Teens can provide information about where they live, go to school, their favorite music, and their interests in hopes of meeting new friends. Unfortunately, they can also meet ill-intentioned strangers at these sites. Many documented encounters involve adults masquerading as teens attempting to meet young people for illicit purposes. Parents are advised to discuss potential dangers, check their children's profiles, and monitor their activities at such Web sites.

Facebook holds a startling amount of information about its more than 900 million members. In addition, many members are not discrete and reveal such information as their health conditions and treatments; where they will be on a certain day (helpful to potential burglars); personal details of members of their family; their sexual, racial, religious, and political affiliations and preferences; and other personal information about their friends and family. Facebook receives a notice every time you visit a Web site with a "Like" button whether or not you click the "Like" button, log on to Facebook, or are a Facebook user.[72] Users and observers have raised concerns about how

Facebook treats this sometimes very personal information. For example, should law enforcement officials file a subpoena for your Facebook information, they could obtain all these details as well as records of your postings, photos you have uploaded, photos in which you have been tagged, and a list of all your Facebook friends.[73] Under what conditions can Facebook provide this information to third parties for marketing or other purposes?

Privacy and Internet Libel Concerns

Libel involves publishing an intentionally false written statement that is damaging to a person's or organization's reputation. Examples of Internet libel include an ex-husband posting lies about his wife on a blog, a disgruntled former employee posting lies about a company on a message board, and a jilted girlfriend posting false statements to her former boyfriend's Facebook account.

Individuals can post information to the Internet using anonymous e-mail accounts or screen names. This anonymity makes it more difficult, but not impossible, to identify the libeler. The offended party can file what is known as a John Doe lawsuit and use the subpoena power it grants to force the ISP to provide whatever information it has about the anonymous poster, including IP address, name, and street address. (Under Section 230 of the Communications Decency Act, ISPs are not usually held accountable for the bad behavior of their subscribers).

Individuals, too, must be careful what they post on the Internet to avoid libel charges. A Georgia jury awarded $404,000 to a plaintiff who could prove he had been falsely accused of being a pedophile and a drug addict.[74]

In many cases, disgruntled former employees are being sued by their former employers for material posted on the Internet. Employees have even been fired for posting disparaging remarks on their Facebook pages. An Ohio prison guard was fired from his job when he posted the following message on his Facebook account: "OK, we got Bin Laden.... Let's go get Kasich (the Ohio governor) next. Who's with me?"[75]

Privacy and Fairness in Information Use

Selling information to other companies can be so lucrative that many companies will continue to store and sell the data they collect on customers, employees, and others. When is this information storage and use fair and reasonable to the people whose data is stored and sold? Do people have a right to know about data stored about them and to decide what data is stored and used? As shown in Table 14.3, these questions can be broken down into four issues that should be addressed: knowledge, control, notice, and consent.

TABLE 14.3 The right to know and the ability to decide federal privacy laws and regulations

Fairness Issues	Database Storage	Database Usage
The right to know	Knowledge	Notice
The ability to decide	Control	Consent

Knowledge. Should people know what data is stored about them? In some cases, people are informed that information about them is stored in a corporate database. In others, they do not know that their personal information is stored in corporate databases.

Control. Should people be able to correct errors in corporate database systems? This ability is possible with most organizations, although it can be difficult in some cases.

Notice. Should an organization that uses personal data for a purpose other than the original purpose notify individuals in advance? Most companies don't do this.

Consent. If information on people is to be used for other purposes, should these people be asked to give their consent before data on them is used? Many companies do not give people the ability to decide if such information will be sold or used for other purposes.

Privacy and Filtering and Classifying Internet Content

To help parents control what their children see on the Internet, some companies provide *filtering software* to help screen Internet content. Many of these screening programs also prevent children from sending personal information over e-mail or through chat groups. These programs stop children from broadcasting their name, address, phone number, or other personal information over the Internet. According to the 2012 Internet Filter Review, the two top-rated filtering software packages costing less than $50.00 are Net Nanny Parental Controls and McAfee Safe Eyes.[76]

Organizations also implement filtering software to prevent employees from visiting Web sites not related to work, particularly those involving gambling and those containing pornographic or other offensive material. Before implementing Web site blocking, the users must be informed about the company's policies and why they exist. To increase compliance, it is best if the organization's Internet users, management, and IS organization work together to define the policy to be implemented. The policy should be clear about the repercussions to employees who attempt to circumvent the blocking measures.

The U.S. Congress has made several attempts to limit children's exposure to online pornography, including the Communications Decency Act (enacted 1996) and the Child Online Protection Act (enacted 1998). Within two years of being enacted, the U.S. Supreme Court found that both these acts violated the First Amendment (freedom of speech) and ruled them to be unconstitutional. The Children's Internet Protection Act (CIPA) was signed into law in 2000 and later upheld by the Supreme Court in 2003. Under CIPA, schools and libraries subject to CIPA do not receive the discounts offered by the "E-Rate" program unless they certify that they have certain Internet safety measures in place to block or filter "visual depictions that are obscene, child pornography, or are harmful to minors." (The E-Rate program provides many schools and libraries support to purchase Internet access and computers).

In the past few decades, significant laws have been passed regarding a person's right to privacy. Others relate to business privacy rights and the fair use of data and information. The following sections briefly summarize these laws.

The Privacy Act of 1974

The major piece of legislation on privacy is the Privacy Act of 1974 (PA74). PA74 applies only to certain federal agencies. The act, which is about 15 pages long, is straightforward and easy to understand. Its purpose is to provide certain safeguards for people against an invasion of personal privacy by requiring federal agencies (except as otherwise provided by law) to do the following:

- Permit people to determine what records pertaining to them are collected, maintained, used, or disseminated by such agencies.
- Permit people to prevent records pertaining to them from being used or made available for another purpose without their consent.
- Permit people to gain access to information pertaining to them in federal agency records, to have a copy of all or any portion thereof, and to correct or amend such records.
- Ensure that federal agencies collect, maintain, use, or disseminate any record of identifiable personal information in a manner that ensures that such action is for a necessary and lawful purpose, that the information is current and accurate for its intended use, and that adequate safeguards are provided to prevent misuse of such information.
- Permit exemptions from this act only in cases of an important public need for such exemption, as determined by specific law-making authority.
- Be subject to civil suit for any damages that occur as a result of willful or intentional action that violates anyone's rights under this act.

PA74, which applies to all federal agencies except the CIA and law enforcement agencies, established a Privacy Study Commission to study existing databases and to recommend rules and legislation for consideration by Congress. PA74 also requires training for all federal employees who interact with a "system of records" under the act. Most of the training is conducted by the Civil Service Commission and the Department of Defense. Another interesting aspect of PA74 concerns the use of Social Security numbers—federal, state, and local governments and agencies cannot discriminate against people for not disclosing or reporting their Social Security number.

Electronic Communications Privacy Act

This law was enacted in 1986 and deals with three main issues: (1) the protection of communications while in transit from sender to receiver, (2) the protection of communications held in electronic storage, and (3) the prohibition of devices to record dialing, routing, addressing, and signaling information without a search warrant. Under Title I of this law, the government is prohibited from intercepting electronic messages unless it obtains a court order based on probable cause. Title II prohibits access to wire and electronic communications for stored communications not readily accessible to the general public.

Gramm-Leach-Bliley Act

This act was passed in 1999 and requires all financial institutions to protect and secure customers' nonpublic data from unauthorized access or use. Under terms of this act, it is assumed that all customers approve of the financial institutions' collecting and storing their personal information. The institutions are required to contact their customers and inform them of this fact. Customers are required to write separate letters to each of their financial institutions and state in writing that they want to opt out of the data collection and storage process. Most people are overwhelmed with the mass mailings they receive from their financial institutions and simply discard them without understanding their importance.

USA Patriot Act

As discussed previously, the 2001 Uniting and Strengthening America by Providing Appropriate Tools Required to Intercept and Obstruct Terrorism Act (USA Patriot Act) was passed in response to the September 11 terrorism acts. Proponents argue that it gives necessary new powers to both domestic law enforcement and international intelligence agencies. Critics argue that the law removes many of the checks and balances that previously allowed the courts to ensure that law enforcement agencies did not abuse their powers. For example, under this act, Internet service providers and telephone companies must turn over customer information, including numbers called, without a court order if the FBI claims that the records are relevant to a terrorism investigation. Also, the company is forbidden to disclose that the FBI is conducting an investigation.

Table 14.4 lists additional laws related to privacy.

Corporate Privacy Policies

Even though privacy laws for private organizations are not very restrictive, most organizations are sensitive to privacy issues and fairness. They realize that invasions of privacy can hurt their business, turn away customers, and dramatically reduce revenues and profits. Consider a major international credit card company. If the company sold confidential financial information on millions of customers to other companies, the results could be disastrous. In a matter of days, the firm's business and revenues could be reduced

TABLE **14.4** Federal privacy laws and their provisions

Law	Provisions
Fair Credit Reporting Act of 1970 (FCRA)	Regulates operations of credit-reporting bureaus, including how they collect, store, and use credit information
Family Education Privacy Act 1974	Restricts collection and use of data by federally funded educational institutions, including specifications for the type of data collected, access by parents and students to the data, and limitations on disclosure
Tax Reform Act of 1976	Restricts collection and use of certain information by the Internal Revenue Service
Right to Financial Privacy Act of 1978	Restricts government access to certain records held by financial institutions
Electronic Funds Transfer Act of 1979	Outlines the responsibilities of companies that use electronic funds transfer systems, including consumer rights and liability for bank debit cards
Electronic Communications Privacy Act of 1986	Defines provisions for the access, use, disclosure, interception, and privacy protections of electronic communications
Computer Matching and Privacy Act of 1988	Regulates cross-references between federal agencies' computer files (e.g.to verify eligibility for federal programs)
Video Privacy Act of 1988	Prevents retail stores from disclosing video rental records without a court order
Telephone Consumer Protection Act of 1991	Limits telemarketers' practices
Cable Act of 1992	Regulates companies and organizations that provide wireless communications services, including cellular phones
Computer Abuse Amendments Act of 1994	Prohibits transmissions of harmful computer programs and code, including viruses
Gramm-Leach-Bliley Act of 1999	Requires all financial institutions to protect and secure customers' nonpublic data from unauthorized access or use
USA Patriot Act of 2001	Requires Internet service providers and telephone companies to turn over customer information, including numbers called, without a court order, if the FBI claims that the records are relevant to a terrorism investigation
E-Government Act of 2002	Requires federal agencies to post machine-readable privacy policies on their Web sites and to perform privacy impact assessments on all new collections of data of ten or more people
Fair and Accurate Credit Transactions Act of 2003	Designed to combat the growing crime of identity theft; allows consumers to get free credit reports from each of the three major consumer credit reporting agencies every 12 months and to place alerts on their credit histories under certain circumstances

dramatically. Therefore, most organizations maintain privacy policies, even though they are not required by law. Some companies even have a privacy bill of rights that specifies how the privacy of employees, clients, and customers will be protected. Corporate privacy policies should address a customer's knowledge, control, notice, and consent over the storage and use of information. They can also cover who has access to private data and when it can be used.

The BBB Code of Business Practices (BBB Accreditation Standards) requires that BBB Accredited Businesses have some sort of privacy notice on their Web site. Each notice must be based on the following five elements:[77]

- **Notice** (what personal information is being collected on the site)
- **Choice** (what options the customer has about how/whether her or his data is collected and used)
- **Access** (how a customer can see what data has been collected and change/correct it if necessary)
- **Security** (state how any data that is collected is stored/protected)
- **Redress** (what customer can do if privacy policy is not met)

Figure 14.4 shows a sample privacy notice that the BBB provides as a guide to businesses to post on their Web sites.

Privacy Notice

This privacy notice discloses the privacy practices for (Web site address). This privacy notice applies solely to information collected by this web site. It will notify you of the following:

1. What personally identifiable information is collected from you through the web site, how it is used and with whom it may be shared.
2. What choices are available to you regarding the use of your data.
3. The security procedures in place to protect the misuse of your information.
4. How you can correct any inaccuracies in the information.

Information Collection, Use, and Sharing
We are the sole owners of the information collected on this site. We only have access to/collect information that you voluntarily give us via email or other direct contact from you. We will not sell or rent this information to anyone.

We will use your information to respond to you, regarding the reason you contacted us. We will not share your information with any third party outside of our organization, other than as necessary to fulfill your request, e.g. to ship an order.

Unless you ask us not to, we may contact you via email in the future to tell you about specials, new products or services, or changes to this privacy policy.

Your Access to and Control Over Information
You may opt out of any future contacts from us at any time. You can do the following at any time by contacting us via the email address or phone number given on our Web site:

- See what data we have about you, if any.
- Change/correct any data we have about you.
- Have us delete any data we have about you.
- Express any concern you have about our use of your data.

Security
We take precautions to protect your information. When you submit sensitive information via the Web site, your information is protected both online and offline.

Wherever we collect sensitive information (such as credit card data), that information is encrypted and transmitted to us in a secure way. You can verify this by looking for a closed lock icon at the bottom of your web browser, or looking for "https" at the beginning of the address of the web page.

While we use encryption to protect sensitive information transmitted online, we also protect your information offline. Only employees who need the information to perform a specific job (for example, billing or customer service) are granted access to personally identifiable information. The computers/servers in which we store personally identifiable information are kept in a secure environment.

If you feel that we are not abiding by this privacy policy, you should contact us immediately via telephone at XXX YYY-ZZZZ or via email.

FIGURE **14.4**

Sample privacy notice

The BBB provides this sample privacy notice as a guide to businesses to post on their Web sites.

Multinational companies face an extremely difficult challenge in implementing data collection and dissemination processes and policies because of the multitude of differing country or regional statutes. For example, Australia requires companies to destroy customer data (including backup files) or make it anonymous after it's no longer needed. Firms that transfer customer and personnel data out of Europe must comply with European privacy laws that allow customers and employees to access data about them and let them determine how that information can be used.

Web sites for a few corporate privacy policies are shown in Table 14.5.

A good database design practice is to assign a single unique identifier to each customer so each has a single record describing all relationships with the company across all its business units. That way, the organization can apply customer privacy preferences consistently throughout all databases. Failure to do so can expose the organization to legal risks—aside from upsetting customers who opted out of some collection practices. Again, the 1999

TABLE 14.5 Corporate privacy policies

Company	URL
Intel	*www.intel.com/sites/sitewide/en_US/privacy/privacy.htm*
Starwood Hotels & Resorts	*www.starwoodhotels.com/corporate/privacy_policy.html*
TransUnion	*www.transunion.com/corporate/privacyPolicy.page*
United Parcel Service	*www.ups.com/content/corp/privacy_policy.html*
Visa	*www.corporate.visa.com/ut/privacy.jsp*
Walt Disney Internet Group	*http://disney.go.com/corporate/privacy/pp_wdig.html*

Gramm-Leach-Bliley Financial Services Modernization Act requires all financial service institutions to communicate their data privacy rules and honor customer preferences.

Individual Efforts to Protect Privacy

Although numerous state and federal laws deal with privacy, the laws do not completely protect individual privacy. In addition, not all companies have privacy policies. As a result, many people are taking steps to increase their own privacy protection. Some of the steps that you can take to protect personal privacy include the following:

- **Find out what is stored about you in existing databases**. Call the major credit bureaus to get a copy of your credit report. You are entitled to a free credit report every 12 months (see *www.freecreditreport.com*). You can also obtain a free report if you have been denied credit in the last 60 days. The major companies are Equifax (800-685-1111, *www.equifax.com*), TransUnion (800-916-8800, *www.transunion.com*), and Experian (888-397-3742, *www.experian.com*). You can also submit a Freedom of Information Act request to a federal agency that you suspect might have information stored on you.
- **Be careful when you share information about yourself**. Don't share information unless it is absolutely necessary. Every time you give information about yourself through an 800, 888, or 900 call, your privacy is at risk. Be vigilant in insisting that your doctor, bank, or financial institution not share information about you with others without your written consent.
- **Be proactive to protect your privacy**. You can get an unlisted phone number and ask the phone company to block caller ID systems from reading your phone number. If you change your address, don't fill out a change-of-address form with the U.S. Postal Service; you can notify the people and companies that you want to have your new address. Destroy copies of your charge card bills and shred monthly statements before disposing of them in the garbage. Be careful about sending personal e-mail messages over a corporate e-mail system. You can also get help in avoiding junk mail and telemarketing calls by visiting the Direct Marketing Association Web site at *www.the-dma.org*. Go to the site and look under Consumer Help-Remove Name from Lists.
- **Take extra care when purchasing anything from a Web site.** Make sure that you safeguard your credit card numbers, passwords, and personal information. Do not do business with a site unless you know that it handles credit card information securely. (Look for a seal of approval from organizations such as the Better Business Bureau Online or TRUSTe. When you open the Web page where you enter credit card information or other personal data, make sure that the Web address begins with *https*

and check to see if a locked padlock icon appears in the Address bar or status bar). Do not provide personal information without reviewing the site's data privacy policy. Many credit card companies issue single-use credit card numbers on request. Charges appear on your usual bill, but the number is destroyed after a single use, eliminating the risk of stolen credit card numbers.

THE WORK ENVIRONMENT

The use of computer-based information systems has changed the makeup of the workforce. Jobs that require IS literacy have increased, and many less-skilled positions have been eliminated. Corporate programs, such as reengineering and continuous improvement, bring with them the concern that, as business processes are restructured and information systems are integrated within them, the people involved in these processes will be removed. Even the simplest tasks have been aided by computers, making customer checkout faster, streamlining order processing, and allowing people with disabilities to participate more actively in the workforce. As computers and other IS components drop in cost and become easier to use, more workers will benefit from the increased productivity and efficiency provided by computers. Yet, despite these increases in productivity and efficiency, information systems can raise other concerns.

Health Concerns

Organizations can increase employee productivity by paying attention to the health concerns in today's work environment. For some people, working with computers can cause occupational stress. Anxieties about job insecurity, loss of control, incompetence, and demotion are just a few of the fears workers might experience. In some cases, the stress can become so severe that workers avoid taking training to learn how to use new computer systems and equipment. Monitoring employee stress can alert companies to potential problems. Training and counseling can often help the employee and deter problems.

Heavy computer use can affect one's physical health as well. A job that requires sitting at a desk and using a computer for many hours a day qualifies as a sedentary job. Such work can double the risk of seated immobility thromboembolism (SIT), the formation of blood clots in the legs or lungs. People leading a sedentary lifestyle are also likely to experience an undesirable weight gain, which can lead to increased fatigue and greater risk of type 2 diabetes, heart problems, and other serious ailments.

Repetitive strain injury (RSI) is an injury or disorder of the muscles, nerves, tendons, ligaments, or joints caused by repetitive motion. RSI is a very common job-related injury. Tendonitis is inflammation of a tendon due to repetitive motion on that tendon. Carpal tunnel syndrome (CTS) is an inflammation of the nerve that connects the forearm to the palm of the wrist. CTS involves wrist pain, a feeling of tingling and numbness, and difficulty grasping and holding objects.

Avoiding Health and Environmental Problems

Two primary causes of computer-related health problems are a poorly designed work environment and failure to take regular breaks to stretch the muscles and rest the eyes. The computer screen can be hard to read because of glare and poor contrast. Desks and chairs can also be uncomfortable. Keyboards and computer screens might be fixed in place or difficult to move. The hazardous activities associated with these unfavorable conditions are collectively referred to as *work stressors*. Although these problems might not be of major concern to casual users of computer systems, continued stressors such as repetitive motion, awkward posture, and eye strain can cause more serious and long-term injuries. If nothing else, these problems can severely limit productivity and performance.

© David Young-Wolff/PhotoEdit

Research has shown that developing certain ergonomically correct habits can reduce the risk of adverse health effects when using a computer.

ergonomics: The science of designing machines, products, and systems to maximize the safety, comfort, and efficiency of the people who use them.

The science of designing machines, products, and systems to maximize the safety, comfort, and efficiency of the people who use them, called **ergonomics**, has suggested some approaches to reducing these health problems. Ergonomic experts carefully study the slope of the keyboard, the positioning and design of display screens, and the placement and design of computer tables and chairs. Flexibility is a major component of ergonomics and an important feature of computer devices. People come in many sizes, have differing preferences, and require different positioning of equipment for best results. Some people, for example, want to place the keyboard in their laps; others prefer it on a solid table. Because of these individual differences, computer designers are attempting to develop systems that provide a great deal of flexibility.

It is never too soon to stop unhealthy computer work habits. Prolonged computer use under poor working conditions can lead to carpal tunnel syndrome, bursitis, headaches, and permanent eye damage. Strain and poor office conditions cannot be left unchecked. Unfortunately, at times, we are all distracted by pressing issues such as the organization's need to raise productivity, improve quality, meet deadlines, and cut costs. We become complacent and fail to pay attention to the importance of healthy working conditions. Table 14.6 lists some common remedies for heavy computer users.

The following is a useful checklist to help you determine if you are properly seated at a correctly positioned keyboard:[78]

- Your elbows are near your body in an open angle to allow circulation to the lower arms and hands
- Your arms are nearly perpendicular to the floor
- Your wrists are nearly straight
- Your ears are in line with the tops of your shoulders
- Your shoulders are in line with your hips
- The height of the surface holding your keyboard and mouse is 1 or 2 inches above your thighs
- The monitor is about one arm's length (20 to 26 inches) away
- The top of your monitor is at eye level
- Your chair has a backrest that supports the curve of your lower (lumbar) back

TABLE 14.6 Avoiding common discomforts associated with heavy use of computers

Common Discomforts Associated with Heavy Use of Computers	Preventative Action
Red, dry, itchy eyes	Change your focus away from the screen every 20 or 30 minutes by looking into the distance and focusing on an object for 20 to 30 seconds
	Make a conscious effort to blink more often
	Consider the use of artificial tears
	Use an LCD screen that provides a much better viewing experience for your eyes by virtually eliminating flicker while still being bright without harsh incandescence.
Neck and shoulder pain	Use proper posture when working at the computer
	Stand up, stretch, and walk around for a few minutes every hour
	Shrug and rotate your shoulders occasionally
Pain, numbness, or tingling sensation in hands	Use proper posture when working at the computer
	Do not rest your elbows on hard surfaces
	Place a wrist rest between your computer keyboard and the edge of your desk.
	Take an occasional break and spread fingers apart while keeping your wrists straight
	Taken an occasional break with your arms resting at your sides and gently shake your hands

Source: Pekker, Michael, "Long Hours at Computer: Health Risks and Prevention Tips," *http://webfreebies4u.blogspot.com/2011/01/long-hours-at-computer-health-risks-and.html*, January 4, 2011.

ETHICAL ISSUES IN INFORMATION SYSTEMS

code of ethics: A code that states the principles and core values that are essential to a set of people and that, therefore, govern these people's behavior.

As you've seen throughout this book in the "Ethical and Societal Issues" boxes, ethical issues deal with what is generally considered right or wrong. Laws do not provide a complete guide to ethical behavior. Just because an activity is defined as legal does not mean that it is ethical. As a result, practitioners in many professions subscribe to a **code of ethics** that states the principles and core values that are essential to their work and, therefore, govern their behavior. The code can become a reference point for weighing what is legal and what is ethical. For example, doctors adhere to varying versions of the 2000-year-old Hippocratic Oath, which medical schools offer as an affirmation to their graduating classes.

Some IS professionals believe that their field offers many opportunities for unethical behavior. They also believe that unethical behavior can be reduced by top-level managers developing, discussing, and enforcing codes of ethics. Various IS-related organizations and associations promote ethically responsible use of information systems and have developed useful codes of ethics. Founded in 1947, the Association for Computing Machinery (ACM) is the oldest computing society and boasts more than 100,000 members in more than 100 countries.[79] The ACM has a code of ethics and professional conduct that includes eight general moral imperatives that can be used to help guide the actions of IS professionals. These guidelines can also be used for those who employ or hire IS professionals to monitor and guide their work. These imperatives are outlined in the following list: As an ACM member I will …

1. Contribute to society and human well-being.
2. Avoid harm to others.
3. Be honest and trustworthy.
4. Be fair and take action not to discriminate.
5. Honor property rights including copyrights and patents.

6. Give proper credit for intellectual property.
7. Respect the privacy of others.
8. Honor confidentiality.[80]

The mishandling of the social issues discussed in this chapter—including waste and mistakes, crime, privacy, health, and ethics—can devastate an organization. The prevention of these problems and recovery from them are important aspects of managing information and information systems as critical corporate assets. More organizations are recognizing that people are the most important component of a computer-based information system and that long-term competitive advantage can be found in a well-trained, motivated, and knowledgeable workforce that adheres to a set of principles and core values that help guide that workforce's actions.

SUMMARY

Principle:

Policies and procedures must be established to avoid waste and mistakes associated with computer usage.

Computer waste is the inappropriate use of computer technology and resources in both the public and private sectors. Computer mistakes relate to errors, failures, and other problems that result in output that is incorrect and without value. At the corporate level, computer waste and mistakes impose unnecessarily high costs for an information system and drag down profits. Waste often results from poor integration of IS components, leading to duplication of efforts and overcapacity. Inefficient procedures also waste IS resources, as do thoughtless disposal of useful resources and misuse of computer time for games and personal use. Inappropriate processing instructions, inaccurate data entry, mishandling of IS output, and poor systems design all cause computer mistakes.

Preventing waste and mistakes involves establishing, implementing, monitoring, and reviewing effective policies and procedures. Companies should develop manuals and training programs to avoid waste and mistakes. Changes to critical tables, HTML, and URLs should be tightly controlled.

Principle:

Computer crime is a serious and rapidly growing area of concern requiring management attention.

Some crimes use computers as tools. For example, a criminal can use a computer to manipulate records, counterfeit money and documents, commit fraud via telecommunications networks, and make unauthorized electronic transfers of money.

Criminals can gain pieces of information to help break into computer systems by dumpster diving and social engineering techniques.

A cyberterrorist is someone who intimidates or coerces a government or organization to advance his political or social objectives by launching computer-based attacks against computers, networks, and the information stored on them.

Identity theft is a crime in which an imposter obtains key pieces of personal identification information to impersonate someone else. The information is then used to obtain credit, merchandise, and services in the name of the victim or to provide the thief with false credentials.

Although Internet gambling is popular, its legality is questionable within the United States.

The computer is also used as a tool to fight crime. The LeadsOnline Web-based system helps law enforcement officers recover stolen property. JusticeXchange provides law enforcement officials with fast, easy access to information about former and current offenders held in participating jails. Offender Watch tracks registered sex offenders. Law enforcement agencies use GPS tracking devices and software to monitor the movement of registered sex offenders. Law enforcement agencies use crime-related data and powerful analysis techniques, coupled with GIS systems, to better understand and even diminish crime risks.

A criminal hacker, also called a *cracker*, is a computer-savvy person who attempts to gain unauthorized or illegal access to computer systems to steal passwords, corrupt files and programs, and even transfer money. Script bunnies are crackers with little technical savvy. Insiders are employees, disgruntled or otherwise, working solo or in concert with outsiders to compromise corporate systems. The greatest fear of many organizations is the potential harm that can be done by insiders who know system logon IDs, passwords, and company procedures.

Computer crimes target computer systems and include illegal access to computer systems by criminal hackers, alteration and destruction of data and programs by viruses, and simple theft of computer resources.

Malware is a general term for software that is harmful or destructive. There are many forms of malware, including viruses, variants, worms, Trojan horses, logic bombs, and rootkits. Spyware is software installed on a personal computer to intercept or take partial control over the user's interactions with the computer without knowledge or permission of the user. A password sniffer is a small program hidden in a network or computer system that records identification numbers and passwords.

When discarding a computer, people should use disk-wiping utilities to avoid the loss of personal or confidential data even after deleting files and emptying the Recycle Bin (or Trash).

Software piracy might represent the most common computer crime. It is estimated that the software industry lost $59 billion in revenue in 2010 to software piracy. The global recording industry loses as much as $12.5 billion in revenue from music piracy each year. Patent infringement is also a major problem for computer software and hardware manufacturers.

Digital rights management refers to the use of any of several technologies to enforce policies for controlling access to digital media.

Computer-related scams, including phishing and vishing, have cost people and companies thousands of dollars. Computer crime is an international issue.

A fundamental concept of good internal controls is the careful separation of duties associated with key processes so that they are spread among more than one person.

Use of an intrusion detection system (IDS) provides another layer of protection in the event that an intruder gets past the outer security layers—passwords, security procedures, and corporate firewall. An IDS monitors system and network resources and notifies network security personnel when it senses a possible intrusion. Many small and mid-sized organizations are outsourcing their network security operations to managed security service providers (MSSPs), which monitor, manage, and maintain network security hardware and software.

Security measures, such as using passwords, identification numbers, and data encryption, help to guard against illegal computer access, especially when supported by effective control procedures. Virus-scanning software identifies and removes damaging computer programs. Organizations can use a security dashboard to provide a comprehensive display of vital data related to its security defenses and threats. Organizations and individuals can use antivirus software to detect the presence of all sorts of malware.

Balancing the right to privacy versus the need for additional monitoring to protect against cyberattacks is an especially challenging problem. Privacy issues are a concern with e-mail, instant messaging, and personal sensing devices. The Children's Online Privacy Protection Act protects minors using the Internet by requiring operators of Web sites that cater to children to post comprehensive privacy policies and to obtain parental consent before they collect any personal information from children under 13 years of age. The Privacy Act of 1974 establishes straightforward and easily understandable requirements for data collection, use, and distribution by federal agencies; federal law also serves as a nationwide moral guideline for privacy rights and activities by private organizations. The Electronic Communications Privacy Act deals with the protection of communications while in transit from sender to receiver, the protections of communications held in electronic storage, and prohibition of devices to record dialing, routing, addressing, and signaling information without a search warrant. The Gramm-Leach-Bliley Act requires all financial institutions to protect and secure customers' nonpublic data from unauthorized access or use. The USA Patriot Act, passed only five weeks after the September 11, 2001, terrorist attacks, requires Internet service providers and telephone companies to turn over customer information, including numbers called, without a court order if the FBI claims that the records are relevant to a terrorism investigation. Also, the company is forbidden to disclose that the FBI is conducting an investigation. Only time will tell how this act will be applied in the future.

A business should develop a clear and thorough policy about privacy rights for customers, including database access. That policy should also address the rights of employees, including electronic monitoring systems and e-mail. Fairness in information use for privacy rights emphasizes knowledge, control, notice, and consent for people profiled in databases. People should know about the data that is stored about them and be able to correct errors in corporate database systems. If information on people is to be used for other purposes, individuals should be asked to give their consent beforehand. Each person has the right to know and to decide.

Principle:

Jobs, equipment, and working conditions must be designed to avoid negative health effects from computers.

Jobs that involve heavy use of computers contribute to a sedentary lifestyle, which increases the risk of health problems. Some critics blame computer systems for emissions of ozone and electromagnetic radiation. Use of cell phones while driving has been linked to increased car accidents.

The study of designing and positioning computer equipment, called ergonomics, has suggested some approaches to reducing these health problems. Ergonomic design principles help to reduce harmful effects and increase the efficiency of an information system. RSI (repetitive strain injury) prevention includes keeping good posture, not ignoring pain or problems, performing stretching and strengthening exercises, and seeking proper treatment. Although they can cause negative health consequences, information systems can also be used to provide a wealth of information on health topics through the Internet and other sources.

Principle:

Practitioners in many professions subscribe to a code of ethics that states the principles and core values that are essential to their work.

A code of ethics states the principles and core values that are essential to the members of a profession or organization. Ethical computer users define acceptable practices more strictly than just refraining from committing crimes; they also consider the effects of their IS activities, including Internet usage, on other people and organizations. The Association for Computing Machinery developed guidelines and a code of ethics. Many IS professionals join computer-related associations and agree to abide by detailed ethical codes.

CHAPTER 14: SELF-ASSESSMENT TEST

Policies and procedures must be established to avoid waste and mistakes associated with computer usage.

1. The use of nonintegrated information systems that make it difficult for decision makers to collaborate and share information is an example of computer waste. True or False?
2. Employees' demand for videos and music to be delivered to company personal computers can exceed a firm's Internet capacity and create a major ____________.
3. Preventing waste and mistakes involves establishing, implementing, monitoring, and ____________ policies and procedures.

Computer crime is a serious and rapidly growing area of concern requiring management attention.

4. Just over ____________ crime reports were made to the Internet Crime Computer Center in 2010.
   a. 3,000
   b. 30,000
   c. 300,000
   d. 3,000,000
5. According to the Internet Crime Computer Center, the two most common online crimes reported included the nondelivery of payments or merchandise amounting to millions of dollars and ____________.
   a. scams involving people posing as FBI agents
   b. requests for help in moving funds out of Nigeria
   c. requests for donations to bogus charitable organizations
   d. bogus political campaign requests to send money to a candidate
6. Talking a critical computer password out of an individual is an example of ____________.
7. ____________ is any premeditated, politically motivated attack against information, computer systems, computer programs, and data which results in violence against noncombatant targets by subnational groups or clandestine agents.
   a. Hacking
   b. Social engineering
   c. Cyberterrorism
   d. Computer crime
8. Child identity theft is a rapidly growing area of computer crime. True or False?
9. The ____________ Online Web-based service system is used by law enforcement to recover stolen property.
10. The CompStat program uses Geographic Information System software to map crime and identify problem precincts. True or False?
11. ____________ is a derogatory term for inexperienced hackers who download programs called scripts that automate the job of breaking into computers.
12. Deleting files and emptying the Recycle Bin ensures that others cannot view the personal data on your recycled computer. True or False?
13. ____________ is a small program hidden in a network or a computer system that records identification numbers and passwords.
14. ____________ refers to the use of any of several technologies to enforce policies for controlling access to digital media.
    a. Software piracy
    b. Digital rights management
    c. Copyright
    d. Patent
15. The ____________ is a fundamental concept of good internal controls that ensures that the responsibility for the key steps in a process is spread among more than one person.

Jobs, equipment, and working conditions must be designed to avoid negative health effects from computers.

16. Heavy computer use can negatively affect one's physical health. True or False?

Practitioners in many professions subscribe to a code of ethics that states the principles and core values that are essential to their work.

17. The Association of Computing Machinery has a code of ethics and professional conduct that includes eight general moral imperatives that can be used to help guide the actions of IS professionals. True or False?

CHAPTER 14: SELF-ASSESSMENT TEST ANSWERS

1. True
2. bottleneck
3. reviewing
4. c
5. a
6. social engineering
7. c
8. True
9. Leads
10. True
11. Script bunny
12. False
13. Password sniffer
14. d
15. separation of duties
16. True
17. True

REVIEW QUESTIONS

1. What is the Internet Crime Computer Center and what does it do?
2. Define the term cyberterrorism. For how long has the threat of cyberterrorism been considered serious by the U.S. government?
3. What are SOPA and PIPA? What is their common goal? What else do they have in common?
4. What is social engineering? What is dumpster diving?
5. Why might some people consider a contractor to be a serious threat to their organization's information systems?
6. How do you distinguish between a hacker and a criminal hacker?
7. Why are insiders one of the biggest threats for company computer systems?
8. What is a virus? What is a worm? How are they different?
9. What is vishing? What actions can you take to reduce the likelihood that you will be a victim of this crime?
10. What is filtering software? Why would organizations use such software? What issues can arise from the use of this software?
11. What does intrusion detection software do? What are some of the issues with the use of this software?
12. What is the difference between a patent and a copyright? What copyright issues come into play when downloading software or music from a Web site?
13. What is the most common online computer crime reported to the Internet Crime Computer Center?
14. What is ergonomics? How can it be applied to office workers?
15. What is digital rights management?
16. What is a code of ethics? Give an example.

DISCUSSION QUESTIONS

1. Discuss how the use of nonintegrated information systems can lead to waste in an organization.
2. Why is it important for an organization to have a policy covering the appropriate use of information resources? What topics should be covered in such a policy?
3. Identify the risks associated with the disposal of obsolete computers. Discuss the steps that one must take to safely dispose of personal computers.
4. Identify and briefly discuss four specific information systems that are used to fight crime.
5. Briefly discuss software piracy. What is it, how widespread is it, and who is harmed by it?
6. Discuss the legality of online gambling in the United States. Discuss the legality of running an online gambling Web site in the United States.
7. Imagine that your friend regularly downloads copies of newly released, full-length motion pictures for free from the Internet and makes copies for others for a small fee. Do you think that this practice is ethical? Is it legal? Would you express any concerns to him?
8. Outline an approach, including specific techniques (e.g., dumpster diving, phishing, social engineering), that you could employ to gain personal data about the members of your class.
9. Your 12-year-old niece shows you a dozen or so innocent photos of herself and a brief biography, including address and cell phone number that she plans to post on Facebook. What advice might you offer her about posting personal information and photos?
10. Imagine that you are a hacker and have developed a Trojan horse program. What tactics might you use to get unsuspecting victims to load the program onto their computer?

11. Briefly discuss the potential for cyberterrorism to cause a major disruption in your daily life. What are some likely targets of a cyberterrorist? What sort of action could a cyberterrorist take against these targets?
12. What is meant by the separation of duties? When does this concept come into play? Provide a business situation where separation of duties is important.
13. Using information presented in this chapter on federal privacy legislation, identify which federal law regulates the following areas and situations: cross-checking IRS and Social Security files to verify the accuracy of information, customer liability for debit cards, the right to access data contained in federal agency files, the IRS obtaining personal information, the government obtaining financial records, and employers' accessing university transcripts.
14. Briefly discuss the differences between acting ethically and acting legally. Give an example of acting legally and yet unethically.

PROBLEM-SOLVING EXERCISES

1. Use presentation software to create a series of slides explaining what SOPA and PIPA are and your position on these controversial bills.
2. Access the Web site for the Business Software Alliance (BSA) and other Web sites to find estimates of the amount of software piracy worldwide for at least the past five years. Using spreadsheet software and appropriate forecasting routines, develop a forecast for the amount of software piracy for the next three years. Document any assumptions you make in developing your forecast.
3. Do research to determine the current status of the SOPA and PIPA acts. Use presentation software to describe these bills and present your position on the impact and legitimacy of these bills.

TEAM ACTIVITIES

1. Visit one of your school's computer labs and evaluate the setup from the standpoint of ergonomics. Are the chairs, desktops, workstations, screens, keyboards, lighting, etc. designed in a sound ergonomic fashion? If not, what changes should be made? Write a brief report documenting your findings and recommendations.
2. Have each member of your team access six different Web sites and summarize their findings in terms of the existence of data privacy policy statements. Did each site have such a policy? Was it easy to find? Did it seem complete and easy to understand?

WEB EXERCISES

1. Do research on the Web and find six current lawsuits among companies in the information systems industry. Write a brief report summarizing each of these cases.
2. Do research on the Web to find the most current information concerning any connection between cell phone use and cancer.
3. Do research on The Internet Crime Computer Center or the Defense Department's Computer Emergency Response Team. Prepare a report describing its founding, mission, and scope of activities.

CAREER EXERCISES

1. You are a new hire in the marketing organization for a manufacturer of children's toys. A recommendation has been made to develop a Web site to promote and sell your firm's products as well as learn more about what parents and their children are looking for in new toys. Your manager has asked you to develop a list of laws and regulations that could affect the design and operation of the Web site. She has also asked you to describe how these will limit the operation of the organization's new Web site.

2. You have just begun a new position in the Public Relations Department of the RIAA. You have been asked to prepare a one-page summary defending the organization's position against music piracy. What key points would you make?

CASE STUDIES

Case One

Fletcher Allen Health Care Deals with Another Kind of Virus

Fletcher Allen Health Care (FAHC) of Burlington is Vermont's university hospital and medical center. In 2011, it admitted over 50,000 patients and had over 60,000 visits to its emergency room. Many of these people had conditions such as influenza that involved a virus, so Fletcher Allen has had a great deal of experience in dealing with viruses. All that experience, however, left it unprepared for the virus it encountered on March 29, 2011.

A few months earlier, Mac McMillan, co-chair of the Privacy and Security Policy Task force of the Healthcare Information and Management Systems Society (HIMSS), had visited Fletcher Allen to discuss network security with CIO Chuck Podesta. Mac had founded CynergisTek eight years earlier to "make quality IT security consulting available and accessible to organizations across industries and regardless of size or complexity." Now he was at one of the largest hospitals in northern New England, but the issues were much the same as they would be at a smaller facility.

Over the coming months, FAHC implemented some of McMillan's recommendations, but due to resource constraints, it couldn't complete all of them. Unexpectedly, an unknown (and therefore potentially devastating) virus forced the hospital to come up with containment procedures on the fly. FAHC created a Computer Incident Response Team (CIRT). What that team did can serve as a "best practices" recipe for other organizations faced with similar problems:

- **Seal the perimeter**. FAHC unplugged the 1,000 computers (out of about 6,000 at FAHC) that testing found to be affected from communication networks to quarantine them.
- **Know the enemy**. The hospital's security team identified the virus as Pinkslipbot. It probably entered via an e-mail claiming to be from a package delivery service. This virus does not harm infected computers directly, but it shuts down virus-protection software and sends information about them back to its host.
- **Alert users**. FAHC noticed a near-total drop in Internet accesses from all computers due to concern over the virus attack.
- **Clean out the infection**. This action took about 90 minutes per machine for 1,000 computers plus the hospital's 700 laptops, which the team had to assume could be infected until they were brought in and checked. The total cost in the value of the team's time that had to be taken away from other work was in six figures.

The operational impact of this invasion was huge. The backlog of trouble tickets accumulated while resources were focused on cleaning out the virus made for ten times the usual amount of work. The CIRT ran further tests for about two weeks before it felt comfortable bringing everyone back online.

Podesta recommends that other organizations create a CIRT right away before they have an incident. "This can be hard to do," he concedes. "It's behind-the-scenes work that you hope to never use [since a CIRT, by definition, only goes to work after an incident]. It's not getting the user the next app or update they need."

The silver lining? According to Podesta, the incident "put us ahead of the game, because the whole organization is now focused on security." On February 10, 2012, FAHC announced that Heather Roszkowski would be chief information security officer and head the hospital's information security operations. A decorated military veteran with two years' service in Iraq, her previous position was director of IT and communications for a U.S. Army brigade. Someone like that takes security seriously. Her appointment suggests that, now, FAHC does too.

Discussion Questions

1. Suppose the virus entered FAHC as they suspect. What would you recommend to prevent a virus from entering again the same way? Who has to follow this policy? How would you train those people on the policy? Who should be responsible for that training? In the broadest possible sense, what does this virus incident suggest to you as a future manager for ensuring information system security at your company?
2. With 5,000 of its 6,000 computers uninfected based on their testing, FAHC was able to continue patient care, though some areas of the hospital did not function optimally. In a worst-case scenario, the numbers could have been the other way round: 5,000 of 6,000 computers infected or even all 6,000. What do you think the impact on patient care would have been then? How would you go about minimizing it?

Critical Thinking Questions

1. The Pinkslipbot attack probably could have been prevented at lower total cost to FAHC were it not for resource constraints. Why is it so difficult for information systems departments to get funds for security measures *before* something happens? Consider the situation of a CEO making budget allocations.
2. When personal health information is breached for 500 or more hospital patients, the incident must be

reported formally to the U.S. Government. It then posts the information on a public Web site. Should a similar rule apply to colleges and universities? Legislation aside, what should colleges and universities do if they suffer a major security breach?

SOURCES: CynergisTek Web site, *cynergistek.com*, accessed May 21, 2012; HIMSS Web site, *www.himss.org*, accessed May 21, 2012; Fletcher Allen Health Care Web site, *www.fletcherallen.org*, accessed May 21, 2012; Staff, "Virus Profile: W32/Pinkslipbot," McAfee, *home.mcafee.com/virusinfo/virusprofile.aspx?key=141235*, accessed May 21, 2012; Smith, L., "Computer Virus Containment at Fletcher Allen Health Care," HIT Community, *www.thehitcommunity.org/2012/01/case-study-a-virus-attack-at-fletcher-allen-health-care-led-to-best-practices-for-hit-security*, January 27, 2012.

Case Two

Security Monitoring and Big Data

Security Information and Event Management (SIEM) systems centralize the work of monitoring information systems and networks for security threats. They use information from the intrusion detection systems you read about in this chapter.

Nearly anything that happens in a server or network is potentially useful security information. It can be important to know that a component works as it should: that may help pin down the time of a problem and thus help identify its cause. However, millions of events occur in a large network every day—including normal activity, false alarms and real attempts at intrusion. Recording this information, therefore, falls into the category of big data, which you read about in Chapter 5, that is, data that exceeds, in some respect, the capabilities of traditional database systems.

Another reason a lot of data is necessary is that security threats are only a small fraction of all the things that happen in an information system. Therefore, a system must accumulate a lot of data in total before it has enough to draw conclusions about security, train software to recognize threat patterns, and perform statistical analyses.

John Oltsik, senior principal analyst at Enterprise Strategy Group, agrees. "Security intelligence demands more data," he writes. "Early SIEMs collected event and log data, then steadily added other data sources.... Large enterprises now regularly collect gigabytes or even terabytes of data for security intelligence, investigations, and forensics. Many existing tools can't meet these scalability needs."

Zions Bancorporation, a bank holding company based in Salt Lake City, Utah, ran up against this problem. "We'd be bumping our heads against the ceiling with SIEM fairly early on," said Preston Wood, chief security officer at Zions. "The underlying data technology just couldn't handle it."

This inability to handle the required volumes of data changed with the Hadoop distributed file system. Hadoop has made life much different for Zions. "We're doing it in near-real-time fashion," says Mike Fowkes, director of fraud prevention and analytics. "We're pulling in data every five minutes, hourly, every two minutes. It just depends on ... how fresh our data needs to be."

Making sense of the results is where Aaron Caldiero comes in. As senior data scientist at Zions, he plays the roles of "part computer scientist, part statistician, and part graphic designer." Caldiero's job is to collect and centralize the data, design methods of synthesizing it, and then present it in a coherent way.

His approach may be foreign to security professionals. "It's a bottom-up process where you're putting the data first," Caldiero said. That approach of starting with the data and seeing where it led has allowed Zions to draw trends, patterns, or correlations that they might never have found had they put the questions first and sorted through terabytes of data for the answers.

Some experts disagree. Richard Stiennon, chief research analyst at IT-Harvest, a firm devoted to analyzing the information security industry, sees the trend but says "it is time to stop the insanity. If you need Big Data to solve your security problems, you are doing something wrong.... You are looking at data, not intelligence. So, before you make massive investments in Big Data, take another look at your security posture. You do not need better ways to handle data, you need more intelligence."

Most, though, feel that big data is the best route to that intelligence. Who is right? Time will tell.

Discussion Questions

1. Every e-mail message carries about 500 bytes of data on where it came from, what software created it, what route it took to its destination, and more. Suppose a company has 1,000 employees, each of whom receives an average of 50 messages per day. (Some receive hardly any, but others receive many more than 50.) The firm saves this information in case it is needed later to analyze a security threat, though it does not save message text or attachments. How long will it take for the firm to accumulate a gigabyte of e-mail header data? How much data will it have after one year?
2. Industry analysts, such as Jon Oltsik and Richard Stiennon who are quoted in this case, study and report on the state of information system security and SIEM systems. Most large organizations subscribe to at least one such industry analysis service. Suppose you as chief security officer for a large firm subscribe to both Oltsik's and Steinnon's analysis services. You don't know what to do about big data, and these two advisors disagree. How would you proceed?

Critical Thinking Questions

1. Vendors and users are investing a great deal of money in applying big data methods to SIEM. If there is a less expensive way to achieve the same objective, someone will think of it and, if Stiennon is right, will obtain a competitive advantage by using it. What can a SIEM software vendor do to protect itself against this possibility?
2. If the use of big data for SIEM is truly a long-term trend, this is another item on the list of technologies that information security professionals must know about. What effects do you think this need will have on careers in this field? Would using big data make such a career more attractive to you or less?

SOURCES: Chickowski, E., "A Case Study in Security Big Data Analysis," *Dark Reading, darkreading.com/security-monitoring/167901086/security/news/232602339/a-case-study-in-security-big-data-analysis.html*, March 9, 2012; Maxwell, K., "3 Reasons Why Big Data Matters for SIEM," *Threat Thoughts* (blog), *threatthoughts.com/2012/02/15/3-reasons-why-big-data-matters-for-siem*, February 15, 2012; Oltsik, J., "The Intersection of Security and Big Data Analytics," blog, *www.esg-global.com/blogs/the-intersection-of-security-intelligence-and-big-data-analytics*, February 12, 2012; Stiennon, R., "If You Feel You Need Big Data for Security, You Are Doing Something Wrong," Security Bistro, *www.securitybistro.com/blog/?p=1431*, April 23, 2012; Zions Bancorporation Web site, *www.zionsbancorporation.com*, accessed May 24, 2012.

Questions for Web Case

See the Web site for this book to read about the Altitude Online case for this chapter. Following are questions concerning this Web case.

Altitude Online: The Personal and Social Impact of Computers

Discussion Questions

1. Why do you think extending access to a corporate network beyond the business's walls dramatically elevates the risk to information security?
2. What tools and policies can be used to minimize that risk?

Critical Thinking Questions

1. Why does information security usually come at the cost of user convenience?
2. How do proper security measures help ensure information privacy?

NOTES

Sources for the opening vignette: Cluley, G., "Bredolab: Jail for Man Who Masterminded Botnet of 30 Million Computers," *Naked Security, nakedsecurity.sophos.com/2012/05/23/bredolab-jail-botnet*, May 23, 2012; Constantin, L., "Bredolab Botnet Author Sentenced to 4 Years in Prison in Armenia," *PC World, www.pcworld.com/businesscenter/article/256048/bredolab_botnet_author_sentenced_to_4_years_in_prison_in_armenia.html*, May 23, 2012; Gates, S., "Georgy Avanesov, Bredolab Botnet Creator, Found Guilty and Sentenced to Four Years in Prison," *Huffington Post, www.huffingtonpost.com/2012/05/24/georgy-avanesov-found-guilty_n_1543687.html*, May 24, 2012.

1. Staff, "Casa Oliveira Toasts Greater Business Insight with SAP and IBM Global Technology Services," IBM Success Stories, IBM, *www-01.ibm.com/software/success/cssdb.nsf/CS/STRD-8DVGT8?OpenDocument&Site=corp&cty=en_us*, February 8, 2011.
2. Staff, "Departments of Defense and Energy Need to Address Potentially Duplicative Investments," GAO Report, GAO-12-241, Government Accounting Office, February 2012.
3. Holthaus, David, "P&G Tries to Close Online Pandora's Box," *The Cincinnati Enquirer*, April 3, 2012, p. A1.
4. Ibid.
5. Hayden, Maureen, "Indiana Countries to Reap Windfall from State's Costly Error," TriStar.com, *http://tribstar.com/indiana_news/x101437973/Indiana-officials-say-205M-in-local-taxes-mishandled*, April 5, 2012.
6. Totenberg, Nina, "Supreme Court OKs Strip Searches for Minor Offenses," NPR, *www.npr.org/2012/04/02/149866209/high-court-supports-strip-searches-for-minor-offenders*, April 7, 2012.
7. Higgins, A.J., "Partisan Tensions Rise over Maine DHHS Computer Errors," Maine Public Broadcasting Network, *www.mpbn.net/Home/tabid/36/ctl/ViewItem/mid/3478/ItemId/21174/Default.aspx*, April 3, 2012.
8. McCartney, Scott, "United Lists Ongoing Computer System Problems, Fixes Underway," *The Wall Street Journal*, March 12, 2012.
9. "That's Not in The Plan! Computer Error Leaves Astronaut Dangling over Ledge of Space Station 220 Miles above Earth," *Mail Online, www.DailyMail.co.uk/sciencetech/article-1361752/Computer-error-leaves-astronaut-Stephen-Bown-dangling-ISS-ledge-220-miles-Earth.html*, March 1, 2011.
10. Staff, "Course 10: Preventing Human Errors," Process Improvement Institute, *www.process-improvement-institute.com/human-error-prevention-training.html*, accessed May 27, 2012.
11. Tempera, Jackie, "Officer Fired for Alleged Internet Abuse," *The Berkeley Beacon, www.berkeleybeacon.com/news/2012/3/22/officer-fired-for-alleged-internet-abuse*, March 22, 2012.
12. "Human, Computer Errors Add to Colorado's Wildfire Disaster," *Pied Type, http://piedtype.com/2012/03/29/human-computer-errors-add-to-colorados-wildfire-disaster*, March 29, 2012.
13. "Millions Set for Tax Rebates or Bills as HMRC Computer System 'Beds In," *The Guardian*, October 19, 2011.
14. Dolan, Jack, "Computer Errors Allow Violent California Prisoners to Be Released Unsupervised," *Los Angeles Times*, May 26, 2011.
15. Staff, "2010 Internet Crime Report," The Internet Crime Complaint Center, 2011.
16. McCaney, Kevin, "Pay Up: The Most Common Type of Online Crime," *Government Computer News, http://gcn.com/articles/2011/02/28/fbi-internet-crime-report.aspx*, February 26, 2011.
17. Staff, Internet Crime Complaint Center Scam Alerts, Internet Crime Complaint Center, *www.ic3.gov/media/2011.aspx*, accessed April 9, 2012.
18. Staff, "Spam E-mails Continue to Utilize FBI Officials' Names, Titles in Online Fraud Schemes," Intelligence Note, Internet Crime Complaint Center, *www.ic3.gov/media/2011.aspx*, August 9, 2011.
19. Staff, "FBI Dumpster Diving Brigade Coming Soon to Snoop in a Trashcan Near You," *NetworkWorld, www.networkworld.com/community/blog/fbi-dumpster-diving-brigade-coming-soon-snoop*, June 15, 2011.

20. Gordon, Sarah and Ford, Richard, "Cyberterrorism?" Symantec Security Response, *www.symantec.com/avcenter/reference/cyberterrorism.pdf*, accessed June 6, 2012.
21. Coleman, Kevin, "Cyberterrorism Now at the Top of the List of Security Concerns, *Defensetech, http://defensetech.org/2011/09/12/cyber-terrorism-now-at-the-top-of-the-list-of-security-concerns*, September 12, 2011.
22. Gordon, Sarah and Ford, Richard, "Cyberterrorism?" Symantec Security Response, *www.symantec.com/avcenter/reference/cyberterrorism.pdf*, accessed June 6, 2012.
23. Frenkel, Sheera, "Israeli and Arab Hackers Square Off in Cyberbattle," NPR, *http://m.npr.org/news/World/145522079*, January 20, 2012.
24. Mezzofiore, Gianluca, "Teampoison Affiliate Hackers Deface NATO Web Site of Croatia," *International Business Times, www.ibtimes.com/articles/323191/20120403/teamp0ison-hackers-deface-nato-website-croatia.htm*, April 3, 2012.
25. Dilanian, Ken, "Virtual War a Real Threat," *Los Angeles Times*, March 28, 2011.
26. "FBI and Homeland Security Launch Probe as Foreign Cyber Attackers Target U.S. Water Supply," *Daily Mail, www.DailyMail.co.uk/news/article-2063444/Cyber-attack-US-water-supply-traced-Russia-FBI-Homeland-Security-launch-probe.html*, November 20, 2011.
27. "Despite Rise in Child ID Theft, Parents Remain Largely Unaware," *The Sacramento Bee, www.sacbee.com/2012/04/10/4403019/despite-rise-in-child-id-theft.html*, April 12, 2012.
28. Goessi, Leigh, "7-Year-Old Boy's Identity Stolen, Child ID Theft on the Rise," *Digital Journal, www.digitaljournal.com/article/325372*, May 23, 2012.
29. Staff, "Examples of Identity Theft Schemes—Fiscal Year 2012," IRS, *www.irs.gov/compliance/enforcement/article/0,,id=250691,00.html*, accessed April 10, 2011.
30. "Elgin Neighbors Fall Victim to Identity Theft," CBS Chicago, *http://chicago.cbslocal.com/2012/04/02/elgin-neighbors-fall-victim-to-identity-theft*, April 2, 2012.
31. Hurtado, P. and Van Voris, R., "Internet Poker Entrepreneurs Charged with Fraud, U.S. Says," *Businessweek*, April 15, 2011.
32. Richtel, Matt, "U.S. Cracks Down on Online Gambling," *The New York Times*, April 15, 2011.
33. Staff, "About Leads Online," LeadsOnline, *https://www.leadsonline.com/main/about-leadsonline/about-leadsonline.php*, accessed April 11, 2012.
34. Staff, "Success Stories: Property Crime," LeadsOnline, *https://www.leadsonline.com/main/success-stories/property-crime.php*, accessed April 11, 2012.
35. JusticeXchange Web site, *www.nysheriffs.org/justicexchange*, accessed April 11, 2012.
36. Community Notification Web site, *www.communitynotification.com*, accessed April 11, 2012.
37. Staff, "Client Testimonials," Cap Index, *www.capindex.com/Testimonials/Testimonials.aspx*, accessed June 6, 2012.
38. "About NER Operations," NER, *www.ner.net/operations.html*, accessed April 13, 2012.
39. SFPD CompStat, San Francisco Police Department, *www.sf-police.org/index.aspx?page=3254*, accessed April 13, 2012.
40. "Tennessee Highway Patrol Teams Up with Verisk's CargoNet," *Market Watch, www.marketwatch.com/story/tennessee-highway-patrol-teams-up-with-verisks-cargonet-2012-01-30*, January 30, 2012.
41. "Federal Reserve Code Stolen by Insider, FBI Says," *Government Computer News, http://gcn.com/articles/2012/01/26/agg-federal-reserve-code-stolen-insider-threats.aspx*, January 26, 2012.
42. Waugh, Rob, "New PC Virus Doesn't Just Steal Your Money—It Creates Fake Online Bank Statements So You Even Don't Know It's Gone," *Mail Online, www.DailyMail.co.uk/sciencetech/article-2083271/SpyEye-trojan-horse-New-PC-virus-steals-money-creates-fake-online-bank-statements.html*, January 6, 2012.
43. Zielenziger, David, "IBM Says Security Threats Mount Despite More Awareness," *Business & Law, www.ibtimes.com/articles/317755/20120321/ibm-cyber-attack-threat-security-computer-hacker.htm*, March 21, 2012.
44. "How to Remove Win 7 Antispyware 2012," *Dedicated 2 Viruses, www.2-viruses.com/remove-win-7-anti-spyware-2012*, June 7, 2011.
45. "Romney Campaign Computers Stolen," UPI.com, *www.upi.com/Top_News/US/2012/03/27/Romney-campaign-computers-stolen/UPI-74061332894331*, March 27, 2012.
46. Staff, "Data Loss," Lee Miller Rehabilitation, *leemillerrehab.com/news*, accessed April 12, 2012.
47. Lefkow, Chris, "Value of Pirated Software Nearly $59B: Study," *Jakarta Globe, www.thejakartaglobe.com/afp/value-of-pirated-software-nearly-59b-study/440674*, May 12, 2011.
48. Staff, "PC Software Piracy Rates by Region," BSA, *www.bsa.org/country/~/media/D02B5A4B60444B0AAF6CFDD598C72CBC.ashx*, accessed April 14, 2012.
49. Staff, "The Cost of Content Theft by the Numbers," mpaa-infographic, *www.scribd.com/doc/62848402/mpaa-infographic*, accessed February 26, 2012.
50. Staff, "Member of Texas Software Piracy Ring Sentenced to 57 Months in Prison," BSA, *www.bsa.org/country/News%20and%20Events/News%20Archives/en/2012/en-02292012-baxtercase.aspx*, February 29, 2012.
51. Staff, "Who Music Theft Hurts," RIAA Recording Industry Association of America, *www.riaa.com/physicalpiracy.php?content_selector=piracy_details_online*, accessed April 16, 2012.
52. "RIAA Announces New Round of Music Theft Lawsuits," Gold and Platinum News, *www.riaa.com/newsitem.php?news_year_filter=&resultpage=13&id=F7ED251F-6E08-52D9-A805-22662F5E4D4F*, April 2012…
53. Schwartz, Mathew J., "SOPA: 10 Key Facts about Piracy Bill," *InformationWeek*, January 18, 2012.
54. "Congress Changes Its Tune: Senate Delays Scheduled Anti-Piracy Vote Amid Fierce Opposition … and House Follows Suit," *Daily Mail, www.Daily Mail.co.uk/news/article-2089582/Senate-House-delay-PIPA-SOPA-vote-amid-fierce-opposition.html*, January 21, 2012.

55. Fitzpatrick, Alex, "Where Do SOPA and PIPA Stand Now?" *Mashable*, *http://mashable.com/2012/01/17/sopa-pipa*, January 17, 2012.
56. "Apple, Sony Sued in California Patent Infringement Claim over iTunes, Playstation Network Systems," PR Newswire, *www.prnewswire.com/news-releases/apple-sony-sued-in-california-patent-infringement-claim-over-itunes-playstation-network-systems-144684235.html*, March 28, 2012.
57. Clark, Don and Tuna, Cari, "Oracle Sues Google, Saying Android Violates Java Copyrights," *The Wall Street Journal*, August 13, 2010.
58. Maag, Christopher, "The New Breed of Phishing Scams: It's Complicated," Credit.com, *www.credit.com/blog/2012/02/the-new-breed-of-phishing-scams-its-complicated*, February 7, 2012.
59. Chang, Andrea, "Facebook, Google, Other Firms Team to Fight Email Phishing Scams," *Los Angeles Times*, January 30, 2012.
60. Harris, Sheryl, "Smishing, vishing, phishing—Scammers Are after You by Phone," *The Cleveland Plain Dealer*, March 20, 2012.
61. Staff, "Examples of Money Laundering Investigations—Fiscal Year 2012—Massachusetts Pharmacist Sentenced for Illegal Distribution of Drugs via the Internet," IRS, *www.irs.gov/compliance/enforcement/article/0,,id=246537,00.html*, accessed May 23, 2012.
62. Staff, "2011 Global Encryption Trends Study Sponsored by Thales e-Security," Ponemon Institute, *www.globaletm.com/whitepaper/2011-global-encryption-trends-study*, February 2012.
63. Staff, "Case Study–Arch Chemicals," SecurDoc Disk Encryption, *www.winmagic.com/resource-centre/case-studies*, accessed April 20, 2012.
64. Schwartz, Matthew J., "Fired Employee Indicted for Hacking Gucci Network," *InformationWeek*, *www.informationweek.com/news/security/NAC/229400909*, April 5, 2011.
65. Johnson, Auston G., "Utah Department of Health Medicaid Operation Division Finding and Recommendation for the Period July 2010 to June 2011," Office of the Utah State Auditor, *www.sao.utah.gov/_finAudit/rpts/2012/12-GOL-A2.pdf*, February 27, 2012.
66. Norton, Leo M., Lt., "Who Goes There? Mobile Fingerprint Readers in Los Angeles County," *Police Chief Magazine*, *www.policechiefmagazine.org/magazine/index.cfm?fuseaction=display_arch&article_id=1824&issue_id=62009*, April 2012.
67. Staff, "Rackspace Hosting Managed Hosting Solutions," Rackspace, *www.rackspace.com/managed_hosting/services/security/threatmgr*, accessed April 20, 2012.
68. Staff, "Blue Coat Secure Web Gateway Helps Texas Children Hospital Control Web 2.0 Applications and Protect Against Malware," Blue Coat, *www.bluecoat.com/company/customers/blue-coat-secure-web-gateway-helps-texas-children%E2%80%99s-hospital-control-web-20*, accessed April 20, 2012.
69. Staff, "The Coca-Cola Company," Symantec, *www.symantec.com/resources/customer_success/detail.jsp?cid=coca_cola*, accessed April 20, 2012.
70. Miller, Ron, "Employees Have No Reasonable Expectation to Privacy for Materials Viewed or Stored on Employer-Owned Computers or Servers," *CCH Employment Law Daily*, *www.employmentlawdaily.com/index.php/author/ron-miller*, November 24, 2011.
71. Staff, "FTC Enforces COPPA against Web Site That Collected Children's Personal Information without Parental Consent," Epic.org, *epic.org/privacy/kids*, accessed April 21, 2012.
72. Palis, Courtney, "Facebook Privacy Options Ignored by Millions: Consumer Reports," *Huffington Post*, *www.huffingtonpost.com/2012/05/03/facebook-privacy-consumer-reports_n_1473920.html*, May 3, 2012.
73. Golijsn, Rosa, "Consumer Reports: Facebook Privacy Problems Are on the Rise," *Technolog*, *www.technolog.msnbc.msn.com/technology/technolog/consumer-reports-facebook-privacy-problems-are-rise-749990*, accessed June 7, 2012.
74. Tucker, Katheryn Hayes, "Flaming Can Be Defaming," *Daily Report*, *www.dailyreportonline.com/Editorial/New/singleEdit.asp?!=100401119372*, March 12, 2012.
75. "Ohio Prison Worker Fired after Posting Comments about John Kasich," *Huffington Post*, *www.huffingtonpost.com/2012/01/19/ohio-prison-fired-facebook-john-kasich_n_1217125.html*, April 22, 2012.
76. Staff, "2012 Best Internet Filter Software and Reviews," Internet Filter Software Review, *http://internet-filter-review.toptenreviews.com*, accessed April 21, 2012.
77. Staff, "BBB Sample Privacy Policy," BBB, *http://utah.bbb.org/sample-privacy*, accessed May 28, 2012.
78. Staff, "How to Sit at a Computer," American Academy of Orthopedic Surgeons, *http://orthoinfo.aaos.org/topic.cfm?topic=a00261*, accessed April 25, 2012.
79. Staff, "ACM at a Glance," Association of Computing Machinery, *www.acm.org/membership/membership/student/acm-at-a-glance*, accessed April 25, 2012.
80. ACM Code of Ethics and Professional Conduct, *www.acm.org/about/code-of-ethics?searchterm=code+of+ethics*, accessed April 25, 2012.

Glossary

A

acceptance testing Conducting any tests required by the user.

accounting MIS An information system that provides aggregate information on accounts payable, accounts receivable, payroll, and many other applications.

ad hoc DSS A DSS concerned with situations or decisions that come up only a few times during the life of the organization.

alpha testing Testing an incomplete or early version of the system.

antivirus program Software that runs in the background to protect your computer from dangers lurking on the Internet and other possible sources of infected files.

application flowcharts Diagrams that show relationships among applications or systems.

application program interface (API) Tools software developers use to build application software without needing to understand the inner workings of the OS and hardware.

application service provider (ASP) A company that provides the software, support, and computer hardware on which to run the software from the user's facilities over a network.

arithmetic/logic unit (ALU) The part of the CPU that performs mathematical calculations and makes logical comparisons.

ARPANET A project started by the U.S. Department of Defense (DoD) in 1969 as both an experiment in reliable networking and a means to link the DoD and military research contractors, including many universities doing military-funded research.

artificial intelligence (AI) A field in which the computer system takes on the characteristics of human intelligence.

artificial intelligence systems People, procedures, hardware, software, data, and knowledge needed to develop computer systems and machines that demonstrate the characteristics of intelligence.

asking directly An approach to gather data that asks users, stakeholders, and other managers about what they want and expect from the new or modified system.

attribute A characteristic of an entity.

auditing Analyzing the financial condition of an organization and determining whether financial statements and reports produced by the financial MIS are accurate.

B

backbone One of the Internet's highspeed, long-distance communications links.

batch processing system A form of data processing whereby business transactions are accumulated over a period of time and prepared for processing as a single unit or batch.

benchmark test An examination that compares computer systems operating under the same conditions.

best practices The most efficient and effective ways to complete a business process.

beta testing Testing a complete and stable system by end users.

blade server A server that houses many individual computer motherboards that include one or more processors, computer memory, computer storage, and computer network connections.

Bluetooth A wireless communications specification that describes how cell phones, computers, faxes, personal digital assistants, printers, and other electronic devices can be interconnected over distances of 10–30 feet at a rate of about 2 Mbps.

brainstorming A decision-making approach that consists of members offering ideas "off the top of their heads," fostering creativity and free thinking.

bridge A telecommunications device that connects two LANs together using the same telecommunications protocol.

broadband communications A relative term but it generally means a telecommunications system that can transmit data very quickly.

business intelligence (BI) The process of gathering enough of the right information in a timely manner and usable form and analyzing it to have a positive impact on business strategy, tactics, or operations.

business-to-business (B2B) e-commerce A subset of e-commerce in which all the participants are organizations.

business-to-consumer (B2C) e-commerce A form of e-commerce in which customers deal directly with an organization and avoid intermediaries.

byte (B) Eight bits that together represent a single character of data.

C

cache memory A type of highspeed memory that a processor can access more rapidly than main memory.

Cascading Style Sheets (CSS) A markup language for defining the visual design of a Web page or group of pages.

CASE repository A database of system descriptions, parameters, and objectives.

central processing unit (CPU) The part of the computer that consists of three associated elements: the arithmetic/logic unit, the control unit, and the register areas.

centralized processing An approach to processing wherein all processing occurs in a single location or facility.

certificate authority (CA) A trusted third-party organization or company that issues digital certificates.

certification A process for testing skills and knowledge, which results in a statement by the certifying authority that confirms an individual is capable of performing particular tasks.

change model A representation of change theories that identifies the phases of change and the best way to implement them.

channel bandwidth The rate at which data is exchanged, usually measured in bps.

character A basic building block of most information, consisting of uppercase letters, lowercase letters, numeric digits, or special symbols.

chief knowledge officer (CKO) A top-level executive who helps the organization work with a KMS to create, store, and use knowledge to achieve organizational goals.

chip-and-PIN card A type of card that employs a computer chip that communicates with a card reader using radio frequencies; it does not need to be swiped at a terminal.

choice stage The third stage of decision making, which requires selecting a course of action.

circuit-switching network A network that sets up a circuit between the sender and receiver before any communications can occur; this circuit is maintained for the duration of the communication and cannot be used to support any other communications until the circuit is released and a new connection is set up.

client/server architecture An approach to computing wherein multiple computer platforms are dedicated to special functions, such as database management, printing, communications, and program execution.

clock speed A series of electronic pulses produced at a predetermined rate that affects machine cycle time.

cloud computing A computing environment where software and storage are provided as an Internet service and are accessed with a Web browser.

code of ethics A code that states the principles and core values that are essential to a set of people and that, therefore, govern these people's behavior.

cold site A computer environment that includes rooms, electrical service, telecommunications links, data storage devices, and the like; it is also called a shell.

command-based user interface A user interface that requires you to give text commands to the computer to perform basic activities.

compact disc read-only memory (CD-ROM) A common form of optical disc on which data cannot be modified once it has been recorded.

competitive advantage A significant and ideally long-term benefit to a company over its competition.

competitive intelligence One aspect of business intelligence limited to information about competitors and the ways that knowledge affects strategy, tactics, and operations.

compiler A special software program that converts the programmer's source code into the machine-language instructions, which consist of binary digits.

computer literacy Knowledge of computer systems and equipment and the ways they function; it includes the knowledge of equipment and devices (hardware), programs and instructions (software), databases, and telecommunications.

computer network The communications media, devices, and software needed to connect two or more computer systems or devices.

computer programs Sequences of instructions for the computer.

computer-aided software engineering (CASE) Tools that automate many of the tasks required in a systems development effort and encourage adherence to the SDLC.

computer-assisted manufacturing (CAM) A system that directly controls manufacturing equipment.

computer-based information system (CBIS) A single set of hardware, software, databases, telecommunications, people, and procedures that are configured to collect, manipulate, store, and process data into information.

computer-integrated manufacturing (CIM) Using computers to link the components of the production process into an effective system.

concurrency control A method of dealing with a situation in which two or more users or applications need to access the same record at the same time.

consumer-to-consumer (C2C) e-commerce A subset of e-commerce that involves electronic transactions between consumers using a third party to facilitate the process.

contactless card A card with an embedded chip that only needs to be held close to a terminal to transfer its data; no PIN number needs to be entered.

content streaming A method for transferring large media files over the Internet so that the data stream of voice and pictures plays more or less continuously as the file is being downloaded.

continuous improvement Constantly seeking ways to improve business processes and add value to products and services.

control unit The part of the CPU that sequentially accesses program instructions, decodes them, and coordinates the flow of data in and out of the ALU, the registers, the primary storage, and even secondary storage and various output devices.

coprocessor The part of the computer that speeds processing by executing specific types of instructions while the CPU works on another processing activity.

cost center A division within a company that does not directly generate revenue.

cost/benefit analysis An approach that lists the costs and benefits of each proposed system. After they are expressed in monetary terms, all the costs are compared with all the benefits.

counter-intelligence The steps an organization takes to protect information sought by "hostile" intelligence gatherers.

creative analysis The investigation of new approaches to existing problems.

criminal hacker (cracker) A computer-savvy person who attempts to gain unauthorized or illegal access to computer systems to steal passwords, corrupt files and programs, or even transfer money.

critical analysis The unbiased and careful questioning of whether system elements are related in the most effective ways.

critical path Activities that, if delayed, would delay the entire project.

critical success factors (CSFs) Factors that are essential to the success of a functional area of an organization.

cross-platform development A software development technique that allows programmers to develop programs that can run on computer systems having different hardware and operating systems, or platforms.

culture A set of major understandings and assumptions shared by a group, such as within an ethnic group or a country.

cybermall A single Web site that offers many products and services at one Internet location.

cyberterrorism Any premeditated, politically motivated attack against information, computer systems, computer programs, and data which results in violence against noncombatant targets by sub-national groups or clandestine agents.

cyberterrorist Someone who intimidates or coerces a government or organization to advance his or her political or social objectives by launching computer-based attacks against computers, networks, and the information stored on them.

D

data Raw facts, such as an employee number, total hours worked in a week, inventory part numbers, or sales orders.

data administrator A nontechnical position responsible for defining and implementing consistent principles for a variety of data issues.

data analysis The manipulation of collected data so that the development team members who are participating in systems analysis can use the data.

data center A climate-controlled building or a set of buildings that houses database servers and the systems that deliver mission-critical information and services.

data cleanup The process of looking for and fixing inconsistencies to ensure that data is accurate and complete.

data collection Capturing and gathering all data necessary to complete the processing of transactions.

data correction The process of reentering data that was not typed or scanned properly.

data definition language (DDL) A collection of instructions and commands used to define and describe data and relationships in a specific database.

data dictionary A detailed description of all the data used in the database.

data editing The process of checking data for validity and completeness.

data entry Converting human-readable data into a machine-readable form.

data input Transferring machine-readable data into the system.

data item The specific value of an attribute.

data loss prevention (DLP) Systems designed to lock down—to identify, monitor, and protect—data within an organization.

data manipulation The process of performing calculations and other data transformations related to business transactions.

data manipulation language (DML) A specific language, provided with a DBMS, which allows users to access and modify the data, to make queries, and to generate reports.

data mart A subset of a data warehouse that is used by small and medium-sized businesses and departments within large companies to support decision making.

data mining An information-analysis tool that involves the automated discovery of patterns and relationships in a data warehouse.

data model A diagram of data entities and their relationships.

data preparation, or data conversion Making sure that all files and databases are ready to be used with new computer software and systems.

data storage The process of updating one or more databases with new transactions.

data warehouse A large database that collects business information from many sources in the enterprise, covering all aspects of the company's processes, products, and customers, in support of management decision making.

database An organized collection of facts and information, typically consisting of two or more related data files.

database administrator (DBA) A skilled IS professional who directs all activities related to an organization's database.

database approach to data management An approach to data management where multiple information systems share a pool of related data.

database management system (DBMS) A group of programs that manipulate the database and provide an interface between the database and the user of the database and other application programs.

decentralized processing An approach to processing wherein processing devices are placed at various remote locations.

decision-making phase The first part of problem solving, including three stages: intelligence, design, and choice.

decision room A room that supports decision making, with the decision makers in the same building, and that combines face-to-face verbal interaction with technology to make the meeting more effective and efficient.

decision support system (DSS) An organized collection of people, procedures, software, databases, and devices used to support problem-specific decision making.

delphi approach A decision-making approach in which group decision makers are geographically dispersed throughout the country or the world; this approach encourages diversity among group members and fosters creativity and original thinking in decision making.

demand report A report developed to give certain information at someone's request rather than on a schedule.

design report The primary result of systems design, reflecting the decisions made for systems design and preparing the way for systems implementation.

design stage The second stage of decision making in which you develop alternative solutions to the problem and evaluate their feasibility.

desktop computer A nonportable computer that fits on a desktop and that provides sufficient computing power, memory, and storage for most business computing tasks.

dialogue manager A user interface that allows decision makers to easily access and manipulate the DSS and to use common business terms and phrases.

digital audio player A device that can store, organize, and play digital music files.

digital camera An input device used with a PC to record and store images and video in digital form.

digital certificate An attachment to an e-mail message or data embedded in a Web site that verifies the identity of a sender or Web site.

digital rights management (DRM) Refers to the use of any of several technologies to enforce policies for controlling access to digital media, such as movies, music, and software.

digital subscriber line (DSL) A telecommunications service that delivers high-speed Internet access to homes and small businesses over the existing phone lines of the local telephone network.

digital video disc (DVD) A storage medium used to store software, video games, and movies.

direct access A retrieval method in which data can be retrieved without the need to read and discard other data.

direct access storage device (DASD) A device used for direct access of secondary storage data.

direct conversion Stopping the old system and starting the new system on a given date; also called plunge or direct cutover.

direct observation Directly observing the existing system in action by one or more members of the analysis team.

disaster recovery The implementation of the disaster plan.

disk mirroring A process of storing data that provides an exact copy that protects users fully in the event of data loss.

distributed database A database in which the data can be spread across several smaller databases connected through telecommunications devices.

distributed processing An approach to processing wherein processing devices are placed at remote locations but are connected to each other via a network.

document production The process of generating output records, documents, and reports.

documentation Text that describes a program's functions to help the user operate the computer system.

domain The allowable values for data attributes.

domain expert The person or group with the expertise or knowledge the expert system is trying to capture (domain).

downsizing Reducing the number of employees to cut costs.

drill-down report A report providing increasingly detailed data about a situation.

dumpster diving Going through the trash of an organization to find secret or confidential information, including information needed to access an information system or its data.

E

economic feasibility The determination of whether the project makes financial sense and whether predicted benefits offset the cost and time needed to obtain them.

economic order quantity (EOQ) The quantity that should be reordered to minimize total inventory costs.

effectiveness A measure of the extent to which a system achieves its goals; it can be computed by dividing the goals actually achieved by the total of the stated goals.

efficiency A measure of what is produced divided by what is consumed.

e-Government The use of information and communications technology to simplify the sharing of information, speed formerly paper-based processes, and improve the relationship between citizens and government.

electronic business (e-business) Using information systems and the Internet to perform all business-related tasks and functions.

electronic cash An amount of money that is computerized, stored, and used as cash for e-commerce transactions.

electronic commerce Conducting business activities (e.g., distribution, buying, selling, marketing, and servicing of products or services) electronically over computer networks.

electronic-document distribution A process that enables the sending and receiving of documents in a digital form without being printed (although printing is possible).

electronic exchange An electronic forum where manufacturers, suppliers, and competitors buy and sell goods, trade market information, and run back-office operations.

electronic retailing (e-tailing) The direct sale of products or services by businesses to consumers through electronic storefronts, typically designed around the familiar electronic catalog and shopping cart model.

empowerment Giving employees and their managers more responsibility and authority to make decisions, take action, and have more control over their jobs.

encryption The process of converting an original message into a form that can be understood only by the intended receiver.

encryption key A variable value that is applied (using an algorithm) to a set of unencrypted text to produce encrypted text or to decrypt encrypted text.

end-user systems development Any systems development project in which the primary effort is undertaken by a combination of business managers and users.

enterprise data modeling Data modeling done at the level of the entire enterprise.

enterprise resource planning (ERP) system A set of integrated programs that manages the vital business operations for an entire multisite, global organization.

enterprise sphere of influence The sphere of influence that serves the needs of the firm in its interaction with its environment.

enterprise system A system central to the organization that ensures information can be shared across all business functions and all levels of management to support the running and managing of a business.

entity A general class of people, places, or things for which data is collected, stored, and maintained.

entity-relationship (ER) diagrams Data models that use basic graphical symbols to show the organization of and relationships between data.

environmental design Also called green design, it involves systems development efforts that slash power consumption, require less physical space, and result in systems that can be disposed of in a way that doesn't negatively affect the environment.

ergonomics The science of designing machines, products, and systems to maximize the safety, comfort, and efficiency of the people who use them.

event-driven review A review triggered by a problem or opportunity such as an error, a corporate merger, a new market for products, or other causes.

exception report A report automatically produced when a situation is unusual or requires management action.

execution time (E-time) The time it takes to execute an instruction and store the results.

executive support system (ESS) Specialized DSS that includes all hardware, software, data, procedures, and people used to assist senior-level executives within the organization.

expert system A system that gives a computer the ability to make suggestions and function like an expert in a particular field.

explanation facility Component of an expert system that allows a user or decision maker to understand how the expert system arrived at certain conclusions or results.

Extensible Markup Language (XML) The markup language designed to transport and store data on the Web.

external auditing Auditing performed by an outside group.

extranet A network based on Web technologies that allows selected outsiders, such as business partners and customers, to access authorized resources of a company's intranet.

F

feasibility analysis Assessment of the technical, economic, legal, operational, and schedule feasibility of a project.

feedback Information from the system that is used to make changes to input or processing activities.

field Typically a name, number, or combination of characters that describes an aspect of a business object or activity.

file A collection of related records.

File Transfer Protocol (FTP) A protocol that provides a file transfer process between a host and a remote computer and allows users to copy files from one computer to another.

final evaluation A detailed investigation of the proposals offered by the vendors remaining after the preliminary evaluation.

financial MIS An information system that provides financial information for executives and for a broader set of people who need to make better decisions on a daily basis.

five-forces model A widely accepted model that identifies five key factors that can lead to attainment of competitive advantage, including (1) the rivalry among existing competitors, (2) the threat of new entrants, (3) the threat of substitute products and services, (4) the bargaining power of buyers, and (5) the bargaining power of suppliers.

flat organizational structure An organizational structure with a reduced number of management layers.

flexible manufacturing system (FMS) An approach that allows manufacturing facilities to rapidly and efficiently change from making one product to making another.

forecasting Predicting future events to avoid problems.

full-duplex channel A communications channel that permits data transmission in both directions at the same time; a full-duplex channel is like two simplex channels.

G

game theory The use of information systems to develop competitive strategies for people, organizations, or even countries.

Gantt chart A graphical tool used for planning, monitoring, and coordinating projects.

gateway A telecommunications device that serves as an entrance to another network.

genetic algorithm An approach to solving large, complex problems in which many related operations or models change and evolve until the best one emerges; also called a *genetic program*.

geographic information system (GIS) A computer system capable of assembling, storing, manipulating, and displaying geographic information, that is, data identified according to its location.

gigahertz (GHz) Billions of cycles per second, a measure of clock speed.

graphical user interface (GUI) An interface that displays pictures (icons) and menus that people use to send commands to the computer system.

graphics processing unit (GPU) A specialized circuit that is very efficient at manipulating computer graphics and is much faster than the typical CPU chip at performing floating point operations and executing algorithms for which processing of large blocks of data is done in parallel.

green computing A program concerned with the efficient and environmentally responsible design, manufacture, operation, and disposal of IS-related products.

grid chart A table that shows relationships among the various aspects of a systems development effort.

grid computing The use of a collection of computers, often owned by multiple individuals or organizations, to work in a coordinated manner to solve a common problem.

group consensus Decision making by a group that is appointed and given the responsibility of making the final evaluation and selection.

group consensus approach A decision-making approach that forces members in the group to reach a unanimous decision.

group support system (GSS) Software application that consists of most of the elements in a DSS, plus software to provide effective support in group decision-making settings; also called *group support system* or *computerized collaborative work system*.

H

hacker A person who enjoys computer technology and spends time learning and using computer systems.

half-duplex channel A communications channel that can transmit data in either direction but not simultaneously.

handheld computer A single-user computer that provides ease of portability because of its small size.

hardware Computer equipment used to perform input, processing, storage, and output activities.

heuristics Commonly accepted guidelines or procedures that usually find a good solution.

hierarchy of data Bits, characters, fields, records, files, and databases.

highly structured problems Problems that are straightforward and require known facts and relationships.

hotsite A duplicate, operational hardware system or immediate access to one through a specialized vendor.

HSPA+ (Evolved High Speed Packet Access) A broadband telecommunications standard that provides data rates of up to 42 Mbps on the downlink and 22 Mbps on the uplink.

HTML tags Codes that tell the Web browser how to format text—as a heading, as a list, or as body text, for example—and whether images, sound, and other elements should be inserted.

human resource MIS (HRMIS) An information system that is concerned with activities related to previous, current, and potential employees of an organization, also called a personnel MIS.

hyperlink Highlighted text or graphics in a Web document that, when clicked, opens a new Web page containing related content.

Hypertext Markup Language (HTML) The standard page description language for Web pages.

I

identify theft Someone using your personal identification information without your permission to commit fraud or other crimes.

IF-THEN statements Rules that suggest certain conclusions.

image log A separate file that contains only changes to applications.

implementation stage A stage of problem solving in which a solution is put into effect.

incremental backup A backup copy of all files changed during the last few days or the last week.

inference engine Part of the expert system that seeks information and relationships from the knowledge base and provides answers, predictions, and suggestions similar to the way a human expert would.

informatics A specialized system that combines traditional disciplines, such as science and medicine, with computer systems and technology.

information A collection of facts organized and processed so that they have additional value beyond the value of the individual facts.

information center A support function that provides users with assistance, training, application development, documentation, equipment selection and setup, standards, technical assistance, and troubleshooting.

information service unit A miniature IS department attached and directly reporting to a functional area in a large organization.

information system (IS) A set of interrelated components that collect, manipulate, store, and disseminate data and information and provide a feedback mechanism to meet an objective.

information systems literacy Knowledge of how data and information are used by individuals, groups, and organizations.

information systems planning Translating strategic and organizational goals into systems development initiatives.

infrared transmission A form of communications that sends signals at a frequency of 300 GHz and above—higher than those of microwaves but lower than those of visible light.

input The activity of gathering and capturing raw data.

insider An employee, disgruntled or otherwise, working solo or in concert with outsiders to compromise corporate systems.

installation The process of physically placing the computer equipment on the site and making it operational.

instant messaging A method that allows two or more people to communicate online in real time using the Internet.

institutional DSS A DSS that handles situations or decisions that occur more than once, usually several times per year or more. An institutional DSS is used repeatedly and refined over the years.

instruction time (I-time) The time it takes to perform the fetch instruction and decode instruction steps of the instruction phase.

integrated development environments (IDEs) A development approach that combines the tools needed for programming with a programming language in one integrated package.

integration testing Testing all related systems together.

intelligence stage The first stage of decision making in which you identify and define potential problems or opportunities.

intelligent agent Programs and a knowledge base used to perform a specific task for a person, a process, or another program; also called an *intelligent robot* or *bot*.

intelligent behavior The ability to learn from experiences and apply knowledge acquired from those experiences; to handle complex situations; to solve problems when important information is missing; to determine what is important and to react quickly and correctly to a new situation; to understand visual images, process and manipulate symbols, be creative and imaginative; and to use heuristics.

internal auditing Auditing performed by individuals within the organization.

Internet The world's largest computer network, consisting of thousands of interconnected networks, all freely exchanging information.

Internet Protocol (IP) A communication standard that enables computers to route communications traffic from one network to another as needed.

Internet service provider (ISP) Any organization that provides Internet access to people.

intranet An internal network based on Web technologies that allows people within an organization to exchange information and work on projects.

intrusion detection system (IDS) Monitors system and network resources and traffic and notifies network security personnel when it senses a possible intrusion.

IP address A 64-bit number that identifies a computer on the Internet.

J

Java An object-oriented programming language from Sun Microsystems based on the C++ programming language, which allows applets to be embedded within an HTML document.

joining Manipulating data to combine two or more tables.

just-in-time (JIT) inventory An inventory management approach in which inventory and materials are delivered just before they are used in manufacturing a product.

K

kernel The heart of the operating system and controls its most critical processes.

key A field or set of fields in a record that is used to identify the record.

key-indicator report A summary of the previous day's critical activities, typically available at the beginning of each workday.

knowledge The awareness and understanding of a set of information and the ways that information can be made useful to support a specific task or reach a decision.

knowledge acquisition facility Part of the expert system that provides a convenient and efficient means of capturing and storing all the components of the knowledge base.

knowledge base The collection of data, rules, procedures, and relationships that must be followed to achieve value or the proper outcome.

knowledge engineer A person who has training or experience in the design, development, implementation, and maintenance of an expert system.

knowledge user The person or group who uses and benefits from the expert system.

L

laptop computer A personal computer designed for use by mobile users, being small and light enough to sit comfortably on a user's lap.

LCD display Flat display that uses liquid crystals—organic, oil-like material placed between two polarizers—to form characters and graphic images on a backlit screen.

learning systems A combination of software and hardware that allows a computer to change how it functions or how it reacts to situations based on feedback it receives.

legal feasibility The determination of whether laws or regulations may prevent or limit a systems development project.

linking The ability to combine two or more tables through common data attributes to form a new table with only the unique data attributes.

local area network (LAN) A network that connects computer systems and devices within a small area, such as an office, home, or several floors in a building.

logical design A description of the functional requirements of a system.

long term evolution (LTE) A standard for wireless communications for mobile phones based on packet switching.

M

machine cycle The instruction phase followed by the execution phase.

magnetic disk A direct access storage device with bits represented by magnetized areas.

magnetic stripe card A type of card that stores a limited amount of data by modifying the magnetism of tiny iron-based particles contained in a band on the card.

magnetic tape A type of sequential secondary storage medium, now used primarily for storing backups of critical organizational data in the event of a disaster.

mainframe computer A large, powerful computer often shared by hundreds of concurrent users connected to the machine over a network.

maintenance team A special IS team responsible for modifying, fixing, and updating existing software.

make-or-buy decision The decision regarding whether to obtain the necessary software from internal or external sources.

management information system (MIS) An organized collection of people, procedures, software, databases, and devices that provides routine information to managers and decision makers.

market segmentation The identification of specific markets to target them with tailored advertising messages.

marketing MIS An information system that supports managerial activities in product development, distribution, pricing decisions, promotional effectiveness, and sales forecasting.

massively parallel processing systems A form of multiprocessing that speeds processing by linking hundreds or thousands of processors to operate at the same time, or in parallel, with each processor having its own bus, memory, disks, copy of the operating system, and applications.

material requirements planning (MRP) A set of inventory-control techniques that help coordinate thousands of inventory items when the demand of one item is dependent on the demand for another.

megahertz (MHz) Millions of cycles per second, a measure of clock speed.

meta tag An HTML code, not visible on the displayed Web page, that contains keywords representing your site's content, which search engines use to build indexes pointing to your Web site.

metropolitan area network (MAN) A telecommunications network that connects users and their computers in a geographical area that spans a campus or city.

middleware Software that allows various systems to communicate and exchange data.

MIPS Millions of instructions per second, a measure of machine cycle time.

mission-critical systems Systems that play a pivotal role in an organization's continued operations and goal attainment.

mobile commerce (m-commerce) The use of mobile, wireless devices to place orders and conduct business.

model base Part of a DSS that allows managers and decision makers to perform quantitative analysis on both internal and external data.

model management software (MMS) Software that coordinates the use of models in a DSS, including financial, statistical analysis, graphical, and project management models.

modem A telecommunications hardware device that converts (modulates and demodulates) communications signals so they can be transmitted over the communication media.

monitoring stage The final stage of the problem-solving process in which decision makers evaluate the implementation.

Moore's Law A hypothesis stating that transistor densities on a single chip will double every two years.

MP3 A standard format for compressing a sound sequence into a small file.

multicore microprocessor A microprocessor that combines two or more independent processors into a single computer so that they share the workload and improve processing capacity.

multimedia Text, graphics, video, animation, audio, and other media that can be used to help an organization efficiently and effectively achieve its goals.

multiplexer A device that combines data from multiple data sources into a single output signal that carries multiple channels, thus reducing the number of communications links needed and lowering telecommunications costs.

multiprocessing The simultaneous execution of two or more instructions at the same time.

N

natural language processing Processing that allows a computer to understand and react to statements and commands made in a "natural" language such as English.

Near Field Communication (NFC) A very short-range wireless connectivity technology designed for consumer electronics, cell phones, and credit cards.

netbook computer A small, light, inexpensive member of the laptop computer family.

nettop computer An inexpensive desktop computer designed to be smaller, lighter, and consume much less power than a traditional desktop computer.

network Computers and equipment that are connected in a building, around the country, or around the world to enable electronic communications.

network operating system (NOS) Systems software that controls the computer systems and devices on a network and allows them to communicate with each other.

network-attached storage (NAS) Hard disk storage that is set up with its own network address rather than being attached to a computer.

networking protocol A set of rules, algorithms, messages, and other mechanisms that enable software and hardware in networked devices to communicate effectively.

network-management software Software that enables a manager on a networked desktop to monitor the use of individual computers and shared hardware (such as printers), scan for viruses, and ensure compliance with software licenses.

neural network A computer system that can act like or simulate the functioning of a human brain.

nominal group technique A decision-making approach that encourages feedback from individual group members, with the final decision being made by voting, similar to a system for electing public officials.

nonprogrammed decision A decision that deals with unusual or exceptional situations.

notebook computer Smaller than a laptop computer, an extremely lightweight computer that weighs less than 4 pounds and can easily fit in a briefcase.

O

object-oriented database A database that stores both data and its processing instructions.

object-oriented database management system (OODBMS) A group of programs that manipulate an object-oriented database and provide a user interface and connections to other application programs.

object-oriented systems development (OOSD) An approach to systems development that combines the logic of the systems development life cycle with the power of objectoriented modeling and programming.

object-relational database management system (ORDBMS) A DBMS capable of manipulating audio, video, and graphical data.

off-the-shelf software Software mass-produced by software vendors to address needs that are common across businesses, organizations, or individuals.

on-demand computing Contracting for computer resources to rapidly respond to an organization's flow of work as the need for computer resources arises. Also called on-demand business and utility computing.

online analytical processing (OLAP) Software that allows users to explore data from a number of perspectives.

online transaction processing (OLTP) A form of data processing where each transaction is processed immediately without the delay of accumulating transactions into a batch.

open-source software Software that is distributed, typically for free, with the source code also available so that it can be studied, changed, and improved by its users.

operating system (OS) A set of computer programs that controls the computer hardware and acts as an interface with applications.

operational feasibility The measure of whether the project can be put into action or operation.

optical storage device A form of data storage that uses lasers to read and write data.

optimization model A process to find the best solution, usually the one that will best help the organization meet its goals.

organic light-emitting diode (OLED) display Flat display that uses a layer of organic material sandwiched between two conductors which in turn are sandwiched between a glass top plate and a glass bottom plate so that when electric current is applied to the two conductors, a bright, electroluminescent light is produced directly from the organic material.

organization A formal collection of people and other resources established to accomplish a set of goals.

organizational change How for-profit and nonprofit organizations plan for, implement, and handle change.

organizational culture The major understandings and assumptions for a business, corporation, or other organization.

organizational learning The adaptations and adjustments based on experience and ideas over time.

organizational structure Organizational subunits and the way they relate to the overall organization.

output Production of useful information, usually in the form of documents and reports.

outsourcing Contracting with outside professional services to meet specific business needs.

P

packet-switching network A network in which no fixed path is created between the communicating devices and the data is broken into packets, with each packet transmitted individually and capable of taking various paths from sender to recipient.

parallel computing The simultaneous execution of the same task on multiple processors to obtain results faster.

parallel start-up Running both the old and new systems for a period of time and comparing the output of the new system closely with the output of the old system; any differences are reconciled. When users are comfortable that the new system is working correctly, the old system is eliminated.

password sniffer A small program hidden in a network or a computer system that records identification numbers and passwords.

patch A minor change to correct a problem or make a small enhancement. It is usually an addition to an existing program.

p-card (procurement card or purchasing card) A credit card used to streamline the traditional purchase order and invoice payment processes.

perceptive system A system that approximates the way a person sees, hears, and feels objects.

personal area network (PAN) A network that supports the interconnection of information technology close to one person.

personal productivity software The software that enables users to improve their personal effectiveness, increasing the amount of work and quality of work they can do.

personal sphere of influence The sphere of influence that serves the needs of an individual user.

personalization The process of tailoring Web pages to specifically target individual consumers.

phase-in approach Slowly replacing components of the old system with those of the new one; this process is repeated for each application until the new system is running every application and performing as expected; it is also called a *piecemeal approach*.

physical design The specification of the characteristics of the system components necessary to put the logical design into action.

pilot start-up Running the new system for one group of users rather than all users.

pipelining A form of CPU operation in which multiple execution phases are performed in a single machine cycle.

pixel A dot of color on a photo image or a point of light on a display screen.

plasma display A type of display using thousands of smart cells (pixels) consisting of electrodes and neon and xenon gases that are electrically turned into plasma (electrically charged atoms and negatively charged particles) to emit light.

point evaluation system An evaluation process in which each evaluation factor is assigned a weight in percentage points based on importance. Then each proposed information system is evaluated in terms of this factor and given a score, such as one ranging from 0 to 100. The scores are totaled, and the system with the greatest total score is selected.

point-of-sale (POS) device A terminal used to enter data into the computer system.

policy-based storage management Automation of storage using previously defined policies.

portable computer A computer small enough to carry easily.

predictive analysis A form of data mining that combines historical data with assumptions about future conditions to predict outcomes of events, such as future product sales or the probability that a customer will default on a loan.

preliminary evaluation An initial assessment whose purpose is to dismiss the unwanted proposals; it begins after all proposals have been submitted.

primary key A field or set of fields that uniquely identifies the record.

primary storage (main memory; memory) The part of the computer that holds program instructions and data.

private branch exchange (PBX) A telephone switching exchange that serves a single organization.

problem solving A process that goes beyond decision making to include the implementation stage.

procedures The strategies, policies, methods, and rules for using a CBIS.

process A set of logically related tasks performed to achieve a defined outcome.

processing Converting or transforming data into useful outputs.

productivity A measure of the output achieved divided by the input required.

profit center A department within an organization that focuses on generating profits.

Program Evaluation and Review Technique (PERT) A formalized approach for developing a project schedule that creates three time estimates for an activity.

programmed decision A decision made using a rule, procedure, or quantitative method.

programming languages Sets of keywords, commands, symbols, and rules for constructing statements by which humans can communicate instructions to a computer.

project deadline The date the entire project is to be completed and operational.

project milestone A critical date for the completion of a major part of the project.

project organizational structure A structure centered on major products or services.

project schedule A detailed description of what is to be done.

projecting Manipulating data to eliminate columns in a table.

proprietary software One-of-a-kind software designed for a specific application and owned by the company, organization, or person that uses it.

prototyping An iterative approach to the systems development process in which at each iteration requirements and alternative solutions to a problem are identified and analyzed, new solutions are designed, and a portion of the system is implemented.

Q

quality The ability of a product or service to meet or exceed customer expectations.

quality control A process that ensures that the finished product meets the customers' needs.

questionnaires A method of gathering data when the data sources are spread over a wide geographic area.

R

Radio Frequency Identification (RFID) A technology that employs a microchip with an antenna to broadcast its unique identifier and location to receivers.

random access memory (RAM) A form of memory in which instructions or data can be temporarily stored.

rapid application development (RAD) A systems development approach that employs tools, techniques, and methodologies designed to speed application development.

read-only memory (ROM) A nonvolatile form of memory.

record A collection of data fields all related to one object, activity, or individual.

redundant array of independent/inexpensive disks (RAID) A method of storing data that generates extra bits of data from existing data, allowing the system to create a "reconstruction map" so that if a hard drive fails, the system can rebuild lost data.

reengineering (process redesign) The radical redesign of business processes, organizational structures, information systems, and values of the organization to achieve a breakthrough in business results.

register A high-speed storage area in the CPU used to temporarily hold small units of program instructions and data immediately before, during, and after execution by the CPU.

relational model A database model that describes data in which all data elements are placed in two-dimensional tables called relations, which are the logical equivalent of files.

release A significant program change that often requires changes in the documentation of the software.

reorder point (ROP) A critical inventory quantity level.

replicated database A database that holds a duplicate set of frequently used data.

report layout A technique that allows designers to diagram and format printed reports.

request for maintenance form A form authorizing modification of programs.

request for proposal (RFP) A document that specifies in detail required resources such as hardware and software.

requirements analysis The determination of user, stakeholder, and organizational needs.

return on investment (ROI) One measure of IS value that investigates the additional profits or benefits that are generated as a percentage of the investment in IS technology.

revenue center A division within a company that generates sales or revenues.

reverse 911 service A communications solution that delivers emergency notifications to users in a selected geographical area.

rich Internet application (RIA) Software that has the functionality and complexity of traditional application software but that does not require local installation and runs in a Web browser.

robotics Mechanical or computer devices that perform tasks requiring a high degree of precision or that are tedious or hazardous for humans.

router A telecommunications device that forwards data packets across two or more distinct networks toward their destinations through a process known as routing.

rule A conditional statement that links conditions to actions or outcomes.

S

satisficing model A model that will find a good—but not necessarily the best—solution to a problem.

scalability The ability to increase the processing capability of a computer system so that it can handle more users, more data, or more transactions in a given period.

schedule feasibility The determination of whether the project can be completed in a reasonable amount of time.

scheduled report A report produced periodically, such as daily, weekly, or monthly.

schema A description of the entire database.

screen layout A technique that allows a designer to quickly and efficiently design the features, layout, and format of a display screen.

script bunny A derogatory term for inexperienced hackers who download programs called scripts that automate the job of breaking into computers.

search engine A valuable tool that enables you to find information on the Web by specifying words that are key to a topic of interest, known as keywords.

secondary storage Devices that store large amounts of data, instructions, and information more permanently than allowed with main memory.

secure sockets layer (SSL) A communications protocol used to secure sensitive data during e-commerce.

security dashboard Software that provides a comprehensive display on a single computer screen of all the vital data related to an organization's security defenses, including threats, exposures, policy compliance and incident alerts.

selecting Manipulating data to eliminate rows according to certain criteria.

semistructured or unstructured problems More complex problems in which the relationships among the pieces of data are not always clear, the data might be in a variety of formats, and the data is often difficult to manipulate or obtain.

separation of duties The careful division of the tasks and responsibilities associated with a key process so that must be performed by more than one person.

sequential access A retrieval method in which data must be accessed in the order in which it is stored.

sequential access storage device (SASD) A device used to sequentially access secondary storage data.

server A computer employed by many users to perform a specific task, such as running network or Internet applications.

service-oriented architecture (SOA) A modular method of developing software and systems that allows users to interact with systems and systems to interact with each other.

simplex channel A communications channel that can transmit data in only one direction and is seldom used for business telecommunications.

single-user license A software license that permits you to install the software on one or more computers, used by one person.

site preparation Preparation of the location of a new system.

slipstream upgrade A minor upgrade—typically a code adjustment or minor bug fix—not worth announcing. It usually requires recompiling all the code, and in so doing, it can create entirely new bugs.

smart card A credit card–sized device with an embedded microchip to provide electronic memory and processing capability.

smartphone A handheld computer that combines the functionality of a mobile phone, camera, Web browser, e-mail tool, MP3 player, and other devices into a single device.

social engineering Using social skills to get computer users to provide information that allows a hacker to access an information system or its data.

Software as a Service (SaaS) A service that allows businesses to subscribe to Web-delivered application software.

software The computer programs that govern the operation of the computer.

software piracy The act of unauthorized copying, downloading, sharing, selling, or installing of copyrighted software.

software suite A collection of single programs packaged together in a bundle.

source data automation Capturing and editing data where it is initially created and in a form that can be directly entered into a computer, thus ensuring accuracy and timeliness.

speech-recognition technology Input devices that recognize human speech.

spyware Software that is installed on a personal computer to intercept or take partial control over the user's interaction with the computer without knowledge or permission of the user.

stakeholders People who, either themselves or through the organization they represent, ultimately benefit from the systems development project.

start-up The process of making the final tested information system fully operational.

statistical sampling Selecting a random sample of data and applying the characteristics of the sample to the whole group.

steering committee An advisory group consisting of senior management and users from the IS department and other functional areas.

storage area network (SAN) A special-purpose, high-speed network that provides high-speed connections among data storage devices and computers over a network.

storage as a service Storage as a service is a data storage model where a data storage service provider rents space to individuals and organizations.

storefront broker A company that acts as an intermediary between your Web site and online merchants who have the products and retail expertise.

strategic alliance (or strategic partnership) An agreement between two or more companies that involves the joint production and distribution of goods and services.

strategic planning Determining long-term objectives by analyzing the strengths and weaknesses of the organization, predicting future trends, and projecting the development of new product lines.

structured interview An interview in which the questions are written in advance.

supercomputers The most powerful computer systems with the fastest processing speeds.

supply chain management (SCM) A system that includes planning, executing, and controlling all activities involved in raw material sourcing and procurement, converting raw materials to finished products, and warehousing and delivering finished products to customers.

switch A telecommunications device that uses the physical device address in each incoming message on the network to determine to which output port it should forward the message to reach another device on the same network.

syntax A set of rules associated with a programming language.

system A set of elements or components that interact to accomplish goals.

system performance measurement Monitoring the system—the number of errors encountered, the amount of memory required, the amount of processing or CPU time needed, and other problems.

system performance products Software that measures all components of the information system, including hardware, software, database, telecommunications, and network systems.

system performance standard A specific objective of the system.

system testing Testing the entire system of programs.

systems analysis The systems development phase involving the study of existing systems and work processes to identify strengths, weaknesses, and opportunities for improvement.

systems controls Rules and procedures to maintain data security.

systems design The systems development phase that defines how the information system will do what it must do to obtain the solution.

systems development The activity of creating or modifying information systems.

systems implementation The systems development phase involving the creation or acquisition of various system components detailed in the systems design, assembling them, and placing the new or modified system into operation.

systems investigation The systems development phase during which problems and opportunities are identified and considered in light of the goals of the business.

systems investigation report A summary of the results of the systems investigation and the process of feasibility analysis and recommendation of a course of action.

systems maintenance A stage of systems development that involves checking, changing, and enhancing the system to make it more useful in achieving user and organizational goals.

systems maintenance and review The systems development phase that ensures the system operates and modifies the system so that it continues to meet changing business needs.

systems operation Use of a new or modified system in all kinds of operating conditions.

systems request form A document filled out by someone who wants the IS department to initiate systems investigation.

systems review The final step of systems development, involving analyzing systems to make sure that they are operating as intended.

T

tablet computer A portable, lightweight computer with no keyboard that allows you to roam the office, home, or factory floor carrying the device like a clipboard.

team organizational structure A structure centered on work teams or groups.

technical documentation Written details used by computer operators to execute the program and by analysts and programmers to solve problems or modify the program.

technical feasibility Assessment of whether the hardware, software, and other system components can be acquired or developed to solve the problem.

technology acceptance model (TAM) A model that specifies the factors that can lead to better attitudes about an information system, along with higher acceptance and usage of it.

technology diffusion A measure of how widely technology is spread throughout the organization.

technology infrastructure All the hardware, software, databases, telecommunications, people, and procedures that are configured to collect, manipulate, store, and process data into information.

technology infusion The extent to which technology permeates an area or department.

telecommunications The electronic transmission of signals for communications that enables organizations to carry out their processes and tasks through effective computer networks.

telecommunications medium Any material substance that carries an electronic signal to support communications between a sending and receiving device.

telecommuting The use of computing devices and networks so that employees can work effectively away from the office.

thin client A low-cost, centrally managed computer with essential but limited capabilities and no extra drives (such as CD or DVD drives) or expansion slots.

time-driven review Review performed after a specified amount of time.

total cost of ownership (TCO) The sum of all costs over the life of an information system, including the costs to acquire components such as the technology, technical support, administrative costs, and end-user operations.

traditional approach to data management An approach to data management whereby each distinct operational system uses data files dedicated to that system.

traditional organizational structure An organizational structure in which the hierarchy of decision making and authority flows from the strategic management at the top down to operational management and non-management employees.

transaction Any business-related exchange such as payments to employees, sales to customers, and payments to suppliers.

transaction processing cycle The process of data collection, data editing, data correction, data manipulation, data storage, and document production.

transaction processing system (TPS) An organized collection of people, procedures, software, databases, and devices used to perform and record business transactions.

Transmission Control Protocol (TCP) The widely used Transport-layer protocol that most Internet applications use with IP.

tunneling The process by which VPNs transfer information by encapsulating traffic in IP packets over the Internet.

U

ultra wideband (UWB) A form of short-range communications that employs extremely short electromagnetic pulses lasting just 50 to 1,000 picoseconds that are transmitted across a broad range of radio frequencies of several gigahertz.

Uniform Resource Locator (URL) A Web address that specifies the exact location of a Web page using letters and words that map to an IP address and a location on the host.

unit testing Testing of individual programs.

unstructured interview An interview in which the questions are not written in advance.

user acceptance document A formal agreement that the user signs stating that a phase of the installation or the complete system is approved.

user documentation Written descriptions developed for people who use a program; in easy-to-understand language, it shows how the program can and should be used.

user interface The element of the operating system that allows people to access and interact with the computer system.

user preparation The process of readying managers, decision makers, employees, other users, and stakeholders for the new system.

users People who will regularly interact with the system.

utility program Program that helps to perform maintenance or correct problems with a computer system.

V

value chain A series (chain) of activities that includes inbound logistics, warehouse and storage, production and manufacturing, finished product storage, outbound logistics, marketing and sales, and customer service.

version A major program change, typically encompassing many new features.

videoconferencing A set of interactive telecommunications technologies that enable people at multiple locations to communicate using simultaneous two-way video and audio transmissions.

virtual organizational structure A structure that uses individuals, groups, or complete business units in geographically dispersed areas; these groups can last for a few weeks or years, often requiring telecommunications and the Internet.

virtual private network (VPN) A private network that uses a public network (usually the Internet) to connect multiple remote locations.

virtual reality The simulation of a real or imagined environment that can be experienced visually in three dimensions.

virtual reality system A system that enables one or more users to move and react in a computer-simulated environment.

virtual tape A storage device for less frequently needed data so that it appears to be stored entirely on tape cartridges, although some parts of it might actually be located on faster hard disks.

virtual workgroups Teams of people located around the world working on common problems.

vision systems The hardware and software that permit computers to capture, store, and manipulate visual images.

voice mail Technology that enables users to send, receive, and store verbal messages to and from other people around the world.

voice mail-to-text service A service that captures voice mail messages, converts them to text, and sends them to an e-mail account.

volume testing Testing the application with a large amount of data.

W

Web Server and client software, the Hypertext Transfer Protocol (HTTP), standards, and markup languages that combine to deliver information and services over the Internet.

Web 2.0 The Web as a computing platform that supports software applications and the sharing of information among users.

Web application framework Web development software that provides the foundational code—or framework—for a professional interactive Web site, allowing developers to customize the code to specific needs.

Web browser Web client software, such as Internet Explorer, Firefox, Chrome, Safari, and Opera, used to view Web pages.

Web log (blog) A Web site that people can create and use to write about their observations, experiences, and opinions on a wide range of topics.

Web portal A Web page that combines useful information and links and acts as an entry point to the Web; portals typically include a search engine, a subject directory, daily headlines, and other items of interest. Many people choose a Web portal as their browser's home page (the first page you open when you begin browsing the Web).

wide area network (WAN) A telecommunications network that connects large geographic regions.

Wi-Fi A medium-range wireless telecommunications technology brand owned by the Wi-Fi Alliance.

wireless mesh A form of communication that uses multiple Wi-Fi access points to link a series of interconnected local area networks to form a wide area network capable of serving a large campus or an entire city.

workgroup Two or more people who work together to achieve a common goal.

workgroup application software Software that supports teamwork, whether team members are in the same location or dispersed around the world.

workgroup sphere of influence The sphere of influence that helps workgroup members attain their common goals.

workstation A more powerful personal computer used for mathematical computing, computer-assisted design, and other high-end processing but still small enough to fit on a desktop.

Z

ZigBee A form of wireless communications frequently used in security systems and heating and cooling control systems.

Subject Index

Note: A boldface page number indicates a key term and the location of its definition in the text.

A

D

E

F

G

H

I

J

K

L

M

N

Q

R

S

T

U

V

W

X

Y

Z

Company Index